Active Physics

A PROJECT-BASED INQUIRY APPROACH

THIRD EDITION

PHYSICS FOR ALL

TEACHER'S EDITION
VOLUME 1

ARTHUR EISENKRAFT, Ph.D.

NSF

IT's ABOUT TIME
HERFF JONES EDUCATION DIVISION

84 Business Park Drive, Armonk, NY 10504
Phone (914) 273-2233 Fax (914) 273-2227
www.its-about-time.com

Program Components

Student Edition

Teacher's Edition – Three-Volume Set

Durable Kits

Consumable Kits

Multimedia

— Student Edition Online

— Teacher's Edition Online

— Teacher's Resources CD – Blackline Masters and Color Overheads

— Multimedia DVD/CD Set – Content Videos and Spreadsheet DVD

— *ExamView* Test Generator

— Constructing Physics Understanding (CPU)

Printed and bound in the United States of America.

ISBN 978-1-60720-002-4

1 2 3 4 5 CRS 13 12 11 10 09

This project was supported, in part, by the
National Science Foundation under Grant No. 0352516.
Opinions expressed are those of the authors and not necessarily
those of the National Science Foundation.

Contents

Acknowledgements vi

Welcome to Active Physics: Your Guide to Success ix

Features of Active Physics xv

Overview of Meeting the Needs of All Students: Differentiated Instruction xx

Students' Prior Conceptions: Overview and 7E Tools xxi

Pacing Guides for Teachers xxii

Engineering Design Cycle xxiv

Active Physics and the National Science Education Standards xxv

Active Physics: Bridging Research and Practice xxviii

Cooperative Learning xxxii

Active Physics–A Research-Based Curriculum xxxvi

Expanding the 5E Model Article xlii

Safety Contract: Safety in the Physics Classroom xlvi

Chapter 1 *Driving the Roads* 1

Chapter Overview 2

Key Physics Concepts 3

Chapter Concept Map 4

Understanding by Design 5

Pacing Guide 6

Chapter Materials and Equipment 10

Teacher Resources 14

Chapter Challenge 16

Section 1 Reaction Time: Responding to Road Hazards 22

Section 2 Measurement: Errors, Accuracy, and Precision 46

Section 3 Average Speed: Following Distance and Models of Motion 68

Section 4 Graphing Motion: Distance, Velocity, and Acceleration 102

Chapter Mini-Challenge 138

Section 5 Negative Acceleration: Braking Your Automobile 142
Section 6 Using Models: Intersections with a Yellow Light 170
Section 7 Centripetal Force: Driving on Curves 200
Physics You Learned 224
Physics Chapter Challenge 226
Physics Connections to Other Sciences 228
Physics At Work 229
Physics Practice Test 230
Sample Assessment Rubric for Chapter 1 234

Chapter 2 *Physics in Action* 237

Chapter Overview 238
Key Physics Concepts 239
Chapter Concept Map 240
Understanding by Design 241
Pacing Guide 242
Chapter Materials and Equipment 246
Teacher Resources 252
Chapter Challenge 254
Section 1 Newton's First Law: A Running Start 258
Section 2 Constant Speed and Acceleration: Measuring Motion 282
Section 3 Newton's Second Law: Push or Pull 304
Section 4 Projectile Motion: Launching Things into the Air 332
Section 5 The Range of Projectiles: The Shot Put 356
Chapter Mini-Challenge 380
Section 6 Newton's Third Law: Run and Jump 382
Section 7 Frictional Forces: The Mu of the Shoe 408
Section 8 Potential and Kinetic Energy: Energy in the Pole Vault 432
Section 9 Conservation of Energy: Defy Gravity 458
Physics You Learned 484

Physics Chapter Challenge 486
Physics Connections to Other Sciences 488
Physics At Work 489
Physics Practice Test 490
Sample Assessment Rubric for Chapter 2 494

Chapter 3 *Safety*........ 497

Chapter Overview 498
Key Physics Concepts 499
Chapter Concept Map 500
Understanding by Design 501
Pacing Guide 502
Chapter Materials and Equipment 506
Teacher Resources 512
Chapter Challenge 514
Section 1 Accidents 518
Section 2 Newton's First Law of Motion: Life and Death before and after Seat Belts 538
Section 3 Energy and Work: Why Air Bags? 562
Section 4 Newton's Second Law of Motion: The Rear-End Collision 592
Chapter Mini-Challenge 614
Section 5 Momentum: Concentrating on Collisions 616
Section 6 Conservation of Momentum 634
Section 7 Impulse and Changes in Momentum: Crumple Zone 658
Physics You Learned 688
Physics Chapter Challenge 690
Physics Connections to Other Sciences 692
Physics At Work 693
Physics Practice Test 694
Sample Assessment Rubric for Chapter 3 698
Credits 701

Acknowledgments

Project Director

Dr. Arthur Eisenkraft has taught high school physics for over 28 years. He is currently the Distinguished Professor of Science Education at the University of Massachusetts, Boston, where he is also an Adjunct Professor of Physics and the Director of the Center of Science and Math In Context (COSMIC). Dr. Eisenkraft is the author of numerous science and educational publications and holds a patent for a Laser Vision Testing System, which tests visual acuity for spatial frequency.

Dr. Eisenkraft has been recognized with numerous awards, including: Presidential Award for Excellence in Science Teaching, 1986 from President Ronald Reagan; American Association of Physics Teachers (AAPT); Distinguished Service Citation for "excellent contributions to the teaching of physics," 1989; Science Teacher of the Year, Disney American Teacher Awards, 1991; Honorary Doctorate of Science, Rensselaer Polytechnic Institute, New York, 1993; AAPT Excellence in Pre-College Teaching Award, 1999; National Science Teachers Association (NSTA) Distinguished Service Award to Science Education, 2005 and recipient of the Robert A. Millikan Medal for "notable and creative contributions in physics education," 2009.

In 1999, Dr. Eisenkraft was elected to a three-year cycle as the President-Elect, President, and Retiring President of the NSTA, the world's largest organization of science teachers. He has served on numerous committees of the National Academy of Sciences, including the content committee that has helped author the National Science Education Standards, and in 2003 he was elected a fellow of the American Association for the Advancement of Science (AAAS). Dr. Eisenkraft has been involved with a number of projects and chaired many notable competitions, including the Toshiba/NSTA ExploraVisions Awards (1991 to present), which he co-created; the Toyota TAPESTRY Grants (1990 to 2005); and the Duracell/NSTA Scholarship Competition (1984 to 2000). In 1993, he served as Executive Director for the XXIV International Physics Olympiad after being Academic Director for the United States Team for six years.

Dr. Eisenkraft is a frequent presenter and keynote speaker at national conventions. He has published over 100 articles and presented over 200 papers and workshops. *Quantoons*, written with L. Kirkpatrick and featuring illustrations by Tomas Bunk, led to an art exhibition at the New York Hall of Science.

Dr. Eisenkraft has been featured in articles in *The New York Times*, *Education Week*, *Physics Today*, *Scientific American*, *The American Journal of Physics*, and *The Physics Teacher*. He has testified before the United States Congress, appeared on *NBC's The Today Show,* National Public Radio, and many other radio and television broadcasts, including serving as the science consultant to ESPN's *Sports Figures*.

Acknowledgments — Active Physics Team

Active Physics, Third Edition

was developed by a team of leading physicists, university educators, and classroom teachers with financial support from the National Science Foundation.

NSF Program Officer

Gerhard Salinger
National Science Foundation
Arlington, VA

Principal Investigators

Arthur Eisenkraft
University of Massachusetts
Boston, MA

Barbara Zahm
It's About Time
Herff Jones Education Division
Armonk, NY

Project Coordinator

Gary Hickernell
It's About Time
Herff Jones Education Division
Armonk, NY

Field Test Coordinator

George Amann
Rhinebeck, NY

Writers

Peter Collings
Swarthmore College
Swarthmore, PA

Ron DeFronzo
East Bay Educational Collaborative
Warren, RI

Robert Hilborn
University of Texas at Dallas
Richardson, TX

Ramon Lopez
University of Texas at Arlington
Arlington, TX

Dwight Neuenschwander
Southern Nazarene University
Bethany, OK

John Rowe
Cincinnati Public Schools
Cincinnati, OH

Sue Vincent
Turner Falls H.S.
Montague, NH

Bruce Williamson
Delaware Valley Friends School
Paoli, PA

Board of Advisors

Marilyn Decker
Boston Public Schools
Boston, MA

Michael Lach
Chicago Public Schools
Chicago, IL

Jim Nelson
AAPT
Gainesville, FL

John Roeder
The Calhoun School
New York, NY

Frederick Stein (deceased)
American Physical Society
Dillon, CO

James Stith
American Institute of Physics
Mitchellville, MD

Clara Tolbert
Urban Strategic Initiative (retired)
Philadelphia, PA

Consultants

George Amann
Rhinebeck, NY

Matthew Anthes-Washburn
East High School
Denver, CO

Scott Bartholomew
Parkway Academy of Technology and Health
West Roxbury, MA

Pat Callahan
Catasaugua H.S.
Center Valley, PA

Gary Curts
Dublin Jerome H.S.
Dublin, OH

Timothy M. Fitzgibbon

John Hubisz
North Carolina State University
Apex, NC

Elena Kaczorowski
Bedford H.S.
Bedford, NY

Ernest Kuehl
Lawrence H.S.
Cedarhurst, NY

John Koser
University of St. Thomas
St. Paul, MN

Holly Marcus

Desiree Phillips
Learning Specialist
Somerville, MA

Mary Quinlan
Radnor H.S.
Radnor, PA

Patricia Rourke
Educational Consultant
Alexandria, VA

Larry Weathers
The Bromfield School
Harvard, MA

Shari Weaver
Boston Public Schools
Dorchester, MA

David Wright
Tidewater Community College
Norfolk, VA

Evaluation Team

Frances Lawrenz
University of Minnesota
Minneapolis, MN

Nathan Wood
North Dakota State University
Fargo, ND

Safety Reviewer

George Amann
Rhinebeck, NY

Field Test Teachers

Andrea Anderson
Athens H.S.
The Plains, OH

Matthew Anthes-Washburn
East High School
Denver, CO

Joel Aquino
Splendora H.S.
Splendora, TX

Scott Bartholomew
Parkway Academy of Technology and Health
West Roxbury, MA

Jeff Briggs
Commodore Perry H.S.
Hadley, PA

Pat Callahan
Delaware Valley H.S.
Frenchtown, NJ

Debra Cayea
Briarcliff H.S.
Briarcliff Manor, NY

Robin Chisholm
International School of Indiana
Indianapolis, IN

Kenneth Dugan
Deep Creek H.S.
Chesapeake, VA

Darrin Ellsworth
Fillmore C.H.S.
Harmony, MN

Pete Flores
Harlandale H.S.
San Antonio, TX

Jane Frye
Carter Co. H.S.
Ekalaka, MT

Tracey Greeley-Adams
Woodward Career Technical H.S.
Cincinnati, OH

Michelle Greenlee
Lauderdale County C.S.
Florence, AL

Stephanie Harmon
Rockcastle Co. H.S.
Mt. Vernon, KY

Karl Hendrickson
McGill-Toolen Catholic H.S.
Mobile, AL

LaTeise Jones
Bardstown H.S.
Bardstown, KY

Danielle Joslin
Fond du Lac H.S.
Fond du Lac, WI

Dolores Keeley
Forest Hills Eastern H.S.
Ada, MI

Mark Klawiter
Deerfield H.S.
Deerfield, WI

Jim Kyte
Urbandale H.S.
Urbandale, IA

Kathy Lucas
Casey County H.S.
Liberty, KY

Jennifer Lynch
Taft Middle School
Oklahoma City, OK

Keith Magni
Boston Community Leadership Academy
Brighton, MA

Robert Malcolm
Greater Johnstown H.S.
Johnstown, PA

Janie Martin
Southwest H.S.
San Antonio, TX

Maryl McCrary
Thomas-Fay-Custer H.S.
Thomas, OK

Frederick Meshna
The Bromfield School
Harvard, MA

Kathy Naughton
West Delaware H.S.
Manchester, IA

Alan Nauretz
Shell Lake H.S.
Shell Lake, WI

Fred Nelson
Manhattan H.S.
Manhattan, KS

Vivian O'Brien
Plymouth Regional H.S.
Plymouth, NH

Steve Oszust
Gateway School for Environmental Resource and Technology
Stevenson H.S.
Bronx, NY

Jane Pollack
The Rayen School
Youngstown, OH

Marcia Powell
West Delaware H.S.
Manchester, IA

James Prosser
Fond du Lac H.S.
Fond du Lac, WI

Pamela Pulliam
Charleston Catholic H.S.
Charleston, WV

Randolph Reed
St. Albans H.S.
St. Albans, WV

Kristie Reighard
Bowling Green H.S.
Bowling Green, OH

Becky Reynolds
Sonoraville H.S.
Calhoun, GA

Milly Rixey
St. Margaret's School
Tappahannock, VA

Joan Salow
West Delaware H.S.
Manchester, IA

Judy Scheffler
Poth Jr. H.S.
Poth, TX

John Scholtz
Haverford H.S.
Havertown, PA

Paul Shafer
Mirta Ramirez Computer Science Charter School
Chicago, IL

Katheryn Shannon
Highland West Jr. H.S.
Moore, OK

Sushma Sharma
School of Entrepreneurship at Southshore
Chicago, IL

Chip Sheffield, Jr.
Robert E. Lee H.S.
Jacksonville, FL

Sandy Shutey
Butte H.S.
Butte, MT

Amy Stewart
Sonoraville H.S.
Calhoun, GA

Cheryl Schwartzwelder
Northpoint H.S. for Science, Technology, and Industry
Pomfret, MD

Carla Taylor
Callisburg H.S.
Gainsville, TX

John Tracey
High Point Regional H.S.
Sussex, NJ

Jeff Voss
West Delaware H.S.
Manchester, IA

Elizabeth Walker
North Cobb H.S.
Kennesaw, GA

John Whitsett
Fond du Lac H.S.
Fond du Lac, WI

Donna Wolz
Cross County H.S.
Cherry Valley, AR

Daniel Wood
Tombstone H.S.
Tombstone, AZ

Active Physics, First and Second Editions

were developed in association with the American Association of Physics Teachers (AAPT) and the American Institute of Physics (AIP).

Primary and Contributing Authors

Richard Berg
Howard Brody
Chris Chiaverina
Ron DeFronzo
Ruta Demery
Carl Duzen
Jon L. Harkness
Ruth Howes
Douglas A. Johnson
Ernest Kuehl
Robert L. Lehrman
Salvatore Levy
Tom Liao
Charles Payne
Mary Quinlan
Harry Rheam
Bob Ritter
John Roeder
John J. Rusch
Patty Rourke
Ceanne Tzimopoulos
Larry Weathers
David Wright

Consultants

Peter Brancazio
Robert Capen
Carole Escobar
Earl Graf
Jack Hehn
Donald F. Kirwan
Gayle Kirwan
James La Porte
Charles Misner
Robert F. Neff
Ingrid Novodvorsky
John Robson
Mark Sanders
Brian Schwartz
Bruce Seiger
Clifford Swartz
Barbara Tinker
Robert E. Tinker
Joyce Weiskopf
Donna Willis

Safety Reviewer

Gregory Puskar

Equity Reviewer

Leo Edwards

Physics at Work

Alex Strauss
Mekea Hurwitz

First Printing Reviewer

John L. Hubisz

Unit Reviewers

Robert Adams
George A. Amann
Patrick Callahan
Beverly Cannon
Barbara Chauvin
Elizabeth Chesick
Chris Chiaverina
Andria Erzberger
Elizabeth Farrell Ramseyer
Mary Gromko
Thomas Guetzloff
Jon L. Harkness
Dawn Harman
James Hill
Bob Kearny
Claudia Khourey-Bowers
Steve Kliewer
Ernest Kuehl
Jane Nelson
Mary Quinlan
John Roeder
Patty Rourke
Gerhard Salinger
Irene Slater

Pilot Test Teachers

John Agosta
Donald Campbell
John Carlson
Veanna Crawford
Janie Edmonds
Eddie Edwards
Arthur Eisenkraft
Tom Ford
Bill Franklin
Roger Goerke
Tom Gordon
Ariel Hepp
John Herrman
Linda Hodges
Ernest Kuehl
Fran Leary
Harold Lefcourt
Cherie Lehman
Kathy Malone
Bill Metzler
Elizabeth Farrell Ramseyer
Daniel Repogle
Evelyn Restivo
Doug Rich
John Roeder
Tom Senior
John Thayer
Carol-Ann Tripp
Yvette Van Hise
Jan Haarvick
Sandra Walton
Larry Wood

Field Test Coordinator

Marilyn Decker

Field Test Workshop Staff

John Carlson
Marilyn Decker
Arthur Eisenkraft
Douglas Johnson
John Koser
Mary Quinlan
Elizabeth Farrell Ramseyer
John Roeder

Field Test Evaluators

Susan Baker-Cohen
Susan Cloutier
George Hein
Judith Kelley

Field Test Teachers

Rob Adams
Benjamin Allen
Robert Applebaum
Joe Arnett
Bix Baker
Debra Beightol
Patrick Callahan
George Coker
Janice Costabile
Stanley Crum
Russel Davison
Christine K. Deyo
Jim Doller
Jessica Downing
Douglas Fackelman
Rick Forrest
Mark Freeman
Jonathan Gillis
Karen Gruner
Larry Harrison
Alan Haught
Steven Iona
Phil Jowell
Deborah Knight
Thomas Kobilarcik
Sheila Kolb
Todd Lindsay
Malinda Mann
Steve Martin
Nancy McGrory
David Morton
Charles Muller
Fred Muller
Vivian O'Brien
Robin Parkinson
Donald Perry
Francis Poodry
John Potts
Doug Rich
John Roeder
Consuelo Rogers
Lee Rossmaessler
John Rowe
Rebecca Bonner Sanders
David Schilpp
Eric Shackelford
Robert Sorensen
Teresa Stalions
Roberta Tanner
Anthony Umelo
Judy Vondruska
Deborah Waldron
Ken Wester
Susan Willis

Welcome to Active Physics: Your Guide to Success

A Five-Minute Introduction

Active Physics differs from a traditional physics program, in that it contains all the physics content you need to teach, but it is presented in an excitingly innovative and meaningful way. *Active Physics* explores forces, energy, waves, electricity, and magnetism as well as optics and modern physics. The content is always placed in a larger context that emphasizes student learning through inquiry, while motivating students through a problem-based learning approach. Students learn about waves, sound, and light as part of their requirement to create a brief show that will entertain their friends (*Let Us Entertain You*). Students apply what they have learned about Newton's laws to build an improved safety device for a car (*Physics in Action*). Students take ownership of their understanding of atomic and nuclear physics by developing a museum exhibit on the atom and suggesting something to be sold in the museum store (*Atoms on Display*).

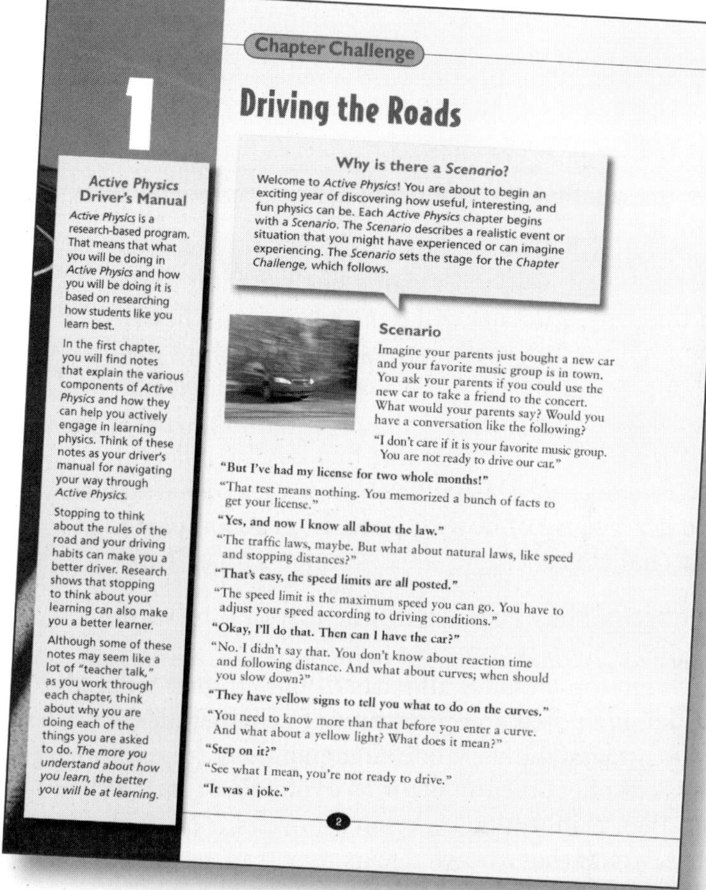

Chapter Challenge

1

Driving the Roads

Active Physics Driver's Manual

Active Physics is a research-based program. That means that what you will be doing in *Active Physics* and how you will be doing it is based on researching how students like you learn best.

In the first chapter, you will find notes that explain the various components of *Active Physics* and how they can help you actively engage in learning physics. Think of these notes as your driver's manual for navigating your way through *Active Physics*.

Stopping to think about the rules of the road and your driving habits can make you a better driver. Research shows that stopping to think about your learning can also make you a better learner.

Although some of these notes may seem like a lot of "teacher talk," as you work through each chapter, think about why you are doing each of the things you are asked to do. *The more you understand about how you learn, the better you will be at learning.*

Why is there a *Scenario*?

Welcome to *Active Physics*! You are about to begin an exciting year of discovering how useful, interesting, and fun physics can be. Each *Active Physics* chapter begins with a *Scenario*. The *Scenario* describes a realistic event or situation that you might have experienced or can imagine experiencing. The *Scenario* sets the stage for the *Chapter Challenge*, which follows.

Scenario

Imagine your parents just bought a new car and your favorite music group is in town. You ask your parents if you could use the new car to take a friend to the concert. What would your parents say? Would you have a conversation like the following?

"I don't care if it is your favorite music group. You are not ready to drive our car."

"But I've had my license for two whole months!"

"That test means nothing. You memorized a bunch of facts to get your license."

"Yes, and now I know all about the law."

"The traffic laws, maybe. But what about natural laws, like speed and stopping distances?"

"That's easy, the speed limits are all posted."

"The speed limit is the maximum speed you can go. You have to adjust your speed according to driving conditions."

"Okay, I'll do that. Then can I have the car?"

"No. I didn't say that. You don't know about reaction time and following distance. And what about curves; when should you slow down?"

"They have yellow signs to tell you what to do on the curves."

"You need to know more than that before you enter a curve. And what about a yellow light? What does it mean?"

"Step on it?"

"See what I mean, you're not ready to drive."

"It was a joke."

2

In this nine-chapter textbook, you will find everything you are required to teach and more. You will also find that the important concepts that your students need are covered in several chapters. This multiple exposure allows students to see the concepts in a variety of contexts and at a variety of depths. You will be able to choose which chapters meet the needs of your students and your state standards. Each section of each chapter also has an *Active Physics Plus* component that provides more mathematics, concept development, inquiry, or depth of treatment.

You may begin the year with any chapter. However, *Driving the Roads* contains the "launcher" material, which introduces the learning components to you and your students and helps provide students with a rationale for why this research-based approach to physics will help them succeed.

For the purpose of understanding the components of all chapters, the *Physics in Action* chapter will be broken down in this Five-Minute Introduction. Your students first learn about the *Chapter Challenge*, which will be the focus of their work. The challenge for this particular chapter is to create a series of voice-over narrations for sporting events. These voice-overs can be considered tryouts for a job as the physics sports commentator. The physics sports commentator will be handed the microphone during a broadcast and will have to explain some of the sports action as examples of physics principles. To actually get this kind of job, physics content knowledge is necessary. The commentator will also have to be entertaining, articulate, and enthusiastic.

How do the students get started on the *Chapter Challenge*? How can they complete such a challenge without the necessary physics knowledge? Students are introduced to the physics they can use to complete the challenge on a need-to-know basis. This is what makes *Active Physics* unique.

Before beginning any of the chapter, have your class discuss the *Criteria for Success*. The class decides what is expected in an excellent physics broadcast and how each of these components will be graded. For instance, they may decide that the rubric for grading will include the following factors:

- the use of physics terms and principles in the narration, including the number of physics principles and the use of equations when appropriate
- the quality of the oral narration, including the entertainment value
- the quality of the written script of the narration

Students will also need to decide whether each factor carries equal weight, or if one has a greater impact. In this way, students will have a sense of what is required for an excellent presentation before they begin, as well as a sense of ownership. Later, they will revisit the criteria before their work on the challenge is finalized. They conclude this first day by reflecting on the *Engineering Design Cycle*, a strategy by which they can get a sense of how to proceed in accomplishing this challenge.

The second day begins with the first of nine sections. As one section is completed, the next one starts. Each section includes all parts of the 7E instructional model, including an opportunity to **elicit** students' prior understanding while **engaging** them, to **explore** a physical phenomenon prior to **explaining** that concept and then **elaborating** on related physics. The students then **extend** that knowledge as they transfer their learning to the *Chapter Challenge*. You and the students will **evaluate** throughout the section. *Active Physics* is an inquiry-based curriculum— students always explore before they explain (ABC = Activity Before Concept).

For example, look at *Section 1 Newton's First Law: A Running Start*. Each section begins with a cartoon, which you can use to introduce some of the aspects that will follow in the section. More importantly, the cartoon is used to begin to engage the interests of the students with a simple *What Do You See?* question. Student answers to this question need not be correct or even relevant to physics, but it is important that they provide some response to the question. Research has found that all students, including those with special needs or English-language learners, can respond to the *What Do You See?* Students become engaged, which provides you the first opportunity to elicit their prior understanding.

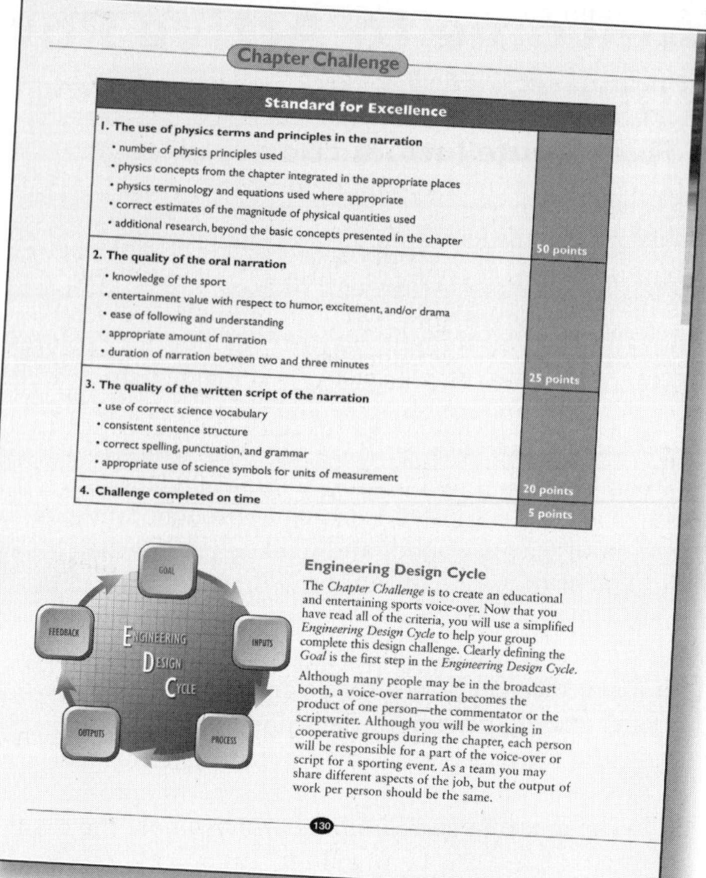

Chapter Challenge

Standard for Excellence	
1. The use of physics terms and principles in the narration • number of physics principles used • physics concepts from the chapter integrated in the appropriate places • physics terminology and equations used where appropriate • correct estimates of the magnitude of physical quantities used • additional research, beyond the basic concepts presented in the chapter	50 points
2. The quality of the oral narration • knowledge of the sport • entertainment value with respect to humor, excitement, and/or drama • ease of following and understanding • appropriate amount of narration • duration of narration between two and three minutes	25 points
3. The quality of the written script of the narration • use of correct science vocabulary • consistent sentence structure • correct spelling, punctuation, and grammar • appropriate use of science symbols for units of measurement	20 points
4. Challenge completed on time	5 points

Engineering Design Cycle

The *Chapter Challenge* is to create an educational and entertaining sports voice-over. Now that you have read all of the criteria, you will use a simplified *Engineering Design Cycle* to help your group complete this design challenge. Clearly defining the *Goal* is the first step in the *Engineering Design Cycle*.

Although many people may be in the broadcast booth, a voice-over narration becomes the product of one person—the commentator or the scriptwriter. Although you will be working in cooperative groups during the chapter, each person will be responsible for a part of the voice-over or script for a sporting event. As a team you may share different aspects of the job, but the output of work per person should be the same.

130

Chapter 2 Physics in Action

Section 1 Newton's First Law: A Running Start

What Do You See?

Learning Outcomes

In this section, you will

- **Describe** Galileo's law of inertia.
- **Apply** Newton's first law of motion.
- **Recognize** inertial mass as a physical property of matter.
- **Use** examples to demonstrate that speed is always relative to some other object.
- **Explain** that the speed of an object depends on the reference frame from which it is being observed.

What Do You Think?

Every sport includes moving objects or people or both. That is what makes sports entertaining.

- **How do figure skaters keep moving across the ice at high speeds for long times while seeming to expend no effort?**
- **Why does a soccer ball continue to roll across the field after it has been kicked?**

Record your ideas about these questions in your *Active Physics* log. Be prepared to discuss your responses with your group and the class.

Investigate

In this *Investigate*, you will use a track and a ball to explore the question, "When a ball is released to roll down a track and up the opposite side of the track, how does the vertical height that the ball reaches on the opposite side of the track relate to the vertical height from which the ball is released?"

1. Make a track that has the same slope on both sides, as shown in the diagram on the next page. Your teacher will suggest how high the ends of the track sections should be.

132

This is followed by the *What Do You Think?* question, "Why does a soccer ball continue to roll across the field after it has been kicked?" This question is intended to further elicit their prior understanding. If you listen intently and ask follow-up questions, students' prior understanding will come to the surface. This is not intended as an opportunity to correct students, but rather an opportunity to find out "what the student thinks." At the end of the section, after the students explore and gather evidence from that investigation to answer questions, you will return to this as a *What Do You Think Now?* question. That is the appropriate time to help students with confusions or inconsistencies in their responses.

The *What Do You Think?* questions are directly related to the physics principles of this section – Newton's first law. Formally, you can say that these questions elicit the students' prior understanding and are part of the constructivist approach. Typically, students write a response for one minute, followed by two minutes of discussion. But you should not try to reach closure here. The questions are designed to open up the conversation.

The students then begin the *Investigate* where they will observe and measure a ball rolling along a u-shaped ramp. They will record how high the ball goes on the far side. They will use the evidence from their observations and, following reasoning similar to that of Galileo's, come to the conclusion that a ball rolling on a flat surface would continue to roll forever if friction were somehow eliminated.

Physics Talk summarizes the physics principles and provides some historical background to enhance the experience. It also presents students with text, illustrations, and photographs that provide greater insight into the physics concepts. Words that may be new or unfamiliar to students are **boldfaced**.

To provide reading support, they are also defined and explained in the margin as *Physics Words*. The *Checking Up* questions at the end of the reading are designed to guide students toward the key concepts of the text. The *Physics Talk* section is most similar to the traditional textbook explanation of physics. A typical text assumes that students have observed a ball rolling across a table, while *Active Physics* ensures that all students have had a common, hands-on experience with which to build conceptual and mathematical understanding. What makes *Physics Talk* unique is that the explanations refer back to investigations and experiences that students assuredly had.

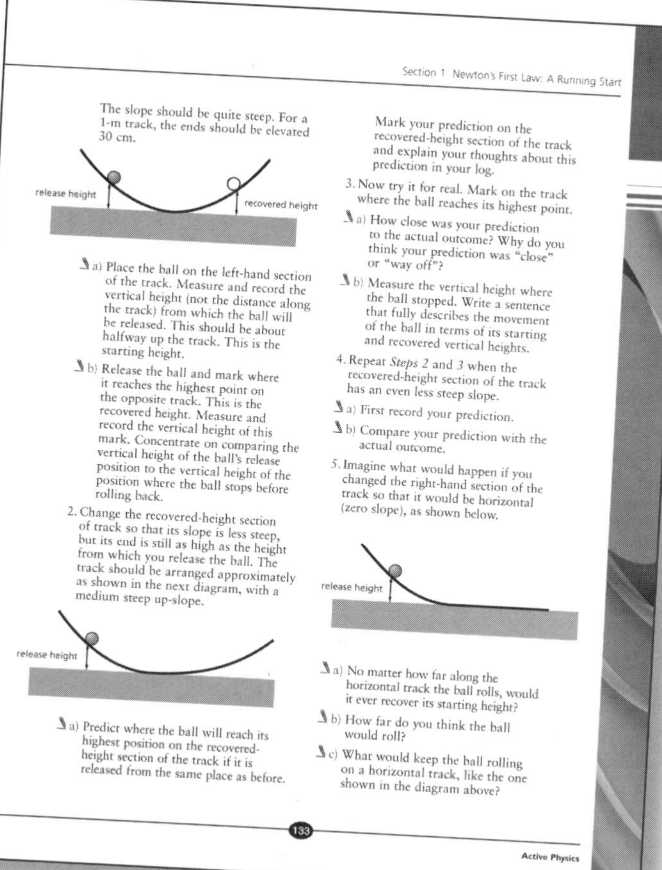
Section 1 Newton's First Law: A Running Start

The slope should be quite steep. For a 1-m track, the ends should be elevated 30 cm.

a) Place the ball on the left-hand section of the track. Measure and record the vertical height (not the distance along the track) from which the ball will be released. This should be about halfway up the track. This is the starting height.

b) Release the ball and mark where it reaches the highest point on the opposite track. This is the recovered height. Measure and record the vertical height of this mark. Concentrate on comparing the vertical height of the ball's release position to the vertical height of the position where the ball stops before rolling back.

2. Change the recovered-height section of track so that its slope is less steep, but its end is still as high as the height from which you release the ball. The track should be arranged approximately as shown in the next diagram, with a medium steep up-slope.

a) Predict where the ball will reach its highest position on the recovered-height section of the track if it is released from the same place as before. Mark your prediction on the recovered-height section of the track and explain your thoughts about this prediction in your log.

3. Now try it for real. Mark on the track where the ball reaches its highest point.

a) How close was your prediction to the actual outcome? Why do you think your prediction was "close" or "way off"?

b) Measure the vertical height where the ball stopped. Write a sentence that fully describes the movement of the ball in terms of its starting and recovered vertical heights.

4. Repeat *Steps 2* and *3* when the recovered-height section of the track has an even less steep slope.

a) First record your prediction.

b) Compare your prediction with the actual outcome.

5. Imagine what would happen if you changed the right-hand section of the track so that it would be horizontal (zero slope), as shown below.

a) No matter how far along the horizontal track the ball rolls, would it ever recover its starting height?

b) How far do you think the ball would roll?

c) What would keep the ball rolling on a horizontal track, like the one shown in the diagram above?

133

Active Physics

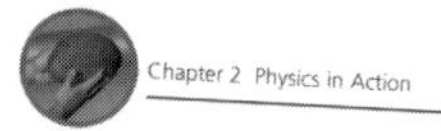
Chapter 2 Physics in Action

Physics Talk

NEWTON'S FIRST LAW OF MOTION

Galileo's Law of Inertia

In the *Investigate*, you observed, measured, and compared the release height of a ball on one side of the track to the recovered height on the other side of the track. You found that they were not exactly equal, but they were close to being equal.

Galileo Galilei (1564–1642) was an Italian physicist, mathematician, astronomer, and philosopher. Galileo is sometimes called the father of modern science. He introduced experimental science to the world. Galileo performed an experiment similar to the one you just completed. He observed that a ball that rolled down one ramp seemed to seek the same height when it rolled up another ramp.

Galileo also did a "thought experiment" in which he imagined a ball made of extremely hard material set into motion on a horizontal, smooth surface, similar to the final track in your investigation. He concluded that the ball would continue its motion on the horizontal surface with constant speed along a straight line "to the horizon" (forever).

From this, and from his observation that an object at rest remains at rest unless something causes it to move, Galileo formed the law of **inertia**: Inertia is the natural tendency of an object to remain at rest or to remain moving with constant speed in a straight line.

Galileo Galilei was a pioneer in the use of precise, quantitative experiments. He insisted on using mathematics to analyze the results of his experiments.

Galileo changed the way in which people viewed motion. Early on, people thought that all moving objects would stop. After Galileo, people thought about how moving objects might continue to move forever unless a **force**, a push or a pull, stopped them. That idea is not easy to understand. Any time you have pushed an object to move it, you have seen it stop. Nobody ever observes an object moving forever. Even when the surface is very, very smooth, the sliding or rolling objects eventually stop. However, Galileo realized that objects do not stop "on their own" but stop because there is a frictional force working that you cannot see and that is the force that stops the object.

Newton's First Law of Motion

Like Galileo, Isaac Newton was a great thinker. He was born in England in 1642, the year of Galileo's death. Newton's achievements brought him a great deal of recognition. Poems were written that honored Newton. Science, government, and philosophy all changed because of Newton's insights about the physics of the world.

Newton used Galileo's law of inertia as the basis for developing his **(Newton's) first law of motion**: In the absence of an unbalanced force, an object at rest remains at rest, and an object already in motion remains in motion with constant speed in a straight-line path.

Physics Words

inertia: the natural tendency of an object to remain at rest or to remain moving with constant speed in a straight line.

force: a push or a pull.

Newton's first law of motion: in the absence of an unbalanced force, an object at rest remains at rest, and an object already in motion remains in motion with constant speed in a straight-line path.

Active Physics 134

In any physics class, there will always be students with a range of interests, mathematical abilities, and motivation. In *Active Physics*, a program which provides access for all high school students, there will be an even wider range of students along these dimensions. This is why we introduce *Active Physics Plus* in each section. This section provides additional explorations, deeper analysis, more mathematics, or more content for some students. You can assign this to students who finish earlier than others, or suggest that students complete this for extra credit. It provides a way to keep all students engaged, while providing some students with more support on the concepts introduced earlier.

What Do You Think Now? revisits the original *What Do You Think?* questions and provides an opportunity for students to reflect on any changes in understanding. You can use this exercise in a similar manner with a written response in the *Active Physics* log, followed by a short classroom discussion.

One of the inherent difficulties in learning physics is connecting the new content to larger concepts that form the skeletal structure of physics, and that distinguish science from other disciplines. *Active Physics* helps students to see both the trees and the forest by focusing on the *Physics Essential Questions*. Students are asked, "What does it mean?" of one of the concepts introduced in the section. This is similar to the questions that are typically asked of students. Students are then asked to respond to "How do you know?" In response to this question, students refer back to some observational or experimental evidence that they have observed in the section. Students are then asked, "Why do we believe?" This takes one of three forms, which help students better understand how physicists view their discipline. Students are shown how the concept introduced in this section connects with other physics content, fits with the "big ideas in science," and meets physics requirements, such as experimental evidence being consistent with models and theories. Finally, students are asked, "Why should I care?" Research has shown that student success is strongly correlated with student engagement. By asking students why they should care, they are required to connect the relevance of this section's physics concepts to the *Chapter Challenge*.

Reflecting on the Activity and the Challenge provides a brief summary of the section and again relates the section to the larger challenge of creating a voice-over narration for a sporting event.

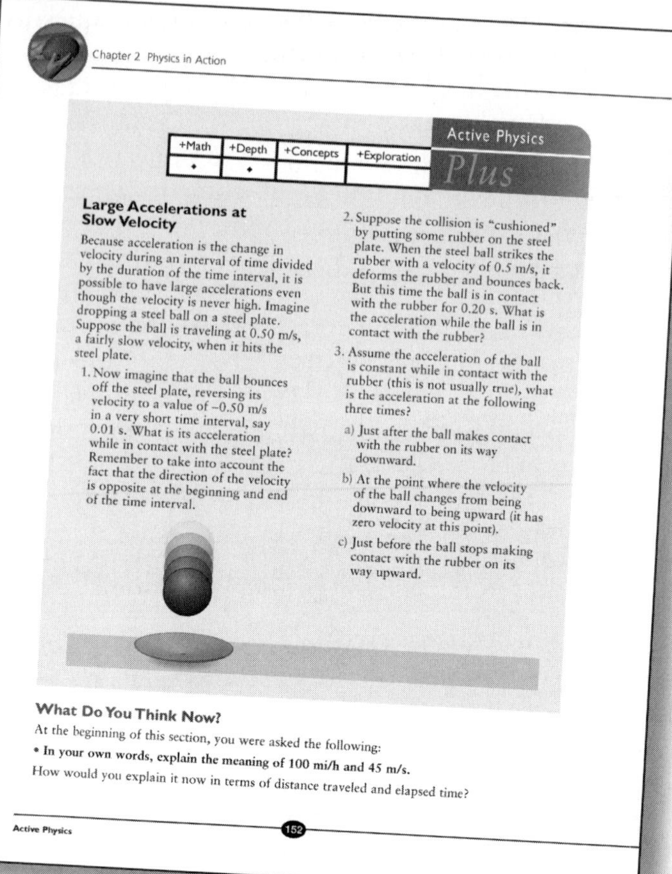
Chapter 2 Physics in Action

Active Physics Plus

+Math	+Depth	+Concepts	+Exploration
•	•		

Large Accelerations at Slow Velocity

Because acceleration is the change in velocity during an interval of time divided by the duration of the time interval, it is possible to have large accelerations even though the velocity is never high. Imagine dropping a steel ball on a steel plate. Suppose the ball is traveling at 0.50 m/s, a fairly slow velocity, when it hits the steel plate.

1. Now imagine that the ball bounces off the steel plate, reversing its velocity to a value of −0.50 m/s in a very short time interval, say 0.01 s. What is its acceleration while in contact with the steel plate? Remember to take into account the fact that the direction of the velocity is opposite at the beginning and end of the time interval.

2. Suppose the collision is "cushioned" by putting some rubber on the steel plate. When the steel ball strikes the rubber with a velocity of 0.5 m/s, it deforms the rubber and bounces back. But this time the ball is in contact with the rubber for 0.20 s. What is the acceleration while the ball is in contact with the rubber?

3. Assume the acceleration of the ball is constant while in contact with the rubber (this is not usually true), what is the acceleration at the following three times?

   a) Just after the ball makes contact with the rubber on its way downward.

   b) At the point where the velocity of the ball changes from being downward to being upward (it has zero velocity at this point).

   c) Just before the ball stops making contact with the rubber on its way upward.

What Do You Think Now?

At the beginning of this section, you were asked the following:

• **In your own words, explain the meaning of 100 mi/h and 45 m/s.**

How would you explain it now in terms of distance traveled and elapsed time?

Active Physics 152

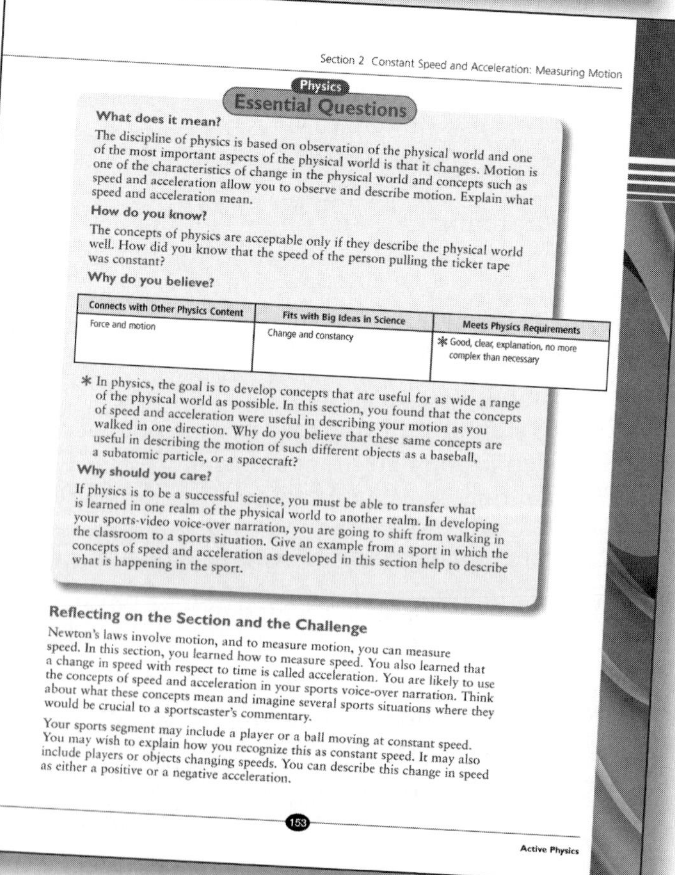
Section 2 Constant Speed and Acceleration: Measuring Motion

Physics Essential Questions

What does it mean?

The discipline of physics is based on observation of the physical world and one of the most important aspects of the physical world is that it changes. Motion is one of the characteristics of change in the physical world and concepts such as speed and acceleration allow you to observe and describe motion. Explain what speed and acceleration mean.

How do you know?

The concepts of physics are acceptable only if they describe the physical world well. How did you know that the speed of the person pulling the ticker tape was constant?

Why do you believe?

Connects with Other Physics Content	Fits with Big Ideas in Science	Meets Physics Requirements
Force and motion	Change and constancy	* Good, clear, explanation, no more complex than necessary

* In physics, the goal is to develop concepts that are useful for as wide a range of the physical world as possible. In this section, you found that the concepts of speed and acceleration were useful in describing your motion as you walked in one direction. Why do you believe that these same concepts are useful in describing the motion of such different objects as a baseball, a subatomic particle, or a spacecraft?

Why should you care?

If physics is to be a successful science, you must be able to transfer what is learned in one realm of the physical world to another realm. In developing your sports-video voice-over narration, you are going to shift from walking in the classroom to a sports situation. Give an example from a sport in which the concepts of speed and acceleration as developed in this section help to describe what is happening in the sport.

Reflecting on the Section and the Challenge

Newton's laws involve motion, and to measure motion, you can measure speed. In this section, you learned how to measure speed. You also learned that a change in speed with respect to time is called acceleration. You are likely to use the concepts of speed and acceleration in your sports voice-over narration. Think about what these concepts mean and imagine several sports situations where they would be crucial to a sportscaster's commentary.

Your sports segment may include a player or a ball moving at constant speed. You may wish to explain how you recognize this as constant speed. It may also include players or objects changing speeds. You can describe this change in speed as either a positive or a negative acceleration.

153 Active Physics

The section continues with a *Physics to Go* homework assignment. Here, students are asked about the specifics of the investigation and required to apply their knowledge of physics principles to new situations. Often, the homework will include an additional investigation, *Preparing for the Chapter Challenge*, which provides students with the chance to do some background work toward their final challenge.

Following the *Physics to Go* homework, many sections have *Inquiring Further* exercises. These exercises often require students to design and carry out an experiment. Typically, the *Inquiring Further* option will be more challenging and can be used for extra credit.

The chapter continues through additional sections where students are introduced to other physics concepts, such as Newton's second law and projectile motion.

In the middle of each chapter, you will find a *Chapter Mini-Challenge*. This provides the students an opportunity to try out one or two of the physics principles that they are considering using for their challenge – the voice-over narration of a sporting event. Students get a first opportunity to gauge the difficulty of the challenge. They also get to see how other teams are approaching the *Chapter Challenge*. They get to receive and provide feedback on their *Mini-Challenge*. Having this chance to see what works and what doesn't work makes the final challenge that much more successful. Such feedback is important for the *Engineering Design Cycle*.

The *Chapter Challenges* that the students complete are often a part of an actual job by real people. *Physics At Work* introduces the students to these people who use the physics in the chapter as part of their careers. Reading about their lives and jobs brings another facet of the importance of physics to students and may get them thinking about their own future career. As student-scientists, they are also encouraged through the *Physics Connections to Other Sciences* to appreciate the connections among various sciences to achieve a more in-depth understanding of nature.

The *Chapter Assessment* is framed around the *Engineering Design Cycle* and begins with a review of the sections and the key concepts. Then, it outlines the steps that the students may follow in completing the challenge. The chapter concludes as the students present one of their voice-over narrations to the class. Before the students present their voice-over narrations

Physics to Go

1. If the launching and landing heights for a projectile are equal, what angle produces the greatest range? Why?
2. Compared to a launch angle of 45°, what happens to the amount of time a projectile is in the air if the launch angle is
   a) greater than 45°?
   b) less than 45°?
3. For a constant launch speed, what angle produces the same range as a launch angle of
   a) 30°?
   b) 15°?
4. Analyses of performances of long jumpers has shown that the typical launch angle is about 18°, far less than the angle needed to produce maximum range. Why do you think this occurs?
5. You might be familiar with Carl Lewis as a medal-winning sprinter. But he is also an Olympic gold medalist in the long jump. Why do you think he was successful in both events?
6. The diagram below shows a ball thrown toward the east and upward at an angle of 30° to the horizontal. Point X represents the ball's highest point.

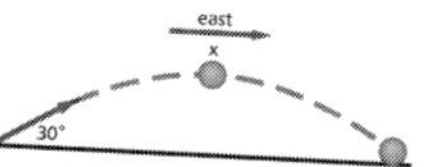

   a) What is the direction of the ball's acceleration at point X? (Ignore friction.)
   b) What is the direction of the ball's velocity at point X?
7. A diver jumps horizontally off a cliff with an initial velocity of 5.0 m/s. The diver strikes the water 3.0 s later.
   a) What is the vertical speed of the diver upon reaching the surface of the water?
   b) What is the horizontal speed of the diver 1.0 s after the diver jumps?
   c) How far from the base of the cliff will the diver strike the water?

Active Physics 194

Chapter Mini-Challenge

Your challenge for this chapter is to create a sports voice-over for the public broadcast service that will engage viewers and introduce physics concepts. Your commentary should be two or three minutes of engaging information that will educate the viewer on the laws of nature governing the sport they are watching.

For your initial design you will need the following:

- A sport that contains physics concepts you have studied
- The physics concepts you have studied linked to the sport
- Appropriate use of physics equations and terminology
- Proper units and approximate values for the magnitude of concepts relating to the sport
- Two to three minutes of live voice-over or recorded voice-over
- A written script of the narration

GOAL · INPUTS · PROCESS · OUTPUTS · FEEDBACK · ENGINEERING DESIGN CYCLE

You still have more to learn before you can complete the challenge but this is a good time to give the *Chapter Challenge* a first try. It will give you a good sense of what the challenge entails and how you and other teams are going to approach your "broadcasting job." Your *Mini-Challenge* for this chapter is to develop and present a one-minute voice-over narration to explain the physics behind the sport that you will be broadcasting. At this point you have a handful of physics topics to choose from and the entire world of sports to apply them to.

Go back and quickly review the *Goal* you established at the start of your chapter. This *Mini-Challenge* offers a unique opportunity because it allows you to complete a full trial-run of the *Chapter Challenge* with the physics information you have learned so far. As you learn additional physics information in the remaining sections you can add that to your *Mini-Challenge* voice-over or you could create an entire second voice-over, even choosing a different sport if you want to.

In the *Engineering Design Cycle*, you are adding a critical *Input* by choosing the sport and the specific sports action that you will be describing. You also have the new physics knowledge you have gained from *Sections 1-5* in this chapter which you should review to help you compose your sports voice-over.

Section 1: You investigated Galileo's principle of inertia and learned about the mass of an object and how mass is related to the concept of inertia. You also read about Newton's first law and reference frames for measuring the speed of an object.

Section 2: You measured speed by making speed vs. time graphs using a ticker timer for objects with constant speeds and objects with changing speeds. You also explored the concept of acceleration, or the rate that speed is increasing or decreasing.

INTRODUCTION

to the class for the *Chapter Challenge*, the *Criteria* should be reviewed by the class and finalized. The final summation is titled *Physics You Learned* and lists key concepts from the chapter in the meaningful context of a sentence. Presenting the *Chapter Challenge* has the following important learning outcomes:

- Students review and increase their learning of the physics concepts in preparation for completing the challenge.
- Students get to review each concept in different contexts as they observe other teams present.
- Students are motivated because they are engaged in the creative aspects of the *Chapter Challenge*.
- Students demonstrate their expertise of how they have found ways to relate physics to their personal challenge execution.
- Students can help achieve equity in that students of different backgrounds and cultures can provide insights into their backgrounds through their *Chapter Challenge*.

In addition to the *Chapter Challenge* assessment, traditional assessment questions and problems are also provided in the *Physics Practice Test*. This is in addition to the hundreds of traditional questions and problems in each section. Students are reminded of these when the *Physics Practice Test* is presented.

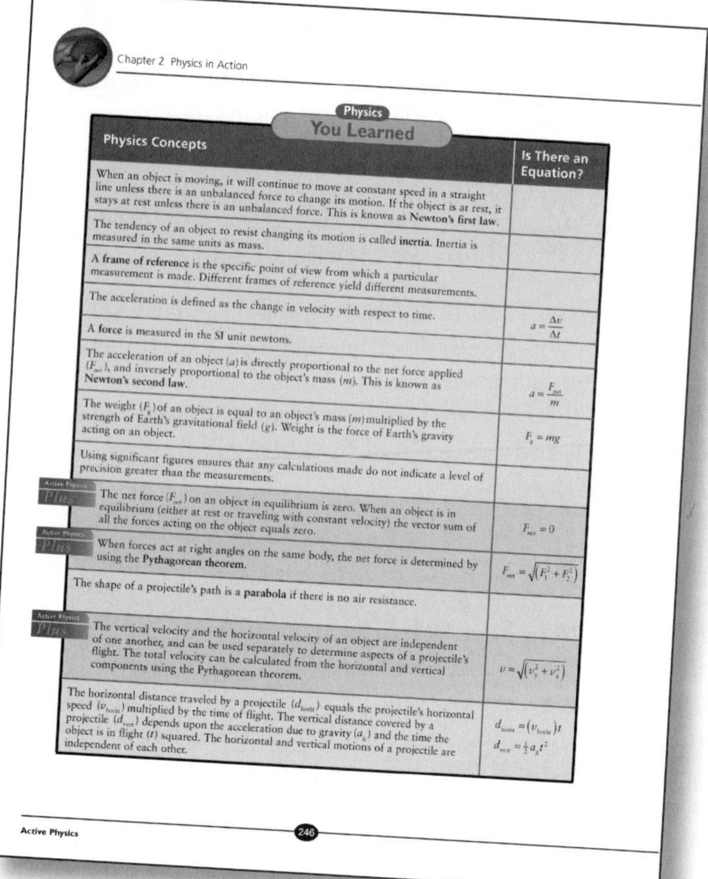

Chapter 2 Physics in Action

Physics You Learned

Physics Concepts	Is There an Equation?
When an object is moving, it will continue to move at constant speed in a straight line unless there is an unbalanced force to change its motion. If the object is at rest, it stays at rest unless there is an unbalanced force. This is known as **Newton's first law.**	
The tendency of an object to resist changing its motion is called **inertia.** Inertia is measured in the same units as mass.	
A **frame of reference** is the specific point of view from which a particular measurement is made. Different frames of reference yield different measurements.	
The acceleration is defined as the change in velocity with respect to time.	$a = \frac{\Delta v}{\Delta t}$
A **force** is measured in the SI unit newtons.	
The acceleration of an object (a) is directly proportional to the net force applied (F_{net}), and inversely proportional to the object's mass (m). This is known as **Newton's second law.**	$a = \frac{F_{net}}{m}$
The weight (F_g) of an object is equal to an object's mass (m) multiplied by the strength of Earth's gravitational field (g). Weight is the force of Earth's gravity acting on an object.	$F_g = mg$
Using significant figures ensures that any calculations made do not indicate a level of precision greater than the measurements.	
Active Physics Plus: The net force (F_{net}) on an object in equilibrium is zero. When an object is in equilibrium (either at rest or traveling with constant velocity) the vector sum of all the forces acting on the object equals zero.	$F_{net} = 0$
Active Physics Plus: When forces act at right angles on the same body, the net force is determined by using the **Pythagorean theorem.**	$F_{net} = \sqrt{(F_1^2 + F_2^2)}$
The shape of a projectile's path is a **parabola** if there is no air resistance.	
Active Physics Plus: The vertical velocity and the horizontal velocity of an object are independent of one another, and can be used separately to determine aspects of a projectile's flight. The total velocity can be calculated from the horizontal and vertical components using the Pythagorean theorem.	$v = \sqrt{(v_y^2 + v_x^2)}$
The horizontal distance traveled by a projectile (d_{horiz}) equals the projectile's horizontal speed (v_{horiz}) multiplied by the time of flight. The vertical distance covered by a projectile (d_{vert}) depends upon the acceleration due to gravity (a_g) and the time the object is in flight (t) squared. The horizontal and vertical motions of a projectile are independent of each other.	$d_{horiz} = (v_{horiz})t$ $d_{vert} = \frac{1}{2}a_g t^2$

Active Physics 246

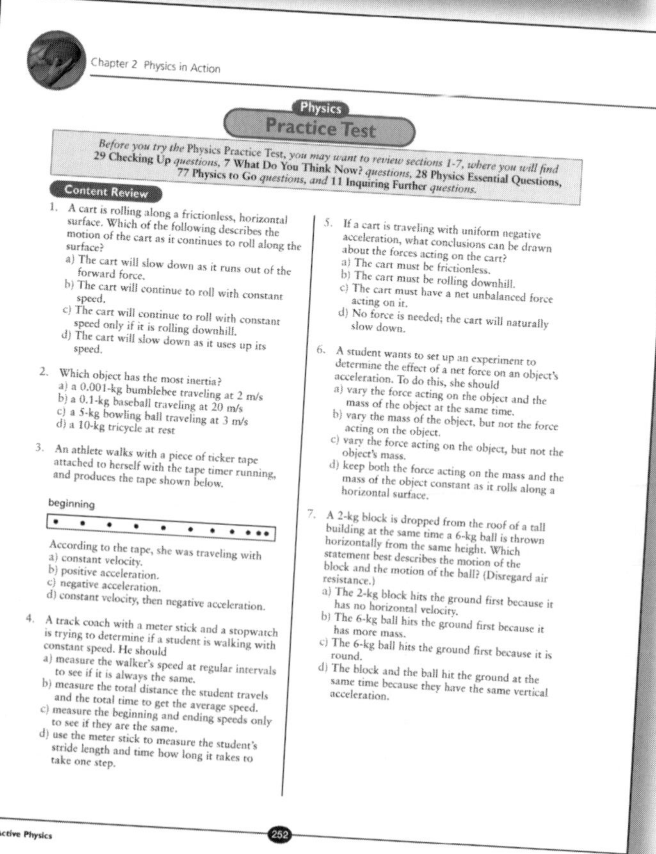

Chapter 2 Physics in Action

Physics Practice Test

Before you try the Physics Practice Test, *you may want to review sections 1-7, where you will find* **29 Checking Up** *questions,* **7 What Do You Think Now?** *questions,* **28 Physics Essential Questions, 77 Physics to Go** *questions, and* **11 Inquiring Further** *questions.*

Content Review

1. A cart is rolling along a frictionless, horizontal surface. Which of the following describes the motion of the cart as it continues to roll along the surface?
   a) The cart will slow down as it runs out of the forward force.
   b) The cart will continue to roll with constant speed.
   c) The cart will continue to roll with constant speed only if it is rolling downhill.
   d) The cart will slow down as it uses up its speed.

2. Which object has the most inertia?
   a) a 0.001-kg bumblebee traveling at 2 m/s
   b) a 0.1-kg baseball traveling at 20 m/s
   c) a 5-kg bowling ball traveling at 3 m/s
   d) a 10-kg tricycle at rest

3. An athlete walks with a piece of ticker tape attached to herself with the tape timer running, and produces the tape shown below.

   beginning

   According to the tape, she was traveling with
   a) constant velocity.
   b) positive acceleration.
   c) negative acceleration.
   d) constant velocity, then negative acceleration.

4. A track coach with a meter stick and a stopwatch is trying to determine if a student is walking with constant speed. He should
   a) measure the walker's speed at regular intervals to see if it is always the same.
   b) measure the total distance the student travels and the total time to get the average speed.
   c) measure the beginning and ending speeds only to see if they are the same.
   d) use the meter stick to measure the student's stride length and time how long it takes to take one step.

5. If a cart is traveling with uniform negative acceleration, what conclusions can be drawn about the forces acting on the cart?
   a) The cart must be frictionless.
   b) The cart must be rolling downhill.
   c) The cart must have a net unbalanced force acting on it.
   d) No force is needed; the cart will naturally slow down.

6. A student wants to set up an experiment to determine the effect of a net force on an object's acceleration. To do this, she should
   a) vary the force acting on the object and the mass of the object at the same time.
   b) vary the mass of the object, but not the force acting on the object.
   c) vary the force acting on the object, but not the object's mass.
   d) keep both the force acting on the mass and the mass of the object constant as it rolls along a horizontal surface.

7. A 2-kg block is dropped from the roof of a tall building at the same time a 6-kg ball is thrown horizontally from the same height. Which statement best describes the motion of the block and the motion of the ball? (Disregard air resistance.)
   a) The 2-kg block hits the ground first because it has no horizontal velocity.
   b) The 6-kg ball hits the ground first because it has more mass.
   c) The 6-kg ball hits the ground first because it is round.
   d) The block and the ball hit the ground at the same time because they have the same vertical acceleration.

Active Physics 252

Features of Active Physics

1. Scenario

Each *Active Physics* chapter opens with an engaging scenario. Students from diverse backgrounds and localities have been interviewed in order to find situations that are not only realistic, but meaningful to the high school population. The scenarios set the stage for the description of the *Chapter Challenge* that immediately follows. You may choose to read the *Scenario* aloud to the class as a way of introducing the new chapter.

2. Your Challenge

The *Chapter Challenges* are the heart and soul of *Active Physics*. They provide students with a purpose for all of the work to follow and the rationale for learning. One of the most common questions that teachers hear from students is, "Why am I learning this?" In *Active Physics*, students do not raise this question. Similarly, you do not have to answer, "Because someday it will be useful to you." The question is already answered because, on day one of the chapter, students are presented with a challenge that, in essence, becomes their job for the next few weeks.

The beauty of the challenges lies in the variety of tasks and opportunities that allows students with different talents and skills to excel. Students who express themselves artistically will have an opportunity to shine in a chapter such as *Atoms on Display*. Students who enjoy sports may find *Physics in Action* a favorite. Other students who enjoy being "on-stage" may find their niche with *Let Us Entertain You*.

The challenges are not contrived situations. For example, engineers who build roller-coasters base their designs on the same physics principles students will learn in *Thrills and Chills*. Students' challenge in *Electricity for Everyone* is to address housing and electricity needs for families in different areas throughout the world. The challenge in *Sports on the Moon* requires students to create, invent, or adapt a sport that can be played on the Moon. This challenge has been successfully completed by ninth-grade high school students, twelfth-grade students, and by NASA engineers. The expectation may be different for each of these audiences, but the challenge is consistent.

3. Criteria for Success

In creating *Active Physics*, the original thought was that the generation of the challenge was enough and that all else would fall in place. Upon reflection, it was soon realized that the *Criteria for Success* must also be included. When students agree to the matrix by which their work will be measured, research has shown that the students will perform better and achieve more. It makes sense. In the simplest situation of cleaning a lab room, the teacher may simply state, "Please clean up the lab." The results are often a minimal cleanup. If the teacher begins by asking, "What does a clean lab room look like?" and students and teacher jointly list the attributes of a clean lab room (e.g., no paper on the floor, all equipment put away, all materials placed on the back of the lab tables, and all power supplies unplugged), the students respond differently and the cleanup is better. When students are asked to include physics principles in the explanation of the challenge, the students should know whether the expectation is for three physics principles or five.

The discussion of grading criteria and the creation of a grading rubric is a crucial ingredient for student success. After the introduction of the challenge, *Active Physics* requires a class discussion about the grading criteria. How much is required? What does an "A" challenge product look like? Should creativity be weighed more than delivery? The criteria will be visited again at the end of the chapter, but at this point it provides clarity to the challenge and helps establish an expectation level that the students will have set for themselves.

4. Engineering Design Cycle

Students are introduced to the *Engineering Design Cycle* to help them solve the problems necessary to successfully complete the *Chapter Challenge*. They will implement the five basic steps of the cycle (Goal, Inputs, Process, Outputs, and Feedback) throughout the chapter, culminating in the completion of the *Chapter Challenge*. Once students are presented with their challenge, they begin the first step of the cycle, which is to establish a clear Goal. Their Goal is established by defining the problem, identifying available resources, drafting potential solutions, and listing constraints and possible actions. As they experience each one of the chapter sections, they will be gaining Inputs to use in the design cycle. At the midway point of the chapter, students will get an opportunity to test their ideas and their approach to the challenge by completing the *Mini-Challenge*. Their *Mini-Challenge* will be their Output for the design cycle, and will also provide them with Feedback about which parts need to be refined. Students will repeat the *Engineering Design Cycle* during the second half of the chapter to help them prepare for the *Chapter Challenge*.

5. Physics Corner

The physics content in each chapter is summarized in a preview cartoon. The cartoon shows the students focused on the investigations and the challenge while the teacher sees the physics that the students will learn. The *Physics Corner* acts as a preview of all the physics concepts the chapter will present.

6. What Do You See?

A cartoon introduces each section and asks the question "*What Do You See?*" The students are given a humorous preview of some of the concepts and activities that the section will provide. You can use this to engage the students' attention and interest in the section. From the discussion, you can also observe what the students already know about the topics. It is not important that student responses are correct or completely relevant.

7. What Do You Think?

During the past few years, much has been written about a constructivist approach to learning. Videos of Harvard graduates, in caps and gowns, show that the students are not able to explain correctly why it is colder in the winter than it is in the summer. These students have previously answered these questions correctly in fourth grade, in middle school, and then again in high school. How else would they have gotten into Harvard? One theory for this is that they never internalized the logic and understanding of the seasons. Part of this problem is that they were never confronted by what they did believe, and were never adequately shown why they should give up that belief system. Certainly, it is worth writing down a "book's perfect answer" on a test to secure a good grade, but to actually believe requires a more thorough examination of competing explanations.

The best way to ascertain a student's prior understanding is through extensive interviewing. Much of the research literature in this area includes the results of such interviews. In a classroom, however, this one-on-one dialogue is rarely possible. The *What Do You Think?* question introduces each section in a way that elicits prior understandings. It gives students an opportunity to verbalize what they think about friction, energy, or light, before they embark on an investigation. The brief discussion of the range of answers brings the students a little closer with that part of their brains which understands friction, energy, or light. The *What Do You Think?* question is not intended to produce a correct answer or a discussion of the features of each question. It is not intended that you bring the discussion to closure. The *Investigate* that follows will provide that discussion as experimental results are analyzed. The *What Do You Think?* question should take no more than a few minutes of class time. It sets the stage for the physics investigation.

Students should be strongly encouraged to write their responses to the questions in their *Active Physics* logs to ensure that they have, in fact, addressed their prior conceptions. After students have discussed their responses in their small groups, activate a class discussion. Ask students to volunteer other students' answers that they found interesting. This may encourage students to exchange ideas without the fear of personally giving a "wrong" answer.

8. Investigate

Active Physics is a hands-on, minds-on curriculum. Students *do* physics; they do not read about doing physics. Each *Investigate* has instructions for each step of the experiment. Students are reminded that data, hypotheses, and conclusions should be recorded in their logs.

Investigates are an opportunity for students to garner the knowledge that they will need to complete the *Chapter Challenge*. Students will understand the physics principle involved because they have investigated it. In *Active Physics*, if a student is asked, "How do you know?" the response is, "Because I did an experiment!"

Recognizing that many students know how to read, but do not like reading, background information is often provided within the context of the section. Students have demonstrated that they will read when the information is required for them to continue with their exploration.

Occasionally, the *Investigate* will require the entire class to participate in a large, single demonstration simultaneously. On other occasions, you may decide that a specific *Investigate* is best done as a teacher demonstration. This would be appropriate if there is limited equipment for that one *Investigate* or the facilities are not available for every student to do the *Investigate*. Viewing demonstrations on an ongoing basis, however, is not what *Active Physics* is about.

Icons throughout the *Investigates* alert students to safety issues that should be given full attention.

⚠ Students are reminded of all safety rules throughout the program. An overview of safety rules for the physics classroom, as well as a safety contract, is available at the end of the *Teacher's Edition* front matter, and is provided as a Blackline Master in your *Teacher Resources CD*.

Most of the *Investigates* require between one and two 45-minute class periods. Considering current trends in class scheduling, there are so many time structures that it is difficult to predict how *Active Physics* will best fit your schedule. The other impact on time is the achievement and preparation level of your students. For example, in a given *Investigate*, students may be required to complete a graph of their data. This is considered one small part of the *Investigate*. If the students have never been exposed to graphing, this could require a two-period lesson to teach the basics of graphing with suitable practice in interpretation. *Active Physics* is accessible to all students. You are in the best position to make accommodations in time that reflect the needs of your students.

9. Physics Talk

Sometimes it is difficult for students to make the conceptual leap from doing an investigation to connecting the ideas into a physics principle. Indeed, if you consider the theory of multiple intelligences, some students grasp concepts more easily by reading. *Physics Talk* summarizes the physics principles and includes physics formulas and equations when appropriate. It presents students with text, illustrations, historical perspective, and photographs that provide greater insight into the physics concepts presented. If sample problems or important laws are integral to the lesson, they can be found in the *Physics Talk*.

10. Physics Words

Science has a language of its own and some concepts are more efficiently communicated when vocabulary is introduced. Certainly, in order to fully participate in science, students need to have an understanding of this language. As a part of the *Physics Talk*, the *Physics Words* highlight the important terms for students. These words are pulled outside the text area and redefined. Such support helps students understand the process of learning as they note important ideas within the text. All *Physics Words* will also be found in the *Glossary*.

11. Checking Up

Another important part of the *Physics Talk* section is the *Checking Up* questions. Each *Physics Talk* ends with several of these questions about the section in the margins outside the text. Students can use these for self-examination to make certain that they understand the most important principles. In addition, you can assign these as homework or *Active Physics* log entries for students to answer as the *Physics Talk* section is read.

12. Active Physics Plus

Active Physics Plus is an opportunity to explore physics in a variety of ways. It provides text, sample problems, and questions for students who want or need more mathematics, depth, concepts, or exploration opportunities. As a teacher, you can use the *Active Physics Plus* to accommodate your state requirements and to guide individual students' instruction to meet appropriate personal challenges so that all students extend their knowledge and skills.

13. What Do You Think Now?

This section offers the students (and you) the opportunity to return to the *What Do You Think?* questions to informally assess changes in knowledge and understanding. This reflection is beneficial to both the students and to you. If you see widespread lapses in understanding, you may spend some time helping the students to understand. Otherwise, you may want to wait for the opportunity to use the spiraling design of the curriculum to foster greater understanding.

14. Physics Essential Questions

Physics can be a difficult study for many students and one of the reasons for this is the lack of organizing principles as it is traditionally taught. For many students, physics is a vast collection of unrelated facts which one needs to memorize for the test. *Active Physics* has developed four questions that capture the organizing principles of physics for the student and make understanding more accessible.

- What does it mean?
- How do you know?
- Why do you believe?
- Why should you care?

If the students can answer these four questions (and they can), then they will likely keep their newly learned principles and concepts forever.

The first essential question, "*What does it mean?*" requires students to describe the content of the section based on what they have learned in their investigation and reading.

The second essential question that the students are asked, "*How do you know?*" is answered by a description of the experimental evidence that they discovered during the *Investigate*. Students "know" because they have done an experiment.

The third essential question, "*Why do you believe?*" emphasizes one of three ideas ("Connects with Other Physics Content," "Fits with Big Ideas in Science," or "Meets Physics Requirements") to help students understand physics as it relates to the world outside the classroom.

Finally, the last question, "*Why should you care?*" requires students to make a direct line from the section to the *Chapter Challenge*. This serves two purposes: the students have an immediate need to know this information, and also to begin planning a response to the *Chapter Challenge*.

15. Reflecting on the Activity and the Challenge

At the close of each section, students are often so involved with the completion of the single experiment that the larger context of the investigation is lost. *Reflecting on the Activity and the Challenge* is an opportunity for students to place their new insights and information into the context of the chapter and the *Chapter Challenge*. If the *Chapter Challenge* is considered a completed picture, each section is a jigsaw-puzzle piece.

By completing half of the sections, the students will be able to fit puzzle pieces together and complete the *Mini-Challenge*. This *Reflecting on the Activity and the Challenge* section ensures that the students do not forget about the larger context and continue their personal momentum toward completion of the *Chapter Challenge*.

16. Physics to Go

This section provides additional questions and problems that can be completed outside of class as homework. Some of the problems are applications of the principles involved in the preceding *Investigate*. Others are replications of the work in the *Investigate*. Still others provide an opportunity to transfer the results of the investigation to the context of the *Chapter Challenge*. The *Physics to Go* questions provide a means by which students can be working on the physics principles behind the larger *Chapter Challenge*. Often, the homework will include an additional activity, *Preparing for the Chapter Challenge*, which provides students with the chance to do some background work toward their final challenge. You may wish to assign these questions to the entire class or to those students who would benefit most from the extra work.

17. Inquiring Further

Following the *Physics to Go* homework, many sections have *Inquiring Further* exercises. These exercises often require students to design and carry out an experiment. Typically, the *Inquiring Further* option will be more challenging and can be used for extra credit.

The outcome of good science instruction should be the ability of students to transfer knowledge to a different problem or task. *Inquiring Further* exercises give students the opportunity to stretch their thinking beyond the textbook and classroom setting. This can be accomplished with a provocative question or problem to solve. Students can put into practice the technique, approaches, and knowledge they have acquired by completing the section and expand upon it to gain new information. The *Inquiring Further* exercises can be assigned as independent study or as a class extension to the investigation.

INTRODUCTION

18. Chapter Mini-Challenge

As a part of a focus on the *Engineering Design Cycle*, you will find a *Chapter Mini-Challenge* in the middle of each chapter. This provides the students an opportunity to set some preliminary goals for their challenge. Using the input from the first four sections of the chapter, they spend some time on the process of integrating the input and goals to design a product for testing. The *Mini-Challenge* presentation serves as the output for their plan. They will receive feedback from the experience, the other students, and you. They will use this feedback later when they refine the design of their final product in the *Chapter Challenge*.

19. Physics You Learned

This section at the end of the chapter provides a list of physics concepts that were studied in the context of the *Investigates* and the *Physics Talk*. It provides students with a sense of accomplishment and serves as a quick review of all that was learned during the preceding weeks. Students can also use this list to select those physics principles which are appropriate for their *Chapter Challenge* project. When describing any physics phenomenon, student physicists should ask themselves, "Is there an equation?" As a reminder, the *Physics You Learned* also provides a checklist of all the equations used in each chapter.

20. Physics Chapter Challenge

The *Physics Chapter Challenge* is the return to the *Your Challenge*, *Criteria for Success*, and the *Engineering Design Cycle*. The students are now ready to complete the challenge. They are able to view the challenge with a clarity that has emerged from the completion of the *Investigates* in each section and the feedback from the *Mini-Challenge*. Students are able to review the chapter as they discuss the synthesis of the information into the required context of the challenge. They should have some class time to work together to complete the challenge and to present their project. In many physics courses, all students are expected to converge on the same solution. In *Active Physics*, each group is expected to have a unique solution. All solutions must have correct physics, but there is ample room for creativity on the students' part. This is one of the features that captures the imagination of students who have often previously chosen not to enroll in physics classes.

Business leaders want to hire people who know how to work effectively in groups and how to complete projects. The *Physics Chapter Challenge* provides guidance on how to begin work on the final *Chapter Challenge* product, to set deadlines, meet all the requirements, and combine the contributions of all members of the team. This section is intended to guide, rather than direct, the student work.
You will find that the best projects will reflect the diverse interests, backgrounds, and cultures of the team members.

21. Physics Connections to Other Sciences

The *Physics Connections to Other Sciences* highlights how the fundamental ideas students learn in each chapter relate to many of the other sciences students will study. By appreciating the connections among various science disciplines, scientists develop a more in-depth understanding of nature.

22. Physics At Work

This section highlights three individuals whose work or hobby is illustrative of the physics in the *Chapter Challenge*. The *Physics At Work* speaks to the authenticity of the challenge. The profiles show how interesting and useful physics is and how important it can be in a variety of careers, and people of all walks of life.

23. Physics Practice Test

The *Physics Practice Test* provides additional assessment opportunities in the format of the traditional state assessments. Students can use the *Physics Practice Test* as a gauge to find out how well they learned the physics contained in each chapter. Before beginning the test, students should be reminded to go back to their *Active Physics* logs and homework to review the hundreds of traditional questions and problems they have answered throughout the chapter in the form of the *Physics to Go*, *Checking Up*, *What Do You Think Now?*, and *Physics Essential Questions*.

Overview of Meeting the Needs of All Students: Differentiated Instruction

Using Augmentation and Accommodations

This feature is designed for you to help struggling students. Every section has its own table with specific information that identifies learning issues in the classroom. The table also gives the location of the issue within each component of a section and recommends instructional and management strategies for helping students compensate for their difficulties. Compensation techniques have been labeled "**Accommodations,**" and the teaching of skills has been labeled "**Augmentation.**"

To understand how augmentations and accommodations can be used as tools by teachers to help struggling students with each section, take an example from *Section 3* in *Safety*. Assume some students in the class find it difficult to follow complex multi-step directions. What will these students learn about energy and work when they cannot comprehend the specific directions embedded in the *Investigate*? Their confusion impedes learning and leads to the distraction of other students. To prevent this from happening, you can either provide the students with directions they can understand (accommodation) or teach them direction-following skills (augmentations).

Augmentation of prerequisite skills involves the direct teaching of the skill. In *Section 3*, *Investigate*, *Step 6*, you might take a group of students and show them how to measure indentations in the landing material for different heights from which an egg is dropped. The advantages of augmentation are that students begin to develop the skills they need to learn independently. The disadvantage is that other students who are allowed to move through an investigation independently will finish sooner, possibly losing interest or motivation as they wait. The best plan is to differentiate the investigation at that point, providing the independent students with a related *extension* or an *anchoring task* while the other students learn to read the directions. Extensions have been provided in the lesson under the heading of *Inquiring Further*. Anchoring tasks are those that can be done independently by everyone. For example, highlighting important aspects of the *Physics Talk* by creating an outline or recording new vocabulary and researching its meaning could serve as anchoring tasks for this lesson.

Accommodations for students have the advantage of allowing them to move along with the group using a teacher-provided scaffold. If students find it difficult to follow directions, you could place students with poor reading skills with other students who are better in that area. You could provide a different set of directions in which the language is simpler, each step is labeled, and a space is provided for the tasks required. The disadvantage of doing this is that a student will most likely need this accommodation for every set of directions until the skill is learned.

Introduction to Strategies for Students with Limited English-Language Proficiency

The *Strategies for Students with Limited English-Language Proficiency* augmentations allow you to adjust your teaching strategies and differentiate them for English-language learners. Language-learning skills that actively engage students in learning new science concepts are provided for each section. These skills are called augmentations because they augment existing frameworks of teaching and learning. To develop proficiency in understanding and using new words, students are given the opportunity to explore words in context. You are provided with language-building strategies that focus on unfamiliar words and help students meet the linguistic challenges of a particular topic. Also, the use of discussion as a means of broadening oral language experiences and bridging content development adds a richer dimension of learning for students and facilitates instruction. When you plan your lessons, you can consult the augmentations to determine where it would be best to incorporate these strategies.

Students' Prior Conceptions: Overview and 7E Tools

Research in cognitive learning and how students establish foundations for understanding science have identified various categories of student conceptions that are alternative frameworks for understanding science. Definitions for these categories are cited below:

Preconceived notions—Popular conceptions rooted in everyday experiences.

Example: Many people believe that water flowing underground must flow in streams because the water they see at Earth's surface flows in streams.

Nonscientific beliefs—Include views learned by students from sources other than scientific education, such as religious or mythical teachings.

Example: Some students have learned through religious instruction about an abbreviated history of Earth and its life forms. The disparity between this widely held belief and the scientific evidence for evolution, dating back to pre-historical times, has led to considerable controversy in the teaching of science.

Conceptual misunderstandings—Arise when students are taught scientific information in a way that does not motivate them to confront paradoxes and conflicts resulting from their own preconceived notions and nonscientific beliefs.

Example: To deal with their confusion, students construct faulty models that usually are so weak that the students themselves are insecure about the concepts.

Vernacular misconceptions—Arise from the use of words that mean one thing in everyday life and another in a scientific context (e.g., "work").

Example: A geology professor noted that students have difficulty with the idea that glaciers retreat, because they picture the glacier stopping, turning around, and moving in the opposite direction. Substitution of the word "melt" for "retreat" helps reinforce the correct interpretation that the front end of the glacier simply melts faster than the ice advances.

Factual misconceptions—False conceptions often learned at an early age and retained unchallenged into adulthood.

Example: The idea that "lightning never strikes twice in the same place" is clearly nonsense, but that notion may be buried somewhere in a student's belief system.

In order for teachers to break down student prior conceptions and align their knowledge to current theory they must:

- Identify students' prior knowledge.
- Provide a forum for students to confront their misconceptions.
- Help students reconstruct and internalize their knowledge, based on scientific models.

The *What Do You Think?* and *Investigates* and sections presented in *Active Physics* are designed so that teachers are informed about basic student prior knowledge and have the tools with which to break down alternative ideas that may be strongly held by students.

Each chapter of *Active Physics* identifies lists of commonly held student preconceptions and offers opportunities for students to test their conceptual ideas. Students continually are asked to explain their reasoning, which enables you to work with students to root out prior conceptions.

Active Physics uses a 7E learning-cycle model. The steps (phases) of the 7E learning cycle are Elicit, Engage, Explore, Explain, Elaborate, Extend, and Evaluate.

The Elicit, Explain, Elaborate, and Extend phases of this learning cycle offer valuable tools for you to interview students to ascertain if prior belief systems leading to misconceptions continue to exist. You have the opportunity to encourage students to compare their preconceptions with their data analysis of the Explore stage of the learning cycle. The critical question, "*Why do you believe?*" and the Evaluate phase enable students to root out personally held misconceptions and to align new conceptual knowledge with current scientific thinking. Constant and open classroom discussions are powerful tools when wielded wisely through the listening and evaluative ear of a classroom educator. The more you understand about how students learn, the more effective you become at crafting classroom interactions to deliver the most effective learning.

Pacing Guides for Teachers

The chart on the next page is intended to show the approximate number of weeks it will take for your class to cover each *Active Physics* chapter in either the standard or accelerated mode. Your pace may be slower or faster depending on a number of variables, including the grade level of the students and your experience in using the *Active Physics* curriculum. In addition, the accelerated column is intended to provide flexibility as a reduced-equipment option, as well as a reduced-time option. Which column you choose will depend upon the time available for your students to complete the work, the academic level of the students, and the equipment available. Columns A and B contain numbers for weeks and days for each chapter in a typical classroom. In general, the average time for each chapter is five and one-half weeks.

This chart is only intended to provide an overview of the time required for the entire year. For a more detailed description of what is suggested for each chapter, see the pacing guide for that chapter. Those guides include an outline of which investigations are recommended for your students to do. It is expected that as you gain experience with the program, fewer of the investigations will be teacher-centered and your students will do more of the investigations independently. It is recommended that you consult these pacing guides prior to making your decision about which chapters to include in your program. Based on an academic year of 180 days, or 36 weeks, you can cover about seven nonaccelerated chapters in a full year. Columns C, D, and E provide three examples using seven chapters in a school year. All of these examples can be completed within a 36-week academic calendar.

To cover all of the national standards, the first eight chapters are recommended. You should make selections based upon chapter content and alignment with state standards and your school's requirements. The table also provides an accelerated schedule covering eight or more chapters in the same 36-week school year. In columns F, G, and H, three examples are shown which use eight or nine chapters. Again, these schedules fit within a school year, allowing time for school events, testing periods, and other interruptions of the schedule.

Mixing accelerated and regular chapters would be another option for a teacher who may not be able to complete the chapters in the recommended time, but still wishes to complete the first eight chapters.

NOTES

Chapter	A Standard	B Accelerated	C Standard	D Standard	E Standard	F Accelerated	G Accelerated	H Accelerated
	# Weeks*	# Weeks*	Weeks (7 Chapters) Option 1	Weeks (7 Chapters) Option 2	Weeks (7 Chapters) Option 3	Weeks (8 Chapters) Option 1	Weeks (8 Chapters) Option 2	Weeks (9 Chapters) Option 3
1 Driving the Roads	5	4.5	5	5	5	4.5	4.5	4.0
2 Physics in Action	5.5	5	5.5	5.5		5.0		4.2
3 Safety	5	4	5		5	4.0	4.0	3.8
4 Thrills and Chills	6	5		5	5	5.0	5.0	4.2
5 Let Us Entertain You	6	5	5	5	5	5.0	5.0	4.4
6 Electricity for Everyone	5.5	4.5	5.5	5.5	5.5	4.5	4.5	4.0
7 Toys for Understanding	4	4	4			4.0	4.0	3.2
8 Atoms on Display	5	4.5	5	5	5	4.0	4.0	4.0
9 Sports on the Moon	5	4		5	5		5.0	3.6
	*Estimated number of weeks including class interruptions							
36 weeks available 180 day/school year		Total Weeks =	35	36	35.5	36	36	35.4

Engineering Design Cycle

Where can you add technology in your physics class? How can you possibly add more content to your cramped curriculum? What if you could add it in the form of a tool that helps students synthesize and present information more efficiently? The *Engineering Design Cycle* will provide a framework for completing the *Mini-Challenge* and the *Chapter Challenge*, while introducing students to the basics of engineering design. The *Engineering Design Cycle* information will show up three times during each chapter, guiding students through the problem-solving design process. The cycle will help to introduce each *Chapter Challenge*, it will be presented with each *Chapter Mini-Challenge*, and then will return to help guide students through their *Chapter Challenge* presentations to conclude each chapter. Consistent formatting for each chapter will help students acclimate quickly to the vocabulary and process of the *Engineering Design Cycle* so that it quickly becomes a tool instead of another requirement. Each iteration through the process will make students more comfortable with a structured way to solve their design challenges and increase the variety of ways they have to assemble the information they have learned.

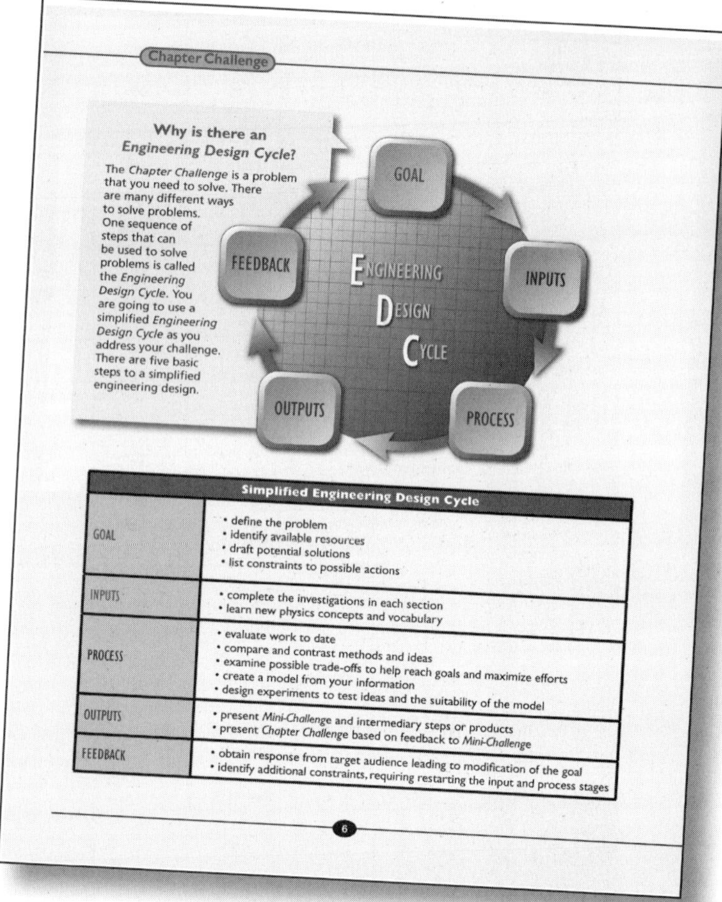

Chapter Challenge

Why is there an *Engineering Design Cycle?*

The *Chapter Challenge* is a problem that you need to solve. There are many different ways to solve problems. One sequence of steps that can be used to solve problems is called the *Engineering Design Cycle*. You are going to use a simplified *Engineering Design Cycle* as you address your challenge. There are five basic steps to a simplified engineering design.

Simplified Engineering Design Cycle	
GOAL	• define the problem • identify available resources • draft potential solutions • list constraints to possible actions
INPUTS	• complete the investigations in each section • learn new physics concepts and vocabulary
PROCESS	• evaluate work to date • compare and contrast methods and ideas • examine possible trade-offs to help reach goals and maximize efforts • create a model from your information • design experiments to test ideas and the suitability of the model
OUTPUTS	• present *Mini-Challenge* and intermediary steps or products • present *Chapter Challenge* based on feedback to *Mini-Challenge*
FEEDBACK	• obtain response from target audience leading to modification of the goal • identify additional constraints, requiring restarting the input and process stages

6

The *Engineering Design Cycle* that is presented uses four simple steps to solve each design challenge in the *Active Physics* text. The four steps form the core of every *Engineering Design Cycle* and therefore will be familiar when students encounter it later in another setting. Each chapter will use INPUTS, PROCESS, OUTPUT, and FEEDBACK steps to help students use the information they have summarized in *What Do You Think Now?* responses and practiced in the *Physics to Go* questions during each chapter section. The steps will guide students to review and include the fundamental information from each section of the chapter. The PROCESS step will highlight a different critical thinking tool for the decision-making process to introduce students to a wide variety of tools that can be used to solve problems and arrive at conclusions. This step is the heart or engine of the *Engineering Design Cycle*. It is where most of the critical thinking must be done. There is not enough time to guide students through each and every decision-making step for each challenge, however by highlighting different tools during each one, students will have been exposed to many useful methods by the end of a typical school year.

Most importantly, the *Engineering Design Cycle* is presented as a lens that can be used to view the chapter challenge function within *Active Physics*. Depending on your personal interest, knowledge, and style, it will have a different function for each classroom. It will likely have a different function for each student within a classroom. For some linear thinkers, the engineering design cycle will provide a map; for more creative thinkers, it will provide an outline or maybe even a launch pad for innovative solutions to a problem. Students can only benefit from considering their design challenges through the lens of a process that is used in similar form in every engineering enterprise today. Students will learn to use a framework to compile their information and solve their chapter challenges. At the same time they will learn that there is a process for solving problems that is independent of the current problem—a process they can practice and use on every problem. Once the students are comfortable with the structure of the design cycle, they will be able to use it to guide them through each *Chapter Mini-Challenge* and final *Chapter Challenge*.

Active Physics and the National Science Education Standards

Active Physics was designed and developed to provide teachers with instructional strategies that model the following from the standards:

Guide and Facilitate Learning

- Focus and support inquiries while interacting with students.
- Orchestrate discourse among students about scientific ideas.
- Challenge students to accept and share responsibility for their own learning.
- Recognize and respond to student diversity; encourage all to participate fully in science learning.
- Encourage and model the skills of scientific inquiry as well as the curiosity, openness to new ideas and data, and skepticism that characterize science.

Engage in Ongoing Assessment of Their Teaching and Student Learning

- Use multiple methods and systematically gather data about student understanding and ability.
- Analyze assessment data to guide teaching.
- Guide students in self-assessment.
- Select teaching and assessment strategies that support the development of student understanding and nurture a community of science learners.

Design and Manage Learning Environments that Provide Students with Time, Space, and Resources needed for Learning Science.

- Structure the time available so students are able to engage in extended investigations.
- Create a setting for student work that is flexible and supportive of science inquiry.
- Make available tools, materials, media, and technological resources accessible to students.
- Identify and use resources outside of school.
- Engage students in designing the learning environment.

Develop Communities of Science Learners that Reflect the Intellectual Rigor of Science Attitudes and Social Values Conducive to Science Learning.

- Display and demand respect for diverse ideas, skills, and experiences of students.
- Enable students to have a significant voice in decisions about content and context of work and require students to take responsibility for the learning of all members of the community.
- Nurture collaboration among students.
- Structure and facilitate ongoing formal and informal discussion based on shared understanding of rules.
- Model and emphasize the skills, attitudes, and values of scientific inquiry.

Assessment Standards

- Features claimed to be measured are actually measured.
- Students have adequate opportunity to demonstrate their achievement and understanding.
- Are deliberately designed with explicitly stated purpose.
- Assessment tasks are authentic and developmentally appropriate, set in familiar context, and are engaging to students with different interests and experiences without regard to gender, ethnicity, or race.
- Assesses student understanding as well as knowledge, the ability to reason scientifically, and communicate effectively about science.
- Improve classroom practice and plan curricula.

Active Physics and the National Science Education Standards

	Driving the Roads	Physics in Action	Safety
Physical Science			
Structure of atoms			
Structure and properties of matter		x	x
Motions and forces	x	x	x
Conservation of energy and increase in disorder		x	
Interactions of energy and matter	x	x	x
Unifying Concepts and Processes			
Systems, order, and organization			
Evidence, models, and explanations	x	x	x
Constancy, change, and measurement	x	x	x
Science as Inquiry			
Identify questions and concepts that guide scientific inquiry	x	x	x
Design and conduct scientific investigations	x	x	x
Use technology and mathematics to improve investigations	x	x	x
Formulate and revise scientific explanations and models using logic and evidence	x	x	x
Communicate and defend a scientific argument	x	x	x
Understand scientific inquiry	x	x	x
Science and Technology			
Identify a problem or a design opportunity	x	x	x
Propose designs and choose between alternate solutions		x	x
Implement a proposed solution	x	x	x
Evaluate the solutions and their consequences		x	x
Communicate the problem, process, and solution	x	x	x
Understand science and technology	x	x	x
Science in Personal and Social Perspectives			
Natural and human induced hazards	x		x
Science and technology in local, national, and global challenges	x		x
History and Nature of Science			
Science as a human endeavor	x	x	x
Nature of scientific knowledge	x	x	x
Historical perspectives		x	

INTRODUCTION

Active Physics and the National Science Education Standards

Thrills and Chills	Let Us Entertain You	Electricity for Everyone	Toys for Understanding	Atoms on Display	Sports on the Moon
		x		x	
x	x	x	x	x	
x		x	x	x	x
x		x	x	x	x
x	x	x	x	x	x
x		x	x	x	
x	x	x	x	x	x
x	x	x	x	x	x
x	x	x	x	x	x
x	x	x	x	x	x
x	x	x	x	x	x
x	x	x	x	x	x
x	x	x	x	x	x
x	x	x	x	x	x
x	x	x	x	x	x
x	x	x	x	x	x
x	x	x	x	x	x
x		x	x	x	x
x	x	x	x	x	x
x	x	x	x	x	x
x		x		x	
		x	x	x	
x	x	x	x	x	x
x	x	x	x	x	x
x		x	x	x	x

Active Physics: Bridging Research and Practice

Over the past 20 years, advances in cognitive science and learning have steered educators to teaching practices that can support more effective learning. No longer must teachers base their practice on instinct and what appears to work. The common wisdom of practice can now be integrated with research results to provide better curriculum.

Active Physics bridges the research with high-quality physics instruction, moving the research results from the cognitive laboratory to the classroom. Below is a brief summary of some approaches in *Active Physics* and the rationale for their use.

The publication by the National Research Council of the National Academy of Sciences entitled *How People Learn* is a must-read for all teachers. It helps provide research and examples in a conversational tone. The three findings emphasized in the book are worthwhile ways to begin explaining how *Active Physics* bridges these findings and classroom instruction.

Key Finding 1: Students come to the classroom with preconceptions about how the world works. If their initial understanding is not engaged, they may fail to grasp new concepts and information, or they may learn them for purposes of a test, but revert to their preconceptions outside the classroom.

All *Active Physics* sections begin by eliciting students' prior understanding. This is first done with a student response to the *What Do You See?* cartoon and is immediately followed by a written *What Do You Think?* prompt. Both set the stage for the content of each section. Each section ends by having students return to confront their initial understandings and the evidence they have generated during the investigation. *What Do You Think Now?* allows students to use evidence to support their arguments and help recognize what they have learned.

Key Finding 2: To develop competence in an area of inquiry, students must: (a) have a deep foundation of factual knowledge, (b) understand facts and ideas in the context of a conceptual framework, and (c) organize knowledge in ways that facilitate retrieval and application.

Active Physics students learn all the required content and factual knowledge in the context of investigations, evidence, and reflection on that evidence. They do not learn the facts in isolation, but as part of the knowledge and information required to complete a *Chapter Challenge*. Through the *Physics Essential Questions* feature, students must also explain:

- What does it mean?
- How do you know?
- Why do you believe?
- Why should you care?

The *Why do you believe?* question requires students to understand the content of each section with respect to (a) connections with other physics content, (b) how it fits with the "big ideas of science," and (c) how it meets physics requirements such as experimental evidence being consistent with models and theories. When students are required to retrieve and apply their content knowledge, they are able to refer to the challenge. For example, when introduced to a new concept, students can ask themselves, "Did I first encounter this concept when I was trying to complete a challenge having to do with sports or a challenge having to do with electrical appliances or a challenge having to do with light and sound?" The next step is for the student to remember which investigation may have had to do with this new topic. Finally, students can retrieve the content that emerged from their investigation. This is in sharp contrast to having a student try to remember a concept because it was in Chapter 1 or Chapter 9.

Key Finding 3: A "metacognitive" approach to instruction can help students learn to take control of their learning by defining learning goals and monitoring their progress in achieving them.

The *What Do You Think Now?* question is one way in which students must reflect on their learning, and come to the understanding that what they knew a few days ago may have to be modified as a result of the evidence from their investigation, analysis, and class discussion. *Active Physics* also requires students to transfer the knowledge from each section to their required execution of the ***Chapter Challenge***. This is not simply recall, but a creative extension of their knowledge. In planning to complete the ***Chapter Challenge***, students must organize their knowledge as well as manage these projects. Both tasks require students to be aware of their own learning processes. From the opening of the chapter, where students are introduced to the ***Chapter Challenge***, they are required to create a set of criteria by which their future project will be evaluated. They also must decide the weighting of each element. This helps students define their learning goals. As they move through the chapter, they monitor their progress in achieving these goals. Additional aspects of *Active Physics* also strengthen the bridge between research and practice.

Problem-based Learning Model

Each chapter begins with an engaging scenario that challenges students and sets the stage for the learning investigations and chapter assessments to follow. Chapter content and *Investigates* are selectively aimed at providing students with the knowledge and skills needed to respond effectively to the challenge. This scenario-challenge framework provides students with a new focus and topic in successive chapters, and covers a wide variety of student interests.

7E Instructional Model

The 7E instructional model used at the section and chapter levels (and described in more detail in this *Teacher's Edition*) ensures that research about learning encourages teaching that supports learning. A table (from the *Student Edition*) listing each component of a section and what phase of the 7E instructional model it addresses, is provided at the bottom of this page.

- **Elicit:** *What Do You See?* and *What Do You Think?* seek to ensure that students' prior knowledge is addressed.
- **Engage:** Students learn when they are motivated.
- **Explore:** Students investigate and have experience with the concept prior to being told, in the abstract, about how things are or how they should be explained.
- **Explain:** The explanation of a concept follows the students' introduction to the phenomenon.
- **Elaborate:** Students can investigate the implications of the new concept and see how it pertains to other situations.
- **Extend:** Students transfer and apply their knowledge from what they have learned to the larger chapter challenge.
- **Evaluate:** There are opportunities for students to evaluate their own learning as well as for teachers to evaluate their learning in all parts of the instructional model and at the end using traditional homework problems.

Phases of the 7E instructional model	Where is it in the section?
Elicit	*What Do You See?, What Do You Think?*
Engage	*What Do You See?, What Do You Think?*
Explore	*Investigate*
Explain	*Physics Talk, Physics Words*
Elaborate	*Physics Talk, What Do You Think Now?* *Checking Up, Physics Essential Questions, Physics to Go*
Extend	*Reflecting on the Section and the Challenge* *Preparing for the Chapter Challenge* *Inquiring Further*
Evaluate	*Formative evaluation — You evaluate your own understanding and the teacher can evaluate your understanding during all components of the chapter. Additional evaluations may include:* Lab reports, *Checking Up*, Quizzes, *What Do You Think Now?, Physics Essential Questions, Physics to Go.*

INTRODUCTION

Multiple Exposure Curriculum

The thematic nature of the course requires students to continually revisit fundamental physics principles throughout the semester, extending and deepening their understanding of these principles as they apply them in fresh contexts and new *Chapter Challenges*. This repeated exposure fosters the retention of content and process, and the transferability of learning. The development of critical-thinking skills also is promoted by multiple exposure and application.

Inquiry-based Learning

All of the six to ten sections per chapter routinely involve inquiry at some level. In some sections, students must design an experiment to explore a concept. At other times, they will collect and interpret data to formulate a new concept. Students are regularly asked to answer questions about what they think and what they do, as well as to support their statements with concrete data. "Activity Before Concept" (ABC) provides a strong base for inquiry-based learning as the students always have a common lab experience upon which to understand the concept. For the same reason, it is important to show the students a concept before exposing them to the vocabulary associated with the concept, "Concept Before Vocabulary" (CBV).

Constructivist Approach

Students are continually asked to explore how they think about certain situations. As they investigate each new situation, they are challenged to either explain observed phenomena using an existing paradigm or develop a more consistent one. This approach can be used to help students recognize previous notions and abandon these in favor of a more powerful idea offered by scientists.

Authentic Assessment

As the culmination of each chapter, students are required to demonstrate their ability to use their newly acquired knowledge by adequately meeting the challenge posed in the chapter introduction. Students are then evaluated on the degree to which they accomplish this performance task using a rubric based on their own input. The curriculum also includes other methods and instruments for authentic assessments as well as nontraditional procedures for evaluating and rewarding desirable achievements.

Cooperative Grouping Assignments

Use of cooperative groups is integral to the curriculum as students work together in small groups to acquire the knowledge and information needed to address the series of challenges presented through the chapter scenarios. Today's business model requires employees to work cooperatively in small groups to accomplish business goals. Students learn to value the abilities and contributions of others while maintaining a focus on the group goal of providing a meaningful response to the *Chapter Challenge*.

Problem Solving

For the curriculum to be meaningful and relevant to students, problem solving related to technological applications is an essential component of the course. Problem solving ranges from simple numerical calculations, to more involved decision-making situations where multiple alternatives must be compared. In some investigations, students are asked to compare class data, determine which data is useful and valid, and suggest reasons for other data falling short of this goal. Seen in its entirety, each chapter is a large problem to be solved, with multiple correct solutions and opportunities for design change before completion.

Challenging Learning Extensions

In every section, an *Active Physics Plus* is provided to encourage further learning through increased mathematical analysis, depth of treatment of the content and concepts, or an additional laboratory exploration. Throughout the text, a variety of *Inquiring Further* and *Preparing for the Chapter Challenge* exercises are provided for more motivated students. These extensions range from more challenging design tasks, to enrichment readings, to intriguing and unusual problems. Many of the extensions provide additional opportunities for the oral and written expression of student ideas.

Research-based Model

Active Physics relies upon the proven NSF developmental process as well as the most up-to-date research in pedagogy and learning theory. The research-based NSF process, with its strict criteria and standard, usually takes three to four years to complete. The curriculum is written and first tested in the classroom in the Pilot Test. Following data collection and evaluation, the curriculum is rewritten and retested in the Field Test classrooms. Following a second evaluation by a nationally recognized independent evaluation team, the curriculum is rewritten again. Expert review panels also guide the process at each stage.

Responds to Equity Concerns

A diverse classroom is a wonderful place for exposure to the various cultures of students and their experiences. However, in most cases, it is impossible for the teacher to become an "expert" in each culture represented by the students and to bring that into the classroom. *Active Physics* honors the diversity of students by allowing them to creatively bring their culture into the classroom in the format of the *Chapter Challenge*, for example, through sports, entertainment, or habitats.

Misconceptions

Students arrive in class knowing a great deal about the world. Researchers in cognitive science have identified various categories of student "alternative" understandings of the world. Each chapter of *Active Physics* identifies lists of commonly held student preconceptions and offers activities for students to test their conceptual ideas about understanding each learning goal. Students continually are asked to explain their reasoning. This process enables the teacher to interview and work with students to root out prior conceptions and establish scientific foundations of understanding of the principles involved.

Understanding by Design

Wiggins and McTighe have introduced the notion that in preparing instruction, teachers should first decide on the enduring understandings and concepts they expect their students to learn. Teachers should then decide what evidence they will accept to show that their students understand these enduring understandings and concepts. Finally, teachers should develop curriculum to ensure that students will be able to present that evidence of learning. In *Active Physics*, the students begin each chapter by identifying and developing a challenge. Next, they develop criteria for excellence for completing this challenge. They then go about learning the concepts in each section while adding their evidence of understanding and creativity to their *Chapter Challenge*.

Student Motivation

- When students are motivated, they learn and remember more. *Active Physics* engages students intellectually by:
- Presenting interesting and meaningful challenges.
- Allowing student creativity in the execution of the challenges.
- Allowing each student group to demonstrate expertise in the physics content and the presentation of their knowledge.
- Providing students with choices about how they will demonstrate their physics knowledge in creative ways.
- Teachers permitting—even encouraging—different forms of expression and respecting student views.
- Students creating original and public products.
- Students sensing that the results of their work were not predetermined or fully predictable.

Integrated Instructional Units

Rather than introducing lab experiments haphazardly, *Active Physics* always incorporates lab investigations into the larger instructional unit to provide the most benefit to the student. As noted in the National Research Council study, *America's Lab Report,* the use of integrated instructional units has been shown to support more learning than do traditional lab approaches.

Cooperative Learning

Benefits of Cooperative Learning

Cooperative learning requires you to organize and structure a lesson so that students work with other students to jointly accomplish a task. Group learning is an essential part of balanced methodology. It should be blended with whole-class instruction and individual study to meet a variety of learning styles and expectations as well as maintain a high level of student involvement.

Cooperative learning has been thoroughly researched and agreement has been reached on a number of results. Cooperative learning:

- promotes trust and risk-taking
- elevates self-esteem
- encourages acceptance of individual differences
- develops social skills
- permits a combination of a wide range of backgrounds and abilities
- provides an inviting atmosphere
- promotes a sense of community
- develops group and individual responsibility
- reduces the time on a task
- results in better attendance
- produces a positive effect on student achievement
- develops key employability skills

As with any learning approach, some students will benefit more than others from cooperative learning. Therefore, you may question as to what extent you should use cooperative learning strategies. It is important to involve the student in helping decide which type of learning approaches they prefer, and to what extent each is used in the classroom. When students have a say in their learning, they will accept to a greater extent any method which you choose to use.

Phases of Cooperative Learning Lessons

Organizational Pre-lesson Decisions

What academic and social objectives will be emphasized? In other words, what content and skills are to be learned and what interaction skills are to be encouraged or practiced?

What will be the group size? Or, what is the most appropriate group size to facilitate the achievement of the academic and social objectives? This will depend on the amount of individual involvement expected (small groups promote more individual involvement), the task (diverse thinking is promoted by larger groups), nature of the task or materials available and the time available (shorter time demands smaller groupings to promote involvement).

Who will make up the different groups? Teacher-selected groups usually have the best mix, but this can only happen after the you get to know your students well enough to know who works well together. Heterogeneous groupings are most successful in that all can learn through active participation. The duration of the groups' existence may have some bearing on deciding the membership of groups.

How should the room be arranged? Practicing routines where students move into their groups quickly and quietly is an important aspect. Having students face-to-face is important. You should still be able to move freely among the groups.

What Materials and/or Rewards Might be Prepared in Advance?

Setting the Lesson

Structure for Positive Interdependence: When students feel they need one another, they are more likely to work together—goal interdependence becomes important. Class interdependence can be promoted by setting class goals which all teams must achieve in order for class success.

Explanation of the Academic Task: Clear explanations and sometimes the use of models can help the students. An explanation of the relevance of the activity is importance. Checks for clear understanding can be done either before the groups form or after, but they are necessary for delimiting frustrations.

Explanation of Criteria for Success: Groups should know how their level of success will be determined. Structure for Individual Accountability: The use of individual follow-up activities for tasks or social skills will provide for individual accountability.

Structure for Individual Accountability: The use of individual follow-up activities for tasks or social skills will provide for individual accountability.

Specification of Desired Social Behaviors: Definition and explanations of the importance of values of social skills will promote student practice and achievement of the different skills.

Monitoring/Intervening During Group Work

Through monitoring students' behaviors, intervention can be used more appropriately. Students can be involved in the monitoring by being "a team observer," but only when the students have a very clear understanding of the behavior being monitored.

Interventions to increase chances for success in completing the task or investigation and for the teaching of collaborative skills should be used as necessary—they should not be interruptions. This means that as the facilitating teacher, you should be moving among the groups as much as possible. During interventions, the problem should be turned back to the students as often as possible, taking care not to frustrate them.

Evaluating the Content and Process of Cooperative Group Work

Assessment of the achievement of content objectives should be completed by both you and the students. Students can go back to their groups after an assignment to review the aspects in which they experienced difficulties.

When assessing the accomplishment of social objectives, two aspects are important: how well things proceeded and where/how improvements might be attempted. Student involvement in this evaluation is a very basic aspect of successful cooperative learning programs.

Organizing and Monitoring Groups

An optimum size of group for most investigations appears to be four; however, for some tasks, two may be more efficient. Heterogeneous groups organized by the teacher are usually the most successful. You will need to decide what factors should be considered in forming the heterogeneous groups. Factors which can be considered are: academic achievement, cultural background, language proficiency, sex, age, learning style, and even personality type.

Level of academic achievement is probably the simplest and initially the best way to form groups. Sort the students on the basis of marks on a particular task or on previous year's achievement. Then choose a student from each quartile to form a group. Once formed, groups should be flexible. Continually monitor groups for compatibility and make adjustments as required.

Students should develop an appreciation that it is a privilege to belong to a group. Remove from group work any student who is a poor participant or one who is repeatedly absent. These individuals can then be assigned the same tasks to be completed in the same timeline as a group. You may also wish to place a ten percent reduction on all group work that is completed individually.

The chart at the end of this Cooperative Learning section presents some possible group structures and their functions.

What Does Cooperative Learning Look Like?

During a cooperative learning situation, students should be assigned a variety of roles related to the particular task at hand. A list of possible roles that students may be given is provided below. It is important that students are given the opportunity of assuming a number of different roles over the course of a semester.

Leader:
Assigns roles for the group. Gets the group started and keeps the group on task.

Organizer:
Helps focus discussion and ensures that all members of the group contribute to the discussion. The organizer ensures that all of the equipment has been gathered and that the group completes all parts of the investigation.

Recorder:
Provides written procedures when required, diagrams where appropriate and records data. The recorder must work closely with the organizer to ensure that all group members contribute.

Researcher:
Seeks written and electronic information to support the findings of the group. In addition, where appropriate, the researcher will develop and test prototypes. The researcher will also exchange information gathered among different groups.

Encourager:
Encourages all group members to participate. Values contributions and supports involvement.

Checker:
Checks that the group has answered all the questions and the group members agree upon and understand the answers.

Diverger:
Seeks alternative explanations and approaches. The task of the diverger is to keep the discussion open. "Are other explanations possible?"

Active Listener:
Repeats or paraphrases what has been said by the different members of the group.

Idea Giver:
Contributes ideas, information, and opinions.

Materials Manager:
Collects and distributes all necessary material for the group.

Observer:
Completes checklists for the group.

Questioner:
Seeks information, opinions, explanations, and justifications from other members of the group.

Reader:
Reads any textual material to the group.

Reporter:
Prepares and/or makes a report on behalf of the group.

Summarizer:
Summarizes the work, conclusions, or results of the group so that they can be presented coherently.

Timekeeper:
Keeps the group members focused on the task and keeps time.

Safety Manager:
Responsible for ensuring that safety measures are being followed, and the equipment is clean prior to and at the end of the investigation.

Group Assessment

Assessment should not end with a group mark. Students and their parents have a right to expect marks to reflect the students' individual contributions to the task. It is impossible for you as the instructor to continuously monitor and record the contribution of each individual student. Therefore, you will need to rely on the students in the group to assign individual marks as merited. There are a number of ways that this can be accomplished. The group mark can be multiplied by the number of students in the group, and then the total mark can be divided among the students.

Some Possible Group Structures and Their Functions*

	Structure	Brief Description	Academic and Social Functions
Team Building	Roundrobin	Each student in turn shares something with his/her teammates.	Expressing ideas and opinions, creating stories. Equal participation, getting acquainted with each other.
Class Building	Corners	Each student moves to a group in a corner or location as determined by the teacher through specified alternatives. Students discuss within groups, then listen to and paraphrase ideas from other groups.	Seeing alternative hypotheses, values, and problem solving approaches. Knowing and respecting differing points of view.
Mastery	Numbered heads together	The teacher asks a question, students consult within their groups to make sure that each member knows the answer. Then one student answers for the group in response to the number called out by the teacher.	Review, checking for knowledge comprehension, analysis, and divergent thinking. Tutoring.
	Color coded co-op cards	Students memorize facts using a flash card game or an adaption. The game is structured so that there is a maximum probability for success at each step, moving from short to long-term memory. Scoring is based on improvement.	Memorizing facts. Helping, praising.
	Pairs check	Students work in pairs within groups of four. Within pairs students alternate — one solves a problem while the other coaches. After every problem or so, the pair checks to see if they have the same answer as the other pair.	Practicing skills. Helping, praising.
Concept Development	Three-step interview	Students interview each other in pairs, first one way, then the other. Each student shares information learned during interviews with the group.	Sharing personal information such as hypotheses, views on an issue, or conclusions from a unit. Participation, involvement.
	Thinkpair-share	Students think to themselves on a topic provided by the teacher; they pair up with another student to discuss it; and then share their thoughts with the class.	Generating and revising hypotheses, inductive and deductive reasoning, and application. Participation and involvement.
	Team wordwebbing	Students write simultaneously on a piece of paper, drawing main concepts, supporting elements, and bridges representing the relation of concepts/ideas.	Analysis of concept into components, understanding multiple relations among ideas, and differentiating concepts. Role-taking.
Multifunctional	Roundtable	Each student in turn writes one answer as a paper and a pencil are passed around the group. With simultaneous roundtables, more than one pencil and paper are used.	Assessing print knowledge, practicing skills, recalling information, and creating designs. Team building, participation of all.
	Partners	Students work in pairs to create or master content. They consult with partners from other teams. Then they share their products or understandings with the other partner pair in their team.	Mastery and presentation of new material, concept development. Presentation and communication skills.
	Jigsaw	Each student from each team becomes an "expert" on one topic area by working with members from other teams assigned to the same topic area. On returning to their own teams, each one teaches the other members of the group and students are assessed on all aspects of the topic.	Acquisition and presentation of new material review and informed debate. Independence, status equalization.

*Adapted from Spencer Kagan (1990), "*The Structural Approach to Cooperative Learning*," Educational Leadership, December 1989/January 1990.

Active Physics—A Research-Based Curriculum

Active Physics has been funded by the National Science Foundation (NSF) and exemplifies the NSF's goals and objectives to improve science, mathematics, and technology education for all students. *Active Physics* is aligned with the National Science Education Standards (NSES) and follows the guidelines of the NSES to improve science-content knowledge, thinking skills, and problem-solving abilities for all students, regardless of background or ability. The NSES were guided by certain principles:

- Science is for all students.
- Learning science is an active process.
- School science reflects the intellectual and cultural traditions that characterize the practice of contemporary science.
- Improving science education is part of systemic education reform.

Adhering to these principles, *Active Physics* relies on the rigorous research of the NSF development process as well as incorporating into its instructional model the most up-to-date research on pedagogy and learning theory. The *Active Physics* curriculum consequently promotes positive student attitudes toward science and positive perceptions of the student as learner. It engages students through its use of real-world contexts and provides a deeper understanding of the role of science and technology in the workplace.

The NSF Research-Based Curriculum Development Process

The NSF Instructional Materials Development Program ensures that each of the curriculum development programs it has funded follows strict *research-based* criteria throughout the development process. The project grants are extremely competitive and only awarded to development teams that have established themselves to be distinguished leaders in science education. The initial development grant for *Active Physics* was awarded to the American Association of Physics Teachers and the American Institute of Physics. The embedded research-based development process with the strict criteria and standards for all NSF-funded programs, usually takes at least three to four years to complete.

Project Evaluation and Research Design

A crucial component to all NSF development projects is the ongoing research and evaluation of the development process and materials by a nationally recognized independent evaluation team. The research and evaluation of these projects is comprehensive and provides both formative and summative information to the development teams as well as to the NSF for review. The formative feedback and information is used to optimize the curricular revision process, and the summative evaluation examines the effectiveness of the curricular materials on teachers and students throughout the three-year process. *Active Physics* initially went through a four-year curriculum development process. An abbreviated description of that process included:

First Year of the Curriculum Development Process—Content Specialists, Master Teachers

- Under the direction of a distinguished, active, and dynamic Advisory Board, the program's Principal Investigators select physicists, physics educators, and high school physics teachers to collaborate on the development of the first draft of the curriculum materials. These teams also serve as part of the Review

Committee to assess each other's works for pedagogical strategies and content accuracy. The curriculum is then reviewed and evaluated by other leading educational specialists for pedagogy, content, safety, equity, readability, cognitive effectiveness, and efficacy, and then the curriculum is revised again based on those results.

- All new materials proceed through the following system for development and revision:
  - Approved by Content Review Committee comprised of leading content experts
  - Approved by the following consultants: science educators, master teachers, and cognitive scientists
  - Micro-tested by the development group (A micro-test is a series of tests of a few students with careful observation and follow-up interviews by the developers.)

Second Year of the Curriculum Development Process—Content Specialists Pilot to Ensure Curriculum Is Correct and Rigorous

- The curriculum is then ready to be pilot tested by a select group of high school teachers from across the country. After an extensive summer training course, these teachers spend the next year piloting the program in their classrooms.
- Pilot-tested by master teachers in their classrooms
- Pilot materials, classes, teachers, and students are studied and evaluated based on an established evaluation and research design model
- Materials are then revised based on the pilot feedback, experts' reviews and evaluation and research reports.

Third Year of the Curriculum Development Process—Diverse Classrooms to Ensure Approach is Appropriate for All Students

- The curriculum is now ready to be field-tested by a broad range of high school teachers from across the country. After an extensive summer training course, these teachers spend the next year field-testing the materials in their classrooms.
- Like the pilot test, the research/evaluation component of the revision process is designed to inform the next iteration and revision of the materials.
- Field-testing of the materials conducted in a wide range of classrooms by teachers with a wide range of experience and expertise
- Field-test materials, classes, teachers, and students are studied and evaluated based on the evaluation and research design model
- Materials are then revised again based on the field-test feedback, experts' reviews and evaluation and research reports.

Fourth Year of the Development Process

- Additional consultant specialists in cognitive psychology, assessment, technology, science education, and equity continue to be brought into the project to review the materials and secure its pedagogical approach and content basis. Finally, the product is turned over to the commercial publisher to mold into a commercial product. Publishing decisions cannot take precedence over content requirements or instructional strategies.

Active Physics Third Edition

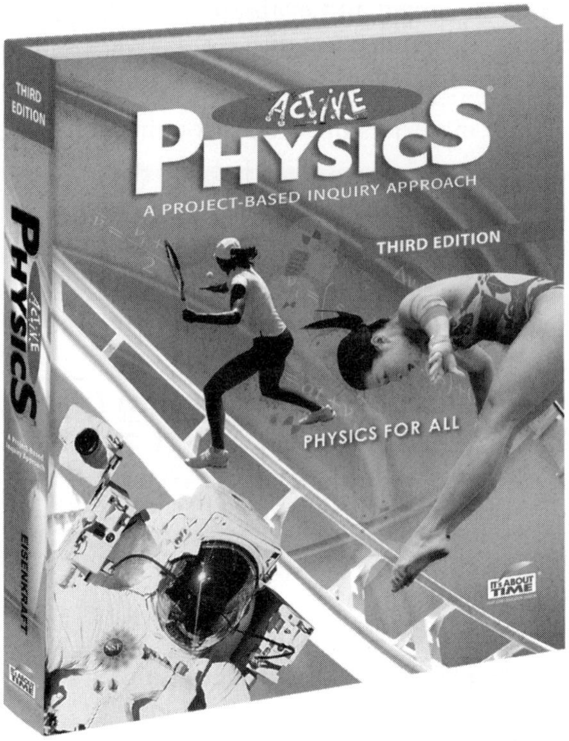

The *Active Physics Third Edition* was also funded by NSF. Revision goals were to revise the original *Active Physics* materials to reflect the feedback, experiences, and research accumulated after several years of implementation in school districts throughout the United States. The original curriculum was designed as six thematic, modular components. Each module contained three independent chapters, thus allowing for the adjustable sequencing of modules and chapters as determined by each school or district. The objectives followed and established by the original *Active Physics* curriculum were to

- Dramatically increase physics enrollment throughout the country
- Significantly decrease the achievement gap in physics
- Allow both male and female students from all ethnic backgrounds to enjoy physics and to succeed in physics
- Allow students to better understand physics-related issues and technologies, and to be able to identify and evaluate personal and societal impacts
- Develop in students an appreciation and understanding of science as a process, develop the skills and attitudes of a scientist, and develop an ability to apply these in realistic and relevant problem-solving and decision-making activities.

Active Physics successfully grew its market share during the eight years it was released, however, there were important lessons learned from the districts implementing the program, and it became clear that a restructuring of the content and format of the materials into a single text book, as well as developing additional content to meet state standards was necessary. To this end, the revision goals and objectives were to

- Restructure and insert additional content appropriate for a "Physics for All" as well as a "Physics First" sequencing
- Provide a stronger emphasis on the organizing principles of physics
- Integrate higher-level mathematics skills into the physics of the curriculum as a separate component to reflect the needs of advanced students, thus creating additional flexibility in the program for use by low- or high-level achieving students.

Active Physics Third Edition also went through a national field test where teachers used the revised materials and reported back to the development group strengths and weaknesses and ways in which the materials could be modified for more effectiveness. Simultaneously, a research study was taking place that compared the physics achievement and attitudes of the more than 3000 students involved in the field test with students taught the prior year by the same teacher using a different curriculum. Some highlights of this research report follow on the next page.

INTRODUCTION

Highlights of the Active Physics Third Edition Research Report

This report compares the outcomes of teachers and students using *Active Physics* Revision to teachers and students using traditional physics curricula. These comparisons were based on a variety of data including several different types of student tests, student surveys, and teacher surveys.

Methodology

Data were collected from both treatment (teachers using *Active Physics* Revision) and comparison (teachers not using *Active Physics* Revision) teachers and their students over the two-year period of this evaluation. The comparison teachers used their "normal" curriculum, while the treatment teachers used the *Active Physics* Revision curriculum. The main comparison used a time-lag design where the outcomes for students in the first year (comparison students) were contrasted to the outcomes for students in the second year (treatment students). This means that the students are not the same in the two years; they are different cohorts, but they are from the same schools and being taught by the same teachers. Therefore, they can be considered a matched comparison group.

Instruments

Several different instruments were used to collect data. Teachers and students completed post-surveys (end of the school year) and students completed one of three physics achievement tests. The teacher survey asked about individual and school characteristics as well as teachers' perceptions of the curriculum they were using that year. The teacher survey also provided information on chapter usage, use of *Chapter Challenges*, and time spent on a given chapter. The student survey asked about individual characteristics, attitudes about science and physics, parental education and involvement in school, and perceptions about the classroom environment. During the field test-year (2005-2006), students and teachers also completed pre-surveys (start of the school year), and students completed pre-tests. These pre-surveys and tests were a subset of items from the post-instruments. The student pre-survey did not ask students about parental involvement or their perceptions of the classroom environment. The teacher pre-survey asked about attitudes toward teaching.

Sample

There were 3657 students overall, with 538 in the comparison group and 3119 in the treatment (*Active Physics* Revision) group. The groups were well matched.

Results

The analyses were designed to determine if any differences existed between *Active Physics* Revision and comparison students and if there were any changes pre-to-post in the teachers and students who used *Active Physics* Revision.

Conclusions

The straightforward, single-variable data analyses show positive results for students studying the *Active Physics* Revision curriculum during the field-test year in comparison to results for students during the comparison year. Although the most conservative analysis did not show any effects for content achievement, less conservative analysis did show that students who participated in *Active*

Physics Revision during the field-test year outperformed students who had studied physics during the comparison year. This was true even though the comparison group students reported higher levels of parental involvement than the *Active Physics* Revision students. Additionally the *Active Physics* Revision students reported being more involved in their physics classes than comparison students.

Pre-to-post comparisons show that the *Active Physics* Revision students gained significantly in content knowledge over the course of the year. Higher post-test scores were also directly related to the amount of time their teachers spent using *Active Physics* Revision.

Three student attitude items show pre-to-post improvement in attitude: *I want to take more science classes; I think about the physics I experience in everyday life* and *to understand physics, I sometimes think about my personal experiences and relate them to the topic being analyzed.* These ideas appear to be directly related to the type of experience provided by *Active Physics* Revision and are good indicators of the positive effect of the *Active Physics* Revision curriculum.

Teacher survey data also show favorable comparative results for *Active Physics* Revision in the field-test year. The teachers in the field-test year reported more favorable attitudes toward the *Active Physics* Revision curriculum than the teachers in the comparison year, although none of the differences were significant. This suggests that the teachers viewed the *Active Physics* Revision curriculum more favorably than the curriculum used during the comparison year. *Active Physics* Revision teachers also reported using three National Standards-based teaching techniques more than the comparison teachers. The *Active Physics* Revision teachers reported that their students spent more time making presentations, completing projects, and writing reflections than the comparison teachers.

In summary, the data indicate that, in general, the students in the field-test year performed as well or better than the students in the comparison year. This suggests that the *Active Physics* Revision curriculum may be superior to the curricula used during the comparison year in some ways and at least comparable in the others investigated.

New Components and Features of Active Physics Third Edition

Active Physics continues to be responsive to the most current research on learning and continues to follow its established and proven successful 7E instructional model that includes the following:

- Eliciting prior understanding through the *What Do You See?* illustrations and the *What Do You Think?* questions.
- Focusing the learning around a *Chapter Challenge* that drives content, motivates students, and builds context.
- Placing investigations before concept introductions and readings.
- Exploring through hands-on inquiry.
- Transferring the learning of content from the investigations and readings to the *Chapter Challenge*.

There are several new components and features in the revised materials. For example, a Launcher Chapter has been created as the first chapter in *Active Physics Third Edition*. In this first chapter, students are introduced to the many features of the program. What is unique about this introduction is that it also explains to the students why each feature is included and how it will help them learn physics. Prefaced by the statement, "The more you understand about how you learn, the better you will be at learning," it encourages students to expand their metacognitive awareness. This unique metacognitive feature will greatly enhance student learning, as research has shown that students learn better when they know how they learn. The first time an *Active Physics* feature is presented, it has a commentary explaining its purpose and begins with 'Why is there a....?' For example, the 7E Instructional Model is fully explained and then referred to later in the textbook.

The knowledge and organizing principles that unite all areas of physics and all science endeavors are overtly emphasized. At the end of each section, students are asked to answer four *Physics Essential Questions* that relate the content in the section the students just completed to the body of science as a whole. *What does it mean?*, *How do you know?*, *Why do you believe?*, and *Why should you care?* are questions that strengthen the scientific inquiry in which students are engaged. The questions assist students in learning to think like a scientist in the classroom, as well as promote their future involvement as responsible citizens in society.

The *Physics Talk* feature has been expanded considerably in *Active Physics Third Edition*. While the *Physics Talk* still uses the students' common experiences during the *Investigate* as the starting point and focusing event for each reading, the new *Physics Talk* readings provide better, more thorough, and more in-depth explanations of the concepts explored in the investigation. The results of the students' investigations are carefully explained in terms of scientific models, laws, and theories. The vocabulary the students need to discuss and explain the physics concepts is boldfaced in the text and also provided in the margin as *Physics Words*.

Active Physics Third Edition takes physics for all a step further than previous editions. An *Active Physics Plus* feature provides text, sample problems, and questions for students who want or need more mathematics, depth, concepts, or exploration. Previous editions were focused on providing sound physics to students who might have ordinarily not considered taking a physics course. The excitement generated in *Active Physics* students through the approach of learning physics concepts as a hands-on, inquiry experience is well documented. Unfortunately, there was a perception that this approach would exclude students that would ordinarily take a physics course. By providing an *Active Physics Plus* feature, *Active Physics Third Edition* is all-inclusive.

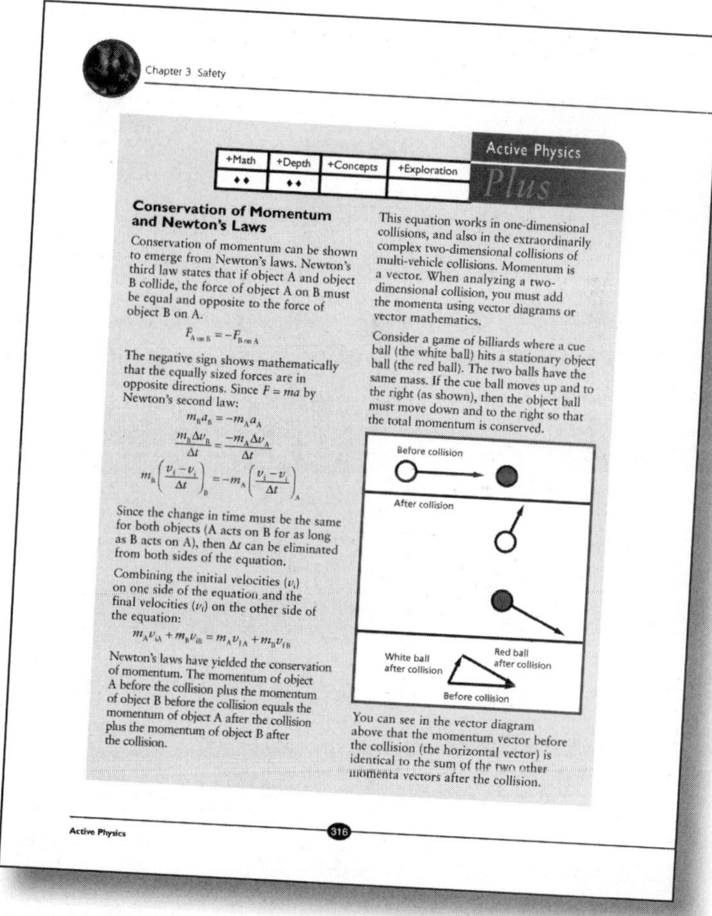

Chapter 3 Safety

Active Physics Plus

+Math	+Depth	+Concepts	+Exploration
◆◆	◆◆		

Conservation of Momentum and Newton's Laws

Conservation of momentum can be shown to emerge from Newton's laws. Newton's third law states that if object A and object B collide, the force of object A on B must be equal and opposite to the force of object B on A.

$$F_{A \text{ on } B} = -F_{B \text{ on } A}$$

The negative sign shows mathematically that the equally sized forces are in opposite directions. Since $F = ma$ by Newton's second law:

$$m_B a_B = -m_A a_A$$

$$\frac{m_B \Delta v_B}{\Delta t} = \frac{-m_A \Delta v_A}{\Delta t}$$

$$m_B\left(\frac{v_f - v_i}{\Delta t}\right)_B = -m_A\left(\frac{v_f - v_i}{\Delta t}\right)_A$$

Since the change in time must be the same for both objects (A acts on B for as long as B acts on A), then Δt can be eliminated from both sides of the equation.

Combining the initial velocities (v_i) on one side of the equation and the final velocities (v_f) on the other side of the equation:

$$m_A v_{iA} + m_B v_{iB} = m_A v_{fA} + m_B v_{fB}$$

Newton's laws have yielded the conservation of momentum. The momentum of object A before the collision plus the momentum of object B before the collision equals the momentum of object A after the collision plus the momentum of object B after the collision.

This equation works in one-dimensional collisions, and also in the extraordinarily complex two-dimensional collisions of multi-vehicle collisions. Momentum is a vector. When analyzing a two-dimensional collision, you must add the momenta using vector diagrams or vector mathematics.

Consider a game of billiards where a cue ball (the white ball) hits a stationary object ball (the red ball). The two balls have the same mass. If the cue ball moves up and to the right (as shown), then the object ball must move down and to the right so that the total momentum is conserved.

You can see in the vector diagram above that the momentum vector before the collision (the horizontal vector) is identical to the sum of the two other momenta vectors after the collision.

Active Physics 316

Expanding the 5E Model

A proposed 7E model emphasizes "transfer of learning" and the importance of eliciting prior understanding

Arthur Eisenkraft

Sometimes a current model must be amended to maintain its value after new information, insights, and knowledge have been gathered. Such is now the case with the highly successful 5E learning cycle and instructional model (Bybee 1997). Research on how people learn and the incorporation of that research into lesson plans and curriculum development demands that the 5E model be expanded to a 7E model.

MIKE OLLIVER

The 5E learning cycle model requires instruction to include the following discrete elements: *engage*, *explore*, *explain*, *elaborate*, and *evaluate*. The proposed 7E model expands the *engage* element into two components—*elicit* and *engage*. Similarly, the 7E model expands the two stages of *elaborate* and *evaluate* into three components— *elaborate*, *evaluate*, and *extend*. The transition from the 5E model to the 7E model is illustrated in Figure 1.

These changes are not suggested to add complexity, but rather to ensure instructors do not omit crucial elements for learning from their lessons while under the incorrect assumption they are meeting the requirements of the learning cycle.

Eliciting prior understandings

Current research in cognitive science has shown that eliciting prior understandings is a necessary component of the learning process. Research also has shown that expert learners are much more adept at the transfer of learning than novices and that practice in the transfer of learning is required in good instruction (Bransford, Brown, and Cocking 2000).

The *engage* component in the 5E model is intended to capture students' attention, get students thinking about the subject matter, raise questions in students' minds, stimulate thinking, and access prior knowledge. For example, teachers may engage students by creating surprise or doubt through a demonstration that shows a piece of steel sinking and a steel toy boat floating. Similarly, a teacher may place an ice cube into a glass of water and have the class observe it float while the same ice cube placed in a second glass of liquid sinks. The corresponding conversation with the students may access their prior learning. The students should have the opportunity to ask and attempt to answer, "Why is it that the toy boat does not sink?"

The engage component includes both accessing prior knowledge and generating enthusiasm for the subject matter. Teachers may excite students, get them interested and ready to learn, and believe they are fulfilling the engage phase of the learning cycle, while ignoring the need to find out what prior knowledge students bring to the topic. The importance of *eliciting* prior understandings in ascertaining what students know prior to a lesson is imperative. Recognizing that students construct knowledge from existing knowledge, teachers need to find out what existing knowledge their students possess. Failure to do so may result in students developing concepts very different from the ones the teacher intends (Bransford, Brown, and Cocking 2000).

A straightforward means by which teachers may elicit prior understandings is by framing a "what do you think" question at the outset of the lesson as is done consistently in some current curricula. For example, a common physics lesson on seat belts might begin with a question about designing seat belts for a racecar traveling at a high rate of speed (Figure 2, p. 58). "How would they be different from ones available on passenger cars?" Students responding to this question communicate what they know about seat belts and inform themselves, their classmates, and the teacher about their prior conceptions and understandings. There is no need to arrive at consensus or closure at this point. Students do not assume the teacher will tell them the "right" answer. The "what do you think" question is intended to begin the conversation.

The proposed expansion of the 5E model does not exchange the *engage* component for the *elicit* component; the engage component is still a necessary element in good instruction. The goal is to continue to excite and interest students in whatever ways possible and to identify prior conceptions. Therefore the elicit component should stand alone as a reminder of its importance in learning and constructing meaning.

Explore and explain

The *explore* phase of the learning cycle provides an opportunity for students to observe, record data, isolate variables, design and plan experiments, create graphs, interpret results, develop hypotheses, and organize their findings. Teachers may frame questions, suggest approaches, provide feedback, and assess understandings. An excellent example of teaching a lesson on the metabolic rate of water fleas (Lawson 2001) illustrates the effectiveness of the learning cycle with varying amounts of teacher and learner ownership and control (Gil 2002).

Students are introduced to models, laws, and theories during the *explain* phase of the learning cycle. Students

FIGURE 1

The proposed 7E learning cycle and instructional model.

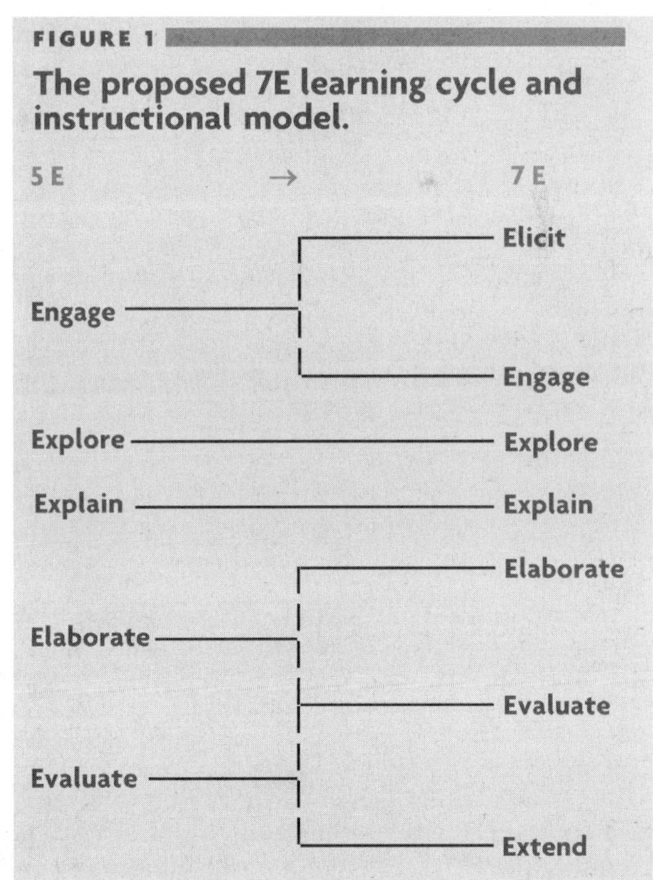

FIGURE 2

Seatbelt lesson using the 7E model.

Elicit prior understandings

- Students are asked, "Suppose you had to design seat belts for a racecar traveling at high speeds. How would they be different from ones available on passenger cars?" The students are required to write a brief response to this "What do you think?" question in their logs and then share with the person sitting next to them. The class then listens to some of the responses. This requires a few minutes of class time.

Engage

- Students relate car accidents they have witnessed in movies or in real life.

Explore

- The first part of the exploration requires students to construct a clay figure they can sit on a cart. The cart is then crashed into a wall. The clay figure hits the wall.

Explain

- Students are given a name for their observations. Newton's first law states, "Objects at rest stay at rest; objects in motion stay in motion unless acted upon by a force."

Engage

- Students view videos of crash test dummies during automobile crashes.

Explore

- Students are asked how they could save the clay figure from injury during the crash into the wall. The suggestion that the clay figure will require a seat belt leads to another experiment. A thin wire is used as a seat belt. The students construct a seat belt from the wire and ram the cart and figure into the wall again. The wire seat belt keeps the clay figure from hitting the wall, but the wire slices halfway through the midsection.

Explain

- Students recognize that a wider seatbelt is needed. The relationship of pressure, force, and area is introduced.

Elaborate

- Students then construct better seat belts and explain their value in terms of Newton's first law and forces.

Evaluate

- Students are asked to design a seat belt for a racing car that travels at 250 km/h. They compare their designs with actual safety belts used by NASCAR.

Extend

- Students are challenged to explore how airbags work and to compare and contrast airbags with seat belts. One of the questions explored is, "How does the airbag get triggered? Why does the airbag not inflate during a small fender-bender but does inflate when the car hits a tree?"

summarize results in terms of these new theories and models. The teacher guides students toward coherent and consistent generalizations, helps students with distinct scientific vocabulary, and provides questions that help students use this vocabulary to explain the results of their explorations. The distinction between the explore and explain components ensures that concepts precede terminology.

Applying knowledge

The *elaborate* phase of the learning cycle provides an opportunity for students to apply their knowledge to new domains, which may include raising new questions and hypotheses to explore. This phase may also include related numerical problems for students to solve. When students explore the heating curve of water and the related heats of fusion and vaporization, they can then perform a similar experiment with another liquid or, using data from a reference table, compare and contrast materials with respect to freezing and boiling points. A further elaboration may ask students to consider the specific heats of metals in comparison to water and to explain why pizza from the oven remains hot but aluminum foil beneath the pizza cools so rapidly.

The elaboration phase ties directly to the psychological construct called "transfer of learning" (Thorndike 1923). Schools are created and supported with the expectation that more general uses of knowledge will be found outside of school and beyond the school years (Hilgard and Bower 1975). Transfer of learning can range from transfer of one concept to another (e.g., Newton's law of gravitation and Coulomb's law of electrostatics); one school subject to another (e.g., math skills, applied scientific investigations); one year to another (e.g., significant figures, graphing, chemistry concepts in physics); and school to nonschool activities (e.g., using a graph to calculate whether it is cost effective to join a video club or pay a higher rate on rentals) (Bransford, Brown, and Cocking 2000).

Too often, the elaboration phase has come to mean an elaboration of the specific concepts. Teachers may provide the specific heat of a second substance and have students perform identical calculations. This practice in transfer of learning seems limited to near transfer as opposed to far or distant transfer (Mayer 1979). Even though teachers expect wonderful results when they limit themselves to near transfer with large similarities between the original task and the transfer task, they know students often find elaborations difficult. And as difficult as near transfer is for students, the distant transfer is usually a much harder road to traverse. Students who are quite able to discuss phase changes of substances and their related freezing points, melting points, and heats of fusion and vaporization may find it exceedingly difficult to transfer the concept of phase change as a means of explaining traffic congestion.

Practicing the transfer of learning

The addition of the *extend* phase to the *elaborate* phase is intended to explicitly remind teachers of the importance for students to practice the transfer of learning. Teachers need to make sure that knowledge is applied in a new context and is not limited to simple elaboration. For instance, in another common activity students may be required to invent a sport that can be played on the moon. An activity on friction informs students that friction increases with weight. Because objects weigh less on the moon, frictional forces are expected to be less on the moon. That elaboration is useful. Students must go one step further and extend this friction concept to the unique sports and corresponding play they are developing for the moon environment.

The *evaluate* phase of the learning cycle continues to include both formative and summative evaluations of student learning. If teachers truly value the learning cycle and experiments that students conduct in the classroom, then teachers should be sure to include aspects of these investigations on tests. Tests should include questions from the lab and should ask students questions about the laboratory activities. Students should be asked to interpret data from a lab similar to the one they completed. Students should also be asked to design experiments as part of their assessment (Colburn and Clough 1997).

Formative evaluation should not be limited to a particular phase of the cycle. The cycle should not be linear. Formative evaluation must take place during all interactions with students. The *elicit* phase is a formative evaluation. The *explore* phase and *explain* phase must always be accompanied by techniques whereby the teacher checks for student understanding.

Replacing *elaborate* and *evaluate* with *elaborate*, *extend*, and *evaluate* as shown in Figure 1, p. 57, is a way to emphasize that the transfer of learning, as required in the extended phase, may also be used as part of the evaluation phase in the learning cycle.

Enhancing the instructional model

Adopting a 7E model ensures that eliciting prior understandings and opportunities for transfer of learning are not omitted. With a 7E model, teachers will *engage* and *elicit* and students will *elaborate* and *extend*. This is not the first enhancement of instructional models, nor will it be the last. Readers should not reject the enhancement because they are used to the traditional 5E model, or worse yet, because they hold the 5E model sacred. The 5E model is itself an enhancement of the three-phrase learning cycle that included exploration, invention, and discovery (Karplus and Thier 1967.) In the 5E model, these phases were initially referred to as explore, explain, and expand. In another learning cycle, they are referred to as exploration, term introduction, and concept application (Lawson 1995).

The 5E learning cycle has been shown to be an extremely effective approach to learning (Lawson 1995; Guzzetti et al. 1993). The goal of the 7E learning model is to emphasize the increasing importance of eliciting prior understandings and the extending, or transfer, of concepts. With this new model, teachers should no longer overlook these essential requirements for student learning. ■

Arthur Eisenkraft is a project director of Active Physics and a past president of NSTA, 60 Stormytown Road, Ossining, NY 10562; e-mail: eisenkraft@att.net.

References

Bransford, J.D., A.L. Brown, and R R. Cocking, eds. 2000. *How People Learn*. Washington, D.C.: National Academy Press.

Bybee, RW. 1997. *Achieving Scientific Literacy*. Portsmouth, N.H.: Heinemann.

Colburn, A., and M.P. Clough. 1997. Implementing the learning cycle. *The Science Teacher* 64(5): 30-33.

Gil, O. 2002. Implications of inquiry curriculum for teaching. Paper presented at National Science Teachers Association Convention, 5- 7 December, in Alburquerque, N.M.

Guzzetti B., T. E. Taylor, G.V. Glass, and W.S. Gammas. 1993. Promoting conceptual change in science: A comparative meta-analysis of instructional interventions from reading education and science education. *Reading Research Quarterly* 28:11 7-1 59.

Hilgard, E.R., and G.H. Bower. 1975. *Theories of Learning*. Englewood Cliffs, N.J.: Prentice Hall.

Karplus, R., and H.D. Thier. 1967. *A New Look at Elementary School Science*. Chicago: Rand McNally.

Lawson, A.E. 1995. *Science Teaching and the Development of Thinking*. Belmont, Calif.: Wadsworth.

Lawson, A.E. 2001.Using the learning cycle to teach biology concepts and reasoning patterns. *Journal of Biological Education* 35(4): 165-169.

Mayer, R.E. 1979. Can advance organizers influence meaningful learning? *Review of Educational Research* 49(2): 371-383.

Thorndike, E.L. 1923, *Educational Psychology, Vol. II: The Psychology of Learning*. New York: Teachers College, Columbia University.

Safety Contract: Safety in the Physics Classroom

Physics is a laboratory science. During this course you will be doing many investigations in which safety is a factor. To ensure the safety of all students, the following safety rules will be followed. You will be responsible for abiding by these rules at all times. After reading the rules, you and a parent or guardian must sign a safety contract acknowledging that you have read and understood the rules and will follow them at all times. The safety contract is also provided as a *Blackline Master* in your *Teacher Resources CD*.

General Rules

1. Never work in the lab unless your teacher or an approved substitute is present.
2. You must follow all directions carefully and use only materials and equipment provided by your teacher. Only experiments approved by your teacher may be carried out in the physics classroom.
3. Do not start work on a lab experiment until told to do so by your teacher.
4. Identify and know the location of a fire extinguisher, fire blanket, emergency shower, eyewash, water shut-offs, and telephone.
5. Eating, drinking, chewing gum, or applying cosmetics is strictly prohibited.
6. All spills and accidents must be reported to your teacher immediately.
7. Be aware of what other groups are doing when you move about in the lab area. Some lab experiments will overlap lab stations at times.
8. If using any kind of flame, use extreme caution. Keep your hands, hair, and clothing away from flames.
9. Long hair must be tied back at all times. No loose clothing/sandals are allowed in the laboratory; long sleeves must be rolled up; bulky jackets, as well as jewelry, must be removed.
10. If any equipment appears defective or breaks while in the lab, report it to your teacher immediately.
11. There will be no running, jumping, pushing, or other behavior considered inappropriate in the science laboratory. You must behave in an orderly and responsible way at all times.

Equipment Rules

1. All equipment must be checked out and returned properly.
2. Do not touch any equipment until you are instructed to do so.
3. When working with electric circuits of any kind, do not plug in or energize the apparatus until the instructor has inspected your circuit and approved the connections and safety considerations.
4. Whenever changing electric connections, always unplug or de-energize the circuit before hand.
5. Assume all electric circuits are dangerous.
6. If any piece of equipment falls on the floor, immediately pick it up and secure it to the table. If working with apparatus that is to be used on the floor, make certain that no other students will walk in that area.
7. Notify your teacher immediately if you observe any unsafe condition.
8. Never stand where a body part is directly below a suspended mass of any kind, or any position where a mass might fall.
9. Some material may become very hot during use. Be aware of these objects and safeguard them so that they can not be touched inadvertently.
10. Safety goggles should be worn in all lab situations unless specifically exempted by your teacher.
11. All equipment is to be used for its intended purpose in the lab only.
12. When doing computer-assisted labs, make no changes to the computer programs without the specific permission of your teacher.
13. Do not start any part of the lab exercise until all members of your group are ready.
14. Never shine a laser in such a manner that the beam or its reflection might strike a person in the eye. Avoid looking at the direct reflection of a laser beam from any shiny surface.
15. If a thermometer or any other glassware breaks, immediately report the breakage to your teacher.
16. Do not handle any radioactive material.
17. No cell phones or other personal electrical devices are to be used in the laboratory space.

Work Area

1. When working in the laboratory, all materials should be removed from the workstation except for instructions, log books, and data tables. Materials should not be placed on the floor, as this is a hazard for someone walking with glassware or equipment.
2. The work area should be kept clean at all times. After completing an experiment, wipe down the area.

Safety Contract

The following contract may be reproduced and must be signed by each student and a parent or guardian before participating in laboratory experiments.

I have read **Safety in the Physics Classroom** and understand the requirements fully. I recognize that there are risks associated with any physics experiment and acknowledge my responsibility in minimizing these risks by abiding by the safety rules at all times.

Please list any known medical conditions or allergies:

____

____

____

____

____

____

I do / do not wear contact lenses. (Circle one)

Emergency phone contact ______________________________

Student signature ______________________ Date __________

Parent or guardian signature ______________________ Date __________

Teacher signature ______________________ Date __________

NOTES

Chapter 1

DRIVING THE ROADS

CHAPTER 1

Driving the Roads

Chapter Overview

Chapter Challenge

Students are challenged to demonstrate some essential knowledge of what they know about the physics of driving in a two-to-three minute presentation to their driving academy's instructors. They are asked to write a report and enhance it using graphs and charts on posters. They are also expected to answer their parents' questions about driving safety. The *Chapter Challenge* is set against the background of insurance companies' policies, which requires drivers to have passed a driving course. The project draws on driving schools that enroll teenage drivers.

In their presentation, students are expected to explain various factors that affect safe driving. They have to show how speed, friction, and the radius of a curve determine the decision to slow down when negotiating a turn on the road. They are required to present the relationship between reaction distance, braking distance, and total stopping distance. Most importantly, students have to explain how these factors affect a driver's decision when approaching an intersection.

As you review the assignments, reassure students, that while they may feel unprepared now, by the end of the chapter they will have the necessary skills and vocabulary to respond adequately. To facilitate cooperative work in groups, have each student take individual responsibility for different tasks that make up the challenge. The academy instructor(s) should be given a rubric of the criteria to evaluate each presentation and write comments to describe how well students met their challenge. The academy instructors could be teachers from your school or other persons you might choose.

Read over the *Chapter Challenge* and the *Criteria for Success* with the class to establish a rubric for assessing student performance. In your grading criteria, include factors such as the number of physics principles referenced, the number of physics terms used properly, clarity of expression, legality of advice, adherence to safety, and credibility.

Chapter Summary

Driving the Roads investigates how reaction time is related to following distance and concepts of velocity and acceleration. Students

- learn why reaction time is critical to avoiding accidents while driving.
- understand the significance of uncertainties in measurements, and distinguish between accurate and precise measurements in order to maintain a safe distance to avoid a collision.
- understand the difference between average speed and instantaneous speed, use graphs of motion to measure velocity, and use equations to calculate average speed and velocity.
- examine how a change in velocity determines the acceleration of an automobile and learn the difference between positive and negative acceleration.
- learn how the speed of an automobile is related to its braking distance. Explore the concept of negative acceleration.
- understand why it is unsafe to stop a vehicle beyond the STOP Zone when the light at an intersection turns yellow.
- learn why it is important to slow down while driving around a curve.

Key Physics Concepts

Section Summaries	Physics Principles
Section 1 Reaction Time: Responding to Road Hazards Using various reaction timers, students explore the time it takes them to react to a situation. This section introduces students to the process of first beginning with their own ideas and predictions, then implementing an investigation that results in both qualitative and quantitative data.	**Reaction time**
Section 2 Measurement: Errors, Accuracy, and Precision Students count the number of strides it takes them to cover a selected distance in an area away from traffic. Students measure the length of their stride using a meter stick and calculate the entire distance by multiplying the total number of strides with the length of each stride. The measurements are then compared by each group. By comparing measurements, students arrive at an understanding of error and the different kinds of errors present in a measurement.	**Errors in measurement** **Accuracy** **Precision**
Section 3 Average Speed: Following Distance and Models of Motion Strobe or multiple-exposure photos of a moving vehicle are used to illustrate speed and acceleration. Students then use a motion detector to measure their walking speed and obtain a computer-generated graph of their motion. Information about speed and velocity is then connected to reaction distance with a discussion on tailgating.	**Average speed** **Instantaneous speed** **Velocity** **Reaction distance**
Section 4 Graphing Motion: Distance, Velocity, and Acceleration Students use sloped tracks to investigate the speed and distance an automobile travels before stopping. They then examine data on time and distance required to stop a vehicle moving at various speeds. This is connected to the total time required to react to a hazard, apply force to the brake, and slow the motion of the vehicle to a complete stop.	**Acceleration** **Positive acceleration** **Negative acceleration** **Vector quantity**
Section 5 Negative Acceleration: Braking Your Automobile The students investigate the relationship between an automobile's speed and the distance required to bring it to a stop. Students draw graphs to study the change in velocity with respect to time. The concept of negative acceleration is explored in this context.	**Negative acceleration** **Braking distance**
Section 6 Using Models: Intersections with a Yellow Light Using a spreadsheet model of an intersection, students explore how reaction time, speed, and stopping distance affect what they should do at a yellow light. This also introduces them to how transportation engineers use a computer simulation to model various factors affecting decisions about speed limits and traffic-light cycles. Students now have the opportunity to apply their understanding of reaction time, distance vs. velocity, and braking distance to identify the STOP, GO, and Dilemma Zones at intersections when they see a yellow light.	**Speed** **Negative acceleration** **Distance vs. time relationships**
Section 7 Centripetal Force: Driving on Curves Students' perceptions and prior learning about the force needed to change the direction of a moving object are challenged in this section. After performing investigations, they reflect on the discrepancy between their perceptions and observed results. Students then read for more information on how forces change the direction.	**Force** **Centripetal force** **Centripetal acceleration**

Chapter Concept Map

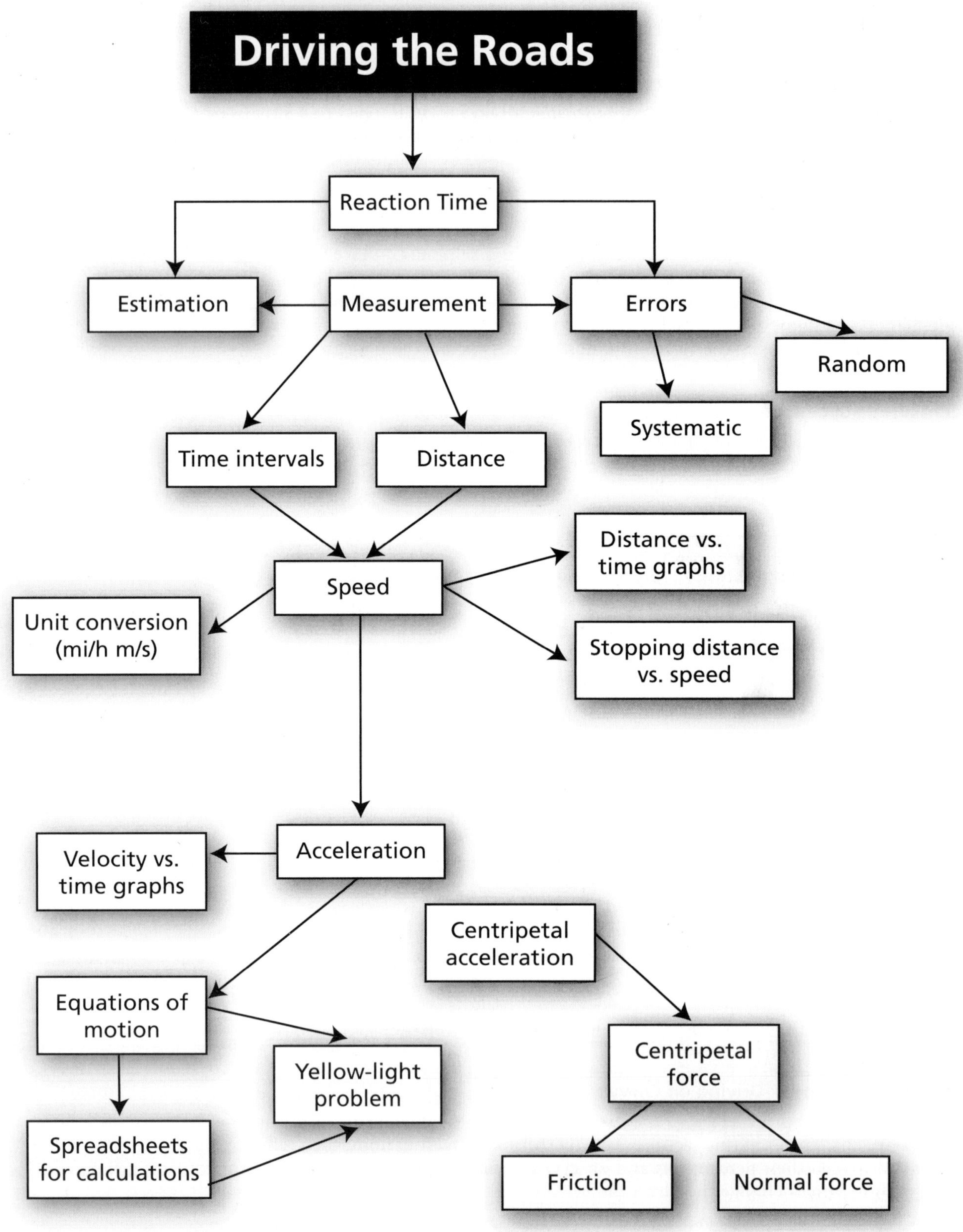

CHAPTER 1

Understanding by Design*

The *Understanding by Design* template focuses on the three stages of backward design:
- **Identify desired results**
- **Determine acceptable evidence**
- **Plan learning experiences**

What overarching understandings are desired?

Traffic safety requires knowledge of physics.
- Motion can be described in terms of position, velocity and acceleration.
- Motion can be described both algebraically and graphically.
- Complex phenomena can be mathematically modeled.
- All measurements have uncertainties (no measurement is exact).

What are the overarching "essential" questions?

- How can physics knowledge help you to be a safe driver?
- How can driving be described using the concepts of position, velocity, and acceleration?
- What is the effect of reaction time on driving?
- How can something as complicated as the traffic at an intersection be modeled using a limited number of variables?
- How are measurements crucial for understanding the motion of a vehicle?

What will students understand as a result of this chapter?

- A person has a measurable reaction time.
- Poor reaction time can lead to more accidents.
- All measurements have uncertainties or random errors.
- Repeated measurements can vary in accuracy and precision.
- Random errors can be attributed to the measurement and/or the measuring instrument.
- Average velocity = total distance traveled over a given time ($v = \Delta d/\Delta t$).
- The slope of a displacement vs. time graph is equal to the velocity.
- Average velocity = total distance traveled over a given time ($v = \Delta d/\Delta t$).
- The slope of a velocity vs. time graph is equal to the acceleration.
- Braking distance is dependent on the negative acceleration of the vehicle (brakes, road surface) and reaction time ($d = v^2/2a$).
- Models can be used to mathematically describe what happens at a yellow light.
- An intersection with a yellow light has a STOP Zone and a GO Zone.
- A safe intersection has an overlap between the GO and STOP Zones.
- An unsafe intersection has positions which are neither in the STOP Zone or the GO Zone.
- Turning a vehicle is an acceleration of the vehicle.
- A force toward the center of a circle is required to turn a vehicle.

What "essential" questions will focus this chapter?

- What is human reaction time and how does it affect driving?
- If five people measure the same event, how do we expect their measurements will differ?
- How is velocity defined and how is it measured?
- How can motion be represented graphically?
- How is braking distance dependent on speed?
- How can an intersection with a yellow light be modeled mathematically?
- What factors describe the motion of a vehicle making a turn?

* Grant Wiggins and Jay McTighe, *Understanding by Design* (Merril/Prentice Hall, 1998), 181.

Pacing Guide

The *Pacing Guide* below is designed so that you have the option to complete the first eight chapters of *Active Physics* during the school year. The *Plan A Pacing Guide* allows the students to complete all the *Investigates*. If you are a new teacher, or unfamiliar with the program, you may have difficulty adhering to *Pacing Guide A*. *Pacing Guide B* suggests places where either time or equipment may be saved if it becomes necessary to complete the chapter in the allotted time. To reach this goal, many of the investigations are whole-class *Investigates* rather than small-group *Investigates*. This will save time and require less equipment than the optimal inquiry-based instruction that the curriculum is intended to provide. In order to choose which plan is better for you, please consult the *Implementation Chart* following this guide.

Note: Each "day" assumes a 45-minute class period, or one half of a 90-minute block.

Day	Plan A (small-group *Investigates*)	Homework (for Plan A and Plan B)	Day	Plan B (combination of whole-class and small-group *Investigates*)	Plan B Equipment Reduction
1	*Scenario, Your Challenge, Criteria for Success*, Scoring Rubric, keeping a log.	Read the material on *Engineering Design Cycle* and give three reasons why you think engineers follow this process.	1	**See Plan A.**	
2	**Section 1** Discuss *What Do You See?* and *What Do You Think?*; students complete *Investigate*.	Read and summarize the *Physics Talk* in your *Active Physics* log, answer *Checking Up* questions.	2	**See Plan A.**	
3	Review *Investigate*, summarize results, discuss *Physics Talk*, review *Checking Up, What Do You Think Now?*, and *Reflecting on the Section and the Challenge*, answer *Physics to Go* Question 3 and discuss with class; students should answer the question in their log.	Answer *Physics to Go* Questions 4-7. Answer the *Preparing for the Chapter Challenge* in your *Active Physics* log.	3	**See Plan A.**	
4	Review *Physics to Go*. Students discuss answers to *Preparing for the Chapter Challenge* in their groups. **Section 2** *What Do You See?, What Do You Think?*; students complete *Investigate* Steps 1-7.	Read *Physics Talk*. Answer the *Checking Up* questions.	4	**See Plan A.**	
5	Review and summarize the *Investigate*; students complete *Investigate* Steps 8 and 9. Discuss *Physics Talk*, SI system, review *Checking Up* questions.	Answer *Physics to Go* Questions 1, 3-8 and record the *Preparing for the Chapter Challenge* in your log.	5	**See Plan A.**	

Day	Plan A (small-group *Investigates*)	Homework (for Plan A and Plan B)	Day	Plan B (combination of whole-class and small-group *Investigates*)	Plan B Equipment Reduction
6	Review *Physics to Go*, discuss *What Do You Think Now?*, *Reflecting on the Section and the Challenge*. **Section 3** *What Do You See?*, *What Do You Think?*; students complete *Investigate* Steps 1-3.	Read *Physics Talk*. Copy Sample Problems 1 and 2 into your *Active Physics* log.	6	See Plan A.	
7	Students complete *Investigate* Steps 4-8.	Read *Physics Talk* and answer the *Checking Up* questions.	7	Complete *Investigate* Steps 4-8 as a teacher-led demonstration. Discuss *Physics Talk* with examples and practice problems for students.	Requires only one computer setup, motion detector, and interface.
8	Discuss *Physics Talk* with examples and practice problems for students.	Answer *Physics to Go* Questions 1-2, 4-7.			
9	Review *Physics to Go*, discuss *Physics Talk*, *What Do You Think Now?*, and *Reflecting on the Section and the Challenge*.	Answer *Physics to Go* Questions 3, 8-12.	8	See Plan A.	
10	Discuss *Physics to Go*. **Section 4** Discuss *What Do You See?*, *What Do You Think?*; students complete *Investigate* Steps 1-7.	Read *Physics Talk* and answer *Checking Up* Questions 1-3, answer *Physics to Go* Questions 1-6.	9	Discuss *Physics to Go*. **Section 4** Discuss *What Do You See?*, *What Do You Think?;* Complete *Investigate* all steps as a teacher-led demonstration. Discuss *Physics Talk*, *What Do You Think Now?* and *Reflecting on the Section and the Challenge*.	Requires only one computer setup, with motion detector and interface, one dynamics cart, ringstand, crossarm, one ramp right-angle holder, and index cards.
11	Review *Physics to Go* and *Checking Up Questions* 1-3. Work with students as they complete *Investigate* Steps 8-12. Discuss *Physics Talk*, *What Do You Think Now?* and *Reflecting on the Section and the Challenge*.	Read *Physics Talk*. Copy graphs on page 64 into your *Active Physics* log, answer *Checking Up* Questions 4-5, and *Physics to Go* Questions 9-12, 15.			
12	Review *Physics to Go*, *Checking Up* Questions 4-5. Review *Preparing for the Chapter Challenge*. **Section 5** Discuss *What Do You See?* and *What Do You Think?;* students complete *Investigate* Steps 1-4.	Read *Physics Talk* and summarize in your log. Answer *Checking Up* questions.	10	Review *Physics to Go*, and *Checking Up* Questions 4-5. Review *Preparing for the Chapter Challenge*. **Section 5** Discuss *What Do You See?* and *What Do You Think?;* complete *Investigate* Steps 1-4 as a teacher-led demonstration.	Requires only one dynamics cart, one ramp, one velocimeter, ringstand, crossarm, right-angle clamp, ruler, meterstick, scissors, index cards, extension clamp, and file folder.

Pacing Guide *(continued)*

Day	Plan A (small-group *Investigates*)	Homework (for Plan A and Plan B)	Day	Plan B (combination of whole-class and small-group *Investigates*)	Plan B Equipment Reduction
13	Students complete *Investigate* Steps 5-8, *What Do You Think Now?* and *Reflecting on the Section and the Challenge.*	Read *Physics Talk*, answer *Physics to Go* Questions 2-4, 8, 10, 11.	11	**See Plan A.**	
14	Discuss *Physics Talk*, review *Physics to Go* and *Preparing for the Chapter Challenge*. Students start preparation for the *Mini-Challenge.*	Read about the *Mini-Challenge.* Choose areas to be addressed in the *Mini-Challenge* and discuss with the group.	12	**See Plan A.**	
15	Students start preparing for the *Mini-Challenge*, discuss with their groups who will complete each part and what will be in the challenge.	Write out response to *Mini-Challenge.*	13	**See Plan A.**	
16	*Mini-Challenge* presentations.	Using the school library or Internet sources, find out how a yellow light works, and record your responses in your *Active Physics* log.	14	**See Plan A.**	
17	***Section 6*** Discuss *What Do You See?, What Do You Think?;* students complete *Investigate* Part A.	Read *Physics Talk* and answer *Physics to Go* Questions 1-4, 6.	15	***Section 6*** Discuss *What Do You See?, What Do You Think?;* complete *Investigate* Parts A and B as a teacher-led demonstration. Discuss *Physics Talk*, review *Physics to Go* homework and *What Do You Think Now?*	Only requires one dynamics cart, computer interface and probes to measure the force and velocity.
18	Do *Investigate* Part B as a teacher-led demonstration. Discuss *Physics Talk* and review *Physics to Go* homework. Discuss *What Do You Think Now?*	Read *Physics Talk*. Answer *Checking Up* questions and *Physics to Go* Question 5.			
19	Review *Physics to Go* and *Checking Up* questions. Ask students what they might write for *Preparing for the Chapter Challenge.* ***Section 7*** Discuss *What Do You See?, What Do You Think?;* students complete *Investigate* Steps 1-10.	Read *Physics Talk* and answer *Checking Up* questions.	16	Review *Physics to Go* and *Checking Up* questions. Ask students what they might write for *Preparing for the Chapter Challenge.* ***Section 7*** Discuss *What Do You See?, What Do You Think?;* Students complete *Investigate* Steps 1-3 and teacher completes Steps 4-10 as a demonstration.	Requires only one turntable, one cork accelerometer, sandpaper, masking tape, and wood piece.
20	Review *Investigate* Steps 1-10. Students complete *Investigate* Steps 11-13. Discuss *Physics Talk, What Do You Think Now?,* students complete *Preparing for the Chapter Challenge.*	Answer *Physics to Go* Questions 1, 3, 4, 7, 9, 10, 11.	17	**See Plan A.**	

Day	Plan A (small-group *Investigates*)	Homework (for Plan A and Plan B)	Day	Plan B (combination of whole-class and small-group *Investigates*)	Plan B Equipment Reduction
21	Review *Physics to Go*. Start preparation for *Chapter Challenge*.	Prepare for *Chapter Challenge*.	18	See Plan A.	
22	Continue preparing for *Chapter Challenge*.	Prepare for *Chapter Challenge*.	19	See Plan A.	
23	Facilitate *Chapter Challenge* presentations.	Study for *Physics Practice Test*.	20	See Plan A.	
24	*Physics Practice Test*.		21	See Plan A.	

Implementation Chart

Hopefully, as you become more experienced and comfortable with the curriculum, you will shift to more small-group *Investigates*. Accordingly, below is an *Implementation Chart* that suggests a three-year timetable to expand the student's role in the chapter by having them do more of the *Investigates*. Although this will require a slightly greater expenditure of time and more equipment, the benefits to the student will be manifest. Eventually, your goal should be to have the students complete almost all the investigations rather than you having to provide the maximum opportunity for inquiry.

	Section 1 Investigate	Section 2 Investigate	Section 3 Investigate	Section 4 Investigate	Section 5 Investigate	Section 6 Investigate	Section 7 Investigate
Year 1	Small group	Small group	Whole class	Whole class	Whole class	Whole class	Small group
Year 2	Small group	Small group	Small group	Whole class	Small group	Whole class	Small group
Year 3	Small group	Small group	Small group	Small group	Small group	Small group	Small group

NOTES

Chapter Materials and Equipment

The following tables contain lists of materials and equipment needed to complete all the experiments. The tables are organized as follows:

- Multimedia Needed
- Durables
- Consumables
- Additional Items Needed Not Supplied

Multimedia contains all the software and content videos that are necessary to complete the sections. All the multimedia items needed will be included in the Multimedia DVD/CD Set. In addition to the Multimedia DVD/CD Set, other multimedia items are available such as the *ExamView*™ Test Generator and the Constructing Physics Understanding simulations.

Durables are items which are not consumed during an experiment. They can be used several times. **Consumables** are items that are used up during each class and must be resupplied for future classes. Both the durables and consumables are broken down by group and by class. A group consists of four students. While the group size will be determined by the teacher based upon logistics and availability of equipment, the information is based upon recommended group size of four students. Items listed per class are based on a class size of 40 students.

Additional Items Needed Not Supplied consists of items that are needed to complete the sections, but are not supplied by **It's About Time**®.

The left-hand column gives the section number(s) in which each item will be used. The next column provides information needed for ordering the particular item. The next column gives the item description. The next columns will indicate quantity, based on either group or class needs. The final column provides a space for you to fill in your total item quantity. You will notice certain rows are shaded and have an asterisk in the last column. This is to indicate a duplicate item listing, based on the needs of different sections. You will not need to figure the items in the shaded rows into your total items needed, as they have already been accounted for in the corresponding non-shaded row.

The *Materials and Equipment* lists are broken down into either PLAN A or PLAN B corresponding to the *Pacing Guide*.

PLAN A

Section	ISBN	Multimedia	Group (4 Students)	Class	Total Items Needed
6	978-1-60720-040-3	Multimedia DVD/CD Set		1 per class	

Section	Item No.	Durables	Group (4 Students)	Class	Total Items Needed
1, 7	SS-7779	Stopwatch	2 per group		
1, 4, 7	RS-2826	Ruler, metric, 30 cm	1 per group		
2	TA-0025	Tape measure, wind-up	1 per group		
2, 5	SM-1676	Meter stick, wood	1 per group		
4, 5	SH-7212	Ring stand, large	1 per group		
4, 5	ST-0002	Rod, aluminum, 12 in. (length) × 3/8 in. (diameter) (to act as crossarm)	1 per group		
4, 5	HO-0002	Holder, right angle (to act as crossarm)	1 per group		

* *Please note duplicate item.*

Section	Item No.	Durables	Group (4 Students)	Class	Total Items Needed
4, 5	SR-6737	Inclined plane ramp for lab cart	1 per group		
4, 5	GC-0001	Dynamics cart	1 per group		
5	SS-1281	Scissors	1 per group		
5	EX-0001	Extension clamp	2 per group		
5	BE-0001	Velocimeter	1 per group		
7	CA-0010	Car, toy, battery operated	1 per group		
7	TT-6833	Turn table, heavy duty	1 per group		
7	WS-6732	Piece, wood, 1 in. × 2 in. × 2 in.	1 per group		
7	AH-9250	Cork, accelerometer	1 per group		
7	BS-7203	Ball, bocce	1 per group		
Section	**Item No.**	**Consumables**	**Group (4 Students)**	**Class**	**Total Items Needed**
7	BS-1608	Battery, AA	1 per group		
7	SS-6078	Sandpaper, 60 grit	1 per group		
2, 7	TS-2662	Tape, masking		1 per class	
4, 5	CS-5011	Index cards, pkg 100		1 per class	
5	FF-7704	Folder, file		10 per class	
5	GS-7706	Paper, graph, pkg of 50		1 per class	
7	SS-7722	String, ball		1 per class	
Section	**Item No.**	**Additional Items Needed Not Supplied**	**Group (4 Students)**	**Class**	**Total Items Needed**
3, 4		Motion detector, probe and interface		1 per class	
3, 4, 6		Computer, station or calculator, CBL or equivalent system	1 per group		
7		Stool, rotating		1 per class	
7		Helmet, safety		1 per class	
7		Pads, knee		1 set per class	
7		Pads, elbow		1 set per class	
7		Rolled up newspaper or magazine	1 per group		
2		Access to a clear hallway		1 per class	

* *Please note duplicate item.*

CHAPTER 1

PLAN B

Section	ISBN	Multimedia Needed	Group (4 Students)	Class	Total Items Needed
6	978-1-60720-040-3	Multimedia DVD/CD Set		1 per class	

Section	Item No.	Durables	Group (4 Students)	Class	Total Items Needed
1, 7	SS-7779	Stopwatch	2 per group		
1, 7	RS-2826	Ruler, metric, 30 cm	1 per group		
4	RS-2826	Ruler, metric, 30 cm		1 per class	*
2	TA-0025	Tape measure, wind-up	1 per group		
2	SM-1676	Meter stick, wood	1 per group		
5	SM-1676	Meter stick, wood		1 per class	*
4, 5	SH-7212	Ring stand, large		1 per class	
4, 5	ST-0002	Rod, aluminum, 12 in. (length) × 3/8 in. (diameter) (to act as crossarm)		1 per class	
4, 5	HO-0002	Holder, right angle (to act as crossarm)		1 per class	
4, 5	SR-6737	Inclined plane ramp for lab cart		1 per class	
4, 5	GC-0001	Dynamics cart		1 per class	
5	SS-1281	Scissors		1 per class	
5	EX-0001	Extension clamp		2 per class	
5	BE-0001	Velocimeter		1 per class	
7	CA-0010	Car, toy, battery operated	1 per group		
7	TT-6833	Turn table, heavy duty		1 per class	
7	WS-6732	Piece, wood, 1 in. × 2 in. × 2 in.		1 per class	
7	AH-9250	Cork, accelerometer		1 per class	
7	BS-7203	Ball, bocce	1 per group		
Section	**Item No.**	**Consumables**	**Group (4 Students)**	**Class**	**Total Items Needed**
7	BS-1608	Battery, AA	1 per group		
7	SS-6078	Sandpaper, 60 grit		1 per class	
2, 7	TS-2662	Tape, masking		1 per class	
4, 5	CS-5011	Index cards, pkg 100		1 per class	
5	FF-7704	Folder, file		10 per class	
5	GS-7706	Paper, graph, pkg of 50		1 per class	
7	SS-7722	String, ball		1 per class	

* *Please note duplicate item.*

Section	Item No.	Additional Items Needed Not Supplied	Group (4 Students)	Class	Total Items Needed
3, 4		Motion detector, probe and interface		1 per class	
3, 4, 6		Computer, station or calculator, CBL or equivalent system		1 per class	
7		Stool, rotating		1 per class	
7		Helmet, safety		1 per class	
7		Pads, knee		1 set per class	
7		Pads, elbow		1 set per class	
7		Rolled up newspaper or magazine	1 per group		
2		Access to a clear hallway		1 per class	

* *Please note duplicate item.*

NOTES

Teacher Resources

Blackline Masters

Available in *Teacher Resources* and on *Color Overheads and BLMs CD.*

Chapter Supports

Title	Point of Use	Blackline Masters
Standard for Excellence	*Chapter Challenge* Introduction	1a
Simplified Engineering Design Cycle	*Chapter Challenge* Introduction	1b

Section Quizzes

Title	Point of Use	Blackline Masters
Section 1 Quiz	Section 1	1-1b
Section 2 Quiz	Section 2	1-2b
Section 3 Quiz	Section 3	1-3b
Section 4 Quiz	Section 4	1-4d
Section 5 Quiz	Section 5	1-5c
Section 6 Quiz	Section 6	1-6b
Section 7 Quiz	Section 7	1-7c

Section Supports

Title	Point of Use	Blackline Masters
Why is there a 7E Instructional Model?	Section 1	1-1a
Table for SI System	Section 2 – *Physics Talk*	1-2a
Distance-Time Graphs A-E	Section 3 – *Physics Talk*	1-3a
Speed Conversion Table	Section 4 – *Investigate*	1-4a
Constant Acceleration Graphs	Section 4 – *Physics Talk*	1-4b
Comparing Motion Graphs	Section 4 – *Physics Talk*	1-4c
Motion of Car with Positive and Negative Acceleration Diagrams A-D	Section 5 – *Physics Talk*	1-5a
Motion of Car with Positive and Negative Acceleration Table	Section 5 – *Physics Talk*	1-5b
Spreadsheet for Input and Output Variables	Section 6 – *Investigate*	1-6a
Cloze Activity	Section 7 – *ELL Strategies*	1-7a
Sports Car and Touring Sedan Data Table	Section 7	1-7b

Chapter Assessment

Title	Point of Use	Blackline Masters
Physics Practice Test	End-of-Chapter	1c
Sample Assesment Rubric	End-of-Chapter	1d

Color Overheads

Available on *Color Overheads* and *BLMs CD.*

Title	Point of Use
Physics Corner	*Chapter Challenge* Introduction
What Do You See?	Section 1
What Do You See?	Section 2
Length of the Hallway	Section 2, *Physics Talk*
Accuracy and Precision	Section 2, *Physics Talk*
What Do You See?	Section 3
What Do You See?	Section 4
What Do You See?	Section 5
What Do You See?	Section 6, *Investigate*
Intersection, Part A	Section 6, *Investigate*
Intersections, Part A, Step 4	Section 6, *Investigate*
Yellow-Light Model	Section 6, *Investigate*
Intersections I, II, and III	Section 6
What Do You See?	Section 7

NOTES

Chapter Challenge

Launcher Material

The first chapter in *Active Physics*, *Driving the Roads*, is a launcher chapter. In the first section of this chapter, students are introduced to the many features of *Active Physics*, as indicated in the manila colored call-out boxes. However, what is unique about this introduction is that it also explains to the students why each feature is included and how it will help them learn physics. Prefaced by the statement "The more you understand about how you learn, the better you will be at learning." encourages students to expand their metacognitive awareness.

Scenario

You may wish to role-play the scene described in the *Scenario* using students as parents and teenage drivers. Ask them how they might try to convince their parents to let them borrow the new car, and how their parents might try to avoid giving it to them. Discuss how their knowledge of physics will help them convince their parents that they are safe drivers. You should encourage students to ask questions in class, and try to create a conversational atmosphere that builds on personal stories and shared experiences.

> 1
>
> **Active Physics Driver's Manual**
>
> *Active Physics* is a research-based program. That means that what you will be doing in *Active Physics* and how you will be doing it is based on researching how students like you learn best.
>
> In the first chapter, you will find notes that explain the various components of *Active Physics* and how they can help you actively engage in learning physics. Think of these notes as your driver's manual for navigating your way through *Active Physics.*
>
> Stopping to think about the rules of the road and your driving habits can make you a better driver. Research shows that stopping to think about your learning can also make you a better learner.
>
> Although some of these notes may seem like a lot of "teacher talk," as you work through each chapter, think about why you are doing each of the things you are asked to do. *The more you understand about how you learn, the better you will be at learning.*
>
> Chapter Challenge
>
> **Driving the Roads**
>
> **Why is there a *Scenario*?**
>
> Welcome to *Active Physics*! You are about to begin an exciting year of discovering how useful, interesting, and fun physics can be. Each *Active Physics* chapter begins with a *Scenario*. The *Scenario* describes a realistic event or situation that you might have experienced or can imagine experiencing. The *Scenario* sets the stage for the *Chapter Challenge*, which follows.
>
> 
>
> **Scenario**
>
> Imagine your parents just bought a new car and your favorite music group is in town. You ask your parents if you could use the new car to take a friend to the concert. What would your parents say? Would you have a conversation like the following?
>
> "I don't care if it is your favorite music group. You are not ready to drive our car."
>
> **"But I've had my license for two whole months!"**
>
> "That test means nothing. You memorized a bunch of facts to get your license."
>
> **"Yes, and now I know all about the law."**
>
> "The traffic laws, maybe. But what about natural laws, like speed and stopping distances?"
>
> **"That's easy, the speed limits are all posted."**
>
> "The speed limit is the maximum speed you can go. You have to adjust your speed according to driving conditions."
>
> **"Okay, I'll do that. Then can I have the car?"**
>
> "No. I didn't say that. You don't know about reaction time and following distance. And what about curves; when should you slow down?"
>
> **"They have yellow signs to tell you what to do on the curves."**
>
> "You need to know more than that before you enter a curve. And what about a yellow light? What does it mean?"
>
> **"Step on it?"**
>
> "See what I mean, you're not ready to drive."
>
> **"It was a joke."**
>
> 2

Keep the *Scenario* focused on the main topic by lightly touching on student answers that broaden their perspective on the importance of safety on the roads, and prompts them to defend their arguments with a reflective mind-set. You may want to invite a professional from a driving school or Department of Motor Vehicles in your area to highlight the importance of why students should be properly informed of the physics behind the rules and regulations of driving. Reassure them that while they may feel ill-prepared now to complete the *Chapter Challenge*, by the end of the chapter they will have the skills and vocabulary to respond adequately. Avoid providing students with too many examples of what could be done, as this may limit their creativity and confidence in seeking their own solutions.

"Driving is no joke. What if you have an accident? What then? What if your friend in the car distracts you?"

"I don't plan on an accident; besides, I'll always wear my seat belt."

"No one plans on an accident—that's why they're called accidents!"

"But I have a valid driver's license."

"Yes, but you still have a lot more to learn."

"You just don't love me."

"I'm doing this *because* I love you."

Why is there a *Chapter Challenge*?

At this point, you are presented with what you and your group members will be required to do as a chapter project. The *Chapter Challenge* may be a problem you are expected to solve or a task you are expected to complete using the knowledge you gain in the chapter. When you first encounter the *Chapter Challenge*, you may find it overwhelming. However, all the physics content in the chapter will help you succeed at the challenge. Each section will provide you with another piece of the puzzle that, when put together, will answer the challenge.

The *Chapter Challenge* is the glue that holds the chapter together. In *Active Physics*, you will never be left wondering, "Why am I learning this?" You will need everything you learn in a chapter to complete the *Chapter Challenge*. You can think of the challenge as the job you need to do over the next few weeks.

Your Challenge

Automobile insurance for most teenage drivers is carried on their parents' policy. Your automobile insurance company says that, if new drivers can pass a course from a certified driving academy, your automobile insurance rates will be reduced. So, before your parents allow you to drive, you must pass a driving course. You find, upon checking with Active Driving Academy—the only driving school in your area—that they enroll students in groups. You are told that, as part of Active Driving Academy's requirements, your group must demonstrate some basic knowledge of the physics of driving. You must demonstrate this knowledge through a two-to three-minute presentation to academy instructors. At a minimum, the presentation must explain the following:

- the relationships among following distance, braking distance, and the total stopping distance, including the factors that affect each;
- how to decide what to do when the light turns yellow as you approach an intersection; and
- the connection among speed, friction, and radius of the curve when turning.

3

CHAPTER 1

Your Challenge

Lead a class discussion about the *Chapter Challenge* and the expectations. You should clarify to the students that the presentation should have a written report with an advertisement, cartoon, or story. Students should be able to explain the main physics concepts in their presentation. It is important for you to emphasize that they enrich their presentation with visual aids. For example, students should be able to show the relationship between reaction distance, braking distance, and stopping distance; should consider what to do when the light turns yellow as they approach an intersection; and should be clear about the connection between speed, friction, and radius of the curve to determine a safe speed limit.

As the class progresses through the chapter, you will need to familiarize students with the content so they can connect the physics principles they learn to their *Chapter Challenge*. You have to keep reminding students that as the chapter unfolds, each new concept will build upon the other, giving them the tools to strengthen their presentation.

Chapter Challenge

Criteria for Success

Take a few moments with your class to develop the criteria for assessing the *Chapter Challenge*. Students should be asked, "What details must your presentation have for you to earn an A?" A way to get started is to make a list of important criteria that are required for the *Chapter Challenge*. List the relevant student suggestions by making frequent references to the physics principles that are likely to be considered.

The criteria should include accurate and clear explanations with original ideas. (It is important that all suggestions be recognized as students volunteer ideas, since this is essentially a brainstorming session.) By soliciting the students' opinions, each criterion should be written in a style that students can understand.

When you have a thorough description of each part of the *Chapter Challenge*, you can have the class vote on how much each part is worth. As students participate in choosing criteria for assessing their own work, they subsequently gain a better understanding of the *Chapter Challenge*.

Chapter Challenge

You will use graphs and charts on posters to enhance your presentation. Using a computer presentation program is permissible but not required. Your group must submit the written report of your presentation to assure academy instructors that you are properly prepared. In addition to the requirements of the driving school, before your parents allow you to drive their car, you must answer their questions about driving safety.

Why are there *Criteria for Success?*

To do well at any job, you need to know what the job expectations are. That applies in the classroom as well as in the workplace. It is essential to define and understand the *Criteria for Success*. Before you begin your job in each chapter, you and your class will discuss and list the criteria that you will be expected to meet. Next, you will determine the relative importance of the assessment criteria. Then, you can assign point values to each component. You will also need to clarify the details of the criteria. For example, you need to know how many physics principles are required to meet a standard of excellence (the best you can do).

Even though physics principles are necessary criteria in addressing the challenge, they are not enough. Each *Chapter Challenge* will also expect you to be imaginative and creative. Each completed project should be unique. It should be a reflection of the interests and talents of the members of your group.

Criteria for Success

For each part of the *Chapter Challenge*, imagine what criteria are required for an excellent presentation and written report. For example, should an excellent presentation have charts or graphs? If you think so, is one chart or graph enough? As a class, list these criteria first. Then read the suggestions given in the *Standard for Excellence* table on the next page and compare them to your class list. By listing your criteria first, you will gain a greater understanding of the challenge. You will also have a chance to revisit the criteria at the end of the chapter before the presentation is due.

Record any notes you have about the *Chapter Challenge* in your *Active Physics* log. You will also need to list the criteria and their point value that your class decided on for assessing the challenge.

4

Read aloud the criteria that you and your students came up with for assessing the *Chapter Challenge*. The reading might bring up other interesting points that you could use to reinforce or modify the criteria.

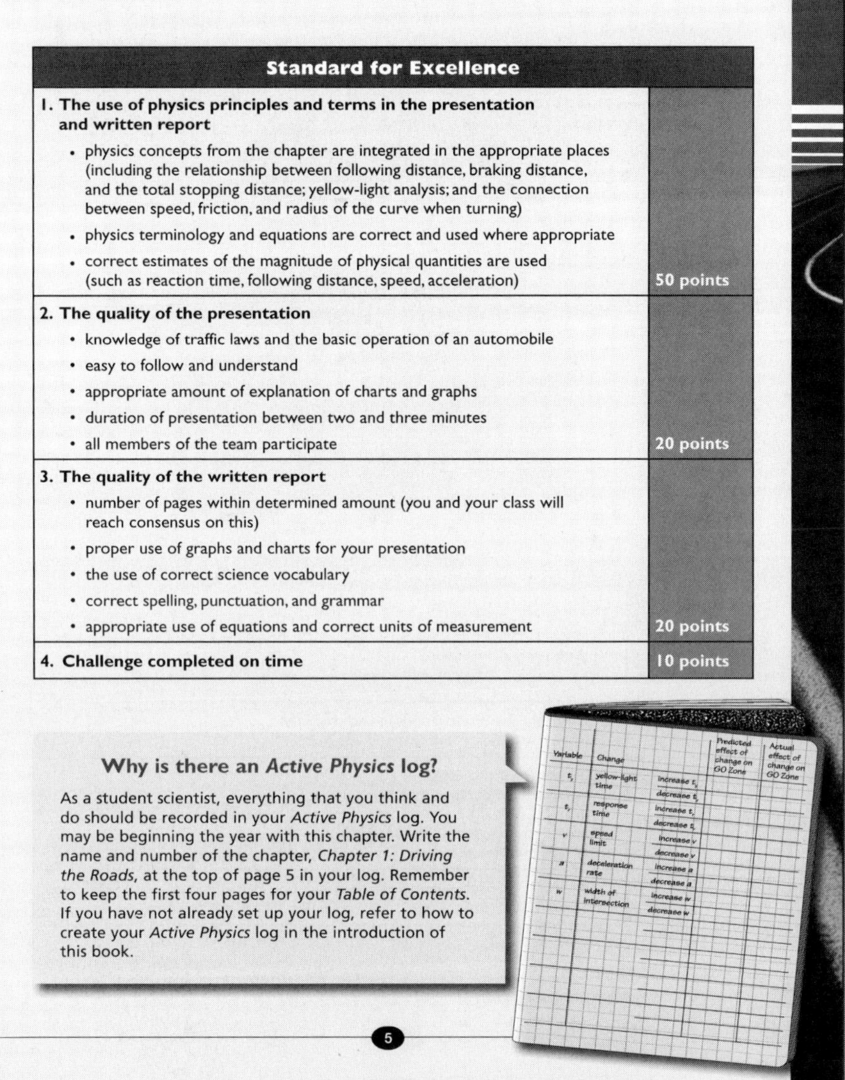

Standard for Excellence	
1. The use of physics principles and terms in the presentation and written report • physics concepts from the chapter are integrated in the appropriate places (including the relationship between following distance, braking distance, and the total stopping distance; yellow-light analysis; and the connection between speed, friction, and radius of the curve when turning) • physics terminology and equations are correct and used where appropriate • correct estimates of the magnitude of physical quantities are used (such as reaction time, following distance, speed, acceleration)	50 points
2. The quality of the presentation • knowledge of traffic laws and the basic operation of an automobile • easy to follow and understand • appropriate amount of explanation of charts and graphs • duration of presentation between two and three minutes • all members of the team participate	20 points
3. The quality of the written report • number of pages within determined amount (you and your class will reach consensus on this) • proper use of graphs and charts for your presentation • the use of correct science vocabulary • correct spelling, punctuation, and grammar • appropriate use of equations and correct units of measurement	20 points
4. Challenge completed on time	10 points

Why is there an *Active Physics* log?

As a student scientist, everything that you think and do should be recorded in your *Active Physics* log. You may be beginning the year with this chapter. Write the name and number of the chapter, *Chapter 1: Driving the Roads*, at the top of page 5 in your log. Remember to keep the first four pages for your *Table of Contents*. If you have not already set up your log, refer to how to create your *Active Physics* log in the introduction of this book.

5

Standard for Excellence

Once the students have compiled a list of criteria, ask them to develop a rough rubric for grading the challenge. This can be as simple as ascribing points to each criterion. You may want to have students assign fewer points to certain components of the rubric. For instance, the quality of the visual presentation would have fewer points than the explanation of physics principles. The rubric should put greater emphasis on how an understanding of physics concepts is utilized in the presentation.

Criteria mentioned in the student text should be included in the rubric. The rubric would therefore cover different aspects of the challenge, such as the use of the physics principles, quality of the written report, and the quality of the visual presentation.

CHAPTER 1

A more comprehensive sample rubric for assessing the *Chapter Challenge* is provided at the end of this chapter in this volume of the *Teacher's Edition*. You could copy and distribute it, or use it as a template for developing scoring guidelines and expectations that suit the needs of your students. For example, you might wish to ensure that core concepts and abilities derived from your local or state science frameworks also appear in the rubric. However, if you decide to evaluate the *Chapter Challenge*, keep in mind that all expectations should be communicated to the students when they begin their work.

Remind students that the rubric can be modified as their understanding of concepts grows. The *Standard for Excellence* table is provided as a Blackline Master in your *Teacher Resources CD*.

1a Blackline Master

Chapter Challenge

Engineering Design Cycle

The *Engineering Design Cycle* is a sequence of steps used to solve problems. Discuss each step of this cycle. Emphasize that this process will continue as students keep learning and testing their understanding of new concepts. Divide students into five groups and give each group the task of determining the purpose of each step in the *Engineering Design Cycle.*

Ask students to discuss and write a brief summary in their *Active Physics* log, explaining the bulleted information given in the chart of the *Simplified Engineering Design Cycle.* Once students have finished writing the summary, have a representative from each group read aloud the explanation that each group member discussed and wrote down. As the student representative is reading his/her group's summary aloud, jot down important ideas on the board and initiate a whole-class discussion to allow all students to collectively participate and share in understanding the purpose of the *Engineering Design Cycle.*

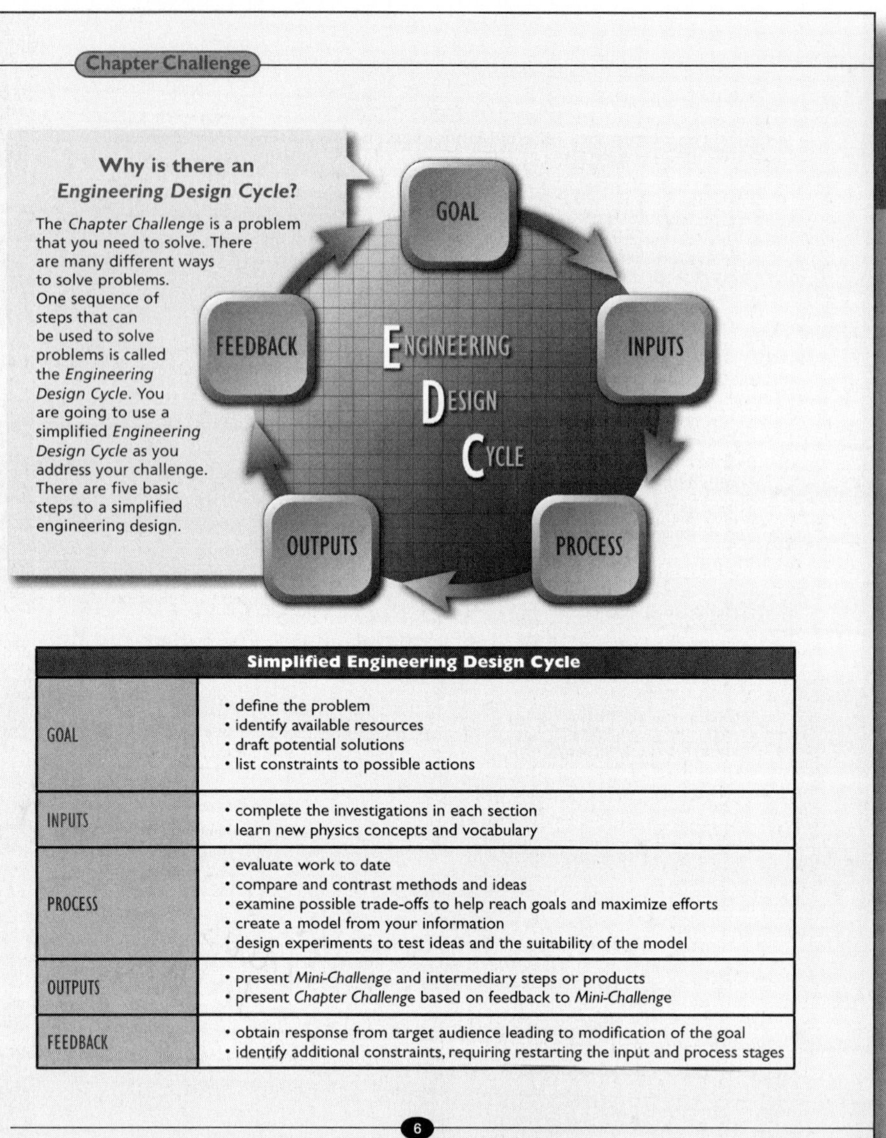
Chapter Challenge

Why is there an *Engineering Design Cycle?*

The *Chapter Challenge* is a problem that you need to solve. There are many different ways to solve problems. One sequence of steps that can be used to solve problems is called the *Engineering Design Cycle*. You are going to use a simplified *Engineering Design Cycle* as you address your challenge. There are five basic steps to a simplified engineering design.

Simplified Engineering Design Cycle	
GOAL	• define the problem • identify available resources • draft potential solutions • list constraints to possible actions
INPUTS	• complete the investigations in each section • learn new physics concepts and vocabulary
PROCESS	• evaluate work to date • compare and contrast methods and ideas • examine possible trade-offs to help reach goals and maximize efforts • create a model from your information • design experiments to test ideas and the suitability of the model
OUTPUTS	• present *Mini-Challenge* and intermediary steps or products • present *Chapter Challenge* based on feedback to *Mini-Challenge*
FEEDBACK	• obtain response from target audience leading to modification of the goal • identify additional constraints, requiring restarting the input and process stages

6

1b **Blackline Master**

You have now heard about your *Chapter Challenge*. You need to give a presentation and write a report that demonstrates your basic knowledge of the physics of driving. You will use a simplified *Engineering Design Cycle* to help your group put together the presentation. Establishing a clear *Goal* is the first step in this process. You have defined the problem you need to solve, identified the *Criteria for Success*, and thought about some of the constraints that you will need to face. You may also already be thinking of some possible ways of giving your presentation. You are well on your way to establishing your *Goal*.

As you experience each one of the chapter sections, you will be gaining *Inputs* to use in the design cycle. These *Inputs* will include new physics concepts, vocabulary, and even equations that will help you with your presentation.

The first *Outputs* of your design cycle will be a short presentation to the driving-academy instructors, along with posters displaying graphs and charts. After several sections, you will work on part of your presentation. Finally, you will receive *Feedback* from your classmates and your instructor about which parts of your presentation are good and which parts need to be refined. You will then repeat the *Engineering Design Cycle* during the second half of the chapter when you gain more *Inputs* and refine your presentation.

The 7E Instructional Model

Active Physics uses a 7E instructional model. The steps (phases) are

• Elicit • Engage • Explore • Explain • Elaborate • Extend • Evaluate

Look for these phases as you work through the first section of this chapter.

Why is there a *Physics Corner*?

The *Physics Corner* lists all the physics principles that the chapter will present. You will learn the physics concepts and master the skills necessary to complete the investigations and the challenge, through active involvement and by engaging your creativity. It will be important for you to understand all the physics principles you are learning, because you will need to apply them at the end of the chapter to complete the *Chapter Challenge*. The *Physics Corner*, along with your teacher, will help you keep track of all the physics concepts you will learn in the chapter.

Physics Corner

Physics in *Driving the Roads*

- Acceleration
- Accuracy in measurement
- Average speed
- Centripetal acceleration
- Centripetal force
- Circular motion
- Distance and time
- Doppler effect
- Friction
- Instantaneous speed
- Momentum
- Positive and negative acceleration
- Precision in measurement
- Reaction time
- Velocity

CHAPTER 1

Physics Corner

The *Physics Corner* illustrates the physics concepts that students are required to learn. This section provides a quick preview of all the physics concepts that will be presented in the chapter. Ask your students if they are familiar with any of the concepts they are about to study and encourage them to provide definitions for the terms they already know. Remind them that the study of these concepts will help them in writing their report, and they will be able to enhance their presentation as they make meaningful connections between concepts.

You will find students motivated by actively engaging in the learning process. As the *Chapter Challenge* approaches, review the *Physics Corner* to help them keep track of all the physics concepts they have learned.

SECTION 1

Reaction Time: Responding To Road Hazards

Section Overview

This section addresses questions of reaction time in relation to bringing a vehicle to rest. Students conduct experiments to measure their reaction time, first in the absence and, later, in the presence of distractions. They estimate how long it would take them to move their foot between imaginary gas and brake pedals. Students also investigate one of two methods to see how reaction time varies in different situations. As lab partners, they record the difference in the time it takes them to turn off their stopwatches or catch a dropped ruler. They then make a list of activities that distract them while driving and simulate those distractions.

Background Information

Background Information for most sections is provided for the interest and insight of the teacher only. It is not intended to be part of the classroom instruction. Reaction time can be understood by grouping physiological processes into three categories: input of sensory information, coordination by the central nervous system, and the response by motor nerves and their effectors, muscles, and/or glands. The simplest reaction pathway is that of a reflex arc. Sensory receptors identify environmental stimuli, causing a sensory nerve cell to become excited. The sensory nerve cell transmits an electrochemical impulse to the spinal cord. Here an intermediary nerve cell transmits a sensory impulse to a motor nerve cell. The impulse is carried by the motor nerve cell to a muscle (or in some cases a gland). The contraction of the muscle signals the response. A knee-jerk response provides an excellent example of this simple nerve pathway. The impulse is triggered through three nerve cells: sensory, interneuron, and motor nerve that travels toward the muscle. Surprisingly, no integration is required by the brain. These reactions occur without thinking.

Reactions that require integration by the central nervous system, such as those that occur when driving, take considerably longer to occur. A moose running in front of a vehicle is identified by visual receptors within the eye. Sensory impulses are carried toward the brain by the optic nerve. Here the information is accumulated and the driver is made aware of the problem. Multiple nerve connections carry the impulses toward the motor area of the brain. A conscious decision is made to lift the foot from the accelerator pedal and push down on the brake. Because the sensory nerves are connected with motor nerves through a maze of circuits within the brain, the reaction time is much longer than that of a reflex arc. Each time an impulse passes between connecting nerve cells, the speed of transmission is slowed.

Conscious decisions, such as braking for a moose, depend upon a number of variables. The time it takes to catch sight of the moose may well be the largest variable. Any distraction or driver fatigue will increase reaction time. Most impulses travel at approximately 100 m/s along a nerve cell, but the time required for the impulse to travel between two different nerve cells varies greatly.

Transmitter chemicals diffuse between connecting nerve cells. Because diffusion takes much more time than the movement of an impulse along a nerve cell, the connections between nerve cells slows reaction time. Not surprisingly, the complexity of integration of sensory impulses by the brain to create a visual image and the number of nerve cells involved also affects reaction time. The greater the number of interconnecting nerves, the slower is the processing time. Moving images require greater time to process and interpret than still images. To accurately determine reaction times, you must consider how the reaction time is measured. The removal of the foot from the driver's pedal takes considerably more time than just pushing down on the brake. The distance the leg moves, the amount of muscle required, and the health of the muscle also affect reaction rates.

The reaction time graph is created by using the following equation for free-fall motion:

$$d = \frac{1}{2}at^2$$

where a is the acceleration due to gravity (9.8 m/s^2), t is the elapsed time, and d is the distance fallen.

Because all objects fall at the same rate, there is no need to be concerned with the mass of the ruler.

Solving the free-fall equation for time:

$$t = \sqrt{\frac{2d}{a}}$$

allows you to compute the reaction time for any given distance. The students will be introduced to this equation later in the course. To provide the equation with no evidence of constant acceleration would not help their understanding at this point. If, on the other hand, they have studied acceleration previously, you may use this equation to provide a reinforcement of this concept.

As people age, reaction rates are said to decline. The buildup of pigmented Nissl bodies within nerve cells slows the transmission of nerve impulses. In addition, the production of transmitter chemicals, the things that allow impulses to travel between nerves, decreases with age. Older people also tend to have less healthy muscles, further increasing the time it takes to respond to a stimulus. But age is not the major factor when considering reaction rates. The alertness of the driver is far more important.

NOTES

Crucial Physics

- A person has measurable reaction time.
  - The reaction time is affected by distractions.
  - Reaction time without distractions is approximately 0.1 s.
- Poor reaction time can lead to more accidents.
- Driving while intoxicated (DWI) is dangerous because of the effect of alcohol on reaction times.

Learning Outcomes	Location in the Section	Evidence of Understanding
Measure reaction time using one of two different methods.	***Investigate*** Method A OR Method B	Students measure the difference between reaction times on each stopwatch or record how long it takes to stop a ruler from falling, or use a reaction-time meter to record reaction times for each method.
Compare the different methods of measuring reaction time.	***Investigate*** Comparing Methods of Reaction Time: Step 1.a) *and* Step 1.c)	Students compare reaction-time measurements from each method to determine which method is most accurate.
Compare the reaction times of your classmates.	***Investigate*** Comparing Methods of Reaction Time: Step 2.a)	Students compare reaction-time measurements with their classmates and record their answers.
Investigate how distractions affect reaction time.	***Investigate*** Reaction Time with Distractions: Steps 1.a)-2.a)	Students explore how distractions affect reaction time when they have to make a decision while driving.

NOTES

Section 1 Materials, Preparation, and Safety

Materials and Equipment

PLAN A		
Materials and Equipment	**Group (4 students)**	**Class**
Stopwatch	2 per group	
Ruler, metric, 30 cm	1 per group	

*Additional items needed not supplied

PLAN B		
Materials and Equipment	**Group (4 students)**	**Class**
Stopwatch	2 per group	
Ruler, metric, 30 cm	1 per group	

*Additional items needed not supplied

Note: Time, Preparation, and Safety requirements are based on Plan A, if using Plan B, please adjust accordingly.

Time Requirement

Approximately 40 minutes are required to complete the experiment.

Teacher Preparation

- Reaction-time software in the form of a spreadsheet game is available from *It's About Time* as an alternative method of testing the students.
- Be sure that stopwatches are counted at the end of the class to ensure that all are returned.

Safety Requirements

- No particular safety precautions are required; however, students should be careful when dropping the rulers to test their reaction times. Rulers should be dropped in areas where they cannot bounce up toward any students' eyes if they are not caught before falling to the floor or table.

Meeting the Needs of All Students

Differentiated Instruction: Augmentation and Accommodations

Learning Issue	Reference	Augmentation and Accommodations
Understanding the purpose of an *Investigate*	***Learning Outcomes***	**Augmentation** • Students often complete investigations in discrete steps that do not seem to be connected to a common goal. This makes it much more difficult to analyze results, draw conclusions, and make meaning from new learning. Explicitly review the *Learning Outcomes* at the beginning of each section to set a purpose and make students aware of the big picture.
Using a textbook as a learning tool	***"Why is there...?"***	**Augmentation** • Some students do not know how to use their textbooks to assist them in their learning process. Read the section explanations in large or small groups. Have a class discussion about the purpose of each section. Small groups could also make posters to represent the purpose of each section and then present them to the whole class. • Display the posters in the classroom.
Following directions and collecting data independently	***Investigate***	**Augmentation** • Many students have a difficult time following the different steps of an *Investigate* because of focus issues and reading comprehension deficits. Instruct students to focus on one step at a time and record their observation. • Set time limits for tasks to give verbal and written time reminders. For example, if students are given 25 minutes to complete the Investigate: Methods A and B, start a timer and give five-minute reminders to keep students on task. Check in with students, especially at the beginning of the investigation, to make sure they understand the directions and are staying on task. **Accommodation** • Provide students with an observation chart to ensure that students record the appropriate data for each step. • Pair students strategically in groups of two to compensate for reading and focus issues. Students who have a difficult time focusing work better when accountable to only one partner.
Estimating reaction times	***Investigate*** Steps 1-2	**Augmentation** • Many students struggle with making sense of numbers in estimating reaction times. Provide students with reasonable examples of reaction times (seconds are more reasonable than minutes).
Visual-motor integration	***Investigate*** Method B	**Augmentation** • Some students will not be able to respond quickly enough to catch a ruler before it falls. Use a meter stick, or something else that is long enough, for students with poor hand-eye coordination.
Calculating an average	***Investigate*** Comparing Methods of Measuring Reaction Time: Step 2.a) ***Investigate*** Reaction Time with Distractions	**Augmentation** • Provide direct instruction on how to calculate an average or mean. Model the procedure for measuring "reaction time with distractions" using a pair of students. Use the data collected from this procedure to review averages.

Learning Issue	Reference	Augmentation and Accommodations
Generalizing data to draw conclusions Number concepts	***Investigate*** Comparing Methods of Measuring Reaction Time	**Augmentation** • Students are asked to compare numbers that may have values with decimals, and the task might be conceptually difficult to determine fastest, slowest, and average reaction times. Provide direct instruction to teach place values of decimal numbers. Using money is a good way to teach this concept. (Five cents = \$0.05, 10 cents = \$0.10, etc.) **Accommodation** • Provide students with a number line that has a range of increasing decimal values.

CHAPTER 1

Strategies for Students with Limited English-Language Proficiency

Learning Issue	Reference	Augmentation
Vocabulary comprehension	***Investigate*** Introduction	ELL students may need support for understanding the grammar of how "the *Investigate*" is used in the textbook. A student may be confused when a term they recognize as a verb is used as a noun. Point out that "the" is one clue, and the italic type is another clue, that this phrase is being used as a noun. Help students with the word "distraction." Explain that it is a combination of the base word "traction," meaning "the condition of being pulled," and the prefix "dis," meaning "apart" or "remove." In context, a distraction is something that "pulls your attention apart" from the task at hand.
Vocabulary comprehension Background knowledge	***Investigate*** Reaction Time with Distractions, Steps 1, 2.b)	Students may need support for using the term "experiment," which can be a verb or a noun. It may help to have students think of simpler terms, such as "clean" or "pump," that can be both verbs and nouns. Encourage ELL students to share their lists of 10 driving distractions with the class. They will benefit from the opportunity to speak, and all students will benefit from hearing of distractions they may not have considered.
Vocabulary comprehension	***Physics Talk*** Reaction Time and Distractions	Students may not be familiar with the nuance of "collision," meaning an accident in which two or more vehicles are involved. In physics, a collision can occur between a moving object and a stationary object, but in common usage a car hitting a tree would not be considered a collision.
Understanding concepts	***Physics Talk*** Other Factors Affecting Reaction Time	It is almost certain that all students will take medication at some point in their lives. Also, many over-the-counter medicines, such as some cough syrups, warn of possible drowsiness, which can interfere with reaction time. Given the relevance to their lives, students may benefit from a discussion of the alcohol and drug paragraph.
Vocabulary comprehension	***Reflecting on the Section and the Challenge***	Students may need help with the slang term "rubbernecking." It may help to have students explain why having a "rubber neck" would aptly describe people taking their eyes off the road to look at an accident or another unusual occurrence.

SECTION 1

Teaching Suggestions and Sample Answers

What Do You See?

Elicit responses from your students on the *What Do You See?* illustration. You might want to ask them about their initial impression of what they see in the illustration, why the artist has chosen to depict an accident, and why it is significant in the context of this section. You are likely to get many different responses. Accept all answers but try to focus on those that provide an opportunity for you to get the students engaged in the physics concepts they are about to learn.

Chapter 1 Driving the Roads

Section 1 Reaction Time: Responding to Road Hazards

Learning Outcomes

In this section, you will

- **Measure** reaction time using one of two different methods.
- **Compare** the different methods of measuring reaction time.
- **Compare** the reaction times of your classmates.
- **Investigate** how distractions affect reaction time.

What Do You See?

Why is there a *What Do You See?* and *What Do You Think?*

The *What Do You See?* and *What Do You Think?* are the **Elicit** and **Engage** phases of learning. You have already spent a number of years at school learning about many different subjects. You watch television, read, or listen to others talk. You have your own ideas about how things work and about what makes things happen. It is very important for you to think about what you already know or what you think you know. That is what you will use to build your understanding. You need to compare what you think you know to what you are learning in the classroom to build a new understanding. The **Elicit** phase of learning is thinking about what you already know.

The **Engage** phase is meant to capture your attention. The *What Do You See?* picture in each section has been drawn by Tomas Bunk. Tomas Bunk is not a physicist but a well-recognized cartoonist. He uses his artistic talent and enjoys drawing humorous illustrations that show real physics concepts in a very personal way. When you look at the illustrations, what do you see? What do you not understand about what is happening in the illustration that you would like to learn more about? How much fun and how personal can you make your encounter with physics?

When you think about the *What Do You Think?* questions, what interests you and what other questions come to your mind that you would like answered? The **Engage** phase of the instructional model is designed to get you interested in what you will be learning.

What Do You Think?

Many deaths that occur on the highway result from the inability of a driver to respond in time to a hazard on the road. The driver could not react quickly enough to avoid being involved in a collision.

- **What factors affect the time you need to react to an emergency situation while driving?**

Begin a new page in your *Active Physics* log. Write *Section 1 Reaction Time* at the top of the new page. Also record the section and page number in your *Table of Contents.* Record your ideas about this question in your log.

8

Students' Prior Conceptions

In this section, students will gain proficiency in applying the analytical skills needed to measure and record data precisely; to gather information necessary to describe objects and events; to characterize relationships among variables; and to argue logically. Mathematics, the precise language of science, is used to study the patterns of motion in the physical world.

1. **Measurement is only linear.** Through observations and careful data collection. students find that measurement is not only linear; this becomes more evident as students work with driving along curves, changing directions, and examine projectiles and other objects traveling through air or space. Time is a quantity that always has a positive or linear measurement. As such, it often becomes the independent variable in motion graphs and is the horizontal or *x*-axis variable.
2. **Any quantity can be measured as accurately as you want.** Students discover any quantity can be measured as accurately as the tools available and that they can only measure to the smallest unit shown on the measuring device. They discover that there is a difference between "precision" and the "accuracy" of measurement.

CHAPTER 1

What Do You Think?

The *What Do You Think?* question is designed to give you a sense of how much your students know about a driver's reaction time when responding to an emergency. It is designed to elicit a discussion on the alertness of a driver while trying to avoid a hazard on the road. You might want to prompt students on the effect of listening to loud music on reaction time, as well as talking while driving, or the effect of fatigue, alcohol, or drugs.

What Do You Think?

A Physicist's Response

The identification of sensory information has the greatest impact on reaction time. The driver's experience may also play an important part in determining reaction time. Experienced drivers are more likely to recognize and prepare for potential dangers before they become emergencies. The amount of time it takes to respond to an emergency depends on the alertness of the driver. An expert driver will recognize and respond to an emergency far more quickly than an inexperienced driver.

The reaction time will also determine the distance it takes to stop an automobile. If an automobile is traveling at a high speed, the time it would take to respond to an emergency would be far more than an automobile traveling slowly because distance is directly proportional to the square of the speed. Reaction time also depends on the distractions at the time of an emergency.

3. **The only "natural" motion is for an object to be at rest.** Recognizing that Earth and all objects on Earth are in motion relative to each other, and that objects only seem at rest relative to each other due to this shared relative motion, is fundamental to students' understanding of what is considered to be a "natural motion"—an object at rest. Motion among these 'objects at rest' must be described relative to an origin or starting point described within the framework of the relative motion of Earth. This "relative motion" is what students perceive as motion. Help students recognize and apply appropriate reference frames for describing the motions observed in everyday life. Establishing origins and the motions of objects relative to these origins enables the students to delve into the physics involved in driving along the roads.
4. **Speed and velocity are the same concept.** It is important for you to lay the groundwork for student understanding of vectors as students measure motions within the "moving" framework. Moving away from the origin is assigned one direction indicated by a positive value, whereas starting away from the origin and moving toward it is assigned a different direction, as indicated by a negative sign.
5. **Objects fall at a constant speed.** Only through careful measurements, made at many different distances in the fall of an object, can students see that the speed of objects increases in a regular manner with the time of fall.
6. **The heavier the object, the faster it falls.** This prior conception persists until students drop various objects that have similar shapes and volumes but different masses from the same height and compare their speeds at the same points along the falling trajectory.

NOTES

Why is there an *Investigate*?

The *Investigate* is the **Explore** phase of the 7E instructional model. Confucius, a Chinese philosopher, said, "I hear and I forget. I see and I remember. I do and I understand." The best way to learn is by doing. In *Active Physics,* whenever possible, you will explore a concept by doing an investigation.

One purpose of the investigation is to "level the playing field" and ensure that everybody has a common experience through which to discuss physics. For example, some students have been in a motor-vehicle accident, while others have not. It would not be sensible for everybody to experience an accident. However, it is possible to provide a classroom experience that everybody can discuss and not limit the discussion to only those students who have been in an accident.

The investigation also provides you with a real dialog with nature. In *Active Physics*, you will not be limited to having to believe what somebody wrote in a book. You will have an opportunity to observe, record data, isolate variables, design and plan experiments, create graphs, interpret results, develop hypotheses, and organize your findings. Sometimes, the entire class will participate in a demonstration.

All scientists value inquiry. The **Explore** phase is part of an inquiry approach to learning. In *Active Physics*, you are not physics students, you are student physicists.

Scientists often record their results in lab books. When you see this symbol ✎, you should record the information required for the *Investigate* in your *Active Physics* log.

Investigate

In this *Investigate*, you will measure reaction time using one of two methods. You will then compare the methods to decide which is the best method of measurement. You will also compare the reaction times of the members of your class, both with and without distractions.

1. To stop an automobile, you must first decide you want to stop. Then you must move your foot from the gas pedal to the brake pedal. The time required to decide to stop and move your foot to the brake is called your *reaction time*. Reaction time includes the time for you to react and the time for you to complete an action.

   Begin by finding how long it takes to move your right foot between imaginary gas and brake pedals.

   a) Estimate how long it takes to move your foot between the imaginary pedals. Record your estimate. (A way in which to estimate time is to count "one one-thousand, two one-thousand, three one-thousand.") Try counting like this until you reach "ten one-thousand" while your partner uses a stopwatch or clock to measure 10 s (seconds). Slow or quicken your counting pace so that it comes close to ten seconds when you finish counting "ten one-thousand." If the time to move your foot from one pedal to the other is less than one second, you can estimate how much time elapsed by how far you got in your counting of one one-thousand [for example, one (¼ s) one- (¼ s) thou- (¼ s) sand (¼ s)].

2. The first step in stopping an automobile occurs even before you move your foot to the brake pedal. It takes time to see or hear something that tells you to move your foot.

Investigate

1.a)

A reasonable estimate would be about half a second. Students usually grossly misjudge the amount of time it takes to move their foot from the gas to the brake pedal, because they are not pressing down on the gas pedal, and are not constrained by things such as the dashboard. Also, it takes approximately 0.12 s for the nerve impulse to travel from the brain to another part of the body.

2.a)

Students typically react to the clap in about 0.25 s if they know it is going to occur, and about 0.5 s if they do not.

Method A: Starting and Stopping Stopwatches

You may choose to have all the groups do all three methods, or have each group only do one method and then repeat their results to the class. The students should then record the result of other groups in their logs.

1.a)

Reaction times will probably be between 0.5 s and 0.8 s.

1.b)

Students calculate average reaction times.

Method B: Catching a Ruler

1.a)

Expect the distances to vary. If students are anticipating the release by closing their fingers periodically, and their fingers are close enough together, results may be as low as 2 cm. Suggest to students that they average a number of trials to obtain a more reasonable result.

1.b)

Students calculate average reaction distances.

1.c)

The students should trace the grid in the graph where the distance intersects the time on the curved line and read down to the horizontal axis to find their reaction time.

Test your reaction time by having a classmate stand behind you and clap. When you hear the sound, move your foot between the imaginary pedals.

a) Estimate how long it took you to react to the sound of the clap. Record your estimate. Your partner can begin counting "one one-thousand" as soon as he or she claps.

Now you will use one of the following methods to measure the time it takes you to react to something you see. Your teacher will assign you one of the following methods.

Method A: Starting and Stopping Stopwatches

1. Obtain two stopwatches. One student starts both stopwatches at the same time, and gives one stopwatch to his/her lab partner. When the first student stops his/her stopwatch, the lab partner stops his/her stopwatch, too. The difference between the times on each stopwatch is the reaction time.

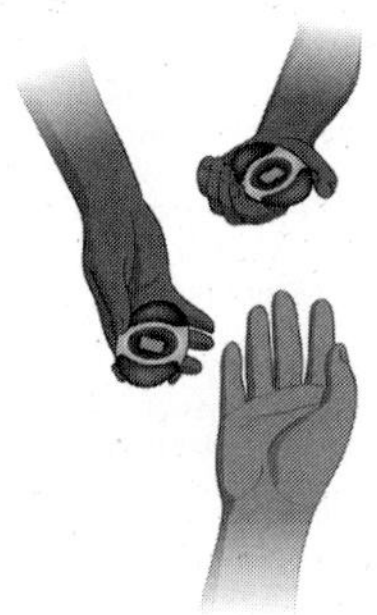

a) Record your reaction time in your *Active Physics* log.

b) Repeat at least three times. Calculate and record your average reaction time.

Method B: Catching a Ruler

1. Obtain a metric ruler. Hold the metric ruler at the top, between thumb and index finger, with the zero centimeter at the bottom. Your lab partner places his/her thumb and index finger at the lower end of the ruler, but does not touch it. Drop the ruler. Your partner must stop the ruler from falling by catching it between his/her thumb and index finger.

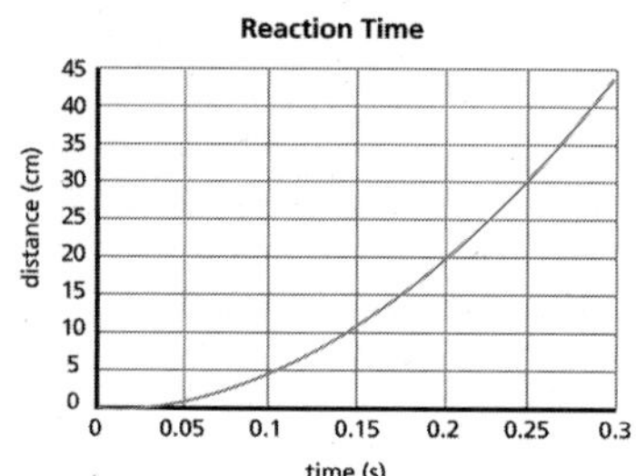

a) The position of your lab partner's fingers on the ruler marks the distance the ruler fell while his/her nervous system was reacting. Record the distance in your log.

b) Repeat at least three times. Calculate and record your average reaction distance.

c) The graph below shows the relationship between the distance the ruler fell and the time it took to catch it. Use the graph to find and record your reaction time.

Reaction Time

distance (cm): 0, 5, 10, 15, 20, 25, 30, 35, 40, 45

time (s): 0, 0.05, 0.1, 0.15, 0.2, 0.25, 0.3

Teaching Tip

For *Method B* to be an accurate measure of reaction time, the student who is dropping the ruler should avoid giving visual cues. This can be accomplished by talking to another person, looking away from the student being tested, and then releasing the ruler unexpectedly. An excellent drop distance would be around 12 cm of fall.

Section 1 Reaction Time: Responding to Road Hazards

Comparing Methods of Measuring Reaction Time

1. Compare your group's average reaction-time measurements with the average reaction-time measurements of other groups using the other method.

   a) Explain why they were not all the same.

   b) Which method do you think most accurately measures reaction time? Explain why.

2. Compare your reaction-time measurements with those of your group and other groups that used the same method.

   a) Record the results for the fastest, slowest, and average reaction times.

   b) Do you think reaction times vary for people of the same age? Discuss this with your group and then record your answer.

Reaction Time with Distractions

1. Before your partner clapped or dropped the ruler, you already knew what you were supposed to do upon receiving that signal. Suppose you had to make a decision after the clap or the ruler drop. Repeat the ruler-catching experiment while being distracted by a decision you have to make.

   The student dropping the ruler now says either "red" at the moment the ruler is dropped, which means you should catch the ruler, or "green" which means you should let the ruler drop. You will have to calculate the average of five reaction times. If you catch the ruler when "green" is called, then you have to do all the trials over. (This is to ensure that you react to the color as well as the ruler dropping.)

   a) How does your reaction time with needing to make a decision compare to your reaction time without needing to make a decision?

   b) How could you apply the difference in reaction time when you need to make a decision to a situation while driving an automobile?

2. Suppose you are talking on a cell phone or changing a CD while driving. How do these distractions affect your reaction time? To find out, repeat the ruler drop with one hand (using the "red" and "green" cues), while at the same time you do one of the following:
   - pretend to change a CD with your other hand.
   - simulate dialing a phone number by entering the phone number on your calculator.

   a) Compare your average reaction time with the distraction to your average reaction time without the distraction.

   b) In your *Active Physics* log, make a list of 10 activities that could distract you from driving safely.

11

Active Physics

Comparing Methods of Measuring Reaction Time

1.a)

Reaction times that require movement of different body parts are expected to be different. Moving fingers to catch a falling ruler is much easier than moving an arm or a leg, which requires more force and more time.

1.b)

Many students will think *Method B* is more accurate.

2.a)

Students should record the fastest, slowest, and average reaction time.

2.b)

Yes, reaction times will vary for different people of the same age. Some of the reasons are genetics, health, and physical conditions.

Reaction Time with Distractions

1.a)

Expect the reaction times to be slower than previous trials. "Dropping the ruler" reaction times should vary from 0.11 s to 0.30 s for some students.

1.b)

Students should realize that because the reaction time while making a decision was greater they should follow at a distance allowing for greater reaction time.

2.a)

Students should find that simulating distractions will slow their reaction time. Point out to them that actually doing the distracting activities while driving would increase their reaction times even more.

2.b)

Some possible distractions while driving might include talking to friends, playing the radio, changing a tape or CD, eating or drinking in the your vehicle, using a cell phone, looking for something dropped on the floor of your vehicle, and so on.

Physics Talk

This *Physics Talk* discusses reaction time. It explains the observations in the *Investigate* and the effect of distractions on reaction time. You might want to point out to your students why taking their eyes off the road or "driving under the influence" increases the possibility of having an accident. Ask students about various factors that affect reaction time. Making a list of responses on the board would be a good way to emphasize how these factors might prevent the driver from responding quickly at the time of an emergency.

As you discuss factors affecting reaction time, quiz students on their knowledge of driving laws in the United States. Use the opportunity to ask specific questions such as why alcohol and drugs are forbidden while driving. Discuss the dangers of using prescribed medication and the hazards of driving under the influence with subtle tact to make teenagers aware of how risky behavior on the roads is the leading cause of accidents in most countries.

Chapter 1 Driving the Roads

Why are there *Physics Words*?

It is easier and more effective to communicate concepts when the appropriate vocabulary is used. In science, a single word is often used to precisely describe a complex idea. *Physics Words* highlight the important terms that you need to know and use. In the *Physics Talk*, these words appear in a **bold-face type** the first time they are used. Sometimes it will be necessary for these words to be used in the *Investigate* first. You will recognize these words because they are printed in *italics* (a slanted type). The best way to learn new vocabulary is to practice using the words frequently and correctly. It is not useful to memorize a lot of terms and definitions.

Why is there a *Physics Talk*?

The *Physics Talk* is the **Explain** phase of the 7E instructional model. Reading the *Physics Talk* and discussing it with other students and your teacher will help you make better sense of the concepts you just explored in the investigation. In the *Physics Talk*, the results of your investigation are explained in terms of scientific models, laws, and theories. You will also be introduced to scientific vocabulary after the concepts are explained. The *Physics Words* highlight the vocabulary you need to know. You will find that using this vocabulary makes it easier to discuss the concepts with your class and answer the *Checking Up* questions. These questions will help you check that you have understood the explanation.

In *Active Physics*, you always **Explore** before you **Explain**. This ensures that you have some experience **(Explore)** with what is being described and discussed **(Explain)**. You can think of this as ABC (Activity Before Concept). You will also be introduced to science vocabulary after you understand the concept. This is what scientists do and how student scientists should learn. You can think of this as CBV (Concept Before Vocabulary).

The *Physics Talk* may also include the **Elaborate** phase of the 7E instructional model. After you are able to explain the physics of the investigation, you will be introduced to additional related physics principles that you will understand based on what you learned in the *Investigate*.

Physics Talk

AVOIDING COLLISIONS

Reaction Time and Distractions

The time taken to respond to a situation is called **reaction time.** Your reaction time while driving can be a matter of life and death. How fast you respond to an emergency could help you avoid an accident. In the *Investigate*, you estimated your reaction time. You found your quickest or best reaction time. You knew something was going to happen, you were ready to respond, and you knew how you were supposed to respond. Then you measured the time of your reaction. You also measured reaction time while you were being distracted in some way.

Physics Words

reaction time: the time it takes to respond to a situation.

12

Active Physics

You probably found you had a slower reaction time when you were distracted. In both situations, your reaction was probably quicker than your reaction time would be while driving, because you knew that you were expected to respond and how you would respond (for example, by catching the ruler, or stopping a stopwatch).

While driving, people are often distracted by conversations, music, or things happening along the road. As you discovered in the *Investigate*, distractions slowed your reaction time. If a decision has to be made suddenly, the slower reaction time may increase the chances of being involved in a collision.

Some distractions cannot be avoided. If you sneeze, your eyes automatically close for a moment and there is nothing you can do about it. However, drivers often consciously decide to take their eyes off the road to look at a passenger in the automobile. Some drivers' reaction time becomes even longer due to eating, changing a CD, or talking on their cell phone while driving.

Other Factors Affecting Reaction Time

Every state in the United States has a law prohibiting driving a vehicle while under the influence of alcohol or drugs. Alcohol and drugs can significantly slow a person's reaction time. Drugs that affect reaction time are not necessarily just illegal drugs. Some medications that are legally prescribed by a doctor instruct the user not to drive after taking the medicine.

There are many other factors that can affect reaction time. Psychologists (scientists who study the human mind) have found that age, gender, practice, fatigue, exercise, attentiveness, and even personality are some of the factors that can increase reaction time. The relationships among these factors and reaction time are complex. You may wish to research some of these relationships further.

Checking Up

1. How do distractions affect reaction time?
2. Why is driving under the influence of alcohol or drugs illegal?
3. Name three factors in addition to distractions and drugs or alcohol that can affect reaction time.

Checking Up

1.

Distractions slow down reaction time, which increases the chances of being involved in a collision.

2.

Driving under the influence of alcohol or drugs is illegal because an intoxicated driver's reaction time is negatively affected, preventing him or her from being able to respond to the challenges of driving.

3.

Age, fatigue, and attentiveness are some of the factors that can affect reaction time.

Active Physics Plus

Encourage students to read the *Why is there an Active Physics Plus*? Discuss how the "diamond notation" indicates the level of intensity. You could have a few student volunteers repeat their experiments to calculate the reaction time according to the equation $d = \frac{1}{2}at^2$. Writing the equation on the board will serve to emphasize its importance.

Why is there an *Active Physics Plus*?

In *Oliver Twist*, a book written by Charles Dickens, the young boy, Oliver, at breakfast declares, "Please sir, I want some more." *Active Physics Plus* is for those students who want more.

Active Physics Plus is more. But you may be wondering, "More what?" The *Active Physics Plus* will be more of one of the following four types of extensions.

Extension 1: More mathematics. Some students appreciate and enjoy the fact that physics content can be expressed efficiently and effectively through mathematics. This type of *Active Physics Plus* extension will provide guidance in how math can help elaborate a topic and add to its understanding.

Extension 2: More depth. Some topics can be elaborated by looking at the content in more depth or by relating it to other topics that have been studied.

Extension 3: More concepts. Sometimes a related concept can be introduced after learning the concept in a particular section.

Extension 4: More exploration. Further investigation of the section concept can include taking additional measurements or performing related investigations.

The *Active Physics Plus* includes the **Elaborate** phase of the 7E instructional model. After you are able to explain the physics of the investigation, you will be introduced to additional physics through the extensions.

Each *Active Physics Plus* component will be noted with a grid, informing you which extension categories will be covered. For example, *Active Physics Plus* for a section may include both a math and a concept extension. The grid at the beginning of the *Active Physics Plus* would look like the following:

+Math	+Depth	+Concepts	+Exploration
••		•	

The diamond notation (•) indicates the level of intensity, with three diamonds ••• signifying an intensive extension. In the example shown above, the depth of the concepts presented is moderate and the math required is more intensive.

Active Physics Plus can be considered optional topics for some students in some schools. For others, your teacher may require you to complete an *Active Physics Plus*. Your teacher is familiar with your state requirements and can guide you to appropriate personal challenges so that all students stretch themselves. Your teacher may also ask you to work as individuals or as teams on the *Active Physics Plus*.

If your teacher decides to skip the *Active Physics Plus*, you can still be sure that you will be able to complete the *Chapter Challenge* and follow the sequence of the sections. *Active Physics Plus* is supplemental and not a required component.

1.a)

Time (s)	Distance (cm)
0.00	0.00
0.01	0.05
0.02	0.20
0.03	0.44
0.04	0.78
0.05	1.23
0.06	1.76

1.b)

The graph the students generate should look like the one below. The axes should be labeled with the correct units, and the graph should have a title of fall distance vs. time.

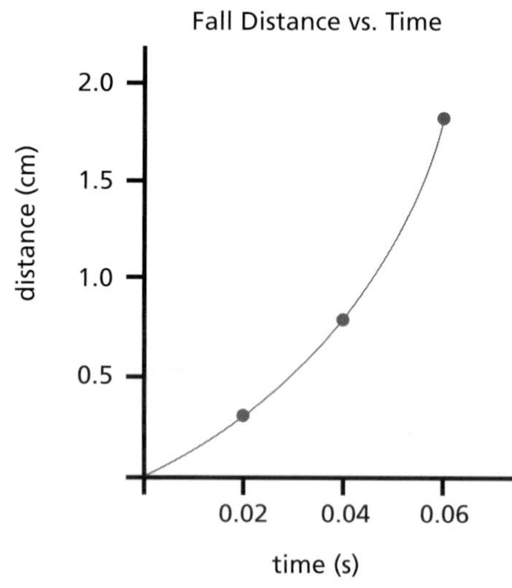

1.c)

The shape of the graph is the same as the graph in the investigation, but this graph concentrates on the lower values of time.

2.a)

Strategy: You know the distance of fall and the acceleration of gravity, so you may use the equation directly.

For the fall distance of 6.0 cm,

$t = \sqrt{2d/a}$

$t = \sqrt{2\left(0.060 \text{ m}/9.8 \text{ m/s}^2\right)}$

$t = 0.11 \text{ s}$

Using the same equation for the 7.5 cm fall gives

$t = \sqrt{2d/a}$

$t = \sqrt{2\left(0.075 \text{ m}/9.8 \text{ m/s}^2\right)}$

$t = 0.12 \text{ s}$.

Active Physics Plus

+Math	+Depth	+Concepts	+Exploration
♦			♦

Calculating Reaction Time

You were able to find your reaction time by dropping a ruler and using a graph that relates the distance a ruler falls to the time it took to catch the ruler. The reaction-time graph was constructed using the following equation:

$$d = \frac{1}{2}at^2$$

where d is the distance the ruler falls (measured in centimeters),
a is the acceleration due to gravity on Earth (980 cm/s^2), and
t is the time of fall (in seconds).

1. Use a computer spreadsheet and graphing program.
   a) Make data for this equation where the time varies from 0 s to 0.6 s in 0.02-s increments.

Time (s)	Distance (cm)
0.00	
0.02	
0.04	
0.06	

   b) Graph this data with time on the x-axis and the distance on the y-axis.
   c) Compare this graph with the one in the investigation.

   To find the time of the fall, the equation can be rewritten as follows:

$$t = \sqrt{\frac{2d}{a}}$$

2. By measuring the distance of fall for the ruler, you can use a calculator to determine the reaction time.
   a) Calculate the reaction time if the ruler is caught at the 6.0-cm mark, and the 7.5-cm mark.
   b) Calculate the reaction time as you repeat the experiment, dropping the ruler and catching it in the following ways:
      - with the thumb and index finger of your right hand
      - with the thumb and index finger of your left hand
      - with the index and middle finger.
   c) Compare the right-hand reaction time with the left-hand reaction time.
   d) Compare the reaction time of catching the ruler with the thumb and index finger to the reaction time of catching the ruler with the index and middle finger.
3. Use the equation to construct a reaction-time ruler with the distance measurement converted to time. You can now read response times directly on the ruler.
4. Do different groups of people have better or worse response times than others? Consider groups such as athletes that need good hand-eye coordination, taxi drivers, video-game players, and so on. Design and carry out an investigation to collect data that will help you find an answer. Include in your plan the number of subjects, how you will test them, and how you will organize and interpret the data collected. Use the response-time ruler to take your measurements. With the approval of your teacher, carry out your investigation. Record your findings and report them to the class.

2.b)

The answer the student obtains for these bullets will depend upon the response distances they record on the ruler.

- The reaction time with the thumb and index finger will be significantly shorter than when the students use their middle and index finger.
- Using their non-dominant hand will slow their response time even further.
- The index finger and middle finger should prove to be the slowest time due to the musculature not being suited for a quick response in this manner.

2.c)

The answer the student obtains for these bullets will depend upon the response distances they record on the ruler. They will find that their non-dominant hand will respond more slowly.

2.d)

Students will make the comparison between the two methods of catching the ruler. The index finger and middle finger should prove to be the slower time.

3.

The students should create a ruler by placing a piece of paper over a standard 30-cm ruler to cover the markings. They should then transpose the readings from the reaction-meter graph from Part B of the *Investigate* onto the paper. At the distances that correspond to the time of fall, times of 0.05 s, 0.10 s, and 0.15 s should be laid out, along with some intermediate times. Note that the intermediate times will not be uniformly spaced between the readings of 0.05 s, 0.10 s, etc., but should be determined by the graph.

4.

This will be the result of the students' investigations with different groups and will vary depending upon which groups are chosen.

What Do You Think Now?

Have students revisit the *What Do You See?* illustration and the *What Do You Think?* question. Encourage them to think of the effect of distractions on reaction time. Discuss how distractions could affect their reaction time at the time of an emergency. Consider sharing *A Physicist's Response* with the class to elicit their opinions. Ask them to return to their findings in the *Investigate* and explain how the alertness of their reaction determined the outcome.

This is a time to clarify previous misconceptions that students might still have. You could cite an example of a person's reaction time in response to a situation, and quiz students on whether the reaction was prompt or delayed. This will serve as a quick formative assessment to help you determine how confident students are in their understanding of reaction time.

Why is there a *What Do You Think Now?*

At the beginning of each section, you are asked to think about one or two questions. At that point, you are not expected to necessarily come up with a correct physics answer, but you are expected to think about what you know. Now that you have completed the investigation, you have learned the physics you need to know to answer the questions. Think about the questions again.

Compare your answers now to the answers you gave initially. Comparing what you think now with what you thought before is a way of "observing your thinking." Remember, research shows that stopping to think about your learning makes you a better learner.

What Do You Think Now?

At the beginning of the section, you were asked the following:

- **What factors affect the time you need to react to an emergency situation while driving?**

How would you answer this question now? Revisit your initial ideas on reaction time, and explain why reaction time is so crucial to avoiding automobile accidents. Explain how distractions can increase the possibility of having an accident in reference to reaction time.

Why are there *Physics Essential Questions?*

As a student physicist, you need to focus on the *Physics Essential Questions* that unite all science endeavors.

- *What does it mean?* (What is the physics content that you are learning?)
- *How do you know?* (What evidence do you have that supports the content?)
- *Why do you believe?* (What are the organizing principles of physics? How is the physics of this section the same as the physics outside this classroom? How does the physics of this section relate to other areas of physics?)
- *Why should you care?* (How is what you learned relevant to your life and/or the *Chapter Challenge?*)

As a student physicist, you are also part of the science community that understands that what you are learning can be placed into the larger context of physics knowledge and organizing principles.

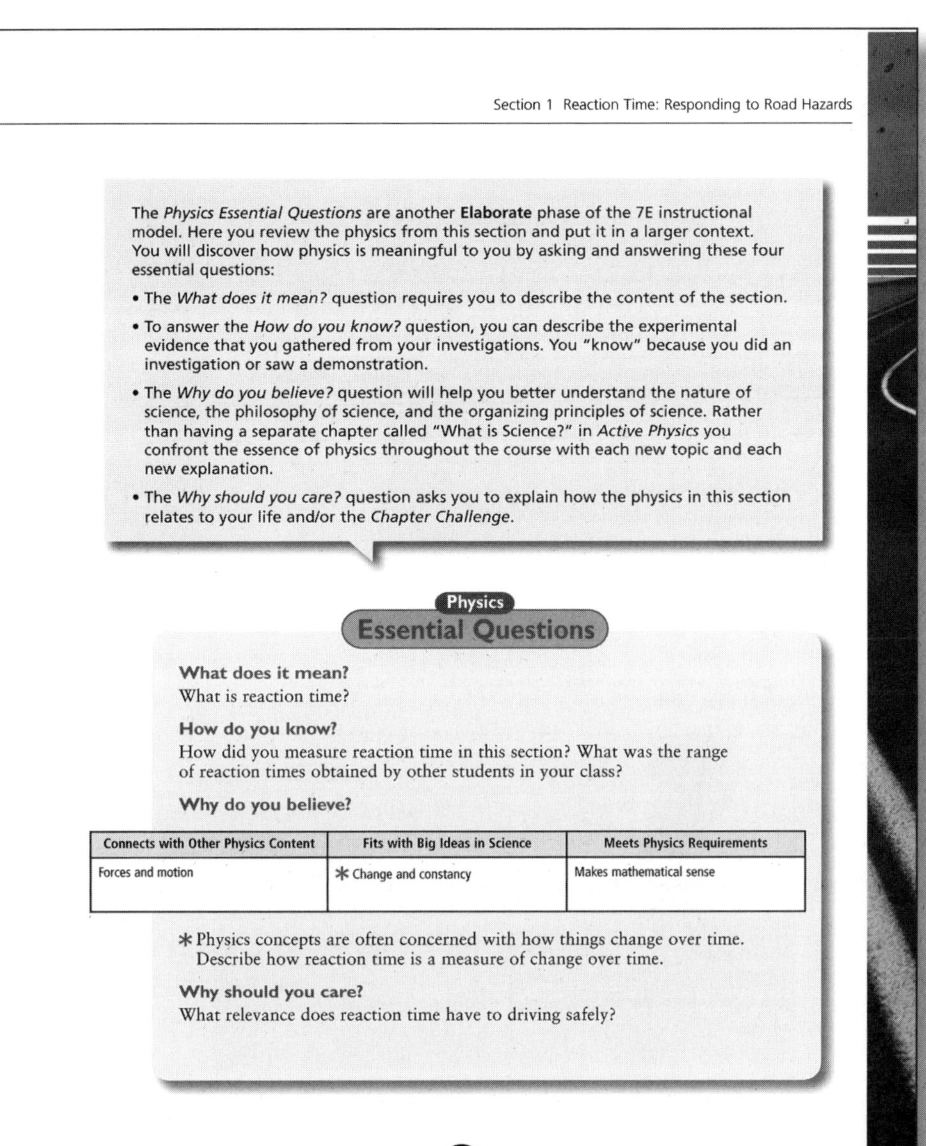
Section 1 Reaction Time: Responding to Road Hazards

The *Physics Essential Questions* are another **Elaborate** phase of the 7E instructional model. Here you review the physics from this section and put it in a larger context. You will discover how physics is meaningful to you by asking and answering these four essential questions:

- The *What does it mean?* question requires you to describe the content of the section.
- To answer the *How do you know?* question, you can describe the experimental evidence that you gathered from your investigations. You "know" because you did an investigation or saw a demonstration.
- The *Why do you believe?* question will help you better understand the nature of science, the philosophy of science, and the organizing principles of science. Rather than having a separate chapter called "What is Science?" in *Active Physics* you confront the essence of physics throughout the course with each new topic and each new explanation.
- The *Why should you care?* question asks you to explain how the physics in this section relates to your life and/or the *Chapter Challenge*.

Physics
Essential Questions

What does it mean?
What is reaction time?

How do you know?
How did you measure reaction time in this section? What was the range of reaction times obtained by other students in your class?

Why do you believe?

Connects with Other Physics Content	Fits with Big Ideas in Science	Meets Physics Requirements
Forces and motion	* Change and constancy	Makes mathematical sense

* Physics concepts are often concerned with how things change over time. Describe how reaction time is a measure of change over time.

Why should you care?
What relevance does reaction time have to driving safely?

17

Active Physics

Physics Essential Questions

What does it mean?

Reaction time is a measure of how long it takes to respond to a stimulus.

How do you know?

Reaction time was measured by catching a ruler and using a graph. It was also measured by stopping a stopwatch after seeing someone else stop theirs. The reaction times of students in the class probably ranged from 0.1 s to 0.2 s.

Why do you believe?

Reaction time is actually a measure of time. To record the reaction time, you must observe somebody doing something like catching a ruler or stepping on the brakes of a vehicle.

Why should you care?

The slower your reaction time, the more chance that you will not have the time to respond to an emergency and being involved in an accident.

Reflecting on the Section and the Challenge

Students can begin to reflect on how this section relates to their *Chapter Challenge*. They should be able to extend their understanding of reaction time to a situation they might want to develop that explains the significance of being alert to potential dangers. Emphasize that they also need to build their presentation with illustrations that depict the effect of driving ability on reaction time, and how being a careful driver reduces the chances of being involved in an accident.

Chapter 1 Driving the Roads

Why is there a *Reflecting on the Section and the Challenge*?

This part of the section is the **Extend** phase of the 7E instructional model. It gives you an opportunity to practice transferring what you learned in a section to another situation. In the case of *Active Physics*, you will need to apply your knowledge to complete the *Chapter Challenge*. Each section of a chapter is like another piece of the puzzle that completes the challenge. Transfer of knowledge is an important element in learning. This component presents a connection between each section and the chapter. It will guide you to producing a better *Chapter Challenge*.

Reflecting on the Section and the Challenge

In a Virginia study reported in 2003, researchers found that traffic, or roadside incidents, caused the largest number of accidents. Rubbernecking was responsible for most of the accidents reported (16%) followed by driver fatigue (12%), looking at scenery or landmarks (10%), passenger or child distractions (9%), adjusting the radio, tape, or CD player (7%), and cell phone use (5%).

The amount of time you need to react to a situation has a direct impact on your driving ability. It takes time to notice a situation and more time to respond. A person who requires more time to respond to what they see or hear is more likely to have an accident than someone who responds in a shorter period of time. One part of your *Chapter Challenge* is to explain the effect of reaction time on driving.

18

Active Physics

Section 1 Reaction Time: Responding to Road Hazards

Why is there a *Physics to Go*?

The *Physics to Go* is another opportunity for you to **Elaborate** on the physics content in the section. It also provides an additional chance to **Extend** your knowledge. Often, you will be assigned *Physics to Go* questions as homework. They are excellent study-guide questions that help you to review and to check your understanding.

The *Physics to Go* is also a part of the **Evaluate** phase. This is one place where you evaluate your learning. However, it is not the only place. You were also evaluating your learning when you asked yourself "What do I see?" and "What do I think?" and "What do I think now?" You also evaluated your learning during the *Investigate* **(Explore)** and the *Physics Talk* **(Explain)**. One difference between beginning and expert learners is that expert learners are more aware of their understanding through a constant evaluation of what they know and do not know.

Physics to Go

1. Test the reaction time of some of your friends and family with the metric ruler by following Method B in the *Investigate*. Obtain results from at least three people of various ages.
2. How did the reaction times you obtained in *Question 1* compare with those you obtained in class? What do you think explains the difference, if any?
3. Cut out a 6 cm × 15 cm rectangle from a sheet of paper. This is about the size of a dollar bill. Fold the paper in half lengthwise. Have your lab partner try to catch the paper between his/her index finger and middle finger.
   a) Explain why it is so difficult to catch the paper. Repeat the paper test, letting people catch it with their thumb and index finger.
   b) Explain why catching the paper with thumb and index finger may have been easier than catching it with index finger and middle finger. Try to include measurements in your answer, such as length of the paper, time for the paper to fall, and average reaction time.
   c) Is there a large range of values for the reaction time? Explain your answer.
   d) How would your reaction time change after repeating the same task several times? Why?
4. Does a race car driver need a faster reaction time than someone driving in a school zone? Explain your answer, giving examples of the dangers each driver encounters.
5. What does alcohol, changing radio stations, or talking on a cell phone do to your reaction time?

19 Active Physics

Physics to Go

1.

Students should test reaction times of family members and friends who are not class members.

2.

Students may find that reaction times for much older and very young family members may be slower. On the other hand, parents and siblings may be more motivated to produce excellent reaction times and therefore, may be more alert than the students tested in class.

3.a)

The length of the dollar bill is 15.7 cm. The free-fall time is under 0.2 s, which makes it nearly impossible to catch the bill unless the hand is lowered or the release is anticipated. The grasping action with the thumb and forefinger is a much more familiar one than between the forefinger and middle finger. Refer the students to the graph in the *Student Edition* following the header Method B: Catching a Ruler.

3.b)

Although the fall time for the "dollar bill" is the same in both cases, catching the paper between the forefinger and middle finger requires the movement of two fingers, while catching the paper between the thumb and middle finger moves only the index finger, hence, a quicker reaction time.

3.c)

The range of reaction times will depend upon physical limitations, age, etc. A range of reaction times that spans a factor of 2-3 would not be uncommon.

3.d)

After several trials, one might expect the reaction time to decrease as the body was trained to respond to the stimulus. Fatigue could start to set in, increasing reaction time.

4.

The speed at which the race car driver is traveling requires a very quick reaction time. Although students will investigate the relationship between distance and time in the next section, most will be able to answer this question from previous experience. Encourage students to become aware of the distractions that the average driver faces, as well as the potential dangers. Compare the focused, alert race car driver encountering oil on the track to a distracted student reacting to an obstacle in the road.

CHAPTER 1

5.

Alcohol, changing radio stations, or talking on the cell phone slow down reaction time.

6.

Some of the consequences drivers face when driving with a slower reaction time include increased chance of accidents as dangerous situations arise, greater tendency toward sudden stops, delayed response to changes such as drifting off the road, possible interaction with police and the courts, including fines, etc.

7.

Auto insurance is more expensive for teenage drivers than older drivers because teenage drivers are less experienced, likely to drive fast, be distracted, and might not take the necessary precautions to avoid hazardous situations. Older drivers are more experienced and careful while driving, taking care to avoid potentially dangerous situations.

8.

Preparing for the Chapter Challenge

Students will probably say that knowing their own reaction time would help them avoid accidents.

Inquiring Further

1. Reaction time of different groups of people

The students may choose to test the reaction times of other students on a school sports team, and possibly compare those times to the math team.

6. What are the consequences of driving if one's reaction time is slow rather than quick?
7. Even though teenagers often have good reaction times, why is auto insurance more expensive for teenage drivers than it is for older, more experienced drivers?

Why is there a *Preparing for the Chapter Challenge*?

This feature serves as a guide to get part of the *Chapter Challenge* completed. As you complete each section or a couple of sections of a chapter, you need to take time to organize the knowledge that you are learning and to try to apply it to the challenge. The *Preparing for the Chapter Challenge* is another **Extend** phase of the 7E instructional model.

8. ***Preparing for the Chapter Challenge***

   Apply what you learned from this section to describe how knowing your own reaction time can help you be a safer driver. You will use this information to meet the *Chapter Challenge*.

Why is there an *Inquiring Further*?

Active Physics uses inquiry as a way of learning. Inquiry lets you think like a scientist. It is the process by which you ask questions, design investigations, gather evidence, formulate answers, and share your answers. You are involved in inquiry during each section of a chapter. However, *Inquiring Further* gives you an additional opportunity to do inquiry on your own. Sometimes you will be asked to design an experiment and with the approval of your teacher, carry out your experiment. Other times, the *Inquiring Further* will ask you to answer questions that require additional sources of information, or to solve more challenging, in-depth problems.

Inquiring Further

1. **Reaction time of different groups of people**

   Do some groups of people have faster or slower reaction times than those of students in your class? Consider groups such as basketball players, video-game players, taxi drivers, older adults, and young children. Plan and carry out an investigation to collect data that will help you find an answer to the question.

2. Red light-green light reaction timer

This assignment may interest students who are enrolled in a technology program and be a way of getting a special reaction timer for use next year. Alternatively, digital circuits for electronic timers can either be purchased or simply constructed on a breadboard circuit. Separate red and green lights may be purchased from electronics hobby suppliers (LEDs would work) as well as the needed switches. Students with an interest in electronics should be able to construct the circuit for under $15 in parts.

1-1a Blackline Master

CHAPTER 1

2. **Red light–green light reaction timer**

Design and build a device with a red light and a green light. If the red light turns on, you must press one button and measure the reaction time. If the green light turns on, you must press a second button and measure the reaction time. Have your teacher approve your design before proceeding. How do reaction times to this decision-making task compare with the reaction times measured earlier?

Why is there a 7E instructional model?

At the beginning of this section, you were introduced to the 7E instructional model. You were also asked to think about why you are asked to do certain things in *Active Physics*. Review the components of this section, and think about what instructional-model phase is addressed by each component.

Phases of the 7E Instructional Model	Where is it in the section?
Elicit	*What Do You See?* *What Do You Think?*
Engage	*What Do You See?* *What Do You Think?*
Explore	*Investigate*
Explain	*Physics Talk* *Physics Words*
Elaborate	*Physics Talk* *What Do You Think Now?* *Checking Up* *Physics Essential Questions* *Physics to Go*
Extend	*Reflecting on the Section and the Challenge* *Preparing for the Chapter Challenge* *Inquiring Further*
Evaluate	*Formative evaluation — You evaluate your own understanding and the teacher can evaluate your understanding during all components of the chapter. Additional evaluations may include:* Lab reports, *Checking Up*, Quizzes, *What Do You Think Now?*, *Physics Essential Questions*, *Physics to Go*.

SECTION 1 QUIZ

1-1b Blackline Master

1. A student is testing her reaction time by trying to keep a light bulb on for exactly 4 s by matching a stopwatch. When she turns the light bulb off, the stopwatch reads 4.13 s. What is her reaction time?

   a) 3.87 s
   b) 0.26 s
   c) 0.13 s
   d) 4.00 s

2. Which of the following will have no effect on the reaction time of a person driving an automobile?

   a) The age of the driver
   b) Distractions in the environment
   c) The speed of the automobile
   d) Talking on a cell phone while driving

3. A race car driver is being tested for reaction time. In several practice trials, the driver's reaction time was recorded as 0.18 s, 0.22 s and 0.17 s. What was the driver's average reaction time?

   a) 0.57 s
   b) 0.17 s
   c) 0.19 s
   d) 0.22 s

4. A student drops a ruler to test her reaction time, and catches it at the 19-cm mark. According to the graph below, her reaction time is closest to

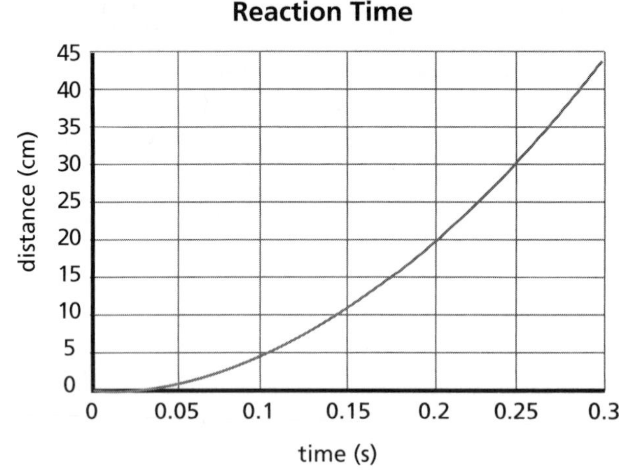

   a) 0.1 s.
   b) 0.1 s.
   c) 0.2 s.
   d) 0.25 s.

5. Two students are having a contest to see who has the quickest reaction time. A third student gives them both stopwatches that were started at the same time and tells them to stop them when he says stop. The student with the quickest reaction time will have the stopwatch with the

   a) nearest time in the chart for *Question 4*.

   b) time nearest the school clock.

   c) highest time recorded.

   d) lowest time recorded.

SECTION 1 QUIZ ANSWERS

1. c) 0.13 s. The student's reaction time is the difference between when the stopwatch hits 4 s and the time it takes for her to respond by stopping the stopwatch. If she would have stopped the stopwatch in exactly 4 s, the answer would have been 1.a). The other answers are not related to the reaction time.

2. c) All the other answers are distractions, which will slow down a driver's reaction time. The automobile's speed does not affect how quickly a person can react, but will affect how far the automobile travels before the reaction starts to take effect.

3. b) The average reaction time is calculated by adding the three reaction times and then dividing by three. The highest or lowest values are reaction times for one trial only, and may be either too high or too low for various reasons.

4. c) According to the chart, the ruler will fall a distance of 19 cm in a time of almost 0.2 s. This is *not* because 0.19 is closest to 0.2!

5. d) The student who reacts in the least amount of time after the stopwatches are started has reacted the quickest.

...TION 2

Measurement: Errors, Accuracy, and Precision

Section Overview

In this section, students study the different sources of errors within measurements. They select an area in the school, away from traffic, and estimate its distance by taking strides along the length of that area. Each stride length is then measured and multiplied by the total number of strides taken to cover the selected distance. The differences in student calculations are recorded each time students measure the distance using the stride method. All measurements are then listed on the board and the sources of error are discussed as random or systematic. The discussion of random and systematic errors leads to the understanding of uncertainty in measurements, which in turn helps students to identify the difference between accuracy and precision.

Background Information

The measurement of a quantity is a comparison to a standard. A meaningful measurement contains both a numerical value and the unit. In the 1790s, the French Academy of Sciences established the first standard unit of length, the meter. Prior to that, different people used different units of length, which varied from place to place. The standard meter was one ten-millionth of the distance from the Earth's equator to either pole, and was represented by a platinum rod. About 100 years later, the meter was defined more precisely as the distance between two engraved marks on a particular bar of platinum-iridium alloy. One of these bars was designated the International Standard and was kept at the International Bureau of Weights and Measures, near Paris, and the others were housed in scientific laboratories all over the world. By 1960, the meter was redefined as 1/1,650,763.73 wavelengths of an orange light emitted by the gas krypton-86. In 1983, the meter was redefined again as the length of a path traveled by light in a vacuum during 1/299,792,458 of a second. The most important system of units today is the Systeme International, or SI. The standard of length is the meter (m); of mass, the kilogram (kg); and of time, the second (s).

Random errors arise because every measurement is uncertain. Any measurement has limited accuracy, and users are unable to read an instrument beyond some fraction of the smallest unit shown. Good measurements include some estimate of the fraction of the smallest unit, and should state the estimated uncertainty. The uncertainty of a measurement is the range within which the actual value is likely to fall, compared to the measured value. The convention for stating uncertainty is to write ± and the smallest measurement that can be made with the instrument after the measurement itself. For example, a length measured as 5.4 cm is approximately written 5.4 cm ± 0.1 cm. Often, the approximation is not written, but understood. The percent error is found by dividing the difference between the measured value and the accepted value by the accepted value, and then multiplying the quotient by 100%.

In addition to their use in measurement, numbers that are defined or counted are used in physics. There is no uncertainty in these numbers. The number of sides of a triangle, for example, is a defined number. The number of pennies in a jar is a counted number.

Crucial Physics

- All measurements have uncertainties or random errors.
  - Some variation in measurement is inevitable.
  - No measurement is exact.
- Repeated measurements can vary in accuracy and precision.
- Random errors can be attributed to the measurement and/or the measuring instrument.

Learning Outcomes	Location in the Section	Evidence of Understanding
Calibrate the length of a stride.	***Investigate*** Step 3	Students measure the length of their stride using a meter stick and record the measurement in their log.
Measure a distance by pacing it off and by using a meter stick.	***Investigate*** Steps 2, 3, and 4	Students pace off the selected distance and record it in their logs. Then students multiply the measured length of one stride by the total number of strides.
Identify sources of error in measurement.	***Investigate*** Step 7.b)	Students list sources of error and give reasons for the sources of error.
Evaluate estimates of measurements as reasonable or unreasonable.	***Investigate*** Step 9)	Students estimate measurements as reasonable or unreasonable.

NOTES

Section 2 Materials, Preparation, and Safety

Materials and Equipment

PLAN A		
Materials and Equipment	**Group (4 students)**	**Class**
Tape measure, wind-up	1 per group	
Meter stick, wood	1 per group	
Tape, masking		1 per class
Access to a clear highway*		1 per class

*Additional items needed not supplied

PLAN B		
Materials and Equipment	**Group (4 students)**	**Class**
Tape measure, wind-up	1 per group	
Meter stick, wood	1 per group	
Tape, masking		1 per class
Access to a clear highway*		1 per class

*Additional items needed not supplied

Note: Time, Preparation, and Safety requirements are based on Plan A, if using Plan B, please adjust accordingly.

Time Requirement

To complete the experiment, 1.5 class periods or 60 minutes are required.

Teacher Preparation

- Notify other teachers in the section of the building where the students will be making measurements in the hallway.
- Mark off the length you wish the students to measure during the *Investigate*. If masking tape is used, be certain that the tape is removed at the end of the class.
- Inform students the day before the *Investigate* that they should wear appropriate clothing because they will be making measurements on the floor.

Safety Requirements

While working on the floor, students doing the measuring must be careful not to have their hands stepped upon by other students.

- All equipment must be picked up and safely returned to the classroom, prior to any change of class that would have non-class students in the halls.
- Students should be cautioned about rulers or other objects on the floor that might present hazards.

NOTES

Meeting the Needs of All Students

Differentiated Instruction: Augmentation and Accommodations

Learning Issue	Reference	Augmentation and Accommodations
Visualizing and estimating measurements	***What Do You Think?***	**Augmentation** • Students with visual-spatial issues often have a difficult time visualizing measurements. Physically show students the difference between 3 m and 10 m and between 3 m and 3.1 m.
Following directions and recording data	***Investigate*** Steps 1-8	**Augmentation** • Students must record data accurately during each step of the *Investigate* to be able to compare measurements and to analyze the sources of error. In each group, assign one person to read directions and one to record data. **Accommodation** • Provide students with an observation chart to record the required data. • Pair students strategically in groups of two to increase their focus on a task.
Making scientific measurements	***Investigate*** Steps 2-4	**Augmentation** • Ask a student to model one or two strides and then show your students how to measure the stride with a meter stick. • Group students intentionally to compensate for measurement difficulties. **Accommodation** • Provide one-on-one or small-group support for students who are really struggling at the beginning of this *Investigate*.
Comparing data	***Investigate*** Step 7	**Augmentation** • Some students struggle to compare numerical data. *Step 7* requires students to compare data and make some generalization about measurement. Have students complete *Step 7* in their logs independently. Then check in with each group to see what ideas they recorded, or facilitate a whole-group discussion to assess if students are making reasonable generalizations.
Estimating	***Investigate*** Step 9	**Augmentation** • Students may not have the prior knowledge or experiences to make reasonable estimations. Students can complete *Step 9* in their logs independently and then compare their answers with a partner. • Tactile and kinesthetic learners may need to manipulate real-life objects for making comparisons. For example, provide 5-kg masses or mark off a distance that is 10 m long, and allow students to move around the room, manipulating objects to estimate measurements.
Reading comprehension	***Physics Talk***	**Augmentation** • Model how to create a Venn diagram to compare these two types of errors. Place the similarities where the circles overlap, and the differences outside. Provide one or two similarities or differences to help students get started. Students may struggle with this tool the first few times, but Venn diagrams will help them to compare concepts and organize information, especially as they become proficient in learning this tool. **Accommodation** • Although the reading is clear and informative, some students may require direct instruction to understand random errors, systematic errors, accuracy, and precision. • Provide students with opportunities to use these words frequently.

Strategies for Students with Limited English-Language Proficiency

Learning Issue	Reference	Augmentation
Understanding concepts Vocabulary comprehension	***What Do You Think?***	Hold a class discussion on what it means to "mentally measure" distances and times. Include "estimation" in your discussion. Students may not know the term "skid marks." Explain how tires can leave rubber on the road when a driver brakes forcefully.
Vocabulary comprehension	***Investigate*** Step 2	You may need to explain the terms "pace" and "stride." A demonstration should help.
Comprehension	***Investigate*** Step 5.b)	Encourage ELL students to share their lists with other students. Have a brief discussion about the reasons for differences in measurements to check student understanding.
Understanding concepts	***Investigate*** Step 8	Discuss "error," "systematic error," and "random error," and clear up any misunderstandings in meaning. To fully understand "random error," students also need to understand "approximation" in context — close to the actual value, but not exact.
Comprehension	***Investigate*** Step 9	ELL students may be unfamiliar with the term "common sense" (good judgment). Discuss and check student understanding of the term.
Understanding prefixes	***Physics Talk*** SI System	Help ELL students use the chart to determine the meaning of the prefixes "kilo-" (thousand), "centi-" (hundredth), and "milli-" (thousandth). Be sure they grasp the difference between "thousand" and "thousandth." To check understanding, have students put these prefixes in front of "meter" and then order the units from smallest to largest. For more practice, repeat this exercise with "gram."
Demonstrating higher-order thinking	***Physics Essential Questions*** Why do you believe?	To give ELL students speaking experience, invite them to explain their thoughts on the question, "How can you trust experiments if all measurements have uncertainties?" Their answers will give you the opportunity to check understanding.

Two important aspects of learning a new language are speaking and writing in that language. Some ELL students will be self-conscious and shy about speaking in front of their peers, while others will be less reluctant to try. Be sure to encourage all ELL students to speak in class, and give them opportunities to write on the board from time to time. Experience will broaden their comfort level.

- To encourage class discussion, make a handout with five targets on a piece of paper. Put dots on one target to represent "accurate," on another to represent "precise," and on another to represent "accurate and precise." Put random dots as distractors on the other two targets. Have students label the targets as "accurate," "precise," and "accurate and precise." Draw the same diagrams on the board and have a volunteer come to the board to label the diagrams. Encourage students to discuss any differences of opinion concerning which term is correctly identified with each target.

SECTION 2

Teaching Suggestions and Sample Answers

What Do You See?

Student reactions will vary. They might state that they see two people walking and another person writing something down. The purpose of this illustration is to ask them what they think people are doing. You should allow them to share their ideas in a discussion as they will have a chance to return to the illustration after completing the *Investigate*.

All responses are valid; however, be alert to misconceptions that tend to distract students from a constructive response that directs them toward the topic they are about to study.

Section 2 **Measurement: Errors, Accuracy, and Precision**

What Do You See?

Learning Outcomes

In this section, you will

- **Calibrate** the length of a stride.
- **Measure** a distance by pacing it off and by using a meter stick.
- **Identify** sources of error in measurement.
- **Evaluate** estimates of measurements as reasonable or unreasonable.

What Do You Think?

When driving a vehicle, you often mentally measure distances and times. When investigating vehicle collisions, police officers take actual measurements at the scene. For example, the length of skid marks help officers to calculate the speed at which a vehicle was traveling.

- **Two students measure the length of the same object. One reports a length of 3 m, the other reports a length of 10 m. Has one of them made a mistake?**
- **If the students reported measurements of 3 m and 3.01 m, do you think one of them has made a mistake?**

Record your ideas about these questions in your *Active Physics* log. Explain your reasoning. Be prepared to discuss your responses with your small group and the class.

Remember to begin a new page in your *Active Physics* log each time you begin a new section. Write *Section 2 Measurement* at the top of the new page. Also record the section and page number in your *Table of Contents*.

Students' Prior Conceptions

Making estimations is a valuable technique in science. Many scientists are known for their ability to estimate results of observations. Naturally, these scientists also seek the most precise results through experimentation. *Section 2* opens the door for the teacher to discuss the value of evaluating estimates of measurements as reasonable or unreasonable.

1. **You can only measure to the smallest unit shown on the measuring device.** This section presents you with the opportunity to encourage students to evaluate their data using the questions "How do you know?" and "Why do you believe?" as they attempt to form reasonable answers from their investigation. The *Elicit*, *Evaluate*, and *Explain* cycles of learning apply well in this section.

2. **You should start at the end of the measuring device when measuring distance.** The human factors involved in this section of measuring and estimating can lead students to recognize how precision can be affected by faulty or worn measuring tools. Students can see the value of employing accurate tools to eliminate sources of error in measurement. Although it is not a standard prior conception, students often confuse the significance of quantitative values with errors in their reading of the tools, rather than an implicit limitation of the tool. It might be helpful for you to talk about percent error in this section.

What Do You Think?

The *What Do You Think?* section is designed to initiate discussion and engage students. You might want to emphasize that there are no "right" answers and that all answers are acceptable. The purpose of these questions is to elicit students' prior knowledge. Ask them to refer to their answers while discovering new physics concepts. Pointing out to them to refer to prior knowledge will help them increase or alter what they already know. It will also help them realize how the scientific process works.

What Do You Think?

A Physicist's Response

- If one person measures 3 m and the other 10 m, both of them cannot be approximately right. A difference in measurement that differs by a factor of three would indicate a serious error. For a measurement of this size, greater accuracy is expected. In some cases, when measuring extremely large or small distances, errors of this proportion would be acceptable. However, for a commonly used measurement of this size, this error would be excessive.
- A difference of 0.1 m out of 3 m is only about a 3 percent difference. Both may be approximately right. A difference in measurement of this size could be acceptable.

NOTES

Investigate

1.

You should choose the distance the students measure. Many schools have either terrazzo blocks (3 ft on a side) or tile squares (1 ft on a side). These may make a convenient measuring device for the students, in which case you should choose a length that does not coincide exactly with the end of one of these markers. However, students should make their measurements using strides.

2.a)

Students record the number of strides. They may be confused when they reach the end and are confronted with a distance that is not a full stride. Tell them to estimate this distance to a reasonable fraction (for example, one-half, one-third, and so on) of a stride.

3.a)

Students find the length of their stride and record it in their logs. Stride lengths will vary greatly, from about 0.4 m to almost a meter for taller students.

Teaching Tip

Students often make the mistake of measuring their stride from the front of one foot to the back of the other. Make certain that students measure from front to front or back to back.

4.

You may choose to set parameters for the students so they will know if they made an error. For example, if the room is 10 m long, you could tell the students that it is between 5 m and 15 m to allow those who are confused to realize when they have made a serious mistake.

4.a)

Students record their calculations.

5.a)

With all the students' measurements on the board, the average measurement should be fairly close to the accepted value.

5.b)

Some reasons for varying results in students' measurements would be inconsistency of the stride, incorrect stride measurement mentioned in *Step 3*, or poor estimation of the fractional stride at the end. Ask if the difference in stride length is compensated for in any manner. Some students should realize the different stride lengths would mean a different number of steps and should be accounted for by the multiplication factor.

5.c)

The students may suggest reporting the measurement several times and averaging the results. They may also suggest the reasons mentioned in *Step 5.b)*. If all students use one method, the range of measurements should be less severe.

6.a)

Students record the measurement. They may find this method does not work well, unless done carefully. You could also give one group a meter stick without metal end caps with 1 cm cut off at the end. The group that uses this meter stick will be shorter than the rest. This can set up your discussion on systematic vs. random errors in *Step 7.b)*.

NOTES

Investigate

In this *Investigate*, you will measure a given distance by various techniques. You will have to determine which technique is best and why it is the best. You will also use estimation to decide if certain measurements are reasonable or not.

1. Your teacher and class will select and agree on a cleared distance along the floor of the cafeteria, corridor, or path around the classroom.

2. Each group will have a member pace off the distance. That is, count the number of strides it takes you to cover the marked-off distance.

   a) Record the number of strides in your log.

3. Have a group member measure the length of your stride using a meter stick. By finding the length of your stride, you are making a calibration, or a scale for a measuring instrument.

   a) Record your measurement in your log.

4. Use the number of strides you took in *Step 2* and the length of your stride to compute the distance in meters.

   a) Record your calculations.

5. List the results of the measurements made by all the groups on the board.

   a) Do all the measurements agree? By how much do the results vary?

   b) Why do you think there are differences among the measurements made by different groups? List as many reasons for the differences in measurements as you can.

   c) Suggest a way of improving your measurements. If all groups try your method, how will the range of measurements change this time?

6. Measure the selected distance with a single meter stick. You will have to move the meter stick over and over.

   a) Record your measurement in your log.

7. List the results of the measurements made by all the groups on the board.

   a) Do all the measurements agree? By how much do the results vary?

   b) Why do you think there are differences among the measurements made by different groups? List as many reasons for the differences in measurements as you can.

   c) Suggest a way of improving your measurements. If all groups try your method, what will the range of measurements be this time?

   d) What do you think would happen if each group were given a very long tape measure? List possible values the different teams may get. Do you think each group would get the exact same value?

   e) Can you develop a system that will produce measurements, all of which agree exactly, or will there always be some difference in measurements? Justify your answers.

8. A difference in measurement close to a certain accepted value is called an error. Physicists identify two kinds of errors in measurement. An error that can be corrected by calculation is called a *systematic error*. For example, if you measured the length of an object starting at the 1 cm mark on a ruler instead of at the end of the ruler, you could correct your measurement by subtracting 1 cm from the final reading on the ruler.

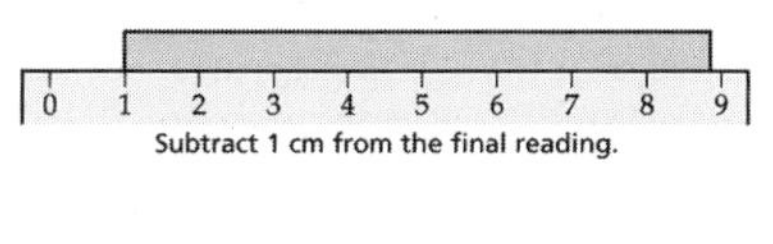

Subtract 1 cm from the final reading.

7.a)

The measurements made by the various groups should now be much closer to one another, unless a group was given the altered meter stick, which should have the largest difference.

7.b)

In measuring the room, there will be some variance about an average. This may be due to inaccurate readings of the meter stick, careless end-to-end placement of meter sticks, not measuring in a straight line, or variations in the length of the meter stick.

7.c)

Suggestions may include making repeated measurements and averaging, using two meter sticks rather than one, so the position on the floor can be accurately determined when moving a meter stick, and possibly drawing a line on the floor to show a path perpendicular to the ends to get the straightest path from one end to the other. If all the groups used these methods, the lengths measured by each group should be closer to each other, reducing both the variance between the readings, and the error size.

7.d)

Tape-measure values should correlate even closer than meterstick values. If each group used a tape measure, variations should be 1–2 cm, assuming the tape measures were identical. This error would be primarily due to not pulling the tape tight when making a measurement.

7.e)

By now the students should start to realize there will never be exact agreement between data sets. "Exact" means accurate to an infinite number of decimal places, and one can never measure anything to an infinite number of decimal places.

Teaching Tip

One estimate of the "size" of the error in measuring with an instrument (such as a meter stick) that employs a graduated scale is to round to the nearest half of the smallest interval on the scale ("least count" estimate of uncertainty).

8.a)

Students' errors would be predominantly random errors. However, if the students initially measured the stride length incorrectly, then used the stride to measure the length, this would be equivalent to having a faulty measuring device, and therefore, a systematic error.

8.b)

The measurement of the room using the stride length would most likely be accurate to approximately 0.5 m. Using the meter sticks to measure the length would probably lead to an error of several centimeters, and the tape measure should be accurate to about 1 cm.

9.a)

100 kg is about 220 lbs (1 kg equals 2.2 lb), so that's a good estimate for a football player.

9.b)

4 m is over 12 ft.

9.c)

1440 min = 24 h. You may feel as if you work 1440 minutes per day!

9.d)

A 20-kg poodle would weigh about 44 lbs, or just over half the weight of a college football player. Some large poodles may come close.

9.e)

If necessary, have the students measure out 1 m^3, then do the estimate.

9.f)

1 km is about 6/10 mi.

Chapter 1 Driving the Roads

An error that cannot be corrected by calculation is called a *random error*. No measurement is perfect. When you measure something, you make an approximation close to a certain accepted value. Random errors exist in any measurement. But you can estimate the amount of uncertainty in measurements that random errors introduce. Scientists provide an estimate of the size of the random errors in their data.

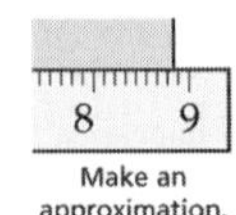

Make an approximation.

a) When measuring the hallway or class, did you have any systematic errors?

b) Estimate the size of your random errors using each technique.

9. Sometimes a precise measurement is not needed. A good estimate will do. What is a good estimate?

Example:

- Suppose one of your friends estimates that a single-serving drink container holds 5 kg (weighing about 11 lb) of liquid.

This is not a good estimate. It is unreasonable. A mass of 5 kg, or a weight of 11 lb, is about the weight of a bowling ball or a turkey. A single-serving drink weighs much less than this.

Use your common sense and prior knowledge to judge if the following measurements are reasonable. Explain your answers.

a) A college football player has a mass of 100 kg (weighing about 220 lb).

b) A high-school basketball player is 4 m (13 ft) tall.

c) Your teacher works 1440 min every day.

d) A poodle has a mass of 60 kg (about 132 lb).

e) Your classroom has a volume of 150 m^3 (about 5300 ft^3).

f) The distance across the school grounds is 1 km (about 0.6 mi).

g) On a rural road, while driving 50 mi/h (about 80 km/h), you encounter a tractor moving very slowly. You are about ¼ mi (0.4 km) away when you see that another automobile is coming toward you. Is it safe to pass the tractor?

h) While driving your pickup truck on a rural road, you approach a narrow bridge and see you will reach it at the same time as a dump truck that is coming from the opposite direction. What must you estimate in order to decide whether to stop and wait for the dump truck to cross the bridge first, or to go ahead and squeeze by the dump truck while on the bridge?

Active Physics 24

9.g)

Assume the oncoming vehicle approaching you is also traveling at 50 mi/h (80 km/h). The speed at which the two vehicles are approaching each other is then 160 km/h. The passing distance available is 0.40 km. Using $v = d/t$, and solving for time gives you $t = d/v$. The time available to pass the slow-moving tractor (assume a very low speed, so that the distance the tractor moves during the passing process is negligible) is $t = 0.40\ \text{km}/(160\ \text{km/h}) = 2.5 \times 10^{-3}$ h. Converting hours to seconds, $(2.5 \times 10^{-3}\ \text{h})(3600\ \text{s/h}) = 9$ s. A 9 s passing time would be sufficient to pass a short vehicle that is moving very slowly. If the tractor is 8 m long and has a velocity of 5 m/s (approx. 10 mi/h), it would only move forward to a distance of 45 m

i) You are driving a motor home with bicycles standing upright in a bicycle rack mounted on the roof. A sign before the entrance to a tunnel states that the maximum height is 21 ft (6.4 m). Will your automobile make it safely through the tunnel?

Physics Talk

ERRORS IN MEASUREMENT

Random Errors

There is no exact measurement. In the *Investigate*, when you used your stride length as the measuring tool, the distance of the hallway was different for many of the groups. If you tried to improve the measurement by using a meter stick, you found that there were still differences in the measurement. Even if you had used a tape measure, there would still have been differences in your measurements.

Physicists know that all measuring tools produce **random errors**, or errors that cannot be corrected by calculating. It is the responsibility of the student scientist to record all the values of a measurement and recognize that the data will include random errors. Every time you measure the length of your desk, you might find that the measurement is different from a previous value by 0.1 cm. This difference could be in either direction (± 0.1 cm). You can use a more precise ruler and that may decrease this random error or uncertainty to only 0.05 cm (± 0.05 cm). However, the uncertainty can never be completely eliminated.

Both the measuring tool and the person doing the measuring are responsible for the uncertainty. A meter stick that has only the centimeters noted would have a greater uncertainty than a meter stick that has the millimeters noted. A meter stick that has millimeters noted may still have a large uncertainty if the person using it is not very careful in aligning the meter stick with the length being measured.

Physics Words

random error: an error that cannot be corrected by calculation.

during the passing time of 9 s. The passing vehicle has a passing distance of 200 m (one-half the 400-m separation), giving it 150 m of distance to pull ahead of the tractor, prior to a collision.

9.h)

You will have to estimate the width of the bridge and compare that to your estimate of the sum of the widths of your pickup and the dump truck. Leave enough space so you and the oncoming truck do not exchange paint. Allow the space for mirrors.

9.i)

Because most bicycles are less than 4 ft tall at the handlebars, allowing for 2 ft for the roof rack means that the motor home would have to be taller than 15 ft to crash the bikes.

Teaching Tip

Students have difficulty making estimates in metric units due to their unfamiliarity. Emphasize that a kilogram is about 2 lb, a liter is about a quart, and a meter is slightly longer that a yard to help them visualize the metric equivalents.

CHAPTER 1

Physics Talk

Students learn that no measurement is perfect. All measurements have errors. These are classified as random and systematic. The *Physics Talk* builds on the *Investigate* where students measured selected distances and recorded their measurements to determine if they were the same or different. You might want to emphasize at this point that differences in measurements are due to an uncertainty that will always be present. Students should be able to clearly distinguish between accurate and precise measurements.

You can enhance their learning experience with a visual focus, drawing students' attention to the bull's-eye diagrams that will reinforce the difference between accuracy and precision. In this section, as students are also introduced to the SI system, it would be useful if they had a table of metric units and the corresponding symbols in their *Active Physics* logs to provide a quick reference.

In your measurement of the distance, you found different distributions of measurement. If you made histograms as shown below, of the length of a hallway using your stride (left figure), the meter stick (middle figure) or a tape measure (right figure), you can get a sense of the uncertainty in each type of measurement. The middle value is probably the "best guess" for the length of the room, but there will always be an uncertainty surrounding that value, as shown by the spread to the left and right of the middle value.

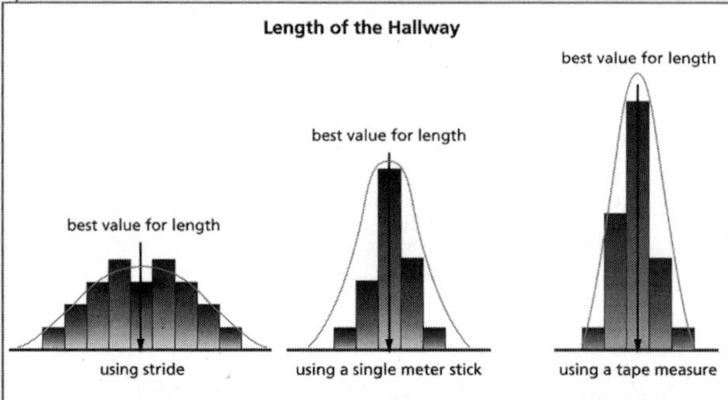

Physics Words

systematic error: an error produced by using the wrong tool or using the tool incorrectly for measurement and can be corrected by calculation.

accuracy: an indication of how close a series of measurements are to an accepted value.

precision: an indication of the frequency with which a measurement produces the same results.

Systematic Errors

There are also **systematic errors.** If you mistake a yardstick for a meter stick and report your measurement as 4 m, when in fact it is 4 yd, that is a systematic error. Every measurement you record with that yardstick will have this error. Systematic errors can be avoided or can be corrected by calculating.

Accuracy and Precision

In shooting arrows at a target, you can have **accuracy** and **precision** by getting all the arrows in the bull's-eye (left figure). You can have precision, but not accuracy by having all the arrows miss the bull's-eye by the same amount (middle figure). You can also have accuracy, but without precision by having all the arrows surrounding the bull's-eye spread out over the area (right figure). Notice that here the average position is the bull's-eye (accuracy), but not one of the arrows actually hit the bull's eye (precision).

Measurements can also vary with accuracy and/or precision just as the arrows in the target.

1-2a Blackline Master

CHAPTER 1

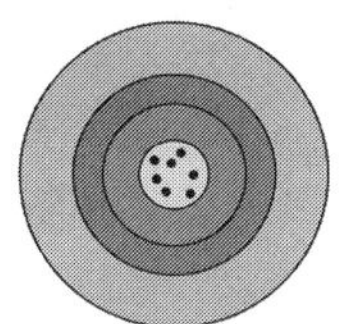
accurate and precise

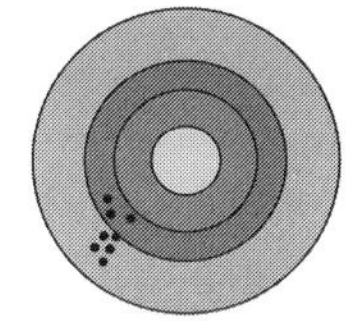
precise not accurate

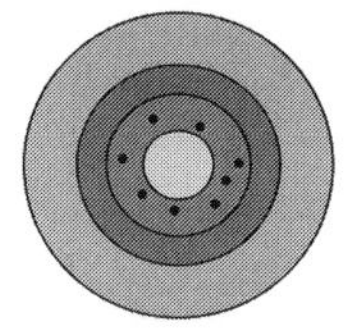
accurate but not precise

People do not always need the same level of precision in their measurements. One decision you must make is how precise a measurement you want. For example, in motor racing, horse racing, or Olympic skiing, time has to be measured to the thousandths or tens-of-thousandths of a second. But when a painter estimates the time required to paint a customer's house, she or he may only need to know the time within a few hours. As you increase the need for precision, the measurement becomes more difficult (and often, more expensive to make).

SI System

In *Active Physics*, you will be using the International System of Units. The units are known as SI units, abbreviated from *Le Système International d'Unités*. This is the system of units that is used by scientists. The system is based on the metric system. All units are related by some multiple of ten. There are seven base units that can be combined to measure all scientific properties. The base units that you will use in *Active Physics* are shown in the table.

Quantity	Unit	Symbol
length	meter	m
mass	kilogram	kg
time	second	s
temperature	kelvin	K
current	ampere	A

You will also be using other units that are a combination of these base units. You will be introduced to these units when you need to use them. The best way to learn units is to use them frequently and correctly. It is not helpful to memorize lots of units.

Checking Up

1.

A systematic error can be corrected by calculation while a random error cannot be corrected by calculation. If you weigh an object with a balance that is not set to the 0-kg mark, but is instead at 1 kg, then you can correct the error by moving the pointer to the 0-kg mark. However, if an object is weighed by different people and the readings are determined to be very close to each other, then the uncertainty in measurement is said to contain a random error.

2.

Uncertainty in measurement can never be eliminated because all measurements are made around a certain value thought to be true. The difference between the true value and the actual value produces an uncertainty in the measurement.

3.

The arrows would be spread out unevenly around the target, illustrating that they were neither accurate nor precise.

In this section, you measured the length of a distance in meters. The meter (m) is the base unit of length. Other units that you will use for measuring and describing length are the kilometer (km), centimeter (cm), and millimeter (mm). These three units are made up of the base unit meter and a prefix.

An important feature of the metric system is that there is a single set of prefixes that relates larger and smaller units. All the prefixes are related by some power (multiple) of ten.

Prefix	Symbol	Multiple of ten by which base unit is multiplied	Example
kilo	k	$10^3 = 1000$	1 km = 1000 m 1 m = 0.001 km
centi	c	$10^{-2} = 0.01$	1 cm = 0.01 m 1 m = 100 cm
milli	m	$10^{-3} = 0.001$	1 mm = 0.001 m 1 m = 1000 mm

Driving the Roads and United States Units of Measurement

The United States does not use the metric system for everyday measurements. Distances along the road are measured in feet, yards, or miles. Speed limits are posted in miles per hour rather than kilometers per hour, as they are in many other countries.

Below are some conversion factors for length in United States measurements.

12 inches (in.) = 1 foot (ft)

3 feet = 1 yard (yd)

5280 feet (1760 yd) =
1 United States statute mile (mi)

In this chapter, *Driving the Roads,* United States measurements will be used to express distances and speeds with respect to driving and traffic. In the classroom, you will use SI units for measuring.

When obeying speed-limit signs, it is important to know what units of measurement are being used.

Checking Up

1. Explain the difference between systematic and random errors.
2. Explain why there will always be uncertainty in measurement.
3. What would the positions of arrows on a target need to be to illustrate measurements that are neither accurate nor precise?

Section 2 Measurement: Errors, Accuracy, and Precision

Active Physics *Plus*

+Math	+Depth	+Concepts	+Exploration
◆	◆		

Precise Measurements and Olympic Records

The events in the Olympic Games depend on precise measurements of distances and times to determine who will receive a gold medal. If one skier has a time that is $^{3}/_{1000}$ of a second better than another skier, this can be the difference between a gold and silver medal. In running and swimming, the athletes compete side by side and whoever is first wins the gold. However, the record times are compared from one Olympics to the next. If a swimmer in one Olympics beats the old record from 4 or 8 years earlier by only $^{1}/_{1000}$ of a second, the swimmer has the new world record. But did that swimmer really swim faster than the prior record holder? Not necessarily. By following a discussion including measurements and uncertainty, you will find that the new world record holder may actually be slower than the old record holder if the time difference was only $^{1}/_{1000}$ of a second.

Every four years, the Summer Olympics are held in a different city, and a new swimming pool must be built. The length of the swimming pool must be 50 m. You know from your investigation that every measurement has an uncertainty associated with it. When a pool is built for the Summer Olympic Games, do you think that pool's length could vary by 1 m? If so, the pool could be 49 m or 51 m in length. This is a huge difference. You can be sure that they build the pools to be closer to 50 m than 49 m.

1. What is the range of lengths for 50-m pools that have an uncertainty of ± 10 cm? ±1 cm? ± 1 mm? (For example, if the uncertainty of the pool were ± 1 m, the range of lengths would be 49–51 m.)

Suppose an Olympic pool is accurate to ± 1 cm. In one Olympics, the pool could be 49.99 m, while in another Olympics it could be 50.01 m. This does not seem to be a big difference and it seems like an accurate 50-m pool. It means that in one Olympic Games in one city, the swimmers may actually swim 50.01 m while in another Olympics in another city, the swimmers may actually swim only 49.99 m.

2. How much extra time does it take to swim 50.01 m than 49.99 m (a difference of 2 cm)? Assume a good swimmer can swim 50 m in 25 s.

The 1500-m race requires a swimmer to swim 30 lengths of the pool. If the pool in one Olympic Games is 50.01 m and in another Olympics the pool is only 49.99 m, then one swimmer will be swimming an extra 2 cm for every lap. In one Olympics, the swimmer may be swimming a total of 60 cm more than the other swimmer.

3. Estimate how long it takes to swim 60 cm. Assume a good time for the 1500-m race is 15 min.

4. In watching the Olympic Games, you hear that someone just broke the record for the 1500-m swim by 1/1000 of a second.

→

29 Active Physics

Active Physics Plus

This *Active Physics Plus* provides an opportunity for students to understand the importance of making precise measurements. Before students begin to answer questions, have them make a list of the reasons that they think are significant to making precise measurements. Ask them how the degree of uncertainty would matter in decisive factors that influence events. Each math problem they solve will emphasize the need for precision in measurements.

1.

The range of uncertainties for these pools would be 49.9 to 50.1 m, 49.99 to 50.01 m, and 49.999 to 50.001 m.

2.

The time to swim an extra 2 cm can be determined from the velocity equation $v = d/t$. Solving for t gives $t = d/v$ so the extra time would be $t = (0.02\text{ m})/(2\text{ m/s}) = 0.01\text{ s}$.

3.

The speed for a 1500 m race in 15 min (900 s) would be $v = d/t = (1500\text{ m})/(900\text{ s}) = 1.67\text{ m/s}$. Using this gives $t = d/v = 0.60\text{ m}/(1.67\text{ m/s}) = 0.36\text{ s}$.

4.

If the pool is short by 1 cm in length, the difference in time would be 0.18 s (half the time for a 2-cm difference), which is 18 hundredths of a second. The swimmer might actually be 0.17 s more than the world record for a pool of the proper length.

5.

Students' letters should indicate what they have discovered in the analysis above from the questions they have answered to explain their case.

What Do You Think Now?

Ask students to review their previous answers. This is an opportunity to identify relevant answers during the discussion and provide them with *A Physicist's Response*. Students should by now have a clear understanding of the sources of systematic and random errors and should be able to explain the difference between these types of errors. You might want to ask them to review the *Physics Talk* before they begin to answer the question on how they could reduce random errors.

Explain how it is possible that this person (the new record holder) may actually be slower than the previous record holder. (Can it be that the new record holder is swimming in a shorter 50 m pool than the prior record holder?)

Comparing records for the 1500-m swim from one Olympics to the next depends on the length of the pools that the swimmers race in. The length of each pool cannot be exactly 50 m.

5. Write a letter to the Olympic commission addressing this issue. Include in your letter a solution to this problem that you have discovered. Including calculations of what would happen if the pools were built with an accuracy of 0.5 cm or 1 mm would make your letter more persuasive. You may also want to include something about the additional cost of making a 50-m pool this much more accurate.

What Do You Think Now?

At the beginning of this section, you were asked to think about the following:

- **Two students measure the length of the same object. One reports a length of 3 m, the other reports a length of 10 m. Has one of them made a mistake?**
- **If the students reported measurements of 3 m and 3.01 m, do you think one of them has made a mistake?**

How would you answer these questions now? Review what you have learned about random errors in measurement. How can you reduce random errors?

Section 2 Measurement: Errors, Accuracy, and Precision

Physics
Essential Questions

What does it mean?
Suppose your friend mistakes a yardstick for a meter stick and measures the length of an intersection in your neighborhood. Is this error random or systematic? Which of these types of errors affect precision or accuracy?

How do you know?
Suppose you want to buy some gold jewelry. The jeweler tells you that the jewelry contains exactly 1 oz of gold. How do you know that the jeweler cannot be sure that it is exactly 1 oz?

Why do you believe?

Connects with Other Physics Content	Fits with Big Ideas in Science	Meets Physics Requirements
All physics includes measurements	Change and constancy	* Experimental evidence is consistent with models and theories

* All physics knowledge is based on experimentation. All experiments require measurements. How can you trust experiments if all measurements have uncertainties?

Why should you care?
What are the consequences of not estimating stopping distances accurately, or the width of a space between your vehicle and other vehicles while driving?

Reflecting on the Section and the Challenge

A measurement is never exact. When you make a measurement, you estimate that measurement. All measurements have systematic and random errors. An example of a systematic error might be using a measuring tape that stretches a little bit when it is pulled tightly. But if you know the amount of stretch, then you can correct the measurement using calculations. In contrast, random errors are part of any measurement process because you can only approximate a mark on a meter stick or the time on a stopwatch, with an accuracy to the closest decimal place. You can try to minimize random errors, but you cannot eliminate them entirely.

When a speed limit is 60 mph (about 100 km/h), you may find that sometimes you drive at 58 mph while other times you drive at 62 mph. These differences are random errors as you try to hold the speed constant. If a police officer stops you because you were driving at 75 mph (about 120 km/h) in a 30 mph zone, you will not be able to convince her that this was just an uncertainty in your measurement. Uncertainties in speeds may be something that you wish to include in your presentation or report.

31 Active Physics

Reflecting on the Section and the Challenge

The concepts students learn in this section establish the presence of random errors in measurements. Because random errors can never be eliminated, they provide a basis for deviating from the accepted value by a small margin. A good example might be variations in the speedometers of vehicles for leeway in speed limits. The students can use this knowledge of random errors to develop their *Chapter Challenge*. They should be able to put forth the connection between driving at a safe speed and reading the speedometer with accuracy.

CHAPTER 1

Physics Essential Questions

What does it mean?

Because a yardstick is smaller than a meter stick, the measured length of the intersection will be incorrect. This is a systematic error and can be corrected once you realize that you used a yardstick instead of a meter stick. Systematic errors affect accuracy. Random errors affect precision.

How do you know?

Every measurement has some uncertainty. There is no such thing as an exact measurement.

Why do you believe?

The uncertainties in measurements place a limitation on what you may be able to conclude. However, you can draw conclusions within these limitations. For example, you may not know your weight to the nearest ounce, but you may know it to the nearest pound. From that knowledge, you can know whether you need to change your diet.

Why should you care?

If you estimate distances poorly, you may find yourself hitting the vehicle in front of you.

Physics to Go

Chapter 1 Driving the Roads

Physics to Go

1. Get a meter stick and centimeter ruler. Find the length of five different-sized objects, such as a door, a tabletop, a large book, a pencil, and a stamp.
   a) Which measuring tool is best for measuring each object?
   b) Estimate the uncertainty in each measurement.
2. Count the number of strides it takes to walk around your classroom and estimate the length of each stride. Calculate the size of the room by multiplying the number of strides taken by the estimated length of each stride. Estimate your accuracy. Then check your accuracy with a meter stick.
3. Give an estimated value of something that you and your friend would agree on. Then, give an estimated value of something that you and your friend would not agree on.
4. An oil tanker is said to hold five million barrels of oil. In your estimate, how accurate is the measurement? Suppose each barrel of oil is worth $100. What is the possible uncertainty in value of the oil tanker's oil?
5. Choose five food products. How accurate are the measurements on labels?
6. Are the following estimates reasonable? Explain your answers.
   a) A 2-L bottle of soft drink is enough to serve 12 people at a meeting.
   b) A mid-sized automobile with a full tank of gas can travel from Boston to New York City without having to refuel.
7. If you are off by 1 m in measuring the width of a room, is that as much as an error as being off by 1 m in measuring the distance between your home and your school?
8. You are driving on a highway that posts a 65 mph (105 km/h) speed limit. The speedometer is accurate within 5 mph (8 km/h).
   a) What speed should you drive as shown on the speedometer to guarantee that you will not exceed the speed limit?
   b) What could a passenger in the vehicle do while you are driving to estimate how accurate the speedometer is? (Hint: The road has mile markers, and the passenger has a wristwatch that shows seconds.)
9. ***Preparing for the Chapter Challenge***

   Many accidents are caused by speeding. To limit the number of collisions, police officers give speeding tickets to drivers. If the speed limit were 30 mph (50 km/h) in a residential neighborhood, a person may get a ticket for driving at 40 mph (65 km/h). Legally, they could also get a ticket for traveling at 31 mph (51 km/h). Given the uncertainties in measurements (the driver has to keep the gas pedal "just right"), you may wish to mention how these uncertainties are a part of safe driving. You may wish to explain why driving 31 mph in a 30 mph zone does or does not warrant a ticket. If you do not think that 31 mph deserves a ticket, you will need to explain what speed should get a ticket and why.

Active Physics 32

1.a)

Use a meter stick for measuring objects that are several meters long; use the ruler for objects that are a few centimeters long. The large book could be measured with either tool.

1.b)

The uncertainty will depend on the smallest markings on the ruler being used.

2.

If the stride is well-calibrated, the students should give an answer within a few centimeters (or tens of centimeters) of the meter-stick method.

3.

As an example, students might agree on the estimated length of a football field at 100 yd, but disagree on the length of a semi-tractor trailer.

4.

Five million barrels is a rough estimate. That probably means "more than four million, nine hundred thousand but less than five million, one hundred thousand."

5.

You may choose to have the students answer this question as part of an in class assignment because it will require them to measure the mass (weight) of the food and compare them with the labeled weight. This will require a balance, and the values and accuracy will depend upon the food choices. Dried foods such as spaghetti are preferable to liquids in cans. Dried food that can be weighed in the box with the weight of the box subtracted will yield the most accurate results. Expect difference from the labeled amounts due to imprecision of the machines filling the boxes, but the errors should be both higher and lower than the listed weight.

6.a)

Two liters is slightly more than 2 qt, so each person would have about 1/6 of a qt. This is a small amount, but reasonable.

6.b)

The distance from New York City to Boston is about 300 mi. A mid-sized vehicle might get 25 mi to the gallon, making its gasoline usage around 12 gal. Most mid-sized vehicles would have a 16-gal tank.

7.

Having the students justify their answers is more important than the answers themselves.

Section 2 Measurement: Errors, Accuracy, and Precision

Inquiring Further

1. **Measurement and national standards**

The National Institute of Standards and Technology (NIST) is the nation's measurement laboratory. It provides companies and other organizations with references to use to check the accuracy of their equipment.

What type of certainty in measurement would you expect if you were buying vegetables by the pound, gas by the gallon, or carpeting by the yard? Investigate what types of measurement standards are regulated by the government. Report to your class the certainty of the measurements that are used in industry and the marketplace.

2. **Random error and number of measurements**

People who study the statistics of measurement have shown that, if you make N independent measurements of the same object, then the size of the random error decreases as:

$$\sqrt{\frac{1}{N}}$$

Plot histograms, graphs that measure frequency distribution, of the measurement of desks in the classroom and see if the size of the random error decreases as you measure more and more identical desks.

8.a)

60 mi/h might really be 65 mi/h, so to be safe, you should drive 60 mi/h.

8.b)

The passenger may watch the mile markers and mark the time when they are passed. A driver traveling at 60 mi/h would travel 1 mi in 60 s (60 mi/h is 60 mi in 60 min), so if the vehicle is traveling 60 mi/h, the next mile marker should be passed in exactly 60 s.

9.

Preparing for the Chapter Challenge

Students may wish to discuss the problems with maintaining an exact speed such as 30 mi/h. Automobile speedometers are legislated to an accuracy of ± 5 percent, but may not actually be that accurate. An accuracy of 5 percent means that at 30 mi/h, the vehicle's true speed is between 28.5 and 31.5 mi/h. If the automobile has tires that are larger than the manufacturer's specifications, the error could be even greater. For reasons such as these, police officers often allow a leeway of ± 5 mph, or for a 30-mi/h speed limit a 15 percent deviation from the speed limit. Allowing for the accuracy of the speedometer, anyone driving at 32 mi/h or higher in a 30-mi/h zone should be liable for a ticket. A driver who spends too much time watching the speedometer to maintain the exact speed limit risks not paying enough attention to other aspects of safe driving. Driving below the speed limit allows the driver to stay within the law and pay more attention to the road.

Inquiring Further

1. Measurement and national standards

Students will have difficulty finding standards for the measurement of volumetric quantities, such as gasoline, because the volume of gas dispensed by a pump varies with temperature. Gasoline pumps are difficult to monitor for a number of reasons, but the industry claims to be accurate to within 1 percent. Carpets likewise will stretch to a certain degree, making exact measurements difficult. Again, accuracy within a few percent is expected. Measurements of the weight of fruits and vegetables are determined by the accuracy of the scales used, and state rules vary as to what is acceptable.

2. Random error and number of measurements

To do this investigation, you may want students to look up "standard deviation" as a measure of random error prior to their making a series of measurements.

SECTION 2 QUIZ

1-2b **Blackline Master**

1. The approximate mass of a high school student is

   a) 1.0 kg. b) 5.0 kg.

   c) 65 kg. d) 250 kg.

2. The approximate length of a baseball bat is

   a) 1 cm. b) 1 m.

   c) 1 km. d) 1 mm.

3. The maximum time allowed for this quiz is approximately

   a) 300 s. b) 2000 s.

   c) 3000 s. d) 10,000 s.

4. Three students measure the length of their classroom and get the following values:

   Student A – 6.54 m Student B – 6.57 m Student C – 6.52 m

   Because all three students received three different measurements, this is an example of

   a) an uncertainty error. b) a predictable error.

   c) a systematic error. d) a random error.

5. Students in a class measure their stride, and use that measurement to determine the length of a school's soccer field. As a result, each group arrives at a different value for the length of the field. The way to achieve the most accurate value for the length of the field would be to

   a) get the class to agree on using one group's value.
   b) pick the highest and the lowest values and average them.
   c) average all the values of each group.
   d) have everyone measure again, and then find out which number is repeated.

SECTION 2 QUIZ ANSWERS

1. c) A high school student might have a weight of 140 lbs. At 2.2 lbs/kg, this is approximately 65 kg. A mass of 5 kg would be 12 lbs. A mass of 250 kg would be 550 lbs.

2. b) A baseball bat is approximately 36 in. long, and a meter stick is about 39 in. long. A length of 1 cm is about $\frac{1}{2}$ in., a millimeter is even smaller, and a kilometer is about 3000 ft long.

3. a) The maximum time allowed for the quiz is 300 s, which is about 5 min. A time of 2000 s is about 35 min (too long for a 5-question quiz) and the other times are even longer.

4. d) Measurements that vary around the correct value would be an example of random error. If all the measurements were on the same side of the accepted value (either higher or lower), the error might be systematic. The other answers are made-up terms.

5. c) Agreeing on one group's value does not improve the accuracy, although choosing the highest and lowest values may work. If most of the values were high and one value was much lower, it would most likely lead to a false result. Having a repeating value does not mean that this is anywhere near the correct value, because it may be due to a systematic error.

SECTION 3

Average Speed: Following Distance and Models of Motion

Section Overview

This section defines speed and reaction distance, and presents different models to analyze speed and motion. Students use strobe photos to analyze constant motion at a slow and then a fast speed. They learn the distinction between speed and velocity, both graphically and mathematically, using a motion detector. They sketch graphs and make predictions to investigate how the graphs vary with a change in direction and speed. These investigations create a backdrop for defining reaction distance. Students then review different position-time graphs with varying slopes that reflect a change in the position of a person traveling at different speeds. By studying equations and comparing their graphs, students learn the difference between instantaneous and average speed and how speed affects reaction distance. Units of speed as tools of measurement are also discussed in relation to driving speeds and collection of data.

Background Information

Kinematics is the study of motion. In physics, the motion of a body is observed, and then, using measurement and graphs, that motion is analyzed. To do this analysis, tools of measurement must be established, which are appropriate for the object in motion and the speed at which it is moving. For example, a geologist who studies the movement of plates within Earth's crust would measure the distances in inches (or cm), and the time in years or even thousands of years. When measuring the speed of an electron in a particle accelerator, distances are measured in meters or kilometers, and time in millionths of seconds.

Motion maps are images of an object at uniform intervals, essentially looking like a strobe photograph of an object as it travels. Strobe photographs are multiple-exposure pictures taken in a darkened room with the shutter of the camera continuously open. In darkness, no image is recorded by the film, but when a bright flash occurs in an instant, an image is recorded. A strobe is a device that emits a bright short-duration flash at regular intervals. Because the flash is so short, the object appears stationary for any given image. If the object is moving during the dark periods between the flashes, a series of separated images will appear on the film. If these images are equally spaced, it is usually assumed that the object is traveling with constant speed in a straight line.

Understanding speed is critical to understanding motion. Suppose one travels a total distance of 100 mi from one city to the next in two hours. If on the return trip 30 minutes are spent fixing a flat tire, the total time of travel changes to 2.5 h, and the average speed is 100 mi/2.5 h or 40 mi/h. Instantaneous speed, on the other hand, is the speed of travel at any given moment. For example, on the return trip, even though the average speed was 40 mi/h, the instantaneous speed may have been 65 mi/h at any particular moment. Instantaneous speed is the speed shown on the speedometer.

A graphical analysis of uniform motion uses collected data plotted onto a graph. Putting the information onto a distance-time graph will produce a straight line as shown below.

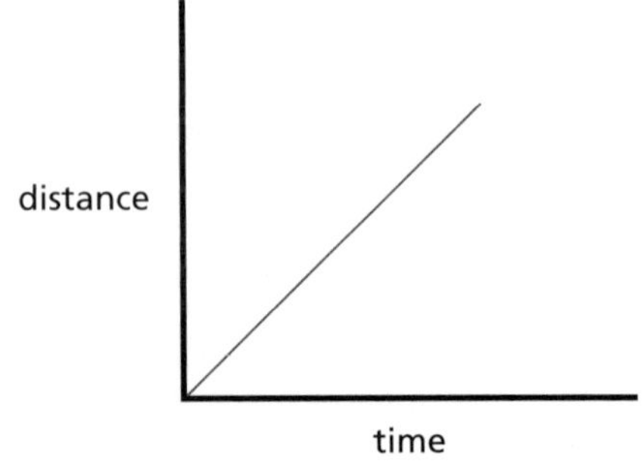

A straight line indicates uniform or constant motion. The slope of that line ($\Delta d/\Delta t$) gives the speed.
$\Delta d = d_2 - d_1$ = meters (m)
(or miles or another unit of distance)
and $\Delta t = t_2 - t_1$ = seconds (s) (or hours).

Therefore, the unit for the slope is mi/s (or mi/h if miles and hours are used).

For distance, d_2 often represents the final distance. In most situations d_1 is the starting point and is often indicated by zero. Similarly, Δt represents change in time, where t_2 is the final or end time and t_1 is the initial time.

The speed-time graph of the same object (traveling with uniform velocity) can be obtained by taking the speed from the distance-time graph, at various times, and plotting it against time on a speed-time graph. The resulting line will be a horizontal straight line, which reinforces the concept of constant motion as shown in the diagram below.

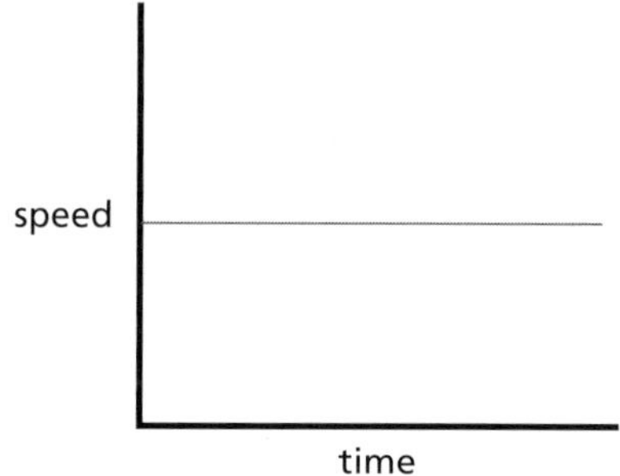

To find the distance an object travels while experiencing constant motion, you find the area under the graph. This will be a rectangle with the top line being the graph, the bottom line being the time axis, the left side the distance axis, and the right side will be a line drawn from the graph to the time axis.

Rectangle Area = length of side × the width of the side or, $A = l \times w$.

Therefore,
$d = v \times t$ since length is speed and width is time.

Mathematically, the equation
$v = d/t$

will give the rearranged equation to solve for d as
$d = v \times t$.

In later sections, students will move from the study of speed to the study of velocity. The concept is essentially the same, except that speed is only the magnitude or how fast something is moving, whereas velocity is the speed and the direction in which the object is moving.

Crucial Physics

- Average velocity = total distance traveled over a given time.
  - $v = \Delta d/\Delta t$
- Average velocity and instantaneous velocity do not have to be equal.
- The slope of a displacement vs. time graph is equal to the velocity.

Learning Outcomes	Location in the Section	Evidence of Understanding
Define and contrast average speed and instantaneous speed.	***Investigate*** Step 7 ***Physics Talk***	Students determine the total distance traveled in each trial to determine average speed. Students define instantaneous speed and contrast it with average speed in a sample problem.
Use strobe photos, graphs, and an equation to describe speed.	***Investigate*** Steps 1-7	Students sketch diagrams of strobe photos to show automobiles traveling at different speeds and arrive at the equation for average speed.
Use a motion detector to measure speed.	***Investigate*** Steps 4-7	Students obtain graphs using a motion detector to measure the velocity of an object.
Construct graphs of your motion.	***Investigate*** Steps 5 and 6	Students produce graphs by walking toward and then away from the motion detector at different speeds and in different directions.
Interpret distance-time graphs.	***Investigate*** Steps 5-7 ***Physics Talk***	Students read graphs to predict and determine distance, time, and average speed. Students study the slope of distance-time graphs to interpret speed.
Calculate speed, distance, and time using the equation for average speed.	***Physics to Go*** Questions 4-7 and 9-10	Students calculate speed, distance, and time in different questions.

NOTES

Section 3 Materials, Preparation, and Safety

Materials and Equipment

PLAN A		
Materials and Equipment	**Group (4 students)**	**Class**
Computer, station or calculator, CBL or equivalent system*		1 per class
Motion detector, probe and interface*	1 per group	

*Additional items needed not supplied

PLAN B		
Materials and Equipment	**Group (4 students)**	**Class**
Computer, station or calculator, CBL or equivalent system*		1 per class
Motion detector, probe and interface*	1 per group	

*Additional items needed not supplied

Note: Time, Preparation, and Safety requirements are based on Plan A, if using Plan B, please adjust accordingly.

Time Requirement

During the class each group of students will require approximately 1.5 class periods or 60 minutes to do the experiment and gather their data.

Teacher Preparation

- Become familiar with the motion detectors and the software that controls their output.
- Instructions come with the equipment, but you should use it yourself before putting it into the classroom. The experiment takes less time if you are aware of the near and far limits of range. Run several trials to be certain that the detector is functioning properly and giving appropriate data.
- Prepare a list of the keystrokes required by your software to erase one graph and get ready to gather new data. Some programs ask several questions before recording data. You should know the best responses to reduce delay.
- Check an appropriate sampling rate for your machine.
- To prepare for the class, set up the detector and computer ahead of time. Put masking tape on the floor to mark the range and perhaps another piece of tape to indicate the lane so that students stay "on track."
- A student may not reflect the sound waves used by the motion detector well enough for the detector to determine his or her position accurately. This occurs because clothing often absorbs more of the sound than it reflects. Give the student a hard surface to hold, such as a section of cardboard, to provide a better reflecting surface.
- Motion detectors will often locate nearby objects that provide good reflections and "lock in" on those objects. Make certain the area that will be used is clear of obstructions and extraneous material to prevent this problem.

Safety Requirements

- Students who are using the equipment will be watching the computer to see the graphs produced and may not watch where they are walking. To prevent a student from accidentally walking into the lab equipment, make certain that the area is clear of any materials that may present a danger.
- The computer (or calculator) that is being used must be placed in a safe position to prevent it from falling and damaging equipment.

Meeting the Needs of All Students

Differentiated Instruction: Augmentation and Accommodations

Learning Issue	Reference	Augmentation and Accommodations
Conceptualizing motion from a still image	***Investigate*** Step 1 ***Physics to Go*** Questions 1 and 2	**Augmentation** • Students who are visual learners will benefit from the strobe photos if they can conceptualize how the still photos represent motion. This concept is a leap for many students. Show students examples of real-life strobe photos. Use the sample in *Physics Talk* or the Internet. • Use a small car to demonstrate how strobe photos are taken. To help students visualize images frozen in motion, say "snapshot" at regular one-second intervals.
Copying sketches from a textbook	***Investigate*** Steps 1 and 2	**Augmentation** • Students with fine-motor issues may have trouble sketching the strobe photos from the textbook. Copying a sketch requires alternating attention between the book and paper, making it difficult for some students to draw an accurate sketch in a timely manner. Direct students to trace the cars by placing their log book page on top of the strobe photos in their textbook. **Accommodation** • For students with more significant fine-motor issues, copy the textbook page for them to tape into their logs.
Following directions Sketching accurate graphs	***Investigate*** Steps 4-6	**Augmentation** • Students with short attention spans have a difficult time following directions for a series of tasks and completing them within a reasonable time limit. • Model the use of the motion detector before you begin the *Investigate* to show students proper usage. • Model a graph that has accurate scales. • Model a table to organize the sketched graphs. **Accommodation** • Provide a worksheet with blank graphs that have an appropriate scale on them. Model how to sketch graphs accurately on the blank grids. • Provide students with a two-column table that already has motion described in words. Require students to sketch the graph of that motion in the other column.
Obtaining information from a graph	***Investigate*** Step 7 ***Physics Talk*** Describing Motion and Speed Using a Distance-Time Graph ***Physics to Go*** Question 3	**Augmentation** • Students with visual-spatial, sequential, or mathematical issues struggle to use graphs for problem solving. This task requires that students copy a graph accurately, read the scales to obtain correct values for distance and time, and then perform a calculation with a new equation. Provide direct instruction to show students how to complete this step. Give a few examples, and then give students an opportunity to practice independently. Remind students that by calculating average speed they are finding the slope of the graph. **Accommodation** • Color-code practice graphs to match the variables in the formula. For example, the distance axis and scale could be red, and the "*d*" in the formula could be red. This strategy provides a visual cue for visual learners and students who may not be consistently focused.

Learning Issue	Reference	Augmentation and Accommodations
Solving word problems	***Investigate*** Step 8 ***Physics to Go*** Questions 4, 5, 6, 7, 9, 10	**Augmentation** • Students with mathematical and executive-function issues struggle to solve multi-step problems. Point out to students that they should have eight answers when they are finished with *Step 8* of the *Investigate*. • Provide direct instruction on how to use the formula before students begin to solve the problems. Use the sections following this investigation as resources. Some students may shut down when faced with a new task that has not been taught. If students learn to solve the problems incorrectly, it might be difficult for them to relearn the skill correctly. **Accommodation** • Provide a worksheet with blank problem-solving boxes for students. (See Problem-Solving Box information and table at the bottom of this page.)
Reading comprehension Extracting important concepts from reading	***Physics Talk***	**Augmentation** • Students with reading, mathematical, and attention issues struggle with longer reading assignments that are laden with math concepts. Use a graphic organizer, or class note-taking strategy, to understand the concept of motion described in different ways. • Provide direct instruction for the new formula and give many opportunities for guided practice, especially if this is the first formula that students learn in the class. • Explain and model how to use the formula circle. This strategy allows students to manipulate formulas, even if they have not yet learned this skill in algebra.
Sketching graphs	***Physics to Go*** Question 11	**Augmentation** • Students struggle to label the correct axes on a graph and set up reasonable scales on each axis. Provide a model graph for students to reference, remind students to use previous work from this section to help them, and/or help students recognize the pattern (+0.25 s) for the reaction-time scale. **Accommodation** • Provide students with a graph that already has labels and scales. Ask them to sketch the coordinates and slopes to check for understanding.
Using academic vocabulary	***Physics to Go*** Question 12	**Augmentation** • Have students use their new academic vocabulary; for example speed, reaction time, reaction distance, and so on. Using vocabulary in their own writing helps students retain more and gain a deeper understanding.

Problem-Solving Box

Students need a way to systematically organize information when solving problems. Use a graphic organizer for problem solving that takes students through the steps. Explicitly teach the way a student can extract necessary information from a word problem by thinking aloud while solving the problem. This organizer can slowly be faded out as students become more independent. An example of a problem-solving graphic organizer is provided to the right.

WANT:	GIVEN:	FORMULA:
Solve:		**Check:**

Strategies for Students with Limited English-Language Proficiency

Learning Issue	Reference	Augmentation
Reading comprehension Identifying parts of speech	***Investigate*** Step 8.f)	Students may be confused by constructions in English in which the part of speech is different from what might be expected. For example, in "automobile lengths" the term "automobile" is used as an adjective. Explain that "automobile lengths" refers to using the length of an automobile as a "yardstick." A driver cannot stop and get out of their vehicle to measure the following distance with a tape measure. Instead, drivers estimate how far ahead the next vehicle is by imagining how many cars could be placed between their vehicle and the next vehicle.
Vocabulary comprehension	***Investigate*** Steps 1 and 2	Certain terms may have cognates in an ELL student's primary language. Ask students to make a list of cognates for technical terms such as "position," "distance," "rapidly," and "velocity." When you describe a physics situation for the class, try to refer back to these cognates while you use a synonym. For example, give the instruction, "Now walk faster, or more rapidly." This extra clarification will help reinforce vocabulary that students need to acquire.
Reading comprehension Identifying parts of speech	***Physics Talk***	In reading the terms "following distance," "braking distance," and "stopping distance," students may read the first word as a verb. Explain that verbs and nouns can be used as adjectives. Other similar terms to look out for in the section include "reaction distance," "reaction time," and "quantity symbol." The terms become even more confusing if an additional adjective is added, for example, "safe following distance." Help students decompose the meaning by starting from the end and working backward. First, they should look at "following distance," and then decide how "safe" further modifies the meaning. Explain that hyphens are an exception to this approach. For example, to understand "multiple-exposure photo," first look at "multiple-exposure" and then see how this modifies "photo." It may help students to illustrate confusing terms with diagrams in their *Active Physics* log. Then they can refer back to the diagrams when they come across one of the words in a word problem.
Reading comprehension	***What Do You Think Now?***	Provide additional practice using technical terms in the section. One way to do this is to ask students to include certain terms in their responses for *What Do You Think Now?* Possible terms in this activity include "elapsed time," "following distance," "average speed," and "instantaneous speed."
Vocabulary comprehension	***Physics to Go*** Question 8	In *Physics to Go, Question 8*, students may be confused when following distance is described using units of time. A historical explanation may help them understand why the term "following distance" is used. Explain that in older driver's education manuals, following distance was recommended to be one car length for every 10-mi/h of speed. Current guidelines recommend a "distance" of 2 s.

NOTES

SECTION 3

Teaching Suggestions and Sample Answers

What Do You See?

Your students will give you a range of responses when you ask them what the illustration depicts. If you want to be more specific, ask them how they think it relates to the title of this section. This approach will help guide them toward the purpose of the *What Do You See?* illustration. Students will begin to take an active interest in the topic, and you might find how quickly the depicted scene can lead to an understanding of physics concepts. At this point, encourage students to answer why the artist has chosen to represent dangerous driving.

Chapter 1 Driving the Roads

Section 3 Average Speed: Following Distance and Models of Motion

What Do You See?

Learning Outcomes

In this section, you will

- **Define** and contrast average speed and instantaneous speed.
- **Use** strobe photos, graphs, and an equation to describe speed.
- **Use** a motion detector to measure speed.
- **Construct** graphs of your motion.
- **Interpret** distance-time graphs.
- **Calculate** speed, distance, and time using the equation for average speed.

What Do You Think?

In a rear-end collision, usually the driver who strikes a vehicle from behind is legally at fault.

- **What is a safe following distance between your automobile and the vehicle in front of you?**
- **How do you decide what a safe following distance is?**

Record your ideas about these questions in your *Active Physics* log. Be prepared to discuss your responses with your small group and the class.

Investigate

In this *Investigate*, you will use strobe photos to observe constant motion at different speeds. You will then use a motion detector to measure velocity and to generate graphs of motion.

1. A "strobe photo" is a combination of photographs taken at regular time intervals. A single picture can then show the position of the object over equal time intervals.

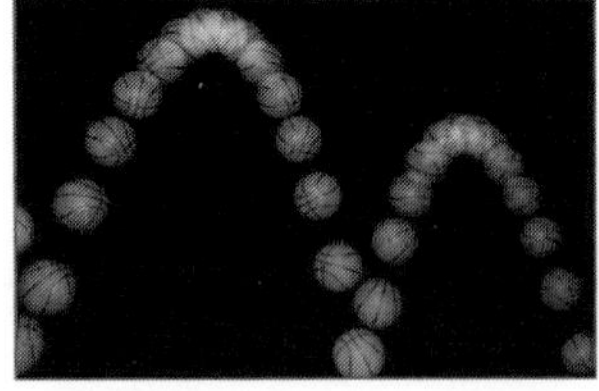

34

Students' Prior Conceptions

The value of this section is introducing students to another language of science and mathematics, and encouraging students to build fundamental skills of interpreting motion through graphical analysis of data.

1. **The location of an object can be described by stating its distance from a given point (ignoring direction).** You must highlight the notion that a student defines a start position within a frame of reference in order to describe motion, either with words or with mathematics. This is the key to support student understating of steady and accelerated motion and to establish the foundation for using vectors. Specific tools are available for students to measure distance and time for various motions and to note the direction in which these motions are occurring along a straight line with reference to an origin. Using their exact measurements of distance, time, and direction, students can produce graphs of motion for analysis.
2. **Time is defined in terms of its measurement.** Time only moves linearly in one direction and subsequently becomes a constant variable in describing motion with distance vs. time or velocity vs. time graphs. Time is the *x*-axis variable.

What Do You Think?

Students will come up with a wide range of answers. Value their responses and encourage them to share their ideas with other students in their group. At this stage, it is important to see what they know of speed in terms of distance. Accepting all answers and posing questions will help generate responses for a few minutes. Ask students to write down their answers. You might want to lead a discussion on the idea of a vehicle following at a safe distance behind another. Point out to your students that they will be returning to the *What Do You Think?* questions for a more comprehensive discussion.

What Do You Think?

A Physicist's Response

A safe distance between your vehicle and the vehicle in front can be determined by the three-second rule. The separation distance provided by this rule allows a driver sufficient time to react to a sudden problem to safely bring a vehicle to a stop and avoid striking the vehicle in front. A physicist would use knowledge of road conditions, vehicle speed, and reaction time to decide on a safe following distance.

Investigate

1.

What is described as a "strobe" sketch in the text is actually equivalent to a time-lapse picture repeatedly exposed with a strobe light. The result is that although the object is moving all the time, the image only appears at regular intervals when the strobe light flashes because the image is taken in the dark.

NOTES

NOTES

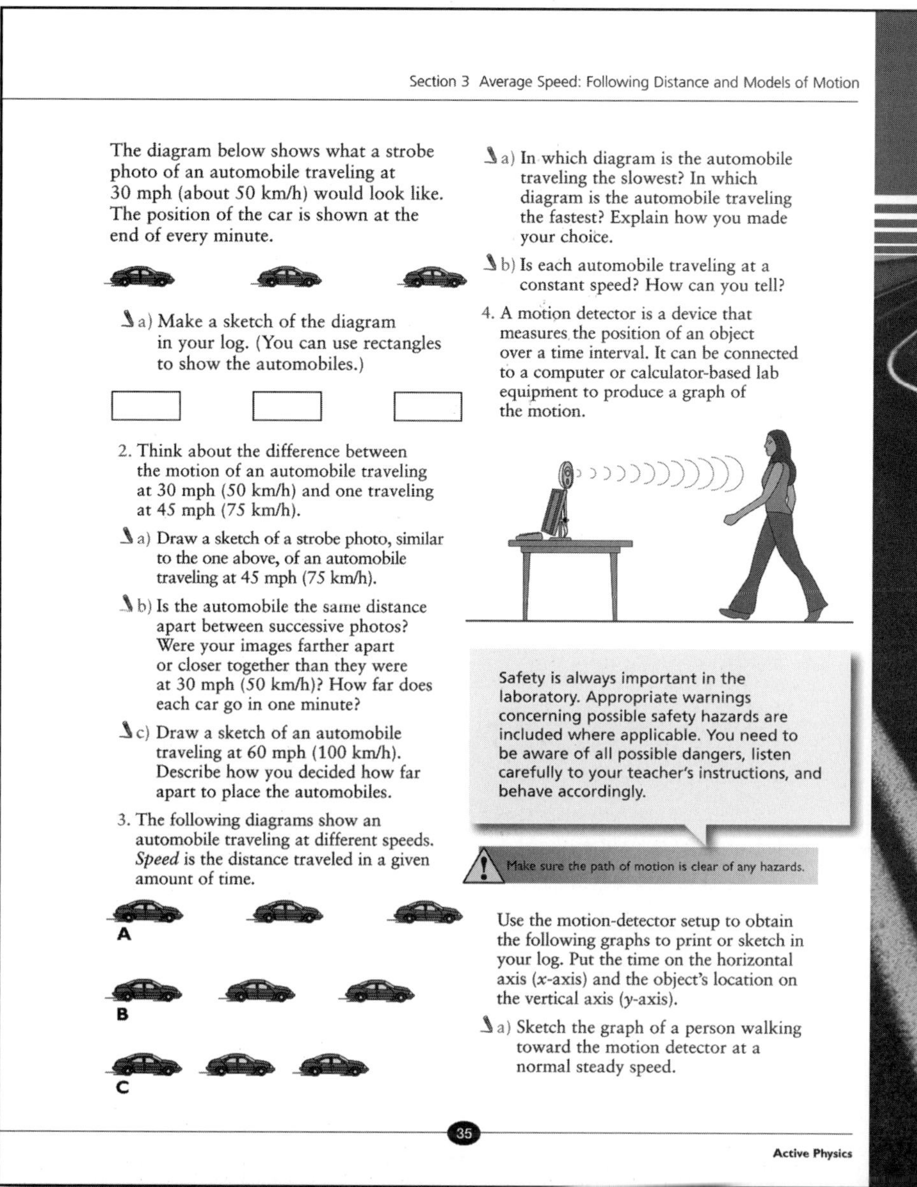

1.a)

If the students do not feel comfortable sketching automobiles, suggest that they draw rectangles instead.

2.

Because the automobile is moving with constant speed in a straight line, the images are equally spaced. Also, the greater the speed, the further apart the images will be.

2.a)

The sketch should show the automobiles with larger, uniform distances between them.

2.b)

Yes, the automobile is the same distance apart in each of the successive photos (indicating constant speed), but further apart than for the automobile going 30 mi/h. In one minute, the 30 mi/h automobile goes $\frac{1}{2}$ mile, and the 45 mi/h automobile goes $\frac{3}{4}$ of a mile. (A car going 60 mi/h goes 1 mile in one minute).

2.c)

The automobiles should be twice the distance apart as they were for 30 mi/h. The distance between images will be doubled because the speed is doubled.

3.a)

The diagram where the automobiles are closest together will represent the slowest speed (diagram C), and furthest upon the fastest (diagram A).

3.b)

Yes, each automobile is traveling with constant speed in the diagrams. This can be determined by measuring the distance between each automobile. If the distance between images is the same for equal time intervals, the automobiles have constant speed, assuming that the flash rate for the pictures shown is constant.

4.

For this *Investigate*, the students should set up the software to give a position vs. time plot. A motion detector typically gives the position of an object by sending out an ultrasonic pulse and then listening for the reflection from the object being measured. This is essentially the same echolocation procedure used by bats to fly in dark caves, or cameras with autofocus to lock onto an object, although cameras typically use an electromagnetic wave rather than sound.

Teaching Tip

Motion detectors from different companies use different software platforms. You should familiarize yourself with the operation of the detector, and check to make certain it is operating correctly before having the students try this *Investigate*. If you or the students are having difficulty getting a good graph of the students' motion, have the students hold a large, solid, flat object such as the cardboard cover from a copy-paper box, or a whiteboard to better reflect the signal back to the detector.

b) Sketch the graph of a person walking away from the motion detector at a normal speed.

c) Sketch the graph of a person walking away from the motion detector then toward it at a very slow speed.

d) Sketch the graph of a person walking in both directions at a fast speed.

e) Describe the similarities and differences among the graphs. Explain how the direction and speed that the person walked contributed to these similarities and differences.

5. Predict what the graph will look like if you walk toward the motion detector at a slow speed and away from it at a fast speed.

a) Sketch a graph of your prediction.

b) Test your prediction. How accurate was your prediction?

6. Do two more trials using the motion detector. In trial 1, walk slowly away from the detector. In trial 2, walk quickly away from the detector.

a) Sketch the lines from the two trials on the same labeled axes. Be sure to record the endpoints for each line.

b) Suppose someone forgot to label the two lines. How can you determine which graph goes with which line?

7. In physics, the total distance traveled by an object during a given time is the *average speed* of the object.

a) From your graph, determine the total distance you walked in each trial.

b) How long did it take you to walk each distance?

c) Divide the distance you walked (your change in position) (d) by the time it took for each trial (t).

This calculation gives you your average speed in meters per second (m/s).

$$v_{av} = \frac{d}{t}$$

d) How could you go about predicting your position after walking for twice the time in trial 2? When you extrapolate data, you make an assumption about the walker. What is the assumption? (Extrapolate means to estimate a value outside the known data points.)

8. An automobile is traveling at 60 ft/s (about 40 mph or 65 km/h).

a) If the reaction time is 0.5 s, how far does the automobile travel in this time?

b) How much farther will the automobile travel if the driver is distracted by talking on a cell phone or unwrapping a sandwich, so that the reaction time increases to 1.5 s?

c) Answer the questions in *Steps 8.a)* and *8.b)* for an automobile moving at 50 ft/s (about 35 mph or 56 km/h).

d) Repeat the calculation for *Step 8.c)* for 70 ft/s (about 48 mph or 77 km/h).

e) Imagine a driver in an automobile in traffic moving at 40 ft/s (about 28 mph or 45 km/h). The driver ahead has collided with another vehicle and has stopped suddenly. How far behind the preceding automobile should a driver be to avoid hitting it, if the reaction time is 0.5 s?

f) An automobile is traveling at 60 ft/s (about 40 mph or 65 km/h). How many automobile lengths does it travel per second? A typical automobile is 15 ft (about 5 m long).

4.a)

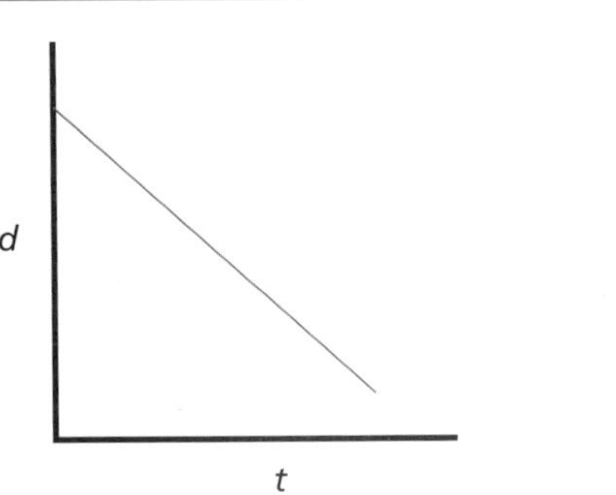

4.b)

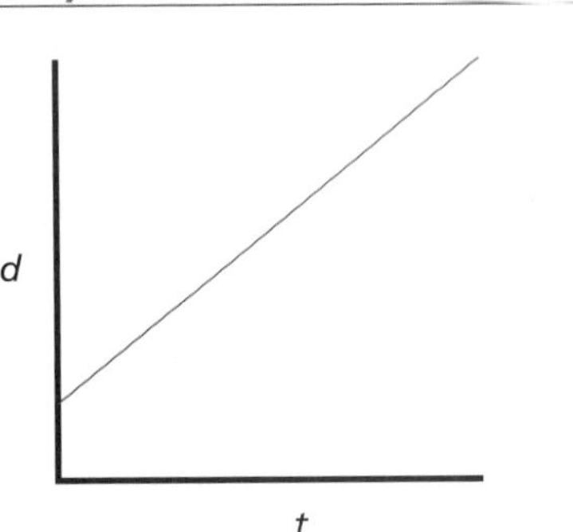

4.c)

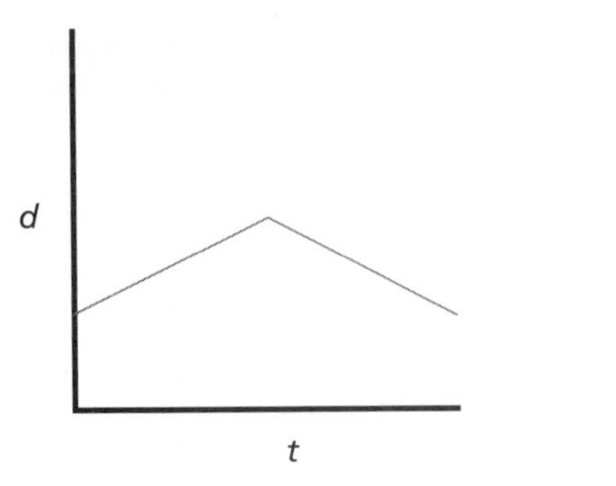

4.d)

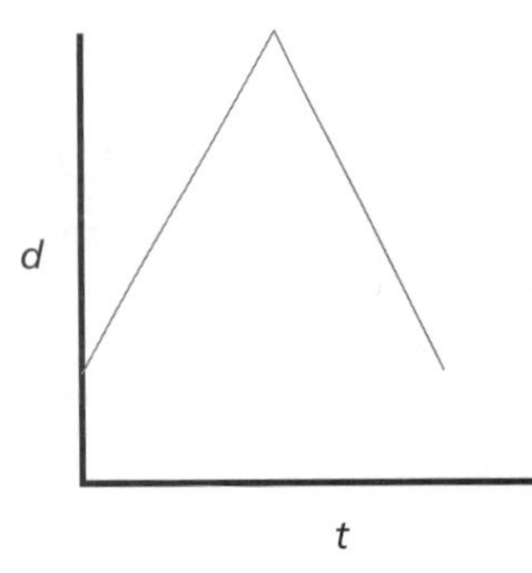

4.e)

When a person walks away from the detector, the graphs have a positive slope. When a person walks toward the detector, the graphs have a negative slope. The graphs have different slopes when a person walks at different speeds and reverses direction, producing peaks in two of them.

5.a)

Students will sketch a graph of their predictions.

5.b)

The prediction of a steeper slope when walking away from the motion detector at a fast speed will be confirmed as accurate.

6.

The slope of the graph will depend upon the direction of motion. Walking toward a detector gives a decreasing distance and a negative slope; walking away, a positive slope. If this is not the case, the software can be adjusted to show this. Older motion detectors do not work well at distances less than 0.50 m or greater than 3 m. Also, the area where the student is walking should be clear of extraneous equipment, people, etc., since the detector may lock onto these objects rather than the person moving.

6.a)

The sketches should look like the graph below. Students' endpoints will vary.

6.b)

Students can determine which line belongs to which trial by looking at the slope. The steeper slope will be the trial with faster motion.

7.

Students can choose distances and times from their data table or graph to use for the calculations in this step, or make their own measurements with a stopwatch and meter stick.

7.a)

Displacement $= x_{(\text{final})} - x_{(\text{initial})}$

7.b)

Elapsed time $= t_{(\text{final})} - t_{(\text{initial})}$

7.c)

Students divide the value they obtained in *7.a)* by the one in *7.b)* to obtain their average speed. Check that the units used are m/s.

7.d)

In twice the time, the walker covers twice the distance, assuming that speed remains constant.

8.a)

Reaction distance (d_{reaction}) equals velocity times reaction time:

$$d_{\text{reaction}} = (v)(t_{\text{reaction}}) =$$
$$(60 \text{ ft/s})(0.5 \text{ s}) = 30 \text{ ft}$$

8.b)

$$d_{\text{reaction}} = (v)(t_{\text{reaction}}) =$$
$$(60 \text{ ft/s})(1.5 \text{ s}) = 90 \text{ ft}$$

or 1.5 times farther.

8.c)

$$d_{\text{reaction}} = (v)(t_{\text{reaction}}) =$$
$$(50 \text{ ft/s})(0.5 \text{ s}) = 25 \text{ ft}$$
$$d_{\text{reaction}} = (v)(t_{\text{reaction}}) =$$
$$(50 \text{ ft/s})(1.5 \text{ s}) = 75 \text{ ft.}$$

8.d)

$$d_{\text{reaction}} = (v)(t_{\text{reaction}}) =$$
$$(70 \text{ ft/s})(0.5 \text{ s}) = 35 \text{ ft}$$
$$d_{\text{reaction}} = (v)(t_{\text{reaction}}) =$$
$$(70 \text{ ft/s})(1.5 \text{ s}) = 105 \text{ ft.}$$

8.e)

$$d_{\text{reaction}} = (v)(t_{\text{reaction}}) =$$
$$(40 \text{ ft/s})(0.5 \text{ s}) = 20 \text{ ft.}$$

However, this does not include the distance required for the brakes to stop the car after they are applied. This is why the suggested following distance is usually 3 s, which would give a distance of 60 ft (20 ft/s)(3s).

8.f)

60 ft/15 ft =

4 automobile-lengths.

NOTES

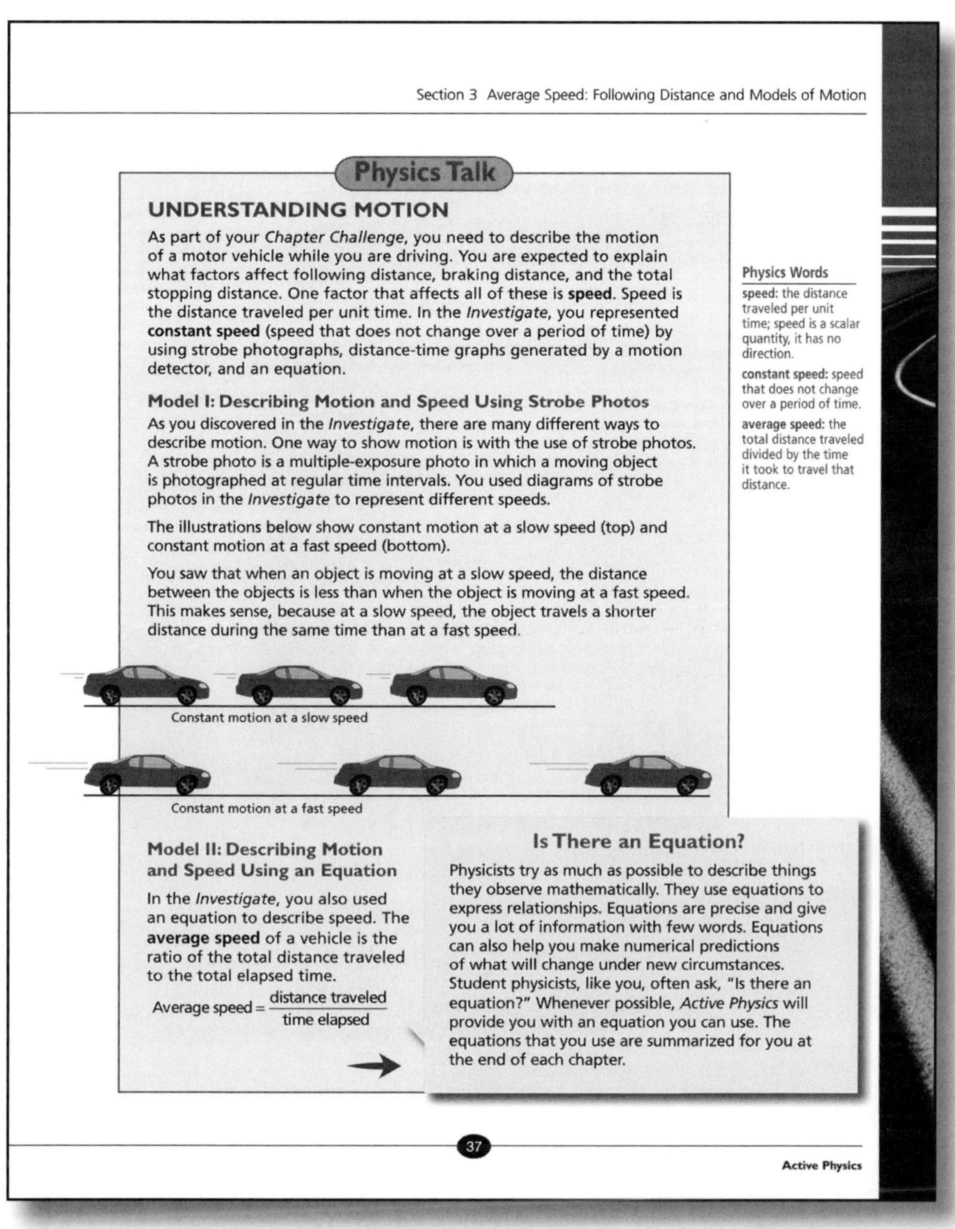

Section 3 Average Speed: Following Distance and Models of Motion

Physics Talk

UNDERSTANDING MOTION

As part of your *Chapter Challenge*, you need to describe the motion of a motor vehicle while you are driving. You are expected to explain what factors affect following distance, braking distance, and the total stopping distance. One factor that affects all of these is **speed**. Speed is the distance traveled per unit time. In the *Investigate*, you represented **constant speed** (speed that does not change over a period of time) by using strobe photographs, distance-time graphs generated by a motion detector, and an equation.

Model I: Describing Motion and Speed Using Strobe Photos

As you discovered in the *Investigate*, there are many different ways to describe motion. One way to show motion is with the use of strobe photos. A strobe photo is a multiple-exposure photo in which a moving object is photographed at regular time intervals. You used diagrams of strobe photos in the *Investigate* to represent different speeds.

The illustrations below show constant motion at a slow speed (top) and constant motion at a fast speed (bottom).

You saw that when an object is moving at a slow speed, the distance between the objects is less than when the object is moving at a fast speed. This makes sense, because at a slow speed, the object travels a shorter distance during the same time than at a fast speed.

Constant motion at a slow speed

Constant motion at a fast speed

Model II: Describing Motion and Speed Using an Equation

In the *Investigate*, you also used an equation to describe speed. The **average speed** of a vehicle is the ratio of the total distance traveled to the total elapsed time.

$$\text{Average speed} = \frac{\text{distance traveled}}{\text{time elapsed}}$$

Physics Words

speed: the distance traveled per unit time; speed is a scalar quantity, it has no direction.

constant speed: speed that does not change over a period of time.

average speed: the total distance traveled divided by the time it took to travel that distance.

Is There an Equation?

Physicists try as much as possible to describe things they observe mathematically. They use equations to express relationships. Equations are precise and give you a lot of information with few words. Equations can also help you make numerical predictions of what will change under new circumstances. Student physicists, like you, often ask, "Is there an equation?" Whenever possible, *Active Physics* will provide you with an equation you can use. The equations that you use are summarized for you at the end of each chapter.

37

Active Physics

Physics Talk

Students review illustrations of a strobe photo showing constant motion at a slow speed and constant motion at a fast speed. Remind them of their sketches in the *Investigate* in which they used strobe photos and motion detectors. Discuss how motion is defined through average and instantaneous speed and have students study the distance-time graphs of an object that represent different types of motion. You will find the distance-time graphs provide strong visuals for data. As students are introduced to the concept of velocity, have them study the slopes of distance-time graphs that reflect the varying speed and direction of an object in motion. By calculating the speed, distance, or time of travel in sample problems, they should rewrite the equation for average speed. You could also use this equation to highlight the importance of reaction distance. While students learn more about the SI system of measurement, it is important to realize that their grasp of this system will gradually improve as they become more familiar with it.

Because maintaining a safe speed is essential to safety, students are introduced to the Doppler effect to explain how an automobile's speed can be accurately determined by another person who is watching the automobile pass from a distance.

You might want to explain the Doppler effect in the *Physics Talk* by citing an example of the frequency shift that is readily observable at racetracks as a race car zooms past the cameras and the stands. Tell students that, in this case, the engine noise replaces the horn as the sound emitter. Mention that exactly the same principle is used in a radar gun which measures the speed of a pitch in a baseball game. Another example you could touch on is the Doppler effect in the "sonic boom," which is heard when a plane breaks the "sound barrier." To check if students understand how police use the Doppler effect in radar guns, ask them to write a brief summary of the Doppler effect in their *Active Physics* logs.

To provide a conceptual basis for the effect, draw students' attention to the two diagrams in the text that show the circular waves being generated by a moving source and a stationary source in the section titled *Speed and the Doppler Effect.* Ask the students to indicate in which direction the moving source is going, where the observed frequency would be higher, where it would be lower, and where there would be no change in frequency.

This equation can be written using quantity symbols.

$$v_{av} = \frac{\Delta d}{\Delta t}$$

where v_{av} is average speed,

Δd is change in position or total distance traveled,

Δt is change in time or elapsed time.

The Greek letter *delta*, "Δ," is often used in science to mean "a change in." In this section, you will be dealing only with situations in which you are given total distance traveled and elapsed time.

Symbols for Physical Quantities

Symbols for SI units are unique and precise. There is only one symbol for each SI unit. For example, "m" stands for meter and "s" stands for second. The same SI symbol is used in every language.

When writing equations in science, there is also a need for other symbols. Symbols are needed for quantities such as distance, time, or speed. These symbols are not a part of SI. They are also not always unique. For example, *V* can stand for volume or voltage. As much as possible, scientists try to use standard symbols for quantities. However the SI symbol V only stands for volts.

When writing equations, you should use the same symbols used in *Active Physics.* To distinguish the two types of symbols in printed materials, sloping *(italic)* type is used for quantity symbols and upright (roman) type is used for SI symbols.

Sample Problem I

If you drive a distance of 400 mi (about 640 km) in 8 h, what is your average speed?

Strategy: You can use the equation for average speed. $v_{av} = \frac{\Delta d}{\Delta t}$

Given: $\Delta d = 400$ mi

$\Delta t = 8$ h

Solution:

$$v_{av} = \frac{\Delta d}{\Delta t} = \frac{400 \text{ mi}}{8 \text{ h}} = 50 \frac{\text{mi}}{\text{h}}$$

Your average speed is 50 mi/h.

The average speed of 50 mi/h (80 km/h) does not tell your fastest speed or your slowest speed. It only tells you that over a period of time, 8 h, you traveled a given distance, 400 mi (80 km).

Section 3 Average Speed: Following Distance and Models of Motion

Before your vehicle started moving, the speed was 0 mi/h. When you stopped your vehicle at the end of the trip, your speed was again 0 mi/h. During the trip you probably slowed down and sped up as you drove along. You may have stopped for a meal. The speedometer reading at any moment during the trip is your **instantaneous speed,** which is the speed at a given moment. Instantaneous speed is the speed measured during an instant.

Physics Words

instantaneous speed: the speed at a given moment.

Using the Equation for Speed to Find Other Quantities

Equations are a powerful mathematical tool. Using the equation for average speed, you are not limited to just solving for speed. If you know the average speed and the time it took you to travel that speed, you can find the distance you traveled in that time. You can also find the time traveled if you know the distance traveled and the average speed.

$$\text{Average speed} = \frac{\text{distance traveled}}{\text{time elapsed}}$$

Using algebra, it follows that

$$\text{Distance} = \text{average speed} \times \text{time}$$

$$\text{Time} = \frac{\text{distance}}{\text{average speed}}$$

Using quantity symbols these equations can be written as

$$v_{av} = \frac{\Delta d}{\Delta t}$$

$$\Delta d = v_{av} \times \Delta t$$

$$\Delta t = \frac{\Delta d}{v_{av}}$$

You can use the following helpful circle to do your calculations:

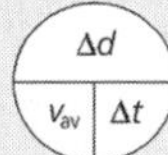

By covering up the variable you wish to find, you can see the equation.

To find average speed (v_{av}), cover up the (v_{av}) and you see $\frac{\Delta d}{\Delta t}$.

To find distance (Δd), cover up (Δd) and you see $v_{av} \times \Delta t$.

To find time (Δt), cover up (Δt) and you see $\frac{\Delta d}{v_{av}}$.

It is important to note that there is one equation for average speed, but you can write it in three equivalent ways.

Sample Problem 2

You are traveling at 35 mph (about 50 ft/s) and your reaction time is 0.2 s. Calculate the distance you travel during your reaction time.

Strategy: You can rewrite the equation for average speed to solve for distance traveled.

$$\Delta d = v_{av} \times \Delta t$$

Remember that ft/s means $\frac{\text{ft}}{\text{s}}$.

Given: $\Delta t = 0.2\ \text{s}$

$v_{av} = 50\ \text{ft/s}$

Solution: $\Delta d = v_{av} \times \Delta t$

$$\Delta d = 50\ \frac{\text{ft}}{\not{\text{s}}} \times 0.2\ \not{\text{s}}$$

$$= 10\ \text{ft}$$

Calculations and Units

In physics, when you do calculations, it is very important to pay close attention to the units in your answer. Notice how in the previous calculation the units for seconds (s) in the top and bottom of the equation cancel out, leaving feet (ft), the unit for distance that you need for your answer. Checking to see if the units make sense is a tool that physicists use to ensure that their calculations make sense and that they have not made a mistake.

Sample Problem 3

In an automobile collision, it was determined that a car traveled 150 ft before the brakes were applied.

a) If the car had been traveling at the speed limit of 40 mph (60 ft/s), what was the driver's reaction-time (time it took to apply the brakes)?

b) Witnesses say that the driver appeared to be under the influence of alcohol. Does your reaction-time data support the witnesses' testimony?

Strategy:

a) You can rewrite the equation for average speed to solve for time elapsed.

$$\Delta t = \frac{\Delta d}{v_{av}}$$

Section 3 Average Speed: Following Distance and Models of Motion

Given: $\Delta d = 150 \text{ ft}$

$v_{av} = 60 \text{ ft/s or } 60 \frac{\text{ft}}{\text{s}}$

Solution: $\Delta t = \frac{\Delta d}{v_{av}}$

$$\Delta t = \frac{150 \text{ ft}}{60 \frac{\text{ft}}{\text{s}}}$$

$$= 2.5 \text{ s}$$

Note, mathematically, $\frac{\text{ft}}{\frac{\text{ft}}{\text{s}}} = \text{ft} \times \frac{\text{s}}{\text{ft}} = \text{s}$

b) Reaction time with distractions was measured in *Section 1*. The reaction time of 2.5 s seems very slow. The driver has the reaction time of someone who could be under the influence of alcohol.

Speed and Velocity

In *Active Physics*, you will often explore the same topic several times. Being exposed to the same topic at different times and in different situations helps you learn and understand the topic better. The difference between speed and velocity will be explored at different times in this book. Velocity will also be explored in greater depth in later chapters.

A term you often hear used when talking about speed is velocity. **Velocity** is speed in a given direction. Velocity always includes both speed and direction.

Physics Words

velocity: the speed in a given direction.

Model III: Describing Motion and Speed Using a Distance-Time Graph

You investigated how to represent motion with strobe photos and with a mathematical equation. A third way to represent motion is with graphs. Graphs are a visual way to represent data. The graph at right shows an automobile traveling at a constant speed of 50 mi/h.

Notice that time is on the *x*-axis and distance is on the *y*-axis. By reading the coordinates on the graph, you can see that the automobile reached the 50 mi position at the end of 1 h; the 100 mi position at the end of 2 h; and the 150 mi position at the end of 3 h.

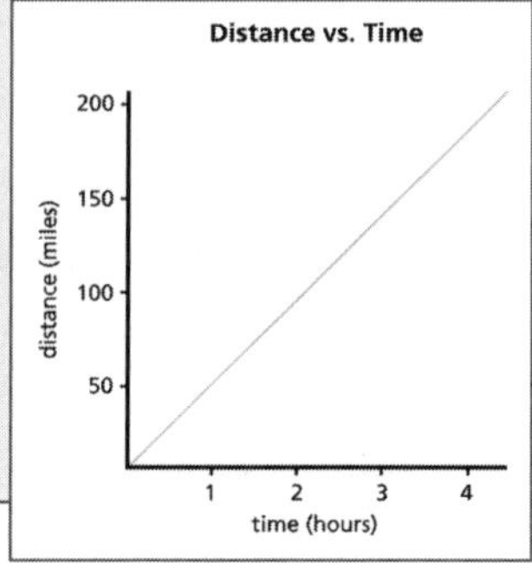

CHAPTER 1

1-3a Blackline Master

It is quite unrealistic to assume that an automobile could keep this speed of 50 mph for a full 3 h. If it did, however, you can see that the distance-time graph forms a straight line. Anytime an object moves at a constant speed, the distance-time graph is a straight line.

In the *Investigate*, you used a motion detector to generate graphs to represent your motion. You can determine the general motion of a person (a vehicle or any object) by reviewing a distance-time graph. Look at the following graphs. All graphs have the same time and distance scales.

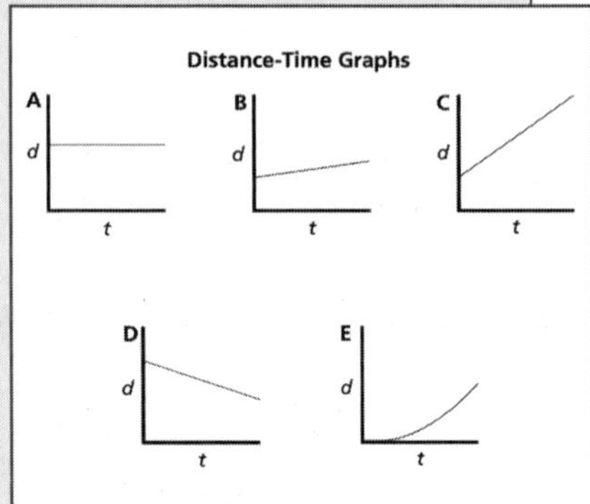

Graph A: A person is at rest. As time increases, there is no change in the position of the person. The person is standing still.

Graph B: A person is traveling at a slow speed. As time increases, there is a small change in the position.

Graph C: A person is traveling at a fast speed. As time increases, there is a greater change in the position.

Notice that the slope of the graph indicates the speed of the person. A slow speed has a gradual slope. A fast speed has a steep slope. No motion has zero slope. The graph of a person at rest is a horizontal line with a slope of zero—representing a speed of zero.

Graph D: A person is traveling in the opposite direction of the person in the previous graphs. As time passes, the change in position is in the opposite direction.

Graph E: A person is changing speed. As time passes, the change in position is increasing for each second. Notice that a changing speed is a curve on a distance-time graph.

In the *Investigate*, you noticed that walking toward the motion detector produced a slope in one direction. Walking away from the motion detector produced a slope in the opposite direction. The slope was zero, or close to zero, when standing still.

Section 3 Average Speed: Following Distance and Models of Motion

Speed and the Slope of a Distance-Time Graph

You compared speeds by looking at the slope of lines on distance-time graphs. You can also use the slopes of distance-time graphs to obtain a quantitative (number) value for speed. The slope of a line is the rise (change along the *y* dimension) divided by the run (change along the *x* dimension). If you look at a distance-time graph, the rise is the distance covered and the run is time taken. Distance divided by time is the equation you used to calculate speed.

$$\text{slope} = \frac{\text{rise}}{\text{run}} \qquad v_{av} = \frac{\Delta d}{\Delta t}$$

The measure of the slope of a *d* vs. *t* graph is equal to the speed of the object.

More about the SI System: Units for Measuring Speed and Velocity

In *Active Physics*, speed and velocity (speed in the direction of motion) in the classroom is measured in meters per second (m/s). Notice that the unit for speed or velocity is made up of a combination of two of the SI base units, meters (m) and seconds (s). These are called derived SI units.

Other units can also be used for speed. For example, highway speeds could be measured in kilometers per hour (km/h). The movement of Earth's crust could be measured in centimeters per year (cm/yr).

Kilometers and Miles

Highway signs and speed limits in the United States of America are given in miles per hour (mi/h or mph). Almost every other country in the world uses kilometers to measure long distances. A kilometer is a little less than two-thirds of a mile (1.0 km ≈ 0.6 mi). Kilometers per hour (km/h) is used to measure highway driving speed. For shorter distances, such as stopping distances and experiments in a science class, speed is measured in meters per second, m/s.

You will use miles per hour when working with driving speeds, but meters per second for data you collect in class. It is important to be able to understand and compare measurements.

There are mathematical conversions that can help you convert from miles per hour to kilometers per hour and meters per second. To help you relate the speed with which you are comfortable to the data you collect in class, the chart at the right gives approximate comparisons. It shows standard speed limits for the United States and Canada.

Speed-Limit Conversion Table

United States (Imperial)		Canada (Metric)	
mph	ft/s	km/h	m/s
20	29	30	8
30	44	50	14
50	73	80	22
70	102	100	28

Speed and the Doppler Effect

Driving safely is everybody's responsibility. Part of safe driving is obeying the speed limit. All roads have a speed limit. This speed limit is posted for some roads but not for all. As a driver, you must know the speed limit of roads without a posting. Most people are aware that going too fast is dangerous. An accident at a high speed results in much greater damage than at slower speeds. However, traveling too slowly can also be dangerous and can result in accidents.

You can find out if you are speeding by reading the speedometer in the car. You can also calculate the speed by using posted mile markers on some roads. If you measure the time it takes your vehicle to travel between two mile-markers, you can find the speed of the vehicle by using the equation for average speed, $v = \frac{\Delta d}{\Delta t}$.

You can also get a sense of the speed of a passing vehicle by listening carefully to the sound of the engine and the tires on the road. Sound travels in waves. The shorter the wavelength, the higher the pitch (frequency). The longer the wavelength, the lower the pitch. As you stand on the side of a street, you can hear the change in pitch of a vehicle as it moves past you. In auto racing, the shift in pitch is quite noticeable. As the race car approaches, you hear a high pitch, and as the car departs, you hear a low pitch. The change in pitch of the sound is indicative of the speed of the car. The driver of the vehicle would not notice any change in pitch, because the driver is traveling along with the source of the sound.

Physics Words

Doppler effect: the change in the pitch, or frequency of a sound (or the frequency of a wave) for an observer that is moving relative to the source of the sound (or source of the wave).

The change in pitch is called the **Doppler effect**. It is named in honor of Christian Doppler, an Austrian physicist and mathematician, who was one of the first to analyze this phenomenon. He noticed that the pitch (frequency) of a train whistle was higher as the train approached and lower as the train departed.

To help you understand why this effect occurs, look at the diagrams on the following page of a stationary source of sound and a moving source. In the first diagram, the sound source is stationary. A person standing at position X and one standing at position Y would hear the same frequency of sound. In the second diagram, the sound source is moving to the right. As the sound waves approach position Y, they are closer together than if the source were stationary. A person at this position would hear a higher frequency sound. The source is moving away from position X. As the sound waves approach that position, they are farther apart. A person standing at X would hear a lower frequency sound. You will learn more about sound waves in a later chapter.

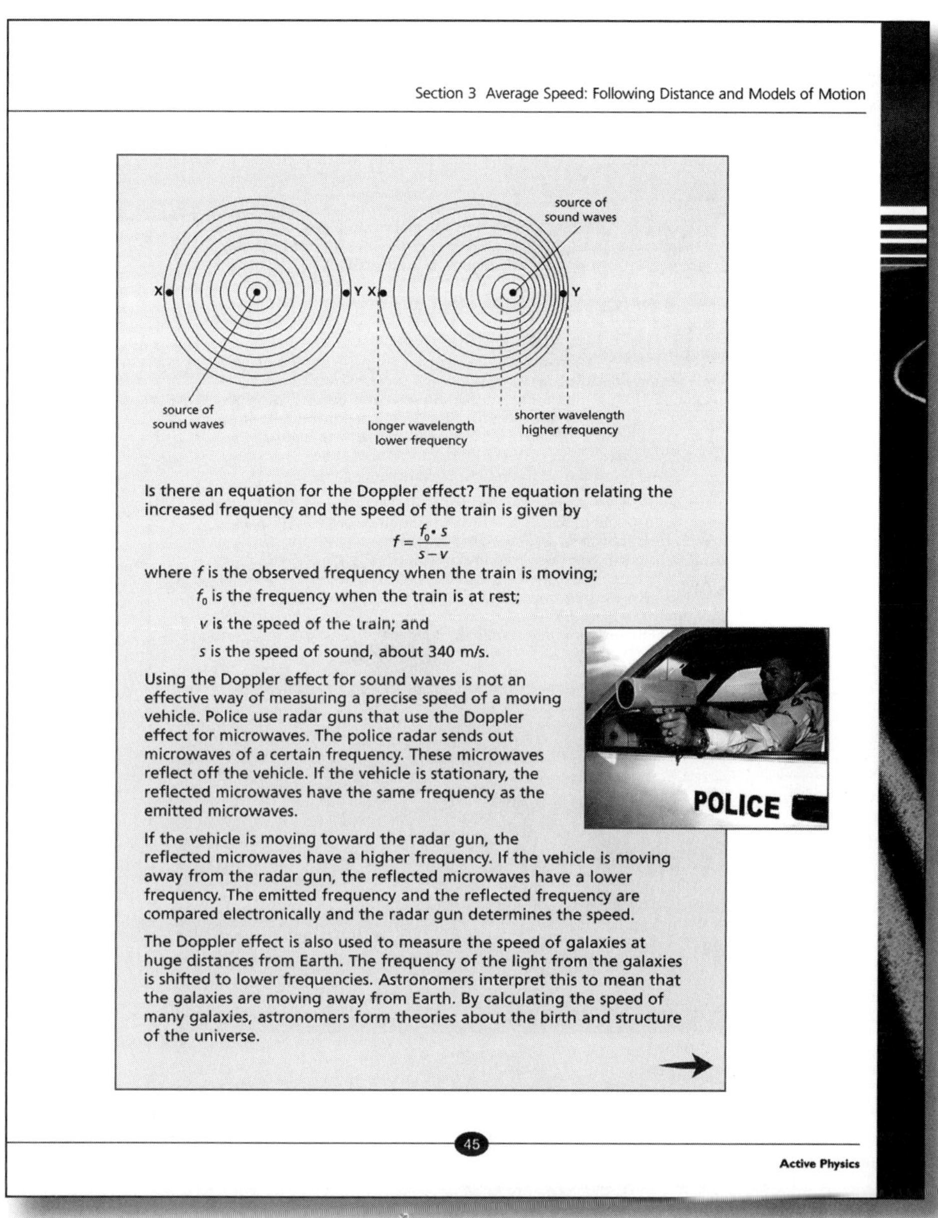

Section 3 Average Speed: Following Distance and Models of Motion

Is there an equation for the Doppler effect? The equation relating the increased frequency and the speed of the train is given by

$$f = \frac{f_0 \cdot s}{s - v}$$

where f is the observed frequency when the train is moving;

f_0 is the frequency when the train is at rest;

v is the speed of the train; and

s is the speed of sound, about 340 m/s.

Using the Doppler effect for sound waves is not an effective way of measuring a precise speed of a moving vehicle. Police use radar guns that use the Doppler effect for microwaves. The police radar sends out microwaves of a certain frequency. These microwaves reflect off the vehicle. If the vehicle is stationary, the reflected microwaves have the same frequency as the emitted microwaves.

If the vehicle is moving toward the radar gun, the reflected microwaves have a higher frequency. If the vehicle is moving away from the radar gun, the reflected microwaves have a lower frequency. The emitted frequency and the reflected frequency are compared electronically and the radar gun determines the speed.

The Doppler effect is also used to measure the speed of galaxies at huge distances from Earth. The frequency of the light from the galaxies is shifted to lower frequencies. Astronomers interpret this to mean that the galaxies are moving away from Earth. By calculating the speed of many galaxies, astronomers form theories about the birth and structure of the universe.

45

Active Physics

Checking Up

1.

Average speed of a vehicle is the total distance it travels divided by the time it took to travel that distance, while instantaneous speed is the speed at any given instant.

2.

Speed is the distance traveled per unit time. Speed is a scalar quantity and has no direction, but velocity is speed in a given direction.

3.

The straight, inclined line on a distance-time graph represents constant speed. The slope of the line is the object's speed.

4.

Reaction time increases or decreases the reaction distance depending on how fast the driver of a vehicle responds to a situation.

Chapter 1 Driving the Roads

The *Physics Talk* may also include the **Elaborate** phase of the 7E learning cycle. The **Elaborate** phase provides an opportunity for you to further your knowledge to new areas.

Reaction Distance

While you are deciding what to do in any given situation, your automobile in the meantime is traveling over the ground, possibly approaching traffic or pedestrians. At a given speed, the time it takes you to respond to a situation corresponds to the distance that the automobile travels. This distance that your automobile travels until you respond is known as the **reaction distance**.

Physics Words

reaction distance: the distance that a vehicle travels in the time it takes the driver to react.

In *Sample Problem 2*, you saw that for a reaction time of 0.2 s, your automobile would move 10 ft if the automobile were traveling at 35 mph (about 50 ft/s). A longer reaction time increases the distance you travel before you even begin to brake or turn. The longer your reaction time, the greater the distance the automobile moves before you begin stopping, swerving, or taking other appropriate action. Your reaction time therefore has a direct effect on the distance your vehicle travels and the possibility of being involved in an accident.

Checking Up

1. Explain how the average speed of a vehicle is different from instantaneous speed.
2. How are the speed and velocity of an object different?
3. If the distance-time graph shows a straight, inclined line, what does the line represent?
4. How does reaction time affect reaction distance?

46

Active Physics

Active Physics Plus

Students understand the difference between taking the average of different speeds and calculating the value of average speed. The concept of average speed is explained in further detail. The *Active Physics Plus* provides an opportunity to emphasize how average speed is calculated and it means that the total distance traveled is divided by the total time. You may want to ask students to share their findings with others in the class.

1.

The distance vs. time graph for the 80-mi trip should look similar to the one at right. The distance vs. time graph for the 100-mi trip should look similar to the one on the following page.

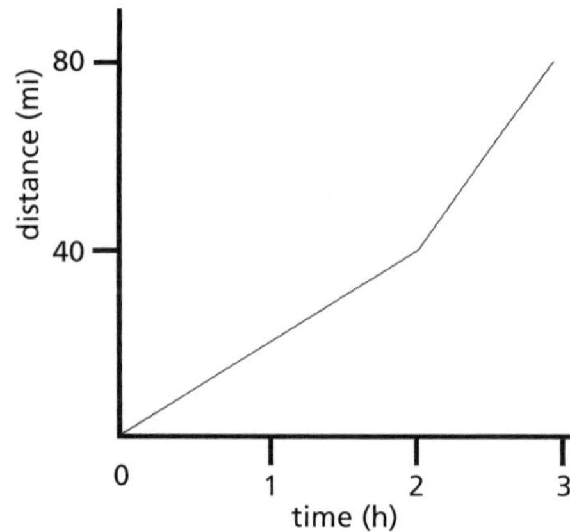

Section 3 Average Speed: Following Distance and Models of Motion

Active Physics *Plus*

+Math	+Depth	+Concepts	+Exploration
◆	◆		

More About Average Speed

An automobile travels the first half of an 80.0 mi trip at 20.0 mi/h and the second half of the trip at 40.0 mi/h. What is the average speed for the entire trip?

A first guess may be 30.0 mi/h because this is the average of the two speeds. However, this is not correct. To find the average speed, you must use the definition of average speed. Average speed is equal to the total distance traveled divided by the total time.

In this problem, you can set up a table like the one below to help you find the average speed.

	Distance	Time
1st half of the trip at 20 mi/h	40.0 mi	2.0 h
2nd half of the trip at 40 mi/h	40.0 mi	1.0 h
Total trip	80.0 mi	3.0 h (1.0 h + 2.0 h)

The average speed of the entire trip is the total distance covered divided by the total time taken.

$$v_{av} = \frac{\Delta d}{\Delta t} = \frac{80.0 \text{ mi}}{3.0 \text{ h}} = 27 \text{ mi/h}$$

Does this answer make sense? Why should the average speed be 27 mi/h instead of 30 mi/h?

To better understand this situation, look at a more extreme case. Imagine an automobile that travels 100 mi. The first 50 mi, the automobile travels at 1 mi/h. The first 50 mi will take 50 h (more than two days of driving). During the last 50 mi, the automobile travels at 50 mi/h. The last half of the trip only requires 1 h. The average speed would not be 25.5 mi/h (the average of 1 mi/h and 50 mi/h). The average would be very close to 1 mi/h because this driver drove at only 1 mi/h for many hours and only got a chance to drive at 50 mi/h for 1 h. Average speed is about distance and time.

1. Draw a distance versus time graph for both situations described above (the 80 mi trip and the 100 mi trip).
2. Draw a strobe sketch for both situations described above (the 80 mi trip and the 100 mi trip).
3. Suppose someone travels 50 mi at 50 mi/h, then travels 50 mi at 25 mi/h, then travels 50 mi at 10 mi/h.
   a) Estimate their average speed.
   b) Calculate the average speed. How close was it to your estimate?
4. If you travel the first half of a trip at 20 mi/h, how fast must you travel the second half of the trip so that your average speed will be 40 mi/h?

47 Active Physics

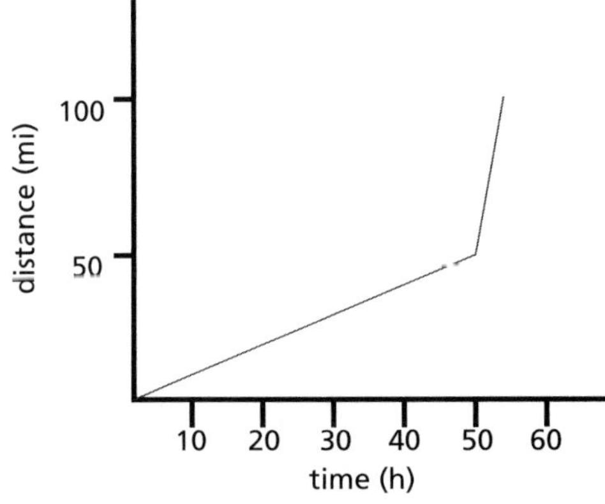

2.

Students' "strobe" picture for the 80-mi trip should look similar to the one below.

20 mi/h
(first half of trip)

40 mi/h
(second half of trip)

For the 100-mi trip, the first part of the trip should show images of a car almost overlapping. The second part of the trip should show images of a car separated 50 times the distance between images in the first part.

3.a)

Students' estimates may vary from an average of the two speeds to many different values.

3.b)

The average speed is the total distance, 150 mi, divided by the total time of 8 h, which gives a speed of 18.75 mi/h.

4.

To get the time traveled for the first half of the trip, $t = \left(\frac{1}{2}d\right)/(20 \text{ mi/h}) = d/(40 \text{ mi/h})$.

For the second half of the trip, $t = \left(\frac{1}{2}d\right)/v = d/2v$, where v is the unknown velocity. For the entire trip, the average speed must be 40 mi/h.

Using $v = d/t$ gives 40 mi/h = total distance/total time
$d/d\left(1/40 \text{ mi/h} + \frac{1}{2}v\right)$
After cancelling, "d" yields
$40 \text{ mi/h} = 1/1\left(40 + \frac{1}{2}v\right) =$
$1/(2a/80v + 40/80v)$.

Solving for v proves impossible because it cancels out of the equation. What has occurred here is that if one travels ½ the trip at half speed, you have already used up the total time needed for the total trip to average 40 mi/h. In other words, this is a trick question with no real answer for the students.

CHAPTER 1

What Do You Think Now?

Students should reflect on their earlier answers to the *What Do You Think?* questions and revise them in light of what they know now about average speed. You might want to provide them with *A Physicist's Response* to give them a better understanding of a safe distance that should be maintained between two automobiles. Students' answers will vary. They should be able to discuss strategies for safe driving.

Reflecting on the Section and the Challenge

Now that students have learned that a driver can change the velocity of an automobile by changing its speed or direction of motion, they can use this information to design their *Chapter Challenge*. Knowing how to follow at a safe distance would help students in making a sound argument against tailgating, and by similar reasoning, against speeding.

Chapter 1 Driving the Roads

What Do You Think Now?

At the beginning of this section, you were asked the following:

- **What is a safe following distance between your automobile and the vehicle in front of you?**
- **How do you decide what a safe following distance is?**

How would you answer these questions now? Now that you know how speed is related to distance and time, why is it important to pay attention to speed while driving? How does speed impact the distance covered when the driver is trying to avoid a rear-end collision?

Physics Essential Questions

What does it mean?

What does it mean to say that the speed of a vehicle is 40 mi/h?

How do you know?

How would you go about measuring the speed of a vehicle? What measurements would you have to take? What calculations would you have to perform?

Why do you believe?

Connects with Other Physics Content	Fits with Big Ideas in Science	Meets Physics Requirements
Forces and motion	* Models	Experimental evidence is consistent with models and theories

* Physicists use models to better understand the world. Speed can be modeled with a strobe photo, an equation, or a graph. How can all three models represent a car moving at 20 m/s?

Why should you care?

Safe driving includes an understanding of speed, reaction time, and reaction distance. Some collisions are difficult to avoid, but any collision would be less severe, if the speed of the vehicles were less. Many highway accidents occur because of tailgating—the practice of leaving very little room between your automobile and the automobile in front of you. Explain how the reaction distance depends on your reaction time and your speed.

Physics Essential Questions

What does it mean?

A speed of 40 mi/h means that the automobile can travel 40 miles in 1 hour. These units can be changed so that you can also find the distance that can be traveled in 1 minute or 1 second.

How do you know?

To measure the speed of a vehicle, you would measure the time it takes for the vehicle to travel a specific distance. You would need a meter stick to measure the distance and a stopwatch to measure the time. The speed of the vehicle would be the distance traveled divided by the time, $v = d/t$.

Why do you believe?

The strobe photo shows the distance the automobile travels every second. The equation allows you to calculate the distance the automobile travels every second. The graph shows the position of the automobile at every time and the steepness of the slope represents the velocity.

Section 3 Average Speed: Following Distance and Models of Motion

Reflecting on the Section and the Challenge

When you drive an automobile, you are controlling its velocity. You change the automobile's velocity by changing its speed (stepping on the gas pedal or brake pedal), and/or by changing its direction of motion (by turning the steering wheel). As you drive, you are continuously monitoring the automobile's velocity (speed and direction). You adjust both speed and direction as necessary.

Based on all the information you have just read, you now have ways to symbolically represent motion. You can use a strobe sketch or a distance–time graph. Also, you can calculate the reaction distance by knowing the speed of the automobile and the driver's reaction time.

You should be able to make a good argument against tailgating as a result of learning about reaction distance as part of the *Chapter Challenge*. Tailgating is when a driver leaves little space between his or her automobile and the automobile in front. You also should be able to make a good argument against excessive speed in any driving situation, especially when approaching an intersection or places where there may be pedestrians.

Physics to Go

1. Describe the motion of each automobile below. The diagrams of strobe photos were taken every 3 s (seconds).

   a) 

   b)  

2. Sketch diagrams of strobe photos of the following:
   a) An automobile starting from rest and reaching a final constant speed.
   b) An automobile traveling at a constant speed then coming to a stop.
3. A race car driver travels at 350 ft/s (that's almost 250 mph) for 20 s. How far has the driver traveled during this time?
4. A salesperson drives the 215 mi from New York City to Washington, DC, in 4.5 h.
   a) What was her average speed?
   b) Do you know how fast she was going when she passed through Baltimore? Explain your answer.
5. If you planned to bike to a park that was five miles away, what average speed would you have to maintain to arrive in about 15 min? (Hint: To compute your speed in miles per hour, consider this: What fraction of an hour is 15 min?)

Why should you care?

The reaction distance is the distance you travel before you brake and while you brake. During the reaction time of deciding to step on the brakes of your automobile, your automobile continues to travel at a constant speed. The longer the reaction time, the further you go. After you move your foot to the brake, an additional distance is traveled as the brakes are applied. This additional distance will also be greater for a greater speed.

Physics to Go

1.a)

The automobile is moving at a constant speed with evenly spaced images every 3 s.

1.b)

Speeds up, and then slows again.

2.a)

Answers will vary, but there should be a measurable increase in the distance of the first few automobiles, and then the same distance while the automobile travels at a constant speed.

2.b)

The automobile moves at a constant speed then slows down. Students' sketches should look similar to the diagram below.

3.

Using the formula
$v = d/t$, $d = vt =$
$350 \text{ ft/s} \times 20 \text{ s} = 7000 \text{ ft}$.

4.a)

$v = d/t = 215 \text{ m}/4.5 \text{ h} \approx 48 \text{ mi/h}$.

Her average speed was about 48 mi/h.

4.b)

This question cannot be answered with the data provided. One can only assume that she was probably going 48 mi/h.

5.

$$v = d/t = \frac{5 \text{ mi}}{0.25 \text{ h}} = 20 \text{ mi/h}.$$

You would need to maintain an average speed of 20 mi/h.

6.a)

The automobile first travels with constant speed away from the origin, then it is at rest.

6.b)

The automobile travels at a high constant speed away from the origin for the first part of the trip, is at rest for the middle part, then travels at a lower constant speed back to the origin.

6.c)

The automobile travels at a low constant speed away from the origin for the first part of the trip, then travels at a higher constant speed away from the origin for the second part of the trip where the slope changes.

6.d)

The automobile is traveling with increasing speed (possibly constant acceleration) away from the origin.

7.a)

At 55 mi/h, the automobile is moving at about 25 m/s; therefore, the distance, $d = vt = 25\text{ m/s} \times \text{reaction time}$ (reaction times will vary according to student). For example, for a reaction time of 0.16 s, $d = vt = 25\text{ m/s} \times 0.16\text{ s} = 4\text{ m}$.

7.b)

Assume a reaction time of 0.8 s. At 35 mi/h, the car will be moving at about 16 m/s, therefore to find the distance $v = d/t$, then $d = vt$
$d = 16\text{ m/s} \times 0.8\text{ s} = 12.8$ or 13 m. A lower speed produces a shorter reaction distance with the same reaction time.

6. For each graph below, describe the motion of the automobile. The vertical axes are labeled with the distance the automobile traveled, denoted *d*.

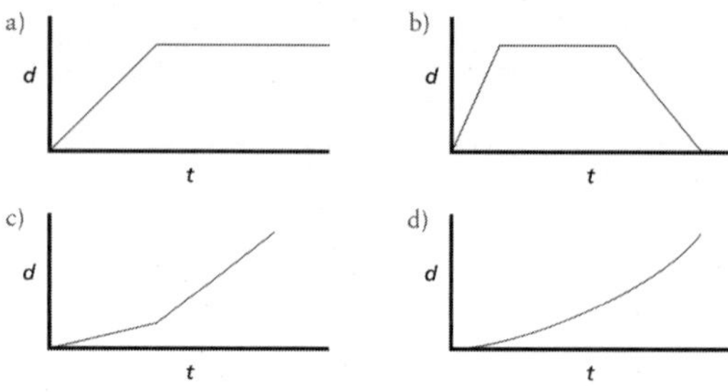

7. Use your average reaction time from *Section 1* to answer the following:
   a) How far does your automobile travel in meters during your reaction time if you are moving at 55 mi/h (25 m/s)?
   b) How far does your automobile travel during your reaction time if you are moving at 35 mi/h (16 m/s)? How does the distance compare with the distance at 55 mi/h?
   c) Suppose you are very tired and your reaction time is doubled. How far would you travel at 55 mi/h during your reaction time?
8. According to traffic experts, the following distance between your automobile and the vehicle in front of you should be three seconds. As the vehicle in front of you passes a fixed point, say to yourself "one thousand one, one thousand two, one thousand three." Your automobile should not reach that point before you complete the phrase.
   a) A second is a unit of time. How can traffic experts be sure this is a safe following distance?
   b) Will three seconds following "distance" be equally as safe on an interstate highway as on a rural road? Explain your answer.
9. A sneeze requires you to close your eyes for one third of a second.
   a) If you are driving at 70 mi/h (100 ft/s), how far will you travel with your eyes closed during a sneeze?
   b) Is this longer than the length of your classroom?

7.c)

You will travel double the distance you calculated for *7.a)*.

8.a)

Because distance traveled is directly proportional to time for constant speed, the three-second rule should work at any speed, and the distance can be described by the travel time. The premise for this rule is that the two automobiles have similar braking ability. The vehicle in front does not "stop on a dime." The three-second distance is not an adequate stopping distance. It is meant to cover the driver's reaction time and the time taken by the lead vehicle to decrease speed before the driver's automobile begins to slow down. Exceptions can be made for poor road, tire, or brake conditions.

Section 3 Average Speed: Following Distance and Models of Motion

10. Imagine you are driving your automobile at 60 mi/h (88 ft/s) moving in a straight line and your reaction time is 0.5 s.
    a) How far does your automobile travel in this time?
    b) How many automobile spaces is this for an automobile that is 15 ft long?
    c) Answer *Questions a)* and *b)* when you travel 30 mi/h.
    d) Answer *Questions a)* and *b)* when you travel 90 mi/h. What fraction of a football field is this distance?
    e) If talking on the cell phone while driving at this speed doubles your reaction time, how do these distance numbers change at 30 mi/h, 60 mi/h, and 90 mi/h?
11. Consider an automobile traveling at 60 mi/h. Sketch a graph showing distance traveled versus reaction time, with reaction times of 0.25 s, 0.50 s, 0.75 s, and 1.00 s.
12. ***Preparing for the Chapter Challenge***

    Apply what you learned in this section to write a convincing argument that describes why tailgating (following an automobile too closely) is dangerous. Include the factors you would use to decide how following too closely counts as tailgating.

Inquiring Further

1. **Calculating speed over a longer distance**

   Measure a distance of about 100 m. You can use a football field or get a long tape measure or trundle wheel to measure a similar distance. You also need a watch capable of measuring seconds. Determine your average speed traveling that distance for each of the following:
   a) a slow walk
   b) a fast walk
   c) running
   d) another method of your choice
2. **Other models for motion**

   In this section, you learned three models that physicists use to describe motion—the strobe photo, the mathematical equation, and the motion graph. Painters, writers, poets, and photographers have also found ways to describe motion. Many people have heard the description of Superman's speed—"faster than a speeding bullet." Investigate and record descriptions of motion by people in the arts. How do artists and writers depict motion? How does one compare the physicist's model with that of the artist?

8.b)

The three-second following rule should prove equally safe on both rural roads and interstates, since the road conditions will be identical for both automobiles. An automobile may not be able to stop as quickly on a rural road due to poor road surface, but the automobile the driver is following will also need more distance to stop, keeping a safe distance between both the automobiles.

9.a)

Using the equation for speed, $v = d/t$ and solving for d, $d = vt = (100\text{ ft/s})(0.33\text{ s}) =$ 33 ft.

9.b)

33 ft would be approximately the length of a laboratory classroom, but more than a typical classroom.

10.a)

$d = vt = (88\text{ ft/s})(0.5\text{ s}) = 44\text{ ft}.$

10.b)

For an automobile that is 15 ft long, a space of 44 ft is almost three automobile-lengths.

10.c)

At 30 mi/h, the reaction distance is 22 ft, and this is almost 1.5 automobile-lengths.

10.d)

At 90 mi/h, the reaction distance would be 66 ft, or a little more than 4 automobile-lengths. This is almost one-fifth of a football field.

10.e)

When the reaction time is doubled, the distance the automobile travels during that reaction time will also double.

11.

The student graph should be similar to the one below:

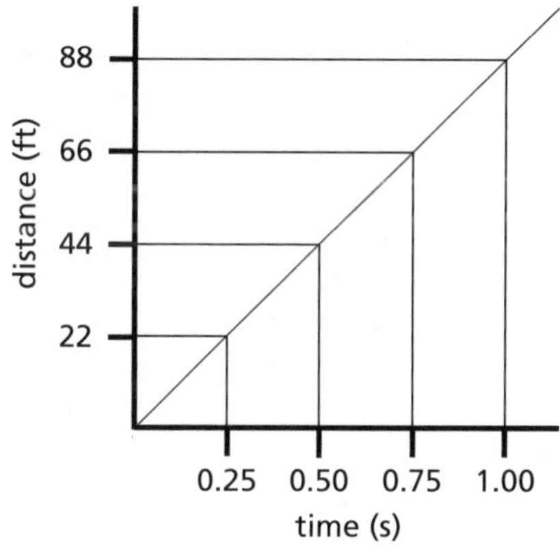

12.

Preparing for the Chapter Challenge

Student answers might include some of the following. Tailgating increases the potential hazards of driving. From the time you see the brakes from the lead automobile flash until the time you hit the brake and begin to slow down, your vehicle is still traveling at top speed. If the two vehicles have equal braking ability, there should be a minimum extra distance between the two automobiles equal to the product of your reaction time and the automobile's speed. Because safe driving means minimizing risks, following distances must be commensurate with speed and reaction distance, in addition to road conditions.

The physical condition of the driver might also affect reaction time. The condition of the vehicle in front, which is not under the control of the person tailgating, must be taken into account as an intangible. For instance, the automobile in front might have faulty tail lights, or very good brakes, or the driver in front might be slowing down using the engine (gearing down) rather than the brakes.

Inquiring Further

1. Calculating speed over a longer distance

The results the students obtain for these activities will depend upon their walking speeds.

a) For a slow walk a speed of 2 m/s or less would be normal.

b) A fast walk will be between 3 and 4 m/s.

c) Running speeds of 5–9 m/s would be normal.

d) Some of the methods students might choose could be walking backward, crawling, wheelbarrow race, three-legged race, and so on. Speeds will vary greatly depending upon the choice and the students who are doing the activity.

2. Other models for motion

Artists, writers, and photographers may all depict speed in different ways. An illustrator for a graphic novel will often show speed by putting lines behind the object or person moving, indicating their motion. The longer the lines, the faster the object is traveling. Photographers will often depict motion by panning the camera to blur the background while keeping the moving object in focus during the exposure, or vice versa, blurring the moving object by keeping the camera fixed. A writer has the most options for describing motion. Often, similes (like "faster than a speeding bullet") are used, or descriptive terms such as "rocketing through the night" or "flying down a long hall" also describe motion.

NOTES

SECTION 3 QUIZ

1-3b Blackline Master

1. An automobile travels between the 100-m and 250-m markers on a highway in 10 s. The average speed of the automobile during this interval is
   a) 10 m/s. b) 15 m/s.
   c) 25 m/s. d) 40 m/s.

2. A baseball pitcher throws a fastball at 42 m/s. If the batter is 18 m from the pitcher, approximately how much time does it take for the ball to reach the batter?
   a) 1.9 s b) 2.3 s
   c) 0.86 s d) 0.43 s

3. The pattern shown below was left by an automobile as it dripped oil at a constant rate on the road while being driven. During which section of the trip was the car traveling with constant velocity?

   a) AB b) BC
   c) CD d) DE

4. The graph to the right shows the distance traveled by two objects, A and B. Compared to the speed of object B, the speed of object A is
   a) the same. b) twice as great.
   c) one half as great. d) three times as great.

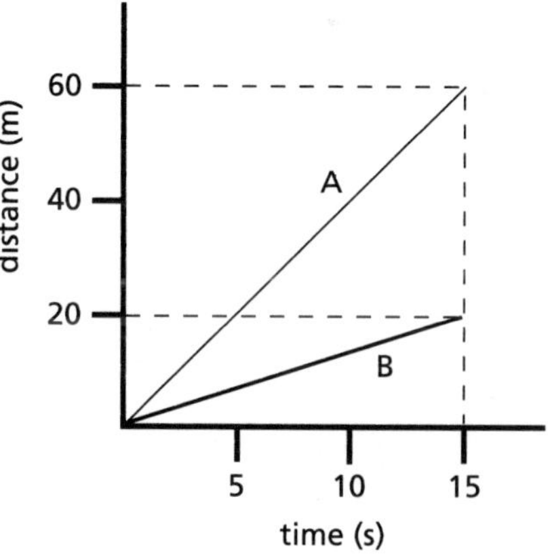

5. The graph to the right represents the relationship between distance and time for an object in straight-line motion. During which interval was the object at rest?
   a) AB only b) BC only
   c) CD only d) both AB and CD

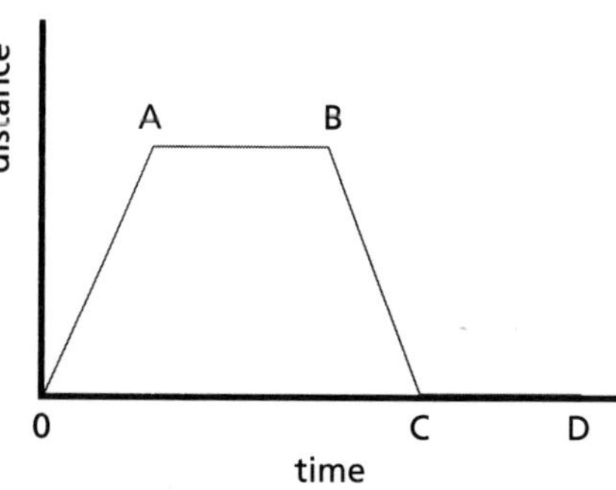

SECTION 3 QUIZ ANSWERS

1. b) The average speed is the distance traveled/time needed to go that distance. The distance between the markers was 150 m (250 m − 100 m), so the speed was 150 m/10 s or 15 m/s. Choices *1.a)* and *1.c)* are wrong because they just use a distance, not the distance traveled, and *1.d)* is wrong because the distances there were added and then divided by the time, rather than subtracted.

2. d) As in the previous question, where the average speed is the distance traveled/time needed to go that distance or $v = d/t$, where $v = 42$ m/s, $d = 18$ m and inserting into the formula gives 42 m/s = 18 m/t. Solving for t gives 0.43 s. If the student mistakenly writes the equation as $v = d \times t$, and then solves, the student will get a wrong answer, 2.3 s. The other answers are double and half the correct value.

3. b) One of the definitions of constant velocity is equal distances in equal times. Because the oil comes out a constant rate, the area where the dots are equally spaced is where the speed will be constant. In section AB, the distance between the dots is increasing, indicating that the car was accelerating, while CD and DE show non-uniform motion with the dots being unevenly spaced.

4. d) The slope of A is three times the slope of B, so the speed must be three times greater. To get the slope, the students can divide the change in the distance by the change in time. Slope A has a change of 60 m in the same time as slope B's change of 20 m.

5. d) During intervals AB and CD, the automobile was at the same distance during the time interval, so its speed must be zero. During interval 0A, the automobile was moving away from the start, and during interval BC, it was moving back toward the starting point.

NOTES

SECTION 4

Graphing Motion: Distance, Velocity, and Acceleration

Section Overview

Students use a motion detector to investigate the motion of a cart as it moves on an inclined plane. They predict how the distance the cart travels changes with respect to time and identify graphs corresponding to various stages of the cart's motion, including when the cart is at rest. Students also predict the changes in velocity with respect to time and sketch graphs to compare the changing slope of distance vs. time with the constant slope of velocity vs. time. They observe how a constantly changing velocity yields a curved distance-time graph as the cart travels on the inclined plane. By observing the changes in velocity, they are given a concrete example of acceleration. Acceleration is finally defined and more velocity-time graphs are sketched to establish how acceleration can be determined from the slope of a velocity-time graph.

Background Information

Acceleration is the rate of change of velocity, or the change in velocity per unit time. A change in velocity (speed with direction) occurs due to a change in speed and/or direction. In the *Investigate*, with the cart being pushed up the ramp, an object is uniformly changing velocity from one value in one direction (called positive) to a value in the opposite direction (called negative), and passes through the zero velocity point where the direction changes. A particularly difficult concept to grasp is that at this zero velocity point, the acceleration is the same as in the decreasing- and increasing-speed phases.

Students should continue to use the same equipment as they study objects with changing velocities (acceleration). Older motion detectors need about 50 cm of leeway in front before they can accurately take measurements. Newer models need only about 5 cm from the detector for proper operation. Small variations in position data can dramatically affect the velocity-time graph. Affixing an index card or some other flat, rigid object to the cart can make for a better surface to reflect the sound waves. If the detector is not properly aimed, or the detection area is not clear, the data may not be indicative of the actual motion.

Acceleration is easier to understand with the use of a graph. Acceleration vs. time and distance vs. time graphs may be easily derived from the velocity-time graph as shown below.

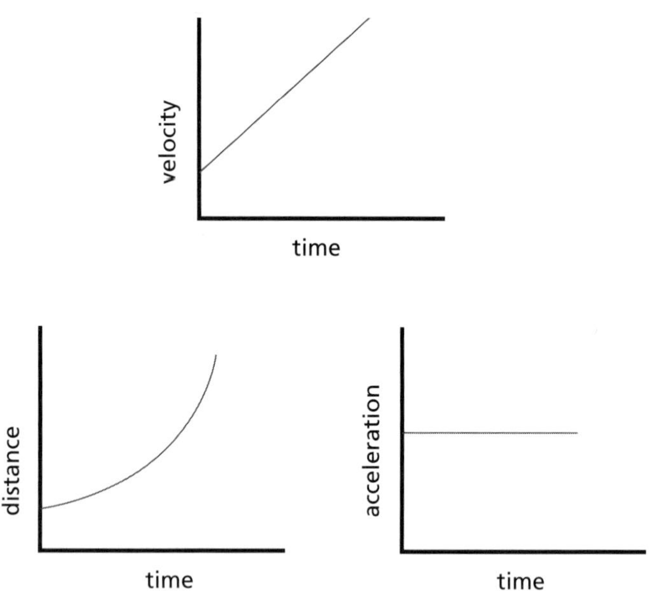

The area underneath the line represented in a velocity-time graph is the total distance traveled by an object during any particular time interval.

The slope of the v vs. t graph is equal to the acceleration.

Crucial Physics

- Acceleration = change in velocity over a given time.
  - $a = \Delta v / \Delta t$
  - There is an acceleration when

    a) There is an increase in speed.

    b) There is a decrease in speed.

    c) There is a turn (a change in direction).
- The slope of a velocity vs. time graph is equal to the acceleration.

Learning Outcomes	Location in the Section	Evidence of Understanding
Measure a change in velocity (acceleration) of a cart on a ramp using a motion detector.	***Investigate*** Step 3	Students use the points on a graph produced by a motion detector to measure a change in velocity.
Construct graphs of the motion of a cart on a ramp.	***Investigate*** Steps 2-7 and 11	Students collect data for the cart's motion on an inclined plane and sketch distance-time and velocity-time graphs.
Define acceleration using words and an equation.	***Investigate*** Step 3 ***Physics Talk***	Students learn about acceleration by studying the slope of the *v-t* graph. They also learn to put it in a word equation.
Calculate speed, distance, and time using the equation for acceleration.	***Physics to Go*** Questions 7-10 and 13-16	Students solve problems to calculate speed, distance, and time by using the equation for acceleration and velocity.
Interpret distance-time and velocity-time graphs for different types of motion.	***Investigate*** Steps 2-7 and 11-12	Students interpret how the slopes of *d-t* and *v-t* graphs vary for the motion of a cart when it changes speed and direction.

Section 4 Materials, Preparation, and Safety

Materials and Equipment

PLAN A		
Materials and Equipment	**Group (4 students)**	**Class**
Ruler, metric, 30 cm	1 per group	
Ring stand, large	1 per group	
Rod, aluminum, 12 in. (length) x 3/8 in. (diameter) (to act as crossarm)	1 per group	
Holder, right angle (to act as crossarm)	1 per group	
Inclined plane ramp for lab cart	1 per group	
Dynamics cart	1 per group	
Index cards, pkg. 100		1 per class
Motion detector, probe and interface*		1 per class
Computer, station or calculator, CBL or equivalent system*	1 per group	

*Additional items needed not supplied

PLAN B		
Materials and Equipment	**Group (4 students)**	**Class**
Ruler, metric, 30 cm		1 per class
Ring stand, large		1 per class
Rod, aluminum, 12 in. (length) x 3/8 in. (diameter) (to act as crossarm)		1 per class
Holder, right angle (to act as crossarm)		1 per class
Inclined plane ramp for lab cart		1 per class
Dynamics cart		1 per class
Index cards, pkg. 100		1 per class
Motion detector, probe and interface*		1 per class
Computer, station or calculator, CBL or equivalent system*		1 per class

*Additional items needed not supplied

Note: Time, Preparation, and Safety requirements are based on Plan A, if using Plan B, please adjust accordingly.

Time Requirement

Approximately 80 minutes are required to complete the experiment.

Teacher Preparation

- See the preparation suggestion for *Section 3* about the use of the motion detector and related equipment.
- Do not allow the motion detector to be struck by a cart when the cart is rolling on the ramp, as this may damage the probe.
- Placing a "sail" or similar surface on the cart when it is rolling on the incline will improve reflection and lead to improved data collection if the detector has trouble "seeing" the cart. As discussed above, make certain the area used for good data collection is free of clutter (students' books, other equipment, etc.).

Safety Requirements

- Many of the dynamics carts being used in labs have wheels with extremely low friction, and will roll off a table with even a slight incline. Instruct the students to invert the carts on the table when they are not in use to keep them from rolling off.
- If the experiment is done on a lab table, make certain that all materials are away from the edge to keep them from falling to the floor. Students should wear appropriate shoes in case an object falls on their feet.

Meeting the Needs of All Students

Differentiated Instruction: Augmentation and Accommodations

Learning Issue	Reference	Augmentation and Accommodations
Following complex directions	***Investigate*** Steps 1-7	**Augmentation** • Students with organization, reading, and attention issues could easily be overwhelmed and give up when following complex directions. Break the task down into smaller steps with reasonable time limits. Model the equipment setup and the data to be recorded. Check in with the class to monitor understanding and frustration level after each step. • Use two different colors to draw *d-t* and *v-t* graphs as a visual cue to show that these graphs represent different aspects of the cart's motion. • Assign a person in each group to be in charge of reading directions orally and making sure that students follow the directions step-by-step. • Focus on one type of motion at a time. Have a whole-group discussion after students collect data and sketches for the cart traveling down the ramp. Then repeat this process for motion up the ramp. **Accommodation** • Students may need direct instruction to review *d-t* graphs and learn about *v-t* graphs before they begin this *Investigate*. Some students are unable to deduce concepts from data and may need to see the big picture before they can make sense of the smaller tasks in this *Investigate*.
Organizing data	***Investigate*** Steps 1-7	**Augmentation** • Students need a method to help them organize all of the predictions, graphs, and comparisons they are asked to record in this *Investigate*. Instruct students to write the number and letter preceding each task before they record their data (*1.a), 1.b), 1.c)*, and so on). Model what an organized log page might look like. **Accommodation** • Give students a table that is numbered and provides adequate space to record data or answer each step.
Learning a new math concept by reading a paragraph	***Investigate*** Step 2.c)	**Augmentation** • This is probably the first time students have been introduced to tangent lines. Use the paragraph description and diagrams of examples and non-examples to teach this concept to students. Point out that tangent lines are used for curved graph sketches that represent changing quantities. **Accommodation** • Understanding tangent lines may impede the progress of students in completing the *Investigate*. Instruct students to skip *2.c)* and come back to it at the end of the *Investigate*.
Sketching graphs	***Investigate*** Steps 2-7, 11.a)	**Augmentation** • Students are asked to compare graphs and do calculations with the data they collect. Remind students that their graphs should have labeled axes and scales, and the lines should be sketched with accuracy. Model two well-drawn graphs that students can refer to as they are working. • Model how to use the "TRACE" function to identify data points on a line.
Applying new learning	***Investigate*** Step 8	**Augmentation** • For students with attention and organization issues, this step could be used as an informal assessment to gauge students' understanding at this point. If students understand the graphs for motion of a cart on an incline, they should be able to answer *Step 8* with accuracy.

Learning Issue	Reference	Augmentation and Accommodations
Locating information in a table	***Investigate*** Step 9	**Augmentation** • Students with visual-spatial and/or memory issues have trouble turning pages to locate and record information from a table. Provide students with a ruler or index card to mark the page that has a table. The ruler or index card can also be used to help students visually scan columns and rows to find information on a table. **Accommodation** • Students will be more successful if they can look at a table or graph and the corresponding questions side-by-side. Provide a copy of tables that are not located on the same page as the questions. On exams, place tables and graphs on the same page as the corresponding questions.
Solving problems with data in a table	***Investigate*** Step 12	**Augmentation** • Students with visual-spatial and math issues struggle to solve problems with numbers in a table because they are being asked to scan a series of numbers, locate the number that corresponds to each quantity in the equation, and then perform the calculation. Provide students with a ruler or index card to single out the row of numbers they are using for each problem. **Accommodation** • Provide students with a copy of the table. Instruct students to write the corresponding symbols next to each value on the table and then solve the problems. It is very important for students who need this accommodation to show their work to check for common calculation errors.
Differentiating scalar and vector quantities	***Physics Talk***	**Augmentation** • Scalar and vector quantities are a recurring topic in physics that students struggle to understand. Create a two-column chart that defines scalar on the left side and vector on the right side. Then post the chart and add examples of each throughout the year as students learn new quantities. This chart can also be copied in student logs.
Vocabulary	***Physics to Go*** Question 8	**Augmentation** • Explain that uniformly means "at a constant rate" or "the same amount for each chunk of time."

NOTES

Strategies for Students with Limited English-Language Proficiency

Learning Issue	Reference	Augmentation
Following complex procedures Vocabulary comprehension	***Investigate***	Break down the *Investigate* into smaller chunks to allow students to comprehend each portion of the *Investigate* before moving on to the next one. This will allow them to get comfortable by following the procedures outlined within each step, and to internalize new concepts and any new vocabulary that is introduced within a step. Lead a brief class discussion after each step to allow students the opportunity to demonstrate knowledge and to use the vocabulary.
Comprehension	***Physics Talk***	Have students read a section of text. Then connect the reading back to the portion of the *Investigate* that addressed that concept. Breaking the reading into smaller portions and providing direct connections with hands-on learning will solidify content for English learners as well as kinesthetic learners.
Reading comprehension	***Physics Talk*** Describing Types of Motion Using Graphs	Some students may have the ability to grasp the concepts of the investigation but may stumble over technical terms. Provide a supplemental vocabulary list and practice using these words in sentences. Possible terms include "slope," "incline," "inclined plane," "horizontal," "vertical," "elapsed," "simultaneously," and "instantaneous." Collaborate with the students' math teachers to determine what level of comprehension students have obtained for reading technical graphs and recognizing the meaning of slope.
Vocabulary comprehension	***Physics Talk*** Vector and Scalar Quantities	Students may have difficulty visualizing the difference between vector quantities and scalar quantities. Point out that the motion detector can record negative velocities when objects are moving toward it. The negative sign is an indication of direction in one dimension. Speed is always zero or positive, so it is a scalar quantity. Acceleration, which was also shown to be positive and negative in the *Investigate*, is another example of a vector quantity. Students may infer incorrectly that negative acceleration means that an object is slowing down. Point out that negative acceleration can also mean that an object is moving faster in a negative direction. In either case, velocity is decreasing, so the acceleration is negative.
Answering higher-order questions	***Physics To Go*** Questions 1–5	ELL students who are visually oriented would benefit from thinking graphically about the first five exercises. Pair up students with different learning styles. Have the students work together to sketch graphs of the situations in each of the first five exercises. Then have them work together to formulate in words an answer to each question.

NOTES

SECTION 4

Teaching Suggestions and Sample Answers

What Do You See?

Students are likely to comment on the automobile passing a stoplight at breakneck speed, while the other automobile stops a short distance away from it. The person running, the hat flying, and the smoke swirling provide a focus for engaging students in the *What Do You See?* illustration.

You might want to ask students what the artist is trying to depict and how the illustration relates to the concept of speed in relation to distance.

Chapter 1 Driving the Roads

Section 4 Graphing Motion: Distance, Velocity, and Acceleration

What Do You See?

Learning Outcomes

In this section, you will

- Measure a change in velocity (acceleration) of a cart on a ramp using a motion detector.
- Construct graphs of the motion of a cart on a ramp.
- Define acceleration using words and an equation.
- Calculate speed, distance, and time using the equation for acceleration.
- Interpret distance-time and velocity-time graphs for different types of motion.

What Do You Think?

Some automobiles can accelerate from 0 to 60 mph (about 100 km/h) in 5 s. Other vehicles can take up to 10 s or more to reach the same speed.

- **An automobile and a bus are stopped at a traffic light. What are some differences and similarities of the motion of these two vehicles as each goes from a stop to the speed limit of 30 mph?**

Record your ideas about this question in your *Active Physics* log. Be prepared to discuss your responses with your small group and the class.

Investigate

In this *Investigate*, you will use a motion detector to explore motion. You will produce distance-time and velocity-time graphs for a cart as it moves down and up an inclined ramp. You will also use the defining equation to calculate acceleration.

1. Set a motion detector at the top of a ramp along with a cart. Before collecting the data, you will make several predictions.

52

Students' Prior Conceptions

This section builds directly on the measurement of distance and time used to create graphs of motion in the previous section. A visual and kinesthetic benefit of this section derives from the use of the motion sensor to obtain and to interpret real-time graphs of motion. Cognitive research indicates that student understanding is enhanced and they learn best about abstract mathematical models when actively using sensors and technology. This is particularly true when students are asked to predict and to interpret velocity vs. time graphs for motions of objects.

1. **Velocity is another word for speed. An object's speed and velocity are always the same.** Encourage students to understand that speed is a scalar quantity; it only has magnitude associated with appropriate dimensional units. Velocity is a vector quantity that has magnitude, units, and direction. It is important to emphasize that a negative direction does not imply slowing down; an object can move with negative velocity and increasing or decreasing speed. A good opening gambits for class discussion is asking students to describe the velocity of a vehicle that moves forward with a given speed and then stops and reverses to move backward along the same path with the same speed. That speed, as read by the speedometer, is constant, but the velocity is positive when the vehicle is moving forward, away from the origin of motion, and negative when the vehicle moves back toward the origin of motion. To complicate matters, the latter is true whether the vehicle reverses gears and merely backs up or

CHAPTER 1

What Do You Think?

To generate enthusiasm, this preliminary question is designed to elicit prior knowledge. Students' initial answers will serve to demonstrate prior learning. It will give students the chance to transfer what they already know to the concepts they will be learning. Because of students' familiarity with driving, most students will be able to come up with at least two correct answers. At this point, the accuracy of their responses should not be of concern.

Encourage students to consider how the information provided in the question can be useful in arriving at an answer.

What Do You Think?

A Physicist's Response

- Both the bus and the automobile have the same initial and final speed, and accelerate by changing their speed from 0 to 30 mi/h (about 50 km/h). However, a bus takes more time to accelerate to 30 mi/h than an automobile. The bus also requires a larger distance than the automobile to reach a speed of 30 mi/h.

turns around, faces the origin with the front of the vehicle, and moves toward the origin.

2. **Acceleration is confused with speed.** Speed is distance divided by time. The value of speed may increase, decrease, or stay the same. Only when there is a change in the speed of an object is there acceleration. By definition, acceleration is the change in the velocity of an object. (Remember that velocity is a vector with magnitude and direction.)
3. **Acceleration always means that an object is speeding up.** Yes, students become mystified when faced with these conditions: speeding up with positive change in velocity and positive acceleration, or speeding up with negative velocity and negative acceleration; and slowing down with a positive velocity and a negative acceleration. An instructional strategy is to remind students that acceleration is the change in the velocity divided by time. Mathematically, they should interpret this to be the second velocity (whether positive or negative) minus the first velocity (whether positive or negative) divided by the change in time, which is always positive. Students can learn to evaluate these various situations as they apply the mathematical principle that subtracting a negative number yields a positive value for that number.
4. **Acceleration always occurs in the same direction as an object is moving.** This alternative notion emerges when students analyze the situation of a vehicle moving forward while slowing down. The motion of the vehicle is forward but the net acceleration of the vehicle is negative, opposite the direction of the moving object.
5. **If an object has a speed of zero (even instantaneously), it has no acceleration.** In this section, it is important for students to understand that a force can act upon a vehicle to stop its forward motion and to bring the speed to zero, but the force may also continue to act upon the vehicle, pushing it in the opposite direction and increasing its instantaneous speed from zero to another value. This preconception may hinder student understanding of what happens in vehicle collisions and what happens to conservation of momentum in later chapters of *Active Physics*.

NOTES

Section 4 Graphing Motion: Distance, Velocity, and Acceleration

a) Predict how the distance the cart travels will change with respect to time. Will it go the first half of the distance in the same amount of time as the last half of the distance?

b) Below are four different distance-time (*d-t*) graphs. In one of them, the cart does not move. In the other three, the more time that elapses, the further the cart has gone. One graph shows that the cart travels at a constant speed. In another, the cart travels faster at the beginning. In another one, the cart travels fastest at the end.

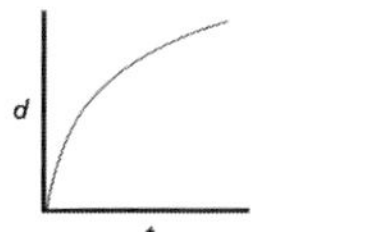

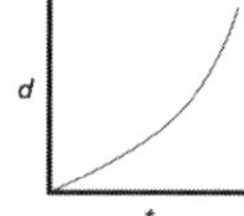

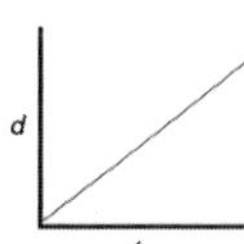

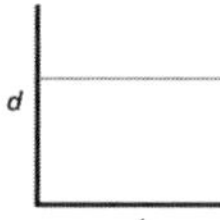

Identify which graph corresponds to which motion. (Hint: Compare the distance traveled in the first few seconds of the trip with the distance traveled in the last few seconds of the trip.)

c) Predict what you think a distance-time graph will look like when your cart is released from the top of the ramp. Sketch your predicted graph in your log along with an explanation.

2. Release the cart and collect the distance-time data. You may need to try this several times to make sure the motion detector collects consistent results.

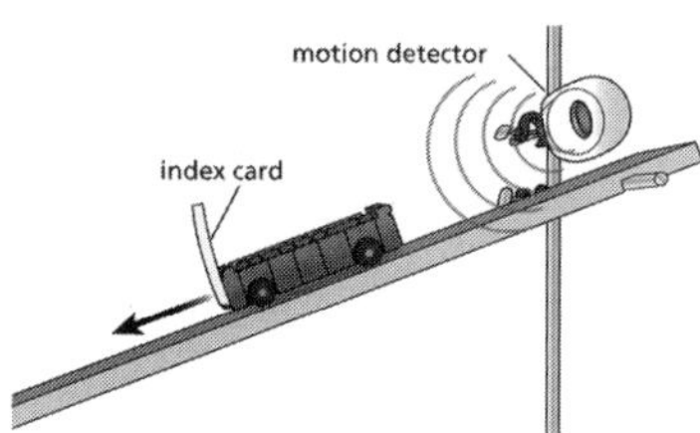

a) Sketch the *d-t* graph from the calculator or computer in your log.

b) Compare your predictions in *Step 1.c)* to what really happened. Explain any differences you find.

c) In *Section 3*, you found that the slope of a distance-time graph represents the speed. If your graph is a curve rather than a straight line, you can still find the slope at a single point on the curve. To do this, choose the point where you want to measure the slope. Then place a ruler so that it intersects the curve at points to the right and left of the point. Slide the ruler so that it finally intersects the curve at a single point. It is now a ***tangent line***. A tangent line is a straight line that touches a curve at only one point.

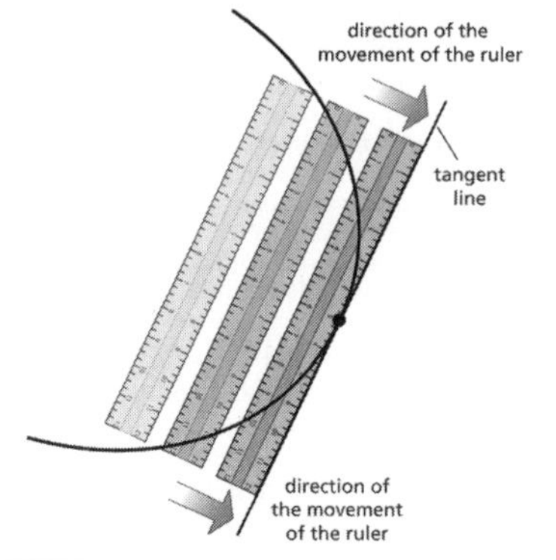

Investigate

1.a)

The cart will cover the last half of the distance in less time because it is traveling at a faster speed.

1.b)

Answers should be (starting from the left graph to the right): decreasing speed, increasing speed, constant speed, and zero speed.

1.c)

Students should realize after *Section 3* that a constant-speed graph will be linear, so the predicted graph should be curved, similar to the one below.

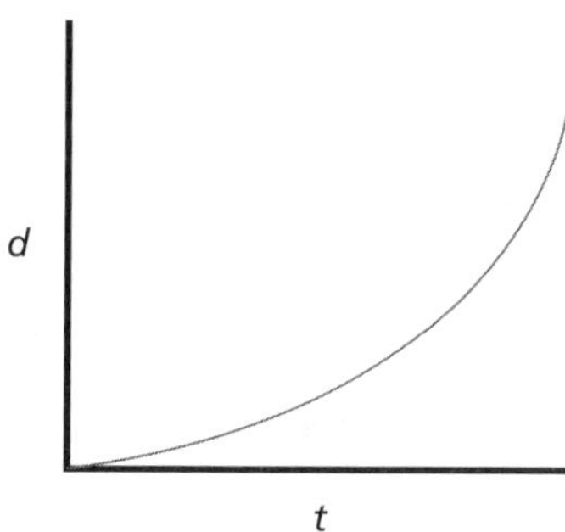

2.a)

The data points should follow a locus that is not scattered. The graph should look like the graph for *Step 1.c)*.

2.b)

Unless there is an exceptional amount of friction, the graphs should not deviate much from the graph of *Step 1.c)*.

2.c)

Starting from the top, the first two lines are tangent to the curve, and the bottom line is not.

2.d)

The slope of the *d-t* graph steadily increases because the velocity (slope) continually increases as the cart accelerates.

2.e)

The *v-t* graph should be a straight line ascending from lower values to higher values. The slope of the *v-t* graph is constant, because the rate of change of velocity (acceleration) is constant.

3.a)

The *v-t* graph should be linear with a positive slope. The graph should appear like the one shown below.

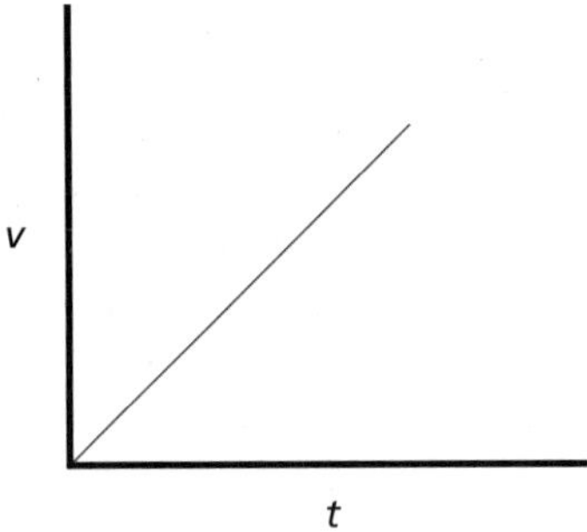

3.b)

The actual graph should look like the prediction made in *Step 2.e)*. If the motion detector started to take data exactly when the cart was released, the graph should start at (0, 0). More likely, the graph will be displaced to the right if the detector is started early, or displaced upward if the cart is moving before the detector starts collecting data.

The cart will start with zero velocity, so if the timer and the cart start at the same time, the graph should begin at (0, 0). The slope is a constant because the cart's speed is constantly increasing as it goes down the plane.

Draw the line, and you can measure the slope. The measure of that slope is equal to the speed of the cart at that point (instantaneous speed).

Look at the following lines. Which lines are tangent lines? If one of the lines is not a tangent line, sketch the curve in your log and draw the correct tangent line.

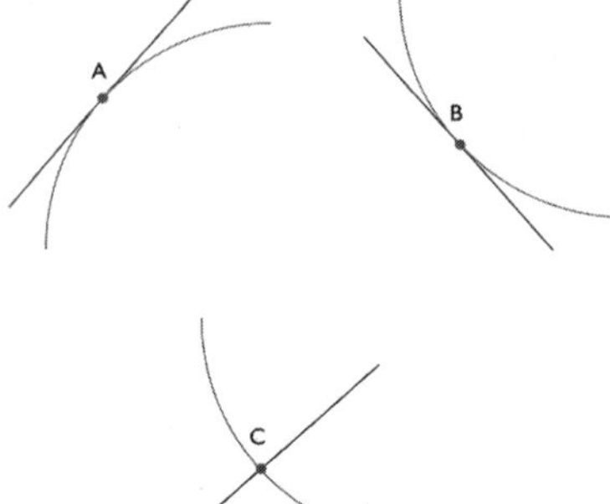

C

d) Returning to your distance-time graph, what happens to the slope of the *d-t* graph as time increases? What does this tell you about the velocity?

e) As you have seen, the motion of the cart can be modeled with a distance-time graph. It could also be modeled with a velocity-time graph. A velocity-time graph shows how the velocity changes as time elapses. Predict what you think a velocity-time (*v-t*) graph will look like for the cart moving down the incline. Sketch it in your log along with an explanation.

3. Replace the cart at the top of the ramp as in *Step 1*. Release the cart and collect the velocity-time data. You may need to try this several times to make sure the motion detector collects accurate data.

a) Sketch the *v-t* graph from the calculator or computer into your log. Use the "TRACE" function to label three to four data points along each line. These data points will assist you in making some calculations.

b) Compare your predictions in *Step 2.e)* to what really happened. Explain any differences you find. Why does the graph start at 0, 0?

c) As time increases, what happens to the slope of the *v-t* graph? Why does this happen?

d) The slope of the *v-t* graph is the *acceleration* of the cart. Acceleration is defined as the change in velocity with respect to a change in time and is expressed as follows:

$$\text{Acceleration} = \frac{\text{change in velocity}}{\text{change in time}}$$

This relationship can be written as an equation using symbols

$$a = \frac{\Delta v}{\Delta t}$$

where a is acceleration,
Δv is change in velocity,
Δt is change in time or elapsed time.

Velocity represents both speed and direction. There is an acceleration:

- if there is a change in speed over a given time,
- if there is a change in direction over a given time, or
- if there is both a change in speed and a change in direction.

Since the cart going down the ramp has no change in direction, you can think of the acceleration as a change in speed with respect to time.

3.c)

As time increases, the slope of the line remains. This happens because the velocity increases at a constant rate for the cart going down the plane.

What happens to the acceleration of the cart as it travels down the ramp?

e) Use pairs of data points from your graph to calculate the acceleration.

4. Prepare to run another trial. This time, move the cart to the bottom of the ramp. Practice giving the cart a push until it nearly reaches the top of the ramp. You can ignore the data for the downward motion. Before taking data, predict the following:

a) What do you think the *d-t* graph will look like? Sketch it in your log along with an explanation.

b) What do you think the *v-t* graph will look like? Sketch it in your log along with an explanation.

5. Give the cart a push and collect the data. Be sure to stop the cart on the way up if it looks like it will hit the motion detector.

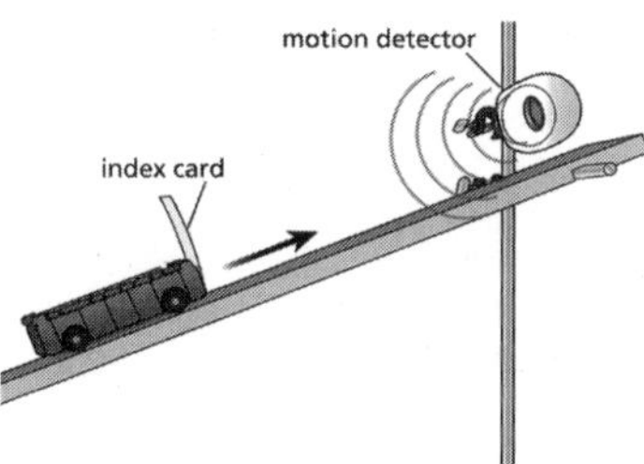

a) Sketch both the *d-t* and *v-t* graphs from the calculator or computer. Use the "TRACE" function to label three to four data points along each line.

b) Compare your predictions in *Steps 4.a)* and *b)* to what really happened. Explain any differences you find.

c) What happens to the slope of the *d-t* graph? Why does this happen?

d) What happens to the slope of the *v-t* graph? Why does this happen?

e) Use pairs of data points from your graph to calculate the acceleration.

6. Prepare to run another trial. This time, move the detector and the cart to the bottom of the ramp, pointing them both toward the top of the ramp. Practice giving the cart a push until it nearly reaches the top of the ramp. Be sure to catch the cart on the way down before it strikes the motion detector. You can ignore the data for the downward motion. Before taking data, predict the following:

a) What do you think the *d-t* graph will look like? Sketch it in your log along with an explanation.

b) What do you think the *v-t* graph will look like? Sketch it in your log along with an explanation.

7. Give the cart a push and collect the data.

a) Sketch both the *d-t* and *v-t* graphs from the calculator or computer. Use the "TRACE" function to label three to four data points along each line.

b) Compare your predictions in *Steps 6.a)* and *b)* to what really happened. Explain any differences you find.

c) What happens to the slope of the *d-t* graph? Why does this happen?

d) What happens to the slope of the *v-t* graph? Why does this happen?

e) Use pairs of data points from your graph to calculate the acceleration.

8. On the next page, you are provided with four graphs. Describe a motion of a cart on an incline that could produce each of these graphs. Include where the motion detector would have to be placed to produce the graph.

3.d)

The acceleration remains constant for the cart traveling down the ramp.

3.e)

Students can choose points by moving the cursor over the graph and recording the associated points. Points farther apart are better than the ones very close together for calculating the acceleration.

4.a)

This experiment mimics (with smaller acceleration) the act of throwing something straight up and letting it fall down. The *d-t* graph should resemble a parabola as shown below (if moving away from the detector is positive).

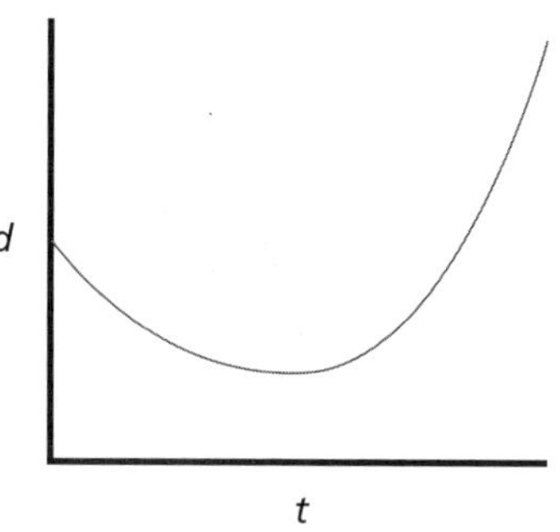

4.b)

The *v-t* graph should appear as shown below. In this graph and the graph for *4.a)*, the students should be concerned only with the left half of the graph.

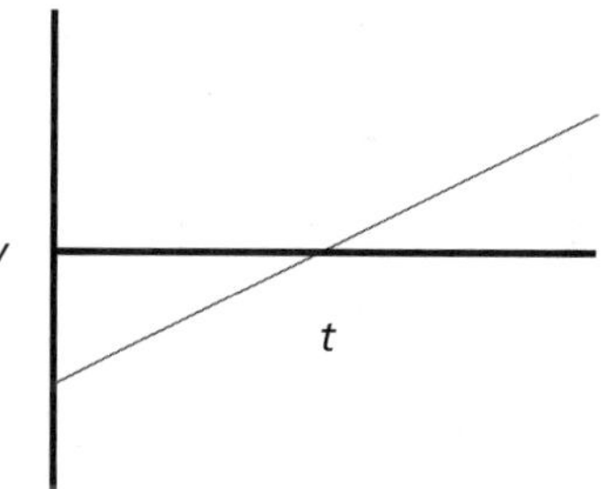

5.a)

The graphs collected should appear similar to those shown above in *4.a)* and *b)*.

5.b)

The actual graphs may differ if the students stop the carts before they begin to come down the plane.

5.c)

The slope of the *d-t* graph starts out negative, and increases to zero, then continues increasing to positive values. That means the velocity started out negative (with the *x*-axis going down the plane, away from the detector), increases to zero, and then becomes positive as the cart goes back down the plane.

5.d)

The slope of the *v-t* graph is positive and constant because the velocity steadily increases (positive acceleration).

5.e)

Students calculate acceleration using pairs of data points from the graphs.

6.a)

The *d-t* graph should look like the one shown below. The cart moves in the + *d* direction with an initial speed away from the detector, slows to a stop, and then reverses direction to come back toward the detector.

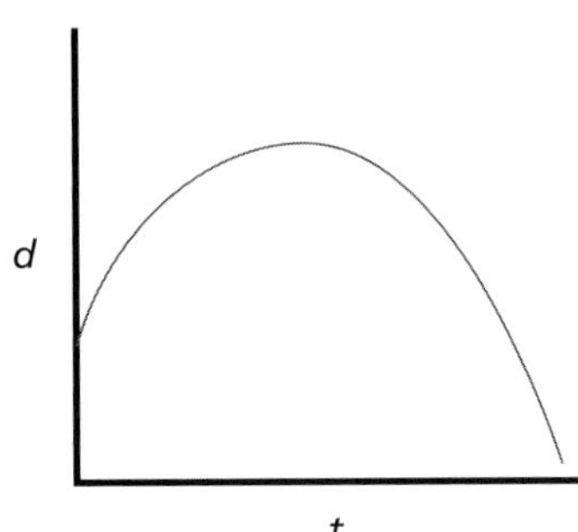

6.b)

The *v-t* graph should look like the one shown below. The cart starts up the plane with a certain velocity, becomes zero at the peak of the rise, and then gains speed on the way down. The slope everywhere is negative (negative acceleration).

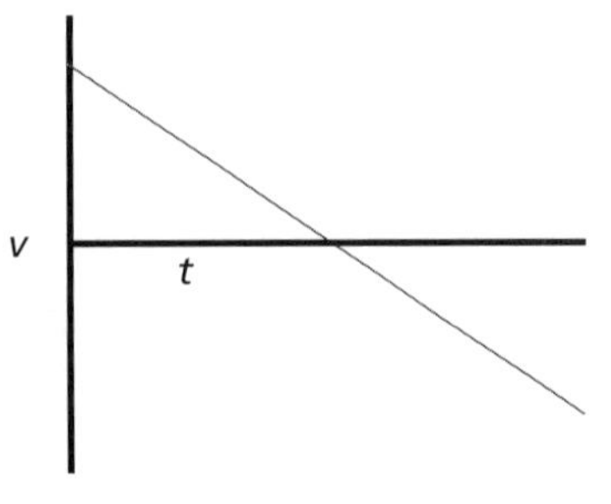

7.a)

The student graphs should appear similar to the ones below.

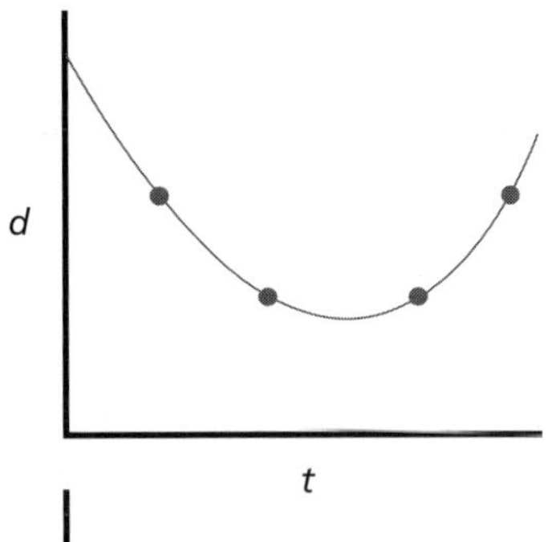

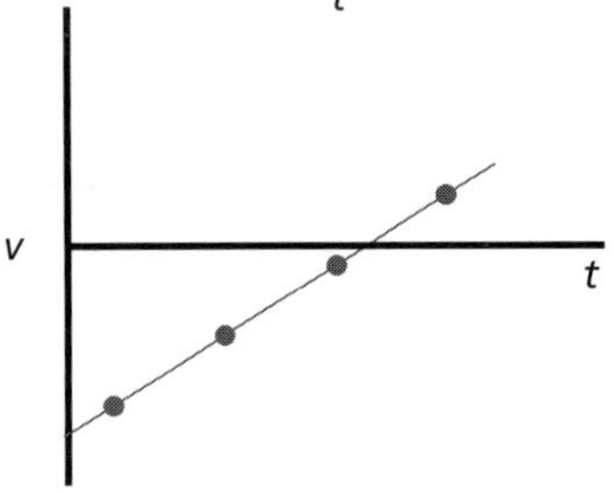

7.b)

Students will discuss their data.

7.c)

The slope change reflects the changing velocity of the cart as it starts and then moves away from the detector again, going down the plane. The value of the slope increases as the cart gains speed.

7.d)

The slope of the *v-t* graph remains the same as the cart rises and then rolls back down the plane. The slope is constant because the acceleration (which is the slope of the *v-t* graph) is a constant, and always in the same direction.

7.e)

Students should choose two data points from the *v-t* graph to calculate the acceleration. Data points near the beginning of the rise, and toward the end of the recorded motion, will most likely provide the most accurate result.

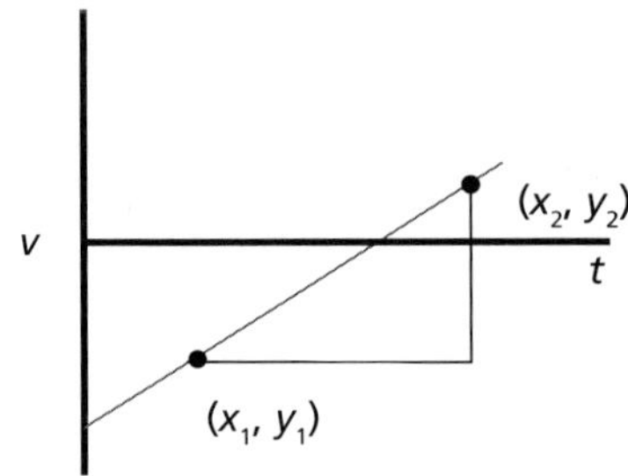

NOTES

8.

Look at the four graphs on the adjacent *Student Edition* page. The two graphs on the left both show increasing speed and would represent a cart moving down an incline, away from a motion detector at the top. The two graphs on the right both show decreasing speed, and would represent a cart moving up an incline with an initial speed when the detector is located at the bottom of the ramp.

9.a)

The acceleration data is located at the top of the third column in the table at the end of *Section 7* of the *Student Edition*.

10.

Students read about conversions.

1-4a Blackline Master

Chapter 1 Driving the Roads

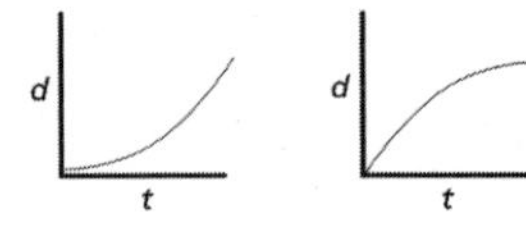

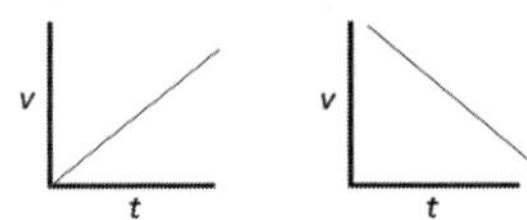

9. You will now take a closer look at acceleration in a straight line. Look at the automobile data provided at the end of this chapter on pages 116-117. The tables contain a lot of information including fuel economy, passenger accommodations, acceleration, and braking. In this section, you will be concerned with acceleration.

a) Record in your log where the acceleration information is located on the automobile table.

10. The speed on the table provided by automobile manufacturers is given in miles per hour (mi/h or mph), but the distances are recorded in feet and the time in seconds. To analyze this data more easily, it is helpful to record the speed in feet per second (ft/s). The table at right converts miles per hour to feet per second. Note that there are 60 min in 1 h and 60 s in 1 min. You should also note that there are 5280 ft in 1 mi. When you convert 60 mi/h to 88 ft/s, the conversion looks like the following:

$$\left(60\frac{\text{mi}}{\text{h}}\right)\left(\frac{1\text{ h}}{60\text{ min}}\right)\left(\frac{1\text{ min}}{60\text{ s}}\right)\left(\frac{5280\text{ ft}}{1\text{ mi}}\right)=88\frac{\text{ft}}{\text{s}}$$

If you deal with the units in the same way that you deal with the numbers, you will see that the miles cancel miles, hours cancel hours, and minutes cancel minutes.

$$\left(60\frac{\cancel{\text{mi}}}{\cancel{\text{h}}}\right)\left(\frac{1\cancel{\text{h}}}{60\cancel{\text{min}}}\right)\left(\frac{1\cancel{\text{min}}}{60\text{ s}}\right)\left(\frac{5280\text{ ft}}{1\cancel{\text{mi}}}\right)=88\frac{\text{ft}}{\text{s}}$$

You should notice that to convert 60 mi/h to 88 ft/s, the 60 mi/hr was multiplied by fractions that always equaled 1 (for example, 1 h and 60 min are the same value of time). Multiplying by 1/1 keeps the value the same.

The following table was constructed on a spreadsheet. You can use the conversions in this table to give you a sense of the different units and to help you answer some of the questions in this chapter.

	A	B	C	D
1	Common Speed Conversions			
2	United States		Canada	
3	mph	ft/s	m/s	km/h
4	0	0	0	0
5	10	15	5	16
6	20	29	9	32
7	30	44	13	49
8	40	59	18	65
9	50	73	23	81
10	60	88	27	97
11	70	103	31	113
12	80	117	36	130
13	90	132	41	146
14	100	147	45	162

11. The sports car's acceleration data from the table at the end of the chapter is shown below with miles per hour changed to feet per second.

Acceleration Data of a Sports Car in Feet per Second

Final speed (ft/s)	Total time (s)
0	0.0
44	2.0
59	2.9
73	4.2
88	5.2
103	6.6
117	8.7
132	10.9
147	13.3

56

Active Physics

Section 4 Graphing Motion: Distance, Velocity, and Acceleration

a) Sketch a graph of speed vs. total time and label it "Velocity-Time Graph." Put the time on the x-axis (horizontal) and the speed on the y-axis (vertical). Plot your points from the table using feet per second (ft/s) units for velocity.

b) During which time interval is the velocity changing the most?

c) During which time interval is the velocity changing the least?

d) Acceleration is defined as the change in velocity for each time interval. Where is acceleration the greatest? Where is acceleration the least?

12. You can now calculate the acceleration for each time interval.

The acceleration is equal to the change in velocity (final speed − initial speed) divided by the change in time.

$$a = \frac{\Delta v}{\Delta t} = \frac{v_f - v_i}{\Delta t}$$

Where a is acceleration,

Δv is change in velocity,

v_f is final velocity,

v_i is initial velocity,

Δt is change in time or elapsed time.

The first acceleration calculation is shown below.

$$a = \frac{\Delta v}{\Delta t} = \frac{v_f - v_i}{\Delta t} = \frac{44 \text{ ft/s} - 0 \text{ ft/s}}{2 \text{ s}} = \frac{22 \text{ ft/s}}{\text{s}}$$

The acceleration is equal to 22 feet per second every second. This is a change in speed (22 ft/s) with respect to time (1 s). This can also be written in the following ways:

22 ft/s every s

22 ft/s per s

22 (ft/s) per s

22 ft/s^2 (feet per second squared)

The last way is the easiest to say, but the first way is the easiest to understand.

If the automobile moved at a constant acceleration of 22 ft/s every second, you would see a constant increase in the speed every second, from 0 ft/s to 22 ft/s, then to 44 ft/s, and then to 66 ft/s. A constant acceleration is what happened to the cart on the ramp. However, this increase is not what usually happens to an automobile. An automobile does not move at a constant acceleration.

a) You can calculate the acceleration for the next time interval by calculating the acceleration of the sports car from 44 ft/s to 59 ft/s. This change in speed required 0.9 s. Complete this calculation. Did you get the value in the table of 16 ft/s every second?

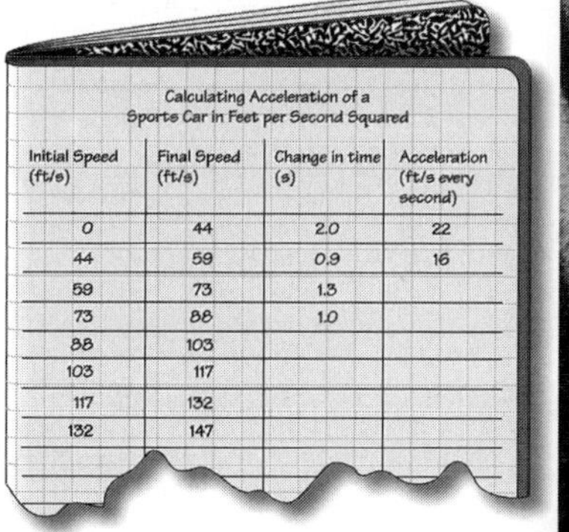

Calculating Acceleration of a Sports Car in Feet per Second Squared

Initial Speed (ft/s)	Final Speed (ft/s)	Change in time (s)	Acceleration (ft/s every second)
0	44	2.0	22
44	59	0.9	16
59	73	1.3	
73	88	1.0	
88	103		
103	117		
117	132		
132	147		

57

Active Physics

11.a)

Students sketch the graph.

11.b)

The velocity changes the most when (change in position)/(time interval) has the maximum value. The greatest change in velocity per unit time is from 0 to 2 s with a change in speed of 22 ft/s.

11.c)

Velocity changes least when (change in position)/(time interval) has the minimum value. The lowest change per time was 10.9 s to 13.3 s with a change of 6.25 ft/s.

11.d)

Acceleration is the greatest from 0 to 2 s (the first time interval) and least for the last time interval in the table of 10.9 – 13.3 s.

12.a)

Yes, 16.3 ft/s.

12.b)

The accelerations for the remaining time intervals (all in ft/second squared) are (59–73) 11, (73–88) 15, (88–103) 11, (103–117) 6.7, (117–132) 6.8, and (132–147) 6.3.

12.c)

The steepest slope for the graph should correspond to the greatest acceleration of the car, so the steepest slope occurs at the beginning of the graph from 0 to 2 s.

Physics Talk

Students read how Galileo applied mathematics to study the change in the speed of falling objects. His technique of using a water clock enabled him to measure small increments of time. He realized that by using an inclined plane he could "slow down" the effects of gravity, which allowed him to investigate how falling objects change speed. Galileo described this change in speed in quantitative terms.

A demonstration of how a water clock works and a discussion with students on Galileo's method of making measurements with precision should give them a clearer picture of how a water clock was used to measure time. This *Physics Talk* illustrates how Galileo arrived at the definition of acceleration with his experiment of balls rolling down an inclined plane. To highlight the concept of acceleration, ask your class to write down the definition of acceleration in their *Active Physics* logs. Draw their attention

b) Work with your group members to complete the Calculating Acceleration of a Sports Car in Feet per Second Squared table in your log.

c) Compare the table with the velocity-time graph you sketched in *Step 11.a)*.

Recall that the slope of the velocity-time graph is equal to the acceleration. Where does the table indicate the greatest acceleration took place? Where does the graph have the steepest slope?

Physics Talk

CHANGING SPEED

Acceleration

When things change speed, it is usually noticeable. The motion of a falling object is a common example of something changing speed. Galileo Galilei, an Italian scientist, was the first person to apply mathematics to the study of the change in the speed of a falling object. To describe the change quantitatively, he needed to make measurements. In the *Investigate*, you used a motion detector to make measurements of **acceleration**. Since Galileo lived in the 1600s, he did not even have access to clocks and had to devise original ways to measure time with accuracy and precision. One technique he used involved a water clock. In a water clock, water flows through a funnel into a bowl. The more time that elapses, the more water is collected.

Physics Words
acceleration: the change in velocity with respect to a change in time.

To help him explore falling objects, he first investigated balls rolling down an incline. He thought that a ball rolling down an incline was like watching a falling object in "slow motion."

Through his experimentation with balls rolling down inclines, Galileo found that if he looked at the change in speed with respect to the change in time, the value remained the same as the ball descended the ramp. He then defined this as acceleration. The definition of acceleration as change in velocity with respect to time is still in use today.

In this section, you observed, just as Galileo did with rolling balls, that a cart traveling down an inclined plane has a constant acceleration. The velocity of the cart changes at a regular rate and is represented by a straight line on the velocity vs. time graph.

to the motion of the cart that traveled down the incline in the *Investigate*, and how its speed changed at a regular rate, in the same manner as Galileo's balls rolling down an inclined plane. Emphasize that constant acceleration is represented by a straight line on the velocity-time graph. It is important for students to understand that the change in speed with respect to time remains the same in the case of constant acceleration.

The distinction between speed and velocity is drawn when students consider an automobile traveling around a curve. As they read why velocity is a vector quantity, your discussion should focus on why speed is considered to be different from velocity and how velocity can be controlled

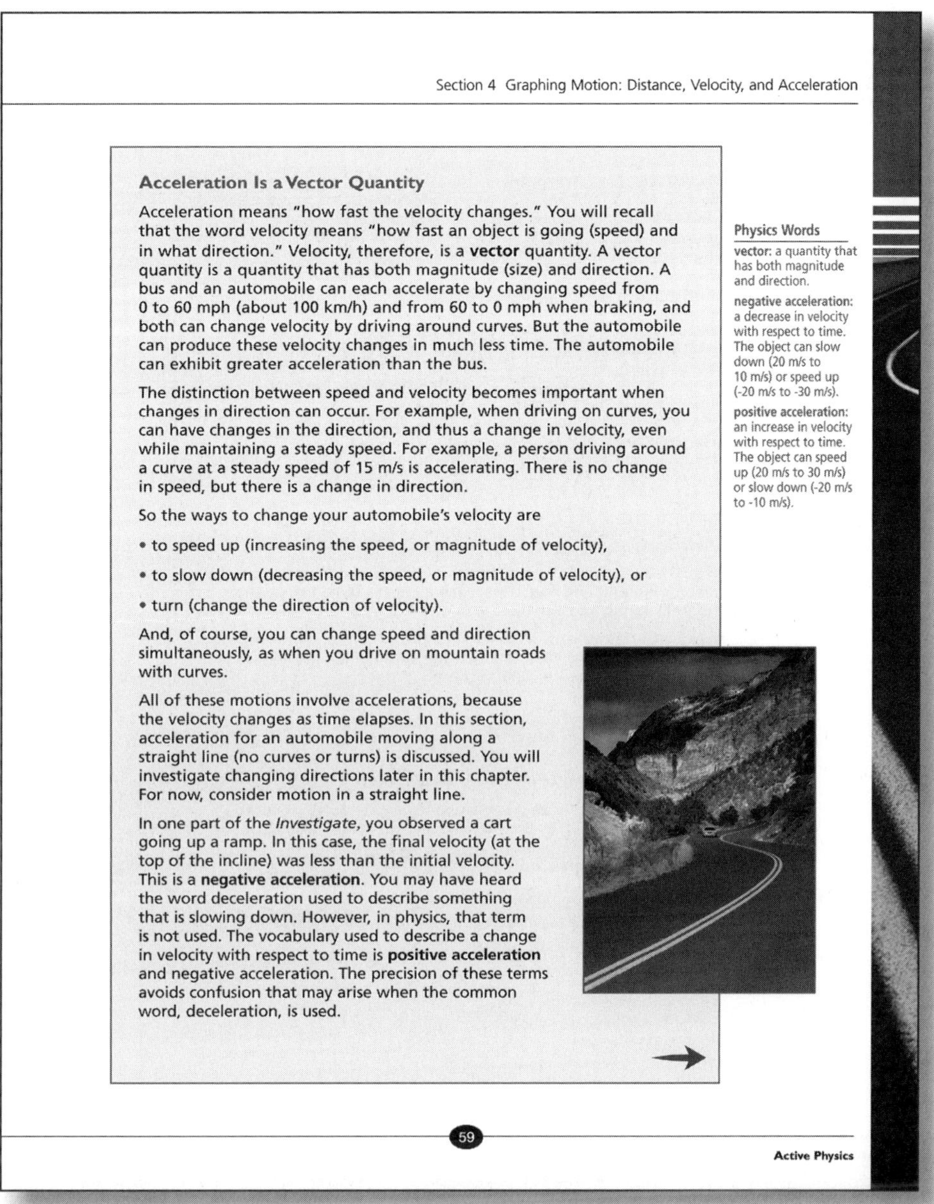

Section 4 Graphing Motion: Distance, Velocity, and Acceleration

Acceleration Is a Vector Quantity

Acceleration means "how fast the velocity changes." You will recall that the word velocity means "how fast an object is going (speed) and in what direction." Velocity, therefore, is a **vector** quantity. A vector quantity is a quantity that has both magnitude (size) and direction. A bus and an automobile can each accelerate by changing speed from 0 to 60 mph (about 100 km/h) and from 60 to 0 mph when braking, and both can change velocity by driving around curves. But the automobile can produce these velocity changes in much less time. The automobile can exhibit greater acceleration than the bus.

The distinction between speed and velocity becomes important when changes in direction can occur. For example, when driving on curves, you can have changes in the direction, and thus a change in velocity, even while maintaining a steady speed. For example, a person driving around a curve at a steady speed of 15 m/s is accelerating. There is no change in speed, but there is a change in direction.

So the ways to change your automobile's velocity are

- to speed up (increasing the speed, or magnitude of velocity),
- to slow down (decreasing the speed, or magnitude of velocity), or
- turn (change the direction of velocity).

And, of course, you can change speed and direction simultaneously, as when you drive on mountain roads with curves.

All of these motions involve accelerations, because the velocity changes as time elapses. In this section, acceleration for an automobile moving along a straight line (no curves or turns) is discussed. You will investigate changing directions later in this chapter. For now, consider motion in a straight line.

In one part of the *Investigate*, you observed a cart going up a ramp. In this case, the final velocity (at the top of the incline) was less than the initial velocity. This is a **negative acceleration**. You may have heard the word deceleration used to describe something that is slowing down. However, in physics, that term is not used. The vocabulary used to describe a change in velocity with respect to time is **positive acceleration** and negative acceleration. The precision of these terms avoids confusion that may arise when the common word, deceleration, is used.

Physics Words

vector: a quantity that has both magnitude and direction.

negative acceleration: a decrease in velocity with respect to time. The object can slow down (20 m/s to 10 m/s) or speed up (-20 m/s to -30 m/s).

positive acceleration: an increase in velocity with respect to time. The object can speed up (20 m/s to 30 m/s) or slow down (-20 m/s to -10 m/s).

59

Active Physics

either by changing the speed, or by changing direction with or without a change in speed. Give the students some examples of scalar and vector quantities, and then ask them to give you some additional examples. Because these concepts might confuse students initially, point out that they will be revisiting scalar and vector quantities while studying other topics in *Active Physics*.

Before introducing the term negative acceleration, recall the steps of the *Investigate* in which the cart goes up an incline. Discuss student observations and ask them why "deceleration" is not used in physics. They should be able to differentiate between negative and positive acceleration. To be able to decide whether acceleration is positive or negative, students must have a coordinate system that determines which direction is positive. For a better perspective, ask them to classify the changes in velocity they have observed, in different situations, as positive or negative acceleration.

As motion is represented in different forms, it is important for students to understand how each description confirms the relationship between acceleration, velocity, and time. Students should sketch models of strobe pictures to show accelerated motion. Have them write out the mathematical equation for each quantity using words and symbols in their log. Encourage students who might want to draw a circle to represent the relationship among variables when giving the equation for acceleration. The sample problem demonstrates how the acceleration of a toy car can be determined. You could ask students to write similar problems, using different values for each quantity, to determine the acceleration of an object. This strategy should make it easier for them to calculate acceleration and manipulate the acceleration equation, further reinforcing student understanding of the concept.

The units for acceleration are often confusing for students. It is preferable to use the term "meters per second per second" in discussions and in written form instead of m/s^2 until the students become more comfortable with the concept of acceleration. When students are learning to represent motion through graphs, make sure they understand that the instantaneous speed can be found at any point on a curve of the distance-time graph, from the slope of the line tangent to the curve at that point. Point out to them that the value of the slope on a *d-t* graph, $\Delta d/\Delta t$, is the same at all points if the object is traveling with constant speed. Similarly, on a velocity-time graph the slope is a straight line when the object is undergoing constant acceleration. While comparing motion graphs, students will see that the graphs of different motions for an automobile will yield different slopes and that the slope of the *d-t* graph is velocity, while the slope of the *v-t* graph is acceleration. By sketching these graphs in their *Active Physics* log, the concept of accelerated motion will be further reinforced.

You will investigate negative acceleration further in *Section 5*. For motion in a straight line, positive acceleration means that the velocity of the object is increasing over time. Negative acceleration means that the velocity of the object is decreasing over time, if the object is moving in a straight line.

Vector and Scalar Quantities

A quantity that involves both direction and size (magnitude) is called a vector quantity. A quantity that has size, but not direction, is called a scalar quantity. Speed is a scalar quantity. It only indicates the change in position over a period of time in a straight line. Velocity is a vector quantity. It can indicate a change in position over a period of time and the direction.

Describing Accelerated Motion Using Strobe Pictures

Recall that you used three different models to describe motion: strobe pictures, graphs, and equations. Each gives the same information, but in different forms. You will use the same models to describe acceleration.

Because the speed is always changing during constant acceleration, the strobe illustration below shows the automobiles moving greater distances during each second of travel.

Describing Acceleration Using an Equation

In the *Investigate*, you used an equation to describe acceleration. You calculated acceleration by finding the change in velocity with respect to time.

$$\text{Acceleration} = \frac{\text{change in velocity}}{\text{change in time}}$$

This relationship can be written as an equation using symbols.

$$a = \frac{\Delta v}{\Delta t}$$

where a is acceleration,
Δv is change in velocity,
Δt is change in time or elapsed time.

Section 4 Graphing Motion: Distance, Velocity, and Acceleration

Units for Measuring Acceleration

To calculate acceleration, you divide change in velocity by change in time $\frac{\Delta v}{\Delta t}$. The units for acceleration are then, by definition, velocity divided by time. Recall from the previous section, the units for velocity can be m/s or km/h. Assume that the time interval is measured in seconds. The units for acceleration would then be (m/s)/s or (km/h)/s. The change in velocity is given in meters per second every second, or kilometers per hour every second.

When writing the units for acceleration, the final units are often simplified. For example, the following all mean the same thing. The simplified units are read as meters per second squared.

$$\frac{\text{m/s}}{\text{s}}, \text{ or (m/s)/s} = \frac{\text{m}}{\text{s}^2} \text{ or m/s}^2$$

In the *Investigate*, you calculated acceleration in feet per second every second, or feet per second squared (ft/s²).

Using the Equation for Acceleration to Find Other Quantities

The defining equation for acceleration shows the relationship between acceleration, velocity, and time. If you know two of these, you can find the third.

$$\text{Acceleration} = \frac{\text{change in velocity}}{\text{change in time}}$$

Using algebra, it follows that

$$\text{Change in velocity} = \text{acceleration} \times \text{time}$$

$$\text{Time} = \frac{\text{change in velocity}}{\text{acceleration}}$$

Using symbols, these equations can be written as

$$a = \frac{\Delta v}{\Delta t}$$

$$\Delta v = a \times \Delta t$$

$$\Delta t = \frac{\Delta v}{a}$$

As you did with the equations for speed in the previous section, you may find it helpful to use a circle, like the following:

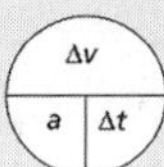

By covering up the variable you wish to find, you can see the equation. To find change in velocity (Δv), cover up the Δv, and you see $a \times \Delta t$.

To find acceleration (a), cover up the a, and you see $\frac{\Delta v}{\Delta t}$.

To find time (Δt), cover up the Δt and you see $\frac{\Delta v}{a}$.

There is only one definition of acceleration. Algebra allows you to write it in different forms.

Sample Problem

At the start of a race, a toy car increases speed from 0 m/s to 5.0 m/s as the clock runs from 0 s to 2.0 s. Find the acceleration of the toy car.

Strategy: Use the definition of acceleration as the change in velocity over a change in time.

Given:

Final velocity $(v_f) = 5.0$ m/s

Initial velocity $(v_i) = 0$ m/s

Final time $(t_f) = 2.0$ s

Initial time $(t_i) = 0$ s

Solution:

$$\begin{aligned} a &= \frac{\Delta v}{\Delta t} \\ &= \frac{v_f - v_i}{t_f - t_i} \\ &= \frac{5.0\ \text{m/s} - 0\ \text{m/s}}{2.0\ \text{s} - 0\ \text{s}} \\ &= \frac{5.0\ \text{m/s}}{2.0\ \text{s}} \\ &= 2.5\ \text{m/s}^2 \end{aligned}$$

The acceleration is 2.5 m/s every second, and can be written and stated in three equivalent ways:

- 2.5 meters per second every second, or
- 2.5 (m/s)/s (meters per second per second), or
- 2.5 m/s^2 (meters per second squared).

1-4b **Blackline Master**

CHAPTER 1

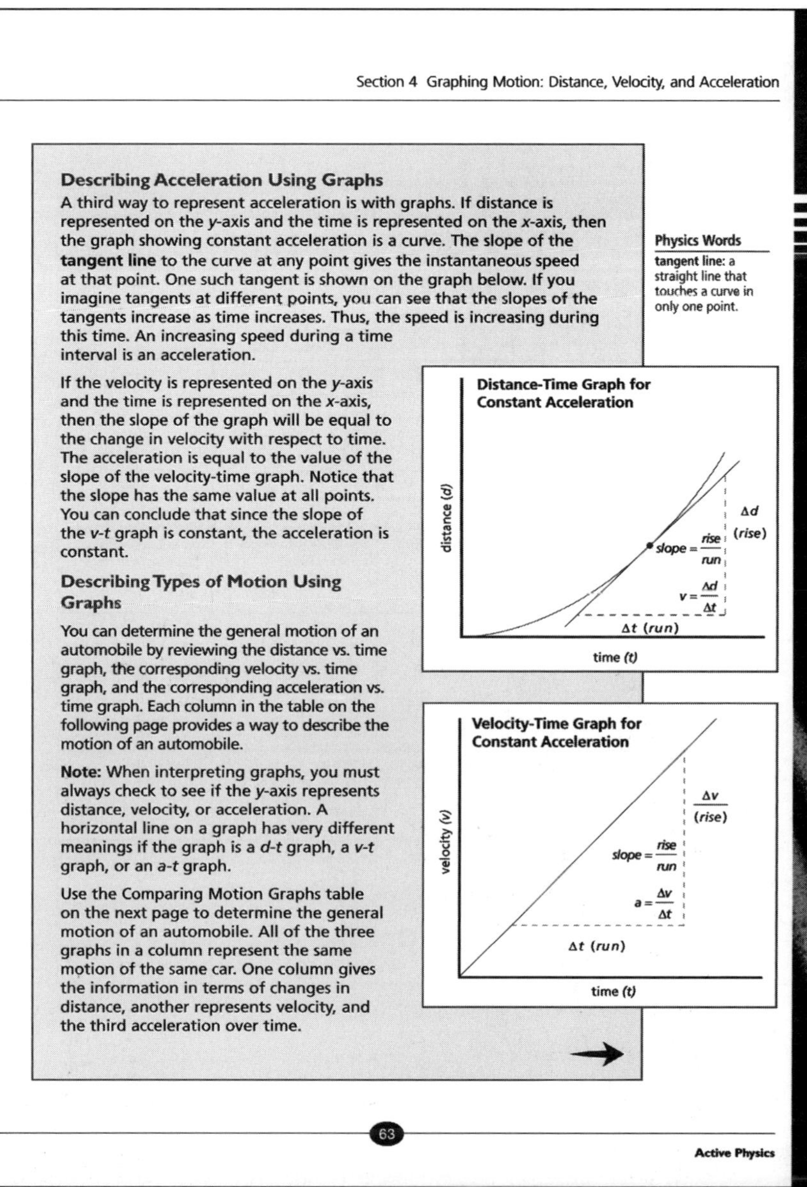

Describing Acceleration Using Graphs

A third way to represent acceleration is with graphs. If distance is represented on the *y*-axis and the time is represented on the *x*-axis, then the graph showing constant acceleration is a curve. The slope of the **tangent line** to the curve at any point gives the instantaneous speed at that point. One such tangent is shown on the graph below. If you imagine tangents at different points, you can see that the slopes of the tangents increase as time increases. Thus, the speed is increasing during this time. An increasing speed during a time interval is an acceleration.

Physics Words

tangent line: a straight line that touches a curve in only one point.

If the velocity is represented on the *y*-axis and the time is represented on the *x*-axis, then the slope of the graph will be equal to the change in velocity with respect to time. The acceleration is equal to the value of the slope of the velocity-time graph. Notice that the slope has the same value at all points. You can conclude that since the slope of the *v-t* graph is constant, the acceleration is constant.

Describing Types of Motion Using Graphs

You can determine the general motion of an automobile by reviewing the distance vs. time graph, the corresponding velocity vs. time graph, and the corresponding acceleration vs. time graph. Each column in the table on the following page provides a way to describe the motion of an automobile.

Note: When interpreting graphs, you must always check to see if the *y*-axis represents distance, velocity, or acceleration. A horizontal line on a graph has very different meanings if the graph is a *d-t* graph, a *v-t* graph, or an *a-t* graph.

Use the Comparing Motion Graphs table on the next page to determine the general motion of an automobile. All of the three graphs in a column represent the same motion of the same car. One column gives the information in terms of changes in distance, another represents velocity, and the third acceleration over time.

Checking Up

1.

Acceleration is the change in velocity with respect to a change in time and is represented using the following symbols:
$a = \Delta v / \Delta t$.

2.

The SI unit for measuring acceleration is m/s^2 (meters per second squared).

3.

A vector quantity is a quantity that involves both size (magnitude) and direction. A scalar quantity involves size, but no direction.

4.a)

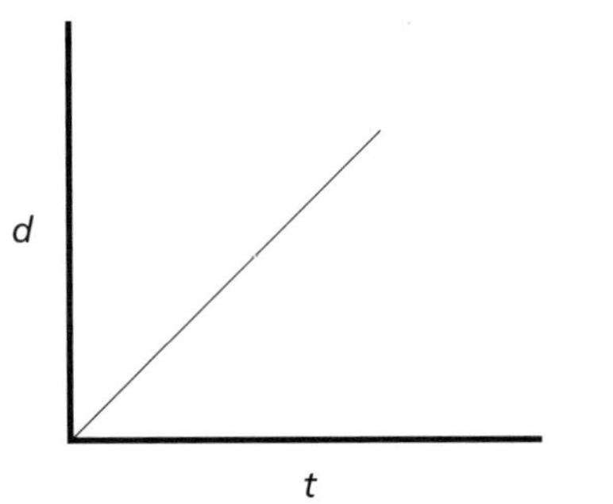

4.b)

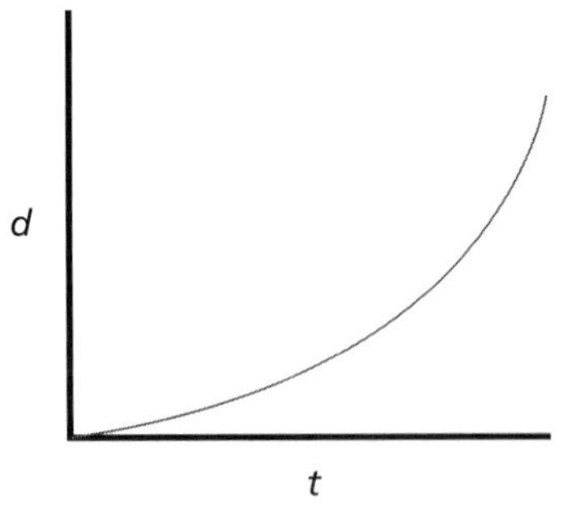

Chapter 1 Driving the Roads

Comparing Motion Graphs

	Automobile at rest	Automobile with constant velocity	Automobile with constant acceleration
d-t graph	Distance (position) does not change.	Distance (position) changes at a constant rate.	Distance (position) changes at a non-constant rate (the *d-t* graph is a curve, not a straight line).
	distance (d) / time (t)	distance (d) / time (t)	distance (d) / time (t)
v-t graph	Velocity is always 0 because the position does not change.	The change in distance vs. time is constant and therefore the velocity is constant rate.	The speed is increasing as can be seen by the increasing slope of the distance vs. time graph.
	velocity (v) / time (t)	velocity (v) / time (t)	velocity (v) / time (t)
a-t graph	Acceleration is always 0 because velocity does not change.	There is a constant velocity. With no change in velocity, there is 0 acceleration.	The velocity is changing at a constant rate. The automobile is moving at a constant acceleration.
	acceleration (a) / time (t)	acceleration (a) / time (t)	acceleration (a) / time (t)

Checking Up

1. Give the defining equation for acceleration in words, and by using symbols.
2. What is an SI unit for measuring acceleration? Use words and unit symbols to describe the unit.
3. What is the difference between a vector and a scalar quantity?
4. Sketch a distance-time graph for
   a) constant velocity
   b) constant acceleration
5. What does the slope of a velocity-time graph represent?

Active Physics 64

5.

The slope of the velocity-time graph represents the change in velocity with respect to time (acceleration).

1-4c Blackline Master

Section 4 Graphing Motion: Distance, Velocity, and Acceleration

Active Physics Plus

+Math	+Depth	+Concepts	+Exploration
•••		••	

Determining Distance Using the Acceleration Equation

The definition of acceleration provides the relationship between velocity, acceleration, and time. If you know the acceleration, you can determine the change in velocity after a given time has elapsed by using the following equation:

$$\Delta v = a\Delta t$$

If an automobile has a constant acceleration, you can also determine the distance traveled after a given time has elapsed.

Knowing the initial and final velocity, you can now determine the average velocity ($\bar{v}$). For a constant acceleration, the average velocity is determined the same way the average of any two numbers is determined.

$$\bar{v} = \frac{v_f + v_i}{2}$$

Once you know the average velocity, you can use the definition of average velocity to determine the distance traveled.

Using some algebra, you can also determine the distance traveled in one step with a newly derived equation.

$$d = \bar{v}t$$

You can now determine how an object's position and velocity depend on the elapsed time from the definition of velocity and acceleration.

$$d = \bar{v}t$$

$$d = \left(\frac{v_f + v_i}{2}\right)t$$

Since $v_f - v_i = at$

then $v_f = v_i + at$

$$d = \left(\frac{(v_i + at) + v_i}{2}\right)t$$

$$d = \left(\frac{(at) + 2v_i}{2}\right)t$$

$$d = \frac{1}{2}at^2 + v_i t$$

Sample Problem

An automobile accelerates from rest at 5.0 m/s every second (5.0 m/s²). How far does it travel after 3.0 s?

Given:

Initial velocity (v_i) = 0 m/s

Acceleration (a) = 5.0 m/s²

Time (Δt) = 3.0 s

Strategy 1:

Find the final velocity using the definition of acceleration; then find the average velocity; and then use the relationship between distance, average velocity, and time.

Solution:

$$v_f = at + v_i$$

$$v_f = \left(5.0\ \frac{\text{m}}{\text{s}^2}\right)(3.0\ \text{s}) + 0$$

$$v_f = 15\ \text{m/s}$$

→

Active Physics Plus

Active Physics Plus draws the curious student to explore physics concepts in more depth and detail. To clarify the purpose of *Active Physics Plus*, make all students read the introduction to this section. You might want to ask a few students to demonstrate the mathematical relationship between acceleration, velocity, and time. When students use the formula $d = (v_i + v_f)(t/2)$ to determine distance using the definition of average velocity, emphasize that the acceleration for the period of time that the distance is being measured remains constant. Highlight the sample problem to show how to use the equation appropriately when the acceleration for which distance is being calculated has a constant value. The students venturing into this section can actively engage other students who are still in the early stages of understanding the concept of acceleration. Through this strategy, all students can eventually participate in developing their understanding of new concepts.

CHAPTER 1

What Do You Think Now?

This is a good time to return to the *What Do You Think?* section and review the answers students gave earlier. It is an opportunity for you to gauge students' prior understanding of velocity. They should now be able to relate velocity to acceleration, and how the driver can adjust velocity to control following distance. You may want to provide answers given in *A Physicist's Response* and allow students to share their opinions. This review will clarify the finer aspects of velocity and acceleration. Students should by now have a more thorough understanding of the $a = \Delta v / \Delta t$ equation. Ask them to share their sketches of velocity-time graphs in small groups to compare and contrast the motion of each vehicle. Encourage them to describe how acceleration is illustrated in their graphs of motion.

Knowing that the final velocity is 15 m/s and the initial velocity equals 0, you can calculate the average velocity.

$$\bar{v} = \frac{v_f + v_i}{2}$$
$$= \frac{15 \text{ m/s} + 0 \text{ m/s}}{2}$$
$$= 7.5 \text{ m/s}$$

Using the definition of average velocity, the distance can be computed:

$$d = \bar{v}t$$
$$= 7.5 \frac{\text{m}}{\cancel{\text{s}}} \times 3.0 \cancel{\text{s}}$$
$$= 22.5 \text{ m}$$

Strategy 2:

Because acceleration, time, and initial velocity are provided, use the derived relationship of distance, acceleration, and time. There is no need to find the final velocity.

Solution:

$$d = \frac{1}{2}at^2 + v_i t$$
$$= \frac{1}{2}\left(5 \frac{\text{m}}{\cancel{\text{s}}^2}\right)(3 \cancel{\text{s}})(3 \cancel{\text{s}}) + 0$$
$$= 22.5 \text{ m}$$

What Do You Think Now?

At the beginning of this section, you were asked the following:

- **An automobile and a bus are stopped at a traffic light. What are some differences and similarities of the motion of these two vehicles as each goes from a stop to the speed limit of 30 mph?**

How would you answer this question now? Now that you have investigated change in velocity over time, compare and contrast the motion of the vehicles using the term acceleration. Sketch a velocity-time graph for each vehicle.

Physics

Essential Questions

What does it mean?

One race car has a greater acceleration than a second race car. But the second race car can reach a higher top speed than the first. How is this possible?

How do you know?

As you enter the highway, your automobile goes from rest to the speed limit. What measurements would you have to take to calculate the acceleration of your automobile as it enters the highway?

Why do you believe?

Connects with Other Physics Content	Fits with Big Ideas in Science	Meets Physics Requirements
Forces and motion	* Models	Experimental evidence is consistent with models and theories

* Physicists use models to better understand the world. How can a distance vs. time graph, a velocity vs. time graph and an acceleration vs. time graph all represent a car moving with a constant acceleration?

Why should you care?

Safe driving saves lives. How can your understanding of velocity and acceleration help you to become a safer driver? In the unfortunate possibility that you are in an accident, how could *d-t*, *v-t*, and *a-t* motion graphs help you explain why the accident was not your fault?

Reflecting on the Section and the Challenge

Driving is all about accelerations. Automobiles accelerate when they speed up, slow down, or make a turn. Drivers depend on negative accelerations to avoid accidents when they apply the brakes. Speeding up too quickly can also lead to accidents.

You now know how to calculate accelerations and to determine accelerations from a velocity-time graph. You may want to use these calculations and/or graphs in your description of safe driving, or perhaps in your presentation.

Reflecting on the Section and the Challenge

Students should be able to reflect on how drivers depend on negative acceleration to avoid accidents. They should be able to understand how velocity-time graphs can help them calculate the rate at which acceleration changes. They can use their calculations to highlight how speeding up in a short interval of time could lead to an accident. The results of their calculations will help the students describe the importance of safe driving in their *Chapter Challenge*.

CHAPTER 1

Physics Essential Questions

What does it mean?

If the second car is able to accelerate for a longer time than the first car, it can continue to gain speed and reach a higher top speed, while the first car is traveling at its maximum speed.

How do you know?

Acceleration is the change in speed during a given amount of time. You know that your beginning speed was 0 mi/h. You can measure your final speed. You can also measure the time required to get to that final speed. This change in speed divided by the time is the acceleration $a = \Delta v / \Delta t$.

Why do you believe?

The acceleration vs. time graph would show a horizontal line. The velocity vs. time graph would show a straight ascending line. The slope of that line would be the acceleration. A distance vs. time graph would show a curved line. The slope of the tangent line at later times would be greater showing an increase in velocity.

Why should you care?

The graphs and your understanding of distance, velocity and acceleration provide you with precision and language to describe what happened. It is more informative to explain that you were traveling at 30 mi/h than it is to say you were going at some intermediate speed.

Physics to Go

1.

An object can have zero acceleration (no change in velocity) and a nonzero velocity. Such an object is called a "free particle" because it moves freely without changing velocity.

2.

An object can have zero velocity and nonzero acceleration. This happened, for example, with the cart rolling up and then down the inclined plane. At the top of the plane, the velocity was instantaneously zero, but it did not stay zero because it was constantly accelerated. The velocity decreased from a positive value, to zero, and through zero to negative values.

3.

If two automobiles have the same acceleration, they do not necessarily have the same velocity. Acceleration measures rate of change of velocity, not velocity itself. The velocity depends not only on the acceleration but also on the time the object has been accelerating, as well as on the initial velocity.

4.

If two automobiles have the same velocity, they do not necessarily have the same acceleration. For example, a ball at rest has zero velocity, and an arrow at the top of its flight when shot straight up has zero velocity (instantaneously). The ball has zero acceleration, but the arrow has a nonzero acceleration. The ball will remain at rest, but the arrow will not.

Physics to Go

1. Can a situation exist in which an object has zero acceleration and nonzero velocity? Explain your answer.
2. Can a situation exist in which an object has zero velocity and nonzero acceleration, even for an instant? Explain your answer.
3. If two automobiles have the same acceleration, do they have the same velocity? Why or why not?
4. If two automobiles have the same velocity, do they have the same acceleration? Why or why not?
5. Can an accelerating automobile be overtaken by an automobile moving with constant velocity?
6. Is it correct to refer to speed-limit signs instead of velocity-limit signs? Why or why not? What units are assumed for speed-limit signs in the United States?
7. Suppose an automobile were accelerating at 2 mi/h every 5 s and could keep accelerating for 2 min at that rate.
   a) How fast would it be going at $t = 2$ min?
   b) How far would it be from the starting line?
8. At an international auto race, a race car leaves the pit after a refueling stop and accelerates uniformly to a speed of 75 m/s in 9 s to rejoin the race.
   a) What is the race car's acceleration during this time?
   b) What was the race car's average speed during the acceleration?
   c) How far does the race car go during the time it is accelerating?
   d) A second race car leaves after its pit stop and accelerates to 75 m/s in 8 s. Compared to the first race car, what is this race car's acceleration, average speed during the acceleration, and distance traveled?
9. During a softball game, a player running from second base to third base reaches a speed of 4.5 m/s before she starts to slide into third base. When she reaches third base 1.3 s after beginning her slide, her speed is reduced to 0.6 m/s.
   a) What is the player's acceleration during the slide?
   b) What was the distance of her slide?
   c) If she had slid for only 1.1 s, how fast would she have been moving when she reached third base? (Assume she had the same acceleration as before.)
   d) Which of these two trials would get her from second base to third base faster?

5.

An automobile that is accelerating may be starting from a lower initial velocity than the automobile that is traveling at constant speed. An example would be an automobile stopped at a stoplight that starts to accelerate when the light turns green. If an automobile rolling through the light has a high velocity, it might easily pass the automobile that accelerates as the light turns green.

6.

You are not really using language incorrectly when talking about "speed limits" because it is assumed that the vehicles are going in a certain direction on the road where the speed limit is posted. Technically, the sign means "the magnitude of your velocity should not exceed this value." In the United States the units are not designated, and "Speed Limit 70" means 70 mi/h. Most other countries assume km/h in their posted speed limits, and they usually do not include the units either.

7.a)

An automobile that accelerates at 2 mi/h/5 s would have an increase in speed of

$\Delta v = a \times \Delta t$.

Thus, $\Delta v = (2 \text{ mi/h})/(5 \text{ s}) \times 120 \text{ s} = 48 \text{ mi/h}$.

7.b)

Starting from rest, the distance covered by the automobile can be given by $d = v_{\text{average}} \Delta t$. The automobile's average velocity would be the

$$\frac{\text{initial speed (0)} - \text{final speed (48)}}{2}$$

$= 24$ mi/h.

The distance covered then would be

$d = (24 \text{ mi/h})(1/30 \text{ h}) = 0.8 \text{ mi}$.

Note: The units of time changed so they would solve both problems in hours to cancel out and leave a unit of distance, miles.

8.a)

The car's acceleration, a, is $a = \Delta v/\Delta t = (75 \text{ m/s} - 0)/(9 \text{ s}) = 8.33 \text{ m/s}^2$.

8.b)

The car's average speed,

$v_{\text{average}} = (v_i + v_f)/2 = (75 \text{ m/s} + 0)/2 = 37.5 \text{ m/s}$.

8.c)

The distance the car goes $d = v_{\text{average}} t = (37.5 \text{ m/s})(9 \text{ s}) = 337.5 \text{ m}$.

8.d)

The second car's acceleration is greater, average speed is the same, but the distance traveled in 8 s is less than the distance covered in 9 s by the first car.

9.a)

The player's acceleration during the slide is

$a = \Delta v/\Delta t = (0.6 \text{ m/s} - 4.5 \text{ m/s})/(1.3 \text{ s}) = -3 \text{ m/s}^2$.

9.b)

The distance of the player's slide is calculated by first determining the average velocity

$v_{\text{average}} = (v_i + v_f)/2 = (4.5 \text{ m/s} + 0.6 \text{ m/s})/2 = 2.55 \text{ m/s}$.

Then the distance,

$d = v_{\text{average}} \times t = 2.55 \text{ m/s} \times 1.3 \text{ s} = 3.315 \text{ m}$.

9.c)

The acceleration of the player's slide

$a = \Delta v/\Delta t = (v_f - v_i)/\Delta t$

$-3 \text{ m/s}^2 = (v_f - 4.5 \text{ m/s})/(1.1 \text{ s})$

$v_f = 1.2 \text{ m/s}$.

9.d)

When the runner slides for 1.1 s, she is running for 0.2 s more than her 1.3 s slide. She has a higher average speed for that time than if she were sliding and slowing down. Also, when she slides for 1.1 s, her average speed during the slide is larger because she doesn't slow down as much. Because her average speed is higher for both parts of the trip, she must arrive in less time than when she slides for 1.3 s.

CHAPTER 1

NOTES

NOTES

10. Suppose an astronaut on an airless planet is trying to determine the acceleration of an object that is falling toward the ground. She has a motion detector in place that records the graph to the right for the falling object until just before it strikes the ground.

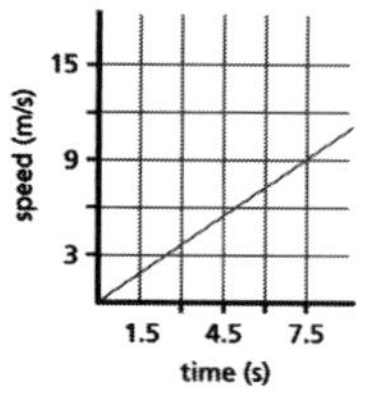

a) From the graph, approximately what was the top speed recorded by the astronaut for the falling object?

b) What is the acceleration of gravity on this planet?

c) If the astronaut had dropped the object from a greater height, what would happen to the object's acceleration as it falls and the object's final velocity before striking the ground?

11. A boy riding a bike with a speed of 5 m/s across level ground comes to a small hill with a constant slope and lets the bike coast up the hill. All graphs have time on the *x*-axis.

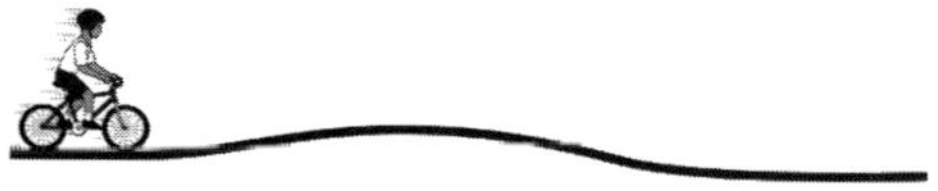

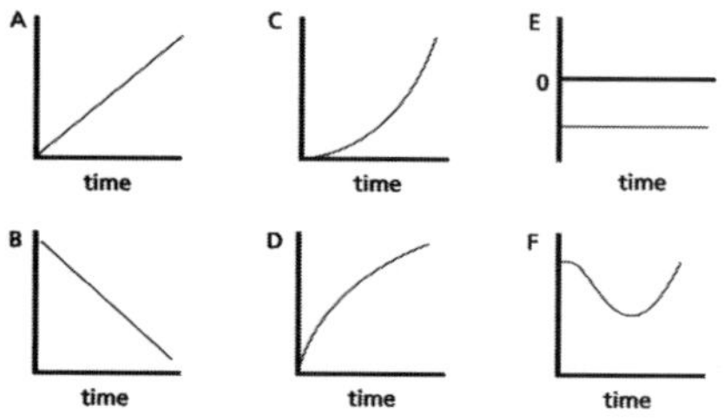

a) Which of the graphs would correctly show the boy's velocity versus time as he coasts up the hill?

b) Which graph shows the distance traveled versus time as he coasts up the hill?

c) Which graph would show the bike's acceleration as it coasts uphill?

d) Which graph shows after reaching the top of the hill, the speed of the boy as he coasts down the hill on the bike?

e) Which graph could show the boy's speed versus time graph as the boy coasts up the hill and then down the hill?

f) Starting from the top of the hill, which graph could correctly show the boy's distance vs. time as he goes down the hill?

11.a)

Graph B would show how the velocity decreases uniformly with time as the bike goes up the hill.

11.b)

Graph D would show how the distance covered is decreasing as the bike's velocity is decreasing as it coasts up the hill.

11.c)

Graph E would indicate a constant negative acceleration as the bike slows down.

11.d)

Graph A shows the speed of the bike increasing as it goes down the hill.

11.e)

Graph F shows decreasing speed as the bike goes up the hill, and increasing speed as the bike goes down the hill.

11.f)

Graph C shows the boy's distance vs. time as he accelerates going down the hill.

10.a)

From the endpoint of the graph, the highest speed would be approximately 15 m/s.

10.b)

Using values from the graph, $v_i = 0$ m/s and $v_f = 9$ m/s.

The change in time for this increase in velocity was 7.5 s.

Using the acceleration formula

$a = \Delta v / \Delta t$

$a = (9 \text{ m/s} - 0 \text{ m/s}) / 7.5 \text{ s} = 1.2 \text{ m/s}^2$.

10.c)

Because the slope of the line is a constant, the acceleration is constant. When the object falls a larger distance the acceleration remains the same, but the final velocity increases.

12.a)

Segment c-d has a constant slope on a *d-t* graph, indicating constant speed.

12.b)

Segments a-b and e-f are showing increasing speed, although e-f shows the object moving back toward the starting point.

12.c)

Segment d-e shows the object at rest.

12.d)

Segments b-c and f-g show decreasing speed.

12.e)

The automobile travels a total distance of 1200 m—600 m away and 600 m returning to the start (0 m) position.

12.f)

The automobile was back at the starting point at a time, *t*, later on. Returning to the starting point means returning to the zero-meter position, but the object cannot return to the zero-time position, since that would mean traveling backward in time!

Chapter 1 Driving the Roads

12. An automobile magazine runs a performance test on a new model car, and records the graph of distance versus time as the car goes around a track. During which segment or segments of the graph is the car

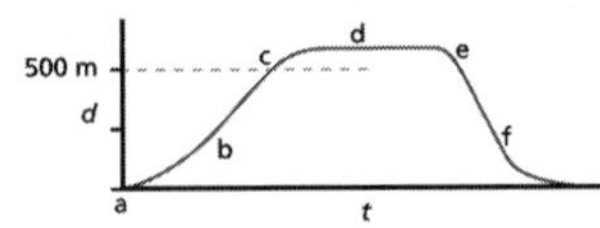

a) traveling with constant speed?
b) increasing speed?
c) at rest?
d) decreasing speed?
e) How far did the car travel during the total test?
f) According to the graph, where was the car when the test was completed?

13. A jet taking off from an aircraft carrier goes from 0 to 250 mi/hr in 30 s.
a) What is the jet's acceleration?
b) If after take-off, the jet continues to accelerate at the same rate for another 15 s, how fast will it be going at that time?
c) How much time does it take for the jet to reach 500 mi/hr?
d) How much distance would it take for that same jet to reach 500 mi/hr?

14. Whenever air resistance can be neglected or eliminated, an object in free-fall near Earth's surface accelerates vertically downward at 9.8 m/s² due to Earth's gravity. This acceleration is also called 1 *g*.
a) If the object falls for 100 m, how fast is it traveling?
b) How much time is required for it to fall this 100 m?
c) If the object falls for 10 s, how fast is it traveling?
d) How far has it fallen in this 10 s?
e) How would your answers to these questions change for an object falling above the Moon, where the acceleration is about ⅙ *g* (1.6 m/s²)?

15. In 1954, in a study of human endurance prior to the manned space program, Colonel John Paul Stapp rode a rocket-powered sled that was boosted to a speed of 632 mi/hr (1017 km/h). The sled and he were then decelerated to a stop in 1.4 s.
a) What was the acceleration of this stop?
b) What is this acceleration in terms of *g*'s?
c) In what distance did the speed of the sled travel as its speed changed from 1017 km/hr to 0?

Active Physics 70

13.a)

$a = \Delta v / \Delta t$

$a = (250 \text{ mi/h})/30 \text{ s} = 8.3 \text{ mi/h/s}$.

13.b)

$a = \Delta v / \Delta t$ or $\Delta v = a\Delta t$

$\Delta v = 8.3 \text{ mi/h/s}(15 \text{ s}) = 125 \text{ mi/h}$

Since $\Delta v = v_f - v_i$,

$125 \text{ mi/h} = v_f - 250 \text{ mi/h}$

and $v_f = 375 \text{ mi/h}$.

13.c)

$a = \Delta v / \Delta t$ or $\Delta t = \Delta v / a = (500 \text{ mi/h})/(8.33 \text{ mi/h/s}) = 60 \text{ s}$.

13.d)

$v_f^2 = v_i^2 + 2ad$

$(500 \text{ mi/h})^2 = 0^2 + 2(8.3 \text{ mi/h/s})d$

$d = 1.5 \times 10^4 (\text{mi/h})(\text{s})$.

Converting seconds to hours using 1 h/3600 s gives

$d = 1.5 \times 10^4 (\text{mi/h})(\text{s}) \times (1 \text{ h}/3600 \text{ s}) = 4.2 \text{ mi}$.

14.a)

$v_f^2 = 2ad$,

$v_f^2 = 2(9.8 \text{ m/s}^2)(100 \text{ m}) = 1960 \text{ m}^2/\text{s}^2$ and $v_f = 44.3 \text{ m/s}$.

Section 4 Graphing Motion: Distance, Velocity, and Acceleration

16. Active Physics Plus An automobile accelerates from rest at 4.0 m/s every second (4.0 m/s^2).

a) How far does it travel after 1.0 s?

b) How far does it travel after 2.0 s?

c) How far does it travel after 3.0 s?

d) How far does it travel after 4.0 s?

e) Complete a *d-t* graph for this automobile.

f) Complete a *v-t* graph for this automobile.

g) How does the motion of this automobile compare with the motion of a real automobile (as you investigated previously)?

17. ***Preparing for the Chapter Challenge***

On highways, you can pass slower-moving vehicles by moving into the left lane and driving past them. You can then return to the right lane, all the while traveling at the speed limit. On a rural road, you must do this by entering the oncoming traffic lane. This can be very dangerous. To pass the vehicle safely and quickly, you may have to accelerate until you get back into your lane. You can describe the motion of both vehicles by creating graphs with two lines on each one—one depicting your vehicle and the other depicting the slower-moving vehicle. Describe how you can safely pass a slower-moving vehicle using *d-t*, *v-t*, and *a-t* graphs to convince the driving academy that you understand safe driving.

Inquiring Further

Speed conversions

Using a spreadsheet program, complete a table like the one below that converts mph to ft/s to m/s to km/hr.

	A	B	C	D
1	mi/hr	ft/s	m/s	km/h
2	0			
3	5			
4	10			
5	15			
6	20			
7	25			
8	30			
9	35			
10	40			
11	45			
12	50			

71 Active Physics

14.b)

$d = \frac{1}{2}a\left(t^2\right)$

$t = \left(2d/a\right)^{1/2} =$

$\left[\left(2 \times 100 \text{ m}\right)/\left(9.8 \text{ m/s}^2\right)\right]^{1/2} = 4.5 \text{ s}.$

14.c)

$v_f = v_i + at;\ v_f =$

$0 \text{ m/s} + \left(9.8 \text{ m/s}^2\right)\left(10 \text{ s}\right) = 98 \text{ m/s}.$

14.d)

$d = \frac{1}{2}a\left(t^2\right)$

$d = \frac{1}{2}\left(9.8 \text{ m/s}^2\right)\left(10 \text{ s}\right)^2 = 490 \text{ m}.$

14.e)

On the Moon, where the acceleration due to gravity is 1/6 g, the answers would be 18 m/s, 11.1 s, 16.3 m/s, and 81.7 m.

15.a)

Converting 1012 km/hr to m/s gives you (1,017,000 m/h)(1 h/3600 s) = 282.5 m/s.
Using $a = (\Delta v)/(\Delta t) =$ (0 m/s – 282.5 m/s)/(1.4 s) = –201.8 m/s^2.

15.b)

To get the acceleration in *g*'s, divide the sled's acceleration (–201.8 m/s^2 by the acceleration of gravity (9.8 m/s^2) to get 20.6 *g*'s.

15.c)

$v_f^2 = v_i^2 + 2ad$ and solving for d gives $d = \left(v_f^2 - v_i^2\right)/\left(2a\right) =$ (0 m/s – 282.5 m/s)/(1.4 s) = –201.8 m/s^2.

16.a)

$v_{\text{average}} = \left(v_i + v_f\right)/2$

$v_{\text{average}} = \left(4 \text{ m/s} + 0 \text{ m/s}\right)/2 =$ 2 m/s, and $d = v_{\text{average}}\left(t\right) = \left(2 \text{ m/s}\right)\left(1 \text{ s}\right) = 2 \text{ m}.$

16.b-d)

Similarly, distances for 2, 3, and 4 would be 8 m, 18 m and 32 m.

16.e)

The *d-t* graph would look like the one below.

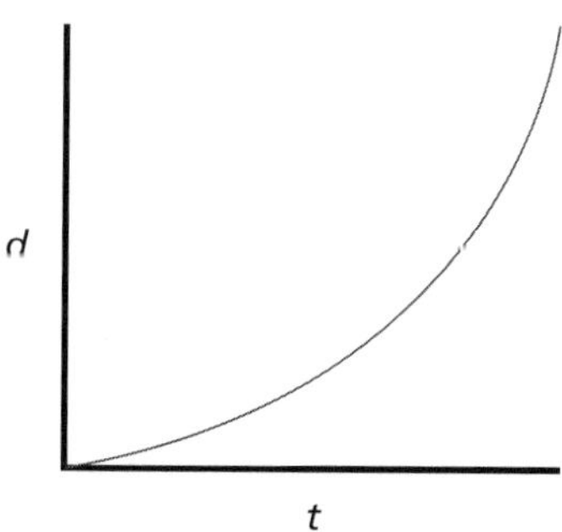

16.f)

The v-t graph would appear as shown below.

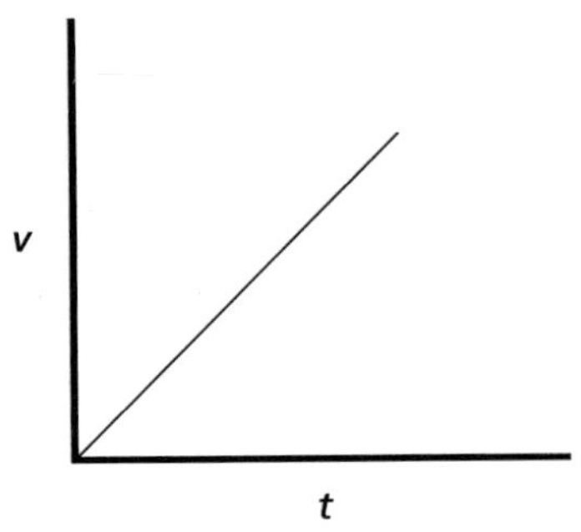

16.g)

The relative motion of this automobile (acceleration of 4 m/s^2) to that of the real automobile shown at the end of *Section 7* is mainly a comparison of accelerations. The "Touring Sedan" has a listed speed of 35 mi/h (approximately 16 m/s) after 2 seconds of acceleration or an acceleration of 8 m/s^2 (8 m/s divided by 2 s). This is significantly higher than the vehicle in *Question 16*. However, by the time the "Touring Sedan" has reached a speed of 60 mi/h, (approximately 27 m/s) in 7.2 seconds, the average acceleration has dropped to 3.7 m/s^2 (27 m/s divided by 7.2 s). This is very close to the value given for the theoretical automobile in the question.

17.

Preparing for the Chapter Challenge

The *Preparing for the Chapter Challenge* graphs should appear as follows:

The d-t graph shows car A will start out initially behind car B, and the two cars have matching velocities as seen from the equal slopes of the lines. At point X, car A starts to accelerate to a higher speed, passing car B at point Y, where each car's distance from the start is equal, and continues at a greater speed to point Z. At point Z, car A begins to slow down to again match the velocity (slope of the line) of car B, but now car A is ahead of car B as shown by its higher position along the distance axis.

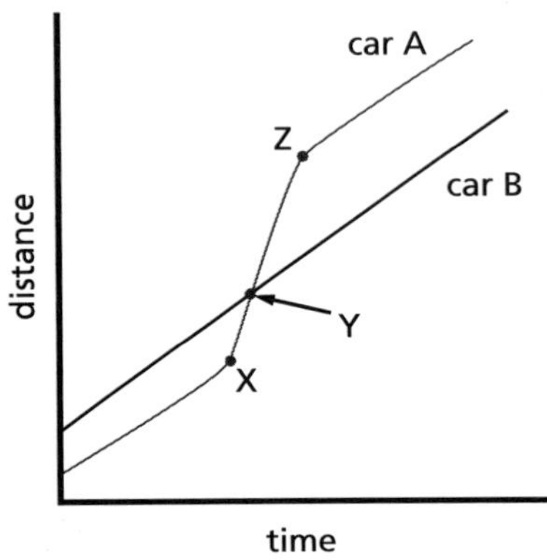

The same process as described in the previous paragraph is also shown on the v-t graph below. Car A increases velocity at point X until it is traveling faster than car B. At some point, where the area under the thin line matches the area under the thick line, near point Y, car A passes car B. Car A continues at a higher speed moving ahead of car B, and then slows down during interval Z. After interval Z, the cars again match speeds, with car A ahead.

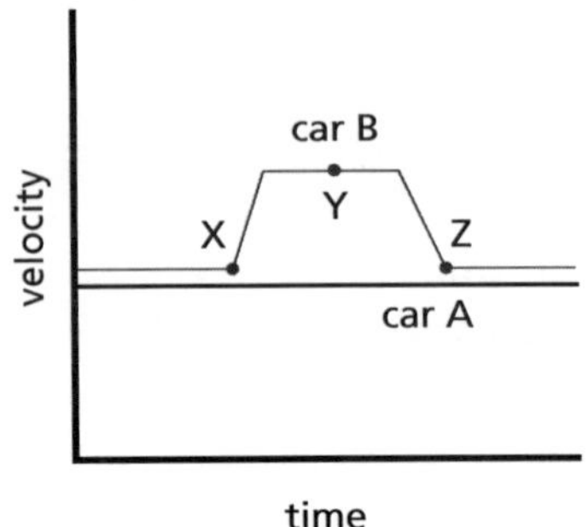

Finally, the a-t graph below shows the cars both traveling with zero acceleration (constant velocity) until car A accelerates at point X to pass car B. Car A then returns to a constant velocity higher than car B, but also zero acceleration, while passing through region Y. After passing car B, car A slows down (negative acceleration) at point Z, until it is again traveling with constant speed.

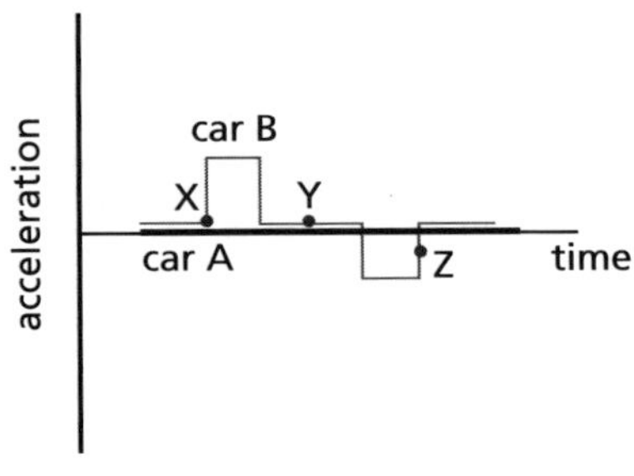

Inquiring Further

The students' table should be the same as follows.

mi/h	ft/s	m/s	km/h
0	0.0	0	0
5	7.3	2.2	8.1
10	14.7	4.5	16.3
15	22.0	6.7	24.4
20	29.3	9.0	32.6
25	36.7	11.2	40.7
30	44.0	13.4	48.8
35	51.3	15.7	57.0
40	58.7	17.9	65.1
45	66.0	20.2	73.3
50	73.3	22.4	81.4

NOTES

SECTION 4 QUIZ

1-4d **Blackline Master**

1. If an automobile's velocity changes from 25 m/s to 15 m/s in 2 s, then what is its acceleration?

   a) $-5\ m/s^2$ b) $7.5\ m/s^2$

   c) $13\ m/s^2$ d) $20\ m/s^2$

2. A strobe photograph is taken of four automobiles undergoing different motions. Which diagram below represents an automobile undergoing constant acceleration?

   a)

   b)

   c)

   d)

3. The graph to the right shows the *d-t* graph for an automobile. According to the graph, the object is

   a) traveling with a constant speed.

   b) undergoing constant acceleration.

   c) traveling with a decreasing velocity.

   d) slowing down and then speeding up.

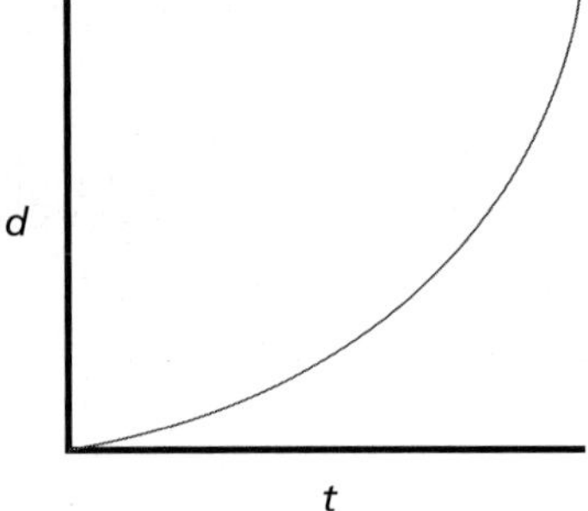

4. A student completes the following two experiments with a cart on an inclined plane.

   Experiment 1: A cart is given a push up an inclined plane and released.

   Experiment 2: A cart is released from the top of an inclined plane.

   Which statement below best describes the velocity and acceleration of the cart in these experiments?

   a) In both experiments, the acceleration is down the plane.

   b) In both experiments, the velocity is down the plane.

   c) In both experiments, the velocity is up the plane and the acceleration is down the plane.

   d) In both experiments, the velocity is down the plane and the acceleration is up the plane.

5. The graph represents the relationship between speed and time for an automobile moving in a straight line. Using any 1-s interval, find the automobile's acceleration.

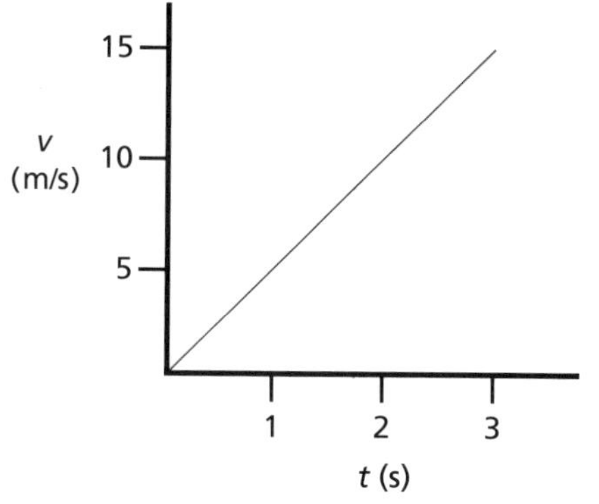

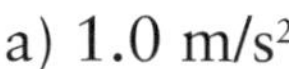

a) 1.0 m/s^2

b) 0.0 m/s^2

c) 10 m/s^2

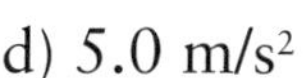

d) 5.0 m/s^2

SECTION 4 QUIZ ANSWERS

1 a) The acceleration is given by $a = \Delta v/\Delta t$. The velocity changes from 25 to 15 (a change of 10 m/s) and the time change is 2 s, giving an acceleration of – 5 m/s^2. Choice *d)* comes from adding the velocities rather than subtracting them, Choice *b)* would be from using the average of the velocities, and Choice *c)* has no justification.

2 b) During acceleration the distance between the images of the automobile would increase at a constant rate. Choice *a)* is constant velocity, Choice *d)* is constant velocity then slowing, and Choice *c)* is random velocities.

3 b) For constant speed, the *d-t* graph would be a straight line. An object that is slowing down would be curved in the opposite direction, and a car that is speeding up and slowing down would be a combination of the constant acceleration graph and one for slowing down.

4 a) The students observed this in the *Investigate*. In both cases, the cart's acceleration is down the plane, although the velocities are in different directions.

5 d) The students should use the values from the graph to calculate the acceleration. For example, using $a = \Delta v/\Delta t$ gives (10 m/s – 5 m/s)/1 s = 5.0 m/s^2. Choice *c)* comes from just using the top speed of 10 m/s for the acceleration, Choice *b)* is unrelated, and Choice *a)* may come from confusing a line at an angle of 45 degrees as always having a slope of one. In this case, since the axes are not equal the 45-degree rule does not hold.

Chapter Mini-Challenge

In each section, students learned new concepts. Now they can recall and review what they have learned and read the *Chapter Mini-Challenge* to address the different phases of the *Engineering Design Cycle*. This is an opportunity for them to reconstruct their learning to meet their *Goal*. Facilitate this process by asking students to reflect on the physics principles of each section. Model a concept map that makes connections between concepts students have learned and the criteria required to complete their *Chapter Challenge*. As students, carry out the process of linking information, point out to them that the *Process* phase is when they decide what information they have that best meets the criteria of their *Goal*. Emphasize that this is the stage of resource analysis, in which students determine what they need to learn to complete their *Chapter Challenge*.

In preparing for their *Mini-Challenge*, students get a chance to work together and share ideas with their class. The purpose behind the *Feedback* phase is to allow time for reflection and explanation. In writing down a first draft of their two-minute presentation, students realize the advantages of sharing information. During this phase of the design cycle, each student should be expected to write two positive comments besides suggestions for improvement.

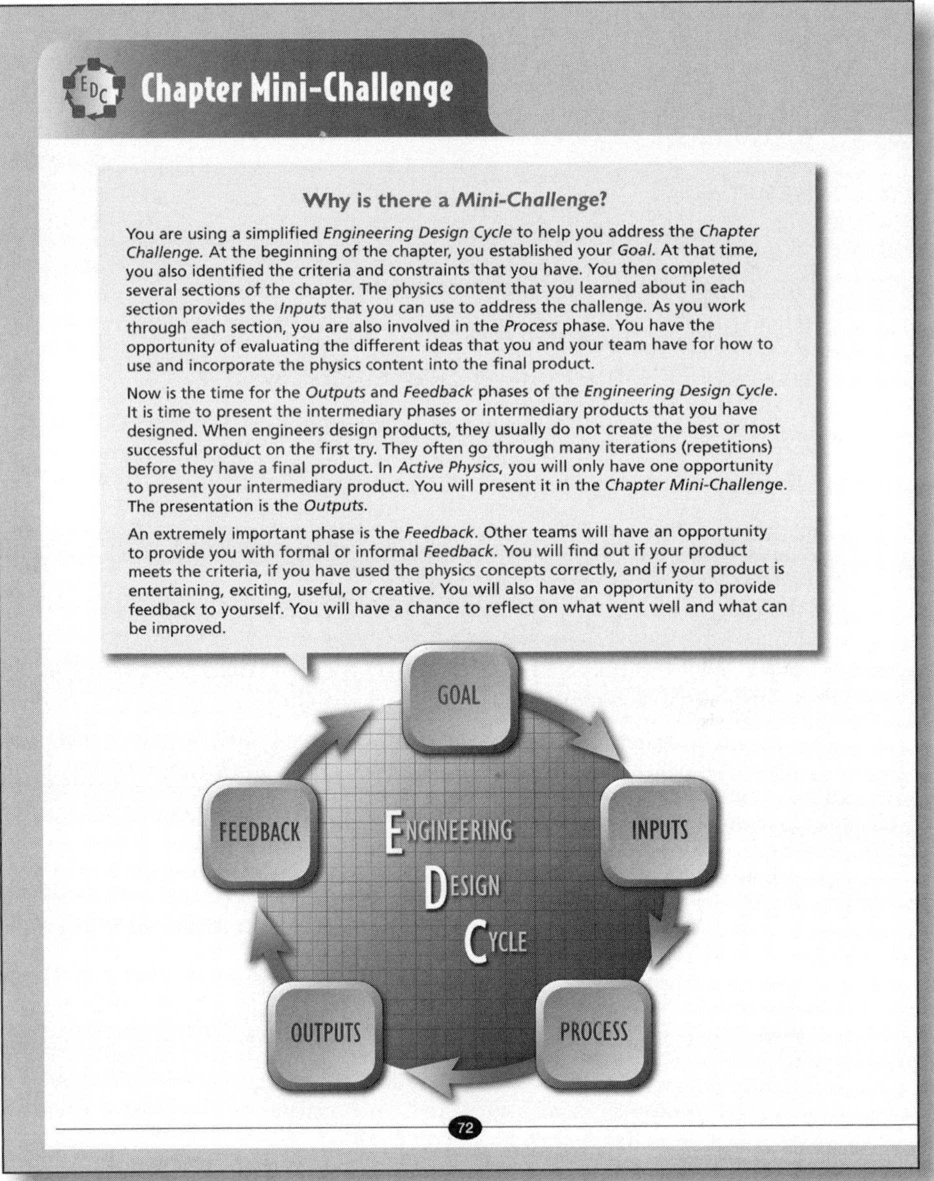

Chapter Mini-Challenge

Why is there a *Mini-Challenge*?

You are using a simplified *Engineering Design Cycle* to help you address the *Chapter Challenge*. At the beginning of the chapter, you established your *Goal*. At that time, you also identified the criteria and constraints that you have. You then completed several sections of the chapter. The physics content that you learned about in each section provides the *Inputs* that you can use to address the challenge. As you work through each section, you are also involved in the *Process* phase. You have the opportunity of evaluating the different ideas that you and your team have for how to use and incorporate the physics content into the final product.

Now is the time for the *Outputs* and *Feedback* phases of the *Engineering Design Cycle*. It is time to present the intermediary phases or intermediary products that you have designed. When engineers design products, they usually do not create the best or most successful product on the first try. They often go through many iterations (repetitions) before they have a final product. In *Active Physics*, you will only have one opportunity to present your intermediary product. You will present it in the *Chapter Mini-Challenge*. The presentation is the *Outputs*.

An extremely important phase is the *Feedback*. Other teams will have an opportunity to provide you with formal or informal *Feedback*. You will find out if your product meets the criteria, if you have used the physics concepts correctly, and if your product is entertaining, exciting, useful, or creative. You will also have an opportunity to provide feedback to yourself. You will have a chance to reflect on what went well and what can be improved.

72

The *Mini-Challenge* is a first attempt at the culminating task. It is the time for students to informally test their understanding of physics principles. For instance, in designing a presentation that highlights the relationship between following distance, braking distance, and the total stopping distance, students will have the opportunity to explain how these principles are a part of safe driving. Their confidence will increase when they are successful in demonstrating their knowledge of the physics of driving, such as why the GO Zone is affected by the yellow-light time, or how road friction and radius of the curve determine the safe speed for negotiating a curve.

Your challenge for this chapter is to create a two-to three-minute group presentation that will convince the Active Driving Academy that you have learned enough about the physics of safe driving to be eligible for graduation. Your group needs to apply the physics concepts you have studied in this chapter to develop that presentation. Your presentation must, at the minimum, include:

- **the relationship between following distance, braking distance, and the total stopping distance, including the factors that affect each;**
- **how to decide what to do when the light turns yellow as you approach an intersection; and**
- **the connection between speed, friction, and radius of the curve when turning.**

Your *Mini-Challenge* for this chapter is a one-to two-minute presentation to the class. The *Mini-Challenge* will help you learn about what you should or should not include in your *Chapter Challenge* presentation. You will not be able to address all of the requirements at this time, but you should do your best to fully address the topics that you have already studied. Anything you create for the *Mini-Challenge* can be used to complete your final *Chapter Challenge*.

Look back at the *Goal* you wrote at the beginning of the chapter. Rewrite your *Goal* so that you are clear on what you will prepare for the *Mini-Challenge*. Review the *Goal* as a class to make sure you have all of the criteria and the necessary constraints.

For the *Inputs* phase of the *Engineering Design Cycle*, you have completed four sections and learned some of the physics content that can be used in your presentation. Your group should review the physics content from these sections to help you begin your safe-driving presentation.

Section 1 You used different methods to measure reaction time and compared the reaction time of different members of your class. You also investigated how distractions affect reaction time.

Section 2 You used a stride and a meter stick to measure distance. You identified the sources of error in measurement and read about units of measurement used in science classrooms and when driving the roads in the United States.

Section 3 You defined average and instantaneous speed and used strobe pictures, graphs, and equations to represent motion. You also used the equation for average speed to calculate speed, distance, and time. You read about how speed can affect following distance when driving on the roads.

Section 4 You learned how changes in speed, direction or acceleration are related to time and distance for a moving vehicle. You also interpreted distance-time and velocity-time graphs for different types of motion.

In addition to the information you learned in the first four sections, your group might also like to look for some statistics about safe stopping distances, safe following distances, and safe braking distances as they relate to vehicle accidents. This information would help make your physics information more impressive to the presentation audience.

73

Chapter Mini-Challenge

The *Process* phase is when you decide what information you have that will help you meet the criteria of the *Goal*. This *Mini-Challenge* requires a thorough evaluation of the physics you have learned so far to help you determine how each piece influences the aspects of time and distance as they relate to driving. You can perform a *Resource Analysis* by creating a list of what you learned in the first four sections of the chapter. For each one, decide if:

- **it can be used to help you measure or determine a safe following distance;**
- **it can be used to help find a braking distance, and**
- **it can be used to find a stopping distance.**

By categorizing the information you already have learned, you can focus your energy on addressing the parts of the challenge that you are prepared to answer at this point.

Your *Resource Analysis* has revealed which of the topics in the first four sections will be helpful for answering each part of the challenge presentation. Your group might assign individuals or teams of two to work on specific answers for your presentation and then put all of the individual answers together to present later. Each person or team will now know which chapter section or sections they can use to help him/her address their part of the presentation.

During your *Resource Analysis* you can also make a list of missing information you still need to complete the *Chapter Challenge*. This list will help you complete the final sections of the chapter and frame your answers to the *Physics Essential Questions*.

The *Outputs* of your *Engineering Design Cycle* is a two-minute presentation. Remember, everyone is working with the same requirements and constraints. You only need to do a good job of meeting the *Goal* requirements to do well.

Presenting your information to the class are your design-cycle *Outputs*. You should have a presentation that addresses factors for safe following distances, safe braking distances, and safe stopping distances. Remember to use graphs or charts from the investigations to help illustrate your explanations. Remember to leave enough time for your group to rehearse or at least agree on who will present which information.

Finally, you will receive *Feedback* from your classmates that will tell you what you have done well according to the criteria from the *Goal*. They might also tell you some things you can improve to get a good grade on the final presentation. To give good *Feedback*, it is important to consider each point of the requirements and the constraints to see how well each different design satisfies them. Your statements should say which parts were satisfied and which, if any, were not. This is an objective process and should focus on the products, not on the engineers who produced them.

This *Feedback* will become an *Input* for your final design in the *Chapter Challenge*. You will have enough time to make corrections and improvements, so you will want to pay attention to the valuable information they provide. Remember to correct any parts of your explanations that you received critical feedback on. You may have also learned something from watching presentations that you want to add to your group's presentation. It will be easier and faster to improve your answers now rather than waiting until the chapter is complete to go back and correct any mistakes. Then, store all of your information in a safe place so that it will be ready to use in the *Chapter Challenge*!

74

NOTES

SECTION 5

Negative Acceleration: Braking Your Automobile

Section Overview

This section addresses the distance traveled during negative acceleration as a function of the initial speed. Students continue the cart-stopping investigation by analyzing the problem of "slowing down" a cart. Students collect data by setting up an experiment in which a toy cart is rolled down a ramp. The initial speed of the cart versus its braking distance is plotted on a graph. The change in speed of the cart as it comes to a stop is explained by using the term negative acceleration. The effect of increasing the initial speed is recorded as data and then plotted on a speed versus time graph. Students study the similarities and differences in their data to find the relationship of initial speed to the braking distance. Braking distance is finally defined and determined to be proportional to the initial velocity squared.

Background Information

The physics of stopping a cart involves reaction time while moving with constant or uniform motion, and braking time while braking with changing speed or nonuniform motion. If students plot the data for distance versus time, for nonuniform motion, they get a parabola.

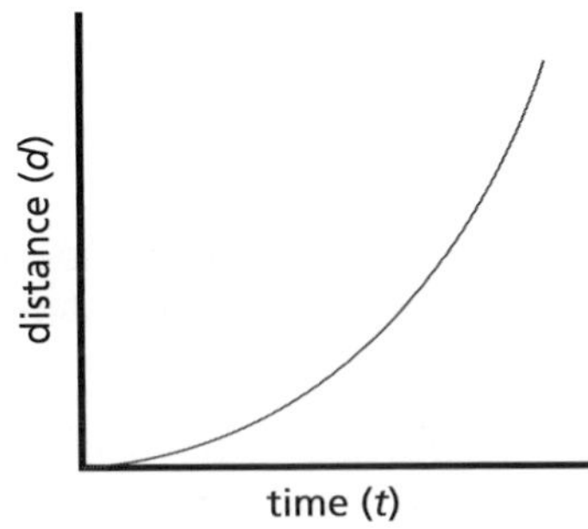

Unfortunately, reading and making predictions from a parabola is very difficult. Students would require calculus to make sense of such a graph. Changing speed or acceleration is the change in the speed of an object in a given time period. Acceleration, $a = \Delta v/\Delta t$ (Δv represents the change in velocity that is measured by recording velocity at two different periods). The initial velocity is often indicated as v_1 and final velocity as v_2. In this section, the students will be solving for the distance required to stop, while traveling at different velocities.

Plotting speed vs. time, results in a straight line, with the slope of that line indicating acceleration.

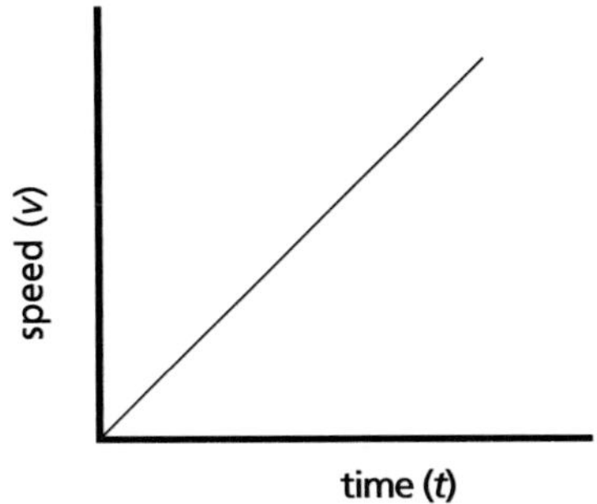

$a = \Delta v/\Delta t$

The unit for v is m/s and for t is s.

Therefore,

$$a = \frac{\Delta v}{\Delta t}$$

and with units,

$$a = \frac{\text{m/s}}{\text{s}} = \text{m/s}^2$$

However, when looking at braking distances, distance traveled is the primary concern. For example, knowing the distance a plane requires to land is important when building a runway. It is also important to know how far an automobile will travel while braking. Therefore, the distance that is required when an automobile is slowing down is considered. From a speed-time graph, finding the area under the graph will give the total distance covered.

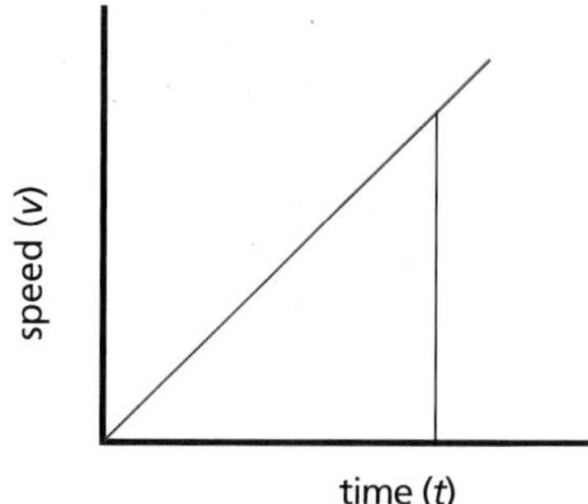

The area under the graph above is expressed as Area = $\frac{1}{2}b \times h$. Using the units from the graph, $d = vt/2$. Going back to the original equation of acceleration, $a = v/t$, and rearranging to solve for t gives $t = v/a$. Combining $d = vt/2$, and gives $d = v^2/2a$, which is the distance an object travels proportional to the square of its speed. Therefore, when the speed the object is traveling is doubled, the distance traveled is quadrupled.

EXAMPLE

An object is traveling at 3 m/s with a negative acceleration of 2 m/s², what is the distance traveled?

Negative acceleration means that the car is slowing down. Therefore,

$$d = \frac{v^2}{2a}$$

or ,

$$d = \frac{(3 \text{ m/s})^2}{2(2 \text{ m/s}^2)} = 2.25 \text{ m.}$$

Now suppose the speed is doubled to 6 m/s,

$$d = \frac{(6 \text{ m/s})^2}{2(2 \text{ m/s}^2)} = 9 \text{ m.}$$

The kinematics of the relationship which relates braking distance and speed is $v^2 = 2ad$.

In this section, expect students to recognize that v^2 is proportional to distance. This can be best understood by looking at changes in the braking distance as it relates to changes in the initial velocity. If the velocity doubles, the braking distance quadruples. If the velocity triples, the braking distance is nine times as great. If the velocity quadruples, the braking distance is sixteen times longer. Similarly, if the velocity is halved, the braking distance is quartered.

This equation can be derived from the definition of velocity and acceleration.

$$a = \frac{\Delta v}{\Delta t} = \frac{v_f - v_i}{t}$$

$$v_f = at + v_i$$

Since

$$v_f = 0 \text{ and } v_i = -at$$

$$v_i^2 = a^2t^2 = \left(\tfrac{1}{2}at^2\right)2a = 2ad.$$

In addition, this section pays particular attention to the concept of negative acceleration, both as an indication of decreasing speed, and also in the context of the vector nature of acceleration. Rather than use the term "deceleration," the term negative acceleration is used.

To clarify the physicist's use of the term negative acceleration, it is necessary to deal with the vector nature of acceleration. Acceleration may occur in any direction. If one direction, for example, east, is arbitrarily chosen to be positive, an increasing speed in the easterly direction would be a positive acceleration, while a decreasing speed in that direction would be considered negative. An increase in speed toward the west (the negative direction) would then be a negative acceleration. If the speed increases, it becomes more negative. If the speed decreases in the negative direction, it gradually becomes less negative and ultimately changes to a positive velocity, indicating a positive acceleration.

As an enrichment activity, the students can run the same investigation on different surfaces. They could then draw analogies to stopping on different road surfaces and how they would drive under those circumstances. An example might be running the cart on sandpaper, wet floor, dirt surface, etc.

Crucial Physics

- Braking distance is dependent on the negative acceleration of the car (brakes, road surface) and reaction time.
  - $d = v^2/2a$.
  - Tripling the speed requires nine times the braking distance.
- Positive acceleration and negative acceleration are not associated uniquely with speeding up or slowing down.
  - One determines if an object speeds up or slows down by considering both the direction of the initial velocity and the direction of the acceleration.

Learning Outcomes	Location in the Section	Evidence of Understanding
Plan and carry out an experiment to relate braking distance to initial speed.	***Investigate*** Steps 2, 3, and 4	Students plan and carry out an experiment to see how the initial speed of a cart affects its braking distance by varying the speed of a cart traveling down a ramp.
Determine braking distance.	***Investigate*** Step 5	Students use data to determine the braking distance of a cart.
Examine accelerated motion.	***Investigate*** Steps 5–8	Students study graphs to see how varying the initial speed changes the braking distance.

NOTES

Section 5 Materials, Preparation, and Safety

CHAPTER 1

Materials and Equipment

PLAN A		
Materials and Equipment	**Group (4 students)**	**Class**
Meter stick, wood	1 per group	
Ring stand, large	1 per group	
Rod, aluminum, 12 in. (length) x 3/8 in. (diameter) (to act as crossarm)	1 per group	
Holder, right angle (to act as crossarm)	1 per group	
Inclined plane ramp for lab cart	1 per group	
Dynamics cart	1 per group	
Scissors	1 per group	
Extension clamp	2 per group	
Velocimeter	1 per group	
Index cards, pkg 100		1 per class
Folder, file		10 per class
Paper, graph, pkg of 50		1 per class

*Additional items needed not supplied

PLAN B		
Materials and Equipment	**Group (4 students)**	**Class**
Meter stick, wood		1 per class
Ring stand, large		1 per class
Rod, aluminum, 12 in. (length) x 3/8 in. (diameter) (to act as crossarm)		1 per class
Holder, right angle (to act as crossarm)		1 per class
Inclined plane ramp for lab cart		1 per class
Dynamics cart		1 per class
Scissors		1 per class
Extension clamp		2 per class
Velocimeter		1 per class
Index cards, pkg 100		1 per class
Folder, file		10 per class
Paper, graph, pkg of 50		1 per class

*Additional items needed not supplied

Note: Time, Preparation, and Safety requirements are based on Plan A, if using Plan B, please adjust accordingly.

Time Requirement

In order for students to run 10 good trials, after proper instruction, they would need at least 45 min. (If there is only one set of equipment for the entire class, then adjust the appropriate amount of time accordingly.) Allow for one more class period to plot the data and analyze the graph.

Teacher Preparation

- Determine the speed-measuring equipment to be used. If a velocimeter or a motion detector is not available, you may use a camcorder, monitor, and a VCR with stop, forward, and advance to show a clear picture of each frozen frame. Practice using the camcorder if you are not familiar with it. Be ready to advise students on how to use the freeze frame advance capability of the tape player or camcorder. Secure a suitable scale behind the path of the cart so that its position is easily observed on the screen.
- If no method is available to measure the speed at the bottom of the ramp, measure the time of descent from rest to the bottom of the ramp. Calculate the average speed on the ramp and double it to get the speed at the bottom. While this method avoids much apparatus, it requires them to trust and understand the mathematics. They are more likely to trust the technology that gives a direct measurement. Be sure that the transition from the ramp to the floor is smooth. A file folder taped to the inclined plane and the surface will suffice.
- If the *Investigate* is to be performed on the floor, inform students the day prior so they may wear appropriate clothing.

Safety Requirements

- If the *Investigate* is done on the floor, students should be aware of the equipment so they do not walk into it. The carts rolling along the floor may present a tripping hazard, and should be collected and returned to the inclined plane as soon as the braking distance has been measured.
- If the *Investigate* is done in a hallway, notify other local teachers that it will be occurring. All the equipment must be picked up and safely stored prior to a change of class.

Meeting the Needs of All Students

Differentiated Instruction: Augmentation and Accommodations

<table>
<tr><th>Learning Issue</th><th>Reference</th><th>Augmentation and Accommodations</th></tr>
<tr><td>Planning an experiment to collect data</td><td>Investigate
Step 2</td><td>Augmentation
• If students are unable to come up with a plan, consider allowing students to ask yes/no questions for feedback.
• Students with reading and processing issues may struggle to organize the information to answer the list of eight planning questions. Provide a worksheet with the experiment's planning questions to give students a place to record their answers.
Accommodation
• Model an experiment setup that could work. Then, as a group, go through the list of planning questions. Students will still have ownership for planning the experiment, but modeling the setup gives students a visual to guide their thinking.</td></tr>
<tr><td>Recording data in a table</td><td>Investigate
Step 4.a)</td><td>Augmentation
• Students are asked to draw a table, to record times and braking distance, and to calculate initial speeds. Students with visual-spatial and graphomotor issues struggle to create tables, especially without a model. Model how to create a four-column table as shown below with a row for each trial. Pair auditory, step-by-step directions with visual cues, and check to see that student tables are adequate for recording data.
<table>
<tr><th>Trial
(n)</th><th>Time
(s)</th><th>Braking Distance
(m)</th><th>Initial Speed
(v_i)</th></tr>
<tr><td></td><td></td><td></td><td></td></tr>
</table>
Accommodation
• Give students a blank table to complete.</td></tr>
<tr><td>Drawing a graph using data from a table</td><td>Investigate
Step 5

Physics to Go
Questions 1 and 6</td><td>Augmentation
• Students struggle to label the correct axes on a graph and to set up reasonable scales on each axis. Check in with students to make sure that initial speed is labeled on the x-axis and braking distance is labeled on the y-axis. Help students recognize the pattern of scales for initial speed and braking distance.
• Review how to plot points on a graph. Provide students with a ruler or index card to aid in the tracking required to plot points on a graph.
• Provide a model graph for students to reference.
Accommodation
• Provide students with a graph that already has labels and scales. Ask them to sketch the lines and check for understanding.</td></tr>
<tr><td>Flipping pages to locate information on a table</td><td>Investigate
Step 8

Physics to Go
Questions 5–7</td><td>Augmentation
• Students with visual-spatial and memory issues have trouble flipping between pages to locate and record information from a table. Provide students with a ruler or index card to mark the page with the table. The ruler or index card can also be used to help students visually scan columns and rows to find information on a table.
Accommodation
• Students will be more successful if they can look at a table or graph and the corresponding questions side-by-side. Provide a copy of any table(s) not located on the same page as the questions. When writing exams, make sure tables and graphs are located on the same page as the corresponding questions.</td></tr>
</table>

Learning Issue	Reference	Augmentation and Accommodations
Understanding positive and negative relative to direction (vectors)	***Physics Talk***	**Augmentation** • Students struggle to understand positive and negative numbers as a general math concept. Adding directions that can be positive and negative is even more confusing. Use the sketch provided in *Physics Talk* to provide direct instruction about this topic.
Vocabulary	***Physics Talk*** Calculating Braking Distance	**Augmentation** • Many students do not understand what "derive" means. Explain the meaning of "derive" and then complete the derivation described in this section as a group. Students may not understand every step of this process, but they will have a much better understanding of what it means to derive an equation.
Choosing correct equation to solve a problem	***Physics to Go*** Question 4	**Augmentation** • Students have learned many motion equations in this chapter, and students with problem-solving or memory issues struggle to figure out which equation to use. Model a problem-solving method or graphic organizer such as the problem-solving box described in the Accommodation for *Section 3*. Make sure the directions are sequential and pair a visual model with auditory cues. Instruct students to identify what question the problem is asking and what information is given. Then model a think-aloud to show students how to choose the correct equation. **Accommodation** • Limit choice to two equations for students who are unable to master this skill with more equations. Increase number of equations from which to choose as the skill is mastered. • Provide a step-by-step checklist with the directions for the problem-solving method. • Provide student with blank problem-solving boxes.

Strategies for Students with Limited English-Language Proficiency

Learning Issue	Reference	Augmentation
Vocabulary comprehension	***Investigate*** Step 2	Students need to have a working definition of "friction": a contact force that opposes the motion of an object. In an automobile, brakes use friction to reduce speed; there is rolling friction between the tires and the road, and there is friction acting on the spinning axle. Check for understanding by asking students where the friction forces act on the cart.
Vocabulary comprehension	***Investigate*** Step 2	Help ELL students with the contextual usage of "trials," meaning "repetitions" or "attempts." In other words, how many times will you do the experiment at each initial speed? Discuss reasons why a single trial at a given initial speed might not give reliable data.
Comprehension	***Investigate*** Step 3	To encourage practice in writing, ask an ELL student from each group to create the flowchart or outline. Check the work for understanding.
Vocabulary comprehension	***Investigate*** Step 6	Have students attempt to determine the meaning of "corresponds to" from context: "goes with" or "matches."
Vocabulary comprehension Answering higher order questions	***Investigate*** Steps 6.a), 7.a)	Be sure students understand the terms "doubling" and "tripling," and that it is clear to them that they should choose actual data. Because students were not directed to double the initial speed in the procedure, they may not have data for initial speeds that are in a 2-to-1 ratio. It may help to teach students how to decode a question with a complex sentence structure. When they read "what is the effect of doubling the initial speed on the distance traveled during braking?" they can simplify by cutting out words: "What is the effect on the distance traveled during braking?"
Understanding complex concepts	***Physics Talk, Negative Acceleration and Positive Acceleration***	Negative acceleration is a difficult concept. Sentences such as "An object could have a negative acceleration by decreasing its speed in the positive direction or increasing its speed in the negative direction" are difficult for all students to unravel. A language barrier can make the task daunting. Spend time reviewing the illustrations of positive and negative acceleration, along with their corresponding descriptions. You may wish to photocopy the page and cut apart the three diagrams illustrations and the three written descriptions, and then have students try to match each illustration with its description. Before moving on, be sure all students understand the conditions of both positive acceleration and negative acceleration. When you think students are comfortable with describing positive and negative acceleration, discuss the equations that represents each condition. Allow students sufficient time to become comfortable with integrating the information in the table.
Understanding graphs	***Physics Talk*** ***Checking Up***	There are three graphs for negative acceleration at the end of *Physics Talk*. Choose one ELL student to interpret each graph orally. You can check their understanding of the graphs and give them speaking practice as well.
Understanding equations	***Active Physics Plus*** Step 3	Check students' understanding of the five motion equations by having them state each equation in words.

NOTES

CHAPTER 1

SECTION 5

Teaching Suggestions and Sample Answers

What Do You See?

The *What Do You See?* illustration will evoke a variety of responses. Students might comment on the tilted vehicle or the unfazed moose. This will be a good opportunity for you to guide them to the topic of this section. You could ask them why the author chose to show a "near miss" collision. A review of the illustration after completing the *Investigate* and the *Physics Talk* would help students in examining how speed affects breaking distance.

What Do You Think?

After posing the *What Do You Think?* question, create a master list and allow the students to give their arguments as they try to justify some factors and dismiss others in order to avoid hitting the animal. The few students who have taken a course in driving may know that the most important factor is speed, and that distance is proportional to the square of the speed. The experiment in the *Investigate* will provide a basis for this relationship.

What Do You Think?

A Physicist's Response

The most important factors in determining if you will stop in time is speed and reaction time. The distance required to stop is proportional to the square of the speed. Other factors include: road conditions, driver's physical health, and environmental conditions.

Students' Prior Conceptions

1. **Force is a property of an object.** An object has force and when it runs out of force it stops moving. This is the traditional view of impetus; an object possesses a force that causes it to move and when that force is used up, the object stops moving. The best way to confront this prior conception is to take a bowling ball or another heavy ball and a billiard or a lacrosse ball and apply the same force to them, or even to try to push them with a finger or a nose. It takes a larger force to propel the larger mass with the same velocity, and each object will continuously roll with a constant velocity on a continuous smooth surface until something impedes the motion.

2. **Friction always hinders motion. Thus, you always want to eliminate friction.** Students will use friction, as applied by the brakes of vehicles, to hinder motion along the road. They also speak about air resistance as a retarding frictional force; however, it will be later in their study of pairs of action-reaction forces that students need to understand that these equal and opposite pairs of objects always act on different objects. It is the friction between the shoe and the surface that enables a walker to move forward.

Section 5 Negative Acceleration: Braking Your Automobile

What Do You See?

Learning Outcomes

In this section, you will

- **Plan and carry out** an experiment to relate braking distance to initial speed.
- **Determine** braking distance.
- **Examine** accelerated motion.

What Do You Think?

In recent years, more than 80 percent of speed-related traffic deaths happened on secondary highways (such as two-lane, rural roads). Imagine you are driving at the speed limit on a secondary highway, and you suddenly see an animal crossing the road ahead of you. Suppose you cannot swerve to miss the animal because of trees on each side of the road.

- **What factors must you consider to determine if you will be able to stop in the distance between you and the animal to avoid hitting it?**

Record your ideas about this question in your *Active Physics* log. Be prepared to discuss your response with your small group and the class.

Investigate

In this *Investigate*, you will plan and carry out an experiment to determine the relationship between the initial speed and the braking distance of an automobile.

1. Knowing how far your automobile will travel after you have stepped on the brake pedal is important. One factor that may have an impact on braking distance is the initial speed of the automobile.

CHAPTER 1

3. **The motion of an object is always in the direction of the net force applied to the object.** This alternative view of how things work in the physical world hinders student appreciation of how the combination of forces results in a net force that opposes the forward motion but that the forward motion continues until the velocity decreases to zero in that direction. This prior conception also appears in student explanations when they encounter moving along curves in *Section 7*, where the net force is perpendicular to the path of the vehicle.

4. **Frictional forces are due to irregularities in surfaces moving past each other.** It puzzles students that larger areas of the same materials do not exhibit larger amounts of friction. Intuitively, students know that rougher surfaces have more static friction and even that you can tilt a surface gently and not have an object immediately move down along the surface. Yet, it is only the mathematical analysis of situations and the analysis of graphs of forces applied over time that help to provide students with the realization that there is an adhesion between the surface molecules, even very smooth ones, that retards motion, just until the applied force equals the force of static friction.

Investigate

1.a)

The students may initially think the graph of braking distance vs. speed will look like graph A. After completion of the *Investigate*, they should have a graph that looks like graph B.

A

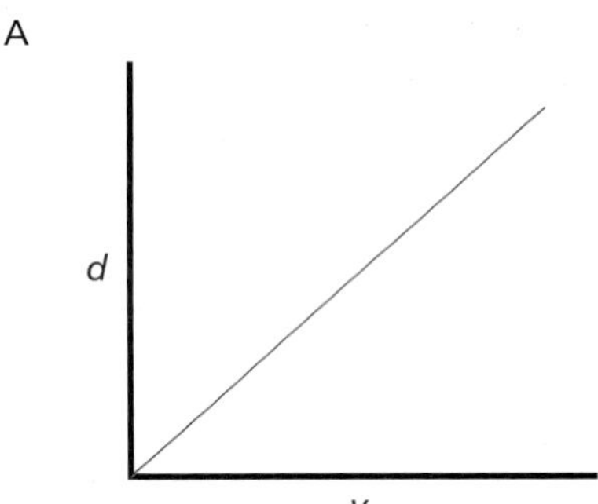

B

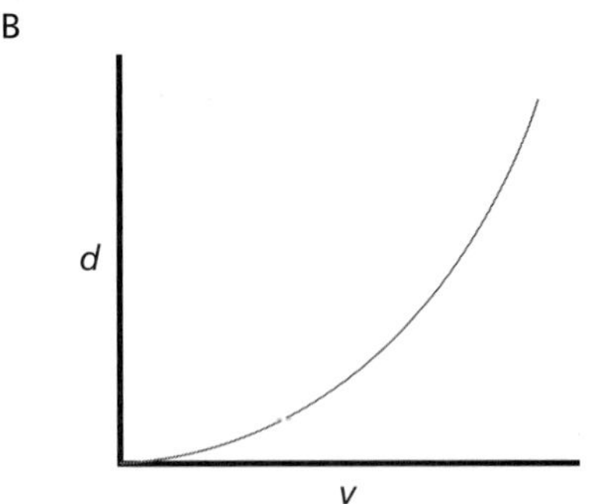

The initial speed is the speed at which you begin to apply the brakes. Braking distance is the distance required to bring the vehicle to rest once the brakes are applied. In your investigation, the initial speed will be the speed at the point at which you begin your measurement of braking distance. You will collect data to study the relationship between initial speed and braking distance.

a) What would a graph of braking distance vs. initial speed look like? Sketch a graph that shows what you think the data would show. (Place the initial speed on the *x*-axis and the braking distance on the *y*-axis.) While sketching the graph, imagine what would happen to the braking distance for a slow-moving vehicle, a faster-moving vehicle, and a very fast-moving vehicle.

b) Provide an explanation for the way you sketched the graph.

2. Your teacher will provide your group with equipment similar to the equipment shown in the illustration below. Discuss with your group how you could use the equipment to study the relationship between initial speed and braking distance.

To plan your experiment, consider the following:

- How will you vary the initial speed of the cart (that is, the velocity the cart has at the bottom of the hill when the brakes are applied)?
- The cart does not really have brakes applied by a driver, but the cart will stop on its own. Friction plays the role of brakes in the cart.
- How will you determine the initial speed of the cart just before it begins braking?
- How will you measure the braking distance? (What tool should you use? Should you measure from the front or the back of the cart? How accurate will you make your measurements?)
- How many different initial speeds will your group need to examine to find a pattern?
- How many trials should you perform at each initial speed?
- What will each group member be responsible for?
- How will you organize your data?

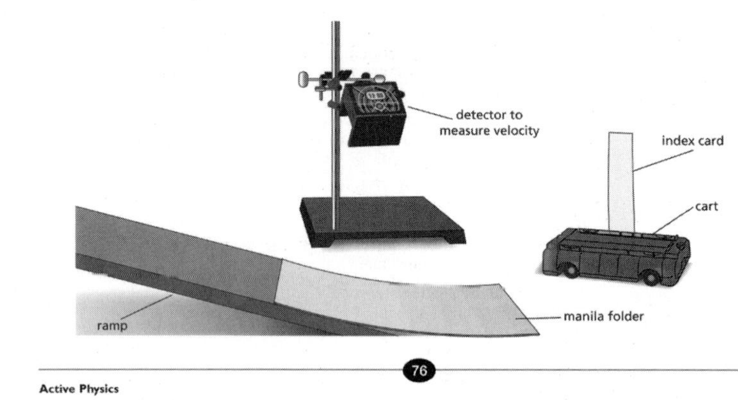

1.b)

The students will probably say that increasing speed implies increasing braking distance.

2.

The students need to measure stopping distance and initial velocity. Things students should consider include:

- Varying the starting height of the cart on the ramp will control speed and allow for repeated accurate trials.
- Because the friction on the rolling wheels is low, long distances may be necessary to stop the cart.
- The velocimeter will be used to measure speed (or another method as described above, in the advanced preparation and setup section.)
- Because braking distances may be large, a tape measure or meter stick will be needed to measure these distances. Measuring from the flag position as it leaves the velocimeter to the flag's final stopped position would give the distance the cart has traveled.
- About six to eight different initial speeds will give a good range, and allow students to find speeds that are triple the beginning speed.
- About four different trials to ensure uniformity would be good.
- One student will be needed for releasing the cart on the ramp, another to measure the velocity, and a third to measure the braking distance.
- The data should be organized in a table.

3. After discussing these questions in your group, develop a plan for what your group will do. Your teacher may ask you to either draw a flowchart or an outline showing the steps you will take.

4. Set up your equipment and perform your experiment.

a) Record both numerical data and observations in your *Active Physics* log.

Place the ramp on the floor in a way that does not obstruct people's ability to walk around the classroom. Do not block the emergency exit.

If you are setting up the ramp on the table, provide some means to contain the cart and prevent it from falling off the table.

5. Use the data you collected to complete the following:

a) Draw a graph showing how the braking distance depends on the initial speed. Place the initial speed on the horizontal axis and the braking distance on the vertical axis.

b) How does the braking distance change with initial speed?

c) How does your graph compare to the graph you sketched in *Step 1.a)*?

d) Compare your graph with those of other groups. What are some similarities and some differences?

e) Does looking at the other groups' graphs make you feel more confident or less confident about your data? Explain your answer.

6. Select two values of initial speed from your graph, with one value approximately twice the value of the other. Note the braking distance which corresponds to each initial speed.

a) What is the effect of doubling the initial speed on the distance traveled during braking?

7. Select two values of initial speed from your graph, with one value approximately three times as fast as the other. Note the braking distance which corresponds to each initial speed.

a) What is the effect of tripling the initial speed on the distance traveled during braking?

b) Predict how going four times faster will affect the braking distance.

8. Use the data on the sports car provided at the end of this chapter on pages 116-117 to answer the following:

a) Where is the braking data located?

b) The braking distance is shown for two speeds. The ratio of the two speeds is 80 mi/hr : 60 mi/hr. This ratio is $^{80}/_{60} = 1.33$. This is an increase of 133 percent. Do you expect the ratio of the braking distances to also be in the ratio of $^{80}/_{60} = 1.33$? What is the ratio of the braking distances? How does it compare with the ratio of the two speeds?

c) How does this data correspond to what you found in your experiment?

Teaching Tip

There may be some confusion about when the speed is to be measured and why. Make sure your students understand that their problem is to find out how far the cart travels on the level surface. The speed you need is its maximum value, which is just as the cart begins to travel horizontally. The purpose of the adjustable ramp is simply to provide a variable series of initial speeds.

3.

Students write out their plan.

4.a)

Students record data and observations.

5.a)

Students graph data in their logs.

5.b)

As you increase the initial speed, the braking distance increases at a much faster pace.

5.c)

The graph will be a parabola, concave side facing upward, which will probably be different from the students' initial graph.

5.d)

All graphs should be roughly parabolic, facing upward. The graphs should differ in values for the velocities and how closely they mimic a parabola.

Note: Unless the students choose a range of values for speed, when the speed doubles and triples, the graph may appear almost linear. If a group does this, have them choose additional values to meet this condition.

5.e)

Looking at the other groups' graphs should improve student confidence.

6.a)

The distance should quadruple.

7.a)

Increasing the speed three times gives a stopping distance about nine times as great.

7.b)

Because this is a quadratic relationship, the braking distance will be 16 times greater.

8.a)

The braking data is located below the data for Fuel Economy.

8.b)

The ratio of the braking distances is approximately 1.69. Students should predict a ratio of $(80)^2$ to $(60)^2$, which would be a 178% increase. The braking ratio listed is 209 ft/118 ft, or a 177% increase.

8.c)

Because the braking distance is proportional to the square of the initial velocity, the ratio of the speeds should be proportional to the square root of 1.69 = 1.3. The ratio 80/60 = 1.33.

Physics Talk

The *Physics Talk* discusses the concept of acceleration as well as the relationship between velocity and braking distance. The change in velocity of an automobile gradually coming to a stop is defined as negative acceleration, if the direction is assumed to be positive. Discuss the concept of negative acceleration in terms of changing speeds in the positive and negative directions. Students should be able to contrast negative acceleration with positive acceleration to highlight the difference between speeding up and slowing down in a particular direction. To further clarify the meaning of positive and negative acceleration, have students define both the terms in their *Active Physics* logs. You might want to emphasize why the use of the term (negative acceleration) is preferred instead of deceleration.

Students should also be reminded that according to the *Investigate*, braking distance is related to the initial velocity of an automobile through the equation $v_i^2 = -2ad$. The negative sign in the equation is the result of the automobile undergoing a negative acceleration while traveling in the positive direction. Students should fully grasp how the knowledge of $v_i^2 = -2ad$ can save lives.

Invite students to discuss how knowledge of the v^2 relationship can save lives when confronted with a situation requiring an emergency stop. Draw their attention to the three graphs that represent negative acceleration in terms of distance vs. time, velocity vs. time, and acceleration vs. time. Ask them why each graph is different and how the slope of each graph shows that the automobile is coming to a rest. Emphasize that the graphs give no information on direction, so the decrease in speed is assumed to be a negative acceleration.

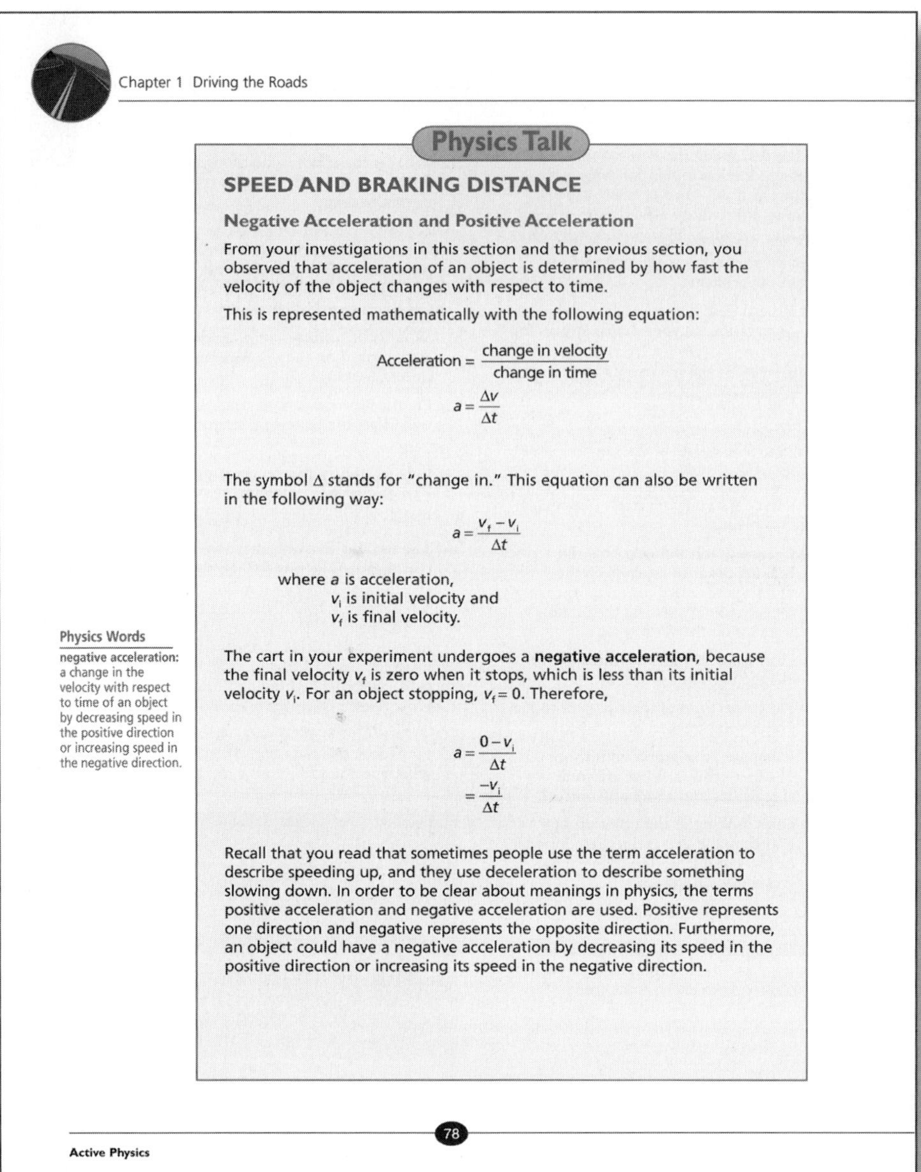
Chapter 1 Driving the Roads

Physics Talk

SPEED AND BRAKING DISTANCE

Negative Acceleration and Positive Acceleration

From your investigations in this section and the previous section, you observed that acceleration of an object is determined by how fast the velocity of the object changes with respect to time.

This is represented mathematically with the following equation:

$$\text{Acceleration} = \frac{\text{change in velocity}}{\text{change in time}}$$

$$a = \frac{\Delta v}{\Delta t}$$

The symbol Δ stands for "change in." This equation can also be written in the following way:

$$a = \frac{v_f - v_i}{\Delta t}$$

where a is acceleration,
v_i is initial velocity and
v_f is final velocity.

Physics Words
negative acceleration: a change in the velocity with respect to time of an object by decreasing speed in the positive direction or increasing speed in the negative direction.

The cart in your experiment undergoes a **negative acceleration,** because the final velocity v_f is zero when it stops, which is less than its initial velocity v_i. For an object stopping, $v_f = 0$. Therefore,

$$a = \frac{0 - v_i}{\Delta t} = \frac{-v_i}{\Delta t}$$

Recall that you read that sometimes people use the term acceleration to describe speeding up, and they use deceleration to describe something slowing down. In order to be clear about meanings in physics, the terms positive acceleration and negative acceleration are used. Positive represents one direction and negative represents the opposite direction. Furthermore, an object could have a negative acceleration by decreasing its speed in the positive direction or increasing its speed in the negative direction.

Active Physics 78

1-5a Blackline Master

CHAPTER 1

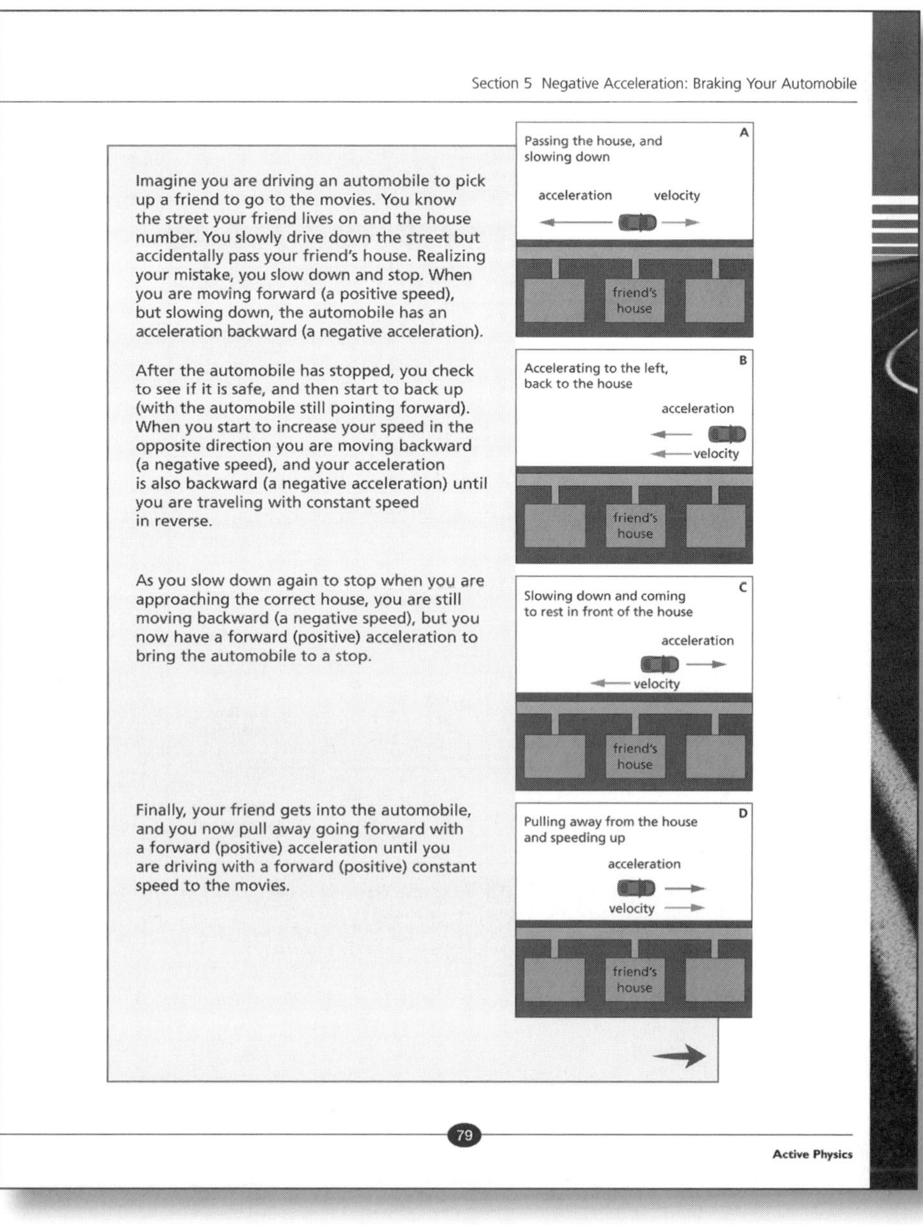

Section 5 Negative Acceleration: Braking Your Automobile

Imagine you are driving an automobile to pick up a friend to go to the movies. You know the street your friend lives on and the house number. You slowly drive down the street but accidentally pass your friend's house. Realizing your mistake, you slow down and stop. When you are moving forward (a positive speed), but slowing down, the automobile has an acceleration backward (a negative acceleration).

After the automobile has stopped, you check to see if it is safe, and then start to back up (with the automobile still pointing forward). When you start to increase your speed in the opposite direction you are moving backward (a negative speed), and your acceleration is also backward (a negative acceleration) until you are traveling with constant speed in reverse.

As you slow down again to stop when you are approaching the correct house, you are still moving backward (a negative speed), but you now have a forward (positive) acceleration to bring the automobile to a stop.

Finally, your friend gets into the automobile, and you now pull away going forward with a forward (positive) acceleration until you are driving with a forward (positive) constant speed to the movies.

A
Passing the house, and slowing down
acceleration
velocity
friend's house

B
Accelerating to the left, back to the house
acceleration
velocity
friend's house

C
Slowing down and coming to rest in front of the house
acceleration
velocity
friend's house

D
Pulling away from the house and speeding up
acceleration
velocity
friend's house

79

Active Physics

1-5b Blackline Master

Motion of Car	Time (s)	Velocity of car (ft/s)	Acceleration of car (ft/s²)	Positive or Negative
Car moving forward and slowing down (Diagram A)	0	+6	$\frac{v_f - v_i}{\Delta t} = \frac{(+4)-(+6)}{1} = -2$	negative acceleration
	1	+4		
Car moving forward, slowing down and car stops	2	+2	$\frac{v_f - v_i}{\Delta t} = \frac{(0)-(+2)}{1} = -2$	negative acceleration
	3	0		
Car moving backward and speeding up (Diagram B)	4	-2	$\frac{v_f - v_i}{\Delta t} = \frac{(-4)-(-2)}{1} = -2$	negative acceleration
	5	-4		
Car moving backward and slowing down (Diagram C)	6	-6	$\frac{v_f - v_i}{\Delta t} = \frac{(-4)-(-6)}{1} = +2$	positive acceleration
	7	-4		
Car moving backward and stopping	8	-2	$\frac{v_f - v_i}{\Delta t} = \frac{(0)-(-2)}{1} = +2$	positive acceleration
	9	0		
Car moving forward and speeding up (Diagram D)	9	0	$\frac{v_f - v_i}{\Delta t} = \frac{(+2)-(0)}{1} = +2$	positive acceleration
	10	2		

In this example, you can see that a negative acceleration can sometimes decrease the speed of an automobile $(t = 0 \text{ to } t = 3)$ or increase the speed of an automobile $(t = 4 \text{ to } t = 5)$, but it always decreases the velocity of the automobile by exactly 2 ft/s every second $(\text{from} +6 \text{ to} +4 \text{ to} +2 \text{ to } 0 \text{ to} -2 \text{ to} -4)$.

Section 5 Negative Acceleration: Braking Your Automobile

Calculating Braking Distance

Using the definition of velocity and acceleration, you can derive an equation for the braking distance when a vehicle comes to rest. The equation is shown below:

$$v_f^2 = 2ad + v_i^2$$

v_f is the final velocity of the car.

v_i is the initial velocity of the vehicle. Notice that it must be squared. This is the same as multiplying it by itself, $v_i^2 = v_i \times v_i$.

a is the acceleration. It is a negative acceleration.

d is the braking distance.

Because the final velocity of a vehicle is zero after the car comes to a stop, you may put a zero in for the final velocity, and of course zero times zero is still zero.

$$0 = 2ad + v_i^2$$

or

$$v_i^2 = -2ad$$

You can use the helpful circle to solve for any of the variables in this equation as well.

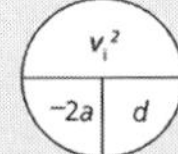

Of all the equations in your first year of physics, this one may have the greatest impact on your safety. Understanding this equation may one day even help to save your life! From this equation, you can see that if you double the initial velocity, then the braking distance d will have to quadruple. If you triple the initial velocity, then the braking distance d will be nine times as great.

You probably found in the *Investigate* that doubling the speed increased the distance traveled while the vehicle was braking by about a factor of four and that tripling the speed increased the braking distance by about a factor of nine. Look at the data for the sports car. The speed increased by 1.33 while the braking distance increased by approximately 1.33 × 1.33 = 1.77. Experiments completed with a great deal of care, ensuring that the braking acceleration is constant between trials, find that this relationship is true.

CHAPTER 1

Checking Up

1.

The braking automobile has undergone negative acceleration because the automobile's velocity decreases to zero as it comes to a sudden stop.

2.

The braking distance is determined by the v^2 equation, which shows that braking distance is raised to the power of two as velocity increases. Thus, if velocity is doubled, braking distance is quadrupled, and if velocity is tripled, braking distance increases by a factor of nine.

3.

Deceleration means slowing down, but does not indicate the direction of velocity as an object slows down. A more precise term, negative acceleration, implies that the body in motion is slowing down in the direction opposite to its velocity, or increasing its velocity in the negative direction.

(The v^2 relationship is derived assuming constant acceleration, which is *approximately* true for real automobiles in everyday stopping situations. The $v^2 = -2ad$ equation models reality very closely, and is therefore useful for describing braking.)

How can knowledge of the v^2 relationship save many lives? If you were to decrease your speed to one third your original speed, you would need only one ninth of the braking distance. If you double the speed you do not require double the distance for the car's brakes to stop the car, but four times the distance. Decreasing your speed can save lives because of the significant effect the slower speed has on braking distance.

You have seen how equations can model the motion of an automobile braking. An automobile with a negative acceleration can also be described using graphs of distance vs. time, velocity vs. time, and acceleration vs. time as shown below.

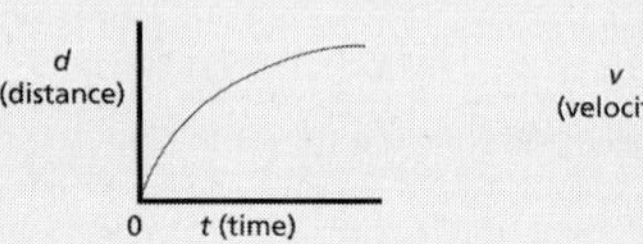

v
(velocity)
0
t (time)

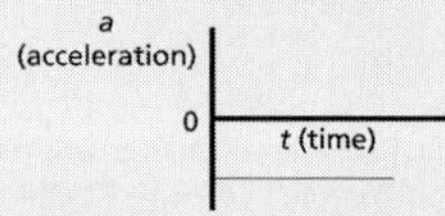

In the velocity vs. time graph, you can see that the velocity is decreasing as the automobile comes to rest. Notice that "at rest" is equivalent to a velocity equal to 0. You should also notice that the slope of the graph is constant. This implies that the acceleration is constant since the slope of the *v-t* graph is equal to the acceleration. Finally, notice that the slope is negative (sloping downward) which implies that the acceleration is negative.

In the acceleration vs. time graph, you can see that the acceleration is constant and negative.

In the distance vs. time graph, you can see how the change in distance for a given time changes as the speed changes. Notice that the slope at the beginning times is very steep, corresponding to a large velocity. Toward the end, the slope becomes 0, corresponding to the car stopping. (Thus, the graph is a curve.)

Checking Up

1. If a vehicle is traveling at constant velocity and then comes to a sudden stop, has it undergone negative acceleration or positive acceleration? Explain your answer.
2. Explain how you know that increasing the velocity of an automobile increases the braking distance.
3. Why is the term negative acceleration used instead of deceleration?

Active Physics *Plus*

+Math	+Depth	+Concepts	+Exploration
◆◆◆	◆◆	◆◆	

Motion Equations

Five motion equations can describe all the relations among position, velocity, and constant acceleration. The equations are all derived from the definitions of velocity and acceleration.

$$d = \bar{v}t$$

$$v_f = at + v_i$$

$$\bar{v} = \frac{v_f + v_i}{2}$$

$$d = \frac{1}{2}at^2 + v_it$$

$$v_f^2 = 2ad + v_i^2$$

The first equation is a restatement of the definition of average velocity. (The v with a bar over the top is a shorthand way of writing $v_{average}$.) The second equation is a restatement of the definition of acceleration. The third equation is for average velocity when there is constant acceleration. The fourth equation helps to determine distance traveled if you know the acceleration and time without the need for first finding the final velocity. The fifth equation relates the stopping distance to the acceleration and velocities without the need for calculating the time.

The fifth equation can be derived from the other four equations using algebra. Assume that the initial velocity equals zero to ease the mathematics.

You may want to try to derive the equation with acceleration not being zero, using the same approach.

These are the variables in the motion equations: d, t, a, v_f, v_i and $v_{average}$.

$v_f = at + v_i$

Assuming $v_i = 0$,

$v_f = at$

Square each side

$v_f^2 = a^2t^2$

$v_f^2 = 2a\left(\frac{1}{2}at^2\right)$

$v_f^2 = 2ad$

If the acceleration is constant and you are able to find or are given any three of these variables, you can use the motion equations to solve for the other two variables and completely describe the motion of the object. The object can be an automobile, an animal, a galaxy, or a cell. The motion equations describe them all.

Sample Problem I

A softball pitcher accelerates a ball from rest to a speed of 25 m/s over a distance of 1.8 m. What is the ball's acceleration?

Strategy:

Using the fifth equation of motion, derived from the other four equations using algebra, and knowing v_i, v_f, and d, you can solve for acceleration.

Given:

$v_i = 0$

$v_f = 25$ m/s

$d = 1.8$ m

Solution:

Knowing that $v_f^2 = 2ad + v_i^2$, and that $v_i = 0$, the equation becomes $v_f^2 = 2ad$.

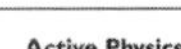

Active Physics Plus

By solving problems and analyzing graphs, students further explore how acceleration affects braking distance. Have the students write the five motion equations in their logs. A discussion of the variables in these equations and how they are derived will help clarify their meaning and how they relate to each other. You might want to ask those students who are able to grasp how the equations are derived to explain the derivations to other students in the class.

1.a)

Negative, since the acceleration is in the opposite direction of the positive motion.

1.b)

$$a = \frac{v}{t} = \frac{-90 \text{ m/s}}{1.5 \text{ s}} = -60 \text{ m/s}^2$$

1.c)

$$v_{\text{average}} = \frac{v_i + v_f}{2} =$$
$$\frac{90 \text{ m/s} + 0 \text{ m/s}}{2} = 45 \text{ m/s}$$

1.d)

$$d = v_i t + \tfrac{1}{2}at^2 = (90 \text{ m/s})(1.5 \text{ s}) +$$
$$\tfrac{1}{2}(-60 \text{ m/s}^2)(1.5 \text{ s})^2 = 67.5 \text{ m}$$

2.a)

$$v_f = v_i + at = 0 + (5 \text{ m/s}^2)(5 \text{ s}) =$$
$$25 \text{ m/s}$$

2.b)

$$d = v_i t + \tfrac{1}{2}at^2 = 0 + \tfrac{1}{2}(8 \text{ m/s})^2 \times$$
$$(6 \text{ s})^2 = 144 \text{ m}$$

2.c)

$$d_{\text{car}} = v_{\text{i car}} t_{\text{car}} + \tfrac{1}{2} a_{\text{car}} t^2_{\text{car}}$$
$$150 \text{ m} = 0 + \tfrac{1}{2}(5 \text{ m/s}^2)t^2_{\text{car}}$$
$$t_{\text{car}} = \sqrt{\frac{150 \text{ m}}{2.5 \text{ m/s}^2}} = 7.7 \text{ s.}$$

$$d_{\text{cycle}} = v_{\text{i cycle}} t_{\text{cycle}} + \tfrac{1}{2} a_{\text{cycle}} t^2_{\text{cycle}}$$
$$200 \text{ m} = 0 + \tfrac{1}{2}(8 \text{ m/s}^2)t^2_{\text{cycle}}$$
$$t_{\text{cycle}} = \sqrt{\frac{200 \text{ m}}{4 \text{ m/s}^2}} = 7.1 \text{ s.}$$

The motorcycle wins!

3.a)

$$v_f^2 = v_i^2 + 2ad$$
$$v_f^2 = (2 \text{ m/s})^2 + 2(0.5 \text{ m/s}^2) \times$$
$$(12 \text{ m}) = 16 \text{ m}^2/\text{s}^2$$
and $v_f = 4$ m/s.

$$v_f^2 = 2ad$$
$$a = \frac{v_f^2}{2d}$$
$$= \frac{(25\text{m/s})^2}{2(1.8 \text{ m})}$$
$$= \frac{25^2(\text{m/s})(\text{m/s})}{2(1.8 \text{ m})}$$
$$= 173.6 \text{ m/s}^2 \text{ or } 170 \text{ m/s}^2$$

Therefore, the acceleration of the softball is 170 m/s^2.

Sample Problem 2

During an auto race, a car with a speed of 75 m/s accelerates past another car at a rate of 3.0 m/s² for 4.0 s. How far does the car travel during this time?

Strategy:

Knowing the car's initial velocity, time, and acceleration you can use the fourth equation

$$d = \frac{1}{2}at^2 + v_i t \text{ to determine}$$

distance traveled without the need for first finding the final velocity.

Given: $v_i = 75\text{m/s}$

$a = 3.0\text{m/s}$

$t = 4.0$

Solution:

Using $d = \frac{1}{2}at^2 + v_i t$ and solving for d gives

$$d = \frac{1}{2}at^2 + v_i t$$
$$d = \frac{1}{2}(3.0\frac{\text{m}}{\text{s}^2})(4.0 \text{ s})^2 + (75\frac{\text{m}}{\text{s}})(4.0 \text{ s})$$
$$d = 324 \text{ m or } 320 \text{ m}$$

Therefore, the car travels 320 m, or almost one quarter of a mi.

1. When a jet lands on an aircraft carrier, its speed goes from 90.0 m/s to zero in 1.5 s as it is stopped by a cable running across the aircraft carrier's deck.
   a) If the direction the jet is traveling is positive, was the jet's acceleration positive or negative?
   b) What is the jet's acceleration during the stopping process?
   c) If the jet undergoes a constant acceleration while stopping, what is the jet's average speed?
   d) How far does the jet travel along the carrier's deck while it is being brought to a stop?
2. A race is held between a sports car and a motorcycle. The sports car can accelerate at 5.0 m/s² and the motorcycle can accelerate at 8.0 m/s². The two vehicles start the race at the same time and accelerate from rest.
   a) After 5.0 s, how fast is the sports car going?
   b) After 6.0 s, what distance will the motorcycle have gone?
   c) To make the race fair, the sports car starts 50.0 m ahead of the motorcycle. If the course is 200.0 m long, which vehicle wins the race? (Hint: The vehicle that covers its distance in the least time wins.)
3. A student on a skateboard pushes off from the top of a small hill with a speed of 2.0 m/s, and then goes down the hill with a constant acceleration of 0.5 m/s².
   a) After traveling a distance 12.0 m, how fast is the student going?
   b) How much time does it take the student to move a distance of 21.0 m while accelerating at this rate?

3.b)

$$d = v_i t + \tfrac{1}{2}at^2$$
$$21 \text{ m} = (2 \text{ m/s})t + \tfrac{1}{2}(0.5 \text{ m/s}^2)t^2$$
$$(0.25 \text{ m/s}^2)t^2 + (2 \text{ m/s})t -$$
$$21 \text{ m} = 0$$

Using the quadratic equation,

$$t = \frac{-(2 \text{ m/s}) \pm \sqrt{(2 \text{ m/s})^2 - 4(0.25 \text{ m/s}^2)(-21 \text{ m})}}{2(0.25 \text{ m/s}^2)}$$
$$= 6 \text{ s.}$$

Section 5 Negative Acceleration: Braking Your Automobile

Graphing Models

You have been using graphs to better understand motion. You have seen that there is a relationship among corresponding *d-t*, *v-t* and *a-t* graphs.

The slope of a *d-t* graph of an automobile is equal to the velocity of the automobile.

The slope of a *v-t* graph of an automobile is equal to the acceleration of the automobile.

Given a *d-t* graph, you can use this information to determine the *v-t* graph and the *a-t* graph as you have seen earlier.

The velocity vs. time graph can also tell you about the distance traveled.

In the following two velocity vs. time graphs, the shaded areas under the velocity vs. time graphs are equal to the distance traveled. This can be proven in the following way.

In the first velocity vs. time graph, the average velocity is constant, because the velocity does not change. With no change in velocity, the acceleration must be zero. The shaded area under the graph is equal to the distance traveled. The shaded area under the graph is the area of a rectangle (A = height × base). This area is (average velocity) × (time), which is the definition of distance traveled.

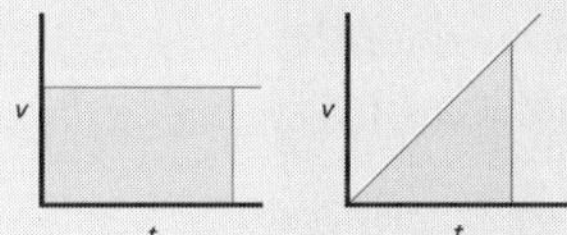

Velocity vs. Time

The second graph shows a constant acceleration. The area under the second graph is identical to the area of a triangle. The area of a triangle is ½ height × base. The base is the time. The height is the final velocity. One half the height is the average of the final velocity and the initial velocity of 0.

$$\frac{1}{2}\text{ height}\times\text{base} = \frac{1}{2}\text{ (final velocity)}\times\text{(time)}$$
$$= \text{(average velocity)}\times\text{(time)}$$

$$\frac{1}{2}h\times b = \frac{1}{2}(v_f)\times(t)$$
$$= (v_{\text{average}})\times(t)$$

Once again, from the definition of average velocity (average velocity = distance/time), there is a way to calculate the distance traveled.

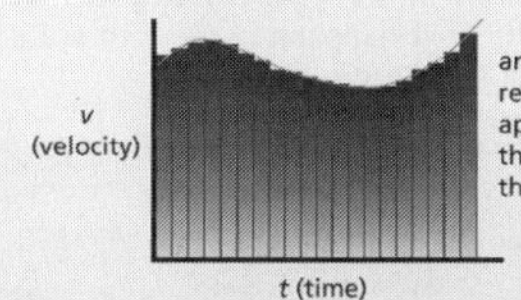

The area under a velocity vs. time graph is always equal to the distance traveled. For non-constant accelerations, the velocity vs. time graph is a curve. You can see in the diagram above how you can break a curve into a series of tiny rectangles that approximates the curve. The total area under the curve is approximately equal to the total area of all the rectangles. This is the beginning of your introduction to calculus—an advanced mathematics invented by Sir Isaac Newton, an English physicist and mathematician, to better understand physics.

1.

The object accelerates forward from rest to a speed of 5 m/s from 0 to 2 s, continues at a constant speed of 5 m/s from 2 to 4 s, accelerates from 5 m/s to 10 m/s from 4 to 6 s, and then slows down from 10 m/s to rest (negative acceleration) from 6 to 8 s.

2.

For 0 to 2 s, $a = \dfrac{(5\text{ m/s} - 0)}{2\text{ s}} = 2.5\text{ m/s}^2$

For 2 to 4 s, $a = \dfrac{(5\text{ m/s} - 5\text{ m/s})}{2\text{ s}} = 0$

For 4 to 6 s, $a = \dfrac{(10\text{ m/s} - 5\text{ m/s})}{2\text{ s}} = 2.5\text{ m/s}^2$

For 6 to 8 s, $a = \dfrac{(0 - 10\text{ m/s})}{2\text{ s}} = -5\text{ m/s}^2$

3.

The distance traveled for each interval is equal to the area under the graph for that interval.

For 0 to 2 s, the area of the triangle is $\text{Area}_t = \frac{1}{2}(\text{base})_t \times (\text{height})_t = \frac{1}{2}(2\text{ s})(5\text{ m/s}) = 5\text{ m}$.

For 2 to 4 s, the area of the rectangle is $\text{Area}_r = (\text{base})_r(\text{height})_r = (2\text{ s})(5\text{ m/s}) = 10\text{ m}$.

For 4 to 6 s, the area of the rectangle plus the triangle is $\text{Area}_r + \text{Area}_t = (\text{base})_r(\text{height})_r + \frac{1}{2}(\text{base})_t(\text{height})_t = (2\text{ s})(5\text{ m/s}) + \frac{1}{2}(2\text{ s})(5\text{ m/s}) = 10\text{ m} + 5\text{ m} = 15\text{ m}$.

Chapter 1 Driving the Roads

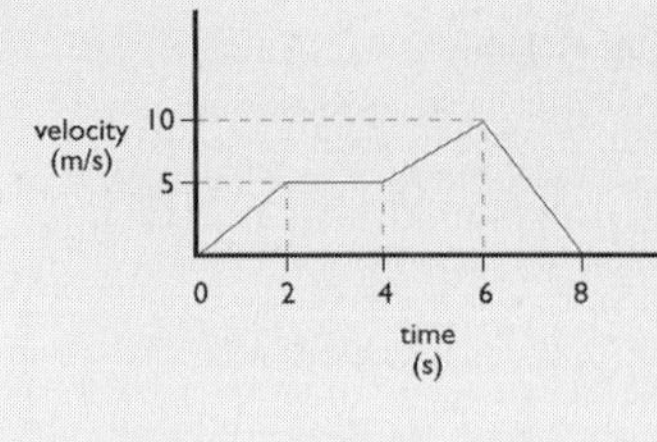

For the velocity vs. time graph shown:

1. Describe the motion from $t = 0$ to $t = 8$ s.
2. Calculate the acceleration of the object for each 2 s.
3. Calculate the distance traveled for each 2 s. (Hint: For $t = 4$ to $t = 6$, the area under the curve is a trapezoid made up of both a rectangle and a triangle.)
4. Calculate the total distance traveled.

What Do You Think Now?

At the beginning of this section, you were asked the following:

- **What factors must you consider to determine if you will be able to stop in the distance between you and the animal to avoid hitting it?**

How would you answer this question now? After studying the equations of motion, how do you think velocity affects the time it takes to suddenly stop an automobile? According to the equation for braking distance, if you double an automobile's speed, what happens to the distance needed for a vehicle's brakes to bring the vehicle to a stop?

86

Active Physics

For 6 to 8 s, the area of the triangle is $\text{Area}_t = \frac{1}{2}(\text{base})_t(\text{height})_t = \frac{1}{2}(2\text{ s})(10\text{ m/s}) = 10\text{ m}$.

4.

The total distance traveled is equal to the sum of all the areas under the graph, or 44 m.

What Do You Think Now?

You were given a *Physicist's Response* to the *What Do You Think?* question for your reference. Provide students with that answer and encourage discussion. Students should be able to articulate the concepts they learned in relation to their responses. Have students respond by pointing out the use of the v^2 equation in determining the braking distance. Write a few responses on the board to draw a connection to concepts they have learned, as these concepts emerge in class discussion.

Physics

Essential Questions

What does it mean?

An automobile safety manual states that the braking distance increases with the square of the velocity of the vehicle. What does this mean? Why is this related to safe driving?

How do you know?

What evidence do you have that tripling the speed of an automobile will increase the braking distance by a factor of $3 \times 3 = 9$?

Why do you believe?

Connects with Other Physics Content	Fits with Big Ideas in Science	Meets Physics Requirements
Forces and motion	Models	* Good, clear, explanation, no more complex than necessary

* Physics tries to use a few simply related principles to describe phenomena. Describing many different things requires a precision in language. In everyday language, you may use the words acceleration and deceleration. In physics, you use only the word acceleration. Describe the difference between positive and negative acceleration.

Why should you care?

Safe driving saves lives. How does knowing about the relationship between speed and braking distance help you to become a safe driver?

Reflecting on the Section and the Challenge

Safe driving requires the ability to stop safely. Some people think that if you triple your speed, the automobile will require triple the braking distance. You now know that it will take more than triple the braking distance – it is closer to nine times the braking distance!

You should be able to explain the importance of braking distance as it relates to speed. You should understand why slowing down is beneficial in terms of braking distance and what will happen to the required braking distance if you decrease your speed by one third.

You should always reduce your speed when driving through a school zone or a parking lot of a crowded supermarket. Slowing down decreases your braking distance and will protect unaware pedestrians.

In your *Chapter Challenge*, you can now demonstrate your understanding of the relationship of speed to braking distance to the Active Driving Academy.

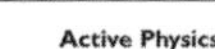

87 Active Physics

Reflecting on the Section and the Challenge

You might wish to read this section aloud and have a class discussion, or have groups of students spend a few minutes in discussion.

Encourage students to reflect on how their knowledge of initial speed versus braking distance contributes to their understanding of driving safely. They should be able to reflect on how speed affects safety and how this can be woven into their *Chapter Challenge*. It is an important opportunity to consolidate the concepts they have learned and relate them to the importance of v^2 in determining braking distances.

CHAPTER 1

Physics Essential Questions

What does it mean?

If you triple your speed, the braking distance will increase by a factor of nine. That's because $3^2 = 9$. Safe drivers realize that dropping their speed can produce a drastic change in braking distance.

How do you know?

When the automobile's speed tripled, the distance along the floor was actually almost nine times as long.

Why do you believe?

If a vehicle is moving to the right (a positive velocity), a positive acceleration will also be to the right and will increase the speed of the vehicle. If the vehicle is moving to the right, a negative acceleration will be to the left and will decrease the speed of the vehicle.

If a vehicle is moving to the left (a negative velocity), a positive acceleration will be to the right and will decrease the speed of the vehicle. If the vehicle is moving to the left, a negative acceleration will also be to the left and will increase the speed of the vehicle.

Why should you care?

It provides guidance about the value and importance of slowing down. If you cut your speed by 1/3, then you will only need 1/9 the distance to stop. Having such a small stopping distance can assist you in places where children may surprise you by running into the road.

Physics to Go

1.

The graph should be a parabola as shown below. Emphasize to students the need for a curved line of best fit. As the speeds increase, the braking distances increase at a much greater rate.

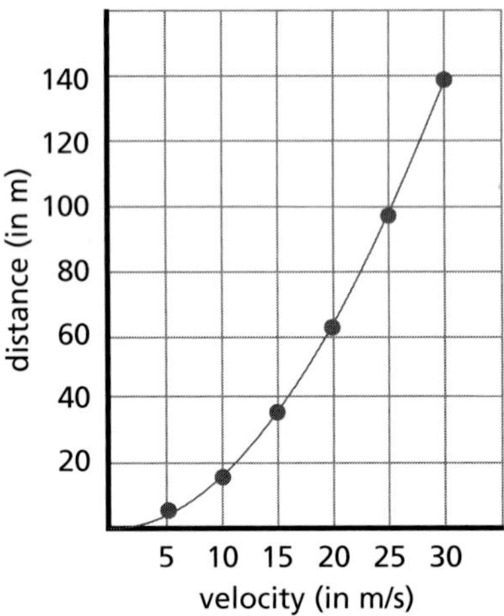

2.

Automobile B has the greater braking distance at slower speeds than automobile A. Using braking distances to determine safety, automobile B is safer.

3.a)

Students should recognize that the braking distance for half the speed will be ¼ of the distance. Therefore, the distance required to stop at 15 mi/h will be 5 m.

3.b)

60 mi/h is twice 30 mi/h; therefore, the distance required to stop will be four times the braking distance, or 100 m.

3.c)

45 mi/h is three times the speed of 15 mi/h; therefore, the distance required to stop will be nine times the braking distance at 15 mi/h or 45 m.

Chapter 1 Driving the Roads

Physics to Go

1. A student measured the braking distance of her automobile and recorded the data in the table. Plot the data on a graph and describe the relationship that exists between initial speed and braking distance.

Initial speed	Braking distance
5 m/s	4 m
10 m/s	15 m
15 m/s	35 m
20 m/s	62 m
25 m/s	98 m
30 m/s	140 m

2. Below is a graph of the braking distances in relation to initial speed for two automobiles. Compare qualitatively (without using numbers) the braking distances when each automobile is going at a slow speed and then again at a higher speed. Which automobile is safer? Why? How did you determine what "safer" means in this question?

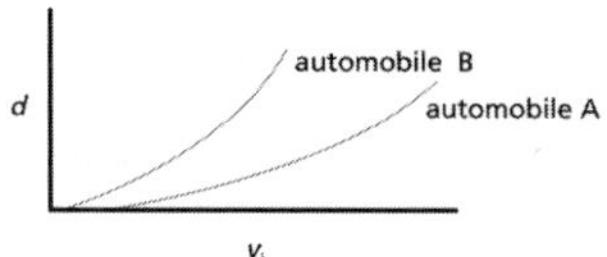

3. An automobile is able to stop in 20 m when traveling at 30 mi/hr. How much distance will it require to stop when traveling at the following:
   a) 15 mi/hr? (half of 30 mi/hr)
   b) 60 mi/hr? (twice 30 mi/hr)
   c) 45 mi/hr? (three times 15 mi/hr)
   d) 75 mi/hr? (five times 15 mi/hr)
4. An automobile traveling at 10 m/s requires a braking distance of 30 m. If the driver requires 0.9 s reaction time, what additional distance will the automobile travel before stopping? What is the total stopping distance, including both the reaction distance and the braking distance?
5. Consult the information for the sports car at the end of this chapter. This shows the stopping distance. How far would you expect this automobile to travel until coming to rest when brakes are applied at a speed of 30 mi/hr?
6. Use the information for the sedan at the end of this chapter. Find the braking distances for 50 mi/hr and 25 mi/hr. Draw a graph using the different braking distances. Plot the speeds on the horizontal axis and the braking distances on the vertical axis.

Active Physics 88

3.d)

75 mi/h is five times the speed of 15 mi/h; therefore, the distance required to stop will be 52 m, or 25 times the braking distance, or 125 m.

4.

The additional distance traveled can be found using $v = d/t$ and solving for d. $d = vt$ gives $d = (10 \text{ m/s})(0.9 \text{ s}) = 9 \text{ m}$ for the distance covered during the driver's reaction time. The total stopping distance covered during the driver's reaction time (9 m) plus the braking distance (30 m) for a total stopping distance of 39 m. The driver's manual will not include this information, since reaction time for each driver varies; therefore, only the braking distance is listed.

7. Does the braking information for the sedan include the driver's reaction time? If it does not, then how much distance is added to the total braking distance, supposing that the driver has a ½ s reaction time? Who should let the consumer know about the ½ s reaction time— the information sheet or a driver training manual?
8. Apply what you learned in this section to write a statement explaining the factors that affect stopping distance. The total stopping distance includes the distance you travel during your reaction time, plus the braking distance. What do you now know about stopping that will make you a safer driver?
9. In a perfect experiment, your data would show that the braking distance is proportional to the square of the velocity. Real data is not perfect. Describe two possible sources of error and explain how they could have impacted your results.
10. How could you revise this experiment to study better/worse braking situations? Predict how your graph might change.
11. ***Preparing for the Chapter Challenge***

    Apply what you have learned in this section to write a convincing argument against excessive speed when approaching an intersection, a traffic light, crosswalk, school zone, or any other driving situation that may require sudden braking on your part. Excessive speed means you cannot stop in the available distance if necessary. What are the consequences of approaching these situations with excessive speed?

Inquiring Further

Reconstructing an accident

Collect newspaper clippings or summarize television news reports of traffic accidents in your city or town that involved automobiles and/or motorcycles. Become an accident investigator and imagine rewinding the events leading into the accident.

- What advice might you have given to the driver(s) involved about speed, reaction time, and braking distances, that would have enabled them to avoid the accident?
- In writing, comment on whether the accident might have been prevented simply by slowing down, or whether there were other contributing factors as well (such as icy roads). If there were other factors, would they be additional reasons for reducing speed?

5.

At 30 mi/h, half of 60 mi/h, the stopping distance will be ¼ of 118 ft, or about 30 ft.

6.

The braking distance for 50 mi/h is about 94 ft. The braking distance for 25 mi/h is about 23 ft. The shape of the graph should be a parabola.

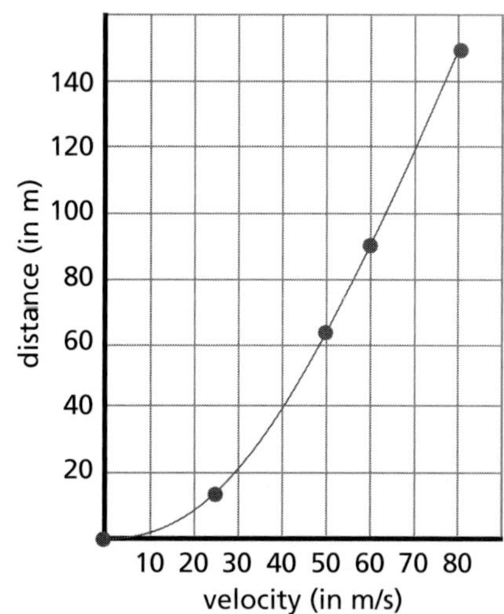

7.

The driver's reaction time is not included on the data sheet. The driver's reaction time would add an additional 44 ft at 60 mi/h. As to who should supply this information, students' answers will vary. Look for sound arguments that can be shared with the class.

8.

Answers will vary. Expect some of the following:

- speed of travel
- road conditions
- brake condition
- reaction times
- tire condition
- weather conditions

9.

Sources of differences between the idealized mathematical model and the real data might include the following:

- Acceleration is not precisely constant.
- Time and distance have measurement errors.

10.

Worse braking situations could use surfaces with less friction (ice, gravel); surfaces with variable friction (blacktop road with some gravel on it); different tires, etc. All the graphs would still remain parabolas, but the curves would be different — steeper for better brakes and less curved for poor brakes.

11.

Preparing for the Chapter Challenge

Students' paragraphs should include information on how the braking distance increases with the square of the automobile's velocity. In areas where the situation may change rapidly and the driver would be required to stop at a short distance, slow speeds and short braking distances are needed.

Inquiring Further

The students should make suggestions detailing their knowledge of how braking distance depends upon the automobile's velocity and how the total braking distance depends upon the reaction distance (and thus reaction time) and the braking distance. If a rear-end collision occurred, they may make suggestions regarding a safe following distance and how this relates to reaction time and braking distances. The students should indicate whether a slower speed would have prevented the collision or if there were other conditions that should have been taken into consideration, such as a sharp bend in the road that might have obscured frontal vision.

The students should also discuss how road conditions might have affected the braking distances involved, and how the drivers might have responded to these conditions to prevent the accident. Conditions such as snow, or rain, will increase braking distances, and fog will increase both reaction time and reaction distance.

If students live in an area with very few automobile accidents, you may wish to provide stored clippings for them from local papers, or download descriptions from the Internet. Many papers have archived online editions that might provide descriptions of multiple vehicle accidents due to conditions such as fog, rain, and snowstorm "white-outs."

NOTES

SECTION 5 QUIZ

1-5c Blackline Master

1. For this question, the direction east is positive, and the direction west is negative. A vehicle is traveling east at 10 m/s when it starts to undergo a negative acceleration of -1 m/s^2 as it comes to rest. Which of the following quantities will be also be negative as the vehicle starts to undergo the negative acceleration coming to rest?

   a) The vehicle's velocity.

   b) The vehicle's change in position.

   c) Both the vehicle's velocity and its change in position.

   d) Neither the vehicle's velocity nor its change in position.

2. A skater uniformly decreases her speed from 6 m/s to zero over a distance of 12 m. Her acceleration would be

   a) 1.5 m/s^2.
   b) -1.5 m/s^2.
   c) 3 m/s^2.
   d) -3 m/s^2.

3. An automobile is going 20 m/s when it applies the brakes and stops after traveling a distance of 20 m. If the automobile was going at 10 m/s, it would be able to stop in a distance of

   a) 10 m.
   b) 20 m.
   c) 80 m.
   d) 5 m.

4. A driver of an automobile with poor brakes finds that if he triples his speed, his stopping distance is 9 times longer. The driver of another car with very good brakes tries the same test. If he triples his speed, he will find that his braking distance has increased by a factor of

   a) 3 times.
   b) between 6 and 9 times.
   c) between 3 and 6 times.
   d) 9 times.

5. Which distance vs. time graph to the right would best represent the velocity of a girl on a bike from the moment she sees the branch fall until the bike comes to rest?

a)

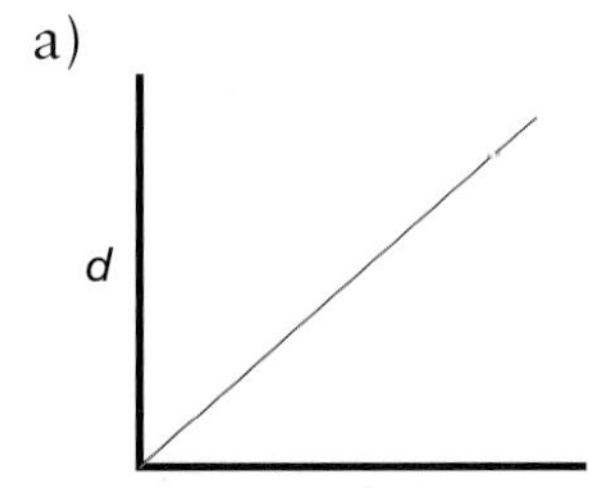

b)

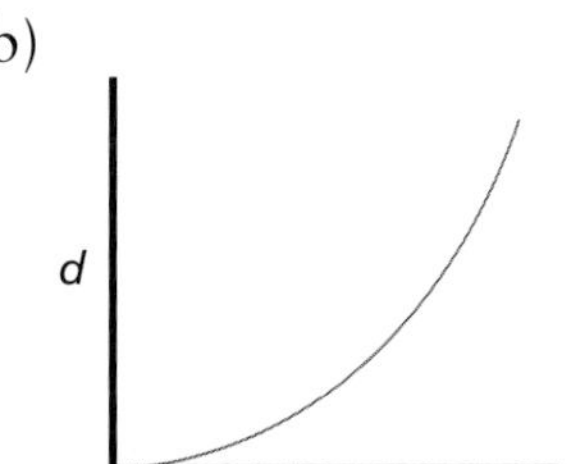

c)

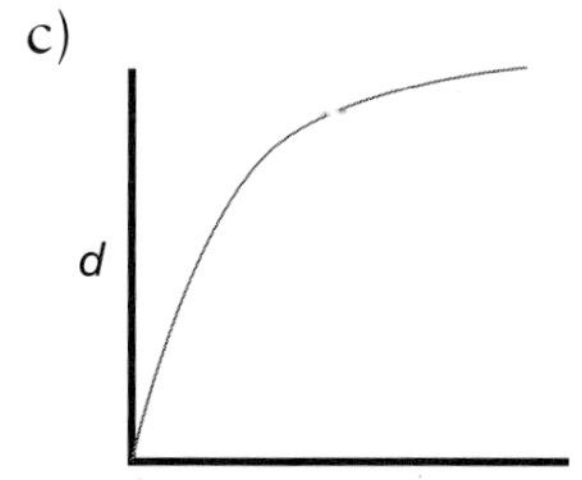

d)

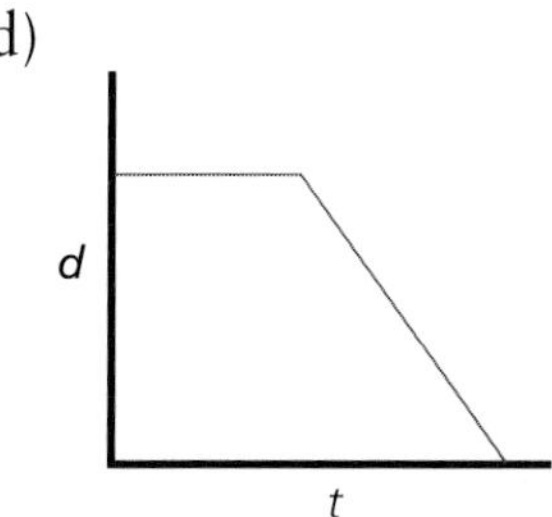

SECTION 5 QUIZ ANSWERS

1. d) Although the vehicle has a negative acceleration, the vehicle will continue traveling east with a reduced velocity and move to the east until it comes to a stop, so both the velocity and change in position are positive even though the acceleration is negative.

2. b) Using the equation $v_i^2 = 2ad$ and solving for a, we have
$a = \frac{v_i^2}{2d} = \frac{(6\text{ m/s})^2}{2(12\text{ m})} = 1.5\text{ m/s}^2$.
Because the skater is slowing down while moving forward, she should undergo a negative acceleration. Students who got the answer 3 m/s^2 did not divide by 2.

3. d) When the speed is cut in half, the braking distance becomes one fourth, or 5 m.

4. d) The driver of the automobile with good brakes would still take nine times the distance to stop when he triples his speed, but the distances would be smaller than those of the automobile with poor brakes.

5. c) The distance vs. time graph shows an object traveling with constant speed (the straight line section), and later with changing speed (the curved section) where the distance does not increase with time and the vehicle gradually comes to a stop.

NOTES

SECTION 6

Using Models: Intersections with a Yellow Light

Section Overview

This section investigates the dilemma a driver faces when encountering a yellow light at an intersection. A mathematical model demonstrates the interplay of different factors that determine whether a driver can safely make it through the intersection. The factors relevant to the driver's decision include reaction time, width of the intersection, speed of the car, and the rate of negative acceleration. The quantitative data for these factors are recorded on a spreadsheet. This data is then inserted into the equations for reaction and braking distance to calculate the actual distance the automobile will cover during the length of the yellow-light time. The equations are then used to determine whether an automobile will be able to go safely through an intersection before the light turns red.

Background Information

While you are driving at a constant speed, you need a certain length of time to react to a stimulus (such as the neighbor's cat running on the road), interpret the stimulus, then tell your muscles to perform an action (push the brake pedal). During the time it takes to react to a yellow light, your vehicle will continue to move with the original speed. It will cover a distance $(d = vt)$.
This distance will be added to the distance that is required to stop once the brakes are applied $(d = v^2/2a)$. Because the acceleration rate is not known for each vehicle, there will be an arbitrary rate chosen to reflect an average. It is usually based on an average car, with good brakes, on clean, dry pavement. The stopping distance, then, is the distance that is required to come to a stop, after the stimulus (the cat) is presented. Therefore, total distance, $d = vt + \frac{v^2}{2a}$.

Computing the GO Zone

Determining the GO Zone is an application of distance = velocity × time. While the yellow light is illuminated, the automobile must be able to go through the intersection. The total distance traveled is the distance from the intersection plus the width of the intersection.

GO Zone + width = velocity × yellow-light time

$GZ + w = vt_y$

$GZ = vt_y - w$

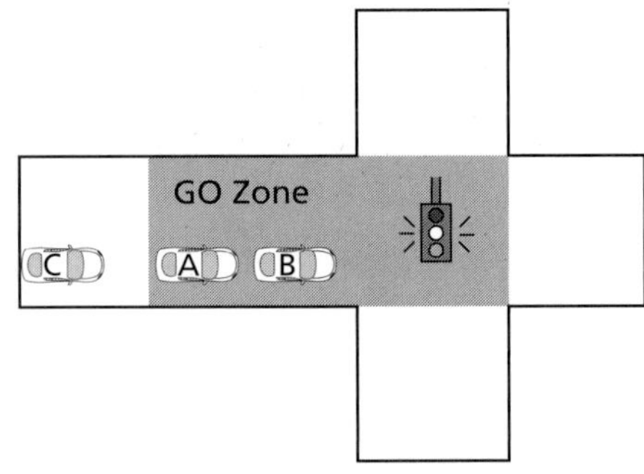

The GO Zone actually includes a range of possible distances. If automobile A goes through the intersection, as shown above, then any automobile closer than automobile A can also make it through the intersection. The correct equation for the GO Zone is $GZ \leq vt_y - w$. It can be shown in a sketch with a shaded area.

Computing the STOP Zone

Determining the STOP Zone requires one to realize that the stopping distance is the sum of the coasting distance (the distance you travel during the reaction time) and the braking distance. These factors were investigated in *Sections 3* and *5*. The coasting distance is an application once again of the equation: distance = velocity × time.

Because time is the reaction time, $d = vt_r$.

The braking distance is an application of the kinematics equation which relates velocity, acceleration, and distance given by $v^2 = 2ad$.

STOP Zone = velocity × reaction time + velocity squared/(2 × negative acceleration rate).

In symbols,

$$SZ = vt_r + \frac{v^2}{2a}.$$

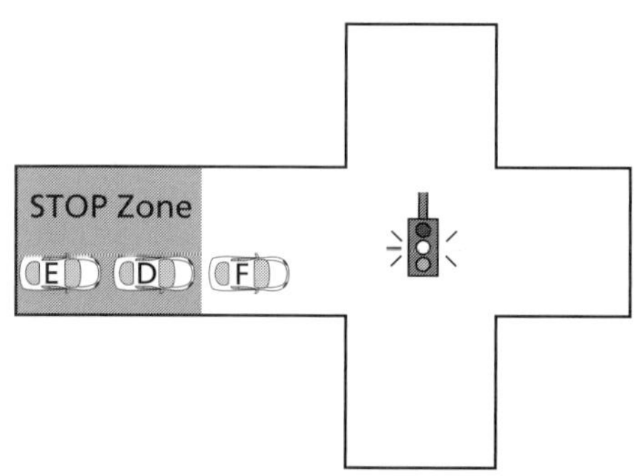

The STOP Zone actually includes a range of possible distances (shown in the shaded area in the diagram above). If automobile D is able to stop before the intersection, as shown above, then any vehicle further than automobile D can also stop before the intersection. The correct equation for the STOP Zone is

$$SZ \geq vt_r + \frac{v^2}{2a}.$$

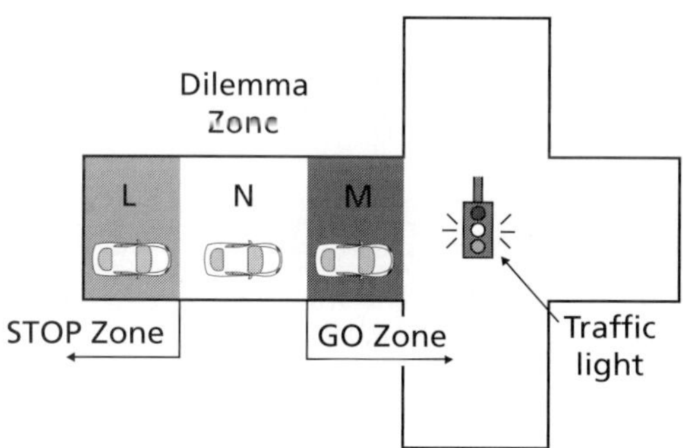

The GO Zone (where it is safe to continue when the light changes), and the STOP Zone (where it is safe to stop) are easily understood. The Dilemma Zone, however, introduces the area or region where the driver must make the more difficult decision of whether to stop or go. Automobile N is in the Dilemma Zone, shown in the diagram above. To calculate the Dilemma Zone, the five input variables must be examined separately.

The sample data below gives a good representation of the Dilemma Zone increasing as the speed of the vehicle increases. This zone is too dangerous to stop (not enough room to stop) and too dangerous to go through—when the light changes to red. There is a good possibility you will either be in the intersection or approaching it at the high speed at which you were initially driving or come to a stop in the intersection after the light has changed.

Intersection data	Velocity	STOP Zone	GO Zone	Dilemma Zone
Yellow-light time = 3 s	10 m/s	20 m	20 m	none
Reaction time = 1 s	15 m/s	37.5 m	35 m	2.5 m
Negative acceleration = 5 m/s²	20 m/s	60 m	50 m	10 m
Insertion width = 10 m				

The intersection will be unsafe if the GO Zone is smaller than the STOP Zone. The Overlap Zone provides a buffer, an area where it is safe to either go or stop. Because there is always some latitude in the way people drive, it is important to provide this Overlap Zone. An Overlap Zone appears when the GO Zone is larger that the STOP Zone, giving the driver time to decide whether to stop before the intersection or proceed through it.

The physics of the GO Zone and the STOP Zone could probably be made clearer if both functions were graphed. In the graph shown, the distances for the corresponding GO Zone and STOP Zone are displayed as a function of velocity.

The GO Zone is a linear graph $\left(GZ = vt_y + w\right)$. Its y-intercept changes with the width of the intersection and its slope changes with the yellow-light time. Either change can alter the Overlap and Dilemma Zones.

The STOP Zone is a parabola $\left(SZ = vt_r + v^2/2a\right)$. Its shape can be altered by varying the reaction time or the negative acceleration rate. The change in the shape or position of the parabola can alter the Overlap and Dilemma Zones. In the graph at the bottom of the next page, the STOP Zone is any distance above the straight line (the solid line), the GO Zone is any distance beneath the parabola (the dotted line), and the Overlap Zone is the area between the parabola and the straight line (shaded area). The Dilemma Zones (cross-hatched area) show where the driver is too close to the intersection to stop safely before the light turns from yellow to red, but does not have enough time to do so.

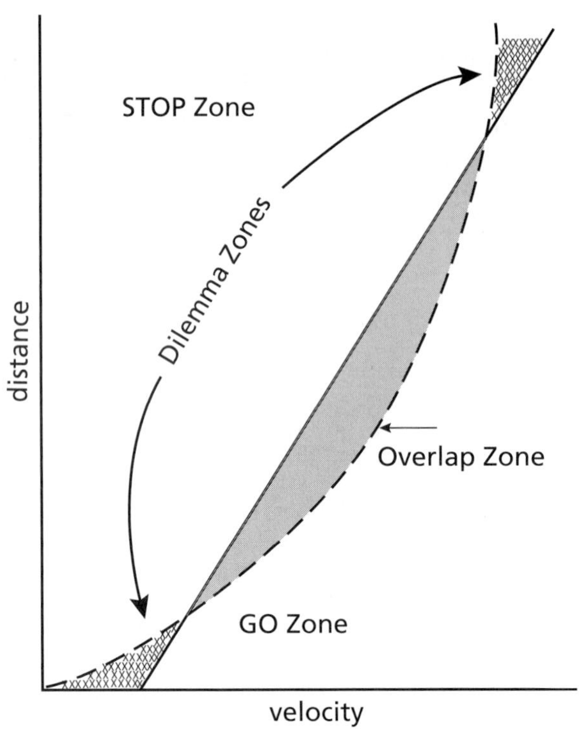

Even though any of the variables can eliminate a Dilemma Zone, it is fairly obvious that a change in the width of an intersection would be an expensive approach. Changing human reaction times, or assuming that people will be more alert, is hardly a worthwhile strategy. Likewise, the negative acceleration rate of cars is so dependent on individual cars and drivers that this is also a poor strategy. This leaves us with lengthening yellow-light time (which can be adjusted easily) or lowering the speed limit.

Crucial Physics

- Models can be used to describe mathematically what happens at a yellow light.
- An intersection with a yellow has a STOP Zone and a GO Zone.
- A safe intersection has an overlap between the GO and STOP Zones.
- An unsafe intersection has positions that are neither in the STOP Zone nor the GO Zone.

Learning Outcomes	Location in the Section	Evidence of Understanding
Investigate the factors that affect the STOP and GO Zones at intersections with traffic lights.	***Investigate*** Part A: Steps 1-9	Students should be able to determine the size of the STOP and GO Zones by using a spreadsheet.
Investigate the factors that result in an Overlap Zone or a Dilemma Zone at intersections with traffic lights.	***Investigate*** Part B: Steps 1-9	Students will be able to define an Overlap and a Dilemma Zone.
Use a computer simulation to mathematically model the situations that can occur at an intersection with traffic lights.	***Investigate*** Part A: Steps 5-9 Part B: Steps 1-9	After using the computer simulations, students will be able to determine the STOP, GO, Overlap, and Dilemma Zones.

Section 6 Materials, Preparation, and Safety

Materials and Equipment

PLAN A		
Materials and Equipment	**Group (4 students)**	**Class**
Multimedia DVD/CD Set		1 per class
Computer, station or calculator, CBL or equivalent system*	1 per group	

*Additional items needed not supplied

PLAN B		
Materials and Equipment	**Group (4 students)**	**Class**
Multimedia DVD/CD Set		1 per class
Computer, station or calculator, CBL or equivalent system*	1 per group	

*Additional items needed not supplied

Note: Time, Preparation, and Safety requirements are based on Plan A, if using Plan B, please adjust accordingly.

Time Requirement

This *Investigate* should take two class periods or 80 minutes.

Teacher Preparation

- Preview the video and note the automobiles that should have stopped for the yellow light and those that did not need to stop.
- A previously prepared spreadsheet is available, but you can easily make your own.
- Input the spreadsheet model into your spreadsheet software. Test your spreadsheet by changing the yellow-light time and see if your samples match those in the text.
- If you are using a local area network, simply download each spreadsheet to the workstations that the student groups will be using, or load the software on the individual stations.
- You need to make sure that specific jobs are assigned to each student in the group and rotate the jobs periodically. If the groups need to share or compare information, assign a communicator to represent each group.
- If you are using a low-tech alternative to this *Investigate*, you will need to copy a set of spreadsheets for each group of students.

Safety Requirements

- If you are in an urban environment, and choose to have the students observe an actual intersection, take the appropriate precautions. Obtain permission from your school administrator prior to taking the class outside the building.
- Follow safe usage for the computers.

Meeting the Needs of All Students

Differentiated Instruction: Augmentation and Accommodations

Learning Issue	Reference	Augmentation and Accommodations
Copying tables	***Investigate*** Part A: Step 5.b)	**Augmentation** • For students with visual-motor, graphomotor, and focus issues, copying tables is a tedious and time-consuming task. Model how to draw a table with 4 columns and 5 rows without any data in it first. Then students can fill in the data from the table in a sequential way (top to bottom or left to right). **Accommodation** • Provide students with a blank table to fill in and tape into their logs.
Tracking information on a spreadsheet	***Investigate*** Part A: Steps 6-9	**Augmentation** • Students with visual integration and focus issues struggle to track and differentiate numbers that are close together in a spreadsheet. Provide students with a ruler or index card to assist in tracking the desired data. **Accommodation** • Provide students with an enlarged copy of the spreadsheet.
Finding patterns in numbers	***Investigate*** Part A: Steps 7.c) and 9	**Augmentation** • Students with math and visual-spatial issues struggle to see patterns in groups of numbers. Limit the amount of numbers in the list to allow students to find the pattern amongst three numbers instead of ten. Give three choices for the pattern and allow students time to find the correct pattern. **Accommodation** • Give students the pattern/formula and ask them to prove why that formula works, given the data in the spreadsheet.
Understanding the differentiation of the four zones	***Investigate*** Part B ***Physics Talk*** Yellow-Light Model ***Physics Essential Questions***	**Augmentation** • Students with reading comprehension issues may be confused by the introduction of two more zones in this section. Draw a diagram to show the GO and STOP Zones with an Overlap Zone and another diagram to show the GO and STOP Zones with a Dilemma Zone. • As a class, create a list of factors that affect and do not affect the size of the zones. This list will help students answer the *Physics Essential Questions.*
Completing a series of calculations	***Physics to Go*** Question 5	**Augmentation** • Remind students to use *Physics to Go,* Question 1, to assist them in performing these calculations. • Students have difficulty switching back and forth between equations. Encourage students to compute all of the GO Zones first and then compute the STOP Zones. **Accommodation** • Limit the number of computations by asking students with more learning issues to complete Questions 5.a), 5.b), and 5.c) only.

Strategies for Students with Limited English-Language Proficiency

Learning Issue	Reference	Augmentation and Accommodations
Vocabulary comprehension	***Investigate*** Introduction	Check students' understanding of "computer simulation." They should know that it is a model of a real-world system by a computer for the purposes of study.
Vocabulary comprehension	***Investigate*** Part A: Step 2.a)	Ask an ELL student to interpret *Step 2.a)* in his or her own words. Listen to be sure the student has correctly interpreted "cutoff" in context as a point on the road behind which a vehicle cannot safely make it through the intersection before the light turns red.
Understanding charts Vocabulary comprehension	***Investigate*** Part A: Step 5	Explain to students that an input-output chart is really just a cause-and-effect chart. A change to any of the input variables (the cause) can affect the size of a safe GO Zone or the STOP Zone (the effect). For example, an increase in speed (a cause) requires a larger GO Zone (the effect) for an automobile to proceed safely through the intersection while the light is yellow. The "boundaries" (real or imaginary lines that mark off an area) need to be moved accordingly.
Comprehension	***Investigate*** Part B: Step 4.d)	To comprehend "Overlap Zone" and "Dilemma Zone," students must have some understanding of the words "overlap," "dilemma," and "zone." An overlap is a space, or zone, held in common by two side-by-side areas. The Overlap Zone, then, is an area from which a vehicle can either safely proceed through the yellow light or stop safely before the light turns red. A dilemma is a situation in which you have a choice between two undesirable options. So, a Dilemma Zone is an area from which a vehicle can neither safely proceed through the yellow light nor stop safely before the light turns red.
Paraphrasing	***Physics Talk*** Checking Up	Have students answer the *Checking Up* questions in writing, so you can assess their ability to express in their own words what they have learned.
Understanding slang	***Active Physics Plus*** Steps 7 and 8	ELL students may not understand the phrase "step on it." Explain that it means press your foot more forcefully on the accelerator pedal to make the vehicle go faster. "Beat the light" means speeding up to try to get through the intersection before the light turns red, even if you were not in the GO Zone when the light turned yellow.

SECTION 6

Teaching Suggestions and Sample Answers

What Do You See?

The function of the *What Do You See?* illustration is to set the stage for further student inquiry. The automobile screeching to a halt or the automobile zooming though the intersection will most likely draw the student's attention. The vivid illustration is designed to stimulate student's curiosity about yellow-light time at traffic intersections.

As students respond in a discussion, direct their attention to different aspects of the visuals. You might want to ask them how the colors that the artist has used also contribute to the meaning of each visual. Ask them if the artist employs an element of humor to convey the main purpose of this illustration.

Chapter 1 Driving the Roads

Section 6 Using Models: Intersections with a Yellow Light

What Do You See?

Learning Outcomes

In this section, you will

- **Investigate** the factors that affect the STOP and GO Zones at intersections with traffic lights.
- **Investigate** the factors that result in an Overlap Zone or a Dilemma Zone at intersections with traffic lights.
- **Use** a computer simulation to mathematically model the situations that can occur at an intersection with traffic lights.

What Do You Think?

Some traffic lights stay yellow for three seconds. Others stay yellow for six seconds.

- **If all traffic lights stayed yellow the same amount of time, how would this affect drivers' decisions at intersections?**
- **How could an intersection with a traffic light be dangerous?**

Record your ideas about these questions in your *Active Physics* log. Be prepared to discuss your responses with your small group and the class.

Investigate

In this *Investigate*, you will use a computer simulation to model the factors that affect the STOP and GO Zones at an intersection with a yellow light. You will then investigate what happens at an intersection when the STOP and GO Zones do not overlap.

Part A: Variables Affecting STOP and GO Zones at Intersections with Traffic Lights

1. Watch the video of an intersection, carefully noting what happens when the light turns yellow.

90

Students' Prior Conceptions

The student preconceptions identified in *Sections 1-6* may still persist, and is vital for the teacher to try to identify prior knowledge that may still not correlate with accepted theory. The prior conceptions focusing on inertia, forces, and friction are pertinent in *Section 6*. Because they were discussed previously, they will not be mentioned again. Students develop new analytical skills as they investigate the factors that result in an Overlap Zone or a Dilemma Zone at intersections with traffic lights. The use of computer simulations to mathematically model the situations addresses different styles of learning.

All seven stages of the 7E Learning Model apply to this section. One student prior conception dealing with friction should be emphasized.

1. **Friction always hinders motion; you always want to eliminate friction.** Frictional forces act to enable a vehicle to travel along the road as well as to enable a vehicle to stop. These forces allow vehicles to move forward, to back up, to negotiate curves, in addition to stopping motion. This phenomenon perplexes students who may consider friction as something that only impedes motion.

Encourage students to ponder the illustration with the title of the section in mind. Remind then that their impressions of the visuals will carry through to what they will be reading subsequently in the section.

What Do You Think?

These are questions that will have varied answers. Do not expect them to be correct. The purpose of these *What Do You Think?* questions is to make students reflect on how the length of the yellow-light time affects a driver's decision. Students should be encouraged to engage in discussing all the factors that drivers might consider when approaching a yellow light at an intersection.

What Do You Think?

A Physicist's Response

If all traffic lights stayed yellow for the same amount of time, numerous problems could develop. For vehicles traveling across very wide intersections, or in areas with a low speed limit, there might not be sufficient time to clear the intersection before the light changed to red. For areas with a higher speed limit, vehicles might not have enough time to stop prior to the intersection if the yellow-light time were short. The advantage would be that drivers would know how long the light would be yellow, and might be able to plan accordingly.

Intersections with traffic lights could be dangerous because an automobile that did not clear the intersection before the light turned red might be struck by another vehicle that is starting to enter the intersection when the light turned green.

As students explore factors that affect the STOP and GO Zones at intersections with traffic lights, it is instructive to guide them to identify all forces acting on the vehicle, the direction in which these forces act, and their magnitude. The focus on reaction time or the time to stop, as implemented by braking (friction) or other interactions in the Dilemma Zone, is important to the student evaluation of the crash worthiness of a situation. What happens to the motion of the vehicle and to the state of the vehicle and its passengers when the interaction time with stopping forces is larger or smaller? How does friction affect the forward motion of the vehicle?

NOTES

a) Are there vehicles that you think should have stopped?

b) Were there vehicles that stopped, but you think should have continued through the intersection?

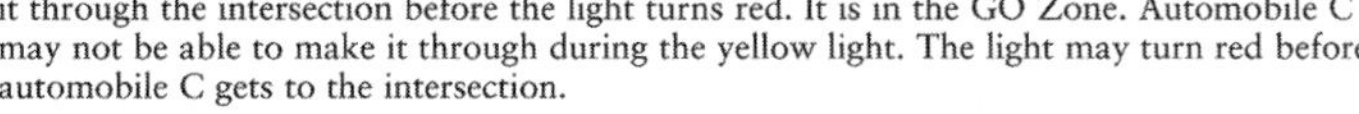

2. Watch the video a second time. This time pay attention to the position of the vehicles at the moment the light turns yellow.

a) Can you identify a "cutoff" point for a vehicle to make it through the intersection before the light turns red?

3. The diagram below shows the position of three automobiles at the moment a light turns yellow. Assume all three automobiles are moving with the same speed. Automobile A is able to make it through the intersection before the light turns red. It is in the GO Zone. Automobile C may not be able to make it through during the yellow light. The light may turn red before automobile C gets to the intersection.

a) Will automobile B be able to make it through during the yellow light?

b) Is automobile B in the GO Zone? Explain your answer.

c) Would any automobile closer to the intersection than automobile A be in the GO Zone?

d) Is automobile C in the GO Zone? What might happen if automobile C decides to continue?

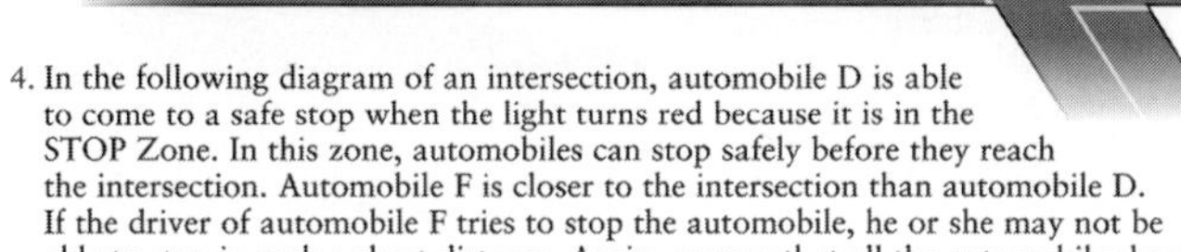

4. In the following diagram of an intersection, automobile D is able to come to a safe stop when the light turns red because it is in the STOP Zone. In this zone, automobiles can stop safely before they reach the intersection. Automobile F is closer to the intersection than automobile D. If the driver of automobile F tries to stop the automobile, he or she may not be able to stop in such a short distance. Again, assume that all the automobiles have the same initial speed.

a) Is automobile E in the STOP Zone? Explain your answer.

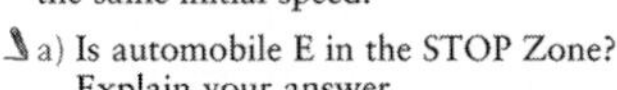

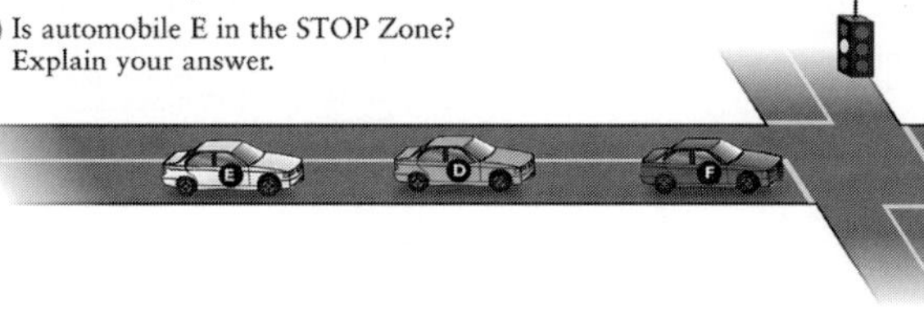

Investigate

Part A: Variables Affecting STOP and GO Zones at Intersections with Traffic Lights

1.a)-b)

Student observations and responses will vary.

Teaching Tip

Have students record their responses to the questions before viewing the video a second time. You may wish to work through the first few examples with the entire class, so you can be sure that the students understand the concept of the STOP and GO Zones.

2.a)

Student observations and responses will vary.

3.a)

Yes.

3.b)

Yes. Because automobile B is closer to the intersection than automobile A, it must be in the GO Zone.

3.c)

Yes.

3.d)

Not sure. It could be, but you can't tell from the data provided. It probably is not. It would be dangerous for the automobile to try.

4.a)

Yes. It is farther away from the intersection than automobile D, and if automobile D can stop, so should automobile E.

CHAPTER 1

4.b)

Not sure. It could be, but you can't tell from the data provided. It appears to be too close to the intersection to stop safely.

4.c)

Student sketches should resemble the diagrams below.

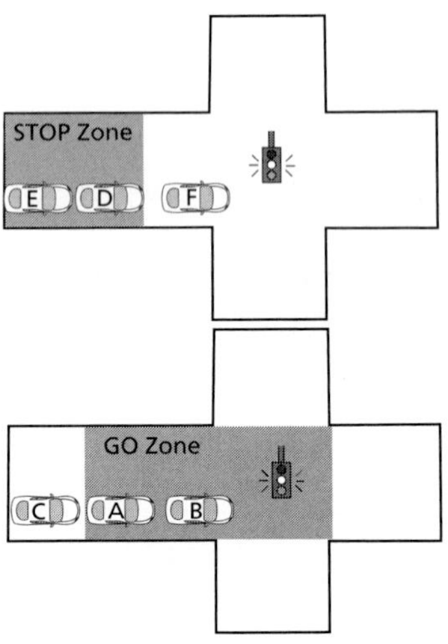

5.a)

Yellow-light time, driver reaction time, speed of vehicle, negative acceleration rate, width of intersection.

5.b)

Students copy table.

5.c)

Student predictions will vary. The predictions should include the following:

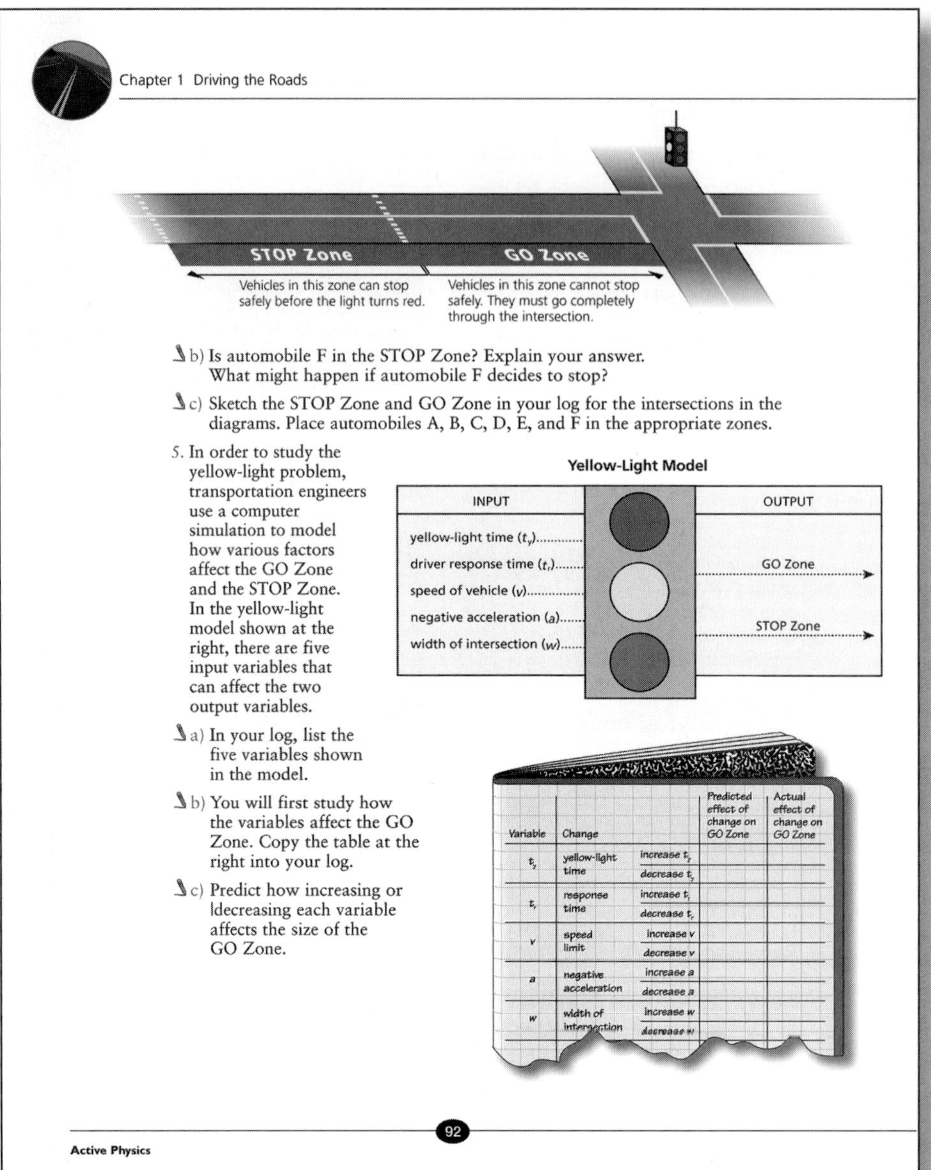

Chapter 1 Driving the Roads

b) Is automobile F in the STOP Zone? Explain your answer. What might happen if automobile F decides to stop?

c) Sketch the STOP Zone and GO Zone in your log for the intersections in the diagrams. Place automobiles A, B, C, D, E, and F in the appropriate zones.

5. In order to study the yellow-light problem, transportation engineers use a computer simulation to model how various factors affect the GO Zone and the STOP Zone. In the yellow-light model shown at the right, there are five input variables that can affect the two output variables.

a) In your log, list the five variables shown in the model.

b) You will first study how the variables affect the GO Zone. Copy the table at the right into your log.

c) Predict how increasing or decreasing each variable affects the size of the GO Zone.

Variable	Change		Predicted effect of change on GO Zone	Actual effect of change on GO Zone
t_y	yellow-light time	increase t_y		
		decrease t_y		
t_r	response time	increase t_r		
		decrease t_r		
v	speed limit	increase v		
		decrease v		
a	negative acceleration	increase a		
		decrease a		
w	width of intersection	increase w		
		decrease w		

Active Physics 92

Variable	Change	Predicted Effect of the change on the GO Zone	Actual Effect of the change on the GO Zone
yellow light time (t_y)	increase t_y decrease t_y	Student prediction	Increase Decrease
response time (t_r)	increase t_r decrease t_r	Student prediction	No change No change
speed (v)	increase v decrease v	Student prediction	Increase Decrease
negative acceleration rate (a)	increase a decrease a	Student prediction	No change No change
width of intersection (w)	increase w decrease w	Student prediction	Decrease Increase

Remember to consider one variable at a time. The other four variables will stay constant. For example, if the time the light is yellow increases from 3 s to 3.5 s, how will the boundaries and size of the GO Zone change? Will the zone increase or decrease? Record your predictions.

6. Look at the copies of the following spreadsheets.

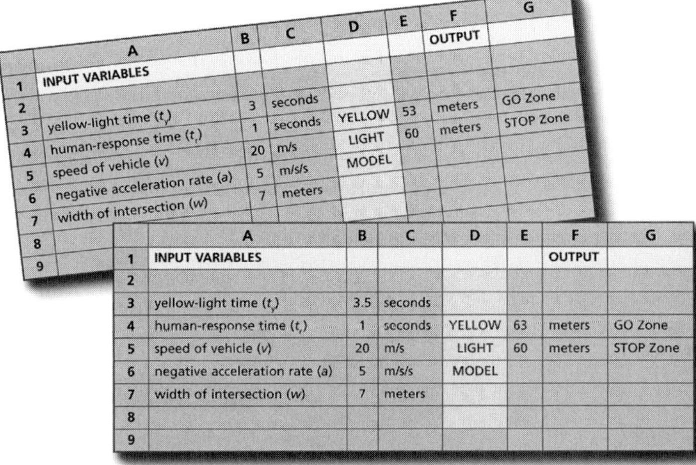

	A	B	C	D	E	F	G
1	INPUT VARIABLES					OUTPUT	
2							
3	yellow-light time (t_y)	3	seconds				
4	human-response time (t_r)	1	seconds	YELLOW	53	meters	GO Zone
5	speed of vehicle (v)	20	m/s	LIGHT	60	meters	STOP Zone
6	negative acceleration rate (a)	5	m/s/s	MODEL			
7	width of intersection (w)	7	meters				
8							
9							

	A	B	C	D	E	F	G
1	INPUT VARIABLES					OUTPUT	
2							
3	yellow-light time (t_y)	3.5	seconds				
4	human-response time (t_r)	1	seconds	YELLOW	63	meters	GO Zone
5	speed of vehicle (v)	20	m/s	LIGHT	60	meters	STOP Zone
6	negative acceleration rate (a)	5	m/s/s	MODEL			
7	width of intersection (w)	7	meters				
8							
9							

a) What is the distance of the GO Zone if the yellow-light time is 3 s?

b) What happens to the GO Zone when the yellow-light time is increased to 3.5 s?

c) Would increasing the yellow-light time allow you to get through the intersection from a further distance away? Explain your answer in your log.

d) Record the effect of changing the yellow-light time in your log.

7. Use a computer spreadsheet program to obtain quantitative data (numbers you can use to test your predictions). Remember to change only one variable at a time. For each variable you investigate, record the following:

a) Did the effect of the change of the variable make sense to you? Explain your answer.

b) Record the effect of changing each variable in your *Active Physics* log. How did the actual effect compare with your predictions in *Step 5.c)*?

6.a)

53 m.

6.b)

The GO Zone increases to 63 m.

6.c)

Yes, increasing the yellow-light time will allow more time to get through the intersection, so the driver will have a larger GO Zone.

6.d)

The students will see from the spreadsheet that increasing the yellow-light time increases the GO Zone.

7.a)

Students record their response.

7.b)

Students will respond either in agreement with their predictions, or with a correction of an incorrect prediction.

7.c)

GO Zone = $vt_y - w$ =
(speed of car × yellow-light time) – width of intersection

7.d)

By increasing the yellow-light time, there is more time to get through the yellow light safely. If there is more time, then at the same speed, you can cover a greater distance. Thus, the GO Zone increases. By increasing the speed, you can cover a greater distance in a shorter time; therefore, the GO Zone increases. The wider the intersection, the greater the distance that needs to be covered; therefore, the GO Zone is smaller.

7.e)

For the GO Zone, the driver does not have a reaction time because it is associated with how long it takes the driver to react to the yellow light and apply the brakes. Likewise, braking (negative acceleration) doesn't occur when the driver is in the GO Zone.

8.a)

Student predictions should be recorded in the STOP Zone chart below.

c) Look for a pattern in your results and try to determine how the GO Zone is calculated by the spreadsheet. Click on the cell that gives the GO Zone value. Look at the formula bar and convert this notation to an equation. Record this equation in your log.

E	F	G
	OUTPUT	
= (B5*B3)-B7	meters	GO Zone
= (B5*B4)+(B5²)/(2*B6)	meters	STOP Zone

d) Discuss with your group and explain why the yellow-light time, speed, and the width of the intersection appear in the equation for the GO Zone. Record this information in your log.

e) Why do the reaction time and negative acceleration not appear in the equation? Why do these two variables have no effect on the GO Zone?

8. Now you will investigate how changing each variable will affect the STOP Zone.

a) Record your predictions in a chart similar to the one you used for the GO Zone.

b) Use the spreadsheet investigation to find the actual effect of each variable on the STOP Zone. Record the effect in your chart.

c) Compare your prediction with the actual effect. Do your results make sense to you? Explain your answer.

9. Look for patterns in your results and try to determine how the STOP Zone is calculated by the spreadsheet.

a) Click on the cell that gives the STOP Zone value. Look at the formula bar and convert this notation to an equation. Record this equation in your log.

b) Discuss the relationship with your group and explain why the yellow-light time and the width of the intersection do not appear in the equation for the STOP Zone.

c) Why do the reaction time, velocity, and negative acceleration appear in the equation? Why do the other two variables have no effect on the STOP Zone?

Variable	Change	Predicted Effect of change on the STOP Zone	Actual Effect of the change on the STOP Zone
yellow light time (t_y)	increase t_y decrease t_y	Student prediction	No change No change
response time (t_r)	increase t_r decrease t_r	Student prediction	Increase Decrease
speed (v)	increase v decrease v	Student prediction	Increase Decrease
negative acceleration rate (a)	increase a decrease a	Student prediction	Decrease Increase
width of intersection (w)	increase w decrease w	Student prediction	No change No change

8.b)

Students will use the spreadsheet to verify their answers and record them in the column for "Actual Effect of the Change on the STOP Zone."

8.c)

While responding, students should look for understanding of why the width of the intersection and the yellow-light time do not affect the STOP Zone.

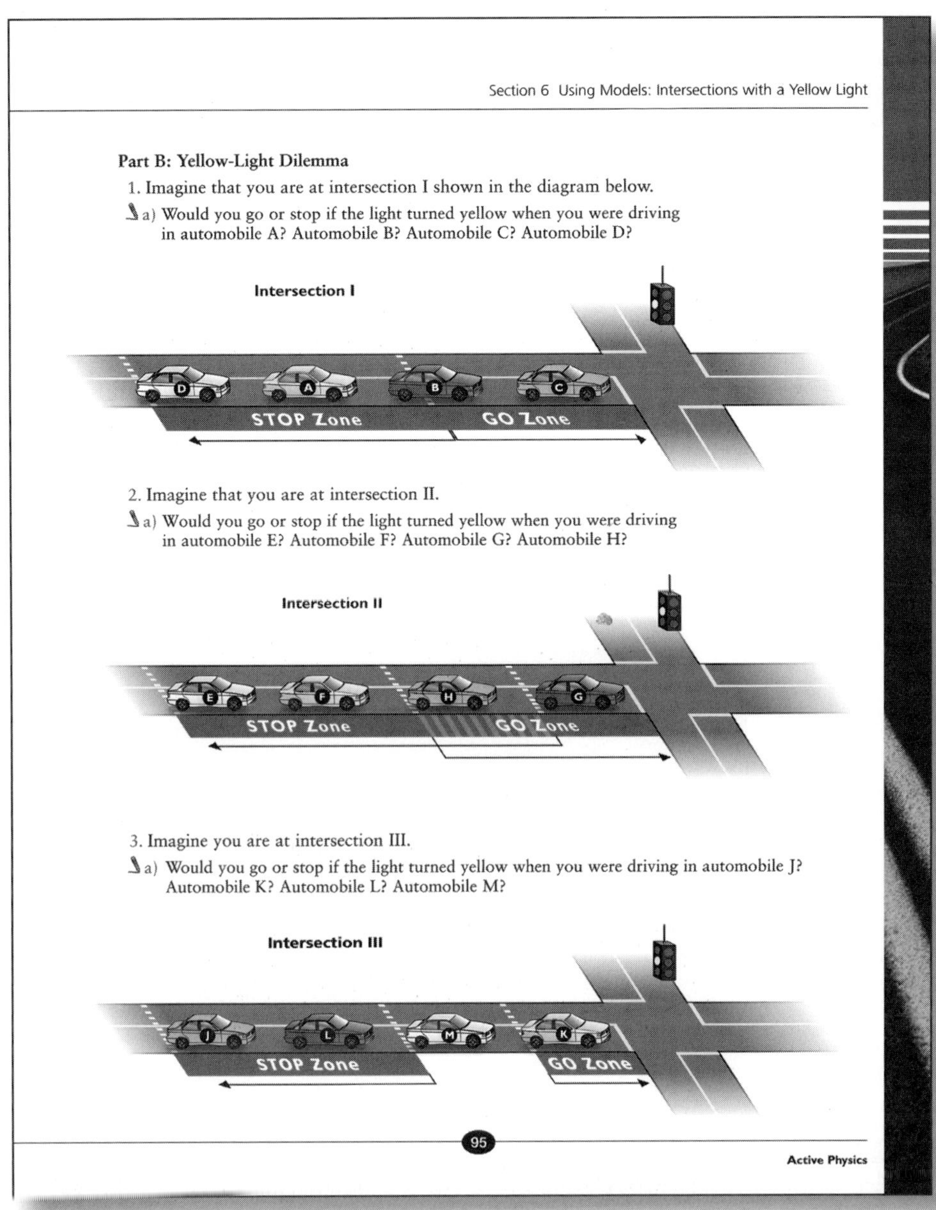
Section 6 Using Models: Intersections with a Yellow Light

Part B: Yellow-Light Dilemma

1. Imagine that you are at intersection I shown in the diagram below.
   a) Would you go or stop if the light turned yellow when you were driving in automobile A? Automobile B? Automobile C? Automobile D?

2. Imagine that you are at intersection II.
   a) Would you go or stop if the light turned yellow when you were driving in automobile E? Automobile F? Automobile G? Automobile H?

3. Imagine you are at intersection III.
   a) Would you go or stop if the light turned yellow when you were driving in automobile J? Automobile K? Automobile L? Automobile M?

95

Active Physics

9.a)

Students record equations from spreadsheet.

9.b)

Students should recognize that the formula for stopping distance is the distance traveled during the reaction time plus the braking distance as shown in the equation below.

(speed of car × reaction time) +
(speed × speed) ÷
(2 × negative accleration)

9.c)

The reaction time and automobile speed determine how far an automobile travels toward the intersection before the brakes are applied, and the speed and the negative acceleration determine the braking distance. The length of time the light stays yellow does not affect the distance in which you can stop. Also, the width of the intersection has no effect on the distance required to stop.

Part B: Yellow-Light Dilemma

1.a)

The driver would stop in automobile A, would go in automobile B, would go in automobile C, and would stop in automobile D.

2.a)

The driver would stop in automobile E, would stop in automobile F, would go in automobile G, and would stop or go in automobile H.

3.a)

The driver would stop in automobile J, would go in automobile K, would stop in automobile L, and would not be able to stop or go safely in automobile M.

4.a)

In Intersection II, the GO Zone and the STOP Zone overlap. In Intersection III, there is a gap between the two zones.

4.b)

The choice would be either to stop or to go. Both choices are safe because the car is in both the STOP and GO Zones.

4.c)

The choice would be either to stop or to go. Neither choice would be safe because the car is in neither the STOP nor the GO Zone.

4.d)

Intersection II has an Overlap Zone, and Intersection III has a Dilemma Zone.

5.a)

The spreadsheet calculates the difference between the GO and STOP Zones. If there is a negative value between the STOP and GO Zones, the intersection is unsafe.

5.b)

No, there is no Overlap or Dilemma Zone at 20 m/s (45 mi/h).

4. Compare the GO Zone and the STOP Zone for intersections I, II, and III.
   a) How are the intersections different?
   b) In intersection II, if the light turned yellow when you were between the GO Zone and the STOP Zone, what would your choices be? Which choice(s) would be safe? Explain your answer.
   c) In intersection III, if the light turned yellow when you were in the space between the STOP Zone and the GO Zone, what would your choices be? Which choice(s) would be safe? Explain your answer.
   d) When both choices are safe, the space between the GO and STOP Zones is called the Overlap Zone. When neither choice is clearly safe, it is called the Dilemma Zone. Intersections with a Dilemma Zone are not safe. Which intersection has an Overlap Zone and which has a Dilemma Zone?
5. Use a computer spreadsheet program, similar to the one you used in *Part A*. There is an additional outcome that tells you whether the intersection is safe and has an Overlap Zone or is unsafe and has a Dilemma Zone. Use the spreadsheet to determine ways in which an unsafe intersection can be made into a safe intersection. Which variables, when adjusted incorrectly, could make the intersection unsafe?

	A	B	C	D	E	F	G
1	INPUT VARIABLES					OUTPUT	
2							
3	yellow-light time (t_y)	3.7	seconds				
4	human-response time (t_r)	1.2	seconds	YELLOW	64	meters	GO Zone
5	speed of vehicle (v)	20	m/s	LIGHT	64	meters	STOP Zone
6	negative acceleration rate (a)	5	m/s/s	MODEL	0	meters	Overlap Zone
7	width of intersection (w)	10	meters				Safe
8							
9							
10							
11	yellow-light time (t_y)	3.7	seconds				
12	human-response time (t_r)	1.2	seconds	YELLOW	101	meters	GO Zone
13	speed of vehicle (v)	30	m/s	LIGHT	126	meters	STOP Zone
14	negative acceleration rate (a)	5	m/s/s	MODEL	-25	meters	UNSAFE
15	width of intersection (w)	10	meters				
16							

   a) How does the spreadsheet determine whether the intersection is safe? What is the relationship between the GO Zone and the STOP Zone at an unsafe intersection?
   b) Use the sample spreadsheet shown above. Is there an Overlap or Dilemma Zone at 20 m/s (45 mi/hr)?

1-6a Blackline Master

c) What happens to the GO Zone and the STOP Zone when the speed is increased to 30 m/s (65 mi/hr)? Is there still an Overlap Zone or Dilemma Zone?

d) Now lower the speed to 10 m/s (20 mi/hr). Is the intersection safer now? Explain your answer.

6. Continue your investigation by resetting the speed to its original value of 20 m/s. Adjust the yellow-light time and determine its effect on the Dilemma and Overlap Zones.

a) Record the results of this investigation in your log. (Include the changes you make to the variable as well as their effect on the zones.)

7. What effect do reaction time, negative acceleration, and width of the intersection have on the safety of the intersection? Does changing any of these variables create a Dilemma Zone? Conduct investigations with your spreadsheet.

a) Record the results in your log. (Include the changes you make to the variable as well as their effect on the zones.)

8. More than one variable change can eliminate a Dilemma Zone and replace it with an Overlap Zone.

a) Of the five variables, explain the ease or difficulty in changing each one to make the intersection safer. For example, why might you suggest changing the yellow-light time rather than changing the width of the intersection?

9. The yellow-light problem is based on a simple model and only provides approximate calculations. It does not include other factors such as whether the road is flat or the length of the automobile.

a) How does the length of the automobile affect the model? Which outputs are affected by the length of the automobile?

5.c)

Both the GO Zone and the STOP Zone increase. There is a Dilemma Zone.

5.d)

The GO Zone becomes 27 m and the STOP Zone is 22 m. The Overlap Zone is 5 m, so the intersection is safer.

6.

Students record their observations. Changing the yellow-light time changes the GO Zone, so when it overlaps the STOP Zone the intersection is safe, but it creates a Dilemma Zone, which is unsafe.

7.a)

Students record the results in their logs. They need to see that increasing the GO Zone to at least the STOP Zone will eliminate the Dilemma Zone. Therefore, decreasing width, decreasing reaction time, and increasing the negative acceleration rate will make the intersection safer by eliminating the Dilemma Zone.

8.a)

Reaction time is hard to change. The width of an intersection is very difficult to change, as it involves rebuilding the road. The speed limit on the road could be changed, but it is difficult to enforce it at all intersections. The negative acceleration rate of vehicles is very difficult to change because different vehicles have different deceleration rates, plus the rate changes depending on the weather, road surface, hills, temperature of tires and road, etc. The easiest variable to change would be the yellow-light time. This might involve only changing a timer or, at a central location, reprogramming a computer.

9.a)

The length of the automobile is affected by the width of the intersection as it takes a longer period to clear the intersection. Therefore, the output which is affected would be the GO Zone.

Physics Talk

The *Physics Talk* highlights the importance of models. Ask students to give you some examples of scientific models that have helped them understand their world. Write up a list of their responses on the board. It is important that they discuss how these models serve to explain scientific ideas. Highlight the significance of the yellow-light model to demonstrate how automobiles can safely pass through an intersection and how engineers use this model to determine if an intersection is safe or how it could be made safer.

Revisit the variables that affect the GO Zone, discuss each variable listed in the *Physics Talk,* and ask students to record how these variables affect the GO Zone in their *Active Physics* logs. Discuss why braking acceleration and a driver's reaction time are variables that don't affect the GO Zone. Determine if students understand the effect of intersection width on the GO Zone.

When discussing the STOP Zone, refer to the spreadsheet model and have students assess which variables would affect the STOP Zone. Ask them to explain why the width of the intersection and the yellow-light time have no effect on the STOP Zone. Consider giving a quiz with different features of the STOP Zone and GO Zone. Make sure they are also aware of the variables that do not affect the STOP Zone.

For Overlap and Dilemma Zones, consider inviting a student to draw a diagram of an Overlap Zone on the board. Then ask each student in the class to draw a diagram that shows an intersection with a Dilemma Zone. Allow students time to focus on the difference between a Dilemma Zone and an Overlap Zone by giving them assignments that require a detailed description of an Overlap Zone and a Dilemma Zone. You could ask them why an Overlap Zone makes an intersection safer, while a Dilemma Zone makes an intersection unsafe. Students will benefit by summarizing the meaning of a Dilemma Zone and an Overlap Zone in their *Active Physics* logs.

While analyzing models, ask students to explain why they developed a yellow-light model and how this model helps traffic engineers. Discuss the limitations of a yellow-light model and have students consider situations in which the yellow-light model is limited in its use. Emphasize that a model is not the same as an actual event; a model describes and analyzes an event.

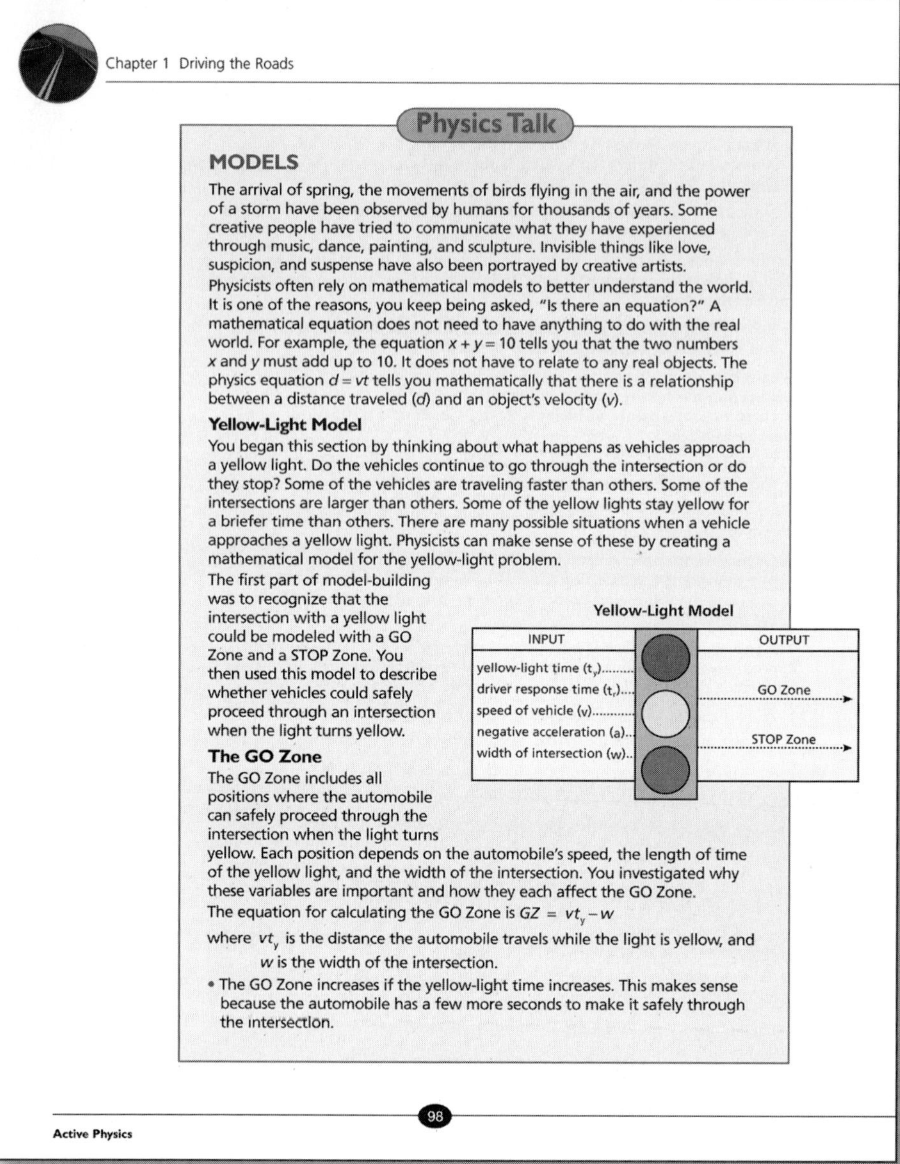

Chapter 1 Driving the Roads

Physics Talk

MODELS

The arrival of spring, the movements of birds flying in the air, and the power of a storm have been observed by humans for thousands of years. Some creative people have tried to communicate what they have experienced through music, dance, painting, and sculpture. Invisible things like love, suspicion, and suspense have also been portrayed by creative artists.

Physicists often rely on mathematical models to better understand the world. It is one of the reasons, you keep being asked, "Is there an equation?" A mathematical equation does not need to have anything to do with the real world. For example, the equation $x + y = 10$ tells you that the two numbers x and y must add up to 10. It does not have to relate to any real objects. The physics equation $d = vt$ tells you mathematically that there is a relationship between a distance traveled (d) and an object's velocity (v).

Yellow-Light Model

You began this section by thinking about what happens as vehicles approach a yellow light. Do the vehicles continue to go through the intersection or do they stop? Some of the vehicles are traveling faster than others. Some of the intersections are larger than others. Some of the yellow lights stay yellow for a briefer time than others. There are many possible situations when a vehicle approaches a yellow light. Physicists can make sense of these by creating a mathematical model for the yellow-light problem.

The first part of model-building was to recognize that the intersection with a yellow light could be modeled with a GO Zone and a STOP Zone. You then used this model to describe whether vehicles could safely proceed through an intersection when the light turns yellow.

The GO Zone

The GO Zone includes all positions where the automobile can safely proceed through the intersection when the light turns yellow. Each position depends on the automobile's speed, the length of time of the yellow light, and the width of the intersection. You investigated why these variables are important and how they each affect the GO Zone.

The equation for calculating the GO Zone is $GZ = vt_y - w$

where vt_y is the distance the automobile travels while the light is yellow, and w is the width of the intersection.

- The GO Zone increases if the yellow-light time increases. This makes sense because the automobile has a few more seconds to make it safely through the intersection.

98

Active Physics

Section 6 Using Models: Intersections with a Yellow Light

- The GO Zone decreases if the width of the intersection increases because the vehicle has to travel a greater distance to safely go through the intersection at the same speed.
- The GO Zone increases with an increase in speed because the vehicle can travel a greater distance during the time that the light is yellow.
- The braking acceleration does not affect the GO Zone because vehicles trying to go through the intersection do not use the brakes.
- The driver's reaction time is also not important because a driver who is going to continue through the intersection does not have to decide to use the brakes and move from the gas to the brake.

The STOP Zone

The STOP Zone includes all positions where the car can safely stop before reaching the intersection. Each position depends on the vehicle's speed, the braking acceleration, and the driver's reaction time. You investigated why these variables are important and how they each affect the STOP Zone.

The equation for calculating the STOP Zone is $SZ = vt_r + \frac{v^2}{2a}$

where vt_r is the distance the vehicle travels as the driver decides to stop (the reaction distance), and

$\frac{v^2}{2a}$ is the braking distance.

- The STOP Zone decreases if the speed decreases. A slower-moving vehicle can be closer to the intersection and still stop safely.
- The STOP Zone decreases if the braking acceleration increases. Better brakes means that a vehicle can stop in a shorter distance.
- The STOP Zone decreases if the human reaction time decreases. The faster someone can react and move the foot from the gas to the brake, the smaller the distance required to stop.
- The width of the intersection does not affect the STOP Zone. Because vehicles in the STOP Zone do not enter the intersection, the size of the intersection does not matter.
- The yellow-light time does not affect the STOP Zone. Because you are stopping your vehicle, it does not matter how long the light stays yellow.

Overlap Zone and Dilemma Zone

A properly designed yellow-light intersection must be safe for drivers when they are obeying the posted speed limits. For this reason, the STOP and GO Zones are adjusted so that a driver can either safely stop the vehicle before the intersection when the light turns yellow, or safely proceed through the intersection before the light turns red. The STOP and GO Zones must form an Overlap Zone. In the Overlap Zone, when the light turns yellow, the driver of the vehicle has the choice of either stopping or going through the intersection safely before the light turns red.

Checking Up

1.

The spreadsheet is referred to as a model because it is used to represent the STOP and GO Zones of an intersection with a yellow light.

2.

The GO Zone is an area in an intersection which includes all positions through which a vehicle can pass safely.

3.

The STOP Zone is an area in an intersection where the automobile can safely stop before reaching the intersection.

4.

The Overlap Zone is an area where the STOP and GO Zones overlap, and the driver has the safe choice of either making it through the intersection or stopping before a red light.

5.

When the STOP and GO Zones are separated by a space, the space represents an area where it is unsafe for vehicles to pass through if the light turns yellow. This zone poses a dilemma between two dangerous choices: to pass through or stop before the light turns red.

Chapter 1 Driving the Roads

In a poorly designed intersection, a Dilemma Zone may exist. The Dilemma Zone is a space between the STOP and GO Zones. In this region, a driver traveling the speed limit cannot safely stop before the intersection or pass through the intersection completely before the light turns red. The driver is faced with the dilemma of choosing between two dangerous options. Fortunately, traffic engineers rarely make errors such as this.

Limitations of the Yellow-Light Model

You began using a simple input/output table based on the model. You input the values for the speed of the vehicle, the width of the intersection, the yellow-light time, the braking acceleration, and the reaction time. The model then provided the output values for the GO Zone and the STOP Zone. You then explored how the model calculated these values. The physics equations of motion were used to determine the values for the GO Zone and the STOP Zone.

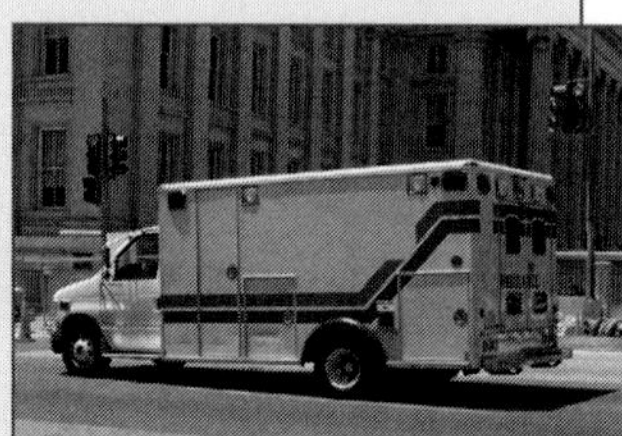

The GO Zone and STOP Zone model lets you analyze an intersection much more precisely than merely watching the motion of cars. Most drivers do not know about the GO Zone and the STOP Zone. They have a sense of what to do when the light turns yellow, but have not developed a model in the way that you have. This model helps traffic engineers to determine if an intersection is safe or if it can be made safer by lowering the speed limit or lengthening the yellow-light time.

All models have limitations. The GO Zone and STOP Zone models that you developed do not take into account the length of the vehicle. You can expand the model by including this variable. The model does not include the colors of the vehicles. A person viewing an intersection may notice that an ambulance is responding to an emergency or a series of vehicles are part of a procession. This information may help explain the motion of the vehicles. The GO Zone and STOP Zone model does not take either of these situations into account. The colors of the vehicles is not an important factor in determining the GO and STOP Zones. You should always ask yourself where the model is useful and where the model is limited. You should also remind yourself that the model is not the same as the actual event. For example, the equations for the GO Zone and STOP Zone are not the same as vehicles approaching a yellow light.

In this section, you worked with a traffic model. In much of physics, you will be working with models that attempt to describe nature. There will be models that describe how forces affect motion, how light behaves and how something called energy can neither be created nor destroyed. The laws of physics are mathematical models that try to accurately predict what will occur and try to give you a glimpse into how nature works.

Checking Up

1. In this section, the spreadsheet is referred to as a model. What makes it a model?
2. In your own words, describe what is meant by the GO Zone.
3. In your own words, describe what is meant by the STOP Zone.
4. Describe what is meant by the Overlap Zone.
5. Describe what is meant by the Dilemma Zone.

Active Physics

Section 6 Using Models: Intersections with a Yellow Light

Active Physics *Plus*

+Math	+Depth	+Concepts	+Exploration
♦♦	♦♦	♦	

Speed and the Yellow-Light Model

As a driver, you have control over the speed of your automobile. How does that speed affect the GO and STOP Zones for a yellow light?

1. Predict how increases or decreases in speed will affect the GO Zone and STOP Zone.
2. Using a spreadsheet program or a graphing calculator, graph the relationship of the GO Zone vs. different speeds.
3. Graph the relationship of the STOP Zone vs. different speeds.
4. Using a graphing calculator or a spreadsheet program, graph the STOP Zone and the GO Zone vs. speed on the same set of axes. Try different values for the other variables from your earlier work.
5. Indicate the Overlap or Dilemma Zones on your graph.
6. In the *Investigate*, you analyzed how decreasing the speed of the automobile can eliminate the Dilemma Zone. Your graph may indicate that there is a new Dilemma Zone at very low speeds. Explain how this can be.
7. In the *Scenario* for this chapter, the teenager jokes that a yellow light means "step on it." Would accelerating help you get through a yellow light? Calculate how much it may help.
8. Revise the spreadsheet model to test your answer to the question above. How would an acceleration of your automobile affect the GO Zone? Try to use realistic values of the acceleration and the time you have to accelerate. If you do increase your speed and find yourself in the intersection when the light has turned red, an accident at the faster speed will be much more severe. Assuming you are traveling at the speed limit, calculate how far above the speed limit you may be going to try to "beat the light."

What Do You Think Now?

At the beginning of this section, you were asked the following:

- **If all traffic lights stayed yellow the same amount of time, how would this affect drivers' decisions at intersections?**
- **How could an intersection with a traffic light be dangerous?**

Now that you have investigated the effects of factors on a GO Zone and STOP Zone, what do you think would be the effect of having all lights stay yellow for the same length of time? Explain what could make an intersection with a traffic light dangerous and what can be done to correct the situation. Include the term Dilemma Zone in your explanation.

Active Physics

Active Physics Plus

Students answer questions posed in *Active Physics Plus* by predicting the effect on STOP and GO Zones and using a spreadsheet program to graph the relationship of the STOP and GO Zones vs. different speeds. Drawing graphs provides a visual understanding of the Overlap and Dilemma Zones. Encourage students to share their graphs with others, and ask them to explain how Dilemma Zones can be changed into Overlap Zones or GO Zones can increase with an increase in acceleration.

1.

As speed increases, the GO Zone also increases in direct proportion, since $GZ \leq vt_y - w$. Likewise, the STOP Zone is given by $SZ \geq vt_r + v^2/2a$ and also increases with velocity, but at a faster rate than the GO Zone.

2.

The GO Zone should resemble the graph below.

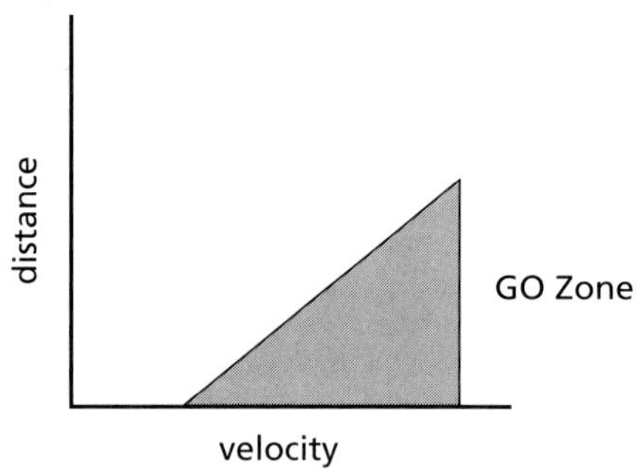

3.

The STOP Zone should resemble the graph below.

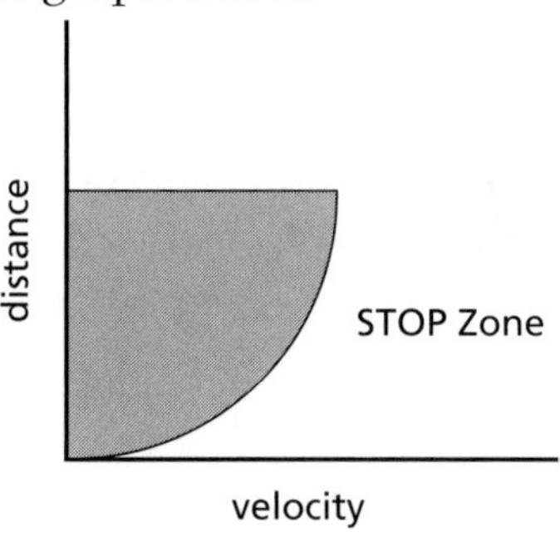

4.

The STOP Zone and GO Zone graphs should be the overlap of the preceding two graphs, and appear like the graph below.

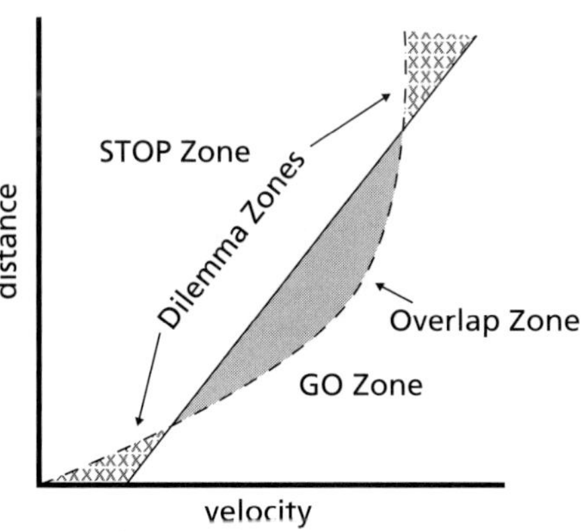

5.

In the graph above, the Overlap Zone is shown by the shaded region, and the Dilemma Zone by the cross-hatched region of the graph.

CHAPTER 1

6.

At very low speeds, the width of the intersection becomes important in determining how long it takes to safely pass through. At low speeds, it may be possible that an automobile is too close to the intersection to stop before entering it, and still the intersection might be too wide for the car to get across before the yellow light turns red.

7.

Yes, accelerating the automobile may help in expanding the size of the GO Zone, possibly enough to eliminate the Dilemma Zone for a driver. How much it will help will depend upon the amount of acceleration.

8.

The speed above the speed limit can be calculated from the formula $v_f = v_i + at$. Here, v_i will be the speed limit, a will be the vehicle's acceleration, and v_f will be the speed above the speed limit.

What Do You Think Now?

Students should reflect on their responses to the *What Do You Think?* section and decide how their original answers could be modified. The effects of various factors on the driver's decision to stop or go through an intersection when approaching a yellow light can be more thoroughly examined at this stage.

You could share the answers found in *A Physicist's Response* and discuss students' opinions. Revisiting prior questions can prompt other queries that you should encourage in order to build a platform that leads to a stimulating discussion. Yellow-light dilemmas and decisions are critical to human safety. Ensure that your students understand why intersections at traffic lights could be dangerous.

NOTES

NOTES

Reflecting on the Section and the Challenge

Reflecting on the *Investigate* and *Physics Talk* can help students make important connections between reaction time, length of yellow-light time, stopping distance, width of an intersection, and rate of deceleration while approaching a yellow light. All these factors that contribute to a driver's decision can be incorporated into the challenge. Students should read the *Reflecting on the Section and the Challenge* to help them compose and present their *Chapter Challenge* project.

Physics Essential Questions

What does it mean?

What factors determine the size of the GO Zone, the STOP Zone, and whether an intersection has a Dilemma Zone?

How do you know?

What measurements can you make to test your understanding of the GO, STOP, Overlap, and Dilemma Zones? In this section, you used equations and calculations on a computer model (spreadsheet) to determine these zones. How could you verify the conclusions from your spreadsheet to determine the zones?

Why do you believe?

Connects with Other Physics Content	Fits with Big Ideas in Science	Meets Physics Requirements
Forces and motion	* Models	Makes mathematical sense

* Physics uses mathematical models to describe physical situations. How do the GO Zone and STOP Zone models help you to improve your understanding of traffic at a yellow-light intersection?

Why should you care?

How can understanding the physics behind the GO, STOP, Overlap, and Dilemma Zones make you a better, more aware, and more informed driver?

Reflecting on the Section and the Challenge

In earlier sections, you learned that an automobile travels a certain distance while you are moving at a constant velocity and deciding whether to stop. You also learned that your automobile travels a certain distance after the brakes have been applied.

In this section, you learned that deciding what to do when you see a yellow light is not a simple decision. It requires a judgment of the distance to the intersection, the width of the intersection, and how much time it will take you to get there at the speed you are traveling. You also need some sense of how well your brakes work and how quickly you can respond. Finally, you need to know the time that the light is yellow before turning red.

You also know that it is important for traffic engineers to make sure that an intersection has an Overlap Zone and not a Dilemma Zone between the STOP and GO Zones.

You now know which factors affect the zones at an intersection. You know any intersection can be made safer by lowering the speed limit or by lengthening the yellow-light time. You also know how these zones may change if your reaction time is slower or if your negative acceleration is being affected by bad weather or road conditions. Part of the *Chapter Challenge* is to explain these factors and how they affect your reaction time based on your investigations and conclusions.

Physics Essential Questions

What does it mean?

The GO Zone is dependent on the speed of the car, the yellow-light time, and the width of the intersection. The STOP Zone is dependent on the speed of the car, the braking acceleration, and the reaction time. The intersection has a Dilemma Zone if there is a gap between the GO Zone and the STOP Zone.

How do you know?

You can measure the width of an intersection, the speed of cars, and the yellow-light time. You can use an approximate value for the braking acceleration and reaction time of drivers. Using motion equations, you can determine the zones, as well as see if cars in the calculated zones travel through or stop at the intersections.

Why do you believe?

The GO and STOP Zones provide a new way to look at the intersection. Without this model, people just guess as to whether they should go or stop at the yellow light.

Why should you care?

Understanding how the speed of your car affects the GO and STOP Zones will encourage you to be more aware of your speed as you approach intersections. You can also conclude that speeding up to make it through a yellow light is particularly dangerous.

Physics to Go

1. An *Active Physics* student group is studying an intersection. The width of the intersection is measured by pacing and is found to be approximately 15-m wide. The yellow-light time for the intersection is 4 s. The speed limit on this road is 30 mi/hr (approximately 15 m/s). The speed of an automobile decreases by 5 m/s every second during negative acceleration. Assume that the people who are driving the automobiles have a reaction time of 1 s.

   a) Calculate the GO Zone using the math equation on the computer spreadsheet. Use a calculator. To guide you, the first two steps are provided for you.

   GO Zone = (velocity × yellow-light time) – width of intersection

   $$GZ = vt_y - w$$

   $$GZ = (15 \text{ m/s})(4 \text{ s}) - 15 \text{ m}$$

   $$GZ =$$

   b) Calculate the STOP Zone using the math equation on the computer spreadsheet. Use a calculator to help you.

   STOP Zone = (velocity × reaction time) + velocity²/(2 × negative acceleration)

   $$SZ = vt_r + \frac{v^2}{2a}$$

   c) Make a sketch of the intersection and label both the GO Zone and the STOP Zone. Include the dimensions of the intersection and each zone.

2. Some people disregard the 30 mi/hr speed limit (15 m/s) and travel at 60 mi/hr (30 m/s) on the road described in *Question 1*.

   a) Use the spreadsheet or calculator to calculate STOP and GO Zones at 60 mi/hr. Sketch the intersection marking both zones. Explain the danger of driving at this speed.

   b) How would a decrease in the speed limit to 20 mi/hr (about 10 m/s) affect the STOP and GO Zones in *Question 1*? Use the spreadsheet or calculator to calculate both, then sketch the intersection, marking both zones.

3. A person is listening to loud music while driving. Explain why the increase in reaction time caused by the music does not affect the GO Zone. Explain how it affects the STOP Zone.

4. An automobile has worn tires and bad brakes. How will this affect the GO Zone and the STOP Zone at a yellow light?

5. Sometimes, when a light turns red at an intersection, the light for the traffic on the cross street does not turn green for a couple of seconds. What is the reason for this delay?

6. In the 1960s, the traffic engineers in a city experimented with a traffic light that featured a clock. As you approached an intersection with a green light, in the space for the yellow light there was a countdown: ..., 5, 4, 3, 2, 1, 0. When the clock reached "0" the light turned yellow.

Physics to Go

1.a)

$$GZ = vt_y - w$$

$$GZ = (15 \text{ m/s})(4.0 \text{ s}) - 15 \text{ m} = 45 \text{ m}.$$

1.b)

$$SZ = vt_r + \frac{v^2}{2a}$$

$$SZ = (15 \text{ m/s})(1.0 \text{ s}) + \frac{(15 \text{ m/s})^2}{2(5 \text{ m/s}^2)} = 37.5 \text{ m}.$$

1.c)

Student sketch. Check that the students include the correct units in the sketch, similar to the one below.

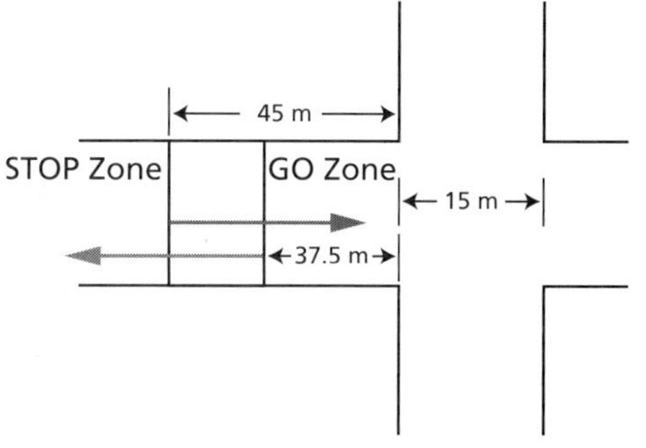

2.a)

Using the same equation provided in *Question 1* gives

$$GZ = vt_y - w$$

$$GZ = (30 \text{ m/s})(4.0 \text{ s}) - 15 \text{ m} = 105 \text{ m}$$

$$SZ = vt_y + \frac{v^2}{2a}$$

$$SZ = (30 \text{ m/s})(1.0 \text{ s}) + \frac{(30 \text{ m/s})^2}{2(5 \text{ m/s}^2)} = 120 \text{ m}.$$

Driving at this speed is dangerous because it leaves a Dilemma Zone between 105 m and 120 m where the automobile is too far away to go through the intersection safely, but too near to stop safely.

CHAPTER 1

2. b)

Reducing the speed limit to 10 m/s would give the following GO and STOP Zones.

$GZ = vt_y - w$

$GZ = (10\text{ m/s})(4.0\text{ s}) - 15\text{ m} =$ 25 m.

$SZ = vt_r + \frac{v^2}{2a}$

$SZ = (10\text{ m/s})(1.0\text{ s}) + \frac{(10\text{ m/s})^2}{2(5\text{ m/s}^2)} =$

20 m.

This would yield an Overlap Zone of 5 m, making the intersection safer.

3.

The STOP Zone includes all positions where the vehicle can safely stop before reaching the intersection. Therefore, the stopping distance is affected by the reaction time, which can be affected by the music. The GO Zone is the area through which you can pass safely in an intersection; therefore, there is no reaction time necessary.

4.

Worn tires and bad brakes will have no effect on the GO Zone because the automobile does not need to stop. However, the STOP Zone will be affected because the worn tires and bad brakes will increase the braking distance, thus increasing the STOP Zone.

5.

Having the intersection light turn green a few seconds after the yellow light turns red for the crossing street, provides an additional safety margin (in effect increasing the yellow-light time). Automobiles that might be in a Dilemma Zone might be able to go through the intersection safely with the extra time.

6.

The countdown was intended to provide extra warning when the light would turn yellow, and possibly increase the STOP Zone. Traffic engineers decided this was not a good idea because the extra warning might tempt drivers to increase their speed and they might still not be able to cross the GO Zone. The countdown would make the intersection more dangerous due to higher traffic speeds.

NOTES

NOTES

7.

Intersection **A**

$GZ = vt_y - w$

$SZ = vt_r + \dfrac{v^2}{2a}$

$GZ = (20 \text{ m/s})(3.0 \text{ s}) - 12 \text{ m} =$ 48 m

$SZ = (20 \text{ m/s})(1.2 \text{ s}) + \dfrac{(20 \text{ m/s})^2}{2(7 \text{ m/s}^2)}$

53 m

This intersection is unsafe, because there is a Dilemma Zone between 48 m and 53 m.

Intersection **B**

$GZ = (20 \text{ m/s})(4.0 \text{ s}) - 8 \text{ m} = 72 \text{ m}$

$SZ = (20 \text{ m/s})(1.2 \text{ s}) + \dfrac{(20 \text{ m/s})^2}{2(7 \text{ m/s}^2)} \approx$

53 m

This intersection is safe. There is an Overlap Zone between 53 m and 72 m.

Intersection **C**

$GZ = (20 \text{ m/s})(3.0 \text{ s}) - 12 \text{ m} =$ 48 m

$SZ = (20 \text{ m/s})(1.0 \text{ s}) +$

$\dfrac{(20 \text{ m/s})^2}{2(7 \text{ m/s}^2)} \approx 49 \text{ m}$

Chapter 1 Driving the Roads

This experiment was never implemented, and not even the city where this experiment was done uses the countdown to the yellow light today. Why do you think the traffic engineers decided this countdown was not a good idea? How does such a countdown affect the STOP and GO Zones of the oncoming traffic?

7. With the grid below, compute the GO and STOP Zones for each intersection. Also, determine if each intersection is safe and describe how you know it is safe.

Intersection	A	B	C	D	E
Yellow-light time	3.0 s	4.0 s	3.0 s	3.0 s	3.5 s
Reaction time	1.2 s	1.2 s	1.0 s	1.8 s	1.2 s
Speed of automobile	20 m/s	20 m/s	20 m/s	20 m/s	15 m/s
Acceleration	-7 m/s^2	-7 m/s^2	-7 m/s^2	-7 m/s^2	-7 m/s^2
Width of intersection	12 m	8 m	12 m	12 m	12 m

8. Do you think it would be a good idea to paint lines at all intersections showing the boundaries of the STOP and GO Zones? Explain your answer.

9. ***Preparing for the Chapter Challenge***

Write a pretend letter to your parents, asking to borrow their car. You must try to convince them based on what you have learned in this section about intersections. Be sure to explain what you have learned about the STOP, GO, Overlap and Dilemma Zones.

Inquiring Further

Study a real intersection

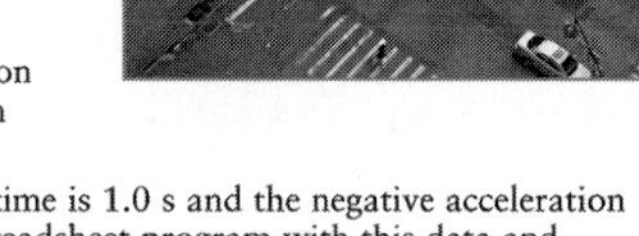

Now that you know how the length of the yellow-light time and other factors affect the safety of a traffic intersection, you are ready to study an actual intersection.

a) Choose a traffic intersection in your neighborhood and measure the yellow-light time.

b) Obtain the width of the intersection from a local police officer or from the municipal government.

c) Assume that the driver's reaction time is 1.0 s and the negative acceleration is 5 m/s every second. Run the spreadsheet program with this data and find the STOP and GO Zones. You may also use a calculator and the appropriate equations.

d) Draw a sketch of the intersection and include the GO Zone and the STOP Zone.

e) From your data, does a Dilemma Zone or an Overlap Zone exist? Is the intersection safe?

Active Physics 

This intersection is unsafe, with a small Dilemma Zone.

Intersection **D**

$GZ = (20 \text{ m/s})(3.0 \text{ s}) - 12 \text{ m} =$ 48 m

$SZ = (20 \text{ m/s})(1.8 \text{ s}) +$
$(20 \text{ m/s})^2/(2 \times 7 \text{ m/s}^2) \approx$
65 m

This intersection is unsafe, with a large Dilemma Zone.

Intersection **E**

$GZ = (15 \text{ m/s})(3.5 \text{ s}) - 12 \text{ m} =$ 41 m

$SZ = (15 \text{ m/s})(1.2 \text{ s}) +$
$(15 \text{ m/s})^2/(2 \times 7 \text{ m/s}^2) \approx 34 \text{ m}$

This intersection is safe, with an Overlap Zone from 34–41 m.

8.

Some students might think it is a good idea, while others might find that they are looking at the painted lines, and not at the stoplight, thus changing their reaction time and increasing the STOP Zone.

9.

Preparing for the Chapter Challenge

Students write letters that show what they have learned about intersections, including the STOP, GO, Overlap, and Dilemma Zones. Be certain that students respond to all four sections of the question, including a description of why the Dilemma Zone is dangerous for drivers. The discussion of the GO Zone should include why reaction time and braking distances are not important in this region, as well as what determines the size of the GO Zone.

Students' responses should include a discussion of how the braking distance depends upon the velocity squared and reaction time to determine the STOP Zone, and how the condition of the car and their concentration on driving will determine these factors. They might also wish to include what they intend to do to ensure these two factors are optimal when they are driving.

Students should discuss the Overlap and Dilemma Zones in their presentation, so that their parents are aware that they understand these regions. The discussion about the Overlap and Dilemma Zones should also include why the Overlap Zone helps them to drive safely, and how the realization of the existence of a Dilemma Zone for a red light will make them even more cautious when they are driving.

Inquiring Further

Depending upon your school's policy, you may wish to determine the yellow-light time and then give it to the students, rather than have them measure it themselves. If the intersection width is not readily available from the police, you or the students may measure it by "pacing off" the distance from curb to curb as you cross the street.

Students' calculations for the GO Zone will depend upon the intersection width, local speed limit, and yellow-light time. The calculations for the STOP Zone can be done from the given values for reaction time, negative acceleration, and the local speed limit.

The drawings for the STOP and GO Zones will be similar to the ones in the *Investigate*, and the location of any Overlap or Dilemma Zones as determined from their calculations, should be clearly marked on their diagrams.

If there are no intersections with a stop light in your local area, you may substitute values from another area without changing the value of the exercise.

CHAPTER 1

SECTION 6 QUIZ

1-6b Blackline Master

1. Which of the following is an important consideration in determining the width of the GO Zone for a yellow light?

   a) The negative acceleration rate of an automobile.

   b) The width of an intersection.

   c) The driver's reaction time.

   d) The condition of an automobile's brakes.

2. An automobile is approaching an intersection with a yellow light. The automobile is traveling with a velocity of 20 m/s, and goes a distance of 15 m before the driver starts to apply the brakes. The automobile stops at a distance of 50 m. What is the driver's reaction time?

   a) 0.75 s
   b) 2.00 s
   c) 0.30 s
   d) 0.40 s

3. For *Question 2*, what would be the automobile's rate of negative acceleration if it comes to a stop in 50 m?

   a) -5.7 m/s^2
   b) -8.0 m/s^2
   c) -13.3 m/s^2
   d) -4.0 m/s^2

4. Which statement below best describes what occurs when the yellow-light time at an intersection is increased while all the other variables remain constant?

   a) The GO Zone decreases, and the STOP Zone remains the same.

   b) The GO Zone decreases, and the STOP Zone decreases.

   c) The GO Zone increases, and the STOP Zone decreases.

   d) The GO Zone increases, and the STOP Zone remains the same.

5. An intersection has an improperly timed yellow light, resulting in a Dilemma Zone for drivers. For automobiles traveling at the legal speed limit, which of the following changes can alter the STOP and GO Zones so that the Dilemma Zone can be eliminated?

   a) Increasing the speed limit.
   b) Increasing the width of the intersection.
   c) Increasing the yellow-light time.
   d) Increasing the driver's reaction time.

SECTION 6 QUIZ ANSWERS

1 b) The GO Zone is determined by the speed of an automobile, the yellow-light time, and the intersection width only. Reaction time, and the negative acceleration rate of an automobile, determine the size of the STOP Zone. The condition of an automobile's brakes determines the negative acceleration rate.

2 a) The distance traveled during the reaction time of the driver is given by $d = vt_r$. To obtain t_r, the distance traveled is divided by the automobile's speed, giving $(15\text{ m})/(20\text{ m/s}) = 0.75\text{ s}$. If the total distance (50 m) is used, the student would get choice *d)*, and if the braking distance is used rather than the reaction distance, choice *c)* is obtained.

3 a) To obtain the automobile's negative acceleration, one needs to know the braking distance. The braking distance is the difference between the stopping distance and the reaction distance, or 50 m – 15 m = 35 m. Using the formula for braking distance, $v^2 = -2ad$ and solving for a gives

$a = \frac{-v^2}{2d} = \frac{-(20\text{ m/s})^2}{2(35\text{ m})} = -5.7\text{ m/s}^2$. Using the total distance for d (50 m), gives choice *d)*.

4 d) The GO Zone is determined by the speed of the automobile, the yellow-light time and the intersection width only. Reaction time, and the negative acceleration rate of the car determine the size of the STOP Zone. An increase in the yellow-light time increases the GO Zone too. The STOP Zone is not affected by yellow-light time.

5 c) Increasing the yellow-light time will increase the GO Zone, but has no effect on the STOP Zone, so the two zones will eventually overlap, eliminating the Dilemma Zone. Increasing the speed limit increases both the STOP and GO Zones, but the STOP Zone increases faster than the GO Zone because the STOP Zone depends upon the velocity squared. Increasing the width of the intersection decreases the GO Zone, without decreasing the STOP Zone, widening the gap, and increasing the driver's reaction time will increase the STOP Zone while leaving the GO Zone unchanged.

SECTION 7

Centripetal Force: Driving on Curves

Section Overview

Students identify and test the nature of centripetal force. A toy car is tied to the end of a string and released. The students then relate factors that control the motion of the toy car in a circular path to the motion of a vehicle traveling around a curve. This section builds on Newton's first law. The formula to calculate the centripetal force is provided at a later stage and sample problems further illustrate how the magnitude of force is determined through the given variables. Students also learn from their investigation what happens to an object moving along a circular path in the absence of a centripetal force.

Background Information

Sir Isaac Newton did more than anyone else to unravel the mysteries of motion. Born on Christmas Day in 1642, this English physicist organized his observations into scientific laws that helped predict motion. Newton noticed objects in motion continued to move until a force acted upon them. A ball rolling would continue to move forever if friction did not eventually slow the ball. A surface, like sandpaper, with great friction would slow the ball faster than a surface with low friction. Newton's first law of motion is often called the law of inertia. The law states that an object in motion or at rest will remain in motion or at rest unless acted upon by an outside, unbalanced force. Therefore, an object moving in a straight path must have an outside unbalanced force acting on it to change its velocity (speed) or its path. Any change in the velocity must involve an acceleration ($a = \Delta v/\Delta t$). The change in velocity can be a change in speed, or a change in direction, as is the case for objects moving in a circular path.

In order for an object to stay in a circular path, there must be a force acting on the object to keep it from moving in a straight line (Newton's first law). This force is the centripetal force (F_c) which always acts toward the center of the circle as shown in the equation $F_c = ma_c$.

A ball on the end of a string stays in a circular path because a force acts on the ball to keep it from going in a straight line. This force is supplied by the tension applied to the string from the center of the circle. Previously, the speed of an object was found by using the formula $v = d/t$.

The same formula is used for an object traveling in a circular path, where the distance is the circumference, $2\pi r$, and the time is the period of one complete rotation (called the period — T). Therefore, the speed of an object is $v = 2\pi r/T$.

To find the period, when you know the frequency, $T = 1/f$.

If a turntable is revolving at 10 revolutions per second, the time (period) for one revolution is

$$T = \frac{1}{f} = \frac{1}{10 \text{ rev/s}} = 0.1 \text{ s/rev} = 0.1 \text{ s}.$$

Centripetal Force

According to Newton's first law of motion, an object in motion will continue in motion unless acted upon by an outside unbalanced force (a push or a pull). Therefore, an object that is moving at a constant speed in a circle must have a force acting on that object to keep it turning. That force is the centripetal force.

The centripetal force is directed toward the center of the circular path. The force is supplied by the tension in the string on a flying ball, friction between the tires and the road of a turning vehicle, and gravity as the space shuttle orbits Earth.

Crucial Physics

- Turning an automobile also produces acceleration.
- A force is required to turn an automobile.
  - The force is toward the center of the circle.
  - The force is always changing direction (since it is always toward the center).

Learning Outcomes	Location in the Section	Evidence of Understanding
Recognize the need for a centripetal force when rounding a curve.	***Investigate*** Steps 2, 5-9	Students make an object go in a curve.
Predict the effect of an inadequate centripetal force.	***Investigate*** Steps 2 and 6	Students calculate speed just before the object slides off the turntable.
Relate speed to centripetal force.	***Investigate*** Step 9	Students change the speed to see a change in the centripetal force.

Section 7 Materials, Preparation, and Safety

Materials and Equipment

PLAN A

Materials and Equipment	Group (4 students)	Class
Stopwatch	2 per group	
Ruler, metric, 30 cm	1 per group	
Car, toy, battery operated	1 per group	
Turn table, heavy duty	1 per group	
Piece, wood, 1 in. x 2 in. x 2 in.	1 per group	
Cork, accelerometer	1 per group	
Ball, bocce	1 per group	
Battery, AA	1 per group	
Sandpaper, 60 grit	1 per group	
Tape, masking		1 per class
String, ball		1 per class
Stool, rotating*		1 per class
Helmet, safety*		1 per class
Pads, knee*		1 set per class
Pads, elbow*		1 set per class
Rolled up newspaper or magazine*	1 per group	

*Additional items needed not supplied

PLAN B

Materials and Equipment	Group (4 students)	Class
Stopwatch	2 per group	
Ruler, metric, 30 cm	1 per group	
Car, toy, battery operated	1 per group	
Turn table, heavy duty		1 per class
Piece, wood, 1 in. x 2 in. x 2 in.		1 per class
Cork, accelerometer		1 per class
Ball, bocce	1 per group	
Battery, AA	1 per group	
Sandpaper, 60 grit		1 per class
Tape, masking		1 per class
String, ball		1 per class
Stool, rotating*		1 per class
Helmet, safety*		1 per class
Pads, knee*		1 set per class
Pads, elbow*		1 set per class
Rolled up newspaper or magazine*	1 per group	

*Additional items needed not supplied

Note: Time, Preparation, and Safety requirements are based on Plan A, if using Plan B, please adjust accordingly.

Time Requirement

The approximate time for this investigation will be 30 minutes. Extra time will be required to perform more tests with the block on the turntable, investigating different surfaces.

Teacher Preparation

- Check to be certain that the toy cars the students will be using have batteries that are in good shape.
- You might wish to tape a card to the outer edge of the turntable to help students in counting the revolutions per minute.
- Accelerometers come in several varieties. Make certain you understand how each works.
- Either a rotating stool or chair may be used for *Step 10*. If one is not available in your classroom, you might be able to obtain one from another lab.

Safety Requirements

- If a student will be sitting in the rotating chair when holding the accelerometer, be certain that the student has the necessary safety gear to prevent injury should he or she fall out of the chair (helmet, kneepads, etc.).
- If toy cars are used on the floor in *Steps 2* and *3*, make certain they are picked up immediately to prevent a tripping hazard.
- When the ball is pushed across the floor with a magazine, make certain the area is clear of any obstructions the student might run into while concentrating on the ball. The ball must be picked up immediately upon completion of this part of the investigation.

NOTES

Meeting the Needs of All Students

Differentiated Instruction: Augmentation and Accommodations

Learning Issue	Reference	Augmentation and Accommodations
Creating a table to record data	***Investigate*** Steps 4-9	**Augmentation** • Students with organizational and graphomotor issues struggle to create a data table without a model. They are asked to record the type of surface, the radius from the block to the center of the turntable, the time required for 10 revolutions, the number of revolutions per minute, the time for one revolution, the circumference of the circle, and the speed of the block. All this data must be recorded for a number of scenarios. Instruct students to turn their page to landscape format. Then model how to draw a table that will fit all of the required information. Make sure that students draw their tables large enough to include all the data. **Accommodation** • Provide students with a table to tape into their *Active Physics* logs. Provide a ruler or index card to assist students in tracking columns and rows to record data.
Completing calculations with data collected in an investigation	***Investigate*** Steps 4-9	**Augmentation** • Model one example of the correct way to collect the data needed for this *Investigate*. Then show students how to complete one row of calculations that they can use as a reference to complete the *Investigate*.
Understanding vocabulary	***Investigate*** Step 11	**Augmentation** • Students are asked to push the ball with a motion that is perpendicular to the motion of the ball. Students may not remember what perpendicular means. Explain the definition and provide an example. • Model how to push the ball with a motion that is perpendicular to the ball's motion.
Using text to find answers	***Checking Up***	**Augmentation** • Students with reading comprehension issues struggle to use texts effectively or efficiently to understand new information. Instruct students to read the *Checking Up* questions before they read the *Physics Talk* section. This creates a purpose for the reading that will help students focus on the main points needed to answer the questions. **Accommodation** • Give students a copy of the *Physics Talk* and tell them to highlight the answers to the *Checking Up* questions. This allows students to focus only on the reading, instead of both reading and writing at the same time. Also, students can see the reading passage and questions side-by-side.
Understanding vocabulary	***Physics Essential Questions*** Why should you care?	**Augmentation** • This question uses many words that may confuse students, such as "consequences," "exceeding," "imposed," and "road-tire interface." Ask students to restate the question in their own words to check for understanding, or assist a confused classmate before students try to answer the question on their own.
Solving word problems	***Physics to Go*** Steps 1-3, 8	**Augmentation** • Remind students to use their work from the *Investigate* to help them solve these problems.

Strategies for Students with Limited English-Language Proficiency

Learning Issue	Reference	Augmentation
Following complex procedures	***Investigate***	Break down the *Investigate* into smaller chunks that allow students to comprehend each portion of the investigation before moving on to the next one. This will allow them to get comfortable following the procedures outlined within each step, and also to internalize new concepts and any new vocabulary that is introduced within a step. Lead a brief class discussion after each step to allow students opportunities to demonstrate knowledge and use the vocabulary.
Calculating with pi (π)	***Investigate,*** Step 7	Collaborate with the students' math teachers to determine what level of familiarity students have with pi (π). Also, decide on the value of π you would like students to use for the calculations in this section.
Inferences, Vocabulary comprehension	***Physics Essential Questions*** Why should you care?	Ask students to infer the meaning of "road-tire interface." Using context clues, students can infer that it means the common boundary of the road and the tires. Explain that "interface" is a combination of the base word "face" and the prefix "inter," meaning "between two faces." It may help to connect with the geometrical meaning of "face": a flat surface of a solid.
Vocabulary comprehension	***Inquiring Further*** Question 1	Students may not know the term "banking" in the context of a curve: to build a road upward from the curve's inside edge. A car on a banked curve is not on a horizontal surface. You can model a car traveling on a banked curve with a toy car or even just your hand.

Cloze Activity

Consider finishing this section with a cloze activity. Cloze activities are useful tools for summarizing material and for giving English-language learners an opportunity to practice vocabulary using science words in context. Write the following paragraph on the board, replacing the underlined words with write-on-lines. Encourage volunteers to fill in the blanks.

1-7a Blackline Master

Centripetal acceleration involves a change in a vehicle's velocity, but not necessarily a change in speed. In the equation, $a = v^2/r$, a is acceleration, v is velocity, and r is the radius of the curve. In the equation for centripetal force, $F = mv^2/r$, F is force and m is mass. Circular motion requires a force toward the center of the circle, called the centripetal force. In the case of a car going around a curve, the force is the friction between the tires and the road surface. The smaller the radius of a curve, the lower the maximum safe speed for driving around that curve.

SECTION 7

Teaching Suggestions and Sample Answers

What Do You See?

A vehicle speeding around a curve conveys the topic of this section in a strong visual to stimulate students' curiosity. You might want to ask them why the illustrator has depicted such a scene and how it might relate to concepts they have learned in the previous sections. Expect a range of responses and prompt students by encouraging them to share their perceptions. Have them focus on the illustration and assure them that the purpose of the *What Do You See?* section is to engage them in an inquiry-based discussion and not to determine the accuracy of their responses.

What Do You Think?

Students will come up with various answers. List all the answers on the board, and emphasize the importance of previous concepts in the chapter. Point out that the new concepts they are about to learn hinge on their understanding of factors influencing the motion of objects. While it is not necessary for all answers to be correct, they should be relevant to the context of the question. *The What Do You Think?* questions should lead to a formative discussion.

What Do You Think?

A Physicist's Response

The sign is indicating to slow down because at the posted speed limit, which is okay for the straight road, there will not be enough friction to make the curve. From the banking angle of the road, and the friction between road and tires, and from Newton's second law, one can find the maximum velocity with which one can negotiate the curve.

Students' Prior Conceptions

1. **A force is needed to keep an object moving with a constant speed.** Students should explore what happens to the motion of the vehicle when the force maintaining motion along the curve ceases to exist. For example, the friction between the tires and the road diminishes drastically, such as in a skid on an icy road. A classroom demonstration to promote brainstorming on this topic is to twirl a foam ball attached to a string overhead and then quickly release the string from the hand. Describe the motion of the foam ball immediately as it is released from the hand.
2. **Students tend to think of force as a property of an object rather than as a relation between objects.** This section offers another opportunity to explore the nature of pairs of action-reaction forces, as they apply to the friction between the wheels of the vehicle and the surface of the road. Encourage students to identify the pair of forces involved and to correlate them with the objects interacting with each other. What forces act on which objects?
3. **Students may confuse friction with inertia.** When explaining the feelings experienced by the human body when riding as a passenger in a vehicle traveling quickly around a tight curve. Inertia wants the passenger's body to continue moving along a straight line tangent to the path of the circular motion, whereas the friction between the seat and the clothes, and the force of the seatbelt maintain the inward centripetal pull on the passenger.

Section 7 Centripetal Force: Driving on Curves

Section 7 Centripetal Force: Driving on Curves

What Do You See?

Learning Outcomes

In this section, you will

- **Recognize** the need for a centripetal force when rounding a curve.
- **Predict** the effect of an inadequate centripetal force.
- **Relate** speed to centripetal force.

What Do You Think?

You are driving along a road at the posted speed limit of 50 mph (80 km/h). A road sign warns that you are approaching a curve and tells you to slow down to 25 mi/hr (40 km/h).

- **Why is the sign indicating to slow down?**
- **How is the amount you should slow down determined?**

Record your ideas about these questions in your *Active Physics* log. Be prepared to discuss your responses with your small group and the class.

Investigate

In this *Investigate*, you will model some of the problems a driver faces when driving around curves. You will investigate how speed and the tightness of a curve can affect what happens to a vehicle on a curve.

1. Driving around a curve produces some unique problems. Physics lets you model some of these problems.

 a) Imagine that you have a toy car at the end of a string, and it is moving in a circle. If you let go of the string, which way would the car travel? The diagrams on the following page show several possibilities.

105 Active Physics

Investigate

1.a)

Student answers will demonstrate their misconceptions. The correct answer is diagram B. The toy car continues on the way it was going (in a straight line after the string was released). Do not expect or offer the correct answer.

CHAPTER 1

Students often say that there is a force pushing them outward, away from the center of the motion along the curve; they believe this force, due to friction, is pressing them against the door. With investigation and analysis, students should alter this prior conception to align with the theory that identifies the vehicle as pushing inward on the passenger's body to maintain circular motion, while inertia wants the body to go straight at every point along the curve.

4. **Students may say that the force required to accelerate an object at rest is working against friction, not inertia.** This section is helpful in reviewing this prior conception of students, and to reinforce scientific explanations on inertia—perhaps to even mention the historical scientific developments on inertia and motion promoted by Galileo Galilei.

5. **Students may believe that the effects of inertia while rounding a curve are due to a "centrifugal force" or one that pushes an object outward, away from the center of the circular motion.** This prior conception correlates with the notion of "feeling" something pushing your body against the door of the vehicle while rounding a curve. As mentioned before, inertia wants the body to continue to move in a straight line, tangent to the curve of the traveling vehicle, while the centripetal forces act to push the body inward along a radial line of the curve. Students interpret this interaction as being pushed outward.

2.a)

The string pulls the car toward the center of the circle.

2.b)

Along the tangent to the circle — note that it is not a radial path.

3.a)

Make sure students realize the car is travelling in a circular curve. Students' diagram should resemble the one below.

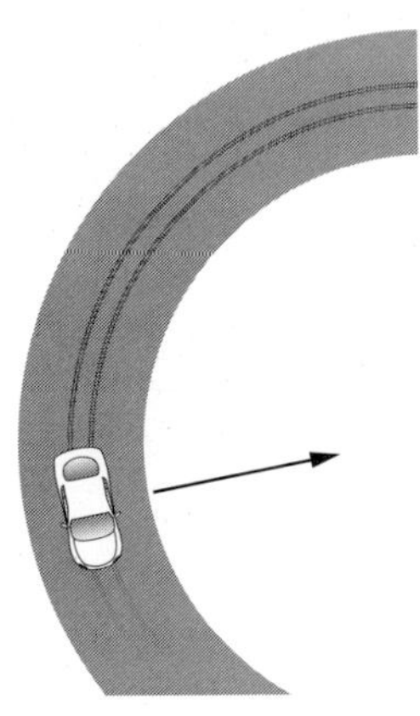

4.a)

Students record the data.

5.

Students spin the turntable to observe the property of friction.

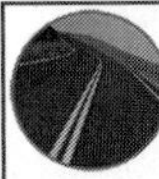

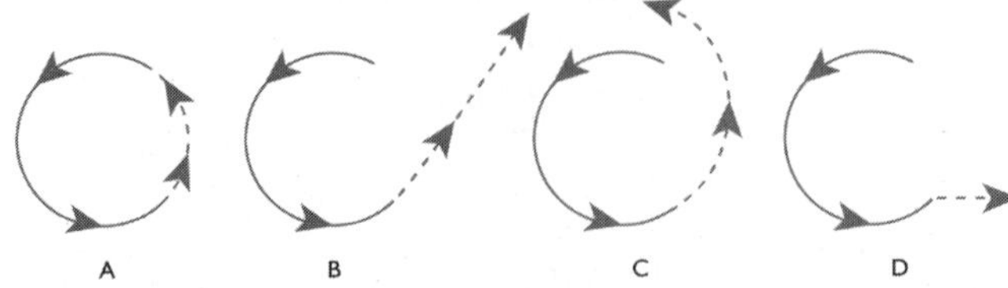

In which direction do you think the car will travel? Write your choice and how you made your decision in your log.

2. The best way to check your answer is to try out the model. That is what it means to do science. Tie a motorized toy car to one end of a string. If you do this on a table, the string should be a little less than half the width of your table. (You can use a longer string if you do this on the floor.) With a finger, hold the other end of the string fixed to the tabletop. Turn on the car's motor, so that the car travels in a circle with your finger at the center, as shown in the diagram.

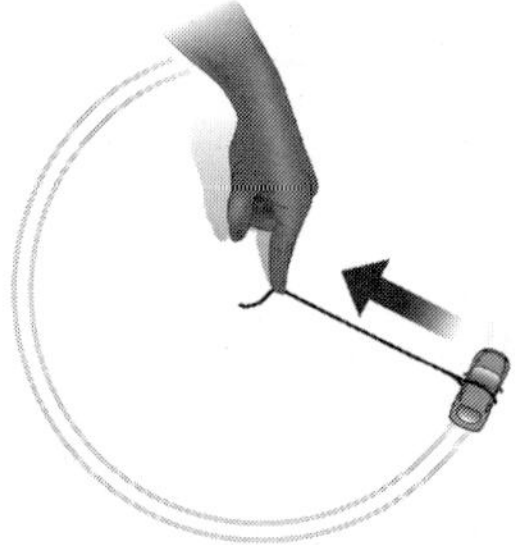

a) The string makes the car travel in a circle. In which direction does the string pull on the car? This pull is referred to as a *force* in physics. Understanding forces is a topic that you will return to many times in this course and every future physics course.

b) Now release the string. Which way does the car travel when it is released?

3. As you know, there is no string attached to a real automobile when it makes a turn, but there must be a force (like the force of the string) that keeps it moving in a circular path. The pull or force toward the center of a circular curve is the friction between the tires and the road. On an icy road, there is very little friction and the automobile cannot move in a curve and continues in a straight line. You can learn more about friction and how to measure friction in another chapter.

a) Draw a diagram of an automobile traveling north and making a right turn. On your diagram, draw the direction of the frictional force that keeps the car moving in a circular curve.

4. To further investigate the factors that determine whether an automobile will stay on the road as it goes around a curve, you will do a second investigation. Place a block of wood near the edge of a turntable (or revolving tray).

a) Record the distance from the block to the center of the turntable. This is the radius of the curve.

5. Spin the turntable. As it spins, the block is held on the spinning surface by friction. (In other words, the block is prevented from sliding off the turntable by friction. If the friction suddenly disappeared while the block was rotating, it would be similar to letting go of the string of the motorized toy car.) This friction between the block and the surface of the turntable

Students' Prior Conceptions *(continued)*

6. **Acceleration is always in a straight line.** In order to apply Newton's second law of motion to motion along a curve, it is helpful for the teacher to interpret the mathematical statement, $\Sigma F = ma$, as the sum of the net forces acting on an object equals the product of its mass and acceleration. Both force and acceleration are vectors, whereas mass is a scalar. The mass of the student remains constant for the short duration of motion; mathematically, this enables the teacher to highlight the direct relationship between the net force and acceleration. They must be in the same direction, parallel to one another. In order to have centripetal motion, a force acts upon an object to push or to pull it in toward the center of the curve; hence, the acceleration also acts inward, toward the center of the curve, perpendicular to the straight-line motion at every point along the curve.

7. **Students may not recognize that a force perpendicular to the straight-line motion of a vehicle is needed at every point along the motion in order for a vehicle to travel along a curve.** Centripetal force, which is a radial force acting toward the center of the curve, confuses students. Prior student knowledge discussed above directly resolves this confusion.

is not identical to the friction that holds an automobile on the road, but it is similar to the friction between the surface of the road and an automobile's tires.

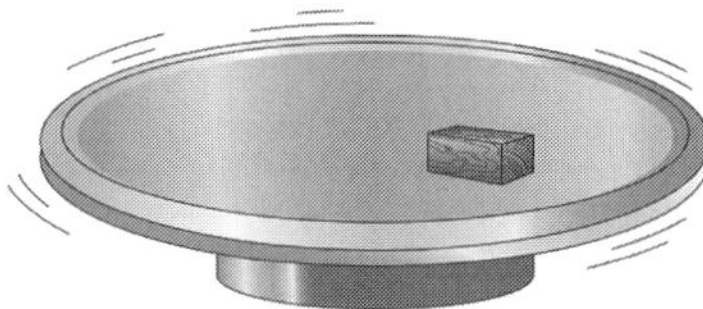

6. Gradually increase the rotational speed of the turntable until the block just begins to slide. To repeat the experiment after the block slides, place it back on the turntable at the original distance from the center. Now practice until you find the fastest speed that allows the block to stay in place.

a) Measure and record the time required for 10 revolutions.

b) Record the number of revolutions per minute made by the block when the friction is strong enough to keep the block going in a circle.

c) How much time goes by during one revolution?

d) How fast (revolutions per second) is the turntable turning when friction can no longer hold the block in place?

7. To calculate the speed of the block, divide the distance traveled by the time needed to go that distance.

When an object moves in a circle, the distance traveled in one revolution is the circumference (distance around the outside) of a circle.

Circumference = $2\pi \times$ radius of circle

$$C = 2\pi r$$

a) What was the speed of the block when it stayed on the turntable?

b) What was the block's speed when it slid off the turntable? Record your results in a table in your log.

Note: You may not be able to find the exact speed at which the block leaves the turntable. You can find a maximum speed at which the block stays in place. You can call this a safe speed. Any speed lower than the safe speed will also be safe. You can also find the minimum speed at which the block is not able to stay on the turntable. You can call this an unsafe speed. Any speed higher than this will also be unsafe.

8. Tape some sandpaper or place a rubber mat on top of the turntable. Place the block on the sandpaper or the mat. Repeat the entire investigation. Keep the distance between the block and the turntable's center the same as it was previously.

a) Record all the necessary data.

b) Calculate the greatest speed at which the block can stay on the sandpaper or rubber mat.

c) Compare the maximum safe speed with the sandpaper or mat to the maximum safe speed without it. How does the surface affect the maximum speed?

9. In addition to the speed and the road surface, you also need to look at the curvature of the road (how tight the turn happens to be). Curves come in many shapes. The arc of a circle is a good approximation for at least a segment of any curve. The arc of a large circle (a circle with a large radius) can represent a gentle curve. The arc of a small circle (a circle with a small radius) can represent a tight curve.

a) Investigate the effect of the amount of curve by placing the block at various distances from the center of the turntable.

6.a)

By timing 10 revolutions, the error in starting and stopping the clock is spread out over a longer time, resulting in less fractional error and higher precision.

6.b)

Write the answer of *6.a)* in rev/s. To convert rev/s into rev/min, multiply by 60 s/min.

6.c)

Time for one revolution or period is equal to 1/frequency. Therefore, from the example above, 1 period = $\frac{1}{2}$ s/rev or 0.5 s.

6.d)

Students observe when the block slides off and records the rpm.

7.a)

Students use $v = 2\pi r/t$ to calculate the speed.

7.b)

Students use $v = 2\pi r/t$.

8.a)

Students record data.

8.b)

Students use $v = 2\pi r/t$ to calculate the speed.

8.c)

The greater the force of friction, supplied by the sandpaper or rubber mat, the greater the speed of the turntable before the block flies off. On a surface with higher friction, an automobile has a higher safe speed while rounding a curve.

9.a)

The students should find that the block will stay on the turntable if the speed remains the same when at a larger radius. If the revolutions per minute are kept the same, the velocity will be greater at the edge, and the block will fly off sooner due to the increased speed and increased radius.

9.b)

As the radius increases, the speed at which the block can travel around the circle without flying off will increase. Make certain the students actually calculate the speed and do not substitute the rpm. The turntable must spin at fewer rpms for the block to have the same linear speed at a larger radius.

9.c)

As the radius of the curve decreases, the cart will fly off at progressively lower speeds.

10.a)

When the automobile brakes, the necklace swings forward; when the automobile speeds up, the necklace swings backward; when the automobile turns, the necklace swings to one side.

10.b)

For a cork-level accelerometer, students should see the cork pointing toward the center of the circle. Have students compare the direction of their acceleration with the indication on the accelerometer.

10.c)

Students' sketches should look like the one below.

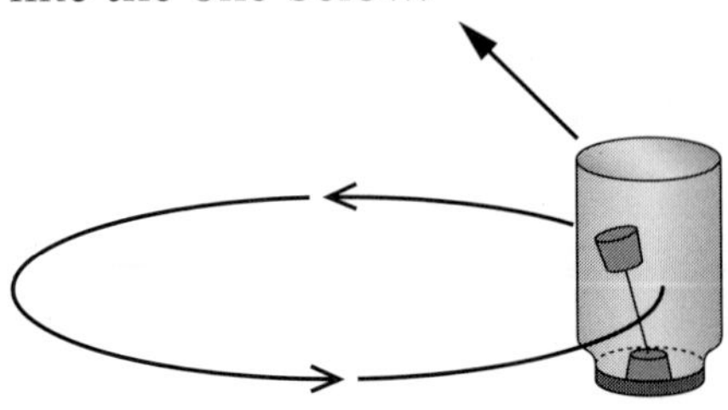

Chapter 1 Driving the Roads

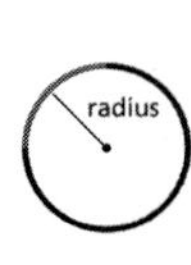

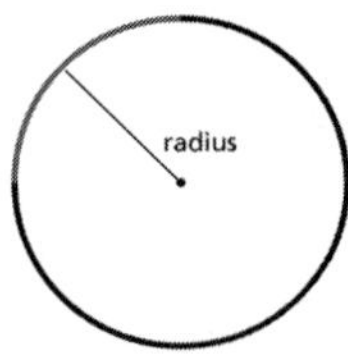

b) At each radius, find the maximum safe speed of the block. Record your data and results in your table.

c) As the radius of the turntable decreases (becomes tighter) what happens to the maximum speed?

10. Get an accelerometer from your teacher. An accelerometer is a device used to measure accelerations.

a) For example, a necklace hanging from the rearview mirror of an automobile acts as a simple accelerometer. How can something such as a hanging necklace tell you if the automobile is accelerating, even if you are not looking out the windows for other clues?

b) Hold an accelerometer in your hands and observe it as you either sit on a rotating stool or spin around while standing. What is the direction of the acceleration indicated by the accelerometer? (For a "cork" accelerometer, you can find out how the cork indicates acceleration by holding it and noting its behavior as you accelerate forward.)

c) In your log, make a sketch that simulates a snapshot photo taken from a horizontal position as the accelerometer was moving along a circular path. Show the circular path, the accelerometer "frozen" at one instant, the cork "frozen" in a leaning position, and an arrow to represent the velocity of the accelerometer at the instant represented by your sketch.

11. Start a ball rolling across the floor. While it is rolling, catch up with the ball and use a rolled-up newspaper or magazine to push the ball sideways or perpendicular to the motion of the ball with a fixed amount of force. Carefully follow alongside the ball and keep adjusting the direction of push so that it is always perpendicular to the motion of the ball.

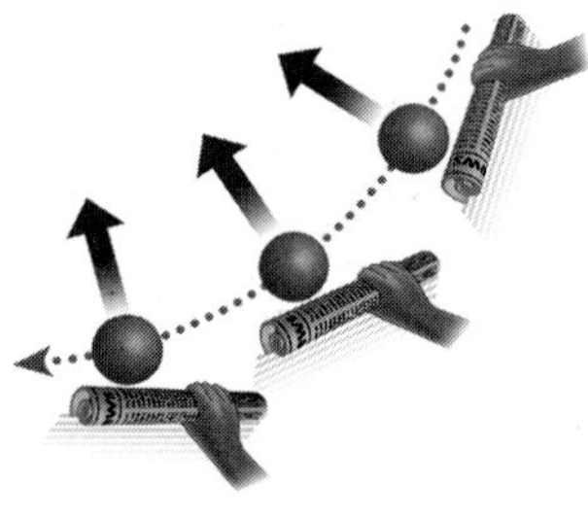

a) Make a top-view sketch in your log that shows:

- A line that represents the straight-line path of the ball before you began pushing sideways on it.
- A dashed line to represent the straight-line path on which the ball would have continued moving if you had not pushed it sideways.

Active Physics 108

11.a)

Students' sketches should look like the one at right.

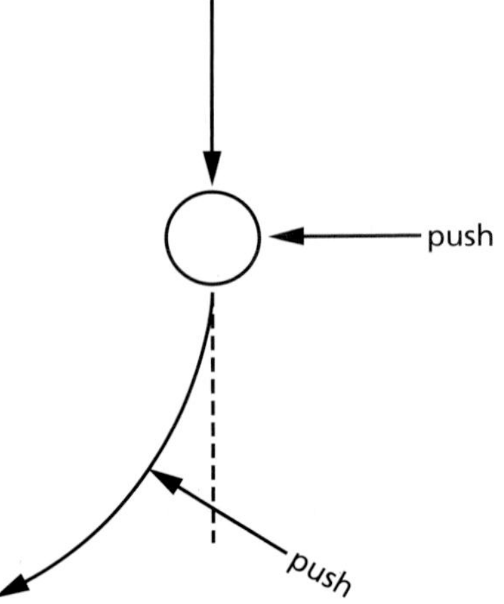

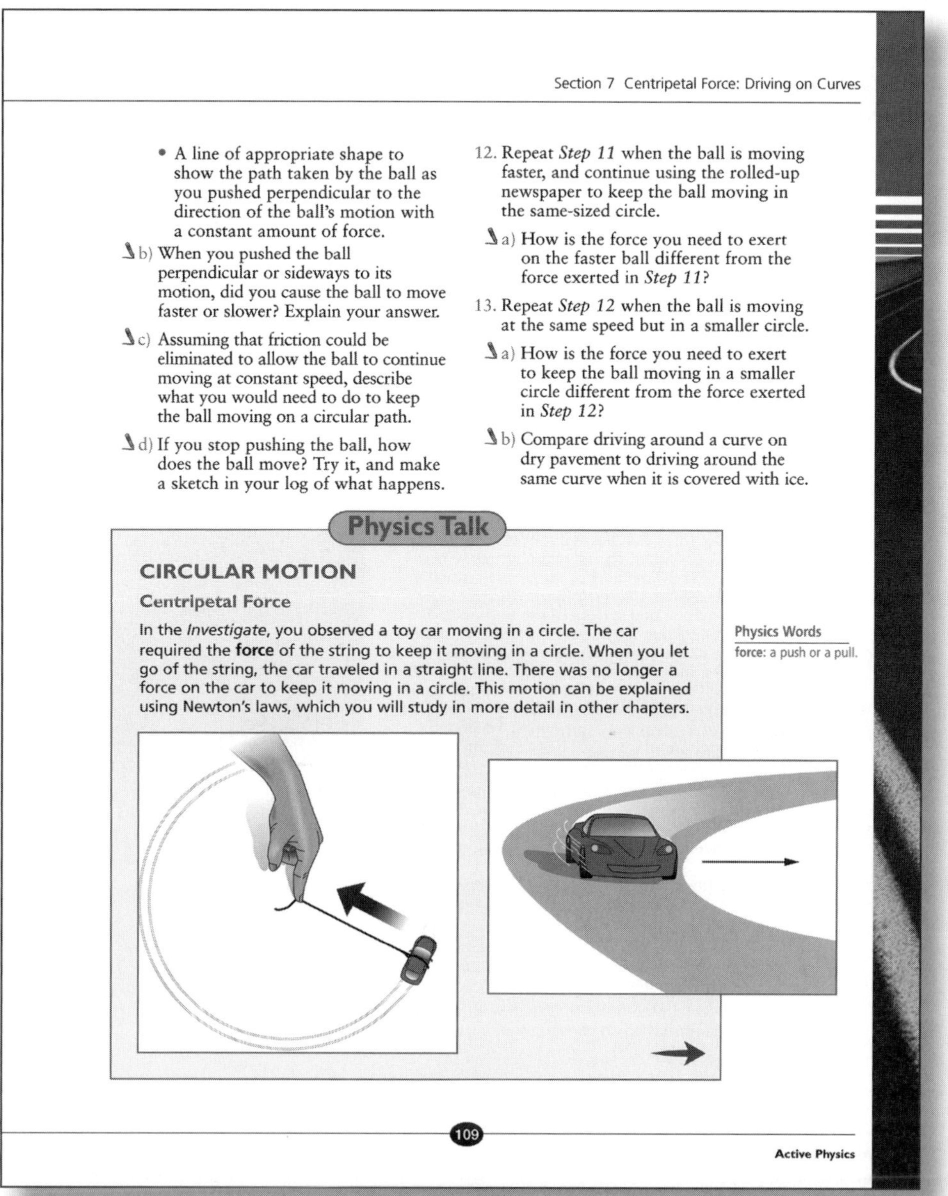

Section 7 Centripetal Force: Driving on Curves

- A line of appropriate shape to show the path taken by the ball as you pushed perpendicular to the direction of the ball's motion with a constant amount of force.

b) When you pushed the ball perpendicular or sideways to its motion, did you cause the ball to move faster or slower? Explain your answer.

c) Assuming that friction could be eliminated to allow the ball to continue moving at constant speed, describe what you would need to do to keep the ball moving on a circular path.

d) If you stop pushing the ball, how does the ball move? Try it, and make a sketch in your log of what happens.

12. Repeat *Step 11* when the ball is moving faster, and continue using the rolled-up newspaper to keep the ball moving in the same-sized circle.

a) How is the force you need to exert on the faster ball different from the force exerted in *Step 11*?

13. Repeat *Step 12* when the ball is moving at the same speed but in a smaller circle.

a) How is the force you need to exert to keep the ball moving in a smaller circle different from the force exerted in *Step 12*?

b) Compare driving around a curve on dry pavement to driving around the same curve when it is covered with ice.

Physics Talk

CIRCULAR MOTION

Centripetal Force

In the *Investigate*, you observed a toy car moving in a circle. The car required the **force** of the string to keep it moving in a circle. When you let go of the string, the car traveled in a straight line. There was no longer a force on the car to keep it moving in a circle. This motion can be explained using Newton's laws, which you will study in more detail in other chapters.

Physics Words

force: a push or a pull.

109

Active Physics

11.b)

The speed will not change because you are never pushing in or against the direction of motion, but the direction of motion changes.

11.c)

Tether the ball to the center somehow, or always push only toward the center to provide the center-seeking force, keeping it on a circular path.

11.d)

With no friction, the ball continues to move with constant speed in a straight line.

12.a)

As speed of the ball increases, students will have to push harder to keep the ball going in the same-size circle.

13.a)

A larger force is needed to make the ball go in a smaller circle at the same speed.

13.b)

When driving on a curve on dry pavement, friction between the tires and the road provides the force toward the center that is needed to make the car change direction. When driving on ice around a curve, the friction is so low that the tires probably will not have enough friction to make the car change direction and the car will go straight off the road.

Physics Talk

Students gain a clearer understanding of centripetal force through examples of different experiments conducted in the *Investigate*. In order to reinforce the concept of centripetal acceleration, try to come up with additional examples of objects traveling in a circular path, such as clothes spinning in a washing machine or amusement park rides. Ask students to indicate the direction of the velocity at any given point, and the direction of the centripetal force and centripetal acceleration. Engage students in discussing how their knowledge of centripetal force would help them in navigating a curved path while driving. Ask students to explain the factors that must be taken into consideration when they are driving to safely negotiate a curve.

Checking Up

1.

The force that is needed to keep an object moving in a circle is always directed toward the center of the circle along the radius.

2.

The force that keeps an object, moving in a circle, directed toward the center, is called the centripetal force.

3.

For an automobile going around a circular path on the road, the force of friction between the tires and the road provides the centripetal force.

4.

Velocity is a vector quantity. All vectors have two components — size and direction. The size of the velocity vector is called the speed. When a moving object changes direction, the object's velocity changes whether the speed changes or not.

5.

Acceleration can occur when an object speeds up, slows down, or changes direction. An automobile has three "accelerators" — the gas pedal to speed up, the brake to slow down, and the steering wheel to change direction. All three will cause the velocity to change, and therefore accelerate the automobile.

Newton's first law of motion states that an object in motion will stay in motion at a constant speed and travel in a straight line unless a force acts on it. When you let go of the string, the car traveled in a straight line, since no force was acting on it. Any time you observe something moving along a curved path, you should recognize that there has to be a force acting on the object.

The force of the string keeps the toy car moving in a circle. This force of the string is always toward the center of the circle. In a similar way, the force of friction between the block and the turntable kept the block moving in a circle. This force of friction is also always toward the center of the circle.

When an automobile makes a turn, it is traveling along part of a circle. There is a force of friction between the tires and the road that keeps the automobile moving in the circle.

Eliminate this friction, which is what happens on an icy road, and the automobile will move in a straight line and will not be able to turn (regardless of what you do with the steering wheel). This force of friction is toward the center of the circular curve.

The force that keeps an object moving in a circular path is called a **centripetal force.** The centripetal force can be the tension in the string, the friction between the block of wood and surface of the turntable, or the friction between an automobile and the road. For Earth moving in a circle around the Sun, the centripetal force is gravity. (You will learn more about the forces of friction and gravity in later chapters.) For a baseball bat moving in a circle during a swing, the centripetal force is the force of the muscles in the batter's arms.

As the toy car moves in a circle, its speed remains the same. It does not appear to go faster or slower. Its velocity does change because the direction is changing. The car is changing its direction. For a moment it is moving east, then it is moving south, then it is moving west, then north, and then east again as it starts its next revolution. Changes in velocity with respect to time are called *accelerations.* In the previous sections, you associated accelerations with changes in speed as a vehicle speeded up or slowed down.

A vehicle that is changing directions is also accelerating. The acceleration associated with an automobile changing directions is referred to as **centripetal acceleration.**

Acceleration is the change in velocity with respect to time. Velocity can change when an object speeds up, slows down, or changes direction.

Physics Words

centripetal force: a force directed toward the center to keep an object in a circular path.

centripetal acceleration: a change in the direction of the velocity with respect to time.

Checking Up

1. What is the direction of the force that keeps an object moving in a circle?
2. What is the name of the force that keeps an object moving in a circle?
3. Name the force that keeps an automobile moving in a circular path on a road.
4. Explain how the velocity of an object can change even if the speed is not changing.
5. Describe three situations in which acceleration can take place.
6. What is the force that keeps Earth moving in a circle around the Sun?

6.

The force that keeps Earth moving in a (roughly) circular path around the Sun is the force of gravitational attraction between Earth and the Sun. This force on Earth is always directed toward the center of the solar system (the Sun).

Active Physics *Plus*

+Math	+Depth	+Concepts	+Exploration
◆	◆		

Calculating Centripetal Acceleration and Centripetal Force

You learned that the force of the string on the toy car or the force of friction between the tires and the road cause an automobile to move in a circle. The direction of the force is toward the center of this circle. The size of the force depends on the mass of the automobile, the speed of the automobile, and the radius of the curve. You can use the following equations to calculate the centripetal force and the corresponding centripetal acceleration.

$$a = \frac{v^2}{r}$$

$$F = \frac{mv^2}{r}$$

where a is acceleration
F is force
v is velocity
m is mass
r is the radius of the curve

Sample Problem

Calculate the centripetal acceleration and centripetal force of a 1000-kg automobile traveling at 27 m/s (60 mi/h) that turns on an unbanked curve having a radius of 100 m.

Strategy: Because the automobile is moving along a circular path, the centripetal acceleration must be directed toward the center of the curve. The magnitude can be found because the speed and radius are given.

Given:

Mass (m) = 1000 kg

Speed (v) = 27 m/s

Radius of curve (r) = 100 m

Solution:

This acceleration changes the direction of the velocity of the automobile.

$$a = \frac{v^2}{r} = \frac{(27 \text{ m/s})^2}{100 \text{ m}} = \frac{729 \text{ m}^2/\text{s}^2}{100 \text{ m}} = 7.29 \text{ m/s}^2 \text{ or } 7.3 \text{ m/s}^2$$

The speed of the automobile remains the same.

Because the automobile is moving along a circular path, the centripetal force must be exerted toward the center of the curve. The magnitude can be found because the mass, speed, and radius are all given.

$$F = \frac{mv^2}{r} = \frac{1000 \text{ kg} \times (27 \text{ m/s})^2}{100.0 \text{ m}} = \frac{1000 \text{ kg} \times 729 \text{ m}^2/\text{s}^2}{100 \text{ m}} = 7290 \text{ kg} \cdot \text{m/s}^2 = 7300 \text{ N}$$

If the centripetal force, which in this case is the force of friction, cannot cause sufficient acceleration, the automobile will not follow the curve and will skid in the direction of its velocity at the instant the tires break loose. Your experiment with the block sliding off the turntable demonstrated this. For a given v and a given r, the centripetal force must be enough to provide the acceleration $\frac{v^2}{r}$.

Active Physics Plus

Have students write in their logs what they know by now about centripetal force. Asking them to consider the direction of centripetal force and how the size of this force is determined will help them solve the problems in this section. You might want one student to come up to the board and show the solution to a problem. You could then use this opportunity to direct questions to the other students in your class so that each student has the opportunity to understand how mathematical calculations explain concepts.

1.a)

The force of friction must be sufficient to provide the necessary centripetal force for the automobile to negotiate the curve. Using the formula for centripetal force,

$$F_c = \frac{mv^2}{r} = \frac{(1000 \text{ kg})(14 \text{ m/s})^2}{40 \text{ m}} = 4900 \text{ N}.$$

1.b)

$$F_c = \frac{mv^2}{r} = \frac{(1000 \text{ kg})(20 \text{ m/s})^2}{40 \text{ m}} = 10{,}000 \text{ N}$$

$F_c = 10{,}000$ N or 5100 N additional frictional force.

1.c)

Reducing the frictional force in *a)* by half would make it 2450 N. Using

$$F_c = \frac{mv^2}{r}$$

$$v = \sqrt{\frac{rF_c}{m}} = \sqrt{\frac{(40 \text{ m})(2450 \text{ N})}{1000 \text{ kg}}} = 9.9 \text{ m/s}.$$

What Do You Think Now?

Ask students to go back to the *What Do You Think?* questions and revise their responses in light of what they have learned about centripetal force. Be sure to insist that speed limits are posted to ensure the safety of drivers and passengers. You should be able to gauge their level of confidence by how comfortable and quick they are in responding to the questions. Toward the end of your discussion, you may provide them with answers in *A Physicist's Response.*

Chapter 1 Driving the Roads

For a given road surface, there is a maximum frictional force that can provide this centripetal force. Hence, there is a maximum $\frac{v^2}{r}$ that a road surface can provide. If your speed is too fast, or the curve too sharp, then the maximum $\frac{v^2}{r}$ will be exceeded, and off the road you go – perhaps to disaster…

A curve with a radius of 40 m has a warning sign that limits the speed to 30 mi/h (14 m/s). Assume that an automobile has a mass of 1000 kg.

a) What is the frictional force of an automobile that is driving the speed limit?

b) How much additional frictional force does the automobile need if the driver decides to exceed the speed limit and travel at 20 m/s?

c) If the frictional force were reduced by half due to wet leaves and water on the road, what speed would you recommend for drivers?

What Do You Think Now?

At the beginning of this section, you were asked the following:

You are driving along a road at the posted speed limit of 50 mi/hr (80 km/h). A road sign warns that you are approaching a curve and tells you to slow down to 25 mi/hr (40 km/h).

- **Why is the sign indicating to slow down?**
- **How is the amount you should slow down determined?**

After having investigated the effect of speed on centripetal force, why should you slow down? Use the results of your investigations to support your answer.

112

Active Physics

Section 7 Centripetal Force: Driving on Curves

Physics
Essential Questions

What does it mean?

What is a centripetal force? Draw a sketch of an automobile making a turn. Show the direction of the velocity and the direction of the centripetal force.

How do you know?

What evidence do you have that circular motion requires a force toward the center of the circle?

Why do you believe?

Connects with Other Physics Content	Fits with Big Ideas in Science	Meets Physics Requirements
Forces and motion	Change and constancy	* Good, clear, explanation, no more complex than necessary

* In physics, a few simply stated principles explain a large variety of phenomena. How can an automobile be accelerating if it does not speed up or slow down?

Why should you care?

What are the consequences of exceeding the physical speed limit imposed by the road-tire interface and the radius of the curve?

Reflecting on the Section and the Challenge

In this section, you learned that friction between the road and the tires helps keep an automobile on the road when it goes around a curve. More friction allows you to move faster and still stay on the road.

A tight turn requires more friction or a slower speed than a wider turn. Because you cannot change or control the friction between the road and tires (other than keeping good tires in good condition), a slower speed will keep the automobile on the road.

Part of your challenge requires you to explain why it is necessary to drive at a slower speed around a curve than on a straight section of the road. You also may want to explain what happens if the road conditions change, if the friction is reduced because the tires are worn out, or if the curve in the road is very tight.

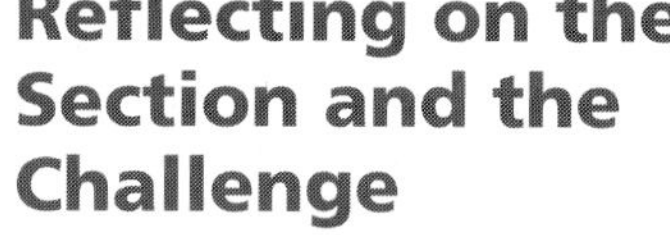

Reflecting on the Section and the Challenge

Ask students to reflect on how they can use their knowledge of friction and centripetal force in their *Chapter Challenge*. They should be able to extend their learning by explaining why it is safer to drive at slower speeds around a curve. In their challenge, students should mention the factors that can be controlled to avoid the possibility of an accident while driving around a curve. They should also mention factors that cannot be controlled while driving around a curve.

Physics Essential Questions

What does it mean?

A centripetal force is a force that keeps an object moving in a circle at constant speed. The centripetal force is the additional name given to the friction between the road and the tires that keeps the automobile moving in a circular path. It is not an additional force. (The sketch should have an arrow from the circumference of the circle pointing to the center along a radial line. The velocity is along a tangent to the circle.)

How do you know?

When the toy car is attached to a fixed string, the car moves in a circle. If the string breaks or is let go, the car travels along a straight line.

Why do you believe?

Acceleration can produce a change in direction of the velocity. Since acceleration is defined as a change in velocity with respect to time, the magnitude or direction of the velocity or both can change during acceleration.

Why should you care?

If you exceed the speed limit, the force of friction may not be large enough to provide the required centripetal force to keep your automobile moving in a circle. The automobile would then slip off the road, hitting a tree or a guard rail, or whatever comes in its way.

Physics to Go

1.

$v = d/t$, where $d = 2\pi r$. Therefore, $d = 2\pi(6,400,000\text{ m}) \approx 40,000,000\text{ m}$.

Speed in meters per second

Using $\pi = 3.14$, $d = 40,000,000\text{ m}$ and $v = (40,000,000\text{ m})/(86,400\text{ s}) = 460\text{ m/s} \approx 1000\text{ mi/h}$.

Speed in kilometers per hour

Using $v = d/t$, where $d = 2\pi r = 2\pi(6,400\text{ km}) = 40,192\text{ km}$ and $t = 24\text{ h}$, yields, $v = 40,000\text{ km}/24\text{ h} \approx 1700\text{ km/h}$.

2.

Speed in meters per second

Using $v = d/t$ and $\pi = 3.14$,

$d = 2\pi r = 2\pi\left(1.5\times10^{11}\text{ m}\right) = 9.42\times10^{11}\text{ m}$ and $t = \left(365\text{ days}\right)\left(24\text{ h/day}\right)\left(60\text{ min/h}\right)\left(60\text{ s/min}\right) = 31,536,000\text{ s}$.

Therefore,

$$v = \frac{9.42\times10^{11}\text{ m}}{31,536,000\text{ s}} = 30,000\text{ m/s} \approx 67,000\text{ mi/h}.$$

Speed in kilometers per hour

Using $v = d/t$, $d = 2\pi r = 2\pi \times \left(1.5\times10^{8}\text{ km}\right) = 9.42\times10^{8}\text{ km}$ and $t = \left(365\text{ days}\right)\left(24\text{ h/day}\right) = 8760\text{ h}$, $v = \dfrac{9.42\times10^{8}\text{ km}}{8760\text{ h}} = 107,500\text{ km/h} \approx 110,000\text{ km/h}$.

Chapter 1 Driving the Roads

Physics to Go

1. A person at the equator travels once around the circumference of Earth in 24 h. The radius of Earth is 6400 km. How fast is the person going? Compute the speed in kilometers per hour (km/h) and in meters per second (m/s). Recall that 1 km is equal to 1000 m.
2. Earth travels in a circular motion around the Sun. The radius of Earth's motion is about 1.5×10^{8} km. What is the speed of Earth around the Sun? Compute the speed in km/h and m/s.
3. A fan turns at a rate of 60 revolutions per second. If the tip of the blade is 15 cm from the center, how fast is the tip moving?
4. Friction can hold an automobile on the road when it is traveling at 20 m/s and the radius of the turn is 15 m. What happens if:
   a) the curve is tighter?
   b) the road surface becomes slippery?
   c) both the curve is tighter and the road is slippery?
5. Think about other examples in which objects travel in curved paths, such as the clothes in a spin dryer, or the Moon traveling around Earth. For each example, explain what produces the force that is constantly being applied to the object toward the center of the curve.
6. Sketch a graph that shows the radial distance and the maximum speed at which the block remains on a turntable for one type of surface.
7. Explain the following statement: "The driver may turn the wheels but it is the road that turns the automobile."
8. Active Physics Plus A jet pilot in level flight at a constant speed of 270 m/s (600 mi/hr) rolls the airplane on its side and executes a tight circular turn that has a radius of 1000 m. What is the pilot's centripetal acceleration? Draw a sketch of the acceleration's direction relative to the ground.
9. Below you will find alternate explanations of the same event given by a person who was not wearing a seat belt when an automobile went around a sharp curve.

   "I was sitting near the middle of the front seat when the automobile turned sharply to the left. A force made my body slide across the seat toward the right, outward from the center of the curve, and then my right shoulder slammed against the door on the passenger side of the automobile."

   "I was sitting near the middle of the front seat when the automobile turned sharply to the left. My body kept going in a straight line while, at the same time due to insufficient friction, the seat slid to the left beneath me, until the door on the passenger side of the automobile had moved far enough to the left to exert a centripetal force against my right shoulder."

   Are both explanations correct? Explain your answer in terms of both explanations.

Active Physics 114

3.

The time for one revolution of the blade is

$$t = \frac{1\text{ rev}}{60\text{ s}} = 0.017\text{ s},$$

and the distance traveled is

$$d = 2\pi r = 2\pi\left(15\text{ cm}\right) = 94.2\text{ cm}.$$

Using $v = d/t$, $v = \dfrac{94.2\text{ cm}}{0.017\text{ s}} \approx \dfrac{1}{n}$ $5500\text{ cm/s} = 55\text{ m/s}$ or 120 mi/h.

4.a)

As the curve gets sharper, the force of friction required to hold the automobile on the road needs to increase; otherwise, the automobile may go off the road.

4.b)

If the road becomes slippery, the friction will be reduced. Therefore, the automobile must travel at a slower speed.

4.c)

If the turn is both tighter and more slippery, the automobile has to go much slower than otherwise.

5.

Many examples will come from the students. Check the physics involved with their explanations, and the centripetal force is in fact acting toward the center of the curve. The clothes in a spin dryer will be getting the centripetal force from the drum of the dryer acting toward the center of the dryer. The Moon gets its centripetal force from the gravitational force exerted by Earth on the Moon.

6.

The graph should show that the radius is directly proportional to the velocity squared. This relationship should be a straight line. If the radius is plotted against only the velocity, then there will still be an increasing relationship — as the radius increases, the velocity increases, but the increase is not linear.

7.

If there were no friction, the automobile would continue in a straight line even if the driver turned the steering wheel. It is the friction of the road on the tire that causes the automobile to turn.

8.

$$a = \frac{v^2}{r} = \frac{(270 \text{ m/s})^2}{1000 \text{ m}} = 73 \text{ m/s}^2$$

or 7.4 g

With the plane of the circle parallel to the ground, the acceleration points horizontally from the plane to the center of the circle. Students' sketches should appear like the one below.

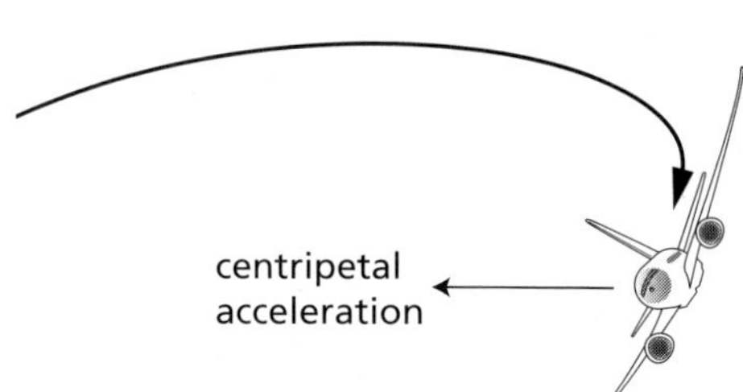

9.

The person will go straight unless there is a centripetal force to make him or her go around the curve with the automobile. This force would be supplied by the door. The person who claimed a force pushed him or her outward is incorrect. There was no real force acting on the person, since he or she was merely going straight as the automobile turned inward and the seat slid underneath. In terms of "force" as it's meant in Newton's laws of motion, the second explanation is correct. The "force" described in the first explanation only appears to the person traveling in a circle and not to anyone who is observing from a stationary point of view.

CHAPTER 1

NOTES

10.

Answers will vary. Frictional forces on the wheels are needed to supply the centripetal force to keep the race car moving in a curved path. Without friction (the outside unbalanced force), the inertia of the object will keep it moving in a straight line.

11.

Vehicles can negotiate a curve with a large radius more quickly than one with a smaller radius. If the curve radius decreases (the curve gets "tighter"), the driver may not lower the vehicle's speed enough to allow friction to apply the necessary centripetal force, and go off the road. A tight curve requires a greater frictional centripetal force than a more open one at the same speed.

12.

When driving on the right-hand side of the road, if the road curves to the right, and the automobile loses friction, the automobile will go into the oncoming lane of traffic. If the road curves to the left under the same circumstances, the automobile would skid off the road.

13.

Preparing for the Chapter Challenge

Students' answers should include a discussion of how the safe speed to travel around a curve depends upon the friction available (dry roads will have greater friction than wet roads, or roads with snow) and the radius of the curve (the larger the radius, the greater the safe speed).

10. Race cars can make turns at 150 mi/hr. What forces act on a race car as it moves along a circular path at constant speed on a flat, horizontal surface?
11. Why are highway curves that have radii that decrease as you go into them especially dangerous? In other words, curves that start out as gentle turns but become tighter and tighter as you get into them.
12. In the United States, vehicles drive on the right-hand side of a two-lane road. If the curve bends to the right and you lose traction in the turn, would you end up in the ditch on your side of the road, or into the lane of oncoming traffic? What if the curve bends to the left?
13. ***Preparing for the Chapter Challenge***

    Write a few sentences telling your parents that you know how to apply the physics from this section to drive safely around curves. You should include information about why you need to slow down around curves in rainy or icy weather.

Inquiring Further

1. **Banking a curve**

   Design an investigation to determine the effect of banking a curve on the speed at which the curve can be safely negotiated. After your teacher approves your procedure, conduct your investigation.
2. **Mass and speed on a curve**

   Design an investigation to determine if the mass of an automobile has an effect on the safe speed around a curve. After your teacher approves your procedure, conduct your investigation.

Inquiring Further

1.

Students may choose to duplicate the exercise they completed with the turntable. The students should suggest using the procedure in the *Investigate*, substituting blocks of different masses to simulate vehicles of varying mass as they negotiate the curves.

2.

As the mass of an automobile increases, the force of friction between the automobile and the road also increases in direct proportion, providing the required centripetal force for the automobile. If the students do not understand fully, point out that this is the reason posted speed limits for curves are the same for all vehicles. It should be pointed out that these posted limits might not apply to large trucks for a different reason. Although the frictional force also increases for trucks, the location of the force (on the tires) may cause a different effect on an automobile than on a truck.

1-7b Blackline Master

0-60 mph	**7.2 s**
0-¼ mi	**15.4 s**
Top speed	**est 143 mph**
Skidpad	**0.83 g**
Slalom	**61.9 mph**
Brake rating	**excellent**

TEST CONDITIONS

Temperature	70°F
Wind	calm
Elevation	1010 ft

ENGINE

Type	aluminum bloc and heads, **V-6**
Valvetrain	doch 4 valve/cyl
Displacement	155 cu in./2544 cc
Bore × stroke	3.24 × 3.11 in./ 82.4 × 79.0 mm
Compression ratio	10.0:1
Horsepower (SAE)	**195 bhp @ 6625 rpm**
Bhp/liter	76.7
Torque	**165 lb-ft @ 5625 rpm**
Maximum engine speed	6750
Fuel injection	elect. sequential port
Fuel	prem unleaded, 91 pump oct

CHASSIS & BODY

Layout	**front engine/front drive**
Body/frame	unit steel
Brakes	
Front	**10.9-in. vented discs**
Rear	**9.9-in. vented discs**
Assist type	vacuum; ABS
Total swept area	366 sq in.
Wheels	cast alloy, **16 × 6½**
Tires	steel-belted touring, **P205/55ZR-16**
Steering	**rack & pinion** power assist
Overall ratio	14.5:1
Turns, lock to lock	2.7
Turning circle	38.4 ft
Suspension	
Front	**struts,** lower A-arms, coil springs, tube shocks, anti-roll bar
Rear	**struts,** trailing links, dual lower lateral links, coil springs, tube shocks, anti-roll bar

DRIVE TRAIN

Transmission .... **5-sp manual**

Gear	Ratio	Overall ratio	(Rpm) Mph
1st	3.42:1	13.89:1	(6750) 34
2nd	2.14:1	8.69:1	(6750) 55
3rd	1.45:1	5.89:1	(6750) 81
4th	1.03:1	4.18:1	(6750) 114
5th	0.77:1	3.13:1	(6750) 143

Final drive ratio 4.06:1

Engine rpm @ 60 mph in 5th 2650

GENERAL DATA

Curb weight	**3055 lb**
Test weight	3180 lb
Weight dist (with driver), f/r, %	63/37
Wheelbase	106.5 in.
Track, f/r	59.2 in./58.5 in.
Length	**183.9 in.**
Width	**69.1 in.**
Height	**54.5 in.**
Ground clearance	8/2 in.
Trunk space	18.0 + 7.0 cu ft

MAINTENANCE

Oil/filter	5000 mi/5000 mi
Tuneup	100,000 mi
Basic warranty	36 mo/36,000 mi

ACCOMMODATIONS

Seating capacity	**5**
Head room, f/r	39.0 in./35.0 in.
Seat width, f/r	2 × 20.5 in./50.0 in.
Front-seat leg room	43.0 in.
Rear-seat leg room	25.0 in.
Seatback adjustment	85 deg
Seat travel	8.5 in.

INTERIOR NOISE

Idle in neutral	54 dBA
Maximum in 1st gear	78 dBA
Constant 50 mph	66 dBA
70 mph	71 dBA

INSTRUMENTATION

160-mph speedometer, 8000-rpm tach, coolant temp, fuel level

ACCELERATION

Time to speed	Seconds
0-35 mph	2.5
0-40 mph	3.8
0-50 mph	5.2
0-60 mph	7.2
0-70 mph	9.3
0-80 mph	11.6
0-90 mph	14.8
0-100 mph	18.8
Time to distance	
0-100 ft	3.3
0-500 ft	8.4
0-1320 ft (¼ mi):	15.4 @ 91.5 mph

FUEL ECONOMY

Normal driving	20.0 mpg
EPA city/highway	20/29 mpg
Cruise range	270 miles
Fuel capacity	14.5 gal

BRAKING

Minimum stopping distance	
From 60 mph	135 ft
From 80 mph	228 ft
Control	excellent
Pedal effort for 0.5 g stop	na
Fade, effort after six 0.5 g stops from 60 mph	na
Brake feel	excellent
Overall brake rating	excellent

HANDLING

Lateral accel (200-ft skidpad)	0.83 g
Balance	moderate understeer
Speed thru 700-ft slalom	61.9 mph
Balance	mild understeer
Lateral seat support	very good

Subjective ratings consists of excellent, very good, good, average, poor; na means information is not available

Section 7 Centripetal Force: Driving on Curves

0-60 mph	**5.2 s**
0-¼ mi	**13.8 s**
Top speed	**est 165 mph**
Skidpad	**na**
Slalom	**62.5 mph**
Brake rating	**excellent**

TEST CONDITIONS

Temperature	86°F
Wind	calm
Elevation	est 700 ft

ENGINE

Type	aluminum bloc and heads, **V-8**
Valvetrain	ohv 2 valve/cyl
Displacement	346 cu in./5666 cc
Bore × stroke	3.90 × 3.62 in./99.0 × 92.0 mm
Compression ratio	10.0:1
Horsepower (SAE)	**345 bhp @ 5600 rpm**
Bhp/liter	60.9
Torque	**350 lb-ft @ 4400 rpm**
Maximum engine speed	6000
Fuel injection	elect. sequential port
Fuel	prem unleaded, 91 pump oct

CHASSIS & BODY

Layout	**front engine/rear drive**
Body/frame	fiberglass/steel unit frame
Brakes	
Front	**12.8-in. vented discs**
Rear	**12.0-in. vented discs**
Assist type	vacuum; ABS
Total swept area	433 sq in.
Swept area/ton	257 sq in.
Wheels	cast magnesium; **17 × 8½ f 18 × 9½ r**
Tires	steel-belted sports; **P245/45ZR-17 f, P275/40ZR-18 R**
Steering	**rack & pinion** variable power assist
Overall ratio	16.1:1
Turns, lock to lock	2.7
Turning circle	38.5 ft
Suspension	
Front	**upper & lower A-arms,** Transverse composite monoleaf spring, tube shocks, anti-roll bar
Rear	**upper & lower A-arms,** toe links, transverse composite monoleaf spring, tube shocks, anti-roll bar

DRIVE TRAIN

Transmission .... **6-sp manual**

Gear	Ratio	Overall ratio	(Rpm) Mph
1st	2.66:1	9.10:1	(6000) 52
2nd	1.78:1	6.09:1	(6000) 77
3rd	1.30:1	4.45:1	(6000) 106
4th	1.00:1	3.42:1	(6000) 137
5th	0.74:1	2.53:1	est (5365) 165
6th	0.50:1	1.71:1	est (3625) 165

Final drive ratio 3.42:1

Engine rpm @ 60 mph in 6th 1320

GENERAL DATA

Curb weight	**est 3240 lb**
Test weight	**est 3380 lb**
Weight dist (with driver), f/r, %	51/49
Wheelbase	104.5 in.
Track, f/r	62.0 in./62.0 in.
Length	**179.7 in.**
Width	**73.6 in.**
Height	**47.7 in.**
Ground clearance	3/7 in.
Trunk space	13. 5 cu ft (top up)/ 10.8 cu ft (top down)

MAINTENANCE

Oil/filter change	7500 mi/7500 mi
Tuneup	100,000 mi
Basic warranty	36 mo/36,000 mi

ACCOMMODATIONS

Seating capacity	**2**
Head room, f/r	36.5 in.
Seat width	2 × 18.0 in.
Leg room	43.5 in.
Seatback adjustment	45 deg
Seat travel	8.0 in.

INTERIOR NOISE

Idle in neutral	61 dBA
Maximum in 1st gear	78 dBA
Constant 50 mph	73 dBA
70 mph	76 dBA

INSTRUMENTATION

200-mph speedometer, 7500-rpm tach, coolant temp, fuel level, volts, oil press.

ACCELERATION

Time to speed	Seconds
0-30 mph	2.0
0-40 mph	2.9
0-50 mph	4.2
0 60 mph	5.2
0-70 mph	6.6
0-80 mph	8.7
0-90 mph	10.9
0-100 mph	13.3
Time to distance	
0-100 ft	3.0
0-500 ft	7.5
0-1320 ft (¼ mi):	13.8 @ 102.1 mph

FUEL ECONOMY

Normal driving	est 18.5 mpg
EPA city/highway	18/28 mpg
Cruise range	270 miles
Fuel capacity	19.1 gal

BRAKING

Minimum stopping distance	
From 60 mph	118 ft
From 80 mph	209 ft
Control	excellent
Pedal effort for 0.5 g stop	na
Fade, effort after six 0.5 g stops from 60 mph	na
Brake feel	excellent
Overall brake rating	excellent

HANDLING

Lateral accel (200-ft skidpad)	na
Balance	na
Speed thru 700-ft slalom	62.5 mph
Balance	moderate understeer
Lateral seat support	excellent

Subjective ratings consists of excellent, very good, good, average, poor; na means information is not available

SECTION 7 QUIZ

1-7c **Blackline Master**

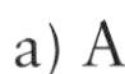

1. The diagram to the right shows a model airplane attached to a wire flying in a horizontal circle in the clockwise direction. If the wire breaks when the plane is in the position shown, the airplane will move toward point

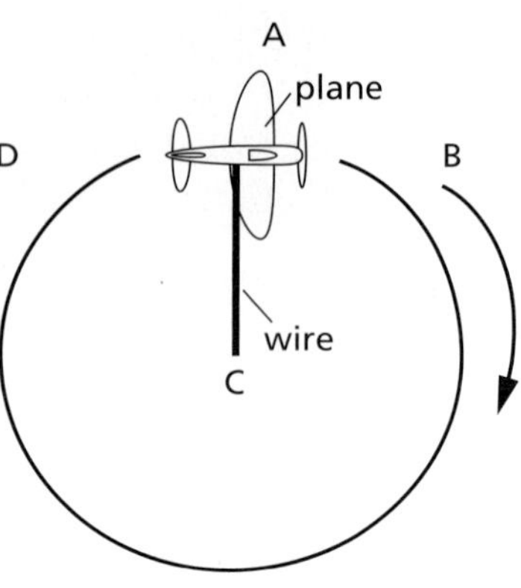

a) A b) B

c) C d) D

2. The centripetal force acting on the plane at the position shown is directed toward point

a) A b) B

c) C d) D

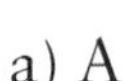

3. A ball attached to a string is moving at a constant speed in a horizontal circular path. A target is near the path as shown in the diagram to the right. At which point along the path should the string be released if the ball is to hit the target?

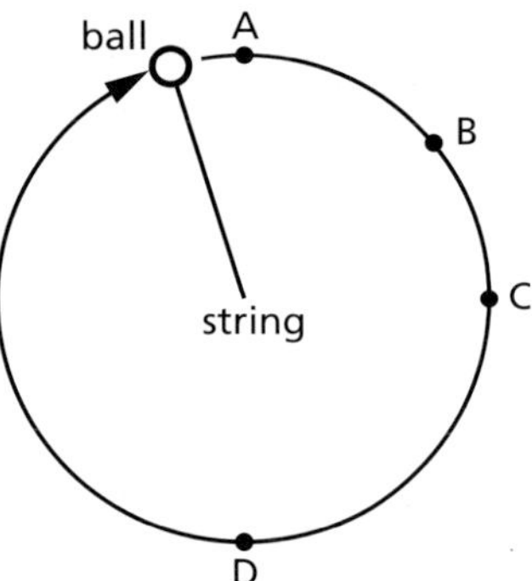

a) A b) B

c) C d) D

4. A student is spinning a turntable with a wooden block resting on it. The block is located 10 cm from the center of the table, and the turntable is rotating 15 times every minute. What is the speed of the wooden block on the turntable?

a) 1.5 cm/s b) 31.4 cm/s

c) 20.7 cm/s d) 15.7 cm/s

5. In *Question 4*, the force that holds the wooden block on the turntable while it is spinning is provided by the force of

a) gravity. b) the student's hand.

c) Earth's magnetic field. d) friction.

SECTION 7 QUIZ ANSWERS

1. b) The velocity of an object traveling in a circular path is always tangent to the circle at that point. If the centripetal force stops (the wire breaks), the plane will continue in a straight-line path, according to Newton's first law.

2. c) The centripetal force is provided by the wire, which is being pulled toward the center of the circle, at point C.

3. b) Because an object's velocity is always tangent to the circle when it is traveling around a circular path, the object must be released at point B to strike the target. This can be verified by drawing a tangent to the circle pointing in the direction of rotation at all four positions. It will be seen that only the tangent from point B intersects the target, giving a hit.

4. d) The speed of the spinning block is given by $v = d/t$, where d is the circumference. The circumference = $2\pi r$ = 2(3.14) × (10 cm) = 62.8 cm. The time it takes the block to rotate around the table, covering a distance equal to the circumference is derived from the rotations/minute – 60 s for 15 rotations means 4 s for 1 rotation. To find the speed, $v = d/t$ = (62.8 cm)/(4 s) = 15.7 cm/s.

5. d) The force of friction is what keeps the block on the spinning turntable. Although the force of gravity helps to determine the size of the frictional force, it is in the wrong direction (downward) to provide the needed horizontal force to keep the block spinning in a circle. The student's hand is not holding the block on the table, but rather spinning the turntable.

Chapter Assessment

Physics You Learned

Consider dividing students into groups and ask them to prepare a collage of important physics concepts and equations to put up on the walls of their classroom. This will provide a visual frame of reference to the physics they have learned. Reviewing a list of physics concepts will help students reinforce what they have learned in *Driving the Roads*. Initiate a discussion on the concepts by referring to the equations that are listed in the *Is There an Equation?* section of the table. You might want to record points of discussion on the board that you can expand upon from time to time. As students summarize and review what they have learned, you must emphasize that the table of physics concepts serves as their reference for the *Physics Practice Test*.

Chapter 1 Driving the Roads

Physics You Learned

Physics Concepts	Is There an Equation?
When driving it takes a certain amount of time, called **reaction time**, to recognize a hazard, decide what to do, and initiate an action such as applying the brakes. During this time, the vehicle is still moving, and the distance traveled is the reaction distance.	
All instruments must be adjusted to read correctly in a process called calibration. This process compares the instrument to a standard to determine its accuracy.	
No measurement is exact (accurate to an infinite number of decimal places). Measurements are often repeated many times to average out uncertainties. Sources of uncertainty include **systematic errors** due to improper calibration, and **random errors**.	
Accuracy refers to the ability of measurements to give an average value close to the accepted standard. **Precision** refers to the ability to repeat a measurement to almost the same value regardless of its accuracy.	
Scientists use the SI system of measurements. Units are related to their sub-units in multiples of ten.	
Average speed is the distance traveled, Δd, in a given interval of time, Δt. By definition, average speed is distance traveled divided by time taken.	$v_{av} = \frac{\Delta d}{\Delta t}$
The equation of average speed can be used to find the time needed for an object to travel a certain distance or the distance traveled during a period of time.	$\Delta d = v_{av} \times \Delta t$ $\Delta t = \frac{\Delta d}{v_{av}}$
The slope of a distance vs. time graph at any point is the object's instantaneous speed. If the object is traveling with constant speed, the graph is a straight line with a constant slope. If the graph is not a straight line, the slope may be found by drawing a tangent to the curve at a point.	
When a source of sound is moving toward or away from an observer, the frequency of the sound detected by the observer is shifted. This shift is referred to as the **Doppler effect.**	$f = \frac{f_0 s}{(s-v)}$
Acceleration is a change in an object's velocity, Δv, with respect to time, Δt. By definition, acceleration is the change in an object's velocity divided by the interval of time. Acceleration can be positive or negative.	$a = \frac{\Delta v}{\Delta t}$
An object's change in velocity with respect to time, or the time that is required for an object to change its velocity, can be found using the definition of acceleration.	$\Delta v = a \times \Delta t$ $\Delta t = \frac{\Delta v}{a}$
The slope of a velocity vs. time graph at any point is the object's acceleration at that time. If the slope of the velocity vs. time graph is constant, the object is traveling with constant acceleration.	

118

Active Physics

When an object is moving, the direction of the movement is as important as the size (magnitude). Quantities that have both size and direction, such as velocity and acceleration, are **vectors**. Those that have only size and no direction, such as mass, are known as **scalars**.	
The equations of motion can be used to predict whether a vehicle is in the STOP, GO, Dilemma, or Overlap Zone when approaching a yellow light.	
Active Physics Plus: The average velocity (v_{av}) of a constantly accelerating object is equal to the quantity initial velocity (v_i) plus the final velocity (v_f) divided by 2. The average velocity is the average of the initial and final velocities for an accelerating object.	$v_{av} = \frac{(v_i + v_f)}{2}$
Active Physics Plus: The distance covered by an accelerating object (d) is equal to the object's initial velocity (v_i) times the time of travel (t) plus one half of the object's acceleration (a) times the square of the travel time (t^2). The distance traveled by an accelerating object depends upon both its initial velocity and its acceleration.	$d = v_i t + \frac{1}{2}at^2$
The distance covered by an object that is undergoing uniform, negative acceleration when coming to rest depends upon the initial velocity squared. The stopping distance for an automobile (d) is equal to the square of the initial velocity $(v_i)^2$ divided by twice the acceleration provided by the brakes (a). The distance covered by an object that is undergoing uniform, negative acceleration when coming to rest depends upon the square of the initial velocity.	$v_i^2 = 2ad$ $d = \frac{v_i^2}{2a}$
Active Physics Plus: The square of the final velocity (v_f^2) of an accelerating object is equal to the square of the initial velocity (v_i^2) plus twice the acceleration times distance traveled while accelerating. When an object accelerates, the final velocity of the object depends upon the initial velocity, the object's acceleration, and the distance traveled during the acceleration.	$v_f^2 = v_i^2 + 2ad$
When an automobile goes around a curve, a centripetal force is needed to cause the direction of the automobile to change so that it can make the turn safely. The force is directed toward the center of the circle.	
Active Physics Plus: The centripetal acceleration (a_c) of an object traveling in a circle at constant speed equals the square of the object's speed (v^2) divided by the radius of the circle (r).	$a_c = \frac{v^2}{r}$
Active Physics Plus: The centripetal force (F_c) on an object traveling in a circular path with constant speed equals the mass of the object (m) multiplied by the square of the object's speed (v^2) divided by the radius (r) of the circle.	$F_c = \frac{mv^2}{r}$

119

Active Physics

Physics Chapter Challenge

Physics

Chapter Challenge

You will now be completing a second cycle of the *Engineering Design Cycle* as you prepare for the *Chapter Challenge*. The goals and criteria remain unchanged. However, your list of *Inputs* has grown.

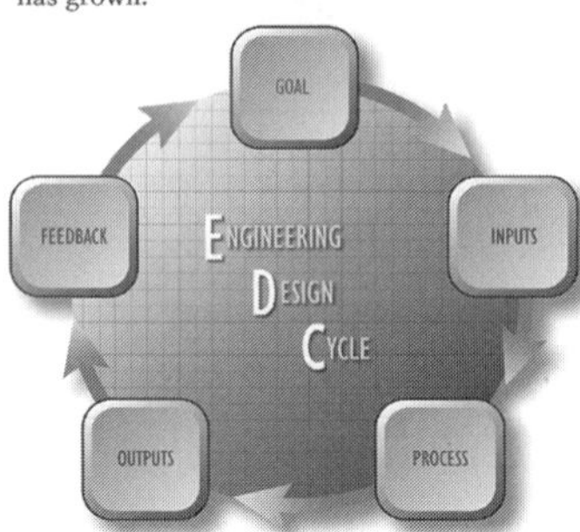

Goal

Your challenge for this chapter is to create a group presentation that will convince the Active Driving Academy that you have learned enough about the physics of safe driving to be eligible for graduation. Review the *Goal* as a class to make sure you are familiar with all the criteria and constraints.

Inputs

You now have additional physics information to help you address the safe-driving topics. You have completed all the sections of this chapter and learned the physics content you need to complete your challenge. This is part of the *Inputs* phase of the *Engineering Design Cycle*. Your group needs to apply these physics concepts to build your presentation. You also have additional *Inputs* of the feedback you received following your *Mini-Challenge* presentation.

Section 1 You used different methods to measure reaction time and compared the reaction time of different members of your class. You also investigated how distractions affect reaction times.

Section 2 You used a stride and a meter stick to measure distance. You identified the sources of error in measurement and read about units of measurement used in science classrooms and when driving the roads in the United States.

Section 3 You defined average and instantaneous speed and used strobe pictures, graphs, and equations to represent motion. You also used the equation for average speed to calculate speed, distance, and time. You read about how speed can affect following distance when driving on the roads.

Section 4 You learned how changes in speed or direction, called accelerations, are related to time and distance for a moving vehicle. You also interpreted distance-time and velocity-time graphs for different types of motion.

Section 5 You designed an experiment to investigate negative acceleration and stopping distance. You used friction as your brakes and measured the stopping distance for different starting speeds to compare speeds and stopping distances.

Section 6 You used models to examine vehicles approaching intersections with yellow lights. You used diagrams and computer models to help isolate locations where drivers are in a safe GO Zone, a safe STOP Zone, a safe Overlap Zone, or an unsafe Dilemma Zone. You also used the computer model to examine which factors influenced the locations of these zones.

Section 7 You explored the force necessary to make a moving object travel in a circle. You also explored the role that friction plays in creating that force for moving vehicles traveling through turns with different radii (sharp or gentle turns).

As students prepare for the *Chapter Challenge*, they go through a second iteration of the *Engineering Design Cycle*. They should now be able to articulate their *Goal* and be clear about the criteria defined in their rubric. The challenge for students is to incorporate the concepts they have learned and build them into their final presentation. For this purpose, a brief summary of each section provides them with the opportunity to reflect on what they have learned. They can now see how additional *Inputs* helped to develop the physics content they knew during their *Mini-Challenge*. Consider pointing out how the yellow-light model applies the concepts of GO and STOP Zones to determine whether or not an intersection has a Dilemma Zone and is safe.

During the *Process* phase, students will have a chance at filtering and refining their information. They will be better equipped during this phase if they are encouraged to share their ideas with the group members to compare and contrast their response to driving situations they might face. Emphasize that once students have decided which model of safe driving they will be describing, they should organize ideas and think of creative ways to present the *Chapter Challenge*. Address readability of text and illustrate how different styles of presentation enhance the audience's interest and focus.

Reiterate that a model similar to the ones used in *Section 6* utilize a fixed reference from which safety requirements can be clearly explained. Point out that the use of a model in their presentation will enable them to bring in the nuances of concepts that they would like to discuss in their presentation. During the *Process* phase, students should refer to the *Goal* and the rubric to guide their work.

As students get ready to present, remind them that should pay careful attention to their presentation skills for their *Outputs* phase, which requires an engaging, well-developed analysis of *Inputs* that have been honed and designed through a process to create a well-developed product.

The *Feedback* stage is designed to help students assess their

Process

In the *Process* phase you need to decide what information you have that you will use to meet the *Goal.* Decide on the format for your presentation. Will your group compare and contrast different driving scenarios and their safe distances, or will you simply address each of the distances individually? You may also have your own excellent idea for organizing your answers. Creativity is encouraged and will help your presentation be memorable to the judges of the Active Driving Academy. Just make sure that every member of your group is included and knows how she or he can contribute to your presentation.

Consider charts, graphs, and diagrams for your presentation. Your presentation could certainly include charts, graphs, and diagrams to illustrate your ideas. Remember to make any materials that you include neat and readable in the presentation format your group will use. Colors and large text can help make your point clearer.

This chapter made extensive use of models in *Section 6* for considering the complex analysis of the yellow light. A similar model would allow you to compare and contrast safe distances for following, braking, and stopping under different driving scenarios. A model for safe driving in curves could also help you compare the different factors that ensure safety for the passengers there. Your group could use models to find and present answers for each of the three safe driving considerations in the challenge. A model will provide a fixed reference point from which you can consider individual changes like a faster speed, more time to react, a sharper curve, and so on. When you describe the result for each change in the model, you enable the audience to easily compare the result to the results for a previous set of design conditions. Models can be very useful for learning and presenting information.

Address each of the three safe-driving topics. Refer back to the goal and constraints for the presentation regularly to make sure you are answering each part. If your class has a grading rubric for the challenge, use that to guide your work so that you don't forget to include any important information.

Outputs

Presenting your information to the class are your design-cycle *Outputs*. You should try to create a very convincing presentation. After all, your freedom is on the line! A combination of good analysis, creativity, well-managed development, and engaging presentation skills will be required to create a successful product.

Feedback

Your classmates will give you *Feedback* on the accuracy and the overall appeal of your presentation based on the criteria of the design challenge. This *Feedback* will likely become part of your grade but could also be useful for additional design iterations. No design is perfect, because there is always room for optimization or improvement, no matter how slight. From your experience with the *Mini-Challenge* you should see how you could continuously rotate through the design cycle to refine almost any idea.

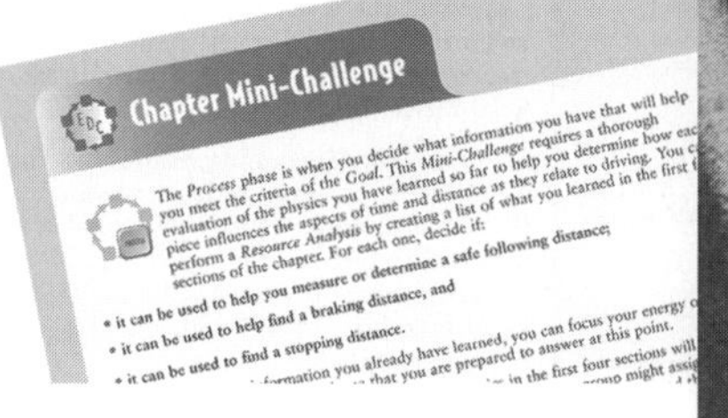

performance by giving it a grade. You should emphasize that the feedback they receive from their classmates will help them in reassessing their *Goal* before its final presentation and it provides an opportunity to include additional ideas or redefine the existing elements of their several design iterations.

Physics Connections to Other Sciences

Physics Connections to Other Sciences is fundamental to how science operates in the real world. This section provides a glimpse of the interconnectedness of different scientific disciplines. The brief descriptions relate students' study of physics concepts to biology, chemistry, and Earth science. Each description deals with an aspect of how various physics principles are intrinsic to the behavior of living beings, chemical substances, and geological phenomena (Earth science).

Discuss the science connections and encourage students to draw analogies with science connections they are familiar with. Students will gain a deep understanding of interdisciplinary interactions if they are actively engaged in the process of thinking about connections between different scientific disciplines. Encourage students to appreciate physics in relation to a broader framework of scientific interactions that are gaining ground as major areas of study— such as geophysics, biochemistry, and biophysics.

If students are unfamiliar with the content of the other branches of science discussed in the examples in their textbooks, explain scientific terms they don't know and ask them to take down notes in their *Active Physics* logs.

Chapter 1 Driving the Roads

Physics
Connections to Other Sciences

The fundamental ideas you have studied in this chapter are also basic to many other sciences that you will study in the future. Appreciating the connections among science disciplines helps scientists achieve a richer understanding of nature. Science research in the twenty-first century depends heavily upon the way these different disciplines interact, with areas such as biophysics and geophysics becoming major areas of study.

Here are some examples of how the concepts you studied in this chapter relate to other sciences.

Response Time

Biology An animal's response time may be critical for its survival. A bird that responds too slowly to a hawk, or a fly that fails to evade a frog's tongue, will not survive long.

Chemistry The time it takes a chemical to respond to light, and then bond with other chemicals, is one of the criteria for determining film speed for cameras.

Earth Science The response of the polar ice caps due to global warming may take hundreds of years to become completely visible.

Circular Motion

Biology Many birds will soar in circles on thermals—rising columns of warm air. The birds bank their wings to provide a centripetal force toward the center of the thermal and the lifting force of the rising air.

Chemistry Magnetic fields cause charged molecular fragments to travel in a circle. The larger the circle, the more massive the charged fragments must be, giving chemists an idea of what the composition of these parts may be.

Earth Science Most of the weather systems on Earth are examples of large masses of air that circle around a central position due to the Coriolis force that is experienced by air masses moving on a rotating Earth. This force is part of the reason for Earth's global winds and ocean currents.

Speed and Velocity

Biology The speeds obtainable by living organisms vary, from that of a diving peregrine falcon (almost 200 mph or 322 km/h) to that of a slime mold (1 mm/hr or 0.04 in./hr).

Chemistry The high speed gained by molecules in a chemical explosion is responsible for the damage they do.

Earth Science The speed of advance of a glacier may be as much as several feet per day.

Acceleration

Biology The fastest land animal, the cheetah, is able to accelerate from rest to a speed of 60 mph in only 3 seconds, or almost 9 m/s^2 (30 ft/s^2)!

Chemistry Electrons are accelerated to a very high speed and collide with molecules in a device known as a mass spectrometer, which is used in forensics for solving crimes.

Earth Science An earthquake may accelerate the floor of the ocean upward for a very short time, causing a tsunami that is capable of damaging large sections of a coastline.

Doppler Effect

Biology The velocity measurement of blood flow in arteries and veins, based on the Doppler effect, is an effective tool for diagnosis of vascular problems.

Chemistry The random motion of atoms of a gas due to their kinetic energy results in a shift in the frequency of the light emitted by the atoms due to the Doppler effect. Thus, the light emitted by a gas has a wider range of frequencies than that of a single atom.

Earth Science The expansion rate of the universe is determined by astronomers who use the Doppler effect to calculate the speed of moving galaxies.

122

Active Physics

This will give you the opportunity to model how you are comfortable explaining scientific phenomena studied in biology, chemistry, or Earth science.

Physics At Work

Christine Lopez

New York State Trooper; Poughkeepsie, NY

On average, Christine Lopez spends 12 hours a day in a car. No, she is not a chauffeur or a professional race car driver. For the past 14 years, Lopez has been a New York State Trooper in Troop K headquarters, which covers Columbia, Dutchess, Putnam and Westchester counties.

Lopez said that State Troopers are responsible for assisting the public in a variety of circumstances, including motor vehicle accidents, burglaries, assaults and larcenies. Lopez said that since a State Trooper's job involves long hours in a vehicle, their risk of being involved in a collision is heightened. "Law enforcement officers are ten times more likely than the average driver to be involved in a collision," said Lopez. "Troopers may be exempt from vehicle and traffic laws while responding to emergencies, but they are bound by the same laws of nature as the average driver."

In 2007 alone, Troop K dealt with 3933 accidents. According to Lopez, 814, or 22 percent of those accidents involved drivers between the ages of 16 and 19. The majority of those accidents were caused by unsafe speed, following too closely, and driver inattentiveness/distraction. "All drivers should adhere to the three-second rule for following distance," said Lopez. "By allowing three seconds between your vehicle and the vehicle in front of you, you will have ample braking distance to allow you to make a complete stop."

Lopez also believes that reaction time plays a crucial role in preventing accidents. "The average reaction time is 1.6 seconds. This varies from individual to individual depending on age, illness, fatigue, and alcohol consumption. A drunk driver will take longer to perceive hazards and will have a slower reaction time," said Lopez.

Dr. Jose Holquin-Veras

Professor, Rensselaer Polytechnic Institute; Troy, NY

Dr. Jose Holquin-Veras is a Professor of Civil and Environmental Engineering at Rensselaer Polytechnic Institute (RPI). Veras believes that an understanding of physics is essential in solving problems and designing roads. "I design traffic signals to deal with the Dilemma Zone, and use braking distance to take reaction time into account." According to Veras, in order to avoid a Dilemma Zone, traffic engineers must use physics to ensure that a yellow light is long enough for a driver to stop or go through the light.

Alyson Coyle

Instructor, Transportation Safety Institute; Oklahoma City, OK

Alyson Coyle is an instructor with the Transportation Safety Institute. Coyle's division is responsible for developing training programs for the National Highway Traffic Safety Administration (NHTSA). Coyle informs participants about everything that happens in a crash, including how seat belts and child safety seats protect occupants. "The best part of my job is helping to save lives every day. According to NHTSA, more than 62,000 lives have been saved by seat belts in the past ten years," said Coyle.

123

Active Physics

Physics At Work

The *Physics At Work* gives students examples of how physics is applied in the real world. Profiles of professionals are provided to help students realize that their *Chapter Challenge* is designed for a practical application of physics, and the investigations they performed were geared to give them a hands-on experience of science.

Each *Investigate* involved in *Active Physics* uses practical procedures that are employed by a variety of people featured in this section. Students read how physics is embedded in many professions and is practiced by a broad range of people. There are no geographic, ethnic, or gender barriers for people who apply physics in their professions. It is not necessary to have an advanced degree or be a scientist to employ physics. The different profiles presented show how people from diverse backgrounds use physics as a part of their job for different purposes.

State Trooper, Christine Lopez's acknowledgment that all drivers are bound by the same laws of nature as an average driver provides an opportunity for yet another meaningful discussion on the speed of a vehicle and the three-second rule for following distance. Point out how Lopez' knowledge of the physics involved in driving improves her chances of avoiding an accident and being safer on the roads. In other words, reaction time while driving plays a crucial role in her life as well, and she has to bear that in mind to help her prevent hazardous situations.

Dr. Jose Holquin-Veras's experience as an engineer helps him in designing roads, and Alyson Coyle's experience in developing training programs to help save lives are both significant profiles that should be emphasized in relation to the physics that students are learning. Consider asking students to write their impression of professional profiles presented to them to explain why they are so fundamental to improving safety on the roads. Students should note how each profile has the potential to contribute to their *Chapter Challenge* by opening up connections between physics and real-life applications that are intrinsic to safety on the roads while driving.

CHAPTER 1

Physics Practice Test

The *Physics Practice Test* is provided as a Blackline Master in your *Teacher Resources CD*.

1c Blackline Master

Content Review

1. c
2. b
3. d
4. a
5. d
6. a
7. b

Physics

Practice Test

Before you try the Physics Practice Test, *you may want to review sections 1-7, where you will find* 29 Checking Up *questions,* 11 What Do You Think Now? *questions,* 28 Physics Essential Questions, 79 Physics to Go *questions, and* 11 Inquiring Further *questions.*

Content Review

1. Many driving experts recommend that novice drivers do not drive with groups of friends in their automobile. The major reason the experts suggest this is because friends may
   a) suggest that the driver exceed the speed limit, increasing risk.
   b) want to drink alcohol in the automobile.
   c) be a distraction that would increase driver reaction time.
   d) urge the driver to go through a yellow light when in the STOP Zone.

2. In a class demonstration, a teacher drops a dollar bill held between the fingers of a student to test how quickly the student can respond by catching the bill. The reason the bill is so difficult to catch is because
   a) the dollar bill is thrown downward.
   b) student's reaction time is too long.
   c) the dollar bill is affected by air resistance.
   d) student's fingers are affected by air resistance.

3. Middle-aged drivers often have better safety records than younger drivers. The most likely reason for this is that middle-aged drivers
   a) have quicker reaction times than teenagers.
   b) are never distracted while driving.
   c) will avoid streets with stoplights to avoid Dilemma Zones.
   d) rely on experience to avoid situations where a short reaction time is important for safety.

4. A friend measures the length of the school soccer field to be sure that it is the correct size. Which measuring device will most likely help your friend get the most accurate answer?
   a) A 50-m tape measure accurate to the nearest cm.
   b) A meter stick accurate to the nearest cm.
   c) A meter stick accurate to the nearest mm.
   d) A 30-cm ruler accurate to the nearest mm.

5. A friend claims that he can measure exactly how much water is in a one-gallon jug after taking a drink from it. You disagree with your friend. Which of the following reason(s) would a scientist give for agreeing with you?

   I. All measurements contain random errors.
   II. All measurements are at best an estimate of the true value.
   III. A perfect measurement requires a very expensive instrument, which your friend cannot afford.

   a) I only
   b) I, II, III
   c) II only
   d) I and II only

6. At a stock car race, you want to check the posted speed for the leading driver in the race. You time how long it takes the driver to make three laps around the track with your stopwatch. What else do you need to know to calculate the race car's average speed for this time?
   a) the length of the track
   b) how many cars the driver passed
   c) the size of the car's wheels
   d) the time that the race began

7. The distance vs. time graph for an automobile is shown below. Which reason below might best explain the automobile's change in motion at point X?

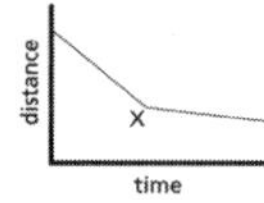

   a) the automobile sped up
   b) the automobile slowed down
   c) the road became less steep
   d) the road was no longer straight

8. Some students decide to take a bike ride. For the first two hours they travel at a speed of 15 mi/hr, they then stop for lunch for an hour. The students then ride for another hour at 10 mi/hr. What was their average speed for the trip?

a) 10 mi/hr c) 13.3 mi/hr
b) 12.5 mi/hr d) 15 mi/hr

9. The graph below shows the velocity of an automobile vs. time as the automobile accelerates on a road. An automobile that has a greater acceleration would have a velocity vs. time graph that has

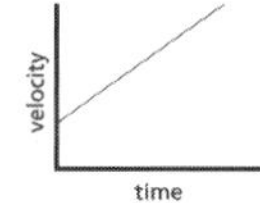

a) a higher velocity at $t = 0$.
b) a lower velocity at $t = 0$.
c) a greater slope.
d) a longer line.

10. A police officer accelerates from rest to catch a speeding motorcycle traveling with constant velocity on a highway. To catch up to the speeder, the police car must
a) have an acceleration less than the speeding motorcycle.
b) match the speeding motorcycle's velocity.
c) match the police car's acceleration to the speeding motorcycle.
d) have a velocity greater than the speeding motorcycle.

11. The velocity vs. time graph for two automobiles is shown below. At time (t), the automobiles have the same

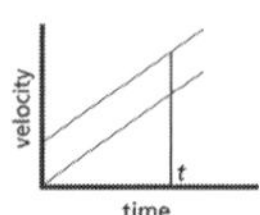

a) velocity.
b) acceleration.
c) velocity and acceleration.
d) velocity, acceleration, and distance traveled

12. Two identical automobiles are approaching a yellow light. The first automobile has a speed 30 mi/hr, and the second automobile has a speed 42 mi/hr. Compared to the 30 mi/hr automobile, the stopping distance of the 42 mi/hr automobile will be

a) the same c) double
b) 1.4 times longer d) 4 times longer

13. Which of the following has no effect on the stopping distance of an automobile approaching a yellow light?
a) the driver's reaction time
b) the automobile's velocity
c) the condition of the automobile's brakes
d) the time the light remains yellow

14. A student is holding an accelerometer made of a ball hanging from a piece of string in her hand. The student is then spun around in a rotating chair as shown. The direction of the hanging string when the chair is rotating indicates

a) the centripetal force is away from the student.
b) the centripetal acceleration is toward the student.
c) the direction of rotation of the chair.
d) the force needed to overcome the friction of the chair.

15. Which of the following objects cannot be used to accelerate an automobile?
a) the gas pedal
b) the brake pedal
c) the steering wheel
d) the rearview mirror

8. a
9. c
10. d
11. b
12. c
13. d
14. b
15. d

CHAPTER 1

Critical Thinking

16.a)

Measuring instruments needed are a stopwatch and velocimeter. Alternative instruments that may be used are a stopwatch and meter stick, depending upon the method a student uses to calculate the acceleration.

16.b)

Possible answers include measuring the speed of the ball at two points using the velocimeter, and the time required for the ball to travel between those two points; or, using the stopwatch and meter stick to determine the distance between the two points, and the time required to travel that distance.

16.c)

Data will be calculated using the equation $a = \Delta v/\Delta t$ or, for the second method, using the equation $d = \frac{1}{2}at^2$.

17.a)

Chapter 1 Driving the Roads

Practice Test *(continued)*

Critical Thinking

16. Your teacher tells you to design an experiment to find the acceleration of a ball rolling down an inclined plane.
    a) What measuring instruments will you need?
    b) What measurements will you take to determine the ball's acceleration?
    c) Show how you will use this data to calculate the ball's acceleration.

17. A ball is rolling across a horizontal table at a constant speed from left to right, then rolls up a ramp where it comes to rest.
    a) Draw a strobe photograph of the ball's motion as it is rolling across the table. Label the first point A and each successive point, B, C, D, and so on.
    b) Draw a velocity vs. time and a distance vs. time graph for the ball as it rolls across the table.
    c) Draw a strobe photograph of the ball's motion as it rolls up the ramp. Label the first point on the ramp 1 and each successive point, 2, 3, 4, and so on.
    d) Draw a velocity vs. time and a distance vs. time graph for the ball as it rolls up the ramp.

18. An automobile is traveling along a smooth, dry road, when the driver suddenly sees a small child run into the street.
    a) If the brakes are applied to the maximum force, what factors other than the condition of the automobile determine how far it takes the automobile to stop?
    b) How would changing each of the factors change the stopping distance?
    c) Which factor would increase the stopping distance the most if it was doubled?

19. Compute the GO Zones and STOP Zones for the intersection described below.
    a) yellow-light time 4.0 s
    reaction time 1.2 s
    speed of automobile 25 m/s
    acceleration $-7\ \text{m/s}^2$
    width of intersection 12 m
    b) Determine if the intersection is safe and describe how you know if it is safe or not.

20. You are driving in an automobile going around a curve at constant speed as shown in the diagram below.

    a) On the diagram, draw the direction of the automobile's acceleration at the position shown.
    b) On the diagram, show the direction of the net force on the automobile at the position shown.
    c) Explain why a passenger in the automobile feels as if he is being pushed outward from the center of the circle.

Active Physics *Plus*

21. A race car accelerates from 50 m/s to 75 m/s over a distance of 400 m. What is the race car's acceleration?

22. An automobile with a mass of 1200 kg is rounding a curve with a radius of 200 meters. If the maximum force of friction the road can provide to the automobile's tires is 2400 newtons, what is the maximum speed at which the automobile can safely take the turn?

23. The graph below shows the acceleration vs. time graph for a jet taking off from the catapult of an aircraft carrier. Draw the graph for the jet's velocity vs. time graph.

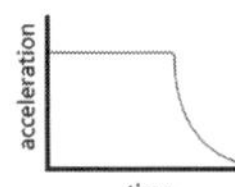

Active Physics 126

17.b)

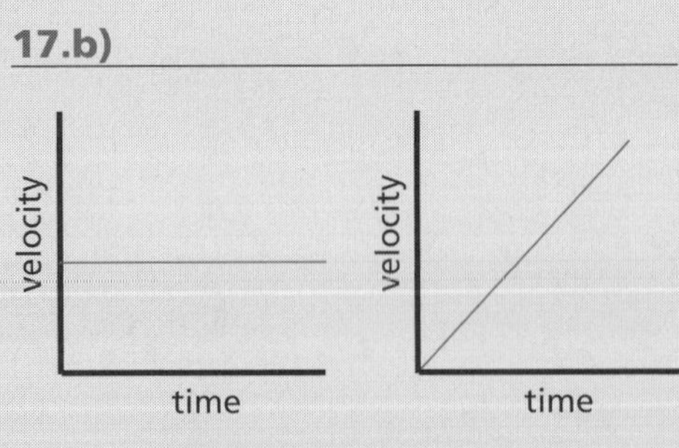

17.c)

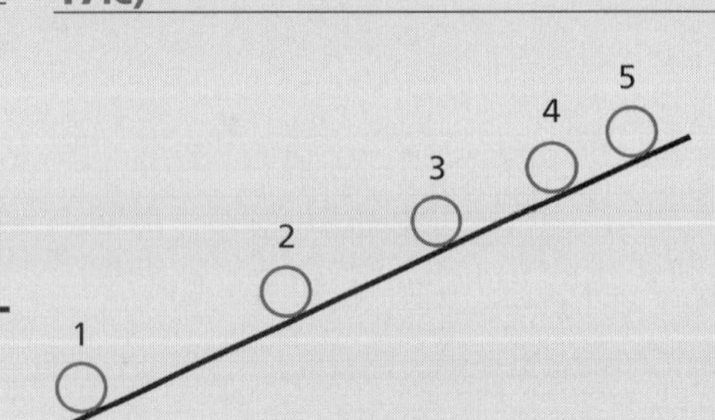

17.d)

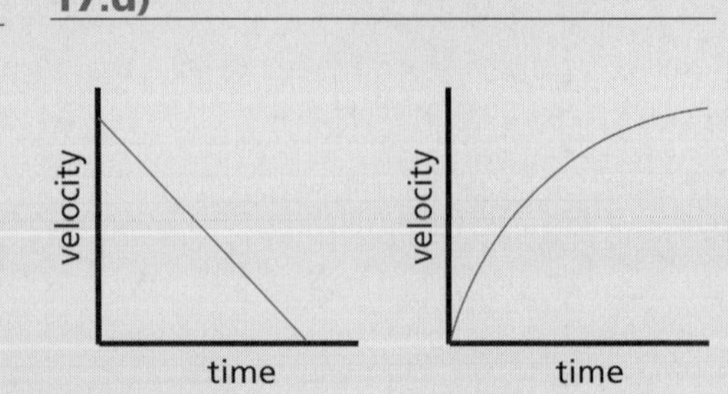

18.a)

The factors that would determine the stopping distance would be reaction time and speed. The road surface also determines the friction, which has an effect on the stopping distance.

18.b)

Increasing the reaction time and the automobile's speed would increase the stopping distance. If the road was wet or snow-covered, the stopping distance would also increase.

18.c)

Doubling the speed would quadruple the stopping distance, which would double the effect of increasing the stopping distance. Doubling the road friction would decrease the stopping distance.

19.a)

GO Zone = (v)(yellow-light time) – intersection width = 88 m

STOP Zone = (v)(reaction time) + $v^2/2a$ = 105 m

19.b)

There is a 17-m Dilemma Zone, so the intersection is unsafe.

20.a)

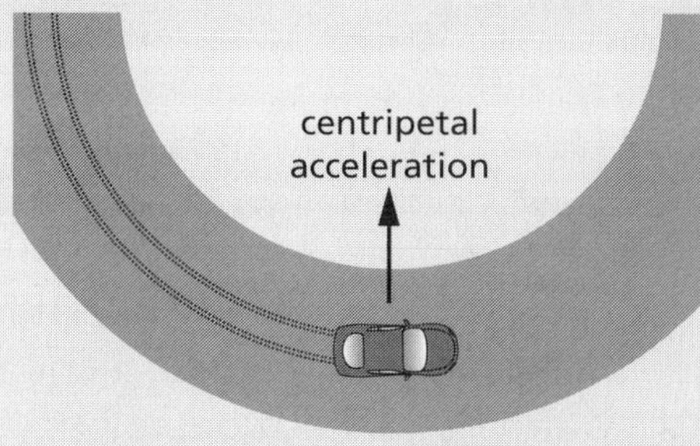

20.b)

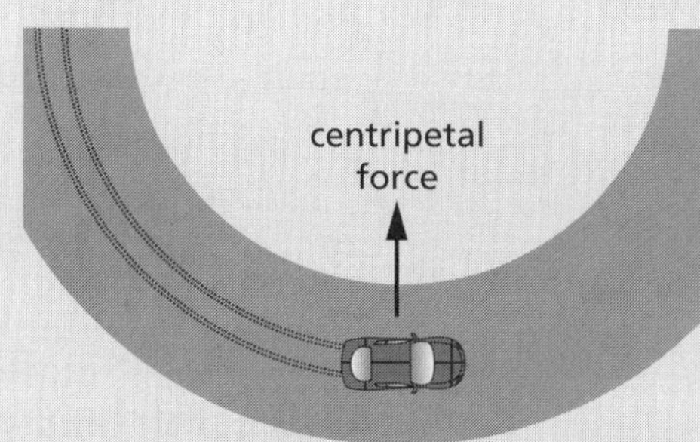

20.c)

The reason the passenger feels as if they are being pushed outward toward the door is that they are going straight, while the automobile is turning into them. The automobile door ultimately will provide the centripetal force that makes the passenger go around the circle with the automobile.

21.

3.9 m/s^2

22.

20 m/s

23.

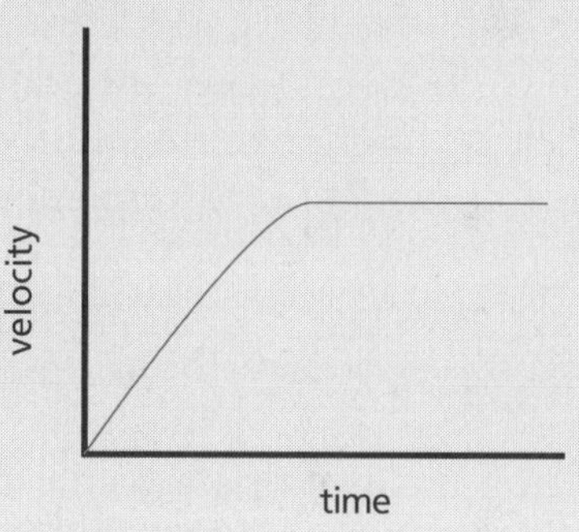

Sample Assessment Rubric

Introduction

Development of rubrics as a template for identifying performance criteria has been shown to increase student achievement. A typical rubric clearly denotes each category to be evaluated and provides specific, required criteria for defining excellence, proficiency, and below-proficiency levels of performance. The sample rubrics for each chapter are intended to serve as guidelines. It should be understood that assessment is more effective when students and teachers tailor it to fit their needs. You are encouraged to work with your colleagues and especially with students to customize the rubric and the criteria. Decisions should be made together with respect to the curricular goals of the project within the particular context. For example, a class may choose to add one requirement in lieu of another, or to change the relative weighting of categories. It is helpful to remember the following recommendations:

1. Assessment should directly address the goals of the *Chapter Challenge*.

   Attention has been paid to the suggested rubrics in addressing the goals of the chapter, and the *Physics You Learned* section should serve as a guide for students and teachers working with the challenge. You may choose to make changes to the rubric in order to emphasize goals important to their context.

2. Students should participate in the assessment of their own performance.

   Students submit their rubric along with the grade they have given themselves. This not only encourages students to take ownership of the project, but it becomes a useful assessment tool for the teacher. If a student earns a "C" and gives himself or herself a "C," the conversation is very different than if a student were to earn a "C" and give himself or herself an "A." The question you might have for the first student is, "Why didn't you choose to do more?" While you might need to review the criteria with the second student and help understand what it takes to get an "A." After the teacher has graded the assignment, students have an opportunity to revise their work, and resubmit it for the "revision" grade. Emphasis should be placed not only on the finished project, but on progress with the rubric during revision.

3. Assessment should begin from a foundation of a proficient level of performance, providing ladders for students to achieve higher orders of thinking.

Finally, the rubric is built from a foundation of proficiency (meets standards). An analogy for this level is that in the real world their is a minimum acceptable standard for performance. A CD must play without skipping, and a shirt must have all of its buttons. Anything less, is substandard. This rubric works the same way. To get a "C" or better, the work must meet all the standards, and fall into the "proficient" category. Work not meeting all standards must be revised and resubmitted. Beyond proficiency, students can do work which shows mastery and may therefore earn a "good" or "excellent" rating (a "B" or an "A"). The scoring column of the rubric includes suggested point ranges for each level of mastery.

For further discussion, see: "Assessment of Laboratory Investigations," Eisenkraft, Arthur and Anthes-Washburn, Matthew. Assessment: Research and Practical Approaches, eds. Coffey, Douglas and Stearns. NSTA Press 2008.

Guide to the Sample Assessment Rubric

Assessment via this rubric will assign students to one of three major groups:

Excellent: Work meets all standards and demonstrates extensive evidence of mastery.

Good: Work meets all standards and demonstrates moderate evidence of mastery.

Proficient: Work meets standards without further evidence of mastery.

Please note that these groups are written at the top of the rubric page as a reminder to students.

In the table, there are three main groups of criteria—Mastery, Meets Standards, and Interventions. A student or team of students should achieve all of the criteria in order to satisfactorily complete the project. Anything less than this will require that the student make another attempt using the Interventions listed in the last column. This is the foundation, or floor of expectations. As teachers, we have to beware that our floor of expectations does not become a ceiling for some students.

1. In the first column, there are suggestions for demonstrating mastery. Completing one or more of these may raise a student or team from Proficient, to Good or Excellent.
2. In the second column, the criteria to meet the standards for the assignment are listed.
3. Some students may have trouble meeting the standards in the Meeting Standards column. The last column provides Interventions, or suggestions for how a student might meet the requirements of the project.
4. In the Scoring column, students submit their own grade, and you respond with a grade and feedback. Students receive a final grade after a revision is submitted. The range of scores for Excellent, Good, and Proficient allow you to assign points that match a student or team's demonstrated mastery. Thus, a student who barely meets standards can receive a different score than one who shows a higher level of mastery.

Implementing the Sample Assessment Rubric

- Modify the rubric with discussions from students.
- Hand out the rubric.
- Review the rubric so that you are confident that students understand each component.
- Have students complete the Scoring column for their work in the chapter by placing checks in each of the boxes. Have students assign themselves a point value for each component.
- Have students total their score for the rubric.
- Collect the student self-appraisal of their work.
- Use the rubric to grade the student work.
- Grade students' work after you and the student agree on the grade. Encourage the student to improve their work for the next chapter. If you and a student disagree, have an appropriate conversation with the student about his or her work and how it could be improved.

The *Sample Assessment Rubric* on the following page is provided as a *Blackline Master* on your *Teacher Resources CD*.

1d **Blackline Master**

CHAPTER 1

Sample Assessment Rubric

Mastery (Students may show mastery through these or other ideas provided by students and teachers.)	**Meets Standards**	**Scoring** (To be discussed by students and teacher)	**Interventions** (Guiding questions and instructions for students falling short of the Standards)
• Study the effect of various distractions (such as cell phone use) on reaction time and discuss the safety risk of driving while distracted. • Create a series of safety posters that educate your peers about safe driving. • Create a myth-busting survey for teenagers to self-test their beliefs and misconceptions about safe driving.	**1. Physics Principles** • Use four physics principles from *Physics You Learned*, which may include velocity and acceleration; the relationship between braking distance and velocity; reaction time and following distance; yellow-light analysis; and the factors determining safe negotiation of curves. • Correctly use scientific terminology, including speed, velocity, acceleration, and centripetal force. • Appropriately use scientific symbols for units and formulae appropriately with correct symbols. • Make correct calculations (where appropriate). • Estimate quantities where appropriate, such as reaction time (0.1 s), yellow-light time (2–3 s) and intersection width (10–20 m).	**Maximum:** 50 Points ***Excellent:** 45–50 **Good:** 40–44 **Proficient:** 35–39 *Student Self Grade:* *Teacher Grade:* *Revision:*	• For four of the following topics, describe a class investigation that helped you understand the concept: – Velocity – Acceleration – Braking distance – Reaction time – Following distance – Yellow-light analysis – Centripetal force
• Create a video public service message to convince teenagers to drive safely. • Use presentation software to present tables and graphs to the audience	**2. Quality of the Presentation** • Prepare and practice your presentation with your group. • Cooperate with your group to ensure that all members participate. • Keep your presentation within the agreed-upon time limit. (Recommended—5 minutes) • Answer questions presented by the audience.	**Maximum:** 25 Points ***Excellent***: 23–25 **Good:** 20–22 **Proficient:** 18–19 *Student Self Grade:* *Teacher Grade:* *Revision:*	• Compose your written report. • Prepare tables, graphs, and diagrams. • Practice your presentation with your team.
• Prepare the presentation for sharing via a poster, web page, or other medium.	**3. Quality of Written Report** • Explain at least two traffic laws and the basic operation of an automobile. • Use tables and graphs where appropriate. • Organize the report so it is easy to follow and understand. • Use correct sentence structure. • Use correct spelling, punctuation, and grammar. • Use the correct number of pages (determined by class and teacher). Suggested—2–3 pages and double spaced.	**Maximum:** 25 Points ***Excellent:** 23–25 **Good:** 20–22 **Proficient:** 18–19 *Student Self Grade:* *Teacher Grade:* *Revision:*	• Follow the suggestions of your teacher and submit a revised script.
		TOTAL: ***Excellent:** 90–100 **Good:** 80–89 **Proficient:** 70-79	

* **Excellent:** Work meets all standards and demonstrates extensive evidence of mastery.
Good: Work meets all standards and demonstrates moderate evidence of mastery.
Proficient: Work meets standards without further evidence of mastery.

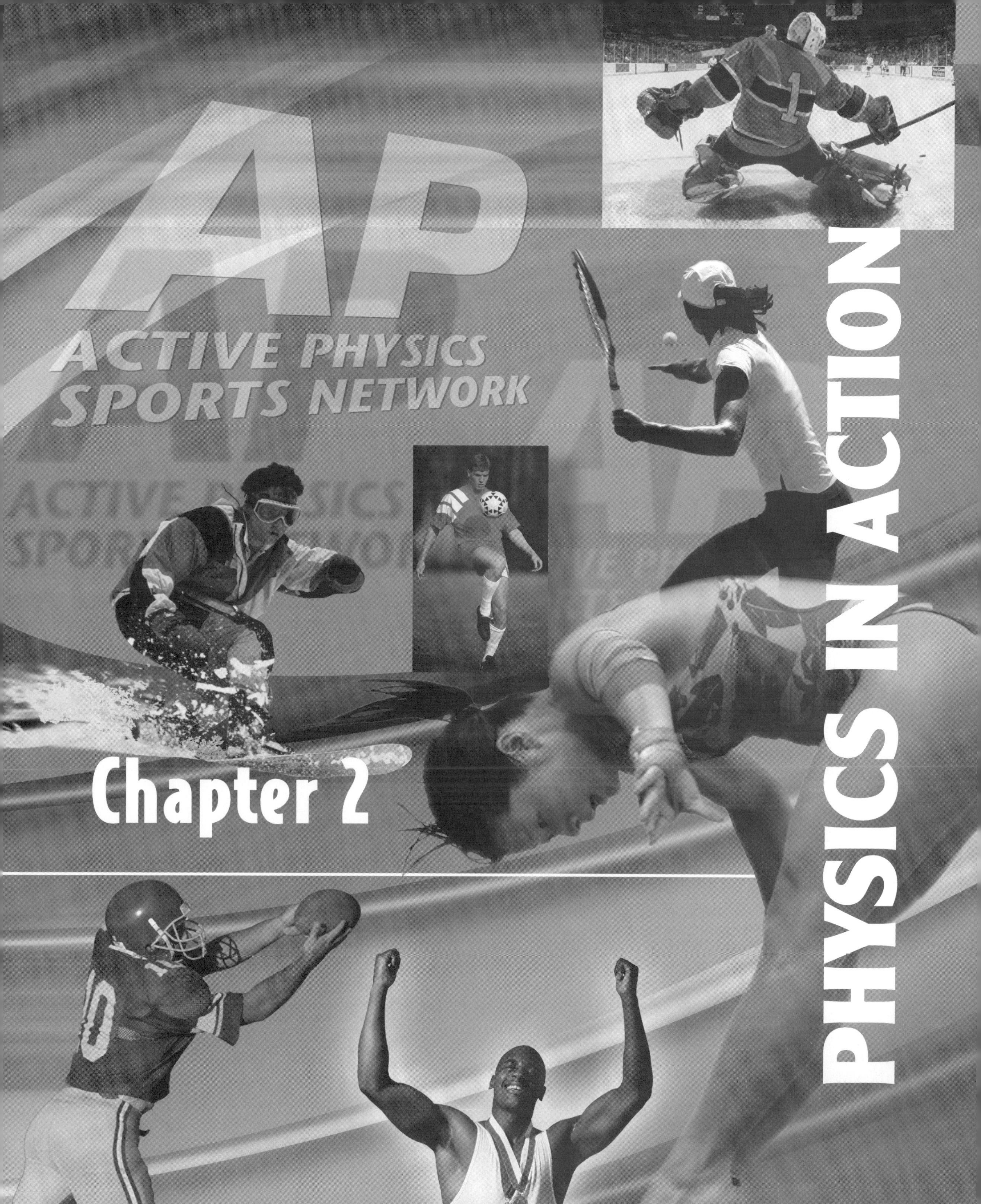
AP
ACTIVE PHYSICS
SPORTS NETWORK
Chapter 2
PHYSICS IN ACTION

CHAPTER 2

Physics in Action

Chapter Overview

Chapter Challenge

This chapter introduces Newton's laws of motion and the concepts of force, inertia (mass), friction, center of mass, work, and energy. The *Chapter Challenge* asks students to produce a two-to-three minute voice-over narration for a sports video explaining the physics behind a sporting event. The students are expected to go beyond the usual description of the event to give the viewer a broader perspective on both sports and the physics involved in that sport. They have to explain the rules of nature that describe the event. The narration can be dubbed into the soundtrack, recorded in an audio version, or provided live. They are also required to submit a written script of the narration.

Students might find it hard to meet the challenge of preparing a scientific documentary. Reassure them that as the concepts in the chapter become more familiar they will be better prepared to complete such an assignment. Remind them that this assignment could be their tryout for a broadcasting job. By the end of the chapter, they will have the skills and knowledge necessary to succeed. The *Chapter Challenge* is designed to set the stage for the physics principles that will be analyzed as the chapter progresses.

You should read the requirements of the challenge along with your students, making sure that students understand exactly what is expected of them. You may select several videos for the students to choose from, or if your students are very ambitious, they can either find some footage themselves or shoot some scenes with a camcorder, or record some sporting events from TV. The entire chapter will build toward the *Chapter Challenge*, and the final evaluation of the students' progress will be based on the video voice-over.

Chapter Summary

The students develop and present a voice-over narration of a sports scene describing the physics involved in the athletes' feats to demonstrate their knowledge of the science involved. Students

- investigate Galileo's law of inertia and Newton's first law of motion, and relate how the laws apply to sporting events.
- explore the terms positive acceleration, negative acceleration, and average speed.
- describe the concept of weight and inertia, and apply Newton's second law of motion.
- recognize that a projectile's vertical motion is independent of its horizontal velocity.
- construct models of trajectories launched at various angles.
- explore Newton's third law of motion and learn the concept of center of mass.
- investigate how the coefficient of sliding friction affects motion.
- describe the law of conservation of energy and define work.
- calculate energy at three different positions of an athlete's jump.

Key Physics Concepts

Section Summaries	Physics Principles
Section 1 Newton's First Law: A Running Start Students release a ball to roll down and then up the sides of a track. They first record its starting height and then the recovered height. From this, they are introduced to the concept of inertia.	Inertia and mass Newton's first law of motion Force Velocity and speed Acceleration Frame of reference
Section 2 Constant Speed and Acceleration: Measuring Motion A timer and paper tape is used to record the motion of various objects. Distance, time, instantaneous and average velocities, and accelerations are calculated from the data.	Instantaneous speed Average speed Positive acceleration Negative acceleration
Section 3 Newton's Second Law: Push or Pull Students calibrate and use a simple force meter to explore the variables involved in accelerating an object. They then connect their observations and data to a study of Newton's second law of motion.	Newton's second law of motion Weight Free-body diagrams Gravitational attraction between masses
Section 4 Projectile Motion: Launching Things into the Air Students explore the motion of objects that are projected in a gravitational field. Differences between objects being dropped, launched horizontally, and launched at an angle are explored in relation to the landing position of objects dropped straight down to those with projected motion.	Gravity Independence of vertical acceleration and constant horizontal velocity for objects in free fall Trajectory of a projectile
Section 5 The Range of Projectiles: The Shot Put Students compare mathematical and physical models of projectile motion to that of a shot put. They apply this to describe the vertical and horizontal motion of the projected object, and predict its trajectory.	Acceleration due to gravity Range of a projectile Mathematical versus physical models
Section 6 Newton's Third Law: Run and Jump Thinking about the direction in which they apply force to move in a desired way introduces students to the concept that every force has an equal and opposite force. They test this concept and then apply it to a variety of motions observed in sports.	Normal force Newton's third law Action-reaction pair forces Free-body diagrams Center of mass
Section 7 Frictional Forces: The Mu of the Shoe Students measure the amount of force necessary to slide athletic shoes on a variety of surfaces. From this and the weight of the shoe, they learn to calculate friction coefficients. They then consider the effect of friction on an athlete's performance.	Friction Coefficient of friction Normal force Weight
Section 8 Potential and Kinetic Energy: Energy in the Pole Vault Students use a penny launched from a ruler to model motion during the pole vault. They connect their observations to the concept of energy conservation.	Gravitational potential energy Kinetic energy Energy conversion Law of conservation of energy Work Spring potential energy
Section 9 Conservation of Energy: Defy Gravity Students learn to measure hang time and analyze vertical jumps of athletes using slow-motion videos. This introduces the concept that work when jumping is force applied against gravity.	Gravitational potential energy Kinetic energy Energy conversion Force and weight Law of conservation of energy Work Spring potential energy

CHAPTER 2

Chapter Concept Map

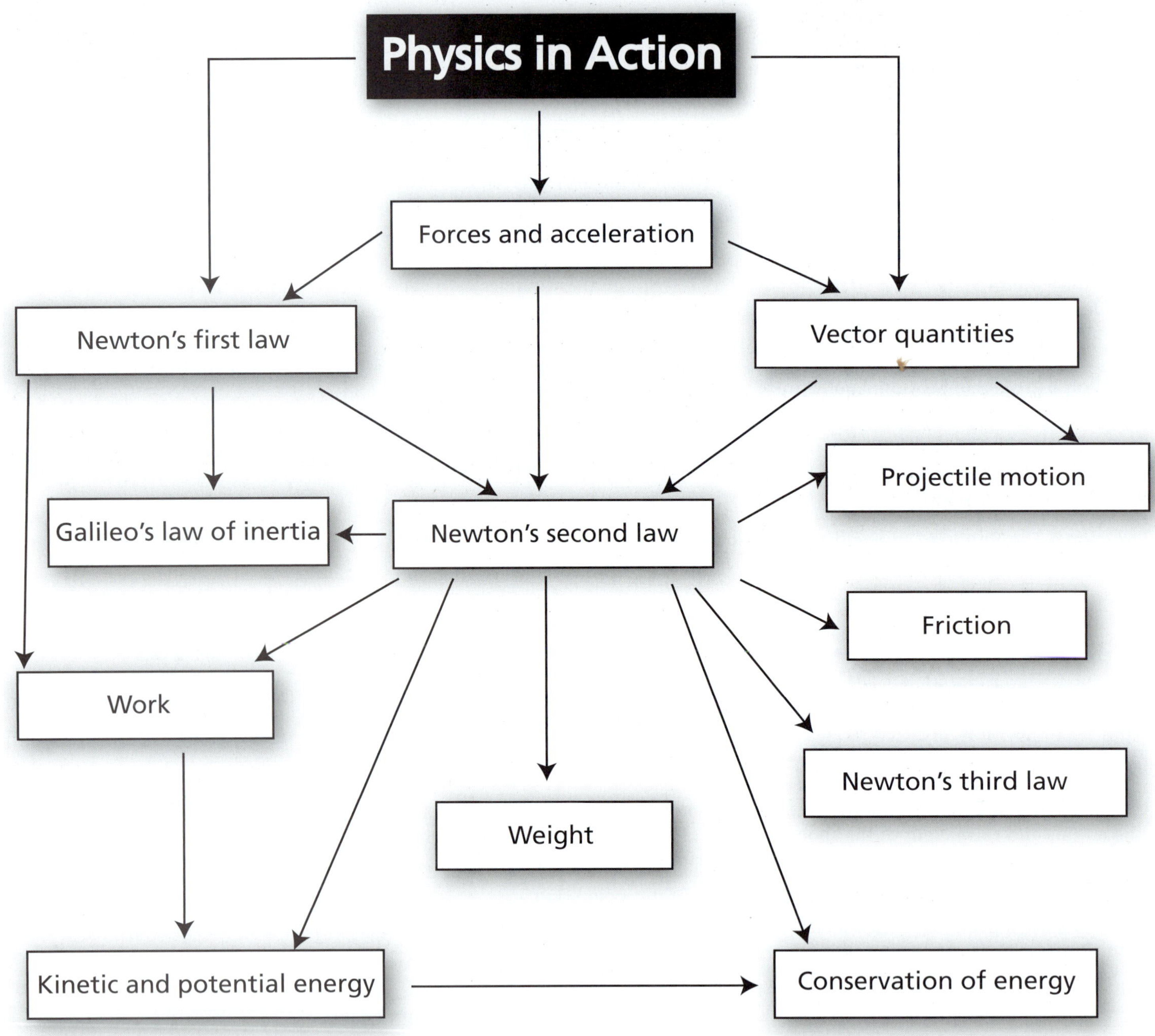

Understanding by Design*

The *Understanding by Design* template focuses on the three stages of backward design:

- **Identify desired results**
- **Determine acceptable evidence**
- **Plan learning experiences**

What overarching understandings are desired?

You can increase your enjoyment of sports by understanding the physics of sports.

- A sportscaster requires knowledge of sports as well as language skills and the ability to clearly articulate thoughts and deliver them in an engaging fashion.
- All sports can be explained with the same laws of physics.
- Knowledge of physics can improve sports performance.
- The motion of people and objects in sports are governed by Newton's laws using mass, position, velocity and acceleration and forces.
- Physics can help explain restrictions to movements in sports.
- Sports can be explained in terms of energy transformations.

What will students understand as a result of this chapter?

- Objects at rest remain at rest and objects in motion remain in motion with a constant velocity along a straight line unless acted upon by an outside force.
- The acceleration of an object is proportional to the net force on it and inversely proportional to its mass. $F = ma$.
- Acceleration is a rate of change of velocity. $a = \Delta v/\Delta t$.
- Velocities and forces add as vectors.
- Weight is the force on an object due to the gravitational attraction between that object and Earth.
- All objects on Earth fall with the same acceleration due to gravity = 9.8 m/s^2 (if air resistance is ignored).
- Newton's third law states that every force has an equal and opposite force. The two forces act on different objects.
- Inanimate objects can exert forces.
- Friction is a force. The coefficient of friction is a property of the two surfaces in contact and is related to the weight of the object. $F_f = \mu F_N$, where F_N is the normal force.
- Kinetic energy, gravitational potential energy, and spring potential energy are three forms of energy.
- Energy can be transformed from one form to another, but the energy of a system not acted on by an outside force is conserved.

What are the overarching "essential" questions?

- What does it mean to say that someone runs faster than someone else?
- How can you increase your speed?
- How can you throw an object further?
- What effect does a shoe have on your sports performance?
- Can you become a world record holder in pole-vaulting by merely purchasing a longer pole?

What "essential" questions will focus this chapter?

- How do objects keep moving after the force on them ceases to act?
- How do velocities add?
- What is inertia (mass)?
- What is acceleration?
- How does acceleration depend on the force on an object and on its mass?
- What is weight?
- What is the acceleration due to gravity?
- What determines the range of a thrown object?
- When an object exerts a force on a second object, what force does the second object exert on the first object?
- What is friction? How is it measured? What does it depend on?
- What does the amount of kinetic, gravitational potential, and elastic energy depend on?
- When is energy conserved and what does it mean that energy is conserved?

* Grant Wiggins and Jay McTighe, *Understanding by Design* (Merril/Prentice Hall, 1998), 181.

CHAPTER 2

Pacing Guide

The *Pacing Guide* below is designed so that you have the option to complete the first eight chapters of *Active Physics* during the school year. The *Plan A Pacing Guide* allows the students to complete all the *Investigates*. If you are a new teacher, or unfamiliar with the program, you may have difficulty adhering to *Pacing Guide A*. *Pacing Guide B* suggests places where either time or equipment may be saved if it becomes necessary to complete the chapter in the allotted time. To reach this goal, many of the investigations are whole-class *Investigates* rather than small-group *Investigates*. This will save time and require less equipment than the optimal inquiry-based instruction that the curriculum is intended to provide. In order to choose which plan is best for you, please consult the *Implementation Chart* following this guide.

Note: Each "day" assumes a 45-minute class period, or one half of a 90-minute block.

Day	Plan A (small-group *Investigates*)	Homework (for Plan A and Plan B)	Day	Plan B (combination of whole-class and small-group *Investigates*)	Plan B Equipment Reduction
1	*Scenario, Chapter Challenge, Chapter Overview, Scoring Rubric,* keeping a log. **Section 1** Have students answer *What Do You See? What Do You Think?*	Obtain a copy of the sport section of a local paper, or a sports magazine. Find one reference to a science and record it in your journal.	1	**See Plan A.**	
2	Students perform the *Investigate* and discuss *Physics Talk.*	Read *Physics Talk* and answer the *Checking Up* questions. Answer *Physics to Go* Questions 1-4 and 10 *(Preparing for the Chapter Challenge).*	2	**Section 1** Teacher does the *Investigate* as a class demonstration. Discuss *Physics Talk*. Answer *What Do You Think Now?* and *Reflecting on the Section and the Challenge.*	Only 1 track with base, steel ball, ruler, c-clamp, masking tape, and felt-tip marker is necessary
3	Review *Checking Up* questions, and *Physics to Go*. Do *What Do You Think Now?* and *Reflecting on the Section and the Challenge.* **Section 2** Have students answer *What Do You See? What Do You Think?* Students perform *Investigate* Steps 1 and 2.	Find and record three sports where there is motion with constant speed, and record when this motion takes place in that sport in their logs.	3	Review *Checking Up* questions, and previous night's *Physics to Go.* **Section 2** answer *What Do You See?* and *What Do You Think?* Students and teacher perform *Investigate* Steps 1 and 11. Teacher does one example of constant speed, increasing speed and decreasing speed with class, then hands out pre-recorded tapes for groups to analyze.	Only 1 tape timer, ruler, scissors, meterstick, and glue stick is required
4	Students perform *Investigate* Steps 3-11.	Students read and summarize *Physics Talk* in their journals. Answer *Checking Up* questions.			
5	Discuss results of the *Investigate*, review students' graphs, and relate to the associated motions. Discuss the *Physics Talk*, and review *Checking Up* questions. Do *What Do You Think Now?* and *Reflecting on the Section and the Challenge.* Students discuss *Physics to Go* Question 14 with their group.	Answer *Physics to Go* Questions 1-4 and 6-11.	4	**See Plan A.**	

Day	Plan A (small-group *Investigates*)	Homework (for Plan A and Plan B)	Day	Plan B (combination of whole-class and small-group *Investigates*)	Plan B Equipment Reduction
6	Go over *Physics to Go* homework. **Section 3** Students answer *What Do You See?* and *What Do You Think?*, perform *Investigate*, all steps, and discuss *Physics Talk*.	Read *Physics Talk* up to "Gravity, Mass, Weight, and Newton's Second Law." Answer *Checking Up* Questions 1 and 2, and *Physics to Go* Questions 1, 3-5, and 9.	5	**See Plan A.**	
7	Review *Checking Up* questions and *Physics to Go* homework, and discuss *Physics Talk* for Section 3, including significant figures.	Answer *Physics to Go* Questions 6-8, and 10-12.	6	**See Plan A.**	
8	Review the *Physics to Go* homework, do *What Do You Think Now?* and *Reflecting on the Section and the Challenge* for Section 3. **Section 4** Answer *What Do You See?* and *What Do You Think?* Students perform *Investigate* Part A.	Answer *Preparing for the Chapter Challenge* (Question 18).	7	**Section 3** Review the *Physics to Go*, homework. Have students answer *What Do You Think Now?* and *Reflecting on the Section and the Challenge*. **Section 4** Have students answer *What Do You See?* and *What Do You Think?* Perform *Investigate* as a teacher-led activity. Discuss *Physics Talk* and have students answer *What Do You Think Now?* and *Reflecting on the Section and the Challenge*.	Requires only one coin launcher and pennies or nickles
9	Discuss students' answers to *Preparing for the Chapter Challenge* and then ask a few students to read what they have written. Review the results of *Investigate* Part A, perform Part B. Discuss *Physics Talk* and have students answer *What Do You Think Now?* and *Reflecting on the Section and the Challenge*	Read *Physics Talk* and answer the *Checking Up* questions. Answer *Physics to Go* Questions 1, 2, 4, and 6.			
10	Review *Physics to Go* from previous day. **Section 5** Students answer *What Do You See?* and *What Do You Think?* Students perform *Investigate* Step 1.	Answer Question 11, *Preparing for the Chapter Challenge*.	8	Review *Physics to Go* from previous day. **Section 5** Students answer *What Do You See?* and *What Do You Think?* Teacher does *Investigate* as a class demonstration to gather data for Step 1, and then hands out pre-recorded tapes for each group. Students and teacher perform *Investigate* Steps 2-7.	Requires only one acceleration of gravity setup (either tape timer or other equipment), scissors, meterstick, ringstand, extension clamp and weight
11	Students perform *Investigate* Parts 2-11 and discuss the Section 5 *Physics Talk*.	Read *Physics Talk* and answer the *Checking Up* questions, Answer *Physics to Go* Questions 2, 3, and 5-10.			
12	Discuss previous day's *Checking Up* questions and *Physics to Go*. Students answer *What Do You Think Now?* and *Reflecting on the Section and the Challenge*. Students read *Chapter Mini-Challenge*.	Record a sports video segment on TV or find one on the Internet to use for the *Mini-Challenge*.	9	**See Plan A.**	

CHAPTER 2

Pacing Guide *(continued)*

Day	Plan A (small-group *Investigates*)	Homework (for Plan A and Plan B)	Day	Plan B (combination of whole-class and small-group *Investigates*)	Plan B Equipment Reduction
13	Students prepare for the *Mini-Challenge.*	Write out the script for the *Mini-Challenge.*	10	**See Plan A.**	
14	*Mini-Challenge* presentations. **Section 6** Students answer *What Do You See?* and *What Do You Think?*	Record any changes they would make in their challenge presentation after seeing the presentations of other groups.	11	**See Plan A.**	
15	Students perform *Investigate* Parts A and B, discuss *Physics Talk*, answer *What Do You Think Now?* and *Reflecting on the Section and the Challenge.*	Read Section 6 *Physics Talk* and answer *Checking Up*. Answer *Physics to Go* Questions 1-7 and *Preparing for the Chapter Challenge* (Question 8).	12	**See Plan A.**	
16	Review previous night's *Physics to Go.* **Section 7** Students answer *What Do You See?* and *What Do You Think?* Students perform the *Investigate* Steps 1-3.	Read *Physics Talk*. Answer *Checking Up* questions and *Physics to Go* Questions 1-5.	13	**See Plan A.**	
17	Review students' results from *Investigate* Steps 1-3. Students finish *Investigate* Steps 4 and 5. Review *Checking Up* and *Physics to Go* questions. Discuss the *Physics Talk*. Discuss *What Do You Think Now?* and *Reflecting on the Section and the Challenge.*	Answer *Physics to Go*, Questions 6-8 and 10-11.	14	**See Plan A.**	
18	Discuss the *Physics to Go* homework. **Section 8** Answer *What Do You See?* and *What Do You Think?* Students do *Investigate* all parts.	Read *Physics Talk* and answer *Checking Up* questions. Answer *Physics to Go* Questions 1, 2, 4, and 5.	15	**See Plan A.**	
19	Review results of Section 8 *Investigate, Checking Up* questions and *Physics to Go.* Discuss *Physics Talk* with special attention to sample problems. Have students answer *What Do You Think Now?* and *Reflecting on the Section and the Challenge.*	Read *Physics to Go*. Answer *Physics to Go* Questions 6, 8, 9, 13 and 16 (*Preparing for the Chapter Challenge*).	16	**See Plan A.**	

Day	Plan A (small-group *Investigates*)	Homework (for Plan A and Plan B)	Day	Plan B (combination of whole-class and small-group *Investigates*)	Plan B Equipment Reduction
20	Discuss the *Physics to Go* homework. **Section 9** Have students answer *What Do You See?* and *What Do You Think?* Students perform *Investigate*, all parts.	Read *Physics to Go* and answer *Checking Up* questions.	17	Discuss the *Physics to Go* homework. **Section 9** Have students answer *What Do You See?* and *What Do You Think?* For Section 9, perform *Investigate* all parts as a teacher-led class demonstration using one boy and one girl for the vertical jump activity. The class should record and analyze the data for each student. Discuss *Physics Talk*. Have students answer *What Do You Think Now?* and *Reflecting on the Section and the Challenge*.	Requires only one meter stick and one computer setup with motion detector and interface
21	Review *Investigate* and *Checking Up* questions. Discuss *Physics Talk*. Do *What Do You Think Now?* and *Reflecting on the Section and the Challenge*.	Answer selected *Physics to Go* questions, plus Question 18.			
22	Review the *Physics to Go* questions. Student groups choose video segments they wish to use for the *Chapter Challenge*. These may either be teacher supplied or teacher approved.	Write first pass at voice-over narration for *Chapter Challenge*.	18	**See Plan A.**	
23	Students compare versions of the voice-over narration with their groups, and refine.	Finish voice-over narration script.	19	**See Plan A.**	
24	Challenge Presentations	Study for *Physics Practice Test*.	20	**See Plan A.**	
25	*Physics Practice Test*		21	**See Plan A.**	

CHAPTER 2

Implementation Chart

Hopefully, as you become more experienced and comfortable with the curriculum, you will shift to more small-group *Investigates*. Accordingly, at the conclusion of the guide is an *Implementation Chart* that suggests a three-year timetable to expand the student's role in the chapter by having them do more of the *Investigates*. Although this will require a slightly greater expenditure of time and more equipment, the benefits to the student will be manifest. Eventually, your goal should be to have the students complete almost all the investigations rather than you having to provide the maximum opportunity for inquiry.

	Section 1 Investigate	Section 2 Investigate	Section 3 Investigate	Section 4 Investigate	Section 5 Investigate	Section 6 Investigate	Section 7 Investigate	Section 8 Investigate	Section 9 Investigate
Year 1	Whole class	Whole class	Small group	Whole class	Whole class	Small group	Small group	Small group	Whole class
Year 2	Small group	Whole class	Small group	Small group	Whole class	Small group	Small group	Small group	Whole class
Year 3	Small group	Small group	Small group	Small group	Small group	Small group	Small group	Small group	Small group

Chapter 2 Materials and Equipment

The following tables contain lists of materials and equipment needed to do all the experiments. The tables are organized as follows:

- Multimedia Needed
- Durables
- Consumables
- Additional Items Needed Not Supplied

Multimedia contains all the software and content videos that are necessary to complete the sections. All the multimedia items needed will be included in the Multimedia DVD/CD Set. In addition to the Multimedia DVD/CD Set, other multimedia items are available such as the *ExamView*™ Test Generator and the Constructing Physics Understanding simulations.

Durables are items which are not consumed during an experiment. They can be used several times. **Consumables** are items that are used up during each class and must be resupplied for future classes. Both the durables and consumables are broken down by group and by class. A group consists of four students. While the group size will be determined by the teacher based upon logistics and availability of equipment, the information is based upon recommended group size of four students. Items listed per class are based on a class size of 40 students.

Additional Items Needed Not Supplied consists of items that are needed to complete the sections, but are not supplied by *It's About Time*®.

The first column gives the section number(s) in which each item will be used. The second column provides information needed for ordering the particular item. The third column gives the item description. The fourth and fifth columns will indicate quantity, based on either group or class needs. The last column provides a space for you to fill in your total item quantity. You will notice certain rows are shaded and have an asterisk in the last column. This is to indicate a duplicate item listing, based on the needs of different sections. You will not need to figure the items in the shaded rows into your total items needed, as they have already been accounted for in the corresponding non-shaded row.

The *Materials and Equipment* lists are broken down into either Plan A or Plan B corresponding to the *Pacing Guide*.

PLAN A					
Section	**ISBN**	**Multimedia**	**Group (4 Students)**	**Class**	**Total Items Needed**
2	978-1-60720-041-3	Multimedia DVD/CD Set		1 per classroom	

Section	Item No.	Durables	Group (4 Students)	Class	Total Items Needed
1, 8	BS-7209	Ball, steel, 1 in. (diameter)	1 per group		
1	TR-0001	Track, plastic (for roller coaster)	1 per group		
1, 2, 6	RS-2826	Ruler, metric, 30 cm	1 per group		*
8	RS-2826	Ruler, metric, 30 cm	2 per group		
2, 5	TT-6100	Timer, ticker tape, AC	1 per group		
2, 5	SS-1281	Scissors	1 per group		
2, 5, 6, 8	SM-1676	Meter stick, wood	1 per group		*
9	SM-1676	Meter stick, wood	2 per group		
3	RS-2723	Ruler, plastic, flexible, 30 cm	1 per group		
3	GC-0001	Cart, dynamics	1 per group		
3	WH-9012	Weights, slotted, set	1 per group		
1, 3, 8	CS-3246	C-clamp, steel, 3 in.	1 per group		
4	PL-0001	Launcher, projectile	1 per group		
5	SH-7212	Ring stand, large	1 per group		
5	EX-0001	Clamp, extension	1 per group		
5	HO-0002	Holder, right angle, cast iron	1 per group		
5, 7, 9	CM-1108	Calculator, basic	1 per group		
5, 6	WS-6910	Washer, 3/4 in. (outside diameter) x 5/16 in. (inside diameter)	2 per group		
5, 6, 7	WS-0376	Weight, slotted, 100 g	10 per group		
6, 7	SS-2304	Scale, spring, 0-20 N	2 per group		
7	SS-7209	Surface, felt, rough, 6 in. x 36 in.	1 per group		
7	BH-9871	Box, friction, 5 3/4 in. (length) x 3 3/4 in. (width) x 3 in. (depth) (with hook)	1 per group		
8	SR-0020	Ramp, starting, 20-degree angle	1 per group		
4, 5	BS-0790	Ball, tennis		1 per class	
5	TH-2143	Model, trajectory		1 per class	
5	FW-0002	Weight, fishing, small		7 per class	
5	PH-6816	Timer, photogate (with electromagnet)		1 per class	
6	SH-7708	Board, roller, Reaction Force (skateboard)		2 per class	

* *Please note duplicate item.*

CHAPTER 2

Section	Item No.	Consumables	Group (4 Students)	Class	Total Items Needed
1, 8	MS-1425	Pen, marking, felt tip	1 per group		
2, 5	TH-2653	Ticker tape, roll	1 per group		
2	GL-1146	Glue stick	1 per group		
1, 3, 8, 9	TS-2662	Tape, masking, 3/4 in. x 60 yds		6 per class	
5	NS-1210	Notes, sticky, pad, 3 in. x 3 in.		3 per class	
5	SS-7722	String, cotton, ball		1 per class	
8	CS-5011	Cards, index, unlined, 3 in. x 5 in., pkg. of 100		1 per class	
Section	**Item No.**	**Additional Items Needed Not Supplied**	**Group (4 Students)**	**Class**	**Total Items Needed**
3, 7		Access to a flat surface (such as a table, floor or other open space)	1 per group		
3		Can or bottle (with less mass then the dynamics cart)	1 per group		
3, 4, 8		Coins (pennies, nickels)	5 per group		
4, 5		Chalkboard		1 per classroom	
4, 5		Chalk		1 per classroom	
4, 6		Chair (with wheels)		1 per classroom	
4, 6		Safety helmet		1 per classroom	
4, 6		Pads, knee		1 per classroom	
4, 6		Pads, elbow		1 per classroom	
4, 6		Stack of books		1 per classroom	
5		Probeware, photogate		1 per classroom	
5		Fence, picket		1 per classroom	
5, 9		MBL or CBL Technology (to record probeware activity)		1 per classroom	
5		Ladder		1 per classroom	
7		Shoe, athletic	1 per group		
9		Probeware, motion detector		1 per classroom	
9		Scale, bathroom		1 per classroom	
9		TV with VCR		1 per classroom	

* *Please note duplicate item.*

PLAN B

Section	ISBN	Multimedia	Group (4 Students)	Class	Total Items Needed
2	978-1-60720-041-3	Multimedia DVD/CD Set		1 per classroom	

Section	Item No.	Durables	Group (4 Students)	Class	Total Items Needed
1	BS-7209	Ball, steel, 1 in. (diameter)		1 per class	*
8	BS-7209	Ball, steel, 1 in. (diameter)	1 per group		
1	TR-0001	Track, plastic (for roller coaster)		1 per class	
1, 2	RS-2826	Ruler, metric, 30 cm		1 per class	*
6	RS-2826	Ruler, metric, 30 cm	1 per group		*
8	RS-2826	Ruler, metric, 30 cm	2 per group		
2, 5	TT-6100	Timer, ticker tape, AC		1 per class	
2, 5	SS-1281	Scissors		1 per class	
2, 5	SM-1676	Meter stick, wood		1 per class	*
6, 8	SM-1676	Meter stick, wood	1 per group		
9	SM-1676	Meter stick, wood		2 per class	*
3	RS-2723	Ruler, plastic, flexible, 30 cm	1 per group		
3	GC-0001	Dynamics cart	1 per group		
3	WH-9012	Weights, slotted, set	1 per group		
1	CS-3246	C-clamp, steel, 3 in.		1 per class	*
3, 8	CS-3246	C-clamp, steel, 3 in.	1 per group		
4	PL-0001	Launcher, projectile		1 per class	
5	SH-7212	Ring stand, large		1 per class	
5	EX-0001	Clamp, extension		1 per class	
5	HO-0002	Holder, right angle, cast iron		1 per class	
5, 7, 9	CM-1108	Calculator, basic	1 per group		
5, 6	WS-6910	Washer, 3/4 in. (outside diameter) x 5/16 in. (inside diameter)	2 per group		
5	WS-0376	Weight, slotted, 100 g		10 per class	*
6, 7	WS-0376	Weight, slotted, 100 g	10 per group		
6, 7	SS-2304	Scale, spring, 0-20 N	2 per group		
7	SS-7209	Surface, felt, rough, 6 in. x 36 in.	1 per group		
7	BH-9871	Box, friction, 5 3/4 in. (length) x 3 3/4 in. (width) x 3 in. (depth) (with hook)	1 per group		
8	SR-0020	Ramp, starting, 20-degree angle	1 per group		
4, 5	BS-0790	Ball, tennis		1 per class	
5	TH-2143	Model, trajectory		1 per class	
5	FW-0002	Weight, fishing, small		7 per class	

* *Please note duplicate item.*

Section	Item No.	Durables	Group (4 Students)	Class	Total Items Needed
5	PH-6816	Timer, photogate (with electromagnet)		1 per class	
6	SH-7708	Board, roller, Reaction Force (skateboard)		2 per class	
Section	**Item No.**	**Consumables**	**Group (4 Students)**	**Class**	**Total Items Needed**
1	MS-1425	Pen, marking, felt tip		1 per class	
8	MS-1425	Pen, marking, felt tip	1 per group		
2, 5	TH-2653	Ticker tape, roll		1 per class	
2	GL-1146	Glue stick		1 per class	
1, 3, 8, 9	TS-2662	Tape, masking, 3/4 in. x 60 yds		6 per class	
5	NS-1210	Notes, sticky, pad, 3 in. x 3 in.		3 per class	
5	SS-7722	String, cotton, ball		1 per class	
8	CS-5011	Cards, index, unlined, 3 in. x 5 in., pkg. of 100		1 per class	
Section	**Item No.**	**Additional Items Needed Not Supplied**	**Group (4 Students)**	**Class**	**Total Items Needed**
3, 7		Access to a flat surface (such as a table, floor or other open space)	1 per group		
3		Can or bottle (with less mass then the dynamics cart)	1 per group		
3, 4, 8		Coins (pennies, nickels)	5 per group		
4, 5		Chalkboard		1 per classroom	
4, 5		Chalk		1 per classroom	
4, 6		Chair with wheels		1 per classroom	
4, 6		Safety helmet		1 per classroom	
4, 6		Pads, knee		1 per classroom	
4, 6		Pads, elbow		1 per classroom	
4, 6		Stack of books		1 per classroom	
5		Probeware, photogate		1 per classroom	
5		Fence, picket		1 per classroom	
5, 9		MBL or CBL Technology (to record probeware activity)		1 per classroom	
5		Ladder		1 per classroom	
7		Shoe, athletic	1 per group		
9		Probeware, motion detector		1 per classroom	
9		Scale, bathroom		1 per classroom	
9		TV with VCR		1 per classroom	

* *Please note duplicate item.*

NOTES

CHAPTER 2

Teacher Resources

Blackline Masters

Available in *Teacher Resources* and on *Color Overheads* and *BLMs* CD.

Chapter Supports

Title	Point of Use	Blackline Masters
Standard for Excellence	*Chapter Challenge* Introduction	2a

Section Quizzes

Title	Point of Use	Blackline Masters
Section 1 Quiz	Section 1	2-1b
Section 2 Quiz	Section 2	2-2c
Section 3 Quiz	Section 3	2-3b
Section 4 Quiz	Section 4	2-4c
Section 5 Quiz	Section 5	2-5b
Section 6 Quiz	Section 6	2-6b
Section 7 Quiz	Section 7	2-7b
Section 8 Quiz	Section 8	2-8a
Section 9 Quiz	Section 9	2-9b

Section Supports

Title	Point of Use	Blackline Masters
Illustrations of Balls Rolling Down Incline	Section 1 – *Investigate*	2-1a
Ticker-Tape Graphs – Examples of Speed	Section 2 – *Physics Talk*	2-2a
Ticker-Tape Graphs – Examples of Acceleration	Section 2 – *Physics Talk*	2-2b
Free-Body Diagrams	Section 3 – *Physics Talk*	2-3a
Coin Drop Diagram	Section 4 – *Physics Talk*	2-4a
Ball Thrown in Air Charts	Section 4 – *Physics Talk*	2-4b
Angling Projectiles Diagram	Section 5 – *Physics Talk*	2-5a
Pulling a Chair Diagram	Section 6 – *Physics Talk*	2-6a
Shoe at Constant Speed Diagram	Section 7 – *Physics Talk*	2-7a
Conservation of Energy Table	Section 9 – *Physics Talk*	2-9a

Chapter Assessment

Title	Point of Use	Blackline Masters
Physics Practice Test	End-of-Chapter	2b
Sample Assessment Rubric	End-of-Chapter	2c

Color Overheads

Available on *Color Overheads* and *BLMs* CD.

Title	Point of Use
Physics Corner	*Chapter Challenge* Introduction
What Do You See?	Section 1
What Do You See?	Section 2
What Do You See?	Section 3
What Do You See?	Section 4
Illustration of Throwing Ball in Moving Chair	Section 4 – *Investigate Part B*
What Do You See?	Section 5
What Do You See?	Section 6
What Do You See?	Section 7
What Do You See?	Section 8
What Do You See?	Section 9

NOTES

Chapter Challenge

Physics in Action

Scenario

Have you ever imagined being a TV sports analyst and having millions of people listen to you describe a football or baseball game? Perhaps you would like to provide the commentary for a sport in the Summer Olympics or an analysis of a figure-skating performance on television?

What qualifications are needed to have a career in sportscasting? Should you major in communication in college or be a retired professional athlete to do this job? Could a physics course be a key to becoming a sports analyst? Perhaps a student with physics knowledge can bring to the TV viewer a different perspective that might provide a new outlook on sporting events.

Your Challenge

A public broadcasting service has decided that it wants to televise a variety of sporting events and wants these programs to be educational as well as entertaining. To test out this idea, you are to provide the voice-over narration for a sports video. The narration will need to explain the physics of the action appearing on the screen. You will do a "science commentary" on a short (two to three minutes) sports video or a series of sports videos that add up to two to three minutes.

The public broadcasting service wants people to understand that the laws of physics deal not only with the things that happen in the laboratory, but also with everyday events in the real world. Your task is not to give a play-by-play description of the sporting event or give the rules of the game, but rather to go a step beyond.

Scenario

Read or have students read the *Scenario* aloud. Have students review the table of *Contents*. You may wish to show them a sports video with a commentary. You might even ask them to describe what they see in a sports video that doesn't have a commentary. In doing so, the students will begin to realize what they need to know about the sport for which they choose to provide a voice-over. Suggest to the students that an analysis of a sport would also help them in their presentation and appreciation of the sport.

To elicit an effective science commentary, remind students that each section corresponds to the physics concepts they will need to understand. Emphasize that the skills and knowledge required to perform the culminating task will accumulate gradually as they go through each section. Suggest strategies, such as using a concept map or keeping an updated record of information in *Active Physics* logs, to incorporate new vocabulary that enriches each student's performance.

Your Challenge

Lead a class discussion to tell your students that there are a number of objectives in this chapter, one of which is that the laws of physics hold true not only in their science class and lab, but also in the outside world. The students should be able to look at a sporting event and realize what physical principle is involved. Hopefully this will carry over to everyday life, and the students will then be able to see the physics in the world around them. Each class might start with a short video segment showing sports bloopers. They are commercially available and many of the students may have their own. After the class has covered some of the material, it is increasingly appropriate to discuss the physics displayed in the blooper. Many of these bloopers are very humorous, and will make students look forward to the beginning of the class.

The main concern of this chapter is to ensure that students become well acquainted with the principle of inertia, Newton's laws, and frames of reference. The *Physics to Go* at the end of each section

You are to educate the audience by describing to them the rules of nature that govern the event. This approach will give the viewer (and you) a different perspective on both sports and physics.

You can think of this narration as a tryout for a broadcasting job. In this tryout, a traditional sports broadcaster will give the play-by-play and then turn the microphone over to you to give the physics of sports overview. You can provide the narration live, dub it onto the video soundtrack, or record an audio version. You will also need to submit a written script of the narration.

Criteria for Success

What criteria should be used to evaluate a voice-over narration or script of a sporting event? Since the intention is to provide an interesting analysis of the physics of sports, the voice-over should include the use of physics terms and physics principles. However, remember that physics principles are not enough. Your voice-over narration will also need to be entertaining.

Work with your class to develop a set of criteria for a successful voice-over narration. When you have decided what is required, compare your list with the list on the following page.

129

often contains more questions than those that are required to be assigned for homework. This section has been written in a way that gives you a choice of how much work, and the nature of work, the students will be expected to do each day out of class. As you work with *Active Physics*, be aware that the same physics concepts appear repeatedly in different contexts. It is not necessary for the students to achieve total understanding the first time they encounter a new concept.

Criteria for Success

Develop a set of criteria for assessment in the *Chapter Challenge* together with your class. You should get students' feedback on the details that should be included in their sports commentary and what criteria would form a relevant assessment. The criteria should be clear and easy to follow and carry frequent references to the physics principles that the students will study. While developing the criteria, be accurate and concise. The students should be able to comprehend the criteria with ease.

If you ask your class to volunteer their suggestions for evaluating the *Chapter Challenge*, they will relate to the assessment with a sense of ownership. Make a list of the criteria on the board and solicit student opinion frequently, so that the evaluation can be written in a style that students understand. Some of the main questions you should ask are—How many physics terms should be included? How much description should the science commentary carry and for how long? Have the physics terms been integrated appropriately? After a discussion of the criteria, you and your class should assign points to each criterion.

Have a 10-minute discussion with your class on the criteria that you and your class select. Once you've decided on the final criteria for assessment, use it as a scoring guide to ensure that student expectations are met consistently. You might wish to modify the final assessment criteria to fit state standards and benchmarks. Keep in mind that all expectations should be communicated to the students before you begin work on this chapter.

Chapter Challenge

Standard for Excellence

Students should be able to decide the criteria for evaluating their *Chapter Challenge*. You can facilitate the process of determining each criterion by providing an outline of standards they will be covering. Students do not have to begin with a final assessment rubric. Emphasize that the class will have a chance to reflect on the rubric as it proceeds further in the chapter. At this stage, students should focus on the broader aspects of assessment relevant to the final project. Discuss the list of sample criteria in the student text of *Active Physics*.

A sample rubric is provided at the end of this chapter. The criteria required to meet the standards of excellence should be clearly defined. Students must understand how each column of the grading rubric, which assigns a grade, differs in its expectations. As the chapter progresses, so should the rubric. At each stage of the rubric's development, students should be given a voice, empowering them to assess their own work. This will instill in them a sense of ownership and self-confidence. To make sure that all important aspects have been covered, introduce a new criterion, then with discussion, establish its significance by asking specific questions. The purpose of the criteria should be highlighted and properly understood so that students know how they can excel in the final project. The *Standard for Excellence* table is provided as a Blackline Master in your *Teacher Resources CD*.

2a **Blackline Master**

Engineering Design Cycle

Consider introducing the *Engineering Design Cycle* with an overhead projector. Draw attention to the different phases of the cycle by examining each phase. As students read this section, ask them to highlight the purpose of each step. Explain how the *Chapter Challenge* is built upon a series of steps

Chapter Challenge

Standard for Excellence	
1. The use of physics terms and principles in the narration • number of physics principles used • physics concepts from the chapter integrated in the appropriate places • physics terminology and equations used where appropriate • correct estimates of the magnitude of physical quantities used • additional research, beyond the basic concepts presented in the chapter	50 points
2. The quality of the oral narration • knowledge of the sport • entertainment value with respect to humor, excitement, and/or drama • ease of following and understanding • appropriate amount of narration • duration of narration between two and three minutes	25 points
3. The quality of the written script of the narration • use of correct science vocabulary • consistent sentence structure • correct spelling, punctuation, and grammar • appropriate use of science symbols for units of measurement	20 points
4. Challenge completed on time	5 points

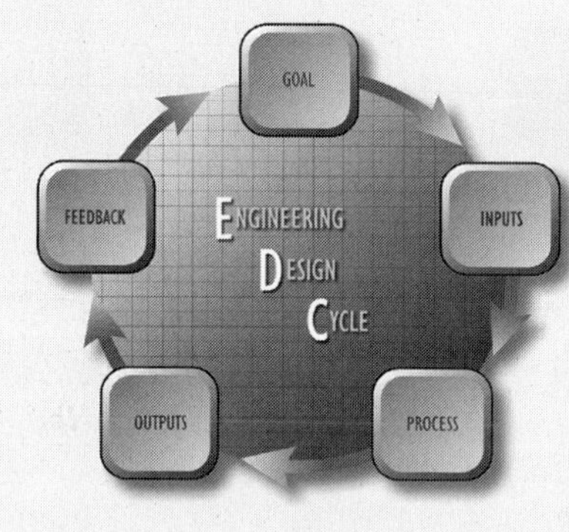

Engineering Design Cycle

The *Chapter Challenge* is to create an educational and entertaining sports voice-over. Now that you have read all of the criteria, you will use a simplified *Engineering Design Cycle* to help your group complete this design challenge. Clearly defining the *Goal* is the first step in the *Engineering Design Cycle.*

Although many people may be in the broadcast booth, a voice-over narration becomes the product of one person—the commentator or the scriptwriter. Although you will be working in cooperative groups during the chapter, each person will be responsible for a part of the voice-over or script for a sporting event. As a team you may share different aspects of the job, but the output of work per person should be the same.

130

As you experience each one of the chapter sections, you will be gaining *Inputs* to use in the design cycle. These *Inputs* will include new physics concepts, vocabulary, and even equations that will help you to educate your sports audience. When your group prepares the *Mini-Challenge* presentation and the *Chapter Challenge*, you will be completing the *Process* step of the *Engineering Design Cycle.* During the *Process* step you will evaluate ideas, consider criteria, compare and contrast potential sports footage, and most importantly, make decisions about what physics principles you will include in your script.

The *Output* of your design cycle will be the sports commentary that your group presents to the class. Finally, you will receive *Feedback* from your classmates and your instructor about what parts of your presentation are good and which parts need to be refined.

Physics Corner

Physics in *Physics in Action*

- Acceleration
- Center of mass
- Coefficient of sliding friction
- Constant speed
- Frames of reference
- Frictional force
- Galileo's law of inertia
- Gravitational potential energy
- Gravity
- Law of conservation of energy
- Law of conservation of momentum
- Momentum = mass × velocity
- Newton's first law of motion
- Newton's second law of motion
- Newton's third law of motion
- Normal force
- Potential and kinetic energy
- Principle of inertia
- Projectile motion
- Relationship of mass and force to acceleration
- Unit of force–newton
- Velocity
- Work

131

that continue to complement each other in a cyclical pattern. You might want to emphasize that each phase of the cycle is experienced as students progress through the chapter. New concepts are developed and understood gradually. The *Inputs* of the *Engineering Design Cycle* incorporate new concepts and vocabulary. The *Process* evaluates, compares, and contrasts the physics principles through the *Chapter Mini-Challenge* and the *Chapter Challenge*. While the *Outputs* take shape in the form of a final product, the *Feedback* continues to be a useful tool in providing a constructive peer analysis of the *Goal.*

Physics Corner

This section previews the concepts the students will be studying. You may wish to know how many of these terms are familiar to your students. When the chapter ends, you might want to revisit the concepts from time to time to chart each student's progress in understanding new terms and principles. What is most important is that students should retain the new concepts they have learned.

Encourage each student to keep a record in their logs of the new terms they learn. Have them chart connections of physics concepts to events that happen in their daily lives, especially in the sports they play. You may want to begin your class with a meaningful concept linked to the daily routine of an individual.

SECTION 1

Newton's First Law: A Running Start

Section Overview

Students investigate the presence of inertia and friction in relation to Newton's first law of motion by letting a ball roll up a tracked slope. They change the height of the slope to see how the change affects the rolling motion of the ball, the height the ball recovers before it comes to a stop, why it stops, and the force that causes the ball to stop. Students explore these questions along with the concept of Frames of Reference.

Background Information

Two major ideas introduced in this section are Galileo's law of inertia (Newton's first law of motion) and frames of reference (relative velocities). Before attempting to identify causes and effects for generating, sustaining, and arresting motion, a pivotal question first must be answered: What kinds of motion require explanation?

Two distinct kinds of motion along a straight line are often encountered in nature. These are motion with constant speed and motion with uniform, or constant, acceleration. Since the contributions of Galileo, physics has operated from the perspective that the first of these kinds of motion, constant speed, has no cause. Galileo devised a number of arguments and demonstrations, some of which are replicated in this section, to support this notion. The cause of all accelerated motion is force; some agent(s) must be pushing or pulling—exerting a force—on any object observed to be accelerating. Sources or kinds of forces abound.

Every situation that involves acceleration has an associated net force. If an orange is dropped, it accelerates because of the downward force due to gravity. When the orange hits the floor, it stops due to another force. The force that stops the orange is provided by the floor, upward. A magnet brought near another magnet will cause the magnets to accelerate; therefore, there must be a magnetic force. Sometimes, forces hiding in constant-speed linear motion can also be discovered. Drop a coffee filter. The filter accelerates downward for a bit, but the amount of acceleration drops to zero, so that the coffee filter falls most of the way at constant speed. Did the force of gravity decrease or disappear? No, a coffee filter seems to weigh (a measure of the force of gravity) the same at every point in the descent path. Therefore, there must be another force, the force of air resistance, acting in the opposite direction to gravity. The force of air resistance eventually balances out the gravitational force.

It is possible for a combination of forces to have a net effect of zero. So, it is the net force on an object that imparts the acceleration. Newton's first law of motion states that an object at rest tends to remain at rest, and an object in motion (in a straight line) tends to remain in motion, unless acted upon by an outside (net, nonzero) force. This statement is more complete than the one provided to the students in *Section 1* of *Active Physics*. Whenever speed, direction, or both speed and direction are observed to change, a net force is the cause. The first law of Newton does not attempt to quantify the relationship between accelerations and the forces that cause them. Establishing the quantitative relationship requires experimental evidence, which is the purpose of the next section.

Crucial Physics

- An object at rest remains at rest.
- An object in motion remains in motion at a constant speed unless acted upon by a force.
- Two people, one on a moving train and one on a platform, can measure different speeds for the same object. Both are correct for their frame of reference.
- Rest is a special case of motion in a straight line with a constant speed (namely, speed equals zero).
- Motion in a straight line with a constant speed is included in Newton's first law of motion and therefore requires no additional explanation.
- The magnitude of an object's velocity depends on the frame of reference in which it is measured.
- Velocities as measured in different frames of reference differ by an amount equal to the relative velocity between the two reference frames.

Learning Outcomes	Location in the Section	Evidence of Understanding
Describe Galileo's law of inertia.	***Investigate*** Steps 1-4 ***Physics Talk***	Students release a ball from a certain height and observe its motion on a tracked slope. This motion is described by Galileo's law of inertia.
Apply Newton's first law of motion.	***Investigate*** Step 5	Students apply their knowledge of Newton's first law to determine what would keep the ball rolling on a horizontal track.
Recognize inertial mass as a physical property of matter.	***Physics Talk***	Students read about Galileo's law of inertia and learn why the mass of an object is the measure of its inertia.
Use examples to demonstrate that speed is always relative to some other object.	***Physics Talk***	Students use different examples to show how the speed of one object is relative to another object's position.
Explain that the speed of an object depends on the reference frame from which it is being observed.	***Physics Talk*** ***Frames of Reference***	Students explain how the speed of an object is relative to the frame of reference from which the motion of the object is being observed.

Section 1 Materials, Preparation, and Safety

Materials and Equipment

PLAN A

Materials and Equipment	Group (4 students)	Class
Ball, steel, 1 in. (diameter)	1 per group	
Track, plastic (for roller coaster)	1 per group	
Ruler, metric, 30 cm	1 per group	
C-clamp, steel, 3 in.	1 per group	
Pen, marking, felt tip	1 per group	
Tape, masking, 3/4 in. x 60 yds		6 per class

*Additional items needed not supplied

PLAN B

Materials and Equipment	Group (4 students)	Class
Ball, steel, 1 in. (diameter)		1 per class
Track, plastic (for roller coaster)		1 per class
Ruler, metric, 30 cm		1 per class
C-clamp, steel, 3 in.		1 per class
Pen, marking, felt tip		1 per class
Tape, masking, 3/4 in. x 60 yds		6 per class

*Additional items needed not supplied

Note: Time, Preparation, and Safety requirements are based on Plan A, if using Plan B, please adjust accordingly.

Time Requirement

This *Investigate* requires 30 minutes.

Teacher Preparation

- If you are using the tracks provided in the kits, you may wish to screw the center of the track onto a board to make it easier for the students to manipulate the ends. Having stacks of books or other material available to prop up the ends of the track would be helpful.

Safety Requirement

- Make sure the students immediately pick up any balls that may have rolled onto the floor, to prevent anyone from slipping on them. One-inch steel balls should be handled carefully and not tossed around the classroom.

NOTES

NOTES

CHAPTER 2

Meeting the Needs of All Students

Differentiated Instruction: Augmentation and Accommodations

Learning Issue	Reference	Augmentation and Accommodations
Making accurate measurements	***Investigate*** Steps 1, 3, and 4	**Augmentation** Students with fine-motor, visual-motor, and/or attention issues often struggle to make accurate measurements. Teach students how to make accurate measurements by modeling proper use of measuring devices. • Draw a larger version of a portion of a meter stick on the board and teach students how to read the scale. • Model how to mark the recovered height for measurement. • Assign group responsibilities for completing this task (recorder, marker, measurer, and roller). **Accommodation** • Use close proximity or hand-over-hand (physically guided) techniques to help students practice measuring. Decrease assistance as accuracy and independence is gained.
Recording organized data	***Investigate***	**Augmentation** • Students with sequential-learning and executive-function issues may struggle to organize collected data in a way that allows comparison. • Ask students to silently read the directions in the *Investigate* and write down the kind of information that needs to be recorded. Then model how to create a table with all of the required information. (See model on the next page.) **Accommodation** • Provide students with a blank copy of the table to complete.
Vocabulary review	***Physics Talk***	**Augmentation** • Students sometimes struggle to commit new vocabulary to their long-term memory. To improve their retention, review the terms vector and scalar in relation to velocity and speed. Continue to add vector and scalar quantities to the class poster created during *Chapter 1.*
Reading comprehension	***Physics Talk*** Frames of Reference	**Augmentation** • The concept of a frame of reference may be difficult for many students to understand. Students with reading issues may not comprehend the examples provided in this section. • Divide the class into groups of four. Assign a frame-of-reference scenario to each of the four groups. Ask them to make a poster and give a 3- to 5-minute presentation to explain their scenario. Students can also create their own scenarios to teach the class.
Using academic vocabulary	***Physics Words*** ***Checking Up*** ***Physics Essential Questions***	**Augmentation** • Students are reluctant to use science vocabulary and often resort to common language that they are comfortable using. Require students to use their *Physics Words* when answering questions orally in class, or when writing answers to *Checking Up* or *Physics Essential Questions.*
Synthesizing information to use in a new way	***Physics to Go*** Question 10	**Augmentation** • Students may not have prior experience of a sports commentary. Without this prior knowledge, they may decide not to complete this task rather than risk an incorrect answer. Provide audio examples of entertaining sportscasts and then generate a class list of traits that would make the commentary, entertaining as well as educational.

Model of a table for the *Investigate*

Scenario	Starting Height (cm)	Predicted Recovered Height (cm)	Measured Recovered Height (cm)
Same slope			
Smaller slope of recovered height			
Smallest slope of recovered height			

Strategies for Students with Limited English-Language Proficiency

Learning Issue	Reference	Augmentation
Accessing prior knowledge	***Investigate*** Step 1	Accessing prior knowledge, which is common in the science classroom, can be particularly helpful with ELL students. It shows that students have some experience with a concept and gives you a way to connect the concept back to what students already know, independent of their grasp of the vocabulary. After students have set up the apparatus for Step 1, before they perform Step 1.a), ask if any students have had experience with sloped tracks (skateboarding, for example). Have these students predict whether the recovered height of the ball will be as high as the starting height, and have them explain their reasoning to the class. Students may have experience with recovered height in other real-world contexts as well. For example, a student who plays basketball could talk about how the ball bounces when you stop dribbling. (The ball's height decreases with each bounce.)
Active learning	***Investigate*** Step 2.a)	When students write and explain their predictions in their *Active Physics* logs, they may benefit from drawing and labeling diagrams to accompany their predictions. This extra step gives students additional experience with the terminology "release point," "starting height," and "recovered height" in context, and helps to demonstrate their understanding of those words. It also allows them to draw arrows from their written predictions to specific points on the drawings to help express their thoughts.
Vocabulary comprehension	***Physics Talk***	The term "frictional force" appears in the *Physics Talk*. Point out that "friction" has been turned into the adjective "frictional" and is modifying "force." Help students understand that "frictional force" is just another way of saying "force of friction."
Vocabulary comprehension Using tools and manipulatives	***Active Physics Plus Physics to Go*** Question 7.c)	When students learn about velocity, they need to understand cardinal directions. Use a map or a globe to point out north, south, east, and west. Show students the word that goes with each direction.
Comprehension	***Active Physics Plus***	Collaborate with the students' math teachers to determine what level of comprehension students have obtained for working with angles.
Vocabulary comprehension Using tools and manipulatives	***Physics to Go*** Question 7.c)	Use two pencils to show the meaning of "perpendicular" and contrast this to the term "parallel." It may help to review other frame-of-reference terms such as "horizontal" and "vertical."

CHAPTER 2

SECTION 1

Teaching Suggestions and Sample Answers

What Do You See?

The *What Do You See?* section gives you an opportunity to catch students' interest. There are vignettes in the illustration that you may want to highlight in relation to Newton's first law. You could query your students on the look of bewilderment/surprise on the players' faces as they see the soccer ball flying over the net. A color overhead would provide a more powerful visual and make it easier for you to steer a probing discussion.

Chapter 2 Physics in Action

Section 1 Newton's First Law: A Running Start

What Do You See?

Learning Outcomes

In this section, you will

- **Describe** Galileo's law of inertia.
- **Apply** Newton's first law of motion.
- **Recognize** inertial mass as a physical property of matter.
- **Use** examples to demonstrate that speed is always relative to some other object.
- **Explain** that the speed of an object depends on the reference frame from which it is being observed.

What Do You Think?

Every sport includes moving objects or people or both. That is what makes sports entertaining.

- **How do figure skaters keep moving across the ice at high speeds for long times while seeming to expend no effort?**
- **Why does a soccer ball continue to roll across the field after it has been kicked?**

Record your ideas about these questions in your *Active Physics* log. Be prepared to discuss your responses with your group and the class.

Investigate

In this *Investigate*, you will use a track and a ball to explore the question, "When a ball is released to roll down a track and up the opposite side of the track, how does the vertical height that the ball reaches on the opposite side of the track relate to the vertical height from which the ball is released?"

1. Make a track that has the same slope on both sides, as shown in the diagram on the next page. Your teacher will suggest how high the ends of the track sections should be.

132

Students' Prior Conceptions

This section establishes the foundation for students to seek consistent explanations for describing an object at rest. Students are led to appreciate both "active" and "passive" forces acting on objects. They read that balanced or unbalanced forces act on an object, and inertial mass and motion are intrinsic to each other. Some preconceptions that may occur are listed below:

1. **The only "natural" motion is for an object to be at rest.** Recognizing that Earth and all objects on Earth are in motion and only seem at rest relative to each other is fundamental to understanding that what is considered to be "natural motion," an object at rest, is at relative motion. The teacher needs to help students to recognize and to apply appropriate frames of reference for motion.
2. **Students tend to consider "true" motion as the motion of an object relative to Earth.** This preconception is directly associated with the previous idea about "natural" motion.
3. **Constant speed needs a cause to sustain it.** Modeling Galileo's historical experiment will enable students to measure distance and time to see that an object in motion can continue to maintain that constant motion without an active force to propel it.

What Do You Think?

Students are not expected to know a "right" answer. These questions are supposed to elicit students' beliefs regarding a very specific prediction or outcome, and students should write a specific answer in their logs. At the same time, they should not shy away from any answer they think is valid. *What Do You Think?* allows the freedom of thinking broadly. Students should not be held back by the notion of being wrong.

What Do You Think?

A Physicist's Response

Skaters maintain speed on ice due to very low friction between the blades and the ice. The skaters' inertia explains the continued motion but in reality no explanation is needed. Uniform motion in a straight line is the starting point for the theory of dynamics. This theory concerns itself with explaining deviations from uniform motion in a straight line. The soccer ball also continues to roll because of the natural tendency of all objects to remain at rest or in motion until a net external force acts on the object.

Investigate

Teaching Tip

As you move about the room during the investigations, you can ask students questions to ascertain their learning and to check on their progress and understanding.

1.a)

Students record the vertical height from which the ball is released.

CHAPTER 2

4. **Friction always hinders motion; you always want to eliminate friction.** The concept of friction is ingrained in students' minds as something that always retards motion. Only with more experience and study will students come to understand that friction is also a vital component to being able to move, for example to walk across very smooth surfaces or even to sit on a chair. *Section 1* is more important in enabling students to recognize relatively steady motion under low friction conditions than in measuring how friction retards motion or causes an object to come to rest.
5. **Objects resist acceleration from the state of rest because of friction.** Students confound the concept of friction with that of inertial mass. The teacher needs to lead students to understand that inertial mass is a property of matter and that objects resist acceleration due to their amount of matter rather than to the friction between the surfaces in contact.
6. **Alternative ideas on measurement.** Through observations and careful data collection, students find that measurement is not only linear. They discover any quantity can be measured as accurately as the tools available and that they can only measure to the smallest unit shown on the measuring device.

NOTES

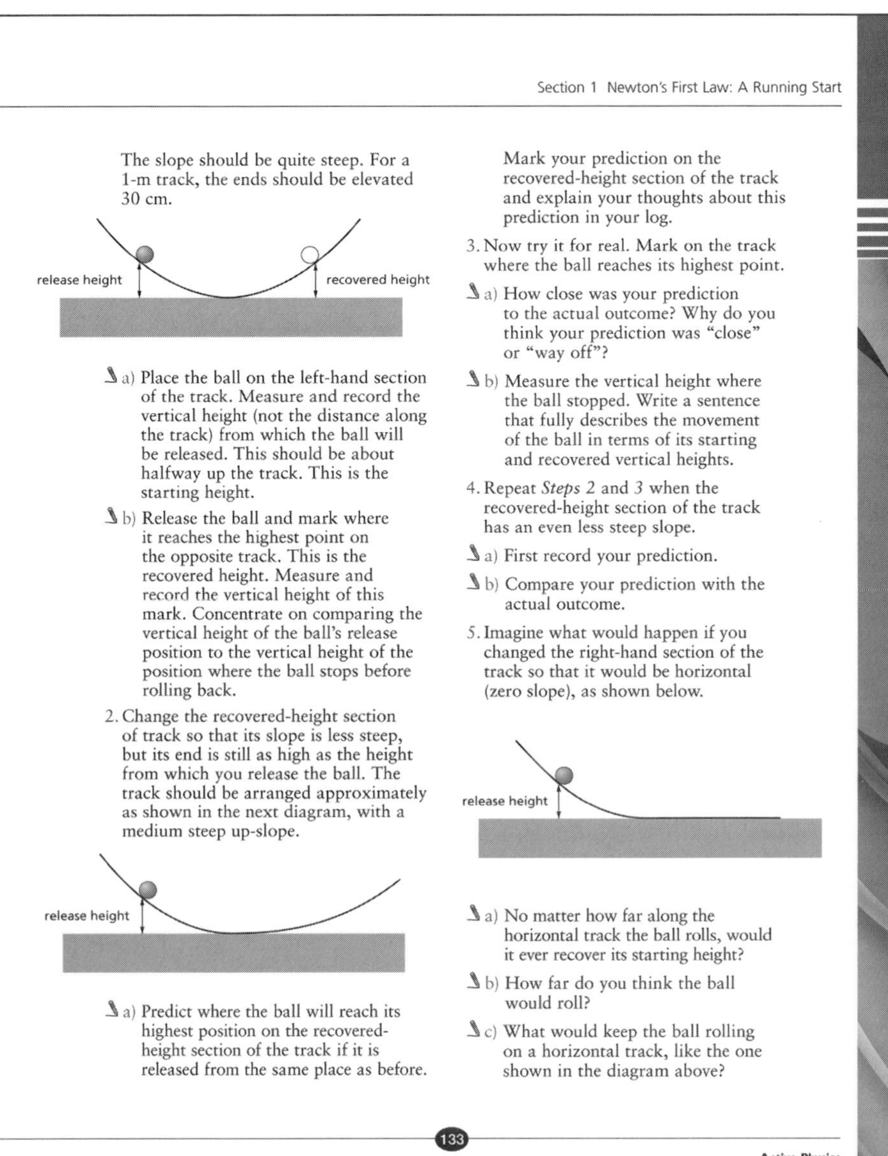

The slope should be quite steep. For a 1-m track, the ends should be elevated 30 cm.

a) Place the ball on the left-hand section of the track. Measure and record the vertical height (not the distance along the track) from which the ball will be released. This should be about halfway up the track. This is the starting height.

b) Release the ball and mark where it reaches the highest point on the opposite track. This is the recovered height. Measure and record the vertical height of this mark. Concentrate on comparing the vertical height of the ball's release position to the vertical height of the position where the ball stops before rolling back.

2. Change the recovered-height section of track so that its slope is less steep, but its end is still as high as the height from which you release the ball. The track should be arranged approximately as shown in the next diagram, with a medium steep up-slope.

a) Predict where the ball will reach its highest position on the recovered-height section of the track if it is released from the same place as before. Mark your prediction on the recovered-height section of the track and explain your thoughts about this prediction in your log.

3. Now try it for real. Mark on the track where the ball reaches its highest point.

a) How close was your prediction to the actual outcome? Why do you think your prediction was "close" or "way off"?

b) Measure the vertical height where the ball stopped. Write a sentence that fully describes the movement of the ball in terms of its starting and recovered vertical heights.

4. Repeat *Steps 2* and *3* when the recovered-height section of the track has an even less steep slope.

a) First record your prediction.

b) Compare your prediction with the actual outcome.

5. Imagine what would happen if you changed the right-hand section of the track so that it would be horizontal (zero slope), as shown below.

a) No matter how far along the horizontal track the ball rolls, would it ever recover its starting height?

b) How far do you think the ball would roll?

c) What would keep the ball rolling on a horizontal track, like the one shown in the diagram above?

1.b)

Students record the vertical height of the ball where it reaches the highest point on the opposite track.

2.a)

Students may predict that they expect the ball to travel the same distance along the track as the ball going down the incline, or to reach the same height as the ball started from.

3.a)

Students record data. Their answers may vary.

3.b)

When rolling the ball within the track, students should find the recovered height to be very nearly equal to the starting height.

4.a)

Students record data.

4.b)

The ratio of recovered distance to starting distance should be only slightly less than 1.00 and typically 0.90 or more. The actual value will, of course, depend on the coefficient of friction for the particular kind of ball and track used. The error of measurement will be nearly as much as the observable difference in distances, indicating nearly complete "conservation" of distance. The ratio should remain essentially constant, regardless of the starting height. Ask the students if there is a pattern that does not depend on the starting distance.

5.a)

The ball can never recover its starting height on a horizontal track.

5.b)

The ball should roll forever in an attempt to recover its starting height.

5.c)

The students may say that the ball's stored motion (inertia) keeps it rolling. More correctly, there is no reason for the ball to stop because there is no force acting on it. The actual reason the ball stops is due to deformation of the track and the ball.

2-1a **Blackline Master**

Physics Talk

After students have read the *Physics Talk*, draw their attention to the similarity between Galileo's analysis of rolling balls down a ramp and the *Investigate*. Students should be able to relate their investigations to the property of inertia. Ask them to write definitions of important physics terms in their logs, so that they can refer to these terms when solving problems later on in the section. Consider giving an assignment that requires them to paraphrase how Galileo arrived at the law of inertia. Students could pull a heavy mass and then a light one to see which mass has greater inertia. Discuss the force of friction and how it relates to Newton's first law.

You might want to use examples of other sports, in addition to lacrosse, to reinforce how "running starts" impact speed. Have students demonstrate their understanding of relative speed by choosing an example of a sport they play. Discuss how their knowledge of velocity would improve their chances of winning a game. Ask them to explain the difference between speed and velocity. Make sure that they also understand the distinction between velocity and acceleration. Assure them that they will revisit these concepts several times in *Active Physics*.

As students become familiar with the term *frames of reference*, introduce various examples of moving objects to point out how a frame of reference would impact the measurement of velocity.

Chapter 2 Physics in Action

Physics Talk

NEWTON'S FIRST LAW OF MOTION

Galileo's Law of Inertia

In the *Investigate*, you observed, measured, and compared the release height of a ball on one side of the track to the recovered height on the other side of the track. You found that they were not exactly equal, but they were close to being equal.

Galileo Galilei was a pioneer in the use of precise, quantitative experiments. He insisted on using mathematics to analyze the results of his experiments.

Galileo Galilei (1564–1642) was an Italian physicist, mathematician, astronomer, and philosopher. Galileo is sometimes called the father of modern science. He introduced experimental science to the world. Galileo performed an experiment similar to the one you just completed. He observed that a ball that rolled down one ramp seemed to seek the same height when it rolled up another ramp.

Galileo also did a "thought experiment" in which he imagined a ball made of extremely hard material set into motion on a horizontal, smooth surface, similar to the final track in your investigation. He concluded that the ball would continue its motion on the horizontal surface with constant speed along a straight line "to the horizon" (forever).

From this, and from his observation that an object at rest remains at rest unless something causes it to move, Galileo formed the law of **inertia**: Inertia is the natural tendency of an object to remain at rest or to remain moving with constant speed in a straight line.

Galileo changed the way in which people viewed motion. Early on, people thought that all moving objects would stop. After Galileo, people thought about how moving objects might continue to move forever unless a **force**, a push or a pull, stopped them. That idea is not easy to understand. Any time you have pushed an object to move it, you have seen it stop. Nobody ever observes an object moving forever. Even when the surface is very, very smooth, the sliding or rolling objects eventually stop. However, Galileo realized that objects do not stop "on their own" but stop because there is a frictional force working that you cannot see and that is the force that stops the object.

Newton's First Law of Motion

Like Galileo, Isaac Newton was a great thinker. He was born in England in 1642, the year of Galileo's death. Newton's achievements brought him a great deal of recognition. Poems were written that honored Newton. Science, government, and philosophy all changed because of Newton's insights about the physics of the world.

Newton used Galileo's law of inertia as the basis for developing his **(Newton's) first law of motion**: In the absence of an unbalanced force, an object at rest remains at rest, and an object already in motion remains in motion with constant speed in a straight-line path.

Physics Words

inertia: the natural tendency of an object to remain at rest or to remain moving with constant speed in a straight line.

force: a push or a pull.

Newton's first law of motion: in the absence of an unbalanced force, an object at rest remains at rest, and an object already in motion remains in motion with constant speed in a straight-line path.

134

Active Physics

Students should be able to explain the examples given in the *Physics Talk* by drawing diagrams that illustrate movement in relative terms.

Section 1 Newton's First Law: A Running Start

Newton also explained that an object's **mass** is a measure of its inertia, or tendency to resist a change in motion. Given different masses moving at the same speed, the one with the greatest mass has the greatest inertia. The tendency of an object at rest to remain at rest appears to be common sense and few people think otherwise. The tendency of an object that is moving to continue moving (forever) unless acted upon by an unbalanced force is very different from what common sense would tell you. The evidence from the investigation you conducted should help to convince you that objects in motion stay in motion unless a force acts upon them. You will have to remind yourself of this many, many times since most people's intuition is that moving objects do not remain in motion, but tend to stop.

Physics Words

mass: the amount of matter in an object.

Isaac Newton credited Galileo and others for their contributions to his thinking. He is quoted as saying, "If I have seen farther than others, it is because I have stood on the shoulders of giants."

Here is an example of how Newton's first law of motion works: An empty grocery cart has a mass of 10 kg and a cart full of groceries has a mass of 30 kg. The cart with the greater mass has greater inertia.

To test your understanding of Newton's first law of motion, decide which of the following carts has the greatest inertia:

a) 1 kg moving at 5 m/s b) 2 kg moving at 3 m/s

c) 3 kg moving at 1 m/s d) 4 kg moving at 1 m/s

The correct response is d) because the 4-kg cart has the most inertia. The speed is not important in determining inertia.

SI System: The Kilogram

In this section, you read that inertia is related to mass. The kilogram is the base unit of mass. This particular base unit is a bit unusual. It is the only base unit that has a prefix. The prefix, kilo (k) placed in front of gram (g) stands for one thousand (10^3). The kilogram is equal to one thousand grams (1 kg = 1000 g).

It might be useful for you to relate the SI units that you will be using in *Active Physics* to the units that you use every day. A two-pound brick has a mass of about one kilogram.

In one of the most important science books of all time, *Principia*, Isaac Newton wrote his first law of motion. It is interesting both historically and in terms of understanding physics to read Newton's first law in his own words:

→

CHAPTER 2

"Every body perseveres in its state of rest, or of uniform motion in a right line, unless it is compelled to change that state by forces impressed thereon."

In Newton's time, "right line" meant "straight line."

Running Starts

You saw how Newton's first law of motion applies to a ball rolling down one track and up another track. Think about how Newton's first law of motion applies to sporting events.

"Running starts" take place in many sporting activities. In sports, where the objective is to maximize the speed of an object or the distance traveled in air, the prior motion of a running start is very important.

For example, in the javelin throw, an athlete is running holding a javelin. At the instance of release, the speed of the javelin is the same as the speed of the hand that is throwing the javelin. Newton's first law of motion tells you that when the athlete releases the javelin, the javelin will continue at the same speed. If the athlete then applies additional force to move the elbow and the shoulder of the arm carrying the javelin forward, the speed of the javelin will be the sum of these speeds.

The hand has a forward speed relative to the elbow, the elbow has a forward speed relative to the shoulder (because the arm is rotating around the elbow and shoulder joints), and the shoulder has a forward speed relative to the ground because the body is rotating and the body is also running forward.

Physics Words

speed: the change in distance per unit of time.

velocity: speed in a given direction.

acceleration: the change in velocity per unit of time.

The **speed** of the javelin is the sum of each of the above speeds. If the thrower is not running forward very fast, then the running speed does not add very much to the javelin's speed relative to the ground.

You can write a velocity equation to show the speeds involved. The letter v stands for velocity.

$$v_{javelin} = v_{hand} + v_{elbow} + v_{shoulder} + v_{body}$$

The term velocity is used in physics more than the term speed. **Velocity** is speed in a given direction. The two terms, speed and velocity, have slightly different meanings, but at this point, you can use them interchangeably.

Motion captures everyone's attention in sports. Sometimes speeds are constant. These motions are examples of Newton's first law: Objects in motion (at constant speed) stay in motion (at constant speed) unless a force acts on them. When a force acts, the speeds change. This change in speed during a specific time is referred to as **acceleration**. Acceleration occurs during starting, stopping, and changing direction.

Section 1 Newton's First Law: A Running Start

Acceleration is definitely an exciting component of many sports. You will be learning about acceleration in other sections of this chapter. However, ordinary, straight-line motion is just as important in sports, but it is easily overlooked.

Speed and Velocity

In *Active Physics*, you will often explore the same topic several times. Being exposed to the same topic at different times and in different situations helps you learn and understand the topic better. The difference between speed and velocity will be explored frequently in this book.

Frames of Reference

In this section, you investigated Newton's first law. In the absence of external forces, an object at rest remains at rest and an object in motion remains in motion. If you were challenged to throw a ball as far as possible, you would now be sure to ask if you could have a running start. If you run with a ball prior to throwing it, the ball gets your speed before you even try to release it.

If you can run at 5 m/s (meters per second), then the ball will get the additional speed of 5 m/s when you throw it. When you do throw the ball, the ball's speed is the sum of your speed before releasing the ball, 5 m/s, and the speed of the release relative to your body.

It may be easier to understand this if you think of a toy cannon that could be placed on a skateboard. The toy cannon always shoots a small ball forward at 7 m/s. This can be checked with multiple trials. The toy cannon is then attached to the skateboard. A release mechanism is set up so that the cannon continues to shoot the ball forward at 7 m/s when the skateboard is held at rest. Now imagine that the skateboard is moved along at a constant speed of 3 m/s. If the cannon releases the ball while the skateboard is being moved at 3 m/s, the ball's speed is now measured to be 10 m/s. From where did the additional speed come? The ball's speed is the sum of the ball's speed from the cannon plus the speed of the skateboard (7 m/s + 3 m/s = 10 m/s).

You may be wondering if the ball is really moving at 7 m/s or 10 m/s. Both values are correct — it depends on your **frame of reference**. The ball is moving at 7 m/s relative to the skateboard. The ball is moving at 10 m/s relative to the ground.

Physics Words

frame of reference: a vantage point with respect to which position and motion may be described.

CHAPTER 2

Checking Up

1.

Inertia is the property of an object to remain at rest or in motion unless something causes it to move. Inertia resists a change in an object's state of motion.

2.

Newton's first law of motion states that in the absence of an external force, an object at rest remains at rest, and an object already in motion remains in motion with constant speed in a straight-line path.

3.

A force needs to act on an object to stop it from moving at a constant speed.

4.

An unbalanced external force stops the ball from moving. This force might be provided by friction or some other force.

5.

The heavier mass will have the greater inertia.

6.

The velocity of the ball measured from the frame of reference of a moving train would differ from the velocity of a ball from the frame of reference of a person standing outside on the ground.

Chapter 2 Physics in Action

Imagine that you are on a train that is stopped at the platform. You begin to walk toward the front of the train at 1 m/s. Everyone in the train will agree that you are moving at 1 m/s toward the front of the train. This is your speed relative to the train. Everyone looking into the train from the platform will also agree that you are moving at 1 m/s toward the front of the train. This is your speed relative to the platform.

Imagine that you are on the same train, but now the train is moving past the platform at 8 m/s. You begin to walk toward the front of the train at 1 m/s. Everyone in the train will agree that you are moving at 1 m/s toward the front of the train. This is your speed relative to the train. Everyone looking into the train from the platform will say that you are moving at 9 m/s (1 m/s + 8 m/s) in the direction the train is moving. This is your speed relative to the platform.

Whenever you describe speed, you must always ask, "Relative to what?" Often, when the speed is relative to the ground, this is not specifically stated and you are expected to assume this fact. If your frame of reference is the ground, then it all seems quite obvious. Frame of reference is a vantage point with respect to which position and motion may be described.

If your frame of reference is the moving train, then more thought is required to figure out the speeds measured by people on the train and by people on the platform.

In sports, where you want to provide the greatest speed to a baseball, lacrosse ball, football, or a tennis ball, that speed could be increased if you were able to get on a moving platform. That being against the rules, an athlete will try to get the body moving with a running start, if allowed. If the running start is not permitted, the athlete tries to move every part of his or her body to get the greatest speed.

Checking Up

1. What is inertia?
2. Describe Newton's first law of motion.
3. What needs to act on an object to stop it from moving at a constant speed?
4. In the real world, a rolling ball does not roll forever. What stops the motion of the ball?
5. Given two different-size masses moving at the same speed, which mass will have the greater inertia?
6. You throw a ball in a moving train. Why is it important to establish a frame of reference when describing the speed of the ball?

138

Active Physics

Active Physics *Plus*

+Math	+Depth	+Concepts	+Exploration
♦♦♦			

Part A: Calculating Velocity for Different Frames of References

You read that when describing speed or velocity it is important to give a frame of reference. You can calculate the velocity relative to a particular frame of reference mathematically by using positive and negative numbers.

Sample Problem 1

A sailboat has a constant velocity of 8.0 m/s east. This is a velocity because it has both a speed and a direction. Someone on the boat prepares to toss a rock into the water.

a) Before being tossed, what is the speed of the rock with respect to the boat?

b) Before being tossed, what is the speed of the rock with respect to the shore?

c) If the rock is tossed with a velocity of 6.0 m/s east, what is the rock's velocity with respect to the shore?

d) If the rock is tossed with a velocity of 6.0 m/s west, what is the rock's velocity with respect to the shore?

Strategy: Before determining a velocity, it is important to check the frame of reference. The rock's velocity with respect to the boat is different from the velocity with respect to the shore.

The direction the rock is thrown also affects the final answer. Let the direction east be a positive value. Use a negative sign to indicate the direction west.

Given:

v_b (velocity of the boat) = 8.0 m/s east

v_r (velocity of the rock) = 6.0 m/s (direction varies)

Solution:

a) With respect to the boat, the rock's velocity is 0 m/s. The rock is moving at the same speed as the boat, but you would not notice this velocity if you were in the boat's frame of reference.

b) With respect to the shore, the rock's velocity is 8.0 m/s east. The rock is on the boat, which is traveling at 8.0 m/s east. Relative to the shore, the boat and everything on it act as a single unit traveling at the same velocity.

c) The relative velocity is the sum of the velocity values. Since each is directed east, the value of each velocity is positive.

$$\begin{aligned} v &= v_b + v_r \\ &= 8.0 \text{ m/s east} + 6.0 \text{ m/s east} \\ &= 8.0 \text{ m/s} + 6.0 \text{ m/s} \\ &= 14.0 \text{ m/s east} \end{aligned}$$

Active Physics Plus

Students doing this section will get the opportunity to extend their understanding of inertia in relation to an object's mass. They will also be able to relate frames of reference to velocity by calculating the velocity of an object thrown from a moving body.

CHAPTER 2

1.

The important physical principle is that the ball will roll up the right-hand ramp to a height equal to its starting height. Therefore, 4.2 m/s + 10.3 m/s = 14.5 m/s and $h = 10$ cm for all the triangles, regardless of the angle of the ramp. If θ is the angle of the ramp, then $\sin\theta = h/d$, so $d = h/\sin\theta$. The distance, d, along the ramp would be for the angles listed.

1.a)

14 cm

1.b)

20 cm

1.c)

29 cm

With respect to the shore, the rock's velocity is now 14.0 m/s east.

d) Since the direction of the rock is the opposite to the direction of the boat, the velocity of the rock has a negative value compared to the velocity of the boat. The relative velocity is the sum of the positive and negative velocities.

$$v = v_b + v_r$$
$$= 8.0 \text{ m/s east} + (6.0 \text{ m/s west})$$
$$= 8.0 \text{ m/s east} + (-6.0 \text{ m/s east})$$
$$= 2.0 \text{ m/s east}$$

With respect to the shore, the rock's velocity is now 2.0 m/s east.

Sample Problem 2

A quarterback on a football team is getting ready to throw a pass. If he is moving backward at 1.5 m/s and he throws the ball forward at 10.0 m/s relative to his body, what is the velocity of the ball relative to the ground?

Strategy: Use a negative sign to indicate the backward direction. Add the two velocities to find the velocity relative to the ground.

Given:

v_q (velocity of the quarterback) = −1.5 m/s

v_f (velocity of the football) = 10.0 m/s

Solution:

Add the velocities.

$$v = v_f + v_q$$
$$= 10.0 \text{ m/s} + (-1.5 \text{ m/s})$$
$$= 8.5 \text{ m/s}$$

The ball is moving forward at 8.5 m/s relative to the ground.

Part B: Calculating Recovered Distance along the Ramp

In the investigation, you predicted and then observed the distance the ball rolled up and along the right-hand slope. Assume that you are using a "perfect ball and ramp" that allows the recovered height to be exactly equal to the starting height. Now that you have completed the investigation, you know that the recovered height is the same as the starting height (in a "perfect" situation). Therefore, you can calculate the distance along the ramp that the ball will roll.

1. Imagine that the ball starts from a point on the left-hand slope with a vertical height of 10 cm. How far up the right-hand slope (measured along the slope) will the ball roll if the angle of the right-hand slope is set at the following angles:

a) 45° b) 30° c) 20°

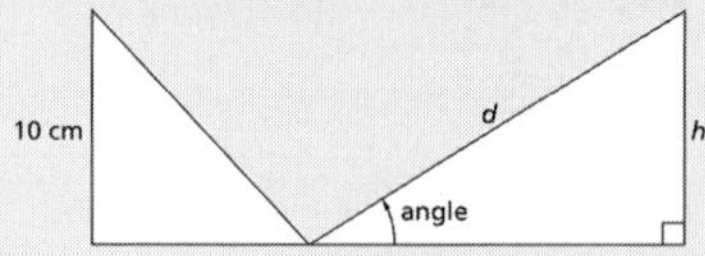

To answer these questions, look at a diagram of the setup above. You can use the right-hand slope and the height of the track to form a right-angled triangle. The hypotenuse of the right-angled triangle is the distance up the ramp the ball rolls (d) and the opposite side is the height of the ball when it stops rolling (h). If you know the angle and you know the height at which the ball was released, you can find the distance along the ramp using a scale diagram. Try this for the three angles given.

You can also use trigonometry to solve this problem by using the value of the sine of the angle of the ramp. The sine of the angle of the right-hand ramp is equal to h/d (opposite/hypotenuse).

The value of the sine of the angle can be found using the "sin" button on your calculator (make sure your calculator is in "degree mode"). Then you may use that value in h/d and solve for d.

2. Use a calculator to check the accuracy of the values of d you obtained using scale diagrams.
3. Use a calculator to find how far up the right-hand slope (measured along the slope) the ball will roll if the angle of the right-hand slope is:
   a) 10°
   b) 1°
   c) 0.1°
   d) 0.01°
4. How in a "perfect frictionless world" would the calculations you did above help you explain Newton's first law of motion?

What Do You Think Now?

At the beginning of this section, you were asked the following:

- **How do figure skaters keep moving across the ice at high speeds for long times while seeming to expend no effort?**
- **Why does a soccer ball continue to roll across the field after it has been kicked?**

The ice skater effortlessly gliding across the ice at high speed and the soccer ball moving across the field are like the ball rolling along the horizontal portion of your track. What determines their horizontal speed and why do they keep moving without someone doing anything to keep them moving?

2.

The calculator values should closely match the scale diagram values.

3.a)

58 cm

3.b)

570 cm

3.c)

5700 cm

3.d)

57,000 cm

4.

The lower the slope, the further the ball will travel. When the slope goes to zero, the ball should roll "forever," which is a statement of Newton's first law.

What Do You Think Now?

Students should by now have a fair understanding of the main concepts discussed in this section. You might want to share *A Physicist's Response* with them, provided at the beginning of this section. Stress the importance of Newton's first law in determining the answers to the *What Do You Think?* questions. Have students revise the answers they originally wrote in their logs. Ask them if their answers have remained the same. Encourage them to discuss any doubts they might have.

Reflecting on the Section and the Challenge

Encourage students to think of the many ways they can incorporate Newton's first law and the properties of matter into their *Chapter Challenge*. Ask them to ponder how they would use the knowledge they now have to improve their "science commentary." Have them read the *Reflecting on the Section and the Challenge* carefully and emphasize that they should watch a variety of sports video segments that illustrate Newton's first law. Tell them to draw links between running starts and inertia and to consider illustrating those links in their voice-over narration.

Physics

Essential Questions

What does it mean?

Even the greatest thinkers may not know why objects have inertia, but your investigations show that they do have inertia. Observation is the basis for all physical concepts. What does it mean when you say that an object with mass has inertia whether it is moving or stationary?

How do you know?

How do you know that the rolling ball you examined in your experiment would keep rolling forever unless some force acted on it? Why is the steady, straight-line motion of an object not "explained" but is simply stated as the way the world works in Newton's first law of motion?

Why do you believe?

Connects with Other Physics Content	Fits with Big Ideas in Science	Meets Physics Requirements
Force and motion	Change and constancy	* Experimental evidence is consistent with models and theories

* In physics, ideal situations are often used to illustrate concepts. Nobody has ever arranged for a rolling ball to roll forever, so why do you believe that it would? Provide examples in which an object might keep moving in a straight line with a constant speed even longer than a rolling ball might.

Why should you care?

In your sports voice-over, you will want to use Newton's first law. Give an example in a sport where an object in motion remains in motion, or where an object at rest remains at rest.

Reflecting on the Section and the Challenge

"Immovable objects," such as defensive linemen in football, illustrate the tendency of highly massive objects to remain at rest and can be observed in many sports. Running starts can also be observed in many sports. Many observers may not realize the important role that inertia plays in preserving the speed already established when an athlete engages in activities such as jumping, throwing, or skating from a running start. For the challenge, you should have no problem finding a great variety of video segments that illustrate Newton's first law.

The segment that you select for your challenge might illustrate:

- That "an object at rest remains at rest."
- That the more massive an object, the more difficult it is to get it to start moving or to stop moving.

Physics Essential Questions

What does it mean?

A massive object has a tendency to keep moving if it is moving and a tendency to stay at rest if it is at rest. This is called inertia.

How do you know?

Newton's first law of motion tells us what happens if there is no external force acting on something. This is an ideal situation—there is always some force acting on an object. The object will keep moving because of the thought experiment that logically shows this as a result.

Why do you believe?

As friction and air resistance are gradually decreased on an object, the object will keep moving further and further. If these forces were completely removed, the ball would keep rolling forever. A wheel with good ball bearings will roll further than a ball.

Why should you care?

A football is thrown and will keep moving until someone catches it or it hits the ground.

A sumo wrestler has enormous inertia and if he is at rest, he will stay at rest unless someone applies a force to move him.

Section 1 Newton's First Law: A Running Start

- How an object will tend to stay in motion until an external force stops it.
- How relative motion depends on the speeds of the player and the ball and the reference frame in which it is measured.

Physics to Go

1. You push a ball to start it rolling along a "perfectly frictionless" surface.
   a) How far will the ball roll?
   b) Explain your answer for a) using Newton's first law of motion.
2. A ball is released from a vertical height of 20 cm. It rolls down a "perfectly frictionless" ramp and up a similar ramp. What vertical height on the second ramp will the ball reach before it starts to roll back down?
3. Do you think it is possible to arrange conditions in the "real world" to have an object move, unassisted, in a straight line at constant speed forever? Explain why or why not.
4. Use what you have learned in this section to describe the motion of a hockey puck between the instant the puck leaves a player's stick and the instant it hits something. (No "slap shot" allowed; the puck must remain in contact with the ice.)
5. Active Physics Plus You are riding your bike and steadily pulling your little brother in his red wagon while someone standing still watches you and your little brother go by. He has a ball, and he throws the ball forward at a velocity of 2.5 m/s relative to his body while you are pulling the wagon at a velocity of 4.5 m/s. At what speed does the person who is standing nearby see the ball go by?
6. Active Physics Plus A track and field athlete is running forward with a javelin at a velocity of 4.2 m/s. If he throws the javelin at a velocity relative to him of 10.3 m/s, what is the velocity of the javelin relative to the ground?
7. Active Physics Plus You are riding in a train. Since the train car is almost empty, you and your friend are pushing a low-friction cart back and forth between the front and rear of the car. The train is moving at a speed of 5.6 m/s. Suppose you push the cart toward each other at 2.4 m/s.
   a) What is the velocity of the cart relative to the ground when the cart is moving toward the front of the car?
   b) What is the velocity of the cart relative to the tracks when it is moving toward the rear of the car?
   c) What if you and your friend push the cart perpendicular to the aisle as the train moves forward? This is a more complicated situation. What is the cart's velocity relative to the ground?

143 Active Physics

Physics to Go

1.a)

The ball keeps rolling if the surface is horizontal.

1.b)

Newton's first law states, in the absence of an external force, an object will continue in its state of rest or motion. When a ball rolls on a frictionless horizontal surface, it will continue to roll because there are no forces to stop it.

2.

The ball will reach a vertical height of 20 cm before it begins to roll down again.

3.

It does not seem possible to eliminate friction to arrive at perpetual motion in the real world, except perhaps in deep space far away from the influence of any source of gravity.

4.

Because the ice exerts almost no frictional force on the hockey puck, the puck will continue to slide with an almost constant speed in a straight line until a force is exerted upon it. If the object it hits exerts a force in the direction opposite its motion, the puck will slow down, stop, or even change direction depending upon the nature of the force. If the force is in another direction, the puck's direction and possibly its speed will change due to the force.

5.

The speed relative to the person watching will be $2.5 \text{ m/s} + 4.5 \text{ m/s} = 7.0 \text{ m/s}$.

6.

The relative velocity will be $4.2 \text{ m/s} + 10.3 \text{ m/s} = 14.5 \text{ m/s}$.

7.a)

The velocity relative to the ground is $5.6 \text{ m/s} + 2.4 \text{ m/s} = 8.0 \text{ m/s}$.

7.b)

The velocity relative to the tracks is $5.6 \text{ m/s} - 2.4 \text{ m/s} = 3.2 \text{ m/s}$.

7.c)

Since the two velocities are perpendicular, we must use the Pythagorean Theorem. So, $(5.6 \text{ m/s})^2 + (2.4 \text{ m/s})^2 = v^2$

$$v = \sqrt{37.12 \text{ m}^2/\text{s}^2}$$

$$v = 6.1 \text{ m/s}.$$

Using the tangent button on the calculator or a vector diagram, the angle is 67°. Students should be able to make the diagram. Some will use the Pythagorean theorem. (More emphasis on this will come in *Section 3*.)

CHAPTER 2

8. Active Physics Plus While riding a horse, a competitor shoots an arrow horizontally toward a target. The speed of the arrow relative to the ground as it reaches the target is 85 m/s. If the horse was traveling at 18 m/s, at what speed did the arrow leave the bow? (Assume the horse and arrow are traveling in the same direction.)

9. Active Physics Plus A ball is released on a ramp at a vertical height of 15 cm. Calculate how far up a second ramp (measured along the slope) the ball will roll if the angle of the second ramp is:

a) 45° b) 20° c) 15° d) 5°

10. ***Preparing for the Chapter Challenge***

a) Provide three examples of Newton's first law in sporting events. Describe the sporting event and which object when at rest stays at rest, or when in motion stays in motion.

b) Describe these same three examples in the manner of a sportscaster.

Inquiring Further

1. **Curling and Newton's first law**

Find out about a sport called curling. It is an Olympic competition that involves some of the oldest Olympians. How can this sport be used to illustrate Newton's first law of motion?

2. **Sliding into base**

Why do baseball players often slide into second base and third base, but they almost never slide into first base after hitting the ball? (Hint: The answer depends on both the rules of baseball and the laws of physics.)

8.

The speed is
85 m/s – 18 m/s = 67 m/s.

9.a)

21.2 cm

9.b)

43.9 cm

9.c)

58 cm

9.d)

172 cm

Preparing for the Chapter Challenge

10.a)

Examples of Newton's first law in sports might include objects in motion that continue in motion in a straight line—such as a hockey puck sliding across the ice, a kicked soccer ball rolling along the ground, a race-car driver traveling with constant speed in a straight line, and an ice skater gliding along the ice. Objects at rest that remain at rest would include a football at the line of scrimmage before it is hiked, race cars at the starting line, and a soccer ball waiting for a corner kick.

10.b)

The students will provide various answers, but the written material should show some of the excitement of the sport as well as the physics involved.

Inquiring Further

1.

Curling uses a heavy, polished, granite stone that is slid along ice toward a target. Once the person who starts the stone sliding releases it, the stone continues to slide with roughly constant speed, illustrating Newton's first law.

2.

Sliding in baseball is a way to both stop the runner and to allow forward velocity to reach a base quickly. Because baseball players can run past first base without penalty, most players will run over the base, relying on Newton's first law to allow them to continue in motion at a high rate of speed, once they are running, and get to the base as quickly as possible. Because a player who overruns second or third base may be tagged out, the players slide into the base to slow themselves down and stop at the base. This method allows them to continue running for a longer time and get to the base quicker.

SECTION 1 QUIZ

2-1b **Blackline Master**

1. A rocket in space can travel without engine power at constant speed in the same direction. This condition is best explained by the concept of

   a) gravity.
   b) inertia.
   c) acceleration.
   d) frames of reference.

2. If the mass of a moving object is doubled, its inertia would be

   a) halved.
   b) the same.
   c) doubled.
   d) four times greater.

3. A coin is resting on a piece of cardboard on a beaker as shown in the diagram. When the cardboard is rapidly removed, the coin falls into the beaker. The two properties of the coin that best explain its fall are its weight and its

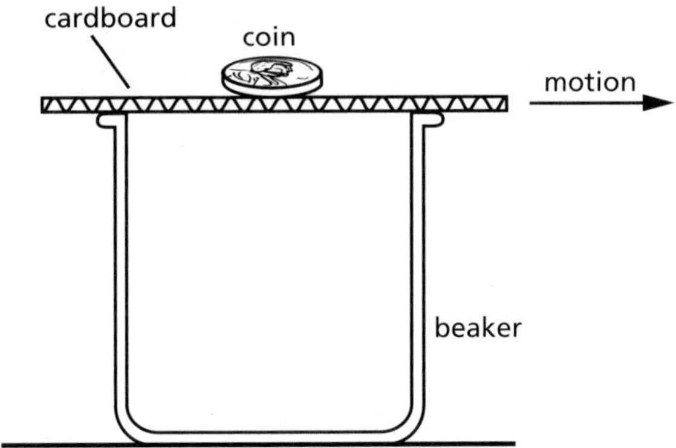

   a) temperature.
   b) inertia.
   c) volume.
   d) shape.

4. If there is no net force acting on an object, the object will

   a) slow down and stop.
   b) change its direction of motion.
   c) accelerate.
   d) continue with constant speed in a straight line.

5. Which person has the greatest inertia?

   a) A 110-kg wrestler resting on a mat
   b) A 90-kg man walking at 2 m/s
   c) A 70-kg long-distance runner traveling at 5 m/s
   d) A 50-kg girl sprinting at 9 m/s

CHAPTER 2

SECTION 1 QUIZ ANSWERS

1. b) Inertia is a measure of mass. The more mass an object has the more inertia, without respect to what the object is composed of. Students may think that feathers would have less inertia than lead, but that is only true if they have less mass. Equal amounts of mass imply equal amounts of inertia.

2. d) The object with the greatest mass has the greatest inertia, irrespective of whether or not the object is in motion.

3. b) Because the object is at rest, it must have no net force acting on it.

4. d) According to the law of inertia, an object at rest or traveling with constant speed in a straight line has no net force acting upon it. A net force would be required to change the direction.

5. a) Inertia is a measure of the object's mass only. Size, shape, volume, and speed do not determine its inertia.

NOTES

NOTES

CHAPTER 2

SECTION 2

Constant Speed and Acceleration: Measuring Motion

Section Overview

In this section, students pull tapes through a dot timer to record motion at three different speeds. They then record and compare motion at a constant speed vs. increasing speed and decreasing speed. By analyzing the spacing of dots on different segments of tape, they compare how the distance between the dots changes as the speed increases, decreases, or remains the same when the velocity is constant. The graph students make with tape segments gives them a visual representation of both constant speed and changing speed, which is defined as acceleration. They use their findings in the *Investigate* to distinguish between average speed, instantaneous speed, and acceleration. Students also reflect on how variations in speed and acceleration affect the outcome of an event. Through the *Investigate* and sample problems in the *Physics Talk*, they learn the strategy for calculating acceleration.

Background Information

An object's motion is described by stating the object's initial and subsequent positions, speed, and acceleration. These quantities basically tell how far and how fast the object's motion is changing. An object's speed is determined by dividing the distance it travels by the time required to traverse that distance. In general, the speed of an object changes as it moves. Consequently, it is necessary to distinguish average speed from instantaneous speed. Average speed is found by dividing the total distance traveled by the total time. Instantaneous speed is the speed at a particular instant. A speedometer provides an automobile's instantaneous speed. In theory, the "instant" is an immeasurably short period of time. In practice, an object's instantaneous speed is actually an average speed calculated over a very short time interval. In this section, the motion of a student is recorded with a dot timer. A tape is pulled through a device that marks a dot on the tape every 1/60 of a second. The faster the person moves, the farther apart the dots. Constant speed results in uniformly spaced dots.

Acceleration is the rate at which the speed or direction of motion of an object changes with respect to time. To determine the acceleration of an object due to speed change, it is necessary to measure the speed of the object at the beginning and the end of a time interval. To determine the acceleration due to the changing direction of an object, a different, slightly more complicated procedure is used. In this section, students will be concerned with an object's acceleration in a straight line. By pulling a tape through a dot timer, a student's motion is displayed and analyzed. First the paper tape is marked and labeled in consecutive segments of 6-dot intervals each. These segments are then cut. Six-dot intervals happen in 1/10 of a second. Pasting them in order on a sheet of paper yields a graph of the average velocity of the student during each 1/10 s interval. Acceleration is revealed by the slope of the graph: a positive slope means the student was speeding up; a negative slope, slowing down; and no slope, constant speed.

Crucial Physics

- Constant speed occurs when objects always cover equal distances in equal times.
- Acceleration is the change in velocity of an object per unit time.
- Positive and negative acceleration are determined by both speed and direction.
- Acceleration units are meters/second per second.
- The formula for calculating an objects average speed is distance traveled divided by the elapsed time or $v_{\text{average}} = \Delta d/\Delta t$. An object may have an average speed without traveling at constant speed.
- The formula for calculating acceleration is change in speed divided by time interval or $a = \Delta v/\Delta t$.
- Scientists refer to increasing speed when traveling in a straight line in one direction as positive acceleration, and when decreasing speed under the same conditions as negative acceleration.
- The units for acceleration are those for the increase in speed for each second. In the
- metric system this would be m/s/s or m/s^2.

Learning Outcomes	Location in the Section	Evidence of Understanding
Give examples of distance, time, speed, and acceleration.	***Investigate*** Steps 8-11	Students pull the tape at a slow speed and then at a faster speed. They measure the length of each strip as the distance covered in 1/10 of a second. They give examples of this change in speed in an interval of time as an example of acceleration.
Differentiate between instantaneous and average speed.	***Physics Talk***	Students learn the difference between instantaneous and average speed by defining the two and studying sample problems.
Recognize when motion is accelerated.	***Investigate*** Steps 8-11 ***Checking Up*** Questions 1 and 4 ***Physics Talk***	Students see how the length of strips change with a change in speed, how the distance between the dots on the ticker tape changes with a change in speed, and solve a problem to calculate average acceleration.
Calculate average speed and acceleration.	***Physics to Go*** Questions 2-6, 8-9, 11-13	Students solve problems to calculate average speed and acceleration.

Section 2 Materials, Preparation, and Safety

Materials and Equipment

PLAN A		
Materials and Equipment	**Group (4 students)**	**Class**
Multimedia DVD/CD Set		1 per class
Ruler, metric, 30 cm	1 per group	
Timer, ticker tape, AC	1 per group	
Scissors	1 per group	
Meter stick, wood	1 per group	
Ticker tape, roll	1 per group	
Glue stick	1 per group	

*Additional items needed not supplied

PLAN B		
Materials and Equipment	**Group (4 students)**	**Class**
Multimedia DVD/CD Set		1 per class
Ruler, metric, 30 cm		1 per class
Timer, ticker tape, AC		1 per class
Scissors		1 per class
Meter stick, wood		1 per class
Ticker tape, roll		1 per class
Glue stick		1 per class

*Additional items needed not supplied

Note: Time, Preparation, and Safety requirements are based on Plan A, if using Plan B, please adjust accordingly.

Teacher Preparation

- Familiarize yourself with the operation of the ticker tape timer. Those that operate on AC power are preset to vibrate at 60 cycles/second. If you have any of the older timers that operate on battery or DC power, you will have to check the vibration rate. This can be done either with a calibrated strobe or by pulling a section of tape through the timer for 5 s, and counting the dots.
- Have spare carbon disks on hand for any timer where the disk is becoming worn out. Replace the disk if the spots on the student's tape become faint. It often is worthwhile for the student to mark faint dots with a pen to make them more visible prior to analyzing the tape.

Safety Requirement

- Make sure the area where the students will walk is clear of any obstructions that may cause them to trip. The students should unplug the timers as soon as they are finished taking measurements to prevent overheating, and to save the carbon disk.

Meeting the Needs of All Students

Differentiated Instruction: Augmentation and Accommodations

Learning Issue	Reference	Augmentation and Accommodations
Following directions	***Investigate***	**Augmentation** • Ask students to read the entire *Investigate* to get a general idea of what is required to complete 11 steps of directions in a timely manner. • Model how to use a ticker timer. • The information in parentheses in *Investigate* 3, is especially important. Make sure students understand how to count six spaces instead of six dots while looking at the illustration below *Step 3*. • Give students a time limit to complete this section. Set a timer and give time updates every 5-10 minutes. **Accommodation** • Provide a checklist that breaks a step into smaller tasks that can be marked off as students complete each one.
Making sketches	***Investigate*** Steps 6.a), 8.a), and 10.a)	**Augmentation** • Students with graphomotor or visual-motor issues may struggle to sketch accurate graphs. Provide graph paper for sketches. • Tell students that the sketch should display general trends or patterns and does not have to be an exact replica of the original bar graph. • Remind students that they should have three bar graphs in their *Active Physics* logs when they are finished.
Understanding trends to interpret graphs	***Physics Talk*** ***Physics to Go***	**Augmentation** • Ask students to describe in their own words the differences between the bar graphs in this section. How is slow speed different from medium speed on a ticker-tape bar graph? How is positive acceleration different from negative acceleration on a ticker-tape bar graph? • To check for understanding, provide bar graphs without labels. Ask students to label the graphs (slow speed, medium speed, etc.) independently. • Understanding the trends in these graphs will help students answer *Physics to Go*, Questions 4, 7, 10, and 14.
Sequential learning and then using two new formulas at the same time	***Physics Talk***	**Augmentation** • Provide direct instruction for the average speed formula and allow opportunities for guided practice. Then provide direct instruction for the acceleration formula and let students practice problems similar to the sample problems.
Differentiating nuances in new vocabulary	***Physics Words*** ***Physics to Go*** Question 1	**Augmentation** • Students may not notice the subtle differences in the definitions for average versus instantaneous speed and positive versus negative acceleration. Ask students to explain the definitions in their own words. • If students are required to copy definitions, ask them to highlight or underline the differences. • Students can also write the words that are different in positive and negative acceleration definitions in capital letters (INCREASE and DECREASE).
Reading vocabulary	***Physics Essential Questions***	**Augmentation** • When reading the *Why do you believe?* section, explain the meaning of *subatomic particle*. Students may know about protons, neutrons, and electrons but may not know the meaning of subatomic particles.

CHAPTER 2

Learning Issue	Reference	Augmentation and Accommodations
Completing long-term projects	***Reflecting on the Section and the Challenge***	**Augmentation** • Assist students in the creation of a calendar, timeline, or checklist for completing the *Chapter Challenge*. • Check in with students each step of the way to make sure they understand and are completing the smaller tasks required to complete the challenge. **Accommodation** • Allow students to work with a partner who has good time-management skills.
Performing calculations with a formula	***Physics to Go*** Question 2	**Augmentation** • Students learned two formulas in this section. Make sure it is clear that students are supposed to use the average speed formula for these problems. • Require that students show their work when solving problems. This allows the teacher and/or student to check for misunderstandings and mistakes. **Accommodation** • Provide students with a sheet of blank problem-solving boxes.
Solving problems with data from a table	***Physics to Go*** Question 5	**Augmentation** • Students with visual-spatial and attention issues may struggle to accurately track numbers in the table. Ask students to use a ruler, note card, or another straight edge to track the numbers in each row that are needed to do the calculation. **Accommodation** • Provide students with a table that has the first two columns reversed so that "Length of 6-tick segment" is column 1 and "Elapsed time" is column 2. Then students can sequentially enter data into their calculators (distance traveled divided by elapsed time).

Strategies for Students With Limited English-Language Proficiency

Learning Issue	Reference	Augmentation
Vocabulary comprehension	***Investigate***	Before beginning this *Investigate*, have each group discuss the word "constant" as used in "constant speed" and "constant acceleration" and share its interpretation with the class. Be sure all students have an appropriate understanding of the word in the context of physics. The word "trend" is used in *Investigate* Step 8.c). Be sure students understand this word in context as well. Try using "pattern" instead.
Comprehension	***Investigate*** Step 1.a)	To help students reinforce their description of the tape, have them draw in their *Active Physics* log what the tape would look like and add labels for positive acceleration and negative acceleration.
Comprehension	***Physics Talk***	To ensure that students understand the *Units of Acceleration* box, collaborate with their math teachers to determine what level of comprehension students have for exponents.
Comprehension	***What Do You Think Now?***	Have students write the answer in their *Active Physics* log. Invite students to share their responses and explain how their understanding may have changed.
Making connections	***Reflecting on the Section and the Challenge***	Remind students of the voice-over narration they are going to write and perform at the end of the chapter. Have them start thinking of the sporting event they want to cover. Ask some guiding questions: What are the rules of nature that govern your event? What physics principles apply to your event? What physics terms can you use to describe your event? How can you make your voice-over lively and engaging?

NOTES

SECTION 2

Teaching Suggestions and Sample Answers

What Do You See?

This illustration brings an array of colorful images together, which deftly connect with the topic of *Section 2*. Consider using an overhead to highlight different aspects of two contrasting visuals. Ask students to note their similarities and differences. Have students recall prior learning to gauge how their understanding of the physics of motion is progressing. You could also touch on the humorous element in the illustration to initiate a lively discussion. Students will invariably respond to the image of the boy running with a bouquet of roses in his hand. The two visuals present ample opportunity to develop ideas and lead students to think about the concepts that students will be investigating in this section. You might want to post an enlarged version of this illustration on the wall of your classroom, so that students can note how their initial perception of the illustration continues to evolve with their investigations and subsequent interpretation of physics concepts.

What Do You Think?

Encourage students to answer the *What Do You Think?* questions and emphasize that all answers will be accepted. You might want to clarify to them at this stage that your prime concern is to prepare the class for physics concepts that will appear later in the section. Stress that the purpose of these questions is to guide your students' understanding. Each question relates to a topic that will be investigated and analyzed in detail.

What Do You Think?

A Physicist's Response

A fastball thrown at 100 mph translates to a speed of 147 ft/s. When objects move at speeds such as this, short time intervals can mean an object travels a significant distance. For a batter who is 1/100 of a second off will miss the ball by 1.47 ft or 18 in.

Students' Prior Conceptions

As they work through *Section 2*, students can root out preconceptions on steady motion, non-steady motion, speed, velocity, and acceleration, either speeding up or slowing down. The preconceptions are as follows:

1. **The location of an object can be described by stating its distance from a given point (ignoring direction).** Students will confront the difference between the total distance an object travels with the displacement of an object relative to a specific point of reference or origin. Students must consider both the distance and the direction from the reference point when describing the location of objects. The concept of magnitude of measurement is distinguishable from magnitude with direction.
2. **The terms *distance* and *displacement* are synonymous and may be used interchangeably.** Thus, the distance an object travels and its displacement are always the same. Student confusion between distance and displacement relates directly to the previous preconception that deals with how the motion of the object changes, and in which direction the change occurs.
3. **Velocity is another word for speed. An object's speed and velocity are always the same.** Teachers need to encourage students to recognize that moving forward, away from the origin with a given speed, differs from starting in front of the origin and moving toward it with the same speed. The direction of the motion, away from the origin (positive velocity) or toward the origin

Section 2 Constant Speed and Acceleration: Measuring Motion

Section 2 Constant Speed and Acceleration: Measuring Motion

What Do You See?

Learning Outcomes

In this section, you will

- **Give** examples of distance, time, speed, and acceleration.
- **Differentiate** between instantaneous speed and average speed.
- **Recognize** when motion is accelerated.
- **Calculate** average speed and acceleration.

What Do You Think?

Some major-league pitchers can throw a baseball 100 mi/h, about 45 m/s. The ball reaches home plate in less than half a second. In that time, the batter must decide whether or not to swing at the pitch. If the batter decides to swing, he then must be able to react quickly enough to the pitch. It is no surprise that so few athletes are capable of competing in the major leagues.

- **In your own words, explain the meaning of 100 mi/h and 45 m/s.**

Record your response in your *Active Physics* log. Be prepared to discuss your response with your group and the class.

Investigate

In this *Investigate*, you will use a measuring device called a ticker timer to explore the concepts of constant speed and acceleration.

1. A ticker timer makes a dot on a paper tape every 1/60 of a second. As you pull the tape through the ticker timer, a dot will be made on the tape every 1/60 of a second. You can use the ticker timer to determine how fast something is moving.

Investigate

1.

Student predictions.

CHAPTER 2

(negative velocity) is vital to student understanding of velocity and to how velocity is interpreted from a graph of displacement vs. time.

4. **Students confuse acceleration with speed; acceleration means that an object is speeding up.** Encourage students to measure the motions of objects so that it is evident that a change in velocity within a given time period is an acceleration. This change in velocity can be positive or negative, leading to a positive acceleration with motion that may either speed up or slow down or a negative acceleration with motion that also may either speed up or slow down. A positive acceleration does not always occur in the same direction as an object is moving.

5. **If an object has a speed of zero, even instantaneously, it has no acceleration.** Because acceleration is the change in velocity, an object can have an instantaneous speed of zero and still be acted upon by a force that gives it acceleration. The correlation between acceleration and force occurs in subsequent activities.

1.a)

At constant speed the dots should be equally spaced.

1.b)

At a faster constant speed, the dots should be equally spaced but further apart.

1.c)

At a slower constant speed, the dots should be equally spaced but closer together.

1.d)

When walking at a constant speed, then faster and faster, the dots at first should be equally spaced and then get progressively farther apart.

2.

Make sure that students have begun to pull the tape at a constant speed before they actually start the timer.

3.

Students will draw a line through the tape at every sixth dot.

4.

Students will number the tape sections.

5.

Students will cut the tape into sections.

6.

Students must be certain to line all the paper strips up along the x-axis. Other students in the group should make sketches of the taped graph in their own logs.

Chapter 2 Physics in Action

Imagine that the end of the tape is attached to your body. Predict what you think the distance between the dots will look like in each of the following situations. Use phrases such as close together, far apart, or evenly/not evenly spaced to describe the distances between the dots.

a) You move at a constant speed.

b) You move at a faster constant speed.

c) You move at a slower constant speed.

d) Predict how you think the distance between the dots will change if you walk at a constant speed, and then walk faster and faster.

2. Your teacher will show your group how to set up the timer. Give the end of a 2-m long piece of the tape to a group member. Let the student begin to pull the tape at a constant speed and then immediately start the timer.

3. The timer makes dots that are separated by equal amounts of time. Call the time interval from one dot to the next a "tick." (A tick is $\frac{1}{60}$ s.) Take the tape from the timer and draw lines across the tape to separate it into segments of 6-tick intervals each. (Count 6 spaces not 6 dots.)

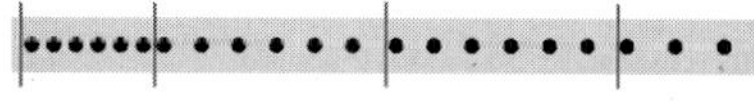

tape with six spaces between vertical lines

4. Number the segments you marked off on the tape. Start by numbering the segment closest to the end your group member held with a "1."

5. Cut the tape along the lines you drew in *Step 3* to make segments.

6. Paste the segments in order and side-by-side on another piece of paper to make a bar graph. Each segment of paper (bar on the graph) is the distance covered by the student during $\frac{1}{10}$ of a second ($6 \times \frac{1}{60}$ s).

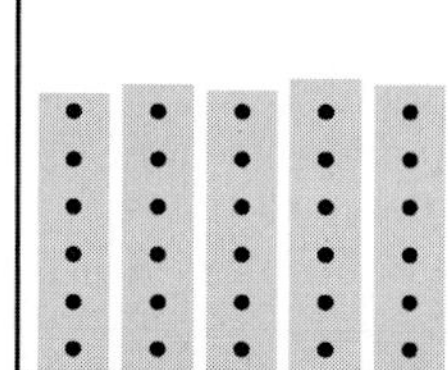

a) One student in the group should paste the piece of paper with the segments into his or her log. The other students should record sketches of the pasted segments in their logs.

7. Interpret the graph you made. The speed is the distance traveled on the tape divided by the time it took the tape to travel that distance. If the speed was constant, all the segments should be approximately equal in length.

Active Physics 146

Section 2 Constant Speed and Acceleration: Measuring Motion

a) Explain why you would expect all the segments to be about equal in length if the speed was constant.

b) Was the speed constant? How could you tell? If the speed was not constant, try again.

8. Use a new 2-m long section of tape. This time, ask the student pulling the tape to start at a slow speed and gradually increase his or her speed. Recall from the previous section, that a change in speed over a given time is called acceleration. Again, mark the tape into segments of 6-tick intervals and number the segments. Cut the segments apart and paste them in order, side-by-side, on a second sheet of paper. The length of each segment (bar on the graph) is the distance covered by the student during 1/10 of a second.

a) One student in the group should paste the piece of paper with the segments into his or her log. The other students should record sketches of the pasted segments in their logs.

b) What does the length of the paper segments (bars on the graph) tell you about the student's speed during each time interval?

c) Is there a trend in the lengths of the paper segments of your graph?

9. Remember that the student pulling the tape started at slow speed (short strip) and then gradually speeded up (increasing the length of subsequent strips). The difference in the length of each successive strip measures the change in the student's speed during that 1/10 of a second. A change in speed is called *acceleration*. Acceleration measures how much an object's speed changes in a given time interval.

a) In your log, measure the acceleration for each time interval on your graph by measuring the difference in length of each strip compared to the previous strip. Do not worry about exact time intervals yet. Use the differences in strip lengths to represent acceleration.

b) Did the student pulling the tape move with a constant acceleration? Was the change in the length of the strips constant?

10. Use another new section of tape. This time, ask the student pulling the tape to start moving at high speed and steadily slow down. Again, mark and cut the tape into 6-tick segments and make a paper-tape bar graph.

a) One student in the group should paste the piece of paper with the segments into his or her log. The other students should record sketches of the pasted segments in their logs.

b) How do you expect the pattern on the graph when decreasing speed to be different from the pattern on the graph when increasing speed?

c) Measure the acceleration for each 1/10 of a second (equal to 6 ticks) by using the difference in strip lengths to represent the change in speed. Did the student travel with a constant acceleration?

11. Compare the graphs for increasing and decreasing speed. Physicists often consider the acceleration with increasing speed positive and the acceleration with decreasing speed negative.

a) Describe the tape of a person who speeds up and then slows down using this +/- acceleration convention.

147

Active Physics

7.a)

Constant speed means covering equal distances in equal times. Because all of the strips are at a time interval of 6 "ticks" (equal times), the strips should all be equal distances for constant speed.

7.b)

If all the strips are approximately the same lengths, the speed is constant. Natural variation in walking will result in some differences.

8.a)

Students should first pull the tape at a slow speed, then gradually increase the speed.

8.b)

The length of each strip increases for the same amount of time by 6 "ticks," which means that the speed increases as the length of each strip increases.

8.c)

Each paper segment is longer than the previous one.

9.a)

Student measure acceleration by measuring the difference in the length of each strip.

9.b)

If the change in length of the strips from one section to the next was always the same (which would be very difficult to do), the acceleration would be constant.

10.a)

Students pull the tape at high speed and steadily slow down. The tape is then cut into 6-tick segments and made into a paper-tape graph.

10.b)

For decreasing speed, each tape strip of 6 "ticks" should be shorter than the previous one.

10.c)

Student calculation using the tape strips.

11.a)

When the person is increasing speed it would be a positive acceleration, and when decreasing speed the acceleration would be negative.

CHAPTER 2

Physics Talk

This *Physics Talk* explores the concepts of speed and acceleration by making a significant connection between the students' experiment in the *Investigate* and the changes in speed and acceleration. Ask students how they could tell from the ticker tape whether a person was traveling at constant speed or accelerating. Have them indicate positive and negative acceleration on the ticker tapes and write down the definitions of these terms. Students should understand the difference between average and instantaneous speed. Discuss the concepts of acceleration and deceleration. Do the sample problems on the board and invite students to ask questions on how to calculate average speed and acceleration. Reemphasize the units for each quantity.

2-2a **Blackline Master**

2-2b **Blackline Master**

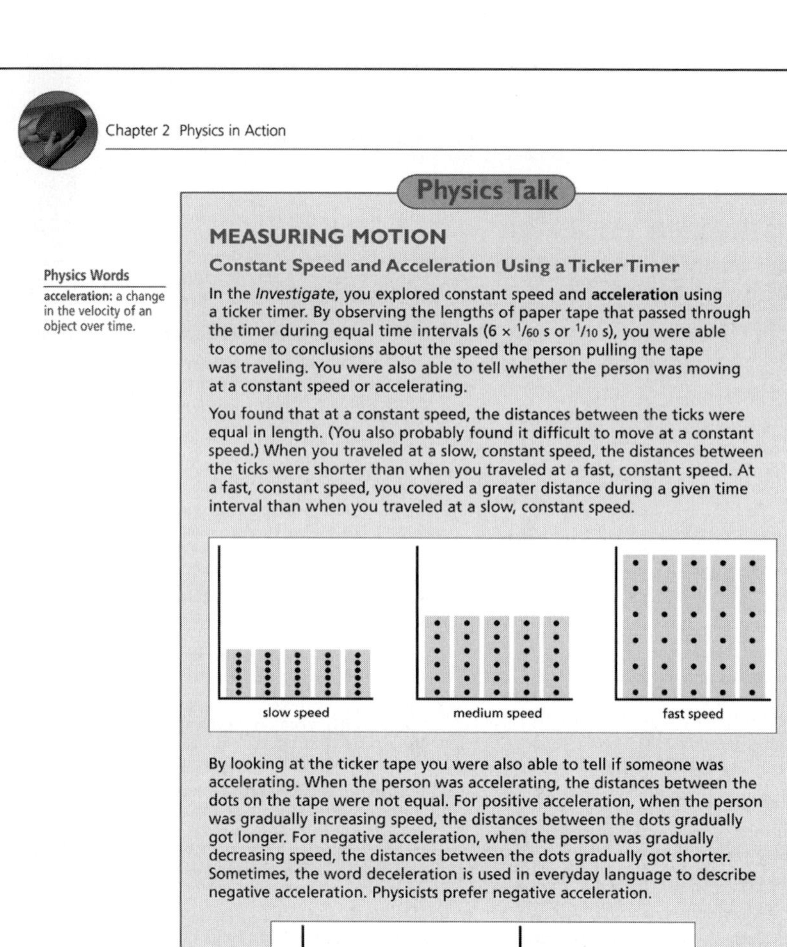

Chapter 2 Physics in Action

Physics Talk

MEASURING MOTION

Constant Speed and Acceleration Using a Ticker Timer

Physics Words

acceleration: a change in the velocity of an object over time.

In the *Investigate*, you explored constant speed and **acceleration** using a ticker timer. By observing the lengths of paper tape that passed through the timer during equal time intervals ($6 \times 1/60$ s or $1/10$ s), you were able to come to conclusions about the speed the person pulling the tape was traveling. You were also able to tell whether the person was moving at a constant speed or accelerating.

You found that at a constant speed, the distances between the ticks were equal in length. (You also probably found it difficult to move at a constant speed.) When you traveled at a slow, constant speed, the distances between the ticks were shorter than when you traveled at a fast, constant speed. At a fast, constant speed, you covered a greater distance during a given time interval than when you traveled at a slow, constant speed.

By looking at the ticker tape you were also able to tell if someone was accelerating. When the person was accelerating, the distances between the dots on the tape were not equal. For positive acceleration, when the person was gradually increasing speed, the distances between the dots gradually got longer. For negative acceleration, when the person was gradually decreasing speed, the distances between the dots gradually got shorter. Sometimes, the word deceleration is used in everyday language to describe negative acceleration. Physicists prefer negative acceleration.

148

Active Physics

Section 2 Constant Speed and Acceleration: Measuring Motion

Calculating Speed

One way to measure motion is to calculate speed. Speed is a ratio of distance traveled to time taken. The unit for speed is always written as a distance per unit of time. **Average speed** is the distance traveled divided by the time taken to travel that distance.

$$\text{Average speed} = \frac{\text{distance traveled}}{\text{time elapsed}}$$

This can be written using symbols.

$$v_{av} = \frac{\Delta d}{\Delta t}$$

where v_{av} is average speed,

Δd is change in distance or total distance,

Δt is change in time or time elapsed.

The Greek letter *delta*, "Δ," is often used in science to mean "a change in."

Sample Problem I

If you drive 140 km in 2 h (hours), calculate your average speed.

Strategy: You can use the equation for average speed.

Given:

$\Delta d = 140$ km

$\Delta t = 2$ h

Solution:

$$v_{av} = \frac{\Delta d}{\Delta t} = \frac{140\ \text{km}}{2\ \text{h}} = 70\ \text{km/h}$$

Your average speed is 70 km/h.

To travel at the average speed of 70 km/h, you might drive 70 km/h throughout the trip. But common sense tells you that your speed changes during a road trip. Before the vehicle starts moving forward, its speed is 0 km/h. At the end of the trip, you slow the vehicle to 0 km/h. And during the trip you probably slow down and speed up as you drive. The speedometer reading at any moment during the trip is your **instantaneous speed**, which is the speed at that moment.

Physics Words

average speed: the distance traveled divided by the time it took to travel that distance.

instantaneous speed: the speed measured during an instant: the speed as the time interval approaches, but does not become zero.

CHAPTER 2

Chapter 2 Physics in Action

Calculating Acceleration

When an object changes its speed, it is accelerating. Acceleration is the change of the speed divided by time. Acceleration is also a change in the direction of motion, but this will be discussed later.

To calculate acceleration in one direction caused by speeding up or slowing down, you divide the change in speed by the time interval during which the change took place.

$$\text{Acceleration} = \frac{\text{change in speed}}{\text{time interval}}$$

$$a = \frac{\Delta v}{\Delta t}$$

where a is acceleration,

Δv is change in speed,

Δt is change in time or time elapsed.

Sample Problem 2

A horse is stopped on a straight path. It begins moving forward and reaches a full gallop along the path in 10 s. The horse gallops at a speed of 20 m/s. What was the horse's acceleration?

Strategy: You can use the equation for acceleration.

Given: $\Delta v = 20$ m/s (20 m/s – 0 m/s)

$\Delta t = 10$ s (10 s – 0 s)

Solution:

$$a = \frac{\Delta v}{\Delta t} = \frac{20\ \frac{\text{m}}{\text{s}}}{10\ \text{s}} = 2\ \frac{\text{m/s}}{\text{s}}$$

The horse accelerated at a rate of two meters per second every second. That is, its speed changed at a rate of + 2 m/s for every second it was moving along the path.

Suppose the horse then came to a stop. In the language of physics, the horse is accelerating! Remember, acceleration is any change in speed or direction. The acceleration can be positive or negative.

Units for Acceleration

To calculate acceleration, you divide change in speed by change in time $\Delta v/\Delta t$. The units for acceleration are then, by definition, speed divided by time. The units for acceleration may be (m/s)/s or (km/h)/s.

When writing acceleration in meters per second per second, the final units are often simplified. For example, the following all mean the same thing.

$$1\ \frac{\frac{\text{m}}{\text{s}}}{\text{s}},\ 1\ \frac{\text{m/s}}{\text{s}},\ \text{or } 1\ (\text{m/s})/\text{s} = 1\ \frac{\text{m}}{\text{s}^2} \text{ or } 1\ \text{m/s}^2$$

The units are read as meter per second squared.

Checking Up

1. Describe the pattern of dots on a ticker tape for each of the following situations:
   a) constant speed
   b) positive acceleration
   c) negative acceleration
2. An athlete runs 400 m in 50 s. What is the runner's average speed?
3. What is the difference between instantaneous speed and average speed?
4. A vehicle accelerates from 0 km/h to 100 km/h in 10 s. What is its average acceleration?

Checking Up

1.a)

Constant speed would have a series of dots equally spaced.

1.b)

The series of dots get progressively further apart as speed gradually increases with positive acceleration.

1.c)

The series of dots come closer together as speed gradually decreases with negative acceleration.

2.

An athlete who runs 400 m in 50 s would have an average speed of

$$v_{\text{average}} = \frac{\text{distance}}{\text{time}} = \frac{400\ \text{m}}{50\ \text{s}} = 8\ \text{m/s}.$$

3.

Instantaneous speed is the measured speed at a given instant while average speed is the distance traveled divided by the time taken to travel that distance. The speedometer of a car registers the instantaneous speed of the car.

4.

The acceleration is

$$a = \frac{\Delta v}{\Delta t} = \frac{100\ \text{km/h}}{10\ \text{s}} = 10\ \frac{\text{km/h}}{\text{s}},$$ or

about 2.8 m/s^2.

Active Physics Plus

The purpose of these problems is for students to understand that although there is small change in velocity it is possible to have a large acceleration for a short interval of time. Therefore, a ball traveling at a slow speed after bouncing off a steel plate will accelerate in the opposite direction of its motion.

Consider the downward direction to be negative and the upward direction to be positive, so the steel ball moving downward would have an initial velocity of –0.5 m/s and, after the bounce when it is moving upward, its final velocity will be +0.5 m/s.

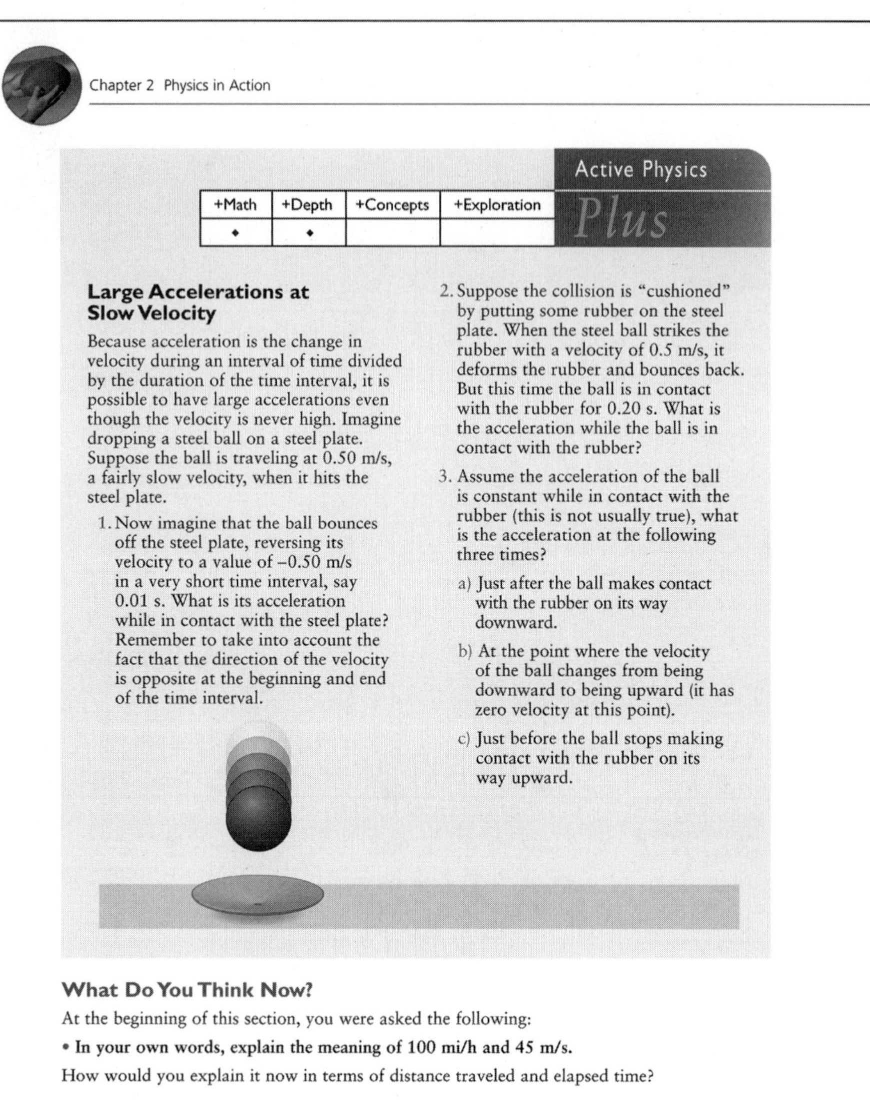

Chapter 2 Physics in Action

Active Physics **Plus**

+Math	+Depth	+Concepts	+Exploration
◆	◆		

Large Accelerations at Slow Velocity

Because acceleration is the change in velocity during an interval of time divided by the duration of the time interval, it is possible to have large accelerations even though the velocity is never high. Imagine dropping a steel ball on a steel plate. Suppose the ball is traveling at 0.50 m/s, a fairly slow velocity, when it hits the steel plate.

1. Now imagine that the ball bounces off the steel plate, reversing its velocity to a value of –0.50 m/s in a very short time interval, say 0.01 s. What is its acceleration while in contact with the steel plate? Remember to take into account the fact that the direction of the velocity is opposite at the beginning and end of the time interval.
2. Suppose the collision is "cushioned" by putting some rubber on the steel plate. When the steel ball strikes the rubber with a velocity of 0.5 m/s, it deforms the rubber and bounces back. But this time the ball is in contact with the rubber for 0.20 s. What is the acceleration while the ball is in contact with the rubber?
3. Assume the acceleration of the ball is constant while in contact with the rubber (this is not usually true), what is the acceleration at the following three times?
   a) Just after the ball makes contact with the rubber on its way downward.
   b) At the point where the velocity of the ball changes from being downward to being upward (it has zero velocity at this point).
   c) Just before the ball stops making contact with the rubber on its way upward.

What Do You Think Now?

At the beginning of this section, you were asked the following:

- **In your own words, explain the meaning of 100 mi/h and 45 m/s.**

How would you explain it now in terms of distance traveled and elapsed time?

Active Physics 152

1.

$$a = \Delta v / \Delta t = (v_f - v_i)/\Delta t =$$

$$\frac{+0.5 \text{ m/s} - (-0.5 \text{ m/s})}{0.01 \text{ s}} = 100 \text{ m/s}^2$$

upward.

2.

$$\frac{+0.5 \text{ m/s} - (-0.5 \text{ m/s})}{0.2 \text{ s}} = 5 \text{ m/s}^2$$

upward.

3.a)

When the object just strikes the rubber, its acceleration is 5 m/s^2 upward.

3.b)

At the lowest point, when it changes direction, its acceleration is still 5 m/s^2 upward.

3.c)

Just before the ball stops making contact with the rubber, its acceleration is 5 m/s^2 upward.

In each case the acceleration is positive and upward. When the ball first strikes the rubber its velocity is negative (downward), but becoming less negative so the change is positive. When the ball momentarily stops at the bottom, its velocity will start to increase upward, so it also has the same acceleration, and also just as it is leaving the rubber. A graph of the ball's velocity would look like the one shown.

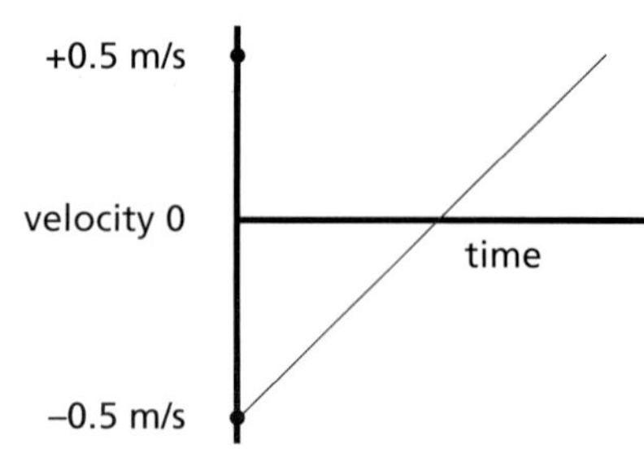

Physics Essential Questions

What does it mean?

The discipline of physics is based on observation of the physical world and one of the most important aspects of the physical world is that it changes. Motion is one of the characteristics of change in the physical world and concepts such as speed and acceleration allow you to observe and describe motion. Explain what speed and acceleration mean.

How do you know?

The concepts of physics are acceptable only if they describe the physical world well. How did you know that the speed of the person pulling the ticker tape was constant?

Why do you believe?

Connects with Other Physics Content	Fits with Big Ideas in Science	Meets Physics Requirements
Force and motion	Change and constancy	* Good, clear, explanation, no more complex than necessary

* In physics, the goal is to develop concepts that are useful for as wide a range of the physical world as possible. In this section, you found that the concepts of speed and acceleration were useful in describing your motion as you walked in one direction. Why do you believe that these same concepts are useful in describing the motion of such different objects as a baseball, a subatomic particle, or a spacecraft?

Why should you care?

If physics is to be a successful science, you must be able to transfer what is learned in one realm of the physical world to another realm. In developing your sports-video voice-over narration, you are going to shift from walking in the classroom to a sports situation. Give an example from a sport in which the concepts of speed and acceleration as developed in this section help to describe what is happening in the sport.

Reflecting on the Section and the Challenge

Newton's laws involve motion, and to measure motion, you can measure speed. In this section, you learned how to measure speed. You also learned that a change in speed with respect to time is called acceleration. You are likely to use the concepts of speed and acceleration in your sports voice-over narration. Think about what these concepts mean and imagine several sports situations where they would be crucial to a sportscaster's commentary.

Your sports segment may include a player or a ball moving at constant speed. You may wish to explain how you recognize this as constant speed. It may also include players or objects changing speeds. You can describe this change in speed as either a positive or a negative acceleration.

What Do You Think Now?

Ask students to revisit their *What Do You Think?* answers. Prompt them to change their answers or add more to what they have written in their logs already. Have a discussion in class to gauge how much students have understood. By now most students should be comfortable answering questions on speed.

Reflecting on the Section and the Challenge

This is the time for students to reflect on the concepts they have explored in this section. Suggest to them that speed, acceleration, and frames of reference are terms that they can incorporate in their voice-over narration for the *Chapter Challenge*. You can have your class brainstorm examples of sports situations where students have an opportunity to discuss their ideas.

Physics Essential Questions

What does it mean?

Speed is a change in distance during an elapsed time. Acceleration is a change in speed during an elapsed time.

How do you know?

Assuming that the dots are created at equal time intervals, we could see that a car was moving at constant speed by measuring whether an equal distance was recorded for each time interval.

Why do you believe?

Although a baseball, a particle and a spacecraft are all very different objects, the definition of speed as a change in distance during an elapsed time can apply to all of them.

Why should you care?

In the sprint, the runner starts from rest and then accelerates to her top speed. For the rest of the race, she travels at a constant speed

Physics to Go

1.

Answers will vary and may include everyday experiences, like driving in a car. Average speed can be calculated for a trip; instantaneous speed is shown on the speedometer.

2.a)

0.06 km/s

2.b)

14 m/s

2.c)

4.8 km/h

2.d)

89 km/h

3.a)

Negative acceleration is occurring as the runner falls to the ground and slows down.

3.b)

Positive acceleration is occurring as the runner gains speed.

3.c)

Acceleration is zero for constant speed in a straight line.

3.d)

Negative acceleration is occurring as the goalie slows down the soccer ball.

3.e)

Acceleration is zero for constant speed in a straight line.

3.f)

Acceleration is zero for constant speed in a straight line.

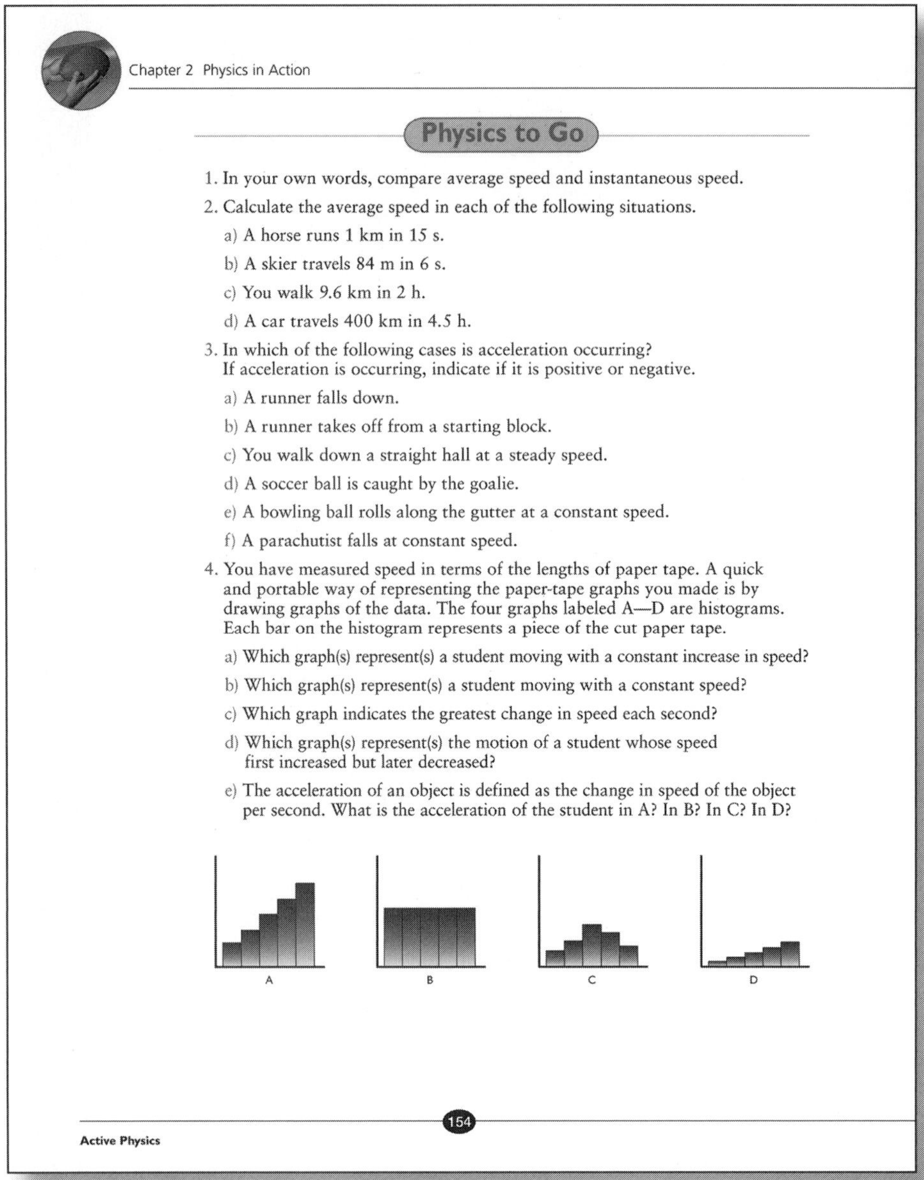

Chapter 2 Physics in Action

Physics to Go

1. In your own words, compare average speed and instantaneous speed.
2. Calculate the average speed in each of the following situations.
   a) A horse runs 1 km in 15 s.
   b) A skier travels 84 m in 6 s.
   c) You walk 9.6 km in 2 h.
   d) A car travels 400 km in 4.5 h.
3. In which of the following cases is acceleration occurring? If acceleration is occurring, indicate if it is positive or negative.
   a) A runner falls down.
   b) A runner takes off from a starting block.
   c) You walk down a straight hall at a steady speed.
   d) A soccer ball is caught by the goalie.
   e) A bowling ball rolls along the gutter at a constant speed.
   f) A parachutist falls at constant speed.
4. You have measured speed in terms of the lengths of paper tape. A quick and portable way of representing the paper-tape graphs you made is by drawing graphs of the data. The four graphs labeled A—D are histograms. Each bar on the histogram represents a piece of the cut paper tape.
   a) Which graph(s) represent(s) a student moving with a constant increase in speed?
   b) Which graph(s) represent(s) a student moving with a constant speed?
   c) Which graph indicates the greatest change in speed each second?
   d) Which graph(s) represent(s) the motion of a student whose speed first increased but later decreased?
   e) The acceleration of an object is defined as the change in speed of the object per second. What is the acceleration of the student in A? In B? In C? In D?

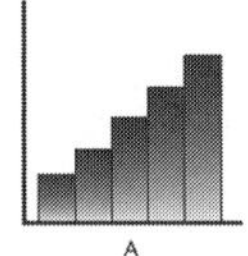

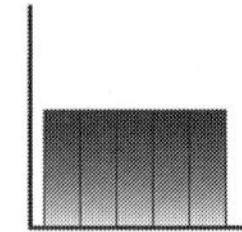

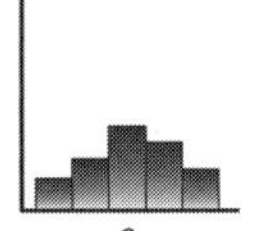

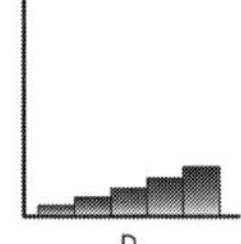

154

Active Physics

4.a)

Graphs A and D

4.b)

Graph B

4.c)

Graph A

4.d)

Graph C

4.e)

Graph A: positive acceleration; Graph B: zero acceleration; Graph C: positive, then negative acceleration.

5.

The completed table should appear as follows on the next page:

Section 2 Constant Speed and Acceleration: Measuring Motion

5. An object's motion was recorded on a ticker-timer tape. The length of each 6-tick segment of tape represents the distance traveled during that 0.1 s interval and is given in the table. Complete the table in your *Active Physics* log by calculating the average speed for each 0.1 s interval (distance traveled divided by elapsed time).

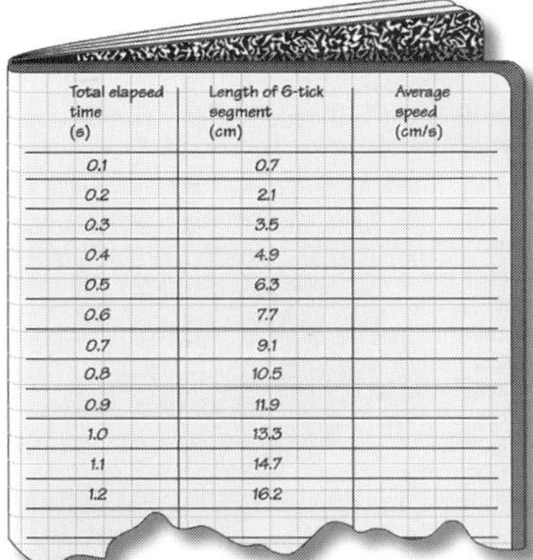

Total elapsed time (s)	Length of 6-tick segment (cm)	Average speed (cm/s)
0.1	0.7	
0.2	2.1	
0.3	3.5	
0.4	4.9	
0.5	6.3	
0.6	7.7	
0.7	9.1	
0.8	10.5	
0.9	11.9	
1.0	13.3	
1.1	14.7	
1.2	16.2	

6. A vehicle traveling at 45 km/h comes to a stop in 9 s. (Hint: When a vehicle comes to a stop, its speed is 0 km/h.)

a) How fast did the vehicle accelerate?

b) Does the acceleration have a positive value or a negative value?

7. Describe the motion of the object that made each of the following ticker tapes:

a)

b)

c)

d)

155

Active Physics

6.a)

$$a = \frac{\Delta v}{\Delta t} = \frac{-45\text{ km/h}}{9\text{ s}} = -5\ \frac{\text{km/h}}{\text{s}}$$

6.b)

The acceleration has a negative acceleration because the speed is decreasing.

7.a)

Constant speed

7.b)

Positive acceleration

7.c)

Slow constant speed, increasing to faster constant speed, slowing down to slow constant speed

7.d)

Negative acceleration to constant speed to positive acceleration

Elapsed time (s)	Length of 6-tick segment (cm)	Average speed (cm/s)
0.1	0.7	7.0
0.2	2.1	21
0.3	3.5	35
0.4	4.9	49
0.5	6.3	63
0.6	7.7	77
0.7	9.1	91
0.8	10.5	105
0.9	11.9	119
1.0	13.3	133
1.1	14.7	147
1.2	16.2	162

8.

50 mi/h

9.

No, this is only the person's average speed, he or she may have gone faster or slower for much of the time, but must have gone 15 mph for at least one instant.

10.

● ● ● ● ● ● ● ●

acceleration constant speed

4 m/s, 8 m/s, 12 m/s, 16 m/s, 20 m/s

Yes, a sprinter running 100 m in 10 s has an average speed of 10 m/s, while the bike is only 6 m/s. However, a sprinter can only run this fast for a short time, while the bicycle rider can travel at this speed for many hours.

13.

The runners in the 400-m relay do not have the advantage of being able to accelerate as quickly as the sprinter due to the starting blocks. In addition, the relay runners have to make certain that the baton is passed from runner to runner. Usually this requires the runners to slow down somewhat to insure that the baton is not dropped.

Chapter 2 Physics in Action

8. A family drives 100 mi in 2 h. What is their average speed?

9. A person drives to work at an average speed of 15 m/s. Does this mean that the person's instantaneous speed was always 15 m/s? Explain.

10. A sprinter (someone who runs short distances very fast) starts from rest and then accelerates to her top speed. If she were pulling on a ticker tape, sketch what the tape might look like.

11. A sports car accelerates at 4 m/s every second. Calculate the speed of the car after each of the first 5 s.

12. The average speed of a bicycle is 6 m/s. If a world-class sprinter can run 100 m in 10 s, can the sprinter move faster than the cyclist?

13. One track event is the 4 × 100-m relay in which each athlete runs 100 m and passes a baton (stick) to the next runner who then runs 100 m, and so on for a total of four runners. The runners receiving the baton can start moving before they receive the baton. They must have the baton in their hand when they begin their 100-m run. The average speed of the 400-m relay is less than the average speed for a 100-m sprint. In the 100-m sprint, the runner begins from a stopped position in starting blocks. Use what you know about acceleration, average speed, and running starts to explain how this is possible.

14. ***Preparing for the Chapter Challenge***
Describe a situation during a sports event that might produce ticker-tape patterns similar to the ones you produced in the *Investigate*.

a) constant motion at an average speed

b) constant motion at a fast speed

c) constant motion at a slow speed

d) positive acceleration

e) negative acceleration

Active Physics 156

Preparing for the Chapter Challenge

14.a)

A long-distance or marathon runner would have constant average velocity, as would be an ice skater in a long race.

14.b)

A football player running for a touchdown or a soccer player racing downfield

14.c)

A race walker traveling at a constant low speed, or a basketball player "walking" the ball up the court

14.d)

A sprinter starting out, or a football player running forward immediately after the ball is snapped, would have positive acceleration.

14.e)

A diver entering the water, or a race-car driver at the end of the race, would have negative acceleration.

SECTION 2 QUIZ

2-2c Blackline Master

1. What is the average velocity of an automobile that travels a distance of 30 km in 0.5 h?

a) 15 km/h
b) 45 km/h
c) 60 km/h
d) 75 km/h

2. A baseball pitcher throws a ball at 42 m/s. If the batter is 18 m from the pitcher, approximately how long does it take the ball to reach the batter?

a) 1.9 s
b) 0.86 s
c) 2.3 s
d) 0.43 s

3. An automobile that is dripping oil leaves a track on the road shown below. The automobile is moving from left to right. According to the oil drops on the pavement, the automobile was traveling

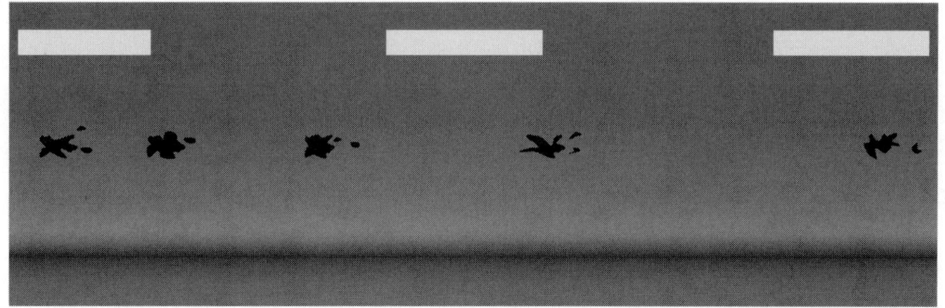

a) with constant speed.
b) with positive acceleration.
c) with negative acceleration.
d) with negative acceleration, then constant speed.

4. An automobile accelerates at 5 m/s^2. How much time is required for the automobile to reach a speed of 30 m/s?

a) 0.17 s
b) 5 s
c) 150 s
d) 6 s

5. A vehicle travels between the 100 m and 250 m highway markers in 10 s. During this time, the vehicle's average speed is

a) 10 m/s.
b) 15 m/s.
c) 25 m/s.
d) 35 m/s.

SECTION 2 QUIZ ANSWERS

1. c) The average speed is given by $v_{average}$ = distance/time = 30 km/0.5 h = 60 km/h. All other answers would be due to mistakes in math.

2. d) The average speed is given by $v_{average}$ = distance/time. Since time is the unknown, we can solve for time as time = distance/$v_{average}$ = (18 m)/(42 m/s) = 0.43 s. Other answers would reflect misapplication of the formula.

3. b) The oil pattern shows the drops are moving successively farther apart indicating positive acceleration since the vehicle started from the left. Negative acceleration would show the drops getting closer together, and for constant velocity the drops would have been equally spaced.

4. d) Using the formula $a = \Delta v/\Delta t$ gives 5 m/s^2 = (30m/s)/t. Solving for t gives t = 6 s. Other answers would reflect incorrect use of the formula.

5. b) The average speed is given by $v_{average}$ = distance/time. The distance traveled is the difference between 250 m and 100 m, which is equal to 150 m. If the students used any other distance from the question, they would have reached one of the incorrect answers.

NOTES

NOTES

SECTION 3

Newton's Second Law: Push or Pull

Section Overview

Students observe how acceleration is directly proportional to the force applied and inversely proportional to an object's mass. By applying a variable force to small and large masses, they obtain a kinesthetic feel for Newton's second law. The *Physics Talk* explains and defines Newton's second law. The use of the newton as a derived unit is emphasized by analyzing the equation for Newton's second law of motion. Students read about the importance of units and significant figures in measurements. Students explore applications of Newton's second law, particularly the force of gravity as a special case. Additionally, they explore scalar and vector quantities, and add forces acting in different directions to measure the resulting unbalanced force. Particular attention is given to the theory and application of significant figures when making experimental calculations.

Background Information

The unit of mass, or quantity of matter, in the International System of Units is the kilogram. One of seven base units from which all other units are derived, the kilogram originally was conceived as the quantity of matter represented by 1 L of water at the temperature of maximum density, 4°C. Today, the kilogram is defined by a carefully protected metal standard called the International Prototype Kilogram. When a balance which employs the force of gravity is used to measure the mass of an object by comparison to prototype masses, the resulting measurement is known as the "gravitational mass" of the object. Mass is also internationally recognized as a measure of the inertial resistance of an object to acceleration. When a standard force is used to compare an object's acceleration to the acceleration of a prototype mass as a means of measuring the mass of the object, the resulting measurement is known as the "inertial mass" of the object.

Extremely accurate measurements indicate that 1 kg of gravitationally determined mass is equivalent to 1 kg of inertial mass. A derived unit of force, the newton, is defined in terms of base units of mass, length, and time using Newton's second law of motion, $F = ma$. One newton is the force which will cause one kilogram to accelerate at one meter per second squared, or $1\ \text{N} = 1\ \text{kg} \cdot \text{m/s}^2$. The word "weight" denotes a force; the weight of an object is the product of its mass and a proportionality constant for objects on the surface of the Earth, 9.81 N/kg. Because weight is the force due to gravity, weight is measured in newtons. 1 N is roughly 1/4 lb, prompting the identification of the familiar 1/4 lb-burger as a "newton burger."

In summary, matter seems to have two distinct properties: it exhibits a resistance to acceleration (a property called "inertia") and it has the property of gravitation, which means that matter is attracted to other matter. All objects, no matter what their mass, have the same free-fall acceleration at a given location. The more mass, the more gravitational force; but the more mass, the more difficult it is to accelerate the object. These two factors exactly compensate to produce the same acceleration for every freely falling object at a given location. This acceleration is equal to the proportionality constant, 9.81 N/kg or 9.81 m/s^2, which is why this constant is often called "the acceleration due to gravity." By emphasizing that weight is equal to the mass times the proper proportionality constant, it will be easier for students to understand that the value of the proportionality constant depends on, for example, whether you are on Earth or the Moon. The acceleration of free-falling objects therefore also depends on whether you are on Earth or the Moon. Students who simply think of 9.81 m/s^2 as the acceleration due to gravity have little basis for understanding why this number is different on Earth and the Moon, because in their minds gravity is involved in both cases.

Crucial Physics

- Forces are "pushes" or "pulls" on an object. Forces on an object add as vectors and it is the total force or net force that determines the acceleration of the object.
- The total or net force on an object is related to its mass and acceleration by Newton's second law, $F = ma$.
- Weight is the force due to gravity on an object and is proportional to the object's mass. On the surface of Earth, the proportionality constant $g = 9.81$ N/kg, so $W = mg$.
- By Newton's second law, the acceleration of an object when the only force acting on it is gravity is $g = 9.81$ N/kg $= 9.81$ m/s^2.
- ***Active Physics Plus:*** Forces add as vectors.

Learning Outcomes	Location in the Section	Evidence of Understanding
Identify the forces on an object.	***Investigate, Physics Talk*** Steps 6-9	Students analyze the *Investigate* to identify forces on an object and learn the definition of a force.
Determine when the forces on an object are either balanced or unbalanced.	***Investigate, Physics Talk*** Steps 7-9	Students keep adding coins to the end of a ruler and learn how an unbalanced force causes the ruler to bend and the coins to drop.
Compare amounts of acceleration semi-quantitatively.	***Physics Talk, Physics to Go***	Students solve problems to compare how the amount of force applied affects acceleration.
Apply the definition of the newton as a unit of force.	***Physics Talk, Physics to Go*** Questions 3, 4, and 10-17	Students write the equivalent form of a newton and solve problems of acceleration and force using the newton as a unit of force.
Describe weight as the force due to gravity on an object.	***Physics Talk***	Students recall the experiment of the ruler and coins in the *Investigate* to describe weight.

Section 3 Materials, Preparation, and Safety

Materials and Equipment

PLAN A		
Materials and Equipment	**Group (4 students)**	**Class**
Ruler, plastic, flexible, 30 cm	1 per group	
Cart, dynamics	1 per group	
Weights, slotted, set	1 per group	
C-clamp, steel, 3 in.	1 per group	
Tape, masking, 3/4 in. x 60 yds		6 per class
Access to a flat surface (such as a table, floor or other open space)*	1 per group	
Can or bottle (with less mass then the dynamics cart)*	1 per group	
Coins (pennies, nickels)*	5 per group	

*Additional items needed not supplied

PLAN B		
Materials and Equipment	**Group (4 students)**	**Class**
Ruler, plastic, flexible, 30 cm	1 per group	
Cart, dynamics	1 per group	
Weights, slotted, set	1 per group	
C-clamp, steel, 3 in.	1 per group	
Tape, masking, 3/4 in. x 60 yds		6 per class
Access to a flat surface (such as a table, floor or other open space)*	1 per group	
Can or bottle (with less mass then the dynamics cart)*	1 per group	
Coins (pennies, nickels)*	5 per group	

*Additional items needed not supplied

Note: Time, Preparation, and Safety requirements are based on Plan A, if using Plan B, please adjust accordingly.

Time Requirement

To complete this requirement, one class period or 40 minutes are required.

Teacher Preparation

For best results, this *Investigate* should be done on a clear area of open floor.

Try to push the cart or can yourself with the ruler to maintain a constant force (bend of the ruler) to get a feel for how difficult this will be for the students. The students will often just give the cart a push to get it going over a short distance rather than chasing after the cart to maintain the force.

The students should discover that they must run faster and faster to keep up with the cart and maintain the same bend of the ruler. The chase is very important because it best demonstrates the acceleration that is taking place. If a large frictional force is present that causes the carts to travel with constant speed with the ruler bent, have the students use the bend of a large ruler until the cart or can accelerates.

If you are using carts, it is helpful if the cart has some additional mass on it for the first trial. When a "smaller" mass is required, mass may be removed from the cart, and later may be added to increase the cart's mass.

Safety Requirements

Make certain that the area where the students will be pushing the carts is free from any obstructions.

Lab partners should watch for any potential problems where the students are pushing the carts so that the students my concentrate on keeping the ruler bent as they chase the cart.

Students should pick up the carts or cans from the floor and place them in a safe location immediately after finishing pushing to prevent anyone from slipping on them.

Meeting the Needs of All Students

Differentiated Instruction: Augmentation and Accommodations

Learning Issue	Reference	Augmentation and Accommodations
Describing motion qualitatively	***Investigate*** Steps 2-5	**Augmentation** • Assist students in brainstorming a list of words that could be used to describe motion (fast, slow, moderate speed, speeding up, slowing down, forward, etc.).
Creating a data table	***Investigate*** Step 5	**Augmentation** • Some students may struggle to create a table that includes all of the required information without seeing a sample. Tell students to decide in groups what kind of information needs to be recorded (object's name, relative mass, and motion). Then ask them how many columns and rows are needed. • Tell students to draw two vertical lines to divide their page into three equal sections. Remind them to record descriptions in the appropriate mass and motion columns, and ask them to make each row a few lines in height. • Pair the oral directions with a visual model of the data table. **Accommodation** • Give students a blank table to tape into their logs and complete.
Understanding qualitative vs. quantitative data	***Investigate***	**Augmentation** • Ask students if they know the meaning of *quantity*. Make an explicit connection between the words *quantity* and *quantitative*. Then ask students to help you create a list of quantities students can measure. Repeat this activity with *quality* and *qualitative*.
Understanding concepts of direct and inverse proportion	***Physics Talk*** ***Physics Essential Questions***	**Augmentation** • Students with reading-comprehension issues may not understand the written explanation. Show the students the relationship in a visual model while explaining the concept orally. Give clear and explicit explanations as in the examples below. • As force increases ($F \Uparrow$), what happens to the acceleration? Acceleration increases ($a \Uparrow$). This is a direct proportion. • This relationship can also be represented visually by increasing the font size of the variables. ($F \Rightarrow F$) causes ($a \Rightarrow a$) • As mass increases ($m \Uparrow$), what happens to acceleration? Acceleration decreases ($a \Downarrow$). This is an inverse proportion. • Provide direct instruction to draw the connection between these relationships to explain $a = F/m$.
Using the helpful circle	***Physics Talk***	**Augmentation** • Students may have forgotten how to use the helpful circle introduced in *Chapter 1*. Provide direct instruction and opportunities for guided practice to use/review the helpful circle.
Understanding key concepts	***Physics Talk***	**Augmentation** • Students with reading-comprehension issues may pass over the key concepts in this section. Students need to know that there are many kinds of forces and that unbalanced forces cause acceleration. • Ask students to brainstorm a list of forces with a partner. Then combine these lists into a class list. This *Investigate* will activate prior knowledge and also show the teacher if there are any misconceptions about kinds of forces. • Ask students if they can think of an example in which unbalanced forces cause acceleration.

Learning Issue	Reference	Augmentation and Accommodations
Understanding the difference between mass and weight	***Physics Talk*** ***Inquiring Further***	**Augmentation** • The average person in society uses mass and weight interchangeably. These concepts can be very confusing. Ask students if they know the difference between mass and weight. Use adequate wait time to give everyone time to think about the answer. • Provide direct instruction to explain the difference. Give an example of the weight of a 1-kg mass on Earth versus the weight of a 1-kg mass on the Moon.
Solving word problems	***Physics to Go*** Questions 3-7, 15	**Augmentation** • Students with reading-comprehension, sequential, and executive-function issues may struggle to extract information from a word problem and follow the steps necessary to solve the problem. • Provide direct instruction to teach key words found in word problems, such as what is, how much, calculate, solve for, etc. • Model a think-aloud to make explicit the mental steps a good problem-solver takes to solve a problem. Pair the think-aloud with visual cues that show each step. **Accommodation** • Provide students with a sheet of blank problem-solving boxes.
Solving word problems that require unit conversions	***Physics Talk*** ***Physics to Go*** Question 14	**Augmentation** • Students with sequential-learning issues may need to see *Sample Problem 2* done as a two-step problem in which the conversion from grams to kilograms is done separately.
Understanding significant figures Using measurement in calculations	***Physics Talk***	**Augmentation** • Students with sequential-learning, executive-function, and attention issues may struggle to understand significant figures because they must follow a list of very specific rules. • Provide direct instruction to teach the rules for determining the number of significant figures. Keep a list of these rules posted in the classroom and provide students with a copy of these rules. • Provide opportunities for guided practice, during which students can receive feedback and ask questions.
Drawing free-body diagrams and finding resultant vectors	***Active Physics Plus*** ***Physics to Go*** Questions 10-13, 16-17	**Augmentation** • This is an opportunity to use differentiated instruction. Provide direct instruction to teach the explicit rules for drawing simple free-body diagrams that only include collinear forces. If students are able to find resultant vectors for collinear forces, introduce finding resultant vectors for forces using the Pythagorean theorem and then the tip-to-tail approach.
Copying tables	***Physics to Go,*** Question 1	**Accommodation** • Students with more significant graphomotor and visual-spatial issues struggle to copy written information accurately. Provide a copy of the table as it appears in the textbook. Then students can tape the table into their notebooks and complete the calculations.

Strategies for Students with Limited English-Language Proficiency

Learning Issue	Reference	Augmentation
Following complex procedures	***Investigate***	Break down the *Investigate* into smaller chunks that allow students to comprehend each portion of the experiment before moving on to the next one. This approach will allow students to get comfortable following the procedures outlined within each step, and also to internalize the new concepts that are introduced. Lead a brief class discussion after each step to allow students the opportunity to demonstrate acquired knowledge and understanding.
Understanding concepts	***Investigate*** Qualitative and Quantitative Observations	It is vital in science that students understand the difference between the similar words "qualitative" and "quantitative." Tell students that qualitative observations discuss the quality of an object by using adjectives to describe the object: green, rough, loud. Quantitative observations involve quantities, or amounts, and often (but not always) are represented with numbers: 89 kg, 47°C, 5.3 m/s.
Vocabulary comprehension	***Physics Talk***	To help all students, especially kinesthetic learners, understand that a force is a push or a pull, and to help ELL students learn the meanings of "push" and "pull," demonstrate the movements of push and pull with your hands and then have students make the movements with you.
Understanding concepts Vocabulary comprehension	***Physics Talk***	All students may need guidance when trying to understand Newton's second law in his own words. To explain the meaning of "impressed" in this context, say: "The root of 'impressed' is press. Pressing on an object is like pushing on the object. When you push on an object, you apply a force to the object. So the force impressed means the force applied." Also, explain that in Newton's time, "right line" meant "straight line." Right after working through Newton's words may be a good time to have students write Newton's second law in their *Active Physics* logs in their own words. You can read the logs to check understanding. Be sure students grasp "unbalanced force" and "inversely proportional."
Understanding concepts	***Physics Talk*** (Gravity, Mass, Weight, and Newton's Second Law)	Students learn that the force of gravity is applied downward on all objects. Be sure students do not think that gravity pushes on objects from above. Gravity is a pulling force from below; Earth pulls objects toward it. Note that the term "on" or "downward on" may confuse ELL students because "on" has a connotation of "on top of" or "from above." In physics, we say "the force on" to mean "the force applied to," regardless of direction. Explain that we can talk about the force of gravity "on" a cinder block, and we also talk about the force the cinder block exerts "on" something under it. Both forces are in the same direction—toward Earth—but you can feel the weight of the block as a push or a pull, depending on where you stand. If you hold a cinder block over your head, you feel it push down on your arms. (Your arms are between the block and Earth.) But if you pick the cinder block up from the ground, it pulls down on your arms. (The block is between your arms and Earth.)
Cooperative learning	***Active Physics Plus*** (Balanced and Unbalanced Forces)	Draw a series of free-body diagrams on the board. Have students work in groups to identify whether the forces in each diagram are balanced or whether there is a net force. If there is a net force, students need to indicate the direction of the net force.
Reading comprehension	***Active Physics Plus***	Collaborate with the students' math teachers to determine what level of comprehension students have obtained for right triangles, the Pythagorean theorem, and tangents.

CHAPTER 2

SECTION 3

Teaching Suggestions and Sample Answers

What Do You See?

Here is yet another skilled representation of how the net force on an object is directly proportional to its acceleration. The three contrasting visuals of the girl pushing the ball with a stick are meant to evoke a response to the different predicaments faced by the girl. Consider asking your students why the girl appears to be so relaxed at first, then puzzled, and eventually so exhausted. Write down responses on the board and highlight key words and phrases that will be used later in the section to develop concepts. Ask questions that initiate a discussion on force and acceleration. This is the time that students might reveal prior misconceptions. Each idea that students present is significant. Keep encouraging them as they discuss the illustration. Emphasize that they will be returning to the *What Do You See?* illustration to discuss how earlier responses get altered with a better understanding of the artist's purpose in relating the physics of a section.

What Do You Think?

The *What Do You Think?* questions are designed to stimulate thinking about the section. Have students share their answers and accept all answers without correction. Encourage them to think of a time when they went bowling or played tennis. Prompt them to think of how a tennis ball is different from a bowling ball in terms of weight. Consider asking them if they can think of how weight and applied force are related. Remind them that they will be returning to these questions later in the section after they have explored the concepts of force and acceleration in relation to weight.

What Do You Think?

A Physicist's Response

In simple terms, a force is a push or a pull. Some forces, such as gravitational and magnetic forces, can act on objects without having to be in contact with them. Many other forces, called mechanical forces, act when particles or objects touch each other. Forces are very important in physics because they determine how matter interacts with other matter.

The same force could be used to move both a bowling ball and a soccer ball, or even a table-tennis ball, as long as there is not much friction or other counteracting forces to interfere. The difference would be the amount by which each is accelerated by the force. The greater the mass, the lesser the acceleration experienced by an object when the same force is applied to it. Mass affects acceleration.

Students' Prior Conceptions

Students recognize that a force is a push or a pull. Forces have both magnitude and direction. When the forces acting on an object are unbalanced in one direction the object moves, either with acceleration or with steady motion, if the force applied equals the force of static friction. Students should grasp the meaning of Newton's second law to establish their foundation for the application of the laws of motion to all subsequent sections in the chapter. Some preconceptions are listed below:

1. **If an object is at rest, no forces are acting on the object.** It is necessary for students to identify all forces acting on an object so that they recognize balanced and unbalanced forces. Often, students overlook the forces acting in the vertical direction, for example, gravity pulling down on an object while the floor, chair, or table push up on the same object. Students might only consider the forces acting in the horizontal direction to affect motion in that direction.
2. **Only animate objects can exert a force.** Students have difficulty recognizing that all interactions involve equal forces acting in opposite directions on the separate interacting bodies. Thus, if an object is at rest on a table, students often say that there are no forces acting upon the object. Students need to identify all of the forces acting on an object, name the forces, discuss their directions, and identify their magnitudes, even if only relative to each other. Recognizing the nature of balanced and unbalanced forces acting on objects is paramount to learning how to apply Newton's second law to the interactions and motions of objects.

Section 3 Newton's Second Law: Push or Pull

Section 3 Newton's Second Law: Push or Pull

What Do You See?

Learning Outcomes

In this section, you will

- **Identify** the forces acting on an object.
- **Determine** when the forces on an object are either balanced or unbalanced.
- **Compare** amounts of acceleration semi-quantitatively.
- **Apply** Newton's second law of motion.
- **Apply** the definition of the newton as a unit of force.
- **Describe** weight as the force due to gravity on an object.

What Do You Think?

Venus Williams is a record holder for one of the fastest serves in the world by a female tennis player. The speed of the serve was almost 208 km/h (129 mi/h). To serve a tennis ball at that speed requires skill, timing, and force.

- **What is a force?**
- **How will the same amount of force affect a tennis ball and a bowling ball differently?**

Record your ideas about these questions in your *Active Physics* log. Be prepared to discuss your responses with your small group and the class.

Investigate

In this *Investigate*, you will use a flexible ruler to continuously push a cart (or a can or plastic bottle) across a table, floor, or open space.

1. Hold one end of the ruler against the table. Push on the other end of the ruler with your finger. Notice that a small force produces a small bend in the ruler and that a large force produces a large bend. You have created a force meter (an instrument that you can use to measure force).

157 Active Physics

Investigate

1.

Explain to the students that the ruler is behaving much like a spring. The bend of the ruler is indicative of the amount of force being applied in the same manner as the stretch of a spring indicates the amount of force. Although this force meter is not calibrated, the students will only be using it to obtain qualitative data on acceleration under an applied force.

CHAPTER 2

3. **Force is a property of an object rather than a relation between objects.** Students may believe that an object has force and when the force runs out the object stops moving. Focus student attention on the concept of inertial mass to recognize that matter moves or stays at rest depending upon the nature of the balanced or unbalanced forces acting upon the object with that amount of matter. More matter requires a larger force to give the same type of motion.

4. **Large objects exert a greater force than small objects.** Students need to understand that force is defined as the product of a mass with its acceleration. A large mass can have a small acceleration and a small force whereas a small mass can have a large acceleration and therefore a large force. The equal nature of pairs of forces in interactions is considered in subsequent activities.

5. **Gravity is a property of an object rather than a force experienced by the object; weight is not a force–rather, the air exerts the force; gravity requires a medium to act through; weight is something that can be "felt"; if an object has no weight then it cannot be "felt"; and mass may not be differentiated from weight.** It is important for students to explore the relationship between mass and weight to recognize that weight is the product of the mass of an object and the acceleration due to the pull of Earth or another larger object or celestial body on an object.

These prior conceptions emerge in subsequent activities, too, as students apply the laws of motion to falling objects.

2.

The students will have to accelerate as they chase the accelerating cart to maintain a constant bend in the ruler (a constant accelerating force).

2.a)

The students should indicate that the cart continually increases in speed as the force is applied. To make the acceleration more obvious to the students, have them apply the force to the cart while it is on the floor, so that the force can be maintained for a longer distance than a lab table, and the students will have to chase after the cart.

Teaching Tip

Try out the acceleration yourself first to determine if the students will have difficulty keeping up with the accelerating cart. If you determine that your carts accelerate too quickly for the students to effectively maintain a constant force, add mass to the cart to slow down the acceleration in these initial steps. Mass will also be added later to show the effect of increasing mass. If masses are added, they can be removed for *Step 4* to simplify that step.

2. Use the ruler to push the cart continuously with only a slight bend in the ruler (a small force) as shown above. Make sure you do not push the cart in spurts. The push (force) must be applied as a continuous motion so the ruler keeps the same amount of bend. You will need to keep up with the cart as it moves and to keep the same amount of bend in the ruler. It may be useful to have another member of your group watch to make sure the ruler keeps the same amount of bend throughout the duration of the push. You may need to practice a few times to be able to do this.

a) Describe the motion of the cart.

3. This time, push the cart continuously with a large amount of bend in the ruler (a large force) as shown above.

Remember, you need to keep up with the cart as you continually push it to keep the large amount of bend in the ruler.

a) Describe the motion of the cart.

b) How was the motion of the object similar with a push from a ruler with a small bend and from a ruler with a large bend?

c) How was the motion of the object different with a push from a ruler with a small bend and from a ruler with a large bend?

d) Remember that acceleration is a measure of the change in speed with respect to time. Write a statement that describes the relationship between the force applied to the cart and the resulting acceleration of the cart.

Begin your statement with: "The greater the constant force pushing on an object, the..."

4. Select an object that has a smaller mass than the cart, can, or bottle. Use the ruler to push the object with a large, steady force (a large bend in the ruler).

a) Record a description of the object (especially its mass) and the motion of the object.

5. Now use the same large amount of force to push objects of greater and greater mass.

a) Record the results for each object in a data table in your log.

b) Complete the statement below that describes the relationship of the mass of an object and its acceleration and write the entire completed statement in your log: "When equal amounts of a constant force are used to push objects having different masses, the more massive object..."

The students should find they have difficulty keeping up with the cart, trying to maintain the constant larger bend in the ruler, as it accelerates at a greater rate.

3.a)

The cart accelerated in both trials.

3.b)

The students should note that the cart accelerates at a greater rate with a larger bend in the ruler (as it produces a larger force) than a smaller bend that produces a smaller force.

3.c)

The acceleration was greater in the second trial.

3.d)

"The greater the constant force pushing on an object, the greater the object's acceleration."

If the students used carts with masses for *Step 2*, they could remove the masses for this step. Otherwise use a smaller mass cart or object.

4.a)

The students should note that the large force on the smaller mass will give an acceleration greater than the previous trial with a large force.

5.a)

The students will see that as the mass becomes larger using the same force, the acceleration of the object will decrease.

6. You conducted two different experiments. You first varied the amount of force on a single object. You then used the same force to push on objects of different mass. By conducting two different experiments, you were able to analyze the effects of changing either the mass or the force.

   a) If you had conducted only one experiment in which you pushed on a large object with a small force and then pushed on a small object with a large force, what conclusions would you have drawn?

7. You noticed earlier that a small bend of the ruler corresponded to a small force and a large bend to a large force.

   You can now check this relationship more precisely. Carefully clamp the flexible ruler to the end of a table.

8. Place one coin on the top surface of the ruler near the outside end. Observe what happens to the ruler.

   a) Record your observations.

9. Repeat by placing two, three, and four coins on the ruler.

   a) What happens to the ruler each time you add a coin?

   b) How many pennies represent a small force? How many pennies represent a large force?

   c) What force is causing the ruler to bend?

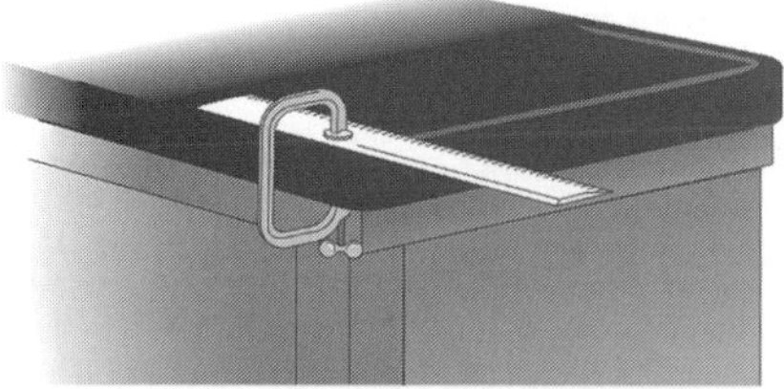

Qualitative and Quantitative Observations

An observation is information that you get through your senses. When you describe the qualities of objects, events, or processes, the observations are qualitative. If you say that something smells spicy, tastes sweet, or feels sticky, you are making qualitative observations.

Observations that are based on measurements or counting are quantitative, because they deal with quantities. The temperature of a sauce cooking on the stove is a quantitative measurement.

In this *Investigate*, you made semi-quantitative observations. The first measurements that you made of the bend of the ruler were small and large. These are semi-quantitative observations. You then calibrated the bend and compared the bend to the number of pennies required to make the ruler bend. The number of pennies that correspond to a small force and a large force are quantitative measurements.

5.b)

"When equal amounts of a constant force are used to push objects having different masses, the more massive object the less the acceleration of that object."

6.

The results of the student's experiments are the essence of Newton's second law.

6.a)

Pushing on a large object with a small force would have produced a small acceleration, and pushing on a small object with a large force would have produced a very large acceleration. The results of these two experiments might have led the students to the correct conclusion about the relationship between force and acceleration. However, they might equally conclude that large forces always mean large accelerations, and small forces always mean small accelerations, regardless of mass. If the experiment had been a large force on a large mass and a small force on a small mass, the conclusion would have been that all masses accelerate at the same rate!

7.

The ruler should be clamped with most of the ruler hanging over the edge of the table.

The coin may need a small piece of tape on the side touching the ruler to keep it from sliding off the ruler.

8.a)

The ruler bends under the weight of the coin.

9.a)

As more coins are added, the bend of the ruler increases.

9.b)

One or two coins would represent a small force, while four coins represent a large force.

9.c)

In this case it is the weight of the coins causing the bend, rather than the force applied by the hand. You might want to point out to the students that equal bends in the ruler imply that equal forces are needed.

Physics Talk

This *Physics Talk* explores Newton's second law of motion qualitatively as well as quantitatively. It would be useful to the students to make connections with the different steps in their *Investigate* and see how each step can be explained through their reading of the *Physics Talk*. To check for student understanding, have students summarize the findings of their *Investigate* as evidence for Newton's second law. Emphasize that an unbalanced force is needed for an object to accelerate, and that although a force may be acting, if it is not unbalanced, the object will not accelerate. In addition, point out that the force of gravity or weight is really only a special case of Newton's second law, where it is the ratio of force to mass that gives the constant acceleration we call "*g*." As students learn about force, make sure they also understand the meaning of derived units. You might want to ask them where they have used these before. You could also ask them to write and discuss the units of force and weight and illustrate Newton's second law with diagrams. While students are working through the sample problems, have them discuss the solution of each problem with their peers. To verify whether students are comfortable determining significant figures in a measurement, ask them to give a few examples of significant figures in their logs. Doing several sample calculations using measurements with different numbers of significant figures will improve their understanding of the role of significant figures in science.

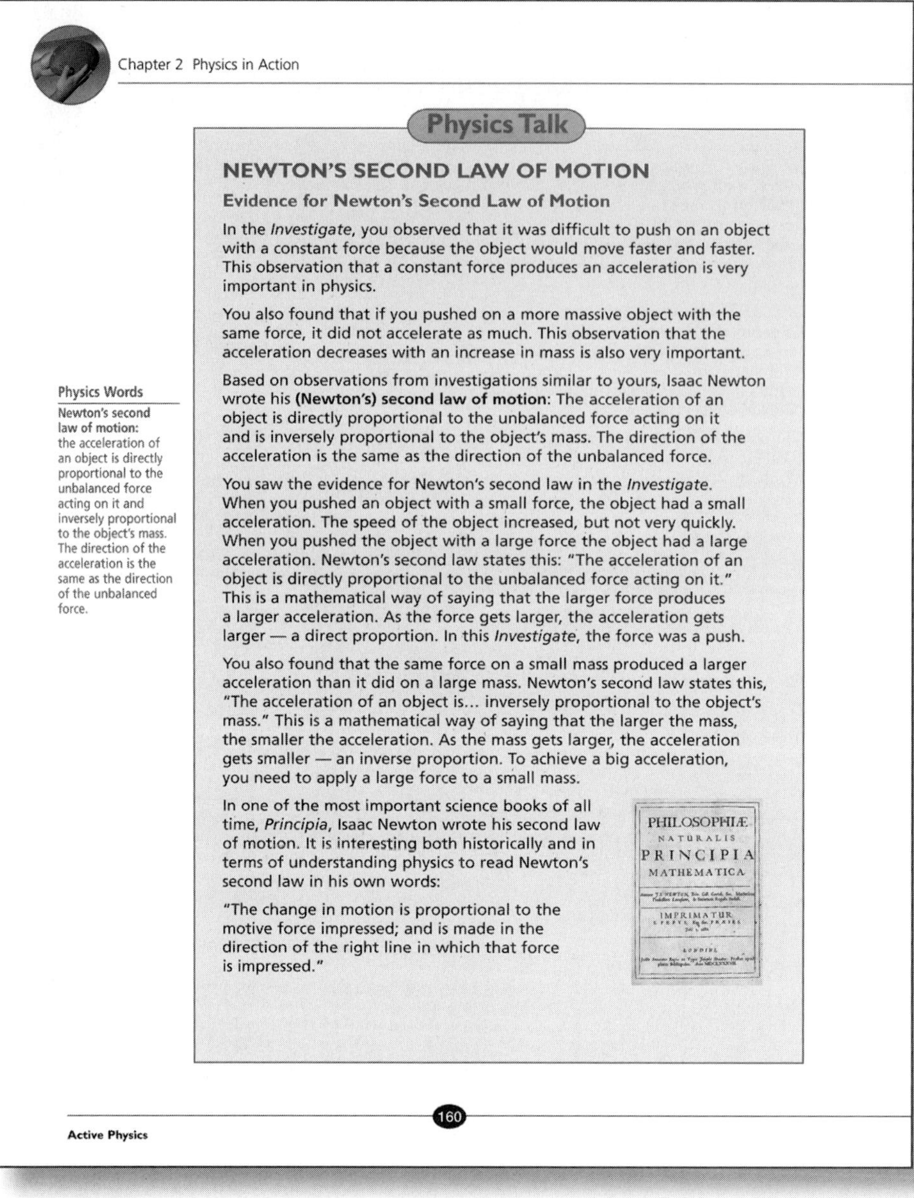
Chapter 2 Physics in Action

Physics Talk

NEWTON'S SECOND LAW OF MOTION

Evidence for Newton's Second Law of Motion

In the *Investigate*, you observed that it was difficult to push on an object with a constant force because the object would move faster and faster. This observation that a constant force produces an acceleration is very important in physics.

You also found that if you pushed on a more massive object with the same force, it did not accelerate as much. This observation that the acceleration decreases with an increase in mass is also very important.

Based on observations from investigations similar to yours, Isaac Newton wrote his **(Newton's) second law of motion**: The acceleration of an object is directly proportional to the unbalanced force acting on it and is inversely proportional to the object's mass. The direction of the acceleration is the same as the direction of the unbalanced force.

Physics Words

Newton's second law of motion: the acceleration of an object is directly proportional to the unbalanced force acting on it and inversely proportional to the object's mass. The direction of the acceleration is the same as the direction of the unbalanced force.

You saw the evidence for Newton's second law in the *Investigate*. When you pushed an object with a small force, the object had a small acceleration. The speed of the object increased, but not very quickly. When you pushed the object with a large force the object had a large acceleration. Newton's second law states this: "The acceleration of an object is directly proportional to the unbalanced force acting on it." This is a mathematical way of saying that the larger force produces a larger acceleration. As the force gets larger, the acceleration gets larger — a direct proportion. In this *Investigate*, the force was a push.

You also found that the same force on a small mass produced a larger acceleration than it did on a large mass. Newton's second law states this, "The acceleration of an object is... inversely proportional to the object's mass." This is a mathematical way of saying that the larger the mass, the smaller the acceleration. As the mass gets larger, the acceleration gets smaller — an inverse proportion. To achieve a big acceleration, you need to apply a large force to a small mass.

In one of the most important science books of all time, *Principia*, Isaac Newton wrote his second law of motion. It is interesting both historically and in terms of understanding physics to read Newton's second law in his own words:

"The change in motion is proportional to the motive force impressed; and is made in the direction of the right line in which that force is impressed."

Active Physics 160

An Equation for Newton's Second Law of Motion

Newton's second law can be written as an equation:

$$\text{Acceleration} = \frac{\text{force}}{\text{mass}}$$

$$a = \frac{F}{m}$$

where a is acceleration expressed in meters per second squared (m/s²),

F is force expressed in newtons (N), and

m is mass expressed in kilograms (kg).

With a bit of algebra, Newton's second law can be arranged so that it is easier to find the unknown quantity of F, m, or a.

$$a = \frac{F}{m} \qquad F = ma \qquad m = \frac{F}{a}$$

Some students like to use a helpful circle:

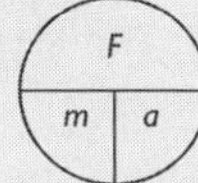

If you want to find:

- force, F, cover it up and you see m next to a (or $F = m \times a$).
- acceleration, a, cover it up and you see F over m (or $a = F \div m$).
- mass, m, cover it up and you see F over a (or $m = F \div a$).

Newton: A Derived SI Unit with a Special Name

When you measured speed and acceleration, you used derived units with compound names. You measured speed in meters per second (symbol, m/s) and acceleration in meters per second per second, or meters per second squared (symbol, (m/s)/s or m/s²). These are derived units. They are made up of one or more base SI units.

In the equation for Newton's second law, force is expressed in newtons (symbol, N). What is a newton? A newton is a derived SI unit with a special name. A newton is the force required to make one kilogram of mass accelerate at one meter per second squared.

CHAPTER 2

With this definition, the unit newton can be written in its equivalent form: $1 \text{ kg} \cdot \text{m/s}^2$.

$$1 \text{ N} = 1 \text{ kg} \cdot \text{m/s}^2$$

Knowing the equivalent form for a newton will be important when using Newton's second law to do calculations to find mass and acceleration. Mathematically, the dot represents multiplication.

$$1 \text{ kg} \cdot \text{m/s}^2 \text{ means } 1 \text{ kg} \times \frac{\text{m}}{\text{s}^2} \text{ or } 1 \text{ kg} \times \frac{\text{m}}{\text{s} \times \text{s}}$$

Where There's Acceleration, There Must Be an Unbalanced Force

There are lots of different everyday forces. There is the force of the bent ruler in this investigation. There is also the force of a spring, the force of a rubber band, the force of a magnet, the force of your hand, the force of a bat hitting a ball, the force of friction, the buoyant force of water, and many more. Newton's second law tells you that accelerations are caused by unbalanced forces. It does not matter what kind of force it is or how it originates. If you observe an acceleration (a change in velocity), then there must be an unbalanced force causing the acceleration. When you apply a force to an object that has a small mass, the acceleration may be quite large. If the object has a large mass, the acceleration will be smaller for the same force. Occasionally, the mass is so large that you cannot measure the acceleration because it is so small.

If you push on a small cart with the largest force you can, the cart will accelerate a great deal. If you push on a car with that same force, the acceleration will be much smaller. If you were to push on a truck, the acceleration would be too small to measure. Can you convince someone that a push on a truck accelerates the truck? Why should you believe something that you cannot measure? If you were to assume that the truck does not accelerate when you push on it, then you would have to believe that Newton's second law stops working when the mass gets too big. If that were so, you would want to determine how big is "too big." When you conduct such experiments, you find that the acceleration gets less and less as the mass gets larger and larger. Eventually, the acceleration gets so small that it is difficult to measure. Your inability to measure it does not mean that it is zero. It just means that it is smaller than your best measurement. In this way, you can assume that Newton's second law is always valid.

Section 3 Newton's Second Law: Push or Pull

Calculations Using Newton's Second Law of Motion

Since Newton's second law relates force, mass, and acceleration, you can use the equations for Newton's second law to solve a variety of problems.

Sample Problem I

As the result of a serve, a tennis ball ($m_t = 58$ g) accelerates at 430 m/s² for the very brief time it is in contact with the racket.

a) What force is responsible for this acceleration?

b) Could an identical force accelerate a 5.0-kg bowling ball at the same rate?

Strategy: Newton's second law states that the acceleration of an object is directly proportional to the applied force and inversely proportional to the mass ($F = ma$).

Given:

$a = 430 \text{ m/s}^2$

$m_t = 58 \text{ g} = 0.058 \text{ kg}$

$m_b = 5.0 \text{ kg}$

Solution:

a)
$$\begin{aligned} F &= m_t a \\ &= (0.058 \text{ kg})(430 \text{ m/s}^2) \\ &= 24.94 \text{ kg} \cdot \text{m/s}^2 \text{ or } 25 \text{ kg} \cdot \text{m/s}^2 \\ &= 25 \text{ N} \end{aligned}$$

Recall that $1 \text{ N} = 1 \text{ kg} \cdot \text{m/s}^2$

b) Since the mass of the bowling ball has a much greater mass than the tennis ball, an identical force will result in a smaller acceleration. (You can calculate the acceleration.)

$$\begin{aligned} a &= \frac{F}{m_b} \\ &= \frac{25 \text{ N}}{5.0 \text{ kg}} \\ &= \frac{25 \cancel{\text{kg}} \cdot \text{m/s}^2}{5.0 \cancel{\text{kg}}} \\ &= 0.5 \text{ m/s}^2 \end{aligned}$$

This is much smaller than the acceleration of the tennis ball.

Calculations and Units

In physics, when you do calculations, it is very important to pay close attention to the units in your answer. Notice how in the calculation above you can write the unit N as kg • m/s². Then the units kg in the top and bottom of the equation cancel out, leaving m/s², the unit for acceleration that you need for your answer.

CHAPTER 2

Sample Problem 2

A tennis racket hits a sand-filled tennis ball with a force of 4.0 N. While the 275-g ball is in contact with the racket, what is its acceleration? (Notice that here "g" stands for grams of mass. You have to really pay attention in physics!)

Strategy: Newton's second law relates the force acting on an object, the mass of the object, and the acceleration given to it by the force. Use the form of the equation that solves for acceleration. The force unit, the newton, is defined as the amount of force needed to give a mass of 1.0 kg an acceleration of 1.0 m/s². Therefore, you will need to change the grams to kilograms.

Given:

$F = 4.0\text{ N}$

$m = 275\text{ g}$

Remember: 1000 g equals 1 kg

Solution:

$$m = (275\text{ g})\left(\frac{1\text{ kg}}{1000\text{ g}}\right)$$

$$m = 0.275\text{ kg}$$

$$a = \frac{F}{m}$$

$$= \frac{4.0\text{ N}}{0.275\text{ kg}}$$

$$= \frac{4.0\text{ kg}\cdot\text{m/s}^2}{0.275\text{ kg}}$$

$$= 14.5\text{ m/s}^2$$

Using Measurements in Calculations

When you perform calculations using measurements, you need to express the result of your calculations in a way that makes sense of the precision of the measurements you used. You must look at the number of significant figures (or digits) in the number. The number of significant figures represents how carefully, and with what level of accuracy or precision, the measurement was taken. A calculation will never add significant figures. If one value from measurements has two significant figures and all your other values were from more precise measurements and had four significant figures, your calculation using these values can have no more than two significant figures.

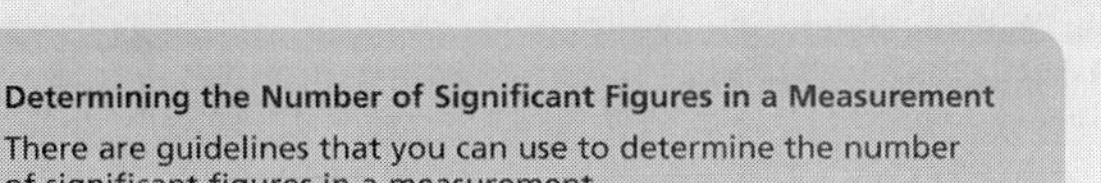

Determining the Number of Significant Figures in a Measurement

There are guidelines that you can use to determine the number of significant figures in a measurement.

All nonzero numbers are considered to be significant figures. In the measurement 152.5 m, all the digits are significant. The measurement has four significant figures.

Zeros may or may not be significant, depending on their place in a number.

- A zero between nonzero digits is a significant figure. In the measurement 308 g, the zero is significant. The measurement has three significant figures.
- A zero at the end of a decimal number is considered significant. In the measurement 1.50 N, the zero is significant. The measurement has three significant figures.
- A zero at the beginning of a decimal number is not significant. In the measurement 0.023 kg, the zeros are not significant. The measurement has two significant figures.
- In a large number without a decimal point, the zeros are not significant. In the measurement 2000 kg, the zeros are not significant. The measurement has one significant figure.

Significant Figures in Calculations

There are also guidelines that you can use when making your calculations.

Adding and Subtracting

When adding or subtracting, the final result should have the same number of decimal places as the measurement with the fewest decimal places.

Multiplying and Dividing

When multiplying or dividing, the result should have no more significant digits than the factor having the fewest number of significant digits.

→

CHAPTER 2

Gravity, Mass, Weight, and Newton's Second Law

In the *Investigate*, you observed another type of force — the force of gravity. As you added coins to the ruler attached to the end of the table, the ruler began to bend. Earlier, you saw the ruler bend when you applied a force from your arm to push the cart. You know that if you observe the ruler bending there must be a force acting. When you drop a ball, you notice that it accelerates to the floor. Newton's second law informs you that if there is an acceleration, there must be an unbalanced force acting.

In both cases, you cannot see the force, but you know it is there from the observations you make. In the case of the force bending the ruler, this is the force due to gravity.

You know that if you apply a force of 1.0 N to a 1.0-kg mass, the mass will accelerate at a rate of 1.0 m/s^2. That means that if you observe a 1.0-kg mass accelerate at 1.0 m/s^2, there must be a 1.0 N force acting on it.

If you drop a 1.0-kg mass on Earth, you will observe that the mass accelerates toward Earth at 9.8 m/s^2. That means that there must be a force of 9.8 N acting downward on the mass. This is the force of gravity acting on the mass. You will have an opportunity to measure the acceleration due to gravity yourself in a later section.

Physics Words

weight: the vertical, downward force exerted on a mass as a result of gravity.

If you put a backpack on your back, you can feel the force you must exert so that the pack does not fall to the ground due to the force of gravity. This force of gravity on the backpack is also called its weight. **Weight** is the force of gravity acting on an object, and it depends on the mass of the object and the acceleration due to gravity.

Using Newton's second law, you can calculate the weight of an object.

$$F_{gravity} = ma_{gravity}$$
$$w = mg$$

where w is the weight (by definition, the force of gravity),

m is the mass in kilograms, and

g is the acceleration due to gravity (9.8 m/s^2)

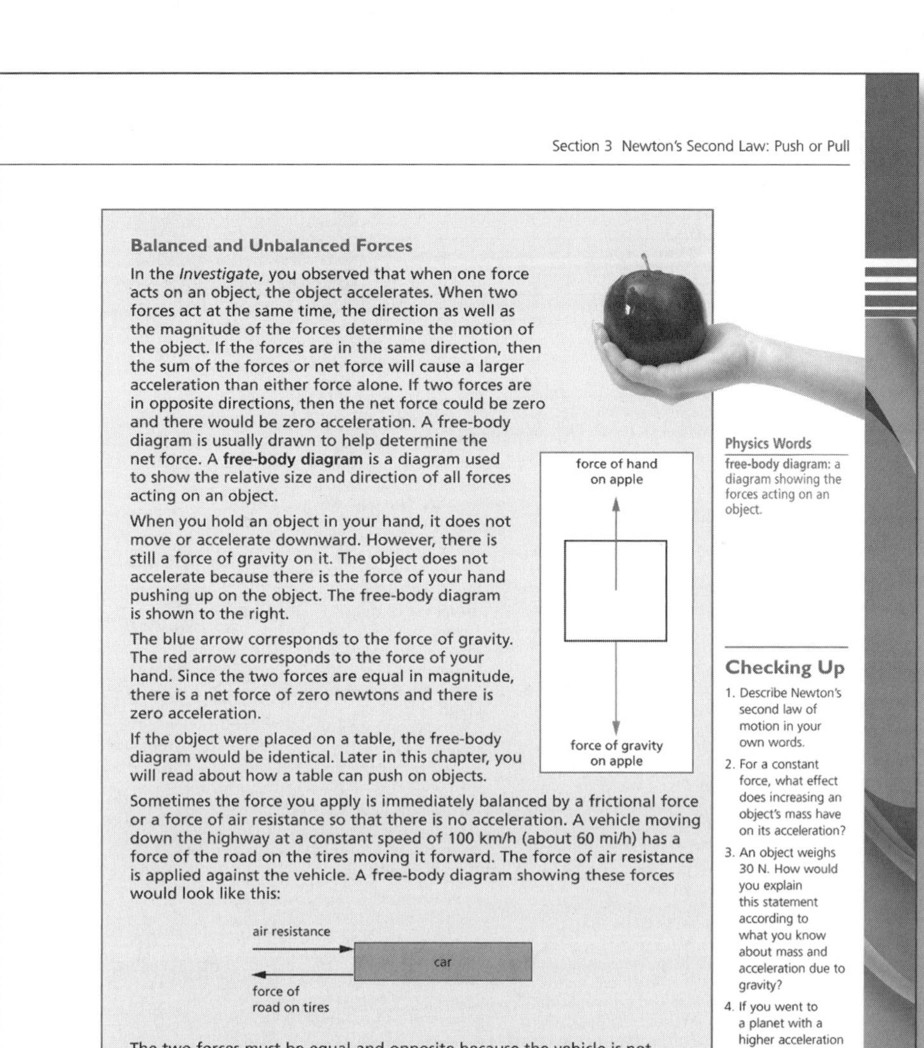

Section 3 Newton's Second Law: Push or Pull

Balanced and Unbalanced Forces

In the *Investigate*, you observed that when one force acts on an object, the object accelerates. When two forces act at the same time, the direction as well as the magnitude of the forces determine the motion of the object. If the forces are in the same direction, then the sum of the forces or net force will cause a larger acceleration than either force alone. If two forces are in opposite directions, then the net force could be zero and there would be zero acceleration. A free-body diagram is usually drawn to help determine the net force. A **free-body diagram** is a diagram used to show the relative size and direction of all forces acting on an object.

When you hold an object in your hand, it does not move or accelerate downward. However, there is still a force of gravity on it. The object does not accelerate because there is the force of your hand pushing up on the object. The free-body diagram is shown to the right.

The blue arrow corresponds to the force of gravity. The red arrow corresponds to the force of your hand. Since the two forces are equal in magnitude, there is a net force of zero newtons and there is zero acceleration.

If the object were placed on a table, the free-body diagram would be identical. Later in this chapter, you will read about how a table can push on objects.

Sometimes the force you apply is immediately balanced by a frictional force or a force of air resistance so that there is no acceleration. A vehicle moving down the highway at a constant speed of 100 km/h (about 60 mi/h) has a force of the road on the tires moving it forward. The force of air resistance is applied against the vehicle. A free-body diagram showing these forces would look like this:

The two forces must be equal and opposite because the vehicle is not accelerating. You know it is not accelerating because the description states that the vehicle is moving at a constant speed in a given direction. No change in speed or direction implies no acceleration.

Physics Words

free-body diagram: a diagram showing the forces acting on an object.

Checking Up

1. Describe Newton's second law of motion in your own words.
2. For a constant force, what effect does increasing an object's mass have on its acceleration?
3. An object weighs 30 N. How would you explain this statement according to what you know about mass and acceleration due to gravity?
4. If you went to a planet with a higher acceleration due to gravity, what would happen to your weight? What would happen to your mass?

167

Active Physics

Checking Up

1.

When an unbalanced force acts on an object, the acceleration of the object is directly proportional to the magnitude of the force and occurs in the same direction that force was applied.

2.

For a constant force, increasing the mass of an object will reduce its acceleration.

3.

The statement means that there will be a force of 30 N acting downward on the mass due to force of Earth's gravity.

4.

Your weight would increase while your mass would remain the same.

2-3a **Blackline Master**

CHAPTER 2

Active Physics Plus

Encourage your students to draw vectors showing the direction of unbalanced forces. Having them define scalar and vector quantities in their logs will help reinforce their learning. To provide a visual focus, drawing examples of vectors on the board will help students to see how two forces create a net resultant force.

+Math	+Depth	+Concepts	+Exploration
◆◆	◆		

Active Physics *Plus*

Adding Vectors

Many of the numbers you use every day are scalars. Scalars are numbers defining quantities that do not have any specific direction associated with them. They only have sizes or magnitudes. Some examples of scalars include temperature, prices, time, mass, lengths, and widths.

Unlike a scalar quantity, a vector is a quantity that has both magnitude and direction. The velocity of an object is a vector. Its magnitude is the speed of the object and its direction is the direction the object is moving.

Force is also a vector because you can measure how big it is (its magnitude) and its direction. Acceleration is also a vector. Newton's second law reminds you that the force and the acceleration must be in the same direction. Mass is not a vector—it has no direction associated with it, so it is a scalar. Weight, however, is a force and does have direction associated with it. All forces are vectors. The direction of the weight vector is down (toward the center of Earth).

Often, more than one force acts on an object. If the two forces are in the same direction, the sum of the forces is simply the algebraic addition of the two forces. A 30-N force by one person and a force of 40 N by a second person (pushing in the same direction) on the same desk provide a 70-N force on the desk. If the two forces are in opposite directions, then you give one of the forces a negative value and one a positive value to show that they act opposite to each other. Then you add them algebraically.

If one student pushes on a desk to the right with a force of 30 N and a second student pushes on the same desk to the left with a force of 40 N, the net force on the desk (also called the total force or the unbalanced force) will be 10 N to the left. Mathematically, you would state that 30 N + (– 40 N) = – 10 N where the negative sign denotes "to the left." Choosing left as the negative direction is an arbitrary choice.

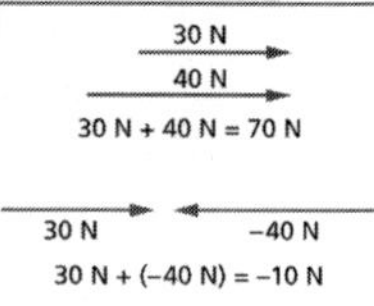

Occasionally, the two forces acting on an object are at right angles. For instance, one student may be kicking a soccer ball with a force of 30 N ahead toward the goal, while the second student kicks the same soccer ball with a force of 40 N toward the sideline. To find the net force on the ball and the direction the ball accelerates, you must use vector addition. You can do this by using a vector diagram or the Pythagorean theorem.

In the actual situation shown on the next page, the two force vectors are shown as arrows acting on the soccer ball. The magnitudes of the vectors are drawn to scale. If you were to draw this using the scale that 10 N = 1.0 cm, then the 30-N force would be 3.0 cm long and the 40-N force would be 4.0 cm long. To add the vectors, slide them so that the tip of the 30-N vector can be placed next to the tail of the 40-N vector (tip to tail method).

Section 3 Newton's Second Law: Push or Pull

The sum of the two vectors is then drawn from the tail of the 30-N vector to the tip of the 40-N vector as shown in the vector diagram below. This resultant vector is measured and is found to be 5.0 cm, which is equivalent to 50 N. The angle is measured with a protractor and is found to be 53°.

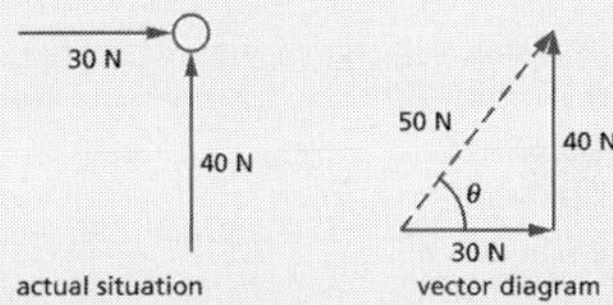

A second method of finding the resultant vector is to recognize that the 30-N and 40-N force vectors form a right triangle. The resultant is the hypotenuse of this triangle. Its length can be found using the Pythagorean theorem.

$$a^2 + b^2 = c^2$$
$$(30\text{ N})^2 + (40\text{ N})^2 = c^2$$
$$900\text{ N}^2 + 1600\text{ N}^2 = c^2$$
$$2500\text{ N}^2 = c^2$$
$$c = \sqrt{2500\text{ N}^2}$$
$$c = 50\text{ N}$$

The angle can be found by using the tangent function.

$$\tan\theta = \frac{\text{opposite}}{\text{adjacent}} = \frac{40\text{ N}}{30\text{ N}} = 1.33$$
$$\theta = 53°$$

Adding vector forces that are not perpendicular is a bit more difficult mathematically, but you can use scale drawings to make vector diagrams. Two other players are kicking a soccer ball in the directions shown in the top right diagram. The resultant vector force can be determined using the tip-to-tail approach.

The two arrows in the left diagram correspond to the actual situation in which two players kick the ball at different angles. The vector diagram at the right shows the two vectors being added "tip to tail." The resultant vector (shown as a dotted line) represents the net force and is the direction of the acceleration of the soccer ball.

1. One player applies a force of 125 N north on a soccer ball. Another player pushes with a force of 125 N west on the ball. What is the magnitude and direction of the resultant force?
2. Three hockey players are fighting for a loose puck. Hockey player A exerts a force of 40 N due north on the puck, while player B exerts a force of 70 N due south. Player C exerts a force of 40 N due west. The forces are shown in the diagram below.

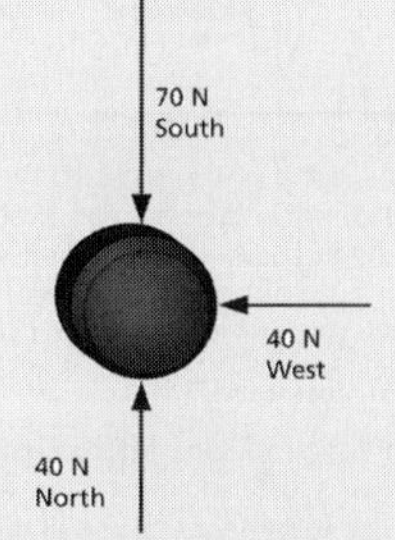

a) What is the resultant force exerted by players A and B on the hockey puck?
b) What is the resultant force of all three players on the hockey puck?
c) What is the direction of the net force on the puck?

1.

Given:

$a = b = 125\text{ N}$

Solution:

Using the Pythagorean theorem, the resultant would be determined by the formula

$a^2 + b^2 = c^2$

$(125\text{ N})^2 + (125\text{ N})^2 = c^2$

$c = \sqrt{(125\text{ N})^2 + (125\text{ N})^2} = 177\text{ N}$

To find the direction use tan

$\theta = \text{opposite}/\text{adjacent} =$

$(125\text{ N})/(125\text{ N}) = 1°$

$\theta = 45°$ North of West

2.a)

Adding the two opposite vertical forces

$40\text{ N} + (-70\text{ N}) = -30\text{ N}$

30 N downward or South

2.b)

$F_{\text{vertical}} = -30\text{ N}$
$F_{\text{horizontal}} = -40\text{ N}$

Using the Pythagorean theorem, the resultant force would be determined by the formula

$F_{\text{resultant}} = \sqrt{(F_{\text{vertical}})^2 + (F_{\text{horizontal}})^2} =$
$\sqrt{(-30\text{ N})^2 + (-40\text{ N})^2} = 50\text{ N}$

2.c)

To find the direction use tan

$$\theta = \left(\frac{F_{\text{vertical}}}{F_{\text{horizontal}}}\right) = \tan^{-1}\left(\frac{-30\text{ N}}{-40\text{ N}}\right) = 37°$$

37° South of West

CHAPTER 2

What Do You Think Now?

Most students will by now have an understanding of Newton's second law and should be well equipped to answer the *What Do You Think Now?* questions with ease. As you discuss their responses, make sure that they have a good understanding of Newton's second law. You might want to touch on certain aspects of the *Investigate* and *Physics Talk*. Recalling previous learning should help them see why it essential for them to revisit questions asked in *What Do You See?*

What Do You Think Now?

At the beginning of this section, you were asked the following:

- **What is a force?**
- **How will the same amount of force affect a tennis ball and a bowling ball differently?**

In the *Investigate*, you changed the force with which you pushed a mass, and the size of the mass you pushed. How would you answer these questions now?

Physics

Essential Questions

What does it mean?

What does it mean when Newton's second law states that acceleration and mass are inversely proportional?

How do you know?

What part of your investigation shows you that stronger forces cause larger accelerations?

Why do you believe?

Connects with Other Physics Content	Fits with Big Ideas in Science	Meets Physics Requirements
Force and motion	* Change and constancy	Good, clear, explanation, no more complex than necessary

* Newton's second law is used to describe and explain motion of large objects and small objects. It helps you better understand the motion of people in sports, cells in the body, colliding atoms, and planets in the Solar System. Entire physics courses in college are based on Newton's second law. Why do you believe that if you push on a truck, the truck has a tiny acceleration?

Why should you care?

All sports involve motion. All accelerated motion involves unbalanced forces. If you identify an acceleration of a person or an object in your sports video, you can discuss the forces that cause that acceleration. What is one way that this idea will come up in your voice-over challenge?

Physics Essential Questions

What does it mean?

If the same force pushed a series of masses, the larger masses would have the smallest accelerations. This is an inverse proportion.

How do you know?

By applying a larger force (more bend in the ruler), you can observe a greater change in speed in a given time interval.

Why do you believe?

Newton's second law states that if the push is the only force acting on the truck, then the truck would have a very tiny acceleration. To assume that there would be no acceleration would require us to then find out why Newton's second law does not apply for either small forces or large masses. It makes more sense to conclude the acceleration is too small to see.

Why should you care?

A baseball leaving the bat has acceleration. Newton's second law states that if there is acceleration, there must be a force. The force is the bat on the ball.

Reflecting on the Section and the Challenge

What you learned in this activity really increases the possibilities for interpreting sports events in terms of physics, particularly when events have motions along straight paths. Now you can explain why accelerations occur in terms of the masses and forces involved. You know that unbalanced forces are the only things that produce accelerations. Therefore, if you see an acceleration occur, you know to look for unbalanced forces. In soccer and baseball, the ball accelerates. In soccer, the force to increase its speed or change the direction of the ball is the player's foot or head. In baseball, one force is the bat hitting the ball. In football, one player tries to accelerate another player by pushing on him. If the player being pushed is small, the acceleration can be quite large. If the player being pushed is quite massive, the acceleration is much smaller. You can apply Newton's second law to the sport you will describe.

You can also discuss the weight of players and objects in the sports by recognizing that weight is a force that is equal to the mass of the object multiplied by g, acceleration due to gravity, which is 9.8 m/s^2 on Earth.

Physics to Go

1. Copy the following table in your log. Use Newton's second law of motion to calculate the missing values in the table. Be sure to include the unit of measurement for each missing item (examples: kg, N, m/s^2).

Newton's second law:	F =	m ×	a
sprinter beginning 100-m dash	?	70 kg	5 m/s^2
long jumper in flight	800 N	?	10 m/s^2
shot-put ball in flight	70 N	7 kg	?
ski jumper going downhill before jumping	400 N	?	5 m/s^2
hockey player "shaving ice" while stopping	–1500 N	100 kg	?
running back being tackled	?	100 kg	–30 m/s^2

2. The following items refer to the table in *Question 1*.
   a) In which cases in the table does the acceleration match g (the acceleration due to gravity, 9.8 m/s^2)? Are the matches to g coincidences or not? Explain your answer.
   b) The force on the hockey player stopping is given in the table as a negative value. Should the player's acceleration also be negative? What do you think it means for a force or an acceleration to be negative?
   c) The acceleration of the running back being tackled also is given as negative. Should the unbalanced force acting on the running back also be negative? Explain your answer.

Newton's second law	F =	m	x a
sprinter beginning 100-meter dash	350 N	70 kg	5 m/s^2
long jumper in flight	800 N	80 kg	10 m/s
shot-put ball in flight	70 N	7 kg	10 m/s^2
ski jumper going down hill before jumping	400 N	80 kg	5 m/s^2
hockey player "shaving ice" while stopping	–1500 N	100 kg	–15 m/s^2
running back being tackled	–3000 N	100 kg	–30 m/s^2

Reflecting on the Section and the Challenge

Students should read this section so that they can begin to form specific connections with the *Chapter Challenge*. A discussion of different sports, especially the ones mentioned in the *Reflecting on the Section and the Challenge*, will help students to see how Newton's second law affects the decisions that are made by players. You might want to provide some time for your students to read this section aloud.

Physics to Go

1.

See chart below.

2.a)

The long jump and the shot put are both cases of free fall; therefore the acceleration is g, the acceleration due to gravity.

2.b)

The negative sign is used to denote that the force and acceleration are in a direction opposite the motion.

2.c)

Because acceleration occurs in the direction of the causal force, yes, the force should be shown as negative.

3.

$42\,\text{N}/0.30\,\text{kg} = 140\,\text{m/s}^2$

4.

$0.040\ \text{kg} \times 20\ \text{m/s}^2 = 0.8\ \text{N}$

5.a)

A bowling ball has greater inertia (mass) than a baseball; therefore, a bowling ball has a greater tendency to either remain at rest or remain in motion than does a baseball.

5.b)

More force is required to cause a bowling ball to accelerate than a baseball; therefore, throwing (accelerating) or catching (decelerating) a bowling ball involves much greater forces than throwing or catching a baseball when equal speeds are involved.

6.

The sandwich would weigh $0.1\ \text{kg} \times 10\ \text{m/s}^2 = 1\ \text{N}$. Names such as "newtonburger" might work.

7.a)

Example:
Weight $= 150\ \text{lb} \times 4.38\ \text{N/lb} = 657\ \text{N}$

7.b)

Mass $= 657\ \text{N}/10\ \text{m/s}^2 = 65.7\ \text{kg}$

8.

Students provide voice-over for tug of war.

9.

No. The force of your hand stops acting on the ball the moment the two are no longer in contact.

Chapter 2 Physics in Action

3. What is the acceleration of a 0.30-kg volleyball when a player uses a force of 42 N to spike the ball?
4. What force would be needed to accelerate a 0.040-kg golf ball at 20.0 m/s²?
5. Most people can throw a baseball farther than a bowling ball, and most people would find it less painful to catch a flying baseball than a bowling ball flying at the same speed as the baseball. Explain these two situations in terms of
   a) Newton's first law of motion.
   b) Newton's second law of motion.
6. Calculate the weight of a new fast-food sandwich that has a mass of 0.1 kg (approximately the mass of a quarter pound). Think of a clever name for the sandwich that would incorporate its weight in newtons.
7. In the United States, people measure body weight in pounds. Imagine a person weighs 150 lb.
   a) Convert the person's weight in pounds to the international unit of force, newtons. To do so, use the following conversion equation:
   Weight in newtons = (weight in pounds) (4.38 newtons per pound)
   b) Use the person's body weight, in newtons, and the equation

   $$\text{Weight} = mg$$

   to calculate the person's body mass (m), in kilograms.
8. If you were doing the voice-over for a tug-of-war competition, how would you explain what was happening? Write a few sentences as if you were the science narrator of that athletic event.
9. You throw a ball. When the ball is many meters away from you, is the force of your hand still acting on the ball? When does the force of your hand stop acting on the ball?
10. Carlo and Sara push on a desk in the same direction. Sara pushes with a force of 50 N, and Carlo pushes with a force of 40 N. What is the unbalanced force acting on the desk? The unbalanced force on an object is sometimes called the total force, or net force, on an object.
11. A vehicle is stuck in the mud. Four adults each push on the back of the vehicle with a force of 200 N. What is the combined force, due to all four adults, on the vehicle?
12. A baseball player throws a ball. While the 700.0-g ball is in the pitcher's hand, there is a force of 125 N on it. What is the acceleration of the ball?
13. Active Physics Plus During a football game, two players try to tackle another player. One player applies a force of 50.0 N to the east. A second player applies a force of 120.0 N to the north. What is the resultant force applied to the player being tackled? (Since force is a vector, you must give both the magnitude and direction of the force.)

172 Active Physics

10.

The resultant is $40\ \text{N} + 50\ \text{N} = 90\ \text{N}$.

11.

The total or combined force is $4 \times 200\ \text{N} = 800\ \text{N}$. But there may also be forces due to the mud, acting in the opposite direction.

12.

$F = ma$

$a = F/m = (125\ \text{N})/(0.7\ \text{kg}) =$

$179\ \text{m/s}^2$

13.

Application of the Pythagorean theorem yields:

$(50\ \text{N})^2 + (120\ \text{N})^2 = F^2$

$F = 130\,\text{N}$

Using the tangent button on the calculator or a vector diagram, the angle is 23° East of North.

14. Active Physics Plus In auto racing, a crash occurs. A red car hits a blue car from the front with a force of 4000 N. A yellow car also hits the blue car from the side with a force of 5000 N. What is the resultant force on the blue car? (Since force is a vector, you must give both the magnitude and direction of the force.)

15. The acceleration due to gravity at the surface of Earth is approximately 9.8 m/s². What force does the gravitational attraction of Earth exert on a 12.8-kg object, such as a toolbox loaded with tools?

16. Active Physics Plus A force of 30.0 N acts on an object. At right angles to this force, another force of 40.0 N acts on the same object.

a) What is the net force on the object?

b) What acceleration would this object have if it is a 5.6-kg wagon?

17. Active Physics Plus Bob exerts a 30.0-N force to the left on a box (m = 100.0 kg). Carol exerts a 20.0-N force on the same box, perpendicular to Bob's force.

a) What is the net force on the box?

b) Determine the acceleration of the box.

c) At what rate would the box accelerate if both forces were to the left instead of perpendicular to each other?

18. ***Preparing for the Chapter Challenge***

Using a sport of your choice, write a script for a voice-over that deals with accelerated motion and forces.

Inquiring Further

Gaining and losing weight

The acceleration due to gravity is different at the surface of the Moon and the other planets in the Solar System. Where would you choose to "live" if you wanted to lose weight? What would be the weight of a 150-lb person on the Moon and on each of the planets?

14.

Application of the Pythagorean theorem yields:

$$(4000\text{ N})^2 + (5000\text{ N})^2 = F^2$$

$$F = 6403\text{ N}$$

Using the tangent button on the calculator or a vector diagram, the angle is 39°.

15.

$$F = ma = (12.8\text{ kg})(9.8\text{ m/s}^2) = 125\text{ N}$$

16.a)

Application of the Pythagorean theorem yields:

$$(40\text{ N})^2 + (30\text{ N})^2 = F^2$$

$$F = \sqrt{(40\text{ N})^2 + (30\text{ N})^2} = 50\text{ N}$$

Using the tangent button on the calculator or a vector diagram, the angle is 53°.

16.b)

$$a = F/m = 50\text{ N}/5.6\text{ kg} = 8.9\text{ m/s}^2$$

17.a)

Application of the Pythagorean theorem yields:

$$(30\text{ N})^2 + (20\text{ N})^2 = F^2$$

$$F = \sqrt{(30\text{ N})^2 + (20\text{ N})^2} = 36\text{ N}$$

Using the tangent button on the calculator or a vector diagram, the angle is 34°.

17.b)

$$a = F/m = 36\text{ N}/100\text{ kg} = 0.36\text{ N/kg} = 0.36\text{ m/s}^2$$

17.c)

If both boxes were pushed toward the right, the new force would be 50 N.

$$(30\text{ N} + 20\text{ N} = 50\text{ N})$$

The acceleration can then be calculated:

$$a = F/m = 50\text{ N}/100\text{ kg} = 0.5\text{ N/kg} = 0.5\text{ m/s}^2$$

18.

Preparing for the Chapter Challenge

Student responses will vary, but should include a description of how Newton's second law will determine the acceleration of an object that is acted upon by one or more forces. The forces should be appropriate for the sports example being illustrated.

Inquiring Further

Because your weight depends upon your mass times the acceleration of gravity, the best place to go to "lose weight" would be the planet with the lowest gravity. Unfortunately, although you would weigh less, your mass would still be the same so you would not look appreciably different.

The acceleration of gravity on the Moon is approximately 1.6 m/s^2 or about one-sixth that of Earth. If you went to the Moon, your new weight would be one-sixth your weight on Earth. A person who weighed 150 lbs on Earth would only weigh 25 lbs on the Moon!

The acceleration of gravity varies greatly from planet to planet, with a high of about 26 m/s^2 for Jupiter, and a low of 0.6 m/s^2 for Pluto. The weight of a 60-kg person would vary from

1560 N (about 350 lbs) on Jupiter to 36 N (about 8 lbs) on Pluto.

NOTES

SECTION 3 QUIZ

2-3b Blackline Master

1. The force required to accelerate a 2-kg mass at 4 m/s^2 is

   a) 6 N. b) 2 N.

   c) 8 N. d) 16 N.

2. A force of *F* newtons will give an object with a mass of *M* an acceleration of *A*. The same force will give a mass of 2 *M* an acceleration of

   a) A/2. b) 2A.

   c) A. d) A/4.

3. On the planet Gamma a 4.0-kg mass experiences a gravitational force of 24 N. What is the acceleration due to gravity on Gamma?

   a) 0.17 m/s^2 b) 6 m/s^2

   c) 9.8 m/s^2 d) 96 m/s^2

4. A cart is uniformly accelerating from rest. The force on the cart must be

   a) decreasing. b) zero.

   c) constant. d) increasing.

5. In the graph below, the acceleration of an object is plotted against the unbalanced force applied. What is the mass of the object?

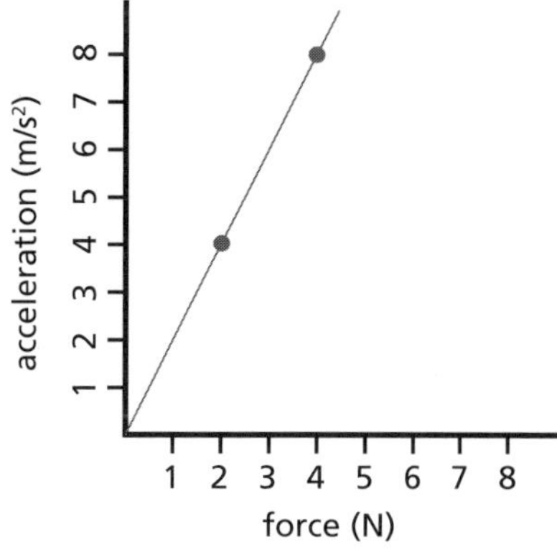

   a) 0.5 kg b) 2 kg

   c) 4 kg d) 0.2 kg

CHAPTER 2

SECTION 3 QUIZ ANSWERS

1. b) Using $F = ma = (2\ \text{kg})(4\ \text{m/s}^2) = 8\ \text{N}$.

2. a) $a = F/M$, then $F/2M = \frac{1}{2}A$

3. b) Using $W = mg$ gives $24\ \text{N} = (4\ \text{kg})(g)$ Solving for $g = 6\ \text{m/s}^2$.

4. c) A constant acceleration requires a constant net force.

5. a) Using values from the graph, when the force is 4 N, the acceleration is 8 m/s^2. Substituting these values in the equation $F = ma$ yields $4\ \text{N} = m(8\ \text{m/s}^2)$, or $m = 0.5\ \text{kg}$.

NOTES

NOTES

CHAPTER 2

SECTION 4

Projectile Motion: Launching Things into the Air

Section Overview

By observing the results of horizontally projected objects, students see how vertical motion of an object in free fall remains unaffected by the horizontal motion of the object and vice versa. In the *Investigate*, students project a coin from across a table to hit another coin placed near the table's edge to observe the rate of fall of each coin. They then compare the motion of the dropped coin and the projected coin to see how they fall relative to each other from a raised surface. Students also observe the combined vertical and horizontal motions of a ball as it is thrown upward at an angle and then allowed to fall as they try to duplicate the trajectory. The upward path of an object and its vertical descent are explained in terms of the rate of change of velocity to demonstrate that all objects fall with the same acceleration.

Background Information

Free Fall: All objects in a state of free fall, regardless of mass, shape, or other characteristics, experience a uniform (constant) acceleration of approximately 10 m/s^2 in the downward direction. (Notice that free fall does not include objects that experience a significant force due to air resistance.)

Projectile Motion: Inertia (an object in motion remains in motion unless a force acts to cause a change in motion) causes a projectile to retain any horizontal motion that it has at the instant it is launched, and it retains that motion in the form of constant horizontal speed.

Gravity causes a projectile to exhibit vertical acceleration that matches free fall, whether or not the projectile has simultaneous horizontal motion. This section does not approach projectile motion quantitatively, but it clearly demonstrates that a coin which simply falls strikes the floor after the same amount of time as a coin launched horizontally from the same height at the same instant. The only way this can happen is if both objects fall downward with identical motions. The fact that one coin simultaneously moves horizontally (at constant speed) as it falls affects where it lands, not when. This section further demonstrates that a ball thrown upward does not have its upward and downward motion affected by any horizontal motion which it may have at the time of launch. The horizontal motion remains one of constant speed and is independent of the vertically accelerated motion. If a student sitting in a stationary chair throws a ball upward, it lands in her hand; if the chair is moving at constant speed throughout the vertical toss, inertia demands that the ball also land in the student's hand. Quantitative treatment of projectile motion, with additional background for the teacher, is presented in the next activity.

Crucial Physics

- Objects in free fall have only the force due to gravity acting on them and move vertically upward or downward with a downward acceleration equal to $g = 9.81\ m/s^2$.
- The horizontal motion and the vertical motion of a trajectory are independent of one another.
- Free fall is a special case in which the horizontal velocity is zero. When the horizontal velocity is not zero, the object's horizontal motion has no acceleration (its velocity is constant) and the object's vertical motion is "free fall" (acceleration equals g). The horizontal and vertical motions operate independently of each other.
- The range of a projectile depends on its initial velocity, the angle its initial velocity makes with the horizontal, and the difference in height of its starting and ending locations.

Learning Outcomes	Location in the Section	Evidence of Understanding
Apply the terms free fall, projectile, trajectory, and range.	***Investigate*** Part B: Step 2.a) ***Physics Talk***	Students write in their logs what they observed for trials in which they varied the speed of the chair and the launching speed of the ball to see how the range and trajectory are affected. The *Physics Talk* applies the term *projectile motion* to help students understand how objects move as observed in the *Investigate*.
Provide evidence concerning projectiles launched horizontally from the same height at different launch speeds (including zero launch speed).	***Investigate*** Part A: Steps 1 and 2.	Students first drop coins from a certain height, then flick a coin across a desk to hit another coin hanging over the edge of a table to observe whether both coins hit the floor at the same time.
Explain the relationship between the vertical and horizontal components of a projectile's motion	***Investigate*** Part B: Steps 1 and 2. ***Physics Talk***	Students observe a ball's trajectory and range as they throw the ball up from a moving chair and try to catch it. The horizontal and vertical movement of a hit or thrown ball is explained in the *Physics Talk*.
Recognize the factors that affect the range of a projectile.	***Physics Talk***	Students recognize that a change in velocity of $9.8\ m/s^2$ acts vertically downwards on the coin while the horizontal velocity of the coin does not change.
Infer the shape of a projectile's trajectory.	***Physics Talk***	Students study the shape of a projectile's trajectory in a graph that represents the motion of an object through different time intervals.

Section 4 Materials, Preparation, and Safety

Materials and Equipment

PLAN A		
Materials and Equipment	**Group (4 students)**	**Class**
Launcher, projectile	1 per group	
Ball, tennis		1 per class
Coins (pennies, nickels)*		1 per class
Chalkboard*		1 per class
Chalk*		1 per class
Chair (with wheels)*		1 per class
Safety helmet*		1 per class
Pads, knee*		1 per class
Pads, elbow*		1 per class
Stack of books*		1 per class

*Additional items needed not supplied

PLAN B		
Materials and Equipment	**Group (4 students)**	**Class**
Launcher, projectile		1 per class
Ball, tennis		1 per class
Coins (pennies, nickels)*		1 per class
Chalkboard*		1 per class
Chalk*		1 per class
Chair (with wheels)*		1 per class
Safety helmet*		1 per class
Pads, knee*		1 per class
Pads, elbow*		1 per class
Stack of books*		1 per class

*Additional items needed not supplied

Note: Time, Preparation, and Safety requirements are based on Plan A, if using Plan B, please adjust accordingly.

Time Requirements

Time required is approximately one period or 40 minutes.

Teacher Preparation

Part A

- Obtain the coins necessary for the *Investigate*. Have extra coins on hand in case some are lost. If you do not have them, coin launchers may be easily built from a piece of scrap metal and a wooden block. This will facilitate the student launching the coins at the same time. If you choose to do the activity with the students "flicking" the coins, the students should do so from a distance of only a few centimeters to increase the chance of a hit.

Part B

- Obtain a sturdy office chair on rollers. It should preferably have five legs, and arms for the student to hold onto while pushed. This *Investigate* should be done as a class demonstration, rather than in individual groups. Before the chair is pushed, have the student who will be sitting in the chair practice throwing the ball straight up and down so that it lands in the same spot from which it was thrown. When the chair is being pushed it is very important that the student be traveling with constant velocity when the ball is thrown upward, and continues with constant velocity while the ball is in the air.

Safety Requirements

- If you use the coin launchers, tell the students not to launch at very high speeds, unless under your direct supervision. Make certain that the students do not leave the coins on the floor, to prevent a slipping hazard.

- For *Part B*, instruct the student pushing the chair on how to push with a uniform velocity that will not endanger the rider. High speed is not necessary to demonstrate the principle. As an extra precaution you may choose to have the rider wear a safety helmet and elbow and knee pads. Do not allow the students to play with the chair, or try to push it at very high velocities, as it may pose a danger to the rider.

NOTES

Meeting the Needs of All Students

Differentiated Instruction: Augmentation and Accommodations

Learning Issue	Reference	Augmentation and Accommodations
Hearing the coins hit the floor	***Investigate*** Part A Steps 1-4	**Augmentation** • Students with hearing, auditory-processing, attention issues may struggle to discriminate the sound of their own coins dropping. Spread groups out as much as possible to do this *Investigate*. • Ask students to set up and practice the procedure for flicking or launching the coins. Then ask one group at a time to flick or launch their coins while the other groups quietly watch and listen. This repetition may help convince more students that both coins hit at the same time and will decrease background noise. **Accommodation** • Perform the procedure as a demonstration with the entire class watching and listening quietly. • Use two of the same objects that are larger and will make a louder sound when hitting the ground.
Understanding vocabulary	***Investigate*** Part B ***Physics Essential Questions***	**Augmentation** • Students may not know the meaning of vertical and horizontal. Ask a student in the class to explain the meaning of both words and show examples.
Maintaining focus during a large group task	***Investigate*** Part B	**Augmentation** • Students with attention issues struggle to focus and follow directions during large-group tasks. Make sure the whole group knows the purpose of the activity and each student knows the task she or he is supposed to complete to assist the large group. • Ask students to repeat directions back to the teacher or a partner to ensure that they have heard and understand their role. **Accommodation** • Ask the students with the most severe attention issues to be the ones sitting in the chair or pushing the chair.
Understanding vocabulary	***Physics Words***	**Augmentation** • Ask students to draw pictures to represent each of the four vocabulary words in this section. This section will aid the students in committing these words to their long-term memories and help the teacher check for understanding.
Understanding vocabulary	***Physics Talk,*** ***Physics to Go*** Question 4	**Augmentation** • Students may not know the meaning of air resistance. Ask a volunteer to explain the meaning of air resistance. • Drop a feather and a coin at the same time to show an example of air resistance.

Strategies for Students with Limited English-Language Proficiency

Learning Issue	Reference	Augmentation
Making inferences Vocabulary comprehension	***Active Physics Plus*** Step 2.d)	To answer the question, students will need to understand the meaning of the word "magnitude." Students may remember that they encountered this word in *Chapter 1* when learning about vectors. Guide students in remembering or inferring the meaning here, which is "size." Be sure students do not think of magnitude in the current context as a synonym for speed.
Making inferences Vocabulary comprehension	***Physics Essential Questions***	To answer the question, "What does it mean to say that the horizontal motion of a projectile is independent of the vertical motion of the same projectile?" students will need to infer the meaning of "independent" in this context. Allow ELL students time to think through and write an answer. Then look at the answers and offer additional guidance if necessary. You may wish to suggest that students think in terms of "separate" or "apart."

Point out new vocabulary words in context and practice using the words as much as possible throughout the section.

air resistance	range
free fall	resultant velocity
horizontal	shot put
javelin	trajectory
launch	variable
projectile	vertical

ELL students benefit from writing in English. Have students write a summary of what they did in this section, using the words in the list above and any other vocabulary words they choose. You may wish to have students work in pairs. Review and correct student work.

Rapid feedback about students' sentences is essential, because the sentences and errors will be fresh in students' minds. A quick and powerful method for providing this feedback is to prepare a list of examples of sentences containing errors from the students' work. Divide examples into the following categories: incorrect science, incorrect usage of vocabulary, incorrect sentence structure, and incorrect grammar. Choose several examples from the collected work to use in each category and edit the sentences until they contain only one or two obvious errors. At the beginning of class the next day, provide each student with a page containing a double-spaced, typed list of the incorrect sentences, with headings for the four categories. Allow students ten minutes to silently make corrections to the sentences. Then place a copy of the list on the overhead projector and ask students to suggest ideas for how to correct the sentences. During the exercise, guide students toward correct science and correct English usage.

SECTION 4

Teaching Suggestions and Sample Answers

What Do You See?

Most students will offer quick responses when they see a visual set in a classroom. Remind students that they will be returning to this illustration further along in this section. Stimulate their interest by asking probing questions that connect to the *Learning Outcomes*. Emphasize that all answers will be considered as a step toward understanding future concepts. You might also want to point out this section provides a backdrop for the concepts they are about to learn. Once they have read this section, they should be able to explain the artist's purpose behind the illustration.

Chapter 2 Physics in Action

Section 4 Projectile Motion: Launching Things into the Air

What Do You See?

Learning Outcomes

In this section, you will

- **Apply** the terms free fall, projectile, trajectory, and range.
- **Provide** evidence concerning projectiles launched horizontally from the same height at different launch speeds (including zero launch speed).
- **Explain** the relationship between the vertical and horizontal components of a projectile's motion.
- **Recognize** the factors that affect the range of a projectile.
- **Infer** the shape of a projectile's trajectory.

What Do You Think?

Some track and field events involve launching things into the air, such as a shot put, a javelin, or even one's body in the case of the long jump. In golf, football, tennis, and baseball, balls move through the air as well.

- **What determines how far an object thrown into the air travels before landing?**

Record your ideas about this question in your *Active Physics* log. Be prepared to discuss your response with your small group and the class.

Investigate

Part A: Observe Two Coins Dropping

In this part of the *Investigate*, you will observe two coins as they fall from a table. One coin will be dropped from the table, and the other will be projected from the table.

1. Hold two coins the same distance above the floor. Drop them at the same time. Listen to the sound they make as they strike the floor.

a) Do they hit the floor at the same time?

174

Students' Prior Conceptions

Students should understand that constant acceleration is the result of a constant force acting on a projectile; forces that act perpendicular to one another act independently on an object; and a projectile can exhibit constant motion in one direction and constant acceleration in another direction, simultaneously. Students should identify what happens to the velocity, both its magnitude and its direction, of an object that is thrown vertically upward in one case and that is projected horizontally in another situation.

1. **The motion of an object is always in the direction of the net force applied to the object.** Measuring and analyzing the velocity of a ball thrown directly upward as a result of the constant downward pull of gravity enables the student to recognize that the ball moves upward until the instantaneous velocity at the top of the trajectory is zero then moves downward with increasing speed; both motions are influenced by the same force of gravity and the same acceleration.
2. **Students may not understand that gravity is a force, believing that the motion of a falling body is natural, therefore needing no further explanation.** Through examining the vertical motions of objects, students learn to recognize that falling is not due to the internal effort of the falling object.
3. **Objects fall with a constant velocity.** Measuring the speed of falling objects at different points in a trajectory is vital for students to align their thinking with the laws governing freely falling objects; the speed of a falling object is directly proportional to the time of fall in the absence of air resistance.

What Do You Think?

You might want to ask your students what happens when they play sports in which they launch things into the air. Invite them to share their answers in class, so that a discussion is started where everybody is encouraged to participate. Your task is to skillfully guide them toward the main question. Make sure they record their answers for another discussion after they have studied this section in more detail.

What Do You Think?

A Physicist's Response

The trajectory of an object is determined by its initial vertical speed and initial horizontal speed, its position at the beginning of the trajectory, and any extra factors (such as friction or fluid speed) that may be important to the specific situation. The initial velocity (speed and direction) determines the range of a projectile; a difference in the elevations of the launch and landing points would also affect the range. A ball with a 100-mph pitch thrown horizontally by a major league player will hit the ground in the same amount of time as a ball with a 10-mph pitch thrown horizontally from the same height by a child. It is a near certainty that many students will not believe this type of fact without some personal experience. The work with coins should present a discrepancy for them which, hopefully, will cause them to confront their misconceptions about projectiles.

Investigate

Part A

1.a)

Both coins will hit the ground at the same time if released at the same time.

CHAPTER 2

4. **Gravity only operates when an object falls.** Measuring the velocity at various points along the trajectory of the path of a ball that is thrown straight up and allowed to return to its origin enables students to see that the acceleration due to gravity slows down the ball in a regular way as the ball travels to the top of its path and speeds up the ball in a similar manner as it falls down again.
5. **An object traveling fast enough in the horizontal direction can defy gravity.** Students may exhibit confusion on this concept as they explore ideas associated with rockets and space travel; they may not recognize that the force of gravity decreases according to the inverse square law as an object moves further away from Earth. They liken breaking away from Earth's pull in the vertical direction with avoiding Earth's pull in the horizontal direction by moving fast enough.
6. **The faster something travels horizontally, the slower it falls.** Propelling two coins or two tennis balls from the same height simultaneously, one vertically and the other horizontally, encourages students to hear and to recognize that the two objects hit the floor simultaneously, regardless of the horizontal speed. Instructional material provides evidence that all projectiles launched horizontally from the same height strike a level surface in equal times, regardless of the launch speed, including a zero launch speed.

NOTES

2.a)

The coins hit the floor at the same time.

Teaching Tip

Having the students make a glancing collision between the two coins at the edge of the table in the first part of the activity may prove frustrating for some students. A device to launch a coin horizontally while simultaneously dropping a second coin may be made from a wood stand, a metal strip like a hacksaw blade, and a folded piece of file folder to hold the coins before they are "launched" as shown in the diagram to the right.

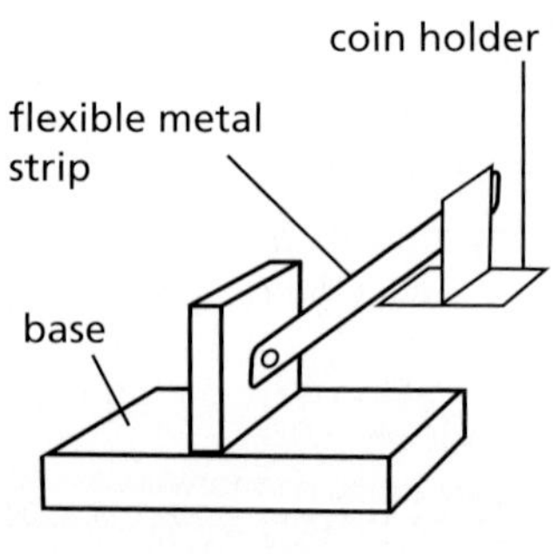

coin launcher

The two coins are placed on opposite sides of the folded file folder at the end of the metal strip. When the strip is then

2. Place one coin at the edge of a table with about half of the coin hanging over the edge. Place another coin flat on the table. Use your fingers to "flick" this coin across the tabletop to strike the first coin. Aim "off center" so that the coin at the edge of the table drops straight down and the projected coin leaves the edge of the table with some horizontal speed.

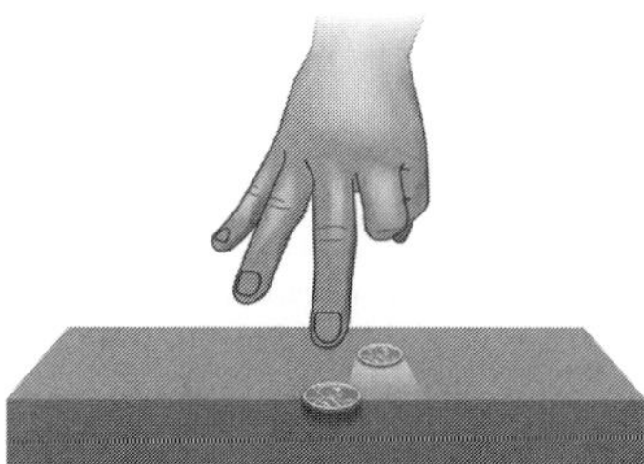

Your teacher may also decide to use a "coin launcher" for this experiment as shown in the diagram below.

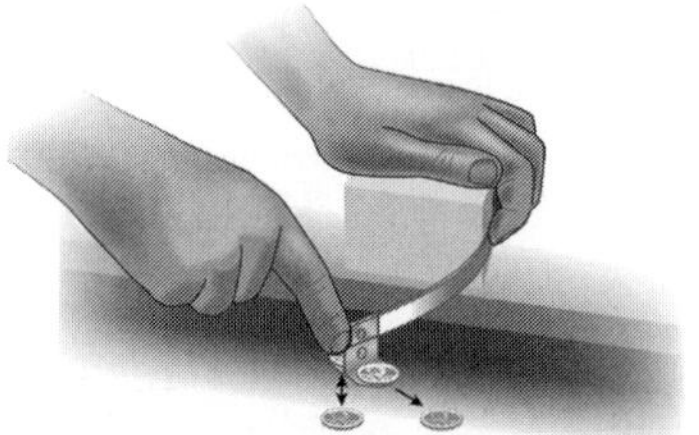

Repeat the event as many times as needed to record your answer to the following question in your log.

a) Do the coins hit the floor at the same or different times? (Hearing is the key to observation here, although you may wish to rely on sight as well.)

3. Vary the speed of the projected coin.

a) Does the speed of the projected coin affect whether the two coins hit the floor at the same time? Explain your answer.

b) Does the speed of the projected coin change how far it lands horizontally from the coin that fell straight down? Explain your answer.

c) Draw a single sketch that includes both the path of the coin that is falling and the coin that is projected. Imagine where each coin is at four identical points in time and note these predicted locations on your sketch. Label these times A, B, C, D.

4. Use a box, chair, or a stack of books to change the height from which you project the coins.

a) Do the coins hit the floor at the same or different times?

b) How does changing the height affect how far the projected coin travels horizontally as it falls?

Part B: Vertical and Horizontal Motion of a Projectile

In this part of the *Investigate*, the class will observe a student throwing a ball into the air while sitting in a moving chair.

pulled back horizontally and released, one coin is propelled horizontally, and the second coin slides off the folder and falls almost vertically to the floor.

3.a)

The initial speed of the coin being "flicked" or launched has no effect on the amount of time it takes for either coin to fall to the floor. Ask the students to go through their reasoning when answering the question. The only way to compare trials with different speeds is to always impart very little velocity to the stationary coin. In other words, a coin falling straight down is the common element in all trials and the basis for comparison.

3.b)

The coin with the greater initial velocity will have a greater range; it will travel further across the floor. Ask the students if this makes sense in light of their answer to *Step 3.a)*. The coin that had the highest initial horizontal speed will maintain that speed as it falls, allowing it to travel farther. The faster the coin is moving, the further it will move.

3.c)

The student's sketch should look similar to the diagram in *Step 2* in the text. Coins labeled with the same time should be the same height from the floor.

4.a)

When the two coins start from a lower elevation, they still hit at the same time. Ask the students what *does* change as the height decreases.

4.b)

With the decreased starting height, the projected coin travels a shorter distance horizontally than it did from a higher starting position. Ask the students if this makes sense, given their observations from the previous parts.

Teaching Tip

This activity may be done by either increasing the time of fall from that of the initial position or decreasing the time of fall. If you are using the coin launcher, increasing the time of fall to the floor is probably easier than trying to have the coins land on a specified target. To do this, just place the coin launcher on a stack of books on the table. The students should make certain they pull the coin launcher back an equal distance for both trials at the different heights to ensure that the difference in range is due to different time of flight, and not a different horizontal velocity.

Chapter 2 Physics in Action

1. To illustrate an object that has both vertical and horizontal motion at the same time, your teacher will supervise a class activity in which one student sits on a chair that is moving at constant speed. While the chair is moving, the student on the chair will throw a ball straight up into the air and try to catch it when it comes down. The class will stand in a line beside the path of the chair to observe the event, prepared to mark the vertical position of the ball as it passes them.

a) In your log, write your prediction of what you think will happen.

2. This activity can be done in several ways. It is ideal if the chair's path can be parallel to the chalkboard. Another option is to take a large roll of paper and tape the paper to a wall parallel to the path of the chair on the opposite side from the observing students. The bottom horizontal side of the board or paper should be at the height the student in the chair launches and catches the ball. Each student observing the event draws a vertical line on the board or paper marking their position beside the path. As the event takes place, each observing student keeps track of the height of the ball as it passes the line representing their position. After the event, each student puts a mark on the board or paper corresponding to the point where the ball passed the line.

a) Write in your log what you observed about the ball's trajectory (shape of the ball's path) and the ball's approximate range (horizontal distance) for trials in which you varied the speed of the chair and the launching speed of the ball. Remember, to see the effect of both the ball's trajectory and approximate range, be sure to only change one variable at a time during your trials.

b) According to your observations, what factors affect the range of the ball?

176

Active Physics

Part B

1.a)

It is important that students record their predictions before the activity is completed. This permits them to confront their misconceptions, if there are any. If students are not encouraged to commit to a prediction, they will often predict the "right answer" after the fact. Although students may learn to respond with the correct answer, they may still not have confronted their misconceptions.

2.

The ball should land in the student's hands as if the chair were not moving when thrown straight upward because, due to inertia, the ball retains its horizontal speed as it independently flies up and down with accelerated vertical motion due to gravity.

2.a)

Students may see that the ball follows a curved path. Hopefully they will measure and record that increased horizontal speed increases range. Increased vertical speed also increases range.

2.b)

Two factors that affect the range of the ball are the vertical speed at which the ball is launched, and the horizontal speed of the chair (and ball) when the ball is launched.

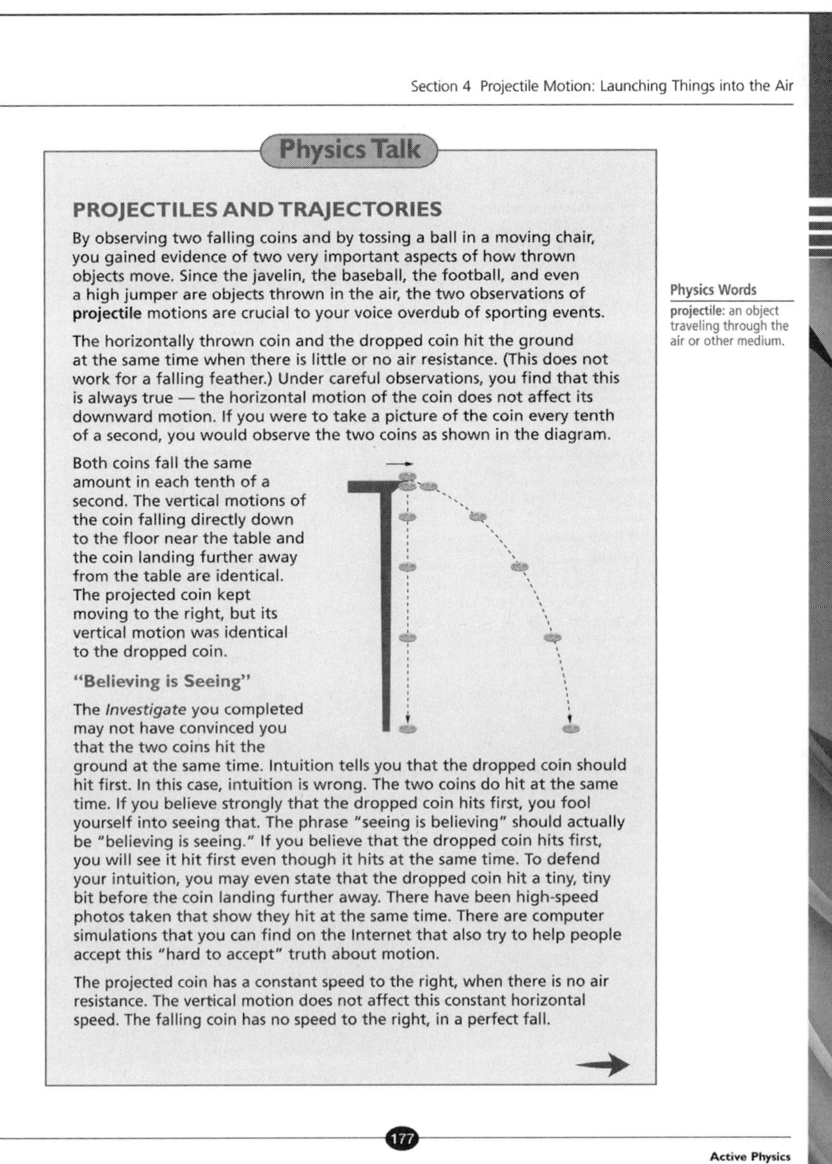

Section 4 Projectile Motion: Launching Things into the Air

Physics Talk

PROJECTILES AND TRAJECTORIES

By observing two falling coins and by tossing a ball in a moving chair, you gained evidence of two very important aspects of how thrown objects move. Since the javelin, the baseball, the football, and even a high jumper are objects thrown in the air, the two observations of **projectile** motions are crucial to your voice overdub of sporting events.

The horizontally thrown coin and the dropped coin hit the ground at the same time when there is little or no air resistance. (This does not work for a falling feather.) Under careful observations, you find that this is always true — the horizontal motion of the coin does not affect its downward motion. If you were to take a picture of the coin every tenth of a second, you would observe the two coins as shown in the diagram.

Both coins fall the same amount in each tenth of a second. The vertical motions of the coin falling directly down to the floor near the table and the coin landing further away from the table are identical. The projected coin kept moving to the right, but its vertical motion was identical to the dropped coin.

"Believing is Seeing"

The *Investigate* you completed may not have convinced you that the two coins hit the ground at the same time. Intuition tells you that the dropped coin should hit first. In this case, intuition is wrong. The two coins do hit at the same time. If you believe strongly that the dropped coin hits first, you fool yourself into seeing that. The phrase "seeing is believing" should actually be "believing is seeing." If you believe that the dropped coin hits first, you will see it hit first even though it hits at the same time. To defend your intuition, you may even state that the dropped coin hit a tiny, tiny bit before the coin landing further away. There have been high-speed photos taken that show they hit at the same time. There are computer simulations that you can find on the Internet that also try to help people accept this "hard to accept" truth about motion.

The projected coin has a constant speed to the right, when there is no air resistance. The vertical motion does not affect this constant horizontal speed. The falling coin has no speed to the right, in a perfect fall.

Physics Words

projectile: an object traveling through the air or other medium.

177

Active Physics

2-4a **Blackline Master**

Physics Talk

Few students have thought much about trajectories. Emphasize that a projectile aimed horizontally will always hit the ground below the point of aim by some amount. In order to hit a target at the same elevation, but at a distance, a projectile must rise first, and then fall. Students who have tried archery might know that they have to aim upward to get an arrow to hit a distant target. Good field and woods archers may even check for obstructions above their line of sight to be sure that the arrow can get to the target without hitting a branch on the way.

Students may take a while to believe that the horizontal motion of an object does not affect its vertical motion, and that the horizontal velocity remains constant throughout the fall. Point out that the value of acceleration of an object thrown upward will be the same as the value of acceleration of that object falling down. However, the object's velocity on its way down will carry a negative sign. Quizzing students on the investigations they made earlier would help them to understand this phenomenon. At the same time, reading the *Physics Talk* should provide focus and explain the *Investigate*.

Checking Up

1.

They will reach the ground at the same time because all objects fall to the ground with the same acceleration, at the rate of 9.8 m/s^2.

2.

As an object falls, its vertical velocity is constantly increasing because it is accelerating due to the unbalanced force of gravity.

3.

At the ball's point of highest rise, the velocity momentarily goes to zero as the ball is changing directions from up to down, and the acceleration is the acceleration of gravity, or 9.8 m/s^2 downward.

2-4b Blackline Master

Chapter 2 Physics in Action

Any hit or thrown ball travels horizontally and vertically. The horizontal velocity remains the same (if there is no air resistance). The vertical velocity is constantly changing. As it rises, the ball slows down. As it falls, the ball speeds up. The change in velocity of the ball is always 9.8 m/s every second or 9.8 m/s^2. For ease of discussion and problem solving, it is sometimes convenient to round this number to be 10 m/s every second or 10 m/s^2. Since the acceleration is always down to Earth, use –10 m/s^2 as the value. Think of any velocity in the "up" direction as + and any velocity in the down direction as –.

If an object is thrown straight up at 40 m/s, its velocity decreases by 10 m/s every second. Its speed at the end of each second is shown in the top diagram.

It comes to rest at the top of the path because its velocity is 0 m/s. Its acceleration will still be –10 m/s^2 because its speed is still changing by –10 m/s every second. Its new speed is –10 m/s one second after it begins its fall.

The horizontal speed of the object will remain constant since no force acts on the ball horizontally.

These two motions can be combined to allow you to mathematically predict the motion of a thrown object. If you space the horizontal position of the ball at equal distances as it rises and falls, you can represent the motion of the ball.

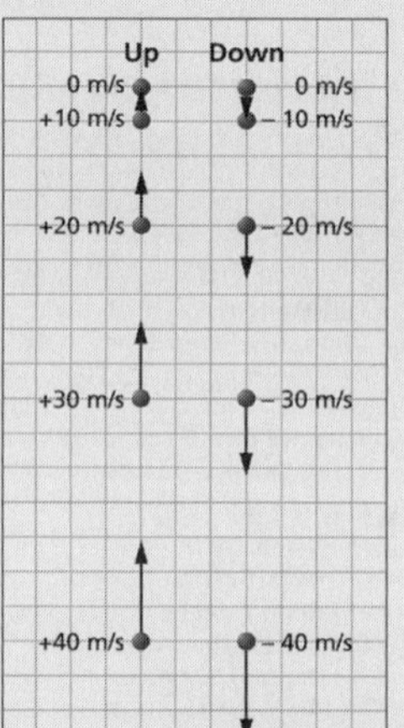

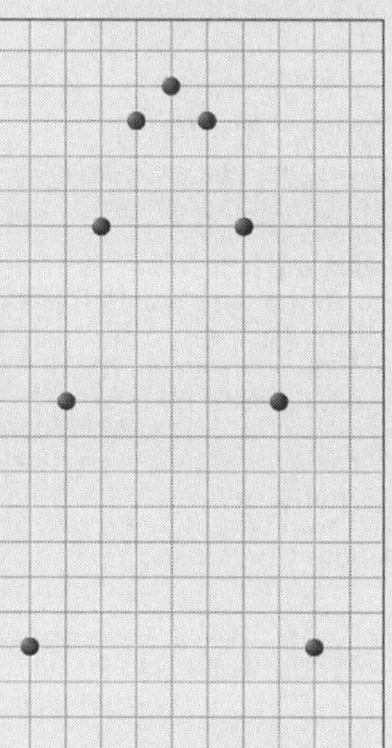

Checking Up

1. If a pen and a ruler are dropped together from the same height, will they reach the ground at the same time? Explain your answer.
2. When an object falls vertically down, does its velocity remain the same? Explain your answer.
3. If a ball is thrown upward, what is the ball's velocity at its point of highest rise? What is the ball's acceleration?

Active Physics 178

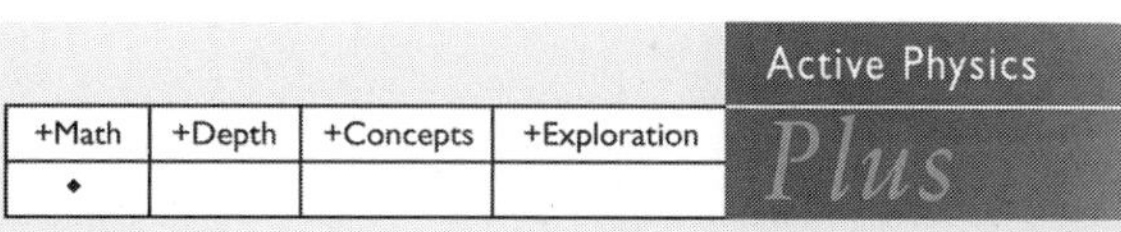

+Math	+Depth	+Concepts	+Exploration
♦			

Vector Components

In the investigation you just completed, the projected coin left the table horizontally. At any point in its motion, the projected coin is moving down and to the right. You can draw its velocity at any time. This velocity has two parts. One part describes the horizontal motion and the second part describes the vertical motion.

A short time after leaving the table, the projected coin has a small vertical speed and a constant horizontal speed.

You can add these parts as vectors. To add these two velocity vectors, use the "tip-to-tail" method. By sliding one vector over (maintaining its length and direction), the resultant is then drawn from the tail of the first vector to the tip of the second vector.

When you look at the coin's velocity some time later, you notice that the coin is moving faster in the vertical direction but continues horizontally at the same speed. If you have the values for the speeds of the vertical and horizontal motions of the coin, you can add the vectors to determine what happens to the total (or resultant) vector.

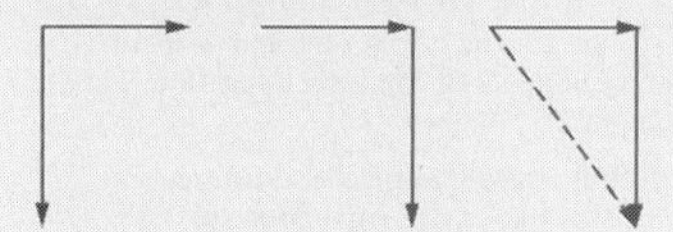

The resultant velocity or total velocity (often simply called the coin's velocity) has become larger and its direction has changed. The coin's resultant or total velocity is pointing in a more vertical direction.

If you measure the velocity at any one point in the path, you could also use that resultant or total velocity vector to find its horizontal and vertical "components." The components of a vector are themselves vectors, namely, the vectors along two perpendicular axes that add up to the vector. First, you draw the total or resultant velocity vector to the correct size and pointing in the correct direction. Second, you draw horizontal and vertical axes from the tail of the vector. Third, you draw lines from the tip of the vector to each axis, making sure the lines are parallel to the other axis. By doing this, you can obtain the horizontal and vertical components of the velocity (the vectors that add together to produce the total or resultant velocity vector).

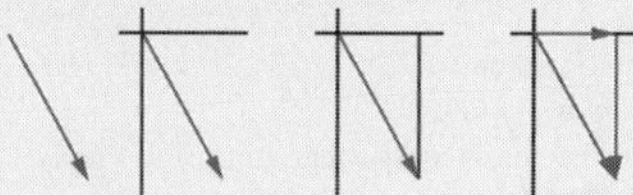

If you were to construct numerous velocity vectors along the path of the object, you would notice two things. First, the horizontal velocity components are always equal. Second, the vertical velocity components increase as time goes on.

Active Physics Plus

As this *Active Physics Plus* focuses mainly on understanding velocity vectors through solving problems, encourage students to draw vectors representing speed and direction to understand the direction of the resultant vector. The students should be able to explain each step of the problem's solution, as vectors can sometimes be confusing. They should also be able to tell when velocity carries a negative sign.

CHAPTER 2

1.a)

The time an object spends in the air depends upon its initial vertical velocity. The force of gravity will provide a downward acceleration of -9.81 m/s^2 (approximated as -10 m/s^2) for each second the object is in the air. The acceleration is negative as it is downward compared to the positive upward initial velocity of the ball. When the football is kicked with a vertical velocity of 30 m/s, its vertical speed will change 10 m/s for each second of flight. For an object that is thrown upward and returns to the same elevation, the rise time is equal to the fall time. This can be shown by using the equation for accelerated motion developed in *Chapter 1, Section 5.*

$$s = v_i t + \tfrac{1}{2} a t^2$$

Because the ball is coming back to the same height, the final position is zero. This gives

$$0 = 30 \text{ m/s}\,(t) + \tfrac{1}{2}\left(-10 \text{ m/s}^2\right)t^2.$$

Solving for t we have

$t = \dfrac{30 \text{ m/s}}{5 \text{ m/s}^2} = 6 \text{ s}$ for the time to return to the ground.

The vertical speed of the ball then will be

$t = 0$	$v = 30$ m/s
$t = 1$	$v = 20$ m/s
$t = 2$	$v = 10$ m/s
$t = 3$	$v = 0$ m/s
$t = 4$	$v = -10$ m/s
$t = 5$	$v = -20$ m/s
$t = 6$	$v = -30$ m/s

The horizontal speed of the ball is always the same while in the air, 10 m/s, and does not change with the changing vertical velocity.

Sample Problem

a) A football is thrown at 20.0 m/s at an angle of 30° with respect to the horizontal. What is its horizontal velocity (often called the x-component of its velocity)?

b) If the football were thrown at 20.0 m/s at an angle of 60°, what is its horizontal velocity?

c) How far does each football travel in the horizontal direction in 3.0 s?

Strategy: You can solve the first two parts by drawing vector diagrams to scale and finding the x-components. In *c)*, you can find how far each football traveled by using the relationship for steady motion:

Distance = (velocity) × (time)

$d = vt$

Solution:

a) The first vector must be 20 units long at an angle of 30°. (The scale is 1 unit = 1 m/s.) Use a protractor to draw the angle accurately.

Measuring the x-component and using the scale, you find the x-component is 17.3 m/s.

b) The second vector is also 20 units long at an angle of 60°.

Measuring the x-component and using the scale, you find the x-component is 10 m/s.

c) The first trajectory has a horizontal velocity component of 17.3 m/s for 3.0 s. Its distance is:

$$\begin{aligned} d &= vt \\ &= (17.3 \text{ m/s})(3.0 \text{ s}) \\ &= (17.3 \times 3.0)\left(\frac{\text{m}}{\cancel{\text{s}}} \times \cancel{\text{s}}\right) \\ &= 51.9 \text{ m or } 52 \text{ m} \end{aligned}$$

The second trajectory has a horizontal velocity of 10 m/s. Its distance is:

$$\begin{aligned} d &= vt \\ &= (10 \text{ m/s})(3.0 \text{ s}) \\ &= 10 \times 3.0\left(\frac{\text{m}}{\cancel{\text{s}}} \times \cancel{\text{s}}\right) \\ &= 30 \text{ m} \end{aligned}$$

1. A football is kicked with a vertical velocity of 30 m/s and a horizontal velocity of 10 m/s.

   a) Calculate the vertical and horizontal velocities for each second that the football is in the air.

   b) Draw a diagram showing the vertical and horizontal positions of the ball after each second.

2. A batted baseball leaves the bat with a velocity of 50 m/s at an angle of 30° from the horizontal.

   a) If the ball leaves the bat at time equal to zero, what are the horizontal and vertical components of the velocity at time equal to zero?

   b) What are the horizontal and vertical components of the velocity at time equal to 1 s?

   c) What are the horizontal and vertical components of the velocity at time equal to 5 s?

   d) What is the magnitude and direction of the velocity at time equal to 5 s?

1.b)

The students would not be expected to go through the solution below, but only to provide the sketch shown. The solution below is for the teacher who may wish to share this with the more interested students.

The vertical position of the ball may be obtained from the same equation as above

$$s = v_i t + \tfrac{1}{2} a t^2$$

and substituting times of 1-6 s into the equation. The vertical and horizontal positions at the various times would be as follows:

Time (t) in seconds	Vertical distance in meters	Horizontal distance in meters
0	0	0
1	25	10
2	40	20
3	45	30
4	40	40
5	25	50
6	0	60

Plotting the points will give a parabola with a trajectory similar to the sketch below.

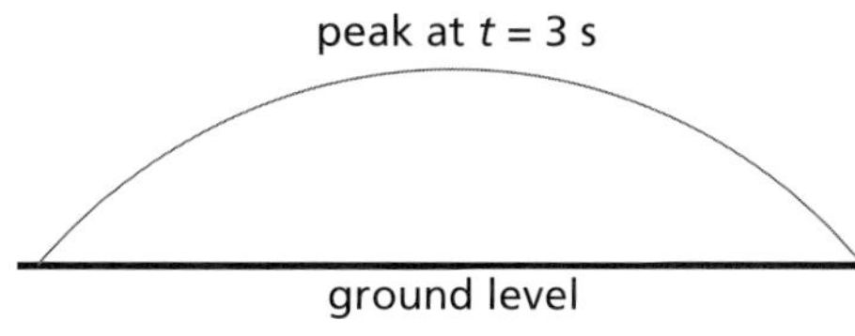

2.a)

To find the vertical and horizontal components of the ball's initial velocity

$v_{\text{vertical}} = v(\sin\theta) =$

$(50\text{ m/s})(\sin 30^\circ) = 25\text{ m/s}$

$v_{\text{horizontal}} = v(\cos\theta) =$

$(50\text{ m/s})(\cos 30^\circ) = 43.3\text{ m/s}$

2.b)

$v_{\text{vertical}} = 15\text{ m/s}$

$v_{\text{horizontal}} = 43\text{ m/s}$

2.c)

$v_{\text{vertical}} = -24\text{ m/s}$

$v_{\text{horizontal}} = 43\text{ m/s}$

2.d)

50 m/s at an angle of 30° below the horizontal (found from using the Pythagorean theorem).

NOTES

CHAPTER 2

NOTES

What Do You Think Now?

At this point, the students should know that the horizontal velocity of a projectile determines the range. It should be clear to them that the vertical velocity does not determine how far an object will travel. You might want to share the answers in the *Physicist's Response* with your class.

Providing a platform for recalling previously learned concepts that will strengthen their confidence and problem-solving skills.

What Do You Think Now?

At the beginning of the section, you were asked the following:

- **What determines how far an object thrown into the air travels before landing?**

From the observations you made in the *Investigate* section, what do you now think determines how far an object travels after it is thrown?

Physics Essential Questions

What does it mean?

A very important principle of physics is that motion in the horizontal direction and motion in the vertical direction are independent of each other. What does it mean to say that the horizontal motion of a projectile is independent of the vertical motion of the same projectile?

How do you know?

What evidence do you have to convince yourself that the horizontal and vertical motions of a projectile are independent of each other?

Why do you believe?

Connects with Other Physics Content	Fits with Big Ideas in Science	Meets Physics Requirements
Force and motion	Models	* Experimental evidence is consistent with models and theories

* Physics attempts to explain as much as possible with a single concept. Both the motion of a projectile and a swimmer crossing a river can be understood if their horizontal and vertical motions are considered independently. Give a reason why you believe that all motion can be examined in this way.

Why should you care?

Many sports involve projectiles. Think of a sport where a projectile is involved and describe how the independence of its horizontal and vertical motions explains its trajectory.

Reflecting on the Section and the Challenge

In *Part A* of this *Investigate* (two falling coins), you observed that the time required for a coin to fall is independent of its horizontal speed. If two long jumpers rise to the same height, they will then remain in the air for identical times.

In *Part B* of the *Investigate* (the rolling chair), you saw that the faster the chair is moving, the farther the ball will travel horizontally. If a long jumper is able to increase horizontal speed, then the jumper will travel farther.

Most sports have objects or people "flying through the air." You can describe how projectile motion relates to a sport you might choose for the challenge.

Reflecting on the Section and the Challenge

Have your students list sporting events that involve projectile motion (long jump, high jump, hurdles, shot put, discus, javelin, hammer throw, baseball, tennis, golf, badminton). Ask them to consider if some of these events have greater dependence on knowledge and technique and less dependence on physical ability than other events. Events involving projectile motion may have high potential to be improved through physics-based help. You may wish to point out that Carl Lewis, one of the fastest men in the world, is one of the world's greatest long jumpers. Jesse Owens' 1936 Olympics performance in the sprints and the long jump also makes a very good story.

CHAPTER 2

Physics Essential Questions

What does it mean?

All objects dropped from a certain height will hit the ground at the same time. This happens regardless of whether the object has a horizontal speed.

How do you know?

When the two coins left the table, the dropped coin and the forward-moving coin hit the ground at the same time. The conclusion is that the forward motion did not affect the vertical motion.

Why do you believe?

A bullet can travel much faster than the coin, but the horizontal motion is still independent of the vertical motion.

Why should you care?

A fast sprinter is also a good broad jumper. The jumper tries to gain maximum height so that the horizontal speed can allow him to travel as far as possible for as long a time as possible.

Physics to Go

1.

Students provide a sketch that should show one coin dropping nearly vertically and the other arcing outward.

2.

Students provide a sketch that should be similar to the first one but with a longer range for the projected coin.

3.

Students provide sketch, similar to the others but with a very long range for the projected bullet. Some students may invoke the curvature of Earth's surface.

Note: Aiming a rifle horizontally is not the same as actually shooting horizontally. Actually shooting horizontally would involve bore sighting, or using a level on the barrel instead of the sight posts.

4.

Answers will vary, but most people would not believe a bullet fired horizontally and one dropped would strike the ground simultaneously on level ground. Encourage students to confront their own misconceptions when answering this question. Flying discs or other airfoil devices contribute to the misconception, but the largest factor is probably that the bullet lands so far away that it is not perceived to have hit the ground until significantly after the dropped bullet.

Physics to Go

1. Draw a sketch of two coins leaving the table. Show where each coin is at the end of each tenth of a second. Remember to emphasize that they both hit the ground at the same time.
2. Repeat the sketch of the two coins leaving the table, but this time have one of the coins moving at a very high speed.
3. It is said that a bullet shot horizontally and a bullet dropped will both hit the ground at the same time if air resistance is neglected. Draw sketches of this (the bullet is like a very, very fast-moving coin).
4. Survey your friends and family members to find out which they think will hit the ground first, a bullet that is dropped, or a horizontally-shot bullet (neglecting air resistance).

   Explain why you think people may believe that the two coins hit the ground at the same time, but that they have a more difficult time believing the same fact about bullets.
5. Use evidence from your observations of the two coins in this section to prove that a 100 mi/h pitch thrown horizontally by a major league player will hit the ground in the same amount of time as a 10 mi/h pitch thrown horizontally from the same height by a child.
6. Use evidence from your observations of the ball and chair in this section to show the truth of the statement, "A projectile's horizontal motion has no effect on its vertical motion, and vice versa."
7. Look at the diagram of an arrow being shot horizontally from a bow and another arrow dropped from the same height. Arrow A is shot horizontally at a speed of 50 m/s. A second arrow, B is dropped from the same height and at the same instant as arrow A is released. Neglecting air friction, how does the time A takes to strike the horizontal plane (ground) compare to the time B takes to strike the horizontal plane?

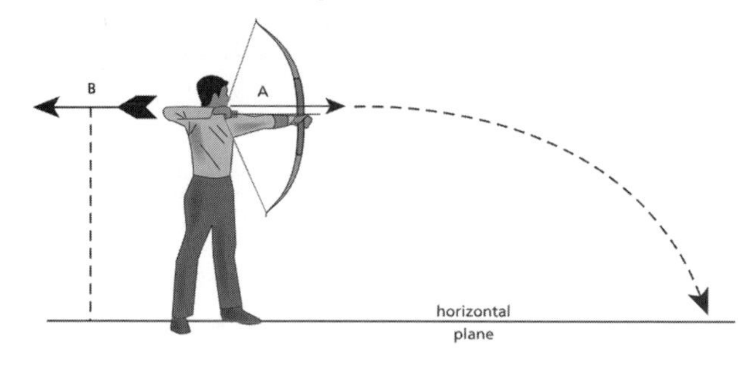

5.

The students should compare the coins that are flicked at different speeds hitting the ground at the same time as the dropped coin to the baseballs thrown at different speeds. They should make the logical jump that not only would a 100-mi/h and a 10-mi/h ball thrown horizontally hit the ground at the same time when released from the same height, but a dropped ball would also strike the ground at the same time.

6.

The ball would not have returned to the hands of the student riding in the chair if the horizontal and vertical motions were not independent of one another.

7.

The times are equal. The dropped arrow and the shot arrow hit the ground at the same time. The vertical motion is independent of the horizontal motion.

Use a protractor for *Questions 8 – 10.*

8. A swimmer jumps into a river and swims directly for the opposite shore at 2.0 km/h as shown in the diagram. The current in the river is 3.0 km/h and flows from left to right in the diagram. What is the swimmer's velocity relative to the shore?

opposite shore

current 3.0 km/h

9. Active Physics Plus A football is thrown at 15 m/s at an angle of 37° in the horizontal direction.
   a) What is its velocity in the horizontal direction?
   b) How far in the horizontal direction has the football traveled in 2.0 s?

10. Active Physics Plus A shot put is released at 12 m/s at an angle of 45° in a horizontal direction.
   a) What is its velocity in the horizontal direction?
   b) How far in the horizontal direction has the shot put traveled in 0.5 s?

11. ***Preparing for the Chapter Challenge***
   Write a script for a sports telecast that describes the motion of a baseball while it is pitched and then hit into the outfield.

Inquiring Further

Investigating more x- and y-components of motion

Ask another student to roll a marble slowly across the table in front of you. As the marble rolls by, apply a momentary force to it with a small object such as a block of wood. Make sure the force you apply is directly away from you and perpendicular to the initial velocity of the marble. If you define the x-direction to be along the initial velocity of the marble, then you can define the y-direction as the direction of the force you are applying. Investigate whether you can change the x-component of the marble's velocity by applying a force in the y-direction. Relate your observation to what you have learned about projectile motion.

8.

Application of the Pythagorean theorem yields:

$(3.0 \text{ km/h})^2 + (2.0 \text{ km/h})^2 = v^2$

Therefore, $v = 3.6$ km/h

Therefore, using the tangent button on the calculator or a vector diagram, the angle is 34°, relative to the shoreline.

9.a)

To find the x-direction component of the velocity, make a scale diagram and measure the x-component. You can also use the cosine function:

$v_x = (15 \text{ m/s})(\cos 37^\circ) = 12.0 \text{ m/s}$

9.b)

If it travels horizontally at 12.0 m/s for 2.0 s, the distance traveled is

$v_x t = d = (12.0 \text{ m/s})(2.0 \text{ s}) = 24 \text{ m}$

10.a)

To find the x-direction component of the velocity, make a scale diagram and measure the x-component. You can also use the cosine function:

$v_x = (12 \text{ m/s})(\cos 45^\circ) = 8.5 \text{ m/s}$

10.b)

If it travels horizontally at 8.5 m/s for 0.5 s, the distance traveled is:

$d = v_x t = (8.5 \text{ m/s})(0.5 \text{ s}) = 4.25 \text{ m}$

Preparing for the Chapter Challenge

11.

Answers will vary. A good answer will include the pitcher throwing the ball with an initial horizontal velocity that is immediately acted upon by gravity as it begins to fall in the vertical direction while maintaining its horizontal velocity. When the ball meets the bat, the force of the bat changes the ball's direction, accelerating it to a new velocity directed upward and horizontally toward the outfield. As the ball rises, it immediately begins to lose its vertical velocity because it is always being accelerated downward by the force of gravity. At the peak of the parabolic trajectory the vertical velocity is zero, after which the downward acceleration of the force of gravity causes the ball to reverse direction, and it begins to fall, gaining vertical speed. All along, the horizontal velocity of the ball has been the same until the ball reaches the level of the bat in the outfield where it will have the same velocity as when it left the bat. (Note: The above description is not descriptive of what occurs to a real baseball in air. The force of air resistance has a tremendous effect on both the ball's speed and its trajectory. For a more detailed description of the ball's true path, a student may want to read *The Physics of Baseball* by Robert Adair.)

Inquiring Further

When doing this section, students will find that they cannot change the x component of the marble's speed with a force applied in the y direction. This is the same as the vertical force of gravity not having an effect on the horizontal motion of a projectile.

NOTES

SECTION 4 QUIZ

2-4c **Blackline Master**

Base your answers to Questions 1 and 2 on the diagram, which shows a ball projected horizontally with an initial velocity of 20 m/s off a cliff 100-m high (neglect air resistance).

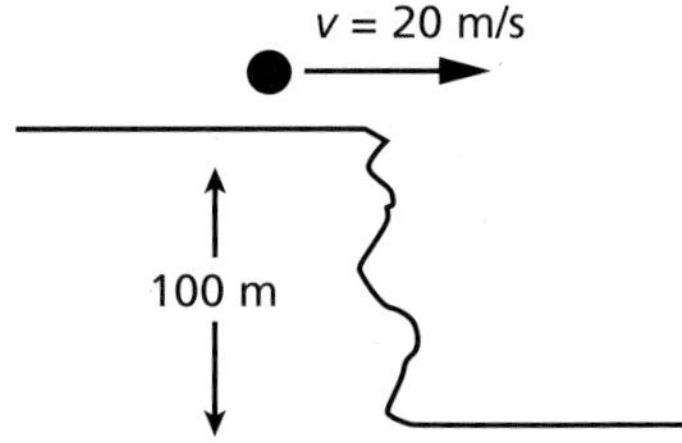

1. Which of the following has no effect on how far the ball travels horizontally before it strikes the ground?

   a) the mass of the ball

   b) the height of the cliff

   c) the speed with which the ball is thrown

   d) the angle at which the ball is thrown

2. During the flight of the ball, what is the direction of its acceleration?

   a) downward
   b) upward

   c) westward
   d) eastward

3. A physics student is doing an experiment at the top of the school football bleachers, where she is going to throw an egg horizontally with a speed of 10 m/s. At the instant she throws the egg, she accidentally knocks over another egg which falls an equal distance to the ground. Which egg will hit the ground first?

   a) the dropped egg

   b) the thrown egg

   c) Both will hit the ground at the same time.

4. A ball is thrown horizontally at a speed of 20 m/s from the top of a cliff. As the speed of the ball is decreased, the time required for the ball to fall 30 m will

   a) decrease.

   b) increase.

   c) remain the same.

5. A ball is tossed straight upward in an airplane doing 500 mi/h in a straight line at constant altitude. When the ball lands in the cabin its position relative to the person who tossed it will be

   a) farther toward the back of the cabin depending upon the plane's speed.

   b) farther toward the front of the cabin depending on the plane's speed.

   c) right back down on the person who tossed it straight upward.

   d) farther toward the back of the plane if the ball is light, but farther toward the front if the ball is heavy.

SECTION 4 QUIZ ANSWERS

1. a) Disregarding air resistance, the acceleration due to gravity is independent of the object's mass for objects that are much smaller than Earth.

2. a) The only force acting on the object is the force of gravity which will cause the object to accelerate downward.

3. c) Both eggs will strike the ground at the same time. The independence of right-angle motions indicates that the horizontal velocity now has an effect on the fall time.

4. c) The horizontal velocity does not affect the time it takes an object to fall a certain distance.

5. c) The ball will land right back in the person's hands. The initial forward motion of the ball was the same as the airplane before it was tossed. The ball retains that initial velocity as it is tossed, and matches the forward motion of the passenger and the plane as it rises and falls to land back in the passenger's hands. Mass has no effect because the ball, regardless of its mass, would have the same velocity as the plane. If this were done in an open convertible where wind resistance was a factor, a light object would be expected to land farther back due to the extra horizontal force of the wind.

NOTES

CHAPTER 2

SECTION 5

The Range of Projectiles: The Shot Put

Section Overview

In this section, students create and compare mathematical and physical models of a projectile's motion to improve performance in events where trajectories occur. Students first measure the acceleration due to gravity, using one of two methods chosen by the teacher, then use this value to calculate an object's position and speed at any time after it enters a state of free fall. In addition, they use this knowledge for a quantitative treatment of the horizontal (constant speed) and vertical (free fall) components of the motion of a projectile. They model the trajectory of a projectile and find some of the variables affecting the height and range of a projectile. Through the *Investigate*, they learn to apply mathematical and physical models to analyze the speed of a falling object. In the mathematical model, students calculate the average speed of an object dropped vertically, using the value of the acceleration due to gravity. In the physical model, they construct a representation of the path of a projectile that is launched at an angle where its velocity has both a horizontal and a vertical component.

Background Information

Acceleration Due to Gravity: All objects in a state of free fall, regardless of mass, shape, or other characteristics, experience a uniform (constant) acceleration, g, of approximately 10 m/s^2 in the downward direction. (Falling objects which experience significant air resistance are not considered to be in a state of free fall.) Earth exerts an amount of downward gravitational pull on objects at or near its surface that is proportional to the mass of each object. The outcome of this is that all objects experience the same amount of force per unit of mass and, therefore, the same acceleration. For example, Earth's pull on a 2-kg rock is twice as much as the pull on a 1-kg rock; the result is that both accelerate at the same rate during free fall. *Active Physics* "rounds off" g to 10 m/s^2, which is within two percent of the average value, about 9.81 m/s^2 on our planet. The acceleration due to gravity depends on the radial distance from the center of Earth; therefore, the value of g varies with location by as much as 0.04 m/s^2 due to terrain differences (mountains, valleys), local rock composition, and the fact that Earth is not a perfect sphere (the equatorial diameter is slightly greater than the polar diameter). Some argue that local variations in g, if taken into account, would sometimes affect performance in a track and field event. For example, a shot-put result at a location with a slightly higher value of g may actually be better than a longer throw made at a location having a lower value of g. Physicists have determined that variations in acceleration due to gravity sometimes could make a difference in records in field events: distance, speed, acceleration, and time of free fall would all be affected by this variation. This section follows arguments originally presented by Galileo regarding free fall.

Galileo hypothesized that free fall involved constant or uniform acceleration. From the definition of constant acceleration, $a = \Delta v/\Delta t$, he reasoned that the speed attained by an object falling from a "rest start" at any time during its fall would be $v = \Delta v = a(\Delta t)$. (The speed, v, at the end of a time of fall, Δt, would be equal to the change in speed, Δv, because the initial speed for a rest start is zero.) Galileo further reasoned that, because the speed of a falling object increases uniformly (or at a constant rate) with time (his hypothesis), the average speed for a time of fall, Δt, would be half of the speed attained at the end of a time of fall. Therefore,

Average speed at the end of a time interval,

$v_{average} = \Delta v/2 = a(\Delta t)/2.$

Finally, Galileo reasoned that the distance of fall and the end of a time of fall, Δt, would be, simply, the average speed multiplied by the time of fall. Therefore, fall distance at the end of time interval, t, is

$$d = \frac{a(\Delta t)}{2} \times \Delta t = \tfrac{1}{2} a\Delta t^2.$$

The above reasoning was used as the strategy for developing the table in *Step 2* of the *Investigate* for this section as an extension of the above for your understanding as the teacher. Galileo was limited in his ability to test his assumptions and reasoning. He did not have adequate instruments to measure acceleration and speed directly, so he devised a test of his thinking which involved what he could measure. From the last line in the above derivation, $d = \tfrac{1}{2} a\Delta t^2$, he solved acceleration as follows:

$$a = \frac{2d}{\Delta t^2}$$

Rolling a sphere down a ramp to "dilute" the effect of gravity (he assumed that a sphere rolling down a ramp also had constant acceleration, but less of it than an object in free fall) he used a crude timing device to measure the amounts of time for the sphere to roll, starting from rest, several measured distances along the ramp. Upon substituting pairs of distance and time measurements into the relation $a = 2d/\Delta t^2$, he obtained a fixed value for the right-hand side of the relation, proving that the acceleration indeed is constant.

The Trajectory of a Projectile

The steps used to develop the model of a trajectory in this section provide a fine quantitative example of the independence of the vertical and horizontal components of the motion of a projectile. While it is not expected that students will predict the range, maximum height, time of flight, and other parameters of a projectile in terms of launch speed and direction, the following equations will prepare you to do so, if needed. For a projectile launched horizontally at speed v from height, h, time of flight, $t = \sqrt{(2h/g)}$, and range (horizontal distance), $R = vt = v\sqrt{(2h/g)}$. For the general case of a projectile launched from ground level at speed v at an angle θ above the horizontal, and traveling over flat ground the following relationships apply:

speed in the horizontal direction is $v_x = v\cos\theta$ (remains constant)

initial speed in the vertical direction is $v_y = v\sin\theta$

speed in the vertical direction at time, t, is $v_y = (v\sin\theta) - gt$

horizontal position at time, t, is $x = (v\cos\theta)t$

vertical position at time, t, is $y = (v\sin\theta)t - \tfrac{1}{2}gt^2$

time to reach maximum height is $t_{max} = (v\sin\theta)/g$

total time of flight is $t = 2t_{max}$

range (horizontal distance) is $R = (v^2 \sin 2\theta)/g$

Note: In the final of the above equations that $\sin 2\theta$ (and, therefore, also the range) has a maximum value of 1 when $\theta = 45^\circ$, $(2\theta = 90^\circ)$. This expression is only true for a projectile returning to the level from which it was launched. If the projectile lands higher or lower than the launch position (for example in the shot put), a more complex relationship is required to predict the optimal angle and range.

CHAPTER 2

Crucial Physics

- Projectiles travel in parabolas (if we ignore the effects of air resistance).
- The horizontal motion and vertical motion of a trajectory are independent of one another.
- A projectile will travel the greatest distance if the angle of release is 45°. At angles less than 45°, the projectile has a greater horizontal speed but does not stay in the air as long. At angles greater than 45°, the projectile stays in the air a longer time, but has a smaller horizontal velocity.
- The acceleration due to gravity can be determined experimentally and is found to be equal to 9.8 m/s^2.
- Mathematical models of projectiles:
    - When the acceleration is constant, the average speed during an interval of time is equal to half the sum of the initial and final velocities for the time interval. If the object is at rest at the beginning of the time interval, the average speed is half the velocity at the end of the time interval.
    - When the acceleration is constant, the distance traveled during an interval of time is equal to the average velocity during the time interval multiplied by the interval of time. If the object is at rest at the beginning of the time interval, the distance traveled is equal to half the velocity at the end of the interval of time multiplied by the interval of time.
    - If the acceleration is constant, the acceleration equals the velocity at the end of the time interval minus the velocity at the beginning of the time interval divided by the time interval. If the object is at rest at the beginning of the time interval, the acceleration is equal to the velocity at the end of the time interval divided by the time interval.
    - If the acceleration is constant and the object starts from rest, then the distance traveled over a time interval is equal to half the acceleration multiplied by the square of the time interval.

Learning Outcomes	Location in the Section	Evidence of Understanding
Measure the acceleration due to gravity.	***Investigate*** Step 1	Students measure acceleration due to gravity, *g*, by using a "picket fence" and a photogate timer attached to a computer. The teacher could also ask students to use an alternative method using a ticker-tape timer and a mass.
Calculate the speed attained by an object that has fallen freely from rest.	***Investigate*** Step 2.b)	Students calculate and record the speed of a falling object at every 0.10 s of its fall. Students complete a table using an example in their *Active Physics* textbook.
Identify the relationship between the average speed of an object that has fallen freely from rest and the final speed attained by the object.	***Investigate*** Step 2.c)	Students calculate the average speed of a falling object by finding the average of the initial and final speeds. Since the initial speed is zero, the average speed comes out to one-half of the final speed.
Calculate the distance traveled by an object that has fallen freely from rest.	***Investigate*** Step 2.c)	Students calculate and record the distance an object falls at the end of each 0.10 s of its fall by using the equation, distance = (Average speed × time).
Use mathematical models of free fall and uniform speed to construct a physical model of the trajectory of a projectile.	***Investigate*** Step 3	Students put together two identical string and mass assemblies, with the string length equal to the distance of fall and the time of fall assigned to each group. The string separation is determined by the horizontal speed of the projectile.
Use the motion of a real projectile to test a physical model of projectile motion.	***Investigate*** Steps 7-10	A student volunteer throws a tennis ball horizontally to match the curve connecting the position of assembly masses on the strings.
Use a physical model of projectile motion to infer the effects of launch speed and launch angle on the range of a projectile.	***Investigate*** Steps 11.a), 11.d)	Students rest the end of the stick corresponding to 0.0 s on the tray at the bottom of the chalkboard and incline it at angles of 30°, 40°, 60°, and 90°. They then predict the greatest range of a projectile.

Section 5 Materials, Preparation, and Safety

Materials and Equipment

PLAN A		
Materials and Equipment	**Group (4 students)**	**Class**
Timer, ticker tape, AC	1 per group	
Scissors	1 per group	
Meter stick, wood	1 per group	
Ring stand, large	1 per group	
Clamp, extension	1 per group	
Holder, right angle, cast iron	1 per group	
Calculator, basic	1 per group	
Washer, 3/4 in. (outside diameter) x 5/16 in. (inside diameter)	2 per group	
Weight, slotted, 100 g	10 per group	
Ball, tennis		1 per class
Model, trajectory		1 per class
Weight, fishing, small		7 per class
Timer, photogate (with electromagnet)		1 per class
Ticker tape, roll	1 per group	
Notes, sticky, pad, 3 in. x 3 in.		3 per class
String, cotton, ball		1 per class
Chalkboard*		1 per class
Chalk*		1 per class
Probeware, photogate*		1 per class
Fence, picket*		1 per class
MBL or CBL Technology (to record probeware activity)*		1 per class
Ladder*		1 per class

*Additional items needed not supplied

PLAN B		
Materials and Equipment	**Group (4 students)**	**Class**
Timer, ticker tape, AC		1 per class
Scissors		1 per class
Meter stick, wood		1 per class
Ring stand, large		1 per class
Clamp, extension		1 per class
Holder, right angle, cast iron		1 per class
Calculator, basic	1 per group	
Washer, 3/4 in. (outside diameter) x 5/16 in. (inside diameter)	2 per group	
Weight, slotted, 100 g		10 per class
Ball, tennis		1 per class
Model, trajectory		1 per class
Weight, fishing, small		7 per class
Timer, photogate (with electromagnet)		1 per class
Ticker tape, roll		1 per class
Notes, sticky, pad, 3 in. x 3 in.		3 per class
String, cotton, ball		1 per class
Chalkboard*		1 per class
Chalk*		1 per class
Probeware, photogate*		1 per class
Fence, picket*		1 per class
MBL or CBL Technology (to record probeware activity)*		1 per class
Ladder*		1 per class

*Additional items needed not supplied

Note: Time, Preparation, and Safety requirements are based on Plan A, if using Plan B, please adjust accordingly.

Time Requirements

This *Investigate* requires two class periods or 80 min.

Teacher Preparation

- Assemble the materials required for the student groups to measure the acceleration due to gravity. The material required will depend upon the method chosen. If you wish to use the ticker-tape timer method, you will need masses to attach to the ticker tape to pull it vertically downward through the timer. Regardless of the method you choose, try several runs yourself prior to the students using the equipment to ensure it is working properly.
- Prepare the "portable" version of the mass and string assembly by attaching masses and strings of the same length as those students will be assembling on the board. These should be attached to a 2.4-m long piece of wood or similar device so that it may be tilted at an angle. If the 2.4-m length proves prohibitive to use in your classroom at the angles suggested, make plans to have the students try to match the trajectories outside or in an area with a higher ceiling.

Safety Requirements

- No particular safety precautions are required for this activity.

CHAPTER 2

NOTES

Meeting the Needs of All Students

Differentiated Instruction: Augmentation and Accommodations

Learning Issue	Reference	Augmentation and Accommodations
Understanding vocabulary	***What Do You Think?***	**Augmentation** • Students with memory and focus issues may struggle to remember the meaning of words including trajectory, projectile, range, and launch. If students drew pictures to represent these words in *Section 4*, ask them to refer back to these pictures before recording their ideas. • If students did not draw pictures in *Section 4*, remind them what these words mean before they record their ideas.
Measuring the acceleration due to gravity	***Investigate*** Step 1	**Augmentation** • Students may need the teacher to model the method the class will use to measure the acceleration due to gravity, especially if students are asked to use the second method with the ticker-tape timer. This method requires students to do calculations they learned earlier in the chapter, and students may not remember how to solve for velocity and acceleration using the ticker tape. **Accommodation** • Model this activity as a whole-group demonstration and ask for volunteers to help with calculations.
Creating a physical model as a group	***Investigate*** Steps 3-8	**Augmentation** • Encourage students to initiate tasks and stay focused, because in this part of the *Investigate* the learning of the whole class depends on everyone doing their task. • Provide pairs of students with a meter stick, string, mass, and some masking tape to construct their string-and-mass assemblies. Then tell students they have 4-7 minutes to construct the string-and-mass assembly. • Pair students strategically to include at least one student who can focus on a task and one who is proficient with measurement. **Accommodation** • Provide students with string-and-mass assemblies that are already constructed and direct them to hang the assemblies on the appropriate pin.
Reading comprehension	***Physics Talk***	**Augmentation** • Instruct students to create a graphic organizer or Venn diagram that names the two kinds of motion involved in projectile motion. Then ask pairs of students to list factors that affect each of these motions. **Accommodation** • Provide students with a list of factors and ask them to sort the factors according to which type of motion they affect.
Understanding key concepts	***Physics Talk*** ***Physics to Go*** Steps 1-3 and 8	**Augmentation** • This diagram of trajectories at different angles summarizes many of the key concepts and vocabulary words about projectiles. Provide students with a copy of the diagram and the statements below it. Then instruct students to use different-colored highlighters to make visual connections between the statements below the diagram and the trajectories in the diagram. Model the first example by highlighting the "45° launch angle" statement in yellow and then tracing that trajectory with the yellow highlighter. **Accommodation** • Provide students with serious visual motor or graphomotor issues with a copy of the diagram and statements that have already been color-coded.

Learning Issue	Reference	Augmentation and Accommodations
Reading comprehension	***Reflecting on the Section and the Challenge***	**Augmentation** • The previous augmentation and accommodation will help students with reading comprehension issues to understand the summary in this reflection. • Students can also use their highlighted diagram to assist them with the voice-over challenge.

Strategies for Students with Limited English-Language Proficiency

Learning Issue	Reference	Augmentation
Following complex procedures	***Investigate***	The steps in this *Investigate* run 4 pages. Break down the *Investigate* into smaller chunks that allow students to comprehend each portion of the activity before moving on to the next one. This approach will allow students to get comfortable following the procedures outlined within each step, to grasp the math, and to internalize new concepts introduced. Lead a brief class discussion after each step to allow students the opportunity to demonstrate acquired knowledge and understanding, and to allow you to correct misunderstandings.
Understanding scientific concepts	***Physics Talk*** ***Physics Essential Questions***	The scientific meaning of the word "model" can be difficult for all students to grasp. Now that students have worked with both a mathematical model and a physical model, explain that in science, a model is anything that accurately represents what we know of how the natural world behaves. This concept is fundamental to an understanding of how scientists work. Revisit the concept of "model" when students answer the "Why do you believe?" question.
Comprehension	***Physics Talk***	The shape of a parabola can be difficult to understand. All students may benefit from looking at and holding a parabola shape. You may wish to cut a few conic sections out of a cone-shaped paper party hat and pass them around for students to examine. Point out that the parabolas are the shapes that are "open," whereas the circles and ellipses are closed shapes.
Vocabulary comprehension	***Active Physics Plus***	ELL students and some other students may have difficulty inferring the meaning of "displacement." Point to the root words "place" and "displace," which give a clue to its meaning. Some students may be familiar with the concept of an object displacing water when it is submerged. Tell students that displacement means how far an object has moved from one place to another. Contrast position and displacement. Position is a scalar, and with this term there is no implicit history of how an object arrived at its position. Displacement, on the other hand, is a vector, and its definition implies specifying an initial position and a final position.
Vocabulary comprehension Comprehension	***Active Physics Plus*** Sample Problem	ELL students may struggle with the word "irrespective" in *Step 1*. Allow them adequate time to think through the possible meaning. Then provide additional guidance if necessary. You may suggest they think back to the meaning of "independent" in *Section 4*. Collaborate with the students' math teachers to determine what level of comprehension students have obtained for working with square roots.

SECTION 5

Teaching Suggestions and Sample Answers

What Do You See?

The *What Do You See?* illustration should give you ample opportunity to hook students' attention to a sport in which guessing the trajectory of a projectile becomes significant to winning or losing. You might want to ask students how familiar they are with soccer and how the illustration makes a connection with the topic. Soccer is a popular game and most students will respond instantly to the illustration. Ask them to use their knowledge of soccer in interpreting the images. Specific questions based on how soccer players are successful in striking goals will guide students towards more meaningful responses.

Chapter 2 Physics in Action

Section 5 The Range of Projectiles: The Shot Put

What Do You See?

Learning Outcomes

In this section, you will

- **Measure** the acceleration due to gravity.
- **Calculate** the speed attained by an object that has fallen freely from rest.
- **Identify** the relationship between the average speed of an object that has fallen freely from rest and the final speed attained by the object.
- **Calculate** the distance traveled by an object that has fallen freely from rest.
- **Use** mathematical models of free fall and uniform speed to construct a physical model of the trajectory of a projectile.
- **Use** the motion of a real projectile to test a physical model of projectile motion.
- **Use** a physical model of projectile motion to infer the effects of launch speed and launch angle on the range of a projectile.

What Do You Think?

A world record in the men's shot put of 23.12 m was set by Randy Barnes of the United States in 1990. In the women's javelin throw, Osleidys Menendez of Cuba broke the world record at 71.70 m in 2005.

- **Describe the trajectories of projectiles launched from the ground at various angles.**
- **Describe how a greater launch speed of a projectile might change the range when the launch angle is the same.**

Record your ideas in your *Active Physics* log. Be prepared to discuss your responses with your small group and the class.

Investigate

In this *Investigate*, you will measure the acceleration due to gravity. You will then use a mathematical model to construct a physical model of the trajectory of a projectile. Finally, you will use a real projectile to test the physical model.

1. Your teacher will provide you with a method of measuring the acceleration caused by Earth's gravity for objects in a condition of free fall.

184

Students' Prior Conceptions

Students need to use mathematical models of free fall and of uniform speed to construct a physical model of the trajectory of a projectile. They then need to analyze the motion of a real projectile to test this physical model. Measuring and analyzing real data encourages students to align prior conceptions to accepted theory and to understand how the constant force of gravity and the launch angle act to determine the range of the projectile.

1. **Students do not understand that the force of gravity is constant as it acts close to the surface of Earth.** This force is proportional to the constant acceleration of a projectile, not to its velocity. The speed and vertical velocity of a projectile change during the flight while the vertical acceleration remains constant. The speed of a projectile is constant in the horizontal direction and the horizontal acceleration is zero in the absence of air resistance.

2. **Students confound constant forces with constant motion.** This preconception may continue from *Sections 1* and *2*. The same is true for the following preconception.

3. **As soon as the propelling force is removed from a projectile, it slows down or stops.**

4. **Gravity simultaneously affects both perpendicular components of projectile motion.** Forces that act perpendicular to projectiles affect trajectories in different ways. Students do not intuitively understand this concept. It is important for students to measure the acceleration due to gravity and to calculate speeds at various points in the trajectory of a projectile in order to confirm that a projectile exhibits constant motion in the horizontal direction of its flight and accelerated motion in the vertical direction.

You might want to ask students what specific intent the artist could have in making the ball bounce over another player's head and why the range of the soccer ball is shown by dotted lines. Check if students know what a projectile is and whether they are familiar with the shot put. Encourage students to discuss their ideas without hesitating to think if they could be trivial or irrelevant. Discuss their answers and streamline their curiosity toward the *What Do You Think?* questions. Students should know that they will have a chance to build on their initial impressions, once they have investigated the concepts presented in this section in more detail.

What Do You Think?

These questions are designed to bring forth a variety of responses. It is important for students to know that you are not looking for correct answers, but only for a discussion of ideas. This is the time when you can find out what your students know about concepts related to this section. Most students will have a limited knowledge of terms like "free fall" and "launch angle." Ask them to record their answers in their *Active Physics* logs.

What Do You Think?

A Physicist's Response

Generally, for a fixed launch velocity (speed and angle), the range will be greater for a higher launch point (as when launching from a tower) and less for a lower launch point (as when launching toward a rising hill). The optimum range across level ground is attained for a 45°-launch angle. The range of a projectile across level ground is governed by the equation: $R = (v^2 \sin 2\theta)/g$ where R is the range, v is the initial speed, θ is the launch angle measured from the horizontal, and g is the acceleration due to gravity. (It is not suggested that this equation be presented to students with the intent of mastery, if at all.)

CHAPTER 2

NOTES

NOTES

Investigate

1.a)

Students may find the procedure difficult if they are not familiar with the ticker-tape timer. The students might initially suggest the activity could be done in a manner similar to that of *Section 2*, where the tape is cut into six-dot strips. Allow the students to drop a mass pulling the tape in order to discover that the time it takes the mass to fall will not provide sufficient dots for an extensive evaluation when the tape is cut into strips this size. Ask the students what size is appropriate, and they may suggest three- or two-dot strips. If the mass falls to the floor from a height of 1 m, there should be approximately 30 dots, or 10 three-dot strips, which should prove sufficient. Remind the students that a three-dot strip would represent ½ the time of a six-dot strip. From the above, the students should be able to obtain the data and do the calculations for finding acceleration due to gravity. Ask the students why the measurement is not done with a meter stick and stopwatch.

Teaching Tip

For the first part of this activity, two methods to measure the acceleration of gravity are detailed. In addition, the use of a motion detector and computer to make a graph of velocity vs. time for a falling object works well. The slope of the line then represents the acceleration. If no equipment, such as a motion detector or those mentioned in the activity, is available, the students may use a stopwatch to time the fall of something (e.g., a softball) over a distance of two meters several times and to obtain the average time. The students can then use the same procedure as discussed in *Step 2* to calculate the final velocity of the ball when it strikes the ground. To calculate the ball's acceleration, the students can use the formula for acceleration:

$$a = \frac{v_f - v_i}{t}$$

Section 5 The Range of Projectiles: The Shot Put

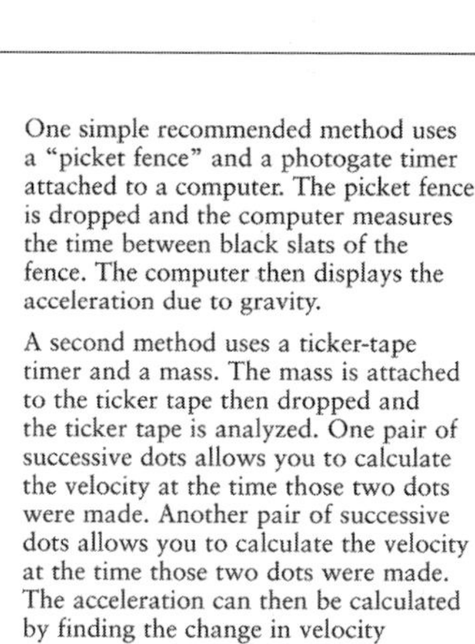

One simple recommended method uses a "picket fence" and a photogate timer attached to a computer. The picket fence is dropped and the computer measures the time between black slats of the fence. The computer then displays the acceleration due to gravity.

A second method uses a ticker-tape timer and a mass. The mass is attached to the ticker tape then dropped and the ticker tape is analyzed. One pair of successive dots allows you to calculate the velocity at the time those two dots were made. Another pair of successive dots allows you to calculate the velocity at the time those two dots were made. The acceleration can then be calculated by finding the change in velocity during the time between the first pair of dots and the second pair of dots. To increase the precision of the calculation, many pairs of dots can be used and an average acceleration can be found.

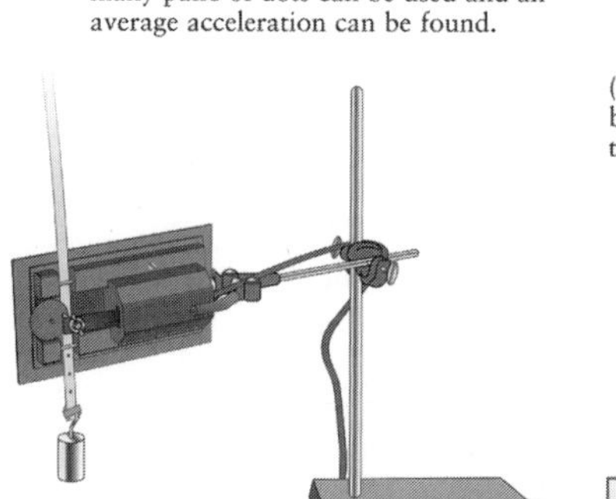

a) In your log, describe the procedure, data, calculations, and the value of the acceleration of gravity obtained. As you have learned, the acceleration due to gravity comes up often and has its own symbol, g.

2. After calculating the acceleration due to gravity (or using the value of $g = 10\ \text{m/s}^2$), you can use this knowledge to analyze the path of a projectile.

a) In your log, make a table similar to the following:

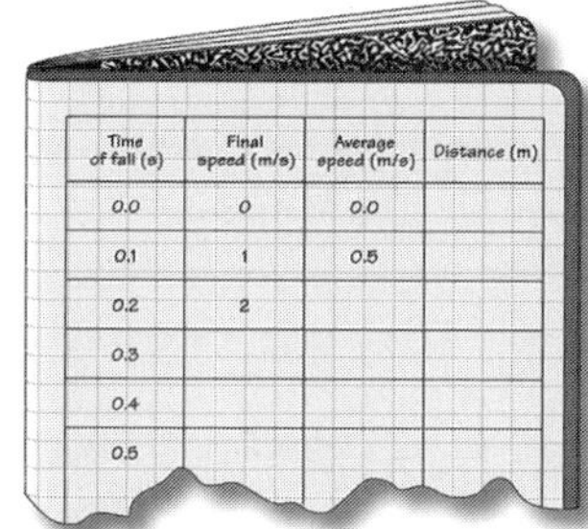

(Some data for a falling object has already been calculated and entered in the table to help you get started.)

b) In the table, calculate and record the speed of a falling object at the end of each 0.10 s of its fall for a total of 0.5 s. To simplify the calculations, use a rounded off value for g of $10\ \text{m/s}^2$. The first three values are provided in the second column. Complete the table using the example below as a guide.

Example:

What you know: $g = 10\ \text{m/s}^2$

Speed = acceleration × time

Speed at the end of 0.2 s = $(10\ \text{m/s}^2) \times (0.2\ \text{s})$

Speed = 2 m/s

185

Active Physics

CHAPTER 2

2.a)-b)

The completed table of calculated values of speed, average speed, and distance at 0.10 s intervals for an object falling from rest is shown to the right. (See chart. For clarity, significant digits have not been used.) Since g is limited to one significant figure, $10\ \text{m/s}^2$, the tabled values calculated using g also should be limited to one significant figure. Strict adherence would suggest that the fall distance of 1.25 m should be rounded to 1 m, but it perhaps is not advisable to distract students with that detail at this time – use your judgment on whether or not to bring this up.

Time of Fall (s)	Speed at End of Fall (m/s)	Average Speed (m/s)	Distance (m)
0.0	0	0.0	0.00
0.1	1	0.5	0.05
0.2	2	1.0	0.20
0.3	3	1.5	0.45
0.4	4	2.0	0.80
0.5	5	2.5	1.25

2.c)

See data table on previous page for values. Ask the students if the velocity increases in regular increments.

2.d)

See data table on the previous page for values. Ask the students if the distance increases in regular increments.

3.-4.

Students will assemble the string-and-mass assemblies assigned by the teacher from the data table. The students should make two identical copies of their assigned lengths of fall, one to be used in *Step* 6 and one for *Step 8*.

5.

From the amount of space you have for the hanging weights, determine the horizontal distance between pins and tell this distance to the students. Here is a schematic drawing of what the string-and-mass assembly should look like in the three cases.

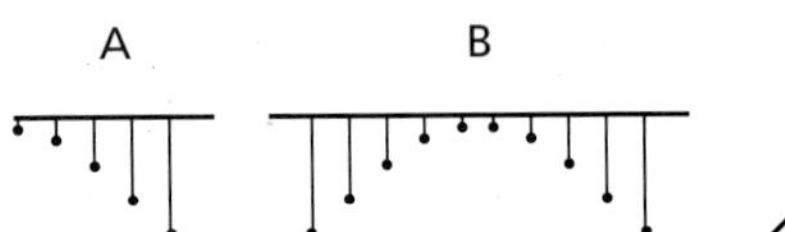

5.a)

If the horizontal distance between pins is 40 cm, the horizontal speed is 400 cm/s or 4 m/s. Ask: What is the force on this object you are modeling traveling horizontally with constant speed?

6)

Students mark end of string assembly on chalkboard.

c) When speeds are changing at a constant rate, then the average speed during a time interval is the average of the speeds at the beginning and the end of the time interval. Calculate and record the average speed for each time interval in the table. The falling object's speed has increased uniformly from zero to the final speed. In each time interval, the average speed will be the average of zero and the final speed reached at the end of each 0.10 s of falling. This average speed will come out to one half of the final speed.

> Example:
>
> Average speed =
>
> $$\frac{\text{zero + speed at the end of time interval}}{2}$$
>
> Average speed during 0.2 s of fall =
>
> $$\frac{(0 \text{ m/s} + 2 \text{ m/s})}{2}$$
>
> $= 1$ m/s

Complete the third column of the chart.

d) Calculate and record the distance the object has fallen at the end of each 0.10 s of its fall. To do this, use the familiar equation:

Distance = average speed × time.

> Example:
>
> The average speed during 0.2 s of falling is 1 m/s.
>
> Distance = average speed × time
>
> = (1 m/s) × (0.2 s)
>
> = 0.2 m

3. The table you have completed is a mathematical model of an object falling freely from rest. Now you will change the mathematical model into a physical model. Your teacher will assign your group a particular row in the data table providing information about the falling object.

   Assemble two identical string and mass assemblies, as shown in the diagram, with an assembly length equal to the distance of fall assigned to your group.

4. Label the mass showing your group's name and the time of fall.

5. Your teacher will place a horizontal row of pins or tape labeled 0.0 s, 0.1 s, 0.2 s, and so on, along the top edge of a chalkboard in your classroom. The times noted on the labels correspond to the instants for which you calculated distances of fall in the table. The horizontal spacing of the pins is a model of the positions an object would have every 0.10 s if it traveled along the horizontal row of pins at a constant speed.

   a) Calculate the horizontal speed by dividing the distance traveled during each 0.1-s time interval by 0.1 s. (Dividing a number by 0.1 is equivalent to multiplying the number by 10.) Show your calculation and the result in your log.

6. Hang one of your string and mass assemblies from the pin corresponding to the time assigned to your group. Place a small mark on the chalkboard at the bottom end of the string and mass assembly.

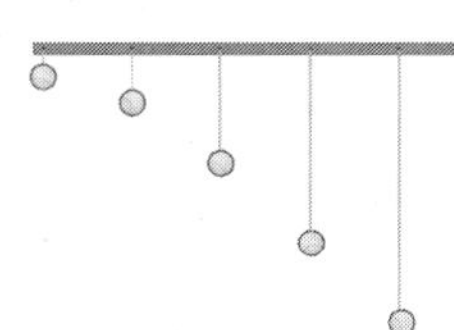

7. A volunteer from the class should draw a smooth curve connecting the marks on the chalkboard. This curve corresponds to the path of an object thrown horizontally. Another volunteer should try to match the path, the trajectory, by throwing a tennis ball horizontally from your starting point (time = 0.0 s). To match the trajectory, the ball will need to be thrown horizontally at the speed calculated in *Step 5.a)*. This may require a few practice tries.

   a) Write your observations in your log.

8. Create the other half of the trajectory by hanging your other mass assembly at the corresponding position to the left of the 0.0 pin. Hang the string and mass assemblies, mark the chalkboard, and connect the points to create the other half of an "arch-shaped" model of a trajectory. The goal is to put the two halves together to produce a single trajectory for an object thrown into the air.

9. If this curve represents the path of a ball, then you should be able to get a thrown ball's path to match this curve. A volunteer should try to throw a ball to match this trajectory. Have another person prepared to catch the ball.

   a) What conditions seem to be necessary to match the trajectory? Write your observations in your log.

   b) When a volunteer is able to match the trajectory, the class should agree upon and give the volunteer instructions to test, one at a time, the effects of launch speed and launch angle on the range of the projectile. Write your observations in your log.

10. Your teacher will show you a "portable" version of the row of pins used in *Step 5*.

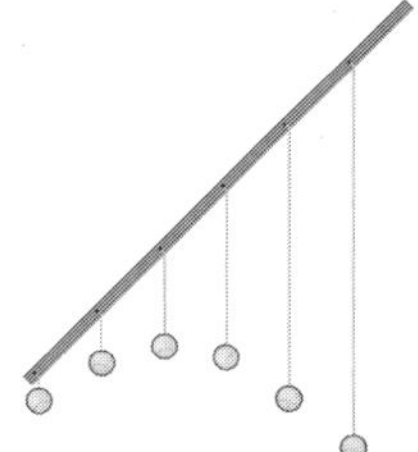

11. Rest the end of the stick corresponding to 0.0 s on the tray at the bottom of the chalkboard while inclining the stick at an angle of 30°.

   a) Is the path indicated by the bottom ends of the string and mass assemblies a "true" trajectory? Have a volunteer try to match it. Record your observations.

   b) Repeat for angles of 45°, 60°, and other angles of interest. Record your observations (it may be necessary to rest the lower end of the model on the floor to prevent the upper end from hitting the ceiling of the room).

   c) What was your observation?

   d) Incline the stick to 90° (straight up). Do this outdoors if the ceiling is not high enough. What is being modeled in this case? Record your thoughts.

7.a)

Students record their observations about a volunteer drawing the curve on the chalkboard and then throwing the ball. The curve on the board and the trajectory of the ball should be parabolas.

8.

The students will assemble the mirror half of the trajectory model to look like diagram B on the previous page.

9.a)

Students throw the ball upward with the angle and speed necessary to duplicate the trajectory in *Step 6*. (Note: This will not be the same speed as in *Step 6*).

9.b)

Students record their observations for a number of different angles and speeds, as suggested by the students.

10.

Teacher activity. You will display portable model of stick and string model of the "Trajectory of a Projectile" with the addition of the string added for 0.6 s that is 1.8 m long.

Teaching Tip

To make the trajectory seem more life like, you can have the students attach tennis or golf balls on the ends of the string at the correct lengths, making what appears to be a time-lapse photograph of the ball's motion as it covers the trajectory.

11.a)

Yes. All are "true trajectories."

11.b)

Students record the height and range of the trajectories for different angles.

11.c)

45°

11.d)

A ball is thrown straight up. Students should be encouraged to realize that the model came from two motions, horizontal motion with a constant velocity and vertical motion due to the force of gravity. That this model agrees with actual projectile motion means that it can be used to calculate what happens during projectile motion.

Physics Talk

It is important to emphasize to students that the path of a projectile has a horizontal as well as a vertical component, both independent of each other. Have them sketch the position of a projectile at regular time intervals to identify these two components of projectile motion. Ask them to label the motion that would be constant speed and the motion that would be downward acceleration.

Students should have a clear grasp of why models are helpful in explaining natural phenomena. You might want to take specific examples from the *Investigate* and ask them how models help in demonstrating the motion of a projectile. Encourage them to use the Internet as a resource for understanding projectile motion. Internet simulations allow students to manipulate variables such as the speed, angle, and height of a launch. Ask them to simulate the path of a projectile at different launch angles by using a computer or graphing calculator. Draw their attention to the diagram in their *Active Physics* textbook to visually reinforce the trajectory of a projectile launched at different angles. Students should summarize their investigations by noting that the symmetry of the projectile's path around a 45°-launch angle means that the projectiles launched at supplementary angles will have the same horizontal range. Supplementary angles result in different times of flight and vertical height of the projectiles at the peak of the trajectory.

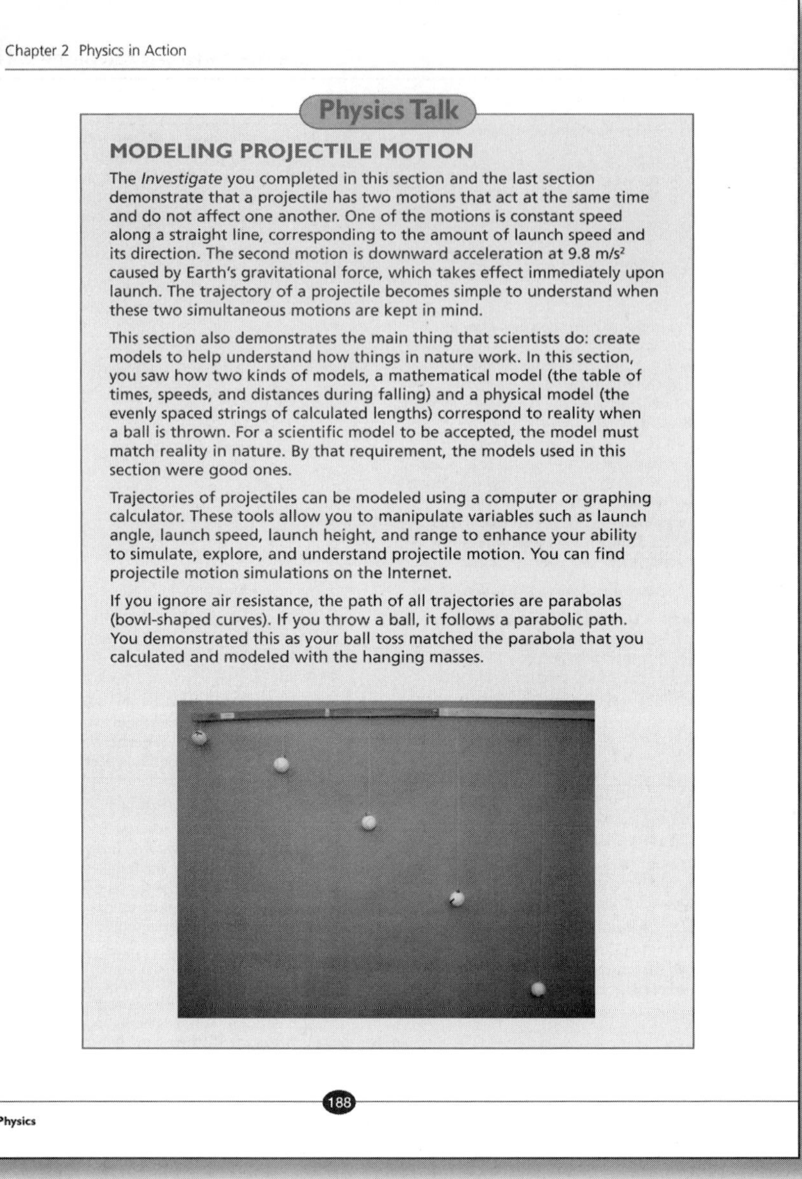

Chapter 2 Physics in Action

Physics Talk

MODELING PROJECTILE MOTION

The *Investigate* you completed in this section and the last section demonstrate that a projectile has two motions that act at the same time and do not affect one another. One of the motions is constant speed along a straight line, corresponding to the amount of launch speed and its direction. The second motion is downward acceleration at 9.8 m/s^2 caused by Earth's gravitational force, which takes effect immediately upon launch. The trajectory of a projectile becomes simple to understand when these two simultaneous motions are kept in mind.

This section also demonstrates the main thing that scientists do: create models to help understand how things in nature work. In this section, you saw how two kinds of models, a mathematical model (the table of times, speeds, and distances during falling) and a physical model (the evenly spaced strings of calculated lengths) correspond to reality when a ball is thrown. For a scientific model to be accepted, the model must match reality in nature. By that requirement, the models used in this section were good ones.

Trajectories of projectiles can be modeled using a computer or graphing calculator. These tools allow you to manipulate variables such as launch angle, launch speed, launch height, and range to enhance your ability to simulate, explore, and understand projectile motion. You can find projectile motion simulations on the Internet.

If you ignore air resistance, the path of all trajectories are parabolas (bowl-shaped curves). If you throw a ball, it follows a parabolic path. You demonstrated this as your ball toss matched the parabola that you calculated and modeled with the hanging masses.

Active Physics 188

The *Physics Talk* ends with the caveat that our model, though useful, still falls short of making accurate predictions in a world of air resistance.

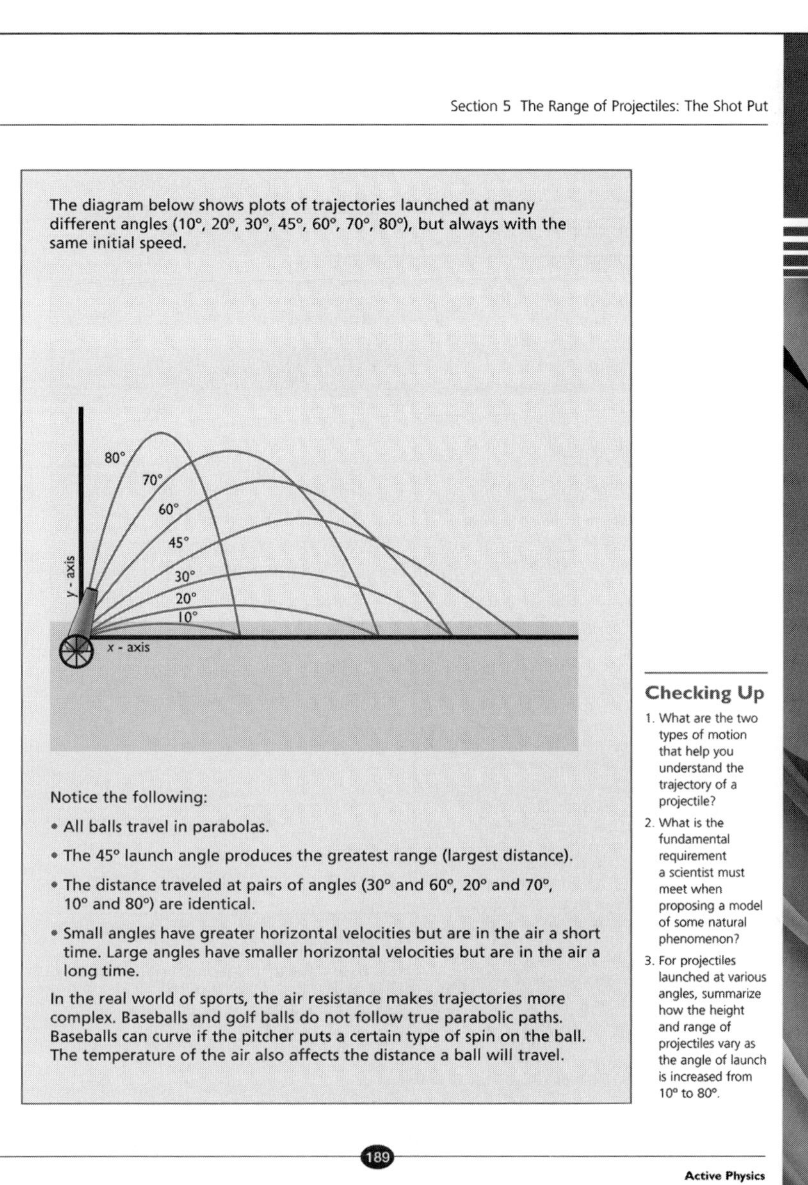

The diagram below shows plots of trajectories launched at many different angles (10°, 20°, 30°, 45°, 60°, 70°, 80°), but always with the same initial speed.

Notice the following:

- All balls travel in parabolas.
- The 45° launch angle produces the greatest range (largest distance).
- The distance traveled at pairs of angles (30° and 60°, 20° and 70°, 10° and 80°) are identical.
- Small angles have greater horizontal velocities but are in the air a short time. Large angles have smaller horizontal velocities but are in the air a long time.

In the real world of sports, the air resistance makes trajectories more complex. Baseballs and golf balls do not follow true parabolic paths. Baseballs can curve if the pitcher puts a certain type of spin on the ball. The temperature of the air also affects the distance a ball will travel.

Checking Up

1. What are the two types of motion that help you understand the trajectory of a projectile?
2. What is the fundamental requirement a scientist must meet when proposing a model of some natural phenomenon?
3. For projectiles launched at various angles, summarize how the height and range of projectiles vary as the angle of launch is increased from 10° to 80°.

Checking Up

1.

The two types of motion that help you understand the trajectory of a projectile are vertical and horizontal motion.

2.

The horizontal speed does not change during the projectile's flight because the vertical descent due to the force of gravity hasn't started.

3.

A projectile launched at an angle that comes back to the same level has its greatest range when launched at 45°.

2-5a **Blackline Master**

CHAPTER 2

Active Physics Plus

Students study sample problems that analyze the horizontal and vertical motion of a long jumper and then calculate the horizontal distance covered by the athlete. They also solve problems to see how far a long jumper and a ball travel horizontally after achieving the maximum height at a certain horizontal velocity.

Active Physics Plus

+Math	+Depth	+Concepts	+Exploration
•			

Analyzing Two-Dimensional Motion Mathematically

You now have a means to analyze two-dimensional motion mathematically. The analysis of two-dimensional motion begins with the recognition that the horizontal and vertical components are independent of one another, as you discovered in this and the previous section. The horizontal speed always remains the same. The vertical speed of a falling object always increases with time as the object descends.

During a long jump the athlete runs and then travels in a parabola. The faster she runs, the faster is her horizontal velocity. She must jump in the air to get height so she can stay in the air longer. She does this without slowing down the horizontal velocity.

If a jumper leaves the ground with the same total velocity but changes the angle, the longest jump occurs when the athlete leaves the ground at an angle of 45°.

Let's see if this makes sense. If the athlete jumps straight up, she maximizes her time in the air but has no horizontal velocity. She will be in the air a long time, but won't go anywhere horizontally. If the athlete jumps straight out at a very small angle, she has a large horizontal component, but is not in the air very long. If she leaves the ground at 45° she is in the air for quite some time and still has a large horizontal velocity. This angle of 45° gives the maximum range.

In physics, you can use mathematical equations to describe the world with accuracy and precision.

Here is a table that describes the horizontal and vertical motion of a trajectory.

	Horizontal Component	Vertical Component
Position	$x = v_x t$ where x is the horizontal displacement v_x is the horizontal component of the velocity t is the time	$y = \frac{1}{2}at^2$ where y is the vertical displacement traveled a is the acceleration due to gravity ($a = 9.8$ m/s^2 on Earth) t is the time
Velocity	The horizontal velocity is constant. There is no net force in the horizontal direction. With no force, there is no acceleration.	$v_y = at$ where v_y is the vertical velocity a is the acceleration due to gravity ($a = 9.8$ m/s^2 on Earth) t is the time
Acceleration	No acceleration in the x-direction.	Acceleration due to gravity in the y-direction = 9.8 m/s^2

Sample Problem

You can analyze a long jumper with the mathematics that you have practiced in this section. Suppose the height that the long jumper achieves is 1.6 m with a horizontal velocity of 6.0 m/s. How far does the jumper move horizontally?

Strategy: Begin by thinking about what will happen if the long jumper jumps horizontally from a ledge with a height of 1.6 m with a horizontal velocity of 6.0 m/s. Where will she land? Jumping from the ledge is identical to the second half of her jump from the maximum height of 1.6 m to the ground.

Solve for the vertical motion and then solve for the horizontal motion.

Step 1: Use the vertical-motion information to determine the time in the air for the second half of the trip. Her vertical fall is 1.6 m irrespective of the horizontal velocity. It is identical to her falling straight down.

If she fell straight down from 1.6 m or jumped horizontally from 1.6 m, her vertical motion would be identical.

You were able to find the vertical distance traveled by first finding the average speed and then multiplying that average speed by the time.

If the vertical speed at the start is zero, the vertical distance traveled can be found in one step by using the equation,

$$y = \frac{1}{2}at^2$$

where a is the acceleration due to gravity (9.8 m/s^2 on Earth).

The value of 9.8 m/s^2 is often rounded up to be 10 m/s^2.

Using the equation $y = \frac{1}{2}at^2$ you can find the time she is in the air.

Given:

$y = 1.6$ m

Solution:

$$y = \frac{1}{2}at^2$$

You can use your calculator to find a value for t, such that:

$$1.6 \text{ m} = \frac{1}{2}at^2$$

or you can practice your algebra skills and rearrange the equation to solve for time.

$$t = \sqrt{\frac{2y}{a}}$$

$$= \sqrt{\frac{2(1.6\text{m})}{9.8\text{ m/s}^2}}$$

$$= 0.57 \text{ s or } 0.6 \text{ s}$$

Strategy:

Step 2: If she has a horizontal velocity of 6.0 m/s and she is in the air for 0.6 s, where will she land? Her horizontal motion can be found by recognizing that distance equals velocity times time.

CHAPTER 2

1.

The long jumper is in the air for 1.18 s.

$$t = \sqrt{\frac{2(1.7\ \text{m})}{9.8\ \text{m/s}^2}} = 0.59\ \text{s, or}$$

$$2t = 1.18\ \text{s}$$

The distance the long jumper achieves is the distance she travels in 1.18 s, which is $d = (7.0\ \text{m/s})(1.18\ \text{s}) = 8.3\ \text{m}$.

2.

To calculate the range of the ball, you must know both the vertical and horizontal velocities. The horizontal velocity is given as 45 m/s. To determine the vertical velocity from the height, use

$$v_f^2 = v_i^2 + 2ad.$$

knowing that the final vertical velocity at the peak is zero gives

$$0^2 = v_i^2 + 2\left(-9.8\ \text{m/s}^2\right)\left(1.5\ \text{m}\right)$$

$$v_i = 5.4\ \text{m/s}.$$

With an initial vertical speed of 5.4 m/s, the ball will take 5.4/9.8 s to reach the peak, and an equal amount of time to return to the ground, giving $t = 1.1$ s.

The horizontal distance traveled will be the horizontal velocity multiplied by the time of flight or

$$d = v_x t = 49.8\ \text{m}.$$

Solution:

$$x = v_x t$$
$$= (6.0\ \text{m/s})(0.6\ \text{s})$$
$$= 3.6\ \text{m}$$

The jumper moves horizontally 3.6 m on the way down for the second half of the trip.

Strategy:

Step 3: If a long jumper achieves a height of 1.6 m and has a horizontal velocity of 6.0 m/s for the second half of the trip, then her horizontal distance is twice the value of the distance for the entire trip, since she moves horizontally on the way up as well. (Remember modeling the other part of the motion in the *Investigate*.)

Solution: $x_{total} = 2(3.6\ \text{m}) = 7.2\ \text{m}$

1. Calculate how far horizontally a long jumper travels if she achieves a height of 1.7 m and a horizontal velocity of 7.0 m/s.

You can solve lots of problems by analyzing half of the motion like this. You can calculate the path of a football or baseball or golf ball. The calculations will not apply to real-life situations as well as you might expect because of the effects of air resistance.

The path of a golf ball should be a parabola. Air resistance changes the shape. A baseball should also travel in a parabola, but when the pitcher puts a certain type of spin on it, the air resistance allows it to curve and therefore change the calculated path of our model.

2. Calculate how far a ball will travel horizontally if the ball reaches a high of 1.5 m above the ground and is thrown at a horizontal velocity of 45.0 m/s. Assume that the ball is caught at the same height it is thrown and that there is no air resistance.

What Do You Think Now?

At the beginning of this section, you were asked the following:

- **Describe the trajectories of projectiles launched from the ground at various angles.**
- **Describe how a greater launch speed of a projectile might change the range when the launch angle is the same.**

You can use evidence from the mathematical model and the physical model of this section to describe the path of the object and to describe how the angle of the trajectory determines the distance the object travels.

What Do You Think Now?

This is a good time to return to the *What Do You Think?* questions and have a student read them aloud. You will find your students more informed as they begin to give their responses. Ask them to refer to the answers they recorded previously in their log books. You might want to emphasize how their grasp of projectile motion increased with their hands-on investigations, and subsequent connections they made in the *Physics Talk*. Expect them to use new terms and concepts introduced in this section with ease. A good way to improve the comfort level in using new terms is to use those terms frequently, yourself, in discussions, so that the unfamiliar is put into different contexts and becomes familiar. A term such as *projectile* may require repeated usage.

Section 5 The Range of Projectiles: The Shot Put

Physics

Essential Questions

What does it mean?

It is said that any thrown object travels in a parabola. Describe three different paths and explain how they each can be a parabola.

How do you know?

What evidence do you have that the mathematics correctly predicted the path that a thrown object would take?

Why do you believe?

Connects with Other Physics Content	Fits with Big Ideas in Science	Meets Physics Requirements
Force and motion	* Models	Experimental evidence is consistent with models and theories

* The use of models is a physicist's way of making sense of the world. Did the mathematical model and the physical model in your investigation adequately describe the path of a trajectory?

Why should you care?

Many sports have objects moving in the air. Baseballs, footballs, and soccer balls all travel in parabolas. Divers and high jumpers also travel in parabolas. As a diver's body twists and turns in the air, how could a television broadcaster show that the path is a parabola?

Reflecting on the Section and the Challenge

The information learned about projectile motion in this section applies not only to the shot put, but to any sporting event that involves throwing things into the air (including the self-launching of a human body, as in the hurdles, long jump, or high jump). It has been reported that one Olympian who competed in the shot put increased his range in that event by nearly 4 m, based on suggestions made by a physicist. You are now a physicist specializing in projectile motion. Imagine what you might say in your voice-over when covering the long jump event or describing a home run ball or a punt in football. You may want to comment on how the vertical motion and horizontal motion are independent of one another. You may wish to mention that the angle will help determine the range of the ball, with 45° producing the longest range. You will certainly want to mention that the curved path of the ball is a parabola. In the real world of sports, the air resistance makes trajectories more complex. Baseballs and golf balls do not follow true parabolic paths. Baseballs can curve if the pitcher puts a certain type of spin on the ball. The temperature of the air affects the distance a ball will travel. Although the details of these are complex to analyze, you may wish to mention them in your voice-over.

Reflecting on the Section and the Challenge

Giving students the time to reflect and ponder is essential to their application of new knowledge. Draw their attention to how an athlete would find the physics behind projectile motion useful, and invite them to discuss how they too might use their present understanding in sports events. This is an opportunity for them to incorporate new information into their *Chapter Challenge*, and make a rough draft of what might be said in a voice-over when describing a sport that involves projectile motion. Also have students reflect on how air resistance might change the course of a trajectory or spinner might put a spin on the ball to change the distance the ball travels, as mentioned in the *Active Physics* textbook.

CHAPTER 2

Physics Essential Questions

What does it mean?

A line drive in baseball; a pop fly in baseball; an outfield hit in baseball. All are parabolas but they all reach different heights.

How do you know?

We showed mathematically that the path would be a parabola. We then tried to toss objects in any shape other than a parabola, but were unsuccessful.

Why do you believe?

The mathematical model predicted a parabola and our investigation of the path showed it to be a parabola.

Why should you care?

A television broadcaster could show how the path of the center of the diver is still a parabola.

Physics to Go

1.

The greatest range is provided by an angle of 45°. This angle provides a lot of time in the air coupled with a lot of horizontal speed, but it does not provide a maximum of either one of these variables.

2.a)

More time

2.b)

Less time

3.a)

The complement of the angle: 60°

3.b)

75°

4.

The horizontal running speed of a long jumper is much greater than the initial vertical speed that the jumper can attain; therefore, the angle is far less than 45°.

5.

Carl Lewis can run fast, which is half of the requirement for a good long jumper. Apparently, he also jumps well vertically.

6.a)

The acceleration at point X is the acceleration due to gravity. Its direction is vertically down.

6.b)

The velocity at point X is horizontal. At the highest point, the ball is neither moving up nor moving down.

Chapter 2 Physics in Action

Physics to Go

1. If the launching and landing heights for a projectile are equal, what angle produces the greatest range? Why?
2. Compared to a launch angle of 45°, what happens to the amount of time a projectile is in the air if the launch angle is
   a) greater than 45°?
   b) less than 45°?
3. For a constant launch speed, what angle produces the same range as a launch angle of
   a) 30°?
   b) 15°?
4. Analyses of performances of long jumpers has shown that the typical launch angle is about 18°, far less than the angle needed to produce maximum range. Why do you think this occurs?
5. You might be familiar with Carl Lewis as a medal-winning sprinter. But he is also an Olympic gold medalist in the long jump. Why do you think he was successful in both events?
6. The diagram below shows a ball thrown toward the east and upward at an angle of 30° to the horizontal. Point X represents the ball's highest point.

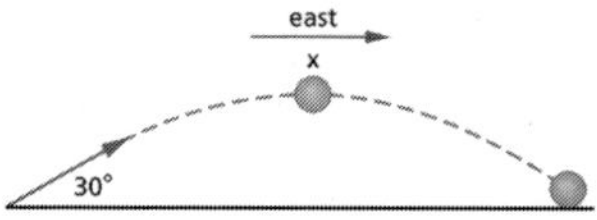

   a) What is the direction of the ball's acceleration at point X? (Ignore friction.)
   b) What is the direction of the ball's velocity at point X?
7. Active Physics Plus A diver jumps horizontally off a cliff with an initial velocity of 5.0 m/s. The diver strikes the water 3.0 s later.
   a) What is the vertical speed of the diver upon reaching the surface of the water?
   b) What is the horizontal speed of the diver 1.0 s after the diver jumps?
   c) How far from the base of the cliff will the diver strike the water?

7.a)

The vertical motion of the horizontally diving person will be identical to the vertical motion of a dropped ball.

$v_y = gt = (9.8 \text{ m/s}^2)(3.0 \text{ s}) =$ 29.4 m/s (or 30 m/s if using $g = 10 \text{ m/s}^2$).

7.b)

It remains 5.0 m/s. The horizontal speed does not change because there is no force in the horizontal direction.

7.c)

$x = v_x t = (5.0 \text{ m/s})(3 \text{ s}) = 15 \text{ m}$

Section 5 The Range of Projectiles: The Shot Put

8. The diagram of the baseball player shows a baseball being hit with a bat. Angle θ represents the angle between the horizontal and the ball's initial direction of motion. Which value of θ would result in the ball traveling the longest horizontal distance if air resistance is neglected?

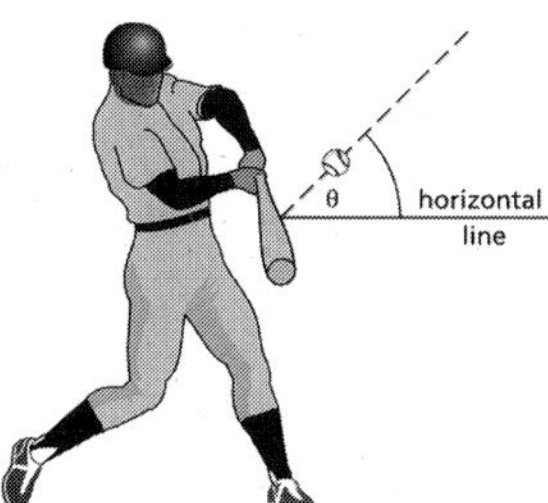

9. Four balls, each with mass (m) and initial velocity (v), are thrown at different angles by a baseball player. Neglecting air friction, which angular direction produces the greatest projectile height?

10. Active Physics Plus The diagram below shows a ball projected horizontally with an initial velocity of 20.0 m/s east, off a cliff 100-m high.

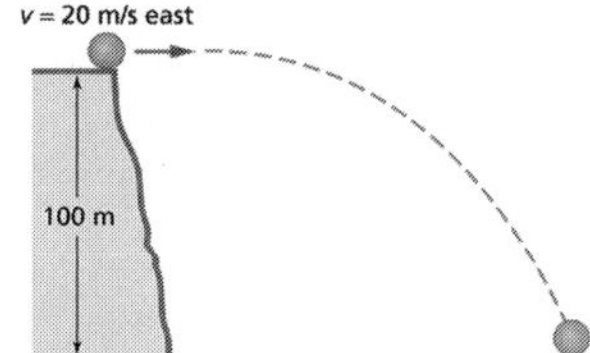

a) During the flight of the ball, what is the direction of its acceleration?

b) How many seconds does the ball take to reach the ground?

c) How far from the base of the cliff does the ball land?

8.

The ball thrown the closest to 45° will travel the farthest. The ones thrown at 35° and 55° will travel identical distances. The 55° ball has less horizontal velocity but is in the air longer than the 35° ball. The ball thrown with the highest angle reaches the greatest height. If it could be thrown 90° (straight up) it would reach the greatest height.

9.

The angle for the longest horizontal distance is 45° when the object leaves the ground and when it returns to the ground. Because the baseball is 1 m above the ground, the optimum angle will be a very small amount less than 45°, as the ball already has some vertical displacement.

10.a)

The acceleration is the acceleration due to gravity. Its direction is vertically down.

10.b)

The ball will take the same amount of time to reach the ground as an object dropped from 100 m.

$$d = \frac{1}{2}gt^2$$

$$100 \text{ m} = \frac{1}{2}\left(10 \text{ m/s}^2\right)t^2$$

$$t = \sqrt{\frac{200 \text{ m}}{10 \text{ m/s}^2}} = 4.5 \text{ s}$$

10.c)

90 m

CHAPTER 2

SECTION 5 QUIZ

2-5b **Blackline Master**

For all the questions in this quiz, air resistance is considered to be zero.

1. A 4.0 kg rock and a 1.0 kg stone fall freely from rest from a height of 100 m. After they fall for 2.0 s, the ratio of the rock's speed to the stone's speed is

   a) 1:1. b) 2:1.

   c) 1:2. d) 4:1.

2. A stone with an initial velocity of zero is dropped from a bridge above a river. After 3 s, the stone strikes the water below the bridge. How fast is the stone traveling when it strikes the water?

   a) 10 m/s b) 20 m/s

   c) 30 m/s d) 45 m/s

3. How far will the stone have fallen in these 3 s?

   a) 10 m b) 20 m

   c) 30 m d) 45 m

4. The diagram below shows a football being kicked. Angle θ represents the angle between the horizontal and the ball's initial direction of motion. Which value of θ would result in the ball traveling the longest distance?

   a) 25° b) 45°

   c) 60° d) 90°

5. As the ball is hit harder at the same angle, its acceleration during its flight after leaving the club will

   a) decrease.

   b) increase.

   c) remain the same.

SECTION 5 QUIZ ANSWERS

1. a) All objects fall at the same rate regardless of their mass if there is no air resistance.

2. c) Objects that are falling freely under the influence of gravity have an acceleration of 10 m/s^2. After 3 s of fall the object's speed will have increased to 30 m/s.

3. d) After 3 s of fall, when the rock started from rest, its average speed would be 15 m/s. The distance traveled in 3 s with an average speed of 15 m/s is 45 m.

4. b) The angle that would provide the greatest range is 45° when the object returns to the same height.

5. c) The acceleration of any object in free fall near Earth's surface is 10 m/s^2, and does not depend upon the object's velocity.

NOTES

CHAPTER 2

Chapter Mini-Challenge

Review the *Goal* with your students and brainstorm strategies to accomplish the *Chapter Challenge*. Establish a framework that students can work from by using the initial design given in the Student Edition of *Active Physics*. Draw attention to key words and phrases (e.g., "engaging information," "educate the viewer," "vocabulary words," etc.) that define the purpose of the *Mini-Challenge*. This is a time for you to gauge students' understanding of the physics concepts they have explored so far.

Pair up students and have them share a quick review of the physics content in *Sections 1-5* of *Chapter 2*. Point out that at this stage they should be able to relate the sport they are about to choose to the physics concepts that they will be highlighting. Refer to the *Engineering Design Cycle* and ask students to identify how they are moving through each stage of the cycle—what *Inputs* they are adding, how they *Process* information, how the *Output* or trial run of the *Chapter Challenge* is presented, and how they are giving and receiving *Feedback* for their presentation.

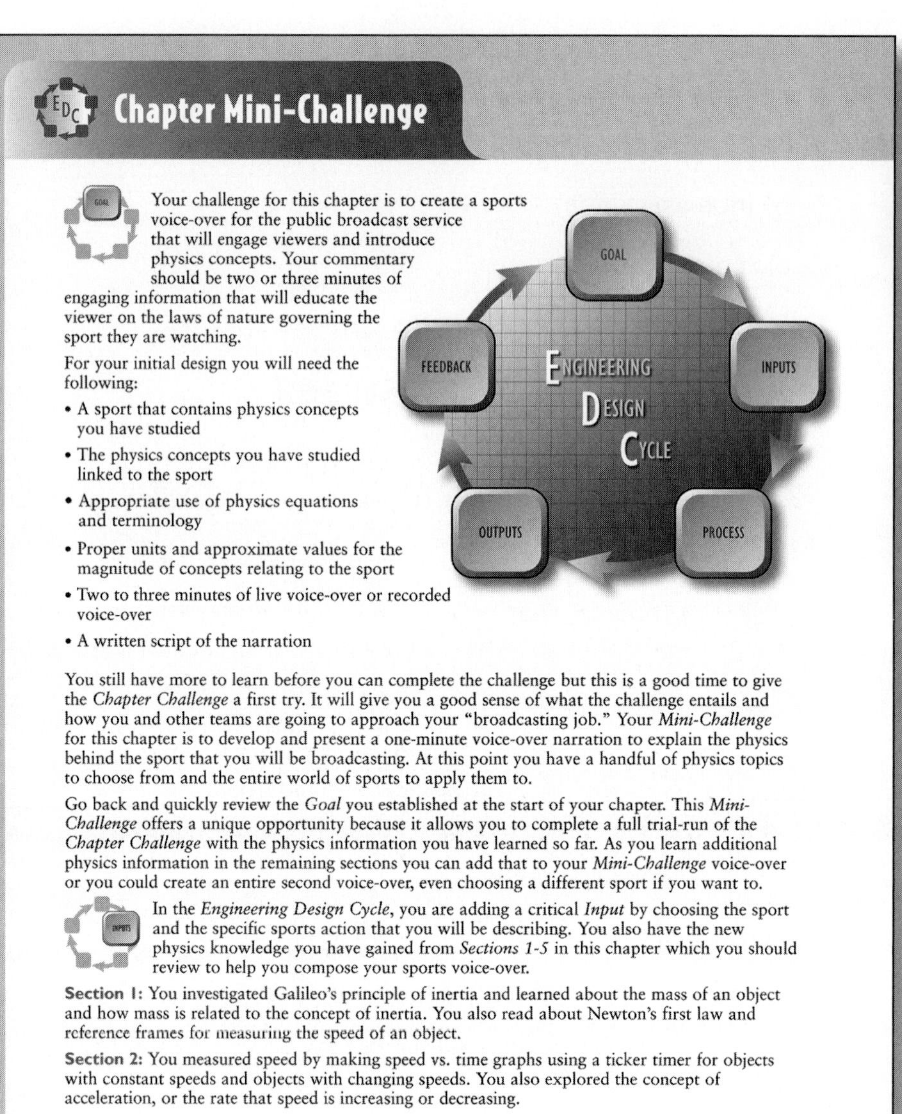

Chapter Mini-Challenge

Your challenge for this chapter is to create a sports voice-over for the public broadcast service that will engage viewers and introduce physics concepts. Your commentary should be two or three minutes of engaging information that will educate the viewer on the laws of nature governing the sport they are watching.

For your initial design you will need the following:

- A sport that contains physics concepts you have studied
- The physics concepts you have studied linked to the sport
- Appropriate use of physics equations and terminology
- Proper units and approximate values for the magnitude of concepts relating to the sport
- Two to three minutes of live voice-over or recorded voice-over
- A written script of the narration

You still have more to learn before you can complete the challenge but this is a good time to give the *Chapter Challenge* a first try. It will give you a good sense of what the challenge entails and how you and other teams are going to approach your "broadcasting job." Your *Mini-Challenge* for this chapter is to develop and present a one-minute voice-over narration to explain the physics behind the sport that you will be broadcasting. At this point you have a handful of physics topics to choose from and the entire world of sports to apply them to.

Go back and quickly review the *Goal* you established at the start of your chapter. This *Mini-Challenge* offers a unique opportunity because it allows you to complete a full trial-run of the *Chapter Challenge* with the physics information you have learned so far. As you learn additional physics information in the remaining sections you can add that to your *Mini-Challenge* voice-over or you could create an entire second voice-over, even choosing a different sport if you want to.

In the *Engineering Design Cycle*, you are adding a critical *Input* by choosing the sport and the specific sports action that you will be describing. You also have the new physics knowledge you have gained from *Sections 1-5* in this chapter which you should review to help you compose your sports voice-over.

Section 1: You investigated Galileo's principle of inertia and learned about the mass of an object and how mass is related to the concept of inertia. You also read about Newton's first law and reference frames for measuring the speed of an object.

Section 2: You measured speed by making speed vs. time graphs using a ticker timer for objects with constant speeds and objects with changing speeds. You also explored the concept of acceleration, or the rate that speed is increasing or decreasing.

196

Section 3: You investigated the relationship between forces and the changes of speed and acceleration of an object. You also used Newton's second law to calculate the unbalanced force, mass, or acceleration for an object when any two of those quantities can be measured.

Section 4: You used models to learn about the horizontal and vertical motion of a projectile. You learned how the horizontal speed and total height of a projectile will affect the horizontal distance it will travel.

Section 5: You measured acceleration due to gravity and discovered how it causes all objects to speed up as they fall toward Earth. You also used calculations and models to describe the trajectory, or path, of a projectile.

The *Process* phase of the *Engineering Design Cycle* is when you decide what information to include to meet the criteria of your *Goal.* It should be easy to select a sport that involves objects with mass, movement, and projectiles. Since almost any sport will work, it might be best to use a "rank list" to help your group decide on a sport. Ask each member of your group to suggest a sport that they would most like to use in the voice-over and list the sports chosen on a small scrap of paper. Each student will then rank the sports in the order in which they would prefer to work on them. One student will then tally the ranking for each sport and the sport with the lowest total is selected. Once your sport is selected you will need a bit of action, something that will be exciting for the audience to hear about.

Presenting your information to the class is your *Output* of the *Engineering Design Cycle*. Your voice-over should describe actual game play and include as many of the physics topics as you can in your one-minute narration. Don't forget that you are also responsible for turning in a written script for your narration. Use your creativity when choosing a character for your voice-over. Sports fans often have a favorite announcer known for his or her distinct voice or personality. The "character" you choose to portray is part of the entertainment value of your presentation.

Your classmates will give you *Feedback* on the accuracy and the overall appeal of your presentation based on the criteria of the *Mini-Challenge*. This feedback will become an *Input* for your final design in the *Chapter Challenge*. You will have enough time to make corrections and improvements, so you will want to pay attention to the valuable information they provide.

Remember to correct any parts of your script that were identified as not correct by your audience. Then, store all of your information in a safe place so that it will be ready to use in the *Chapter Challenge*.

Take another look at your sports action play. Look for pieces of sports action that you did not have a comment for or you felt you could not address completely. Additional information in the remaining sections may help you describe that action. You will study additional physics topics that apply to the general motions in sports, so it is likely you will be able to give a better description later in the chapter.

Your group may also decide that the sport you chose was not as good of a fit as you might have liked. You are welcome to pick a new sport now that you have a better idea of which sports work well with your challenge. You may also find that a different sport fits better with the physics from the remaining chapter sections.

197

Engineering Design Cycle

Ask your students to make a bulleted list of all the physics concepts they have learned so far. Discuss why they should treat these concepts as *Inputs* for their *Engineering Design Cycle*. Have a student read aloud a few concepts from their list. To increase student focus write each concept on the board or use an overhead projector. After students have reviewed the *Inputs* phase, draw their attention to the *Process* phase. Remind them that each time they reflect on how they should meet their *Goal*, they will be moving through the *Process* phase. It is important that students recognize which requirement they have met and what they still need to learn in order to complete their *Mini-Challenge*. During the *Process* phase students should select the physics principles that are related to their sport and work on developing an entertaining narration.

In the *Outputs* phase encourage students to present their *Mini-Challenge. Mini-Challenge* presentations should not take more than 30 s. Make sure that all students work around a set of established requirements and that the environment is relaxed and forthcoming. Refer to the *Criteria* in the *Goal* and ask students to base their *Feedback* on how they see their peers meeting the established requirements. Student engagement in preparation of the challenge is reflected in this phase. Remind students that they should evaluate their peers based on the merits of the presentation alone to be objective and helpful.

SECTION 6

Newton's Third Law: Run and Jump

Section Overview

In this section, students analyze the forces involved in running, stopping, and jumping. To illustrate this concept, a student on a skateboard pushes against a wall and slowly accelerates to roll across the floor with constant speed. To explain movement away from the wall, students identify the force causing the acceleration. Then they perform a thought experiment about the forces involved when running or walking on a horizontal surface, using words and sketches to answer questions on the forces. To assist in identifying the acting forces, students learn about free-body diagrams. Finally, to understand the concept of the normal force that is supplied by an inanimate object, students do an experiment with a meter stick and weights of different masses to see what happens each time a weight is placed on the meter stick. The bending of the meter stick under the force of the weights demonstrates the spring-like nature of surfaces, and helps explain how a normal force is exerted.

Background Information

Before proceeding with this section, it is recommended that you read *Physics Talk: Newton's Third Law of Motion* in the student text. The explanation of forces involved in walking given in the teacher's *Background Information* for *Section 7* will serve to explain the forces involved with walking and running brought up in this section. The pairs of equal and opposite forces identified during earlier sections to explain friction and walking are examples of Newton's third law of motion, often stated as: "For every action there is an equal and opposite reaction." Another equal and opposite pair of forces arises during this *Investigate* when a student standing on a skateboard sets himself into motion by using a leg and foot to push off from the wall.

Inevitably, forces exist in equal and opposite pairs, and often the force that you identify as the force responsible for motion is not the correct one. For example, a person who says, "I pushed down on the trampoline with a mighty force, and my force launched me upward in a high jump," is mistaken; it was the equal and opposite reaction force provided by the trampoline that launched the person upward.

One of the key concepts associated with Newton's third law but often misunderstood is that although action-reaction forces come in pairs, they never act on the same body. This restriction means that these forces never cancel one another out because they are not acting on the same object. In the *Investigate*, the student pushing off the wall while on the skateboard is exerting a force on the wall by trying to make the wall move. The reaction force exerted by the wall on the student is pushing the student in the opposite direction. It is this force that moves the student, not the student's foot. To convince the students, simply have the skateboarder stand on the skateboard when away from the wall and try to push. To further the argument, have some vertical meter sticks brought into position so the student can push off on them. The meter stick's obvious bending should convince the students that the bendable nature of the meter sticks provides the forward force.

Crucial Physics

- Forces come from interactions between objects. This means that forces come in pairs, with each of the forces acting on each of the interacting objects.
- For each pair of forces due to an interaction, the forces are equal but point in opposite directions.

Learning Outcomes	Location in the Section	Evidence of Understanding
Provide evidence that forces come in pairs, with each force acting on a different object.	***Investigate*** Part A: Steps 1-4	Students state reaction force of wall on skateboarder is responsible for the acceleration.
Use Newton's third law to analyze physical situations.	***Investigate*** Part B: Steps 2-4	Students demonstrate knowledge of the normal force to explain the bending of a meter stick when masses are added.
Describe how Newton's third law explains much of the motion in your everyday life.	***Investigate*** Part A: Steps 3-4 Part B: Steps 3-4 ***Physics Talk***	Students correctly identify the action–reaction force pairs in cases such as a student walking across the floor.

NOTES

CHAPTER 2

Section 6 Materials, Preparation, and Safety

Materials and Equipment

PLAN A		
Materials and Equipment	**Group (4 students)**	**Class**
Ruler, metric, 30 cm	1 per group	
Meter stick, wood	1 per class	
Washer, 3/4 in. (outside diameter) x 5/16 in. (inside diameter)	2 per group	
Weight, slotted, 100 g	10 per group	
Scale, spring, 0-20 N	2 per group	
Board, roller, Reaction Force (skateboard)		2 per class
Chair (with wheels)*		1 per class
Safety helmet*		1 per class
Pads, knee*		1 per class
Pads, elbow*		1 per class
Stack of books*		1 per class

*Additional items needed not supplied

PLAN B		
Materials and Equipment	**Group (4 students)**	**Class**
Ruler, metric, 30 cm	1 per group	
Meter stick, wood	1 per class	
Washer, 3/4 in. (outside diameter) x 5/16 in. (inside diameter)	2 per group	
Weight, slotted, 100 g	10 per group	
Scale, spring, 0-20 N	2 per group	
Board, roller, Reaction Force (skateboard)		2 per class
Chair (with wheels)*		1 per class
Safety helmet*		1 per class
Pads, knee*		1 per class
Pads, elbow*		1 per class
Stack of books*		1 per class

*Additional items needed not supplied

Note: Time, Preparation, and Safety requirements are based on Plan A, if using Plan B, please adjust accordingly.

Time Requirements

This *Investigate* will take approximately one class period.

Teacher Preparation

- If you do not have access to skateboards and the associated safety equipment, ask for student volunteers to bring theirs to class the day of the *Investigate*. The first part of this *Investigate* should be done as a class demonstration with one or two students pushing off the wall while on skateboards. Two skateboard riders or two wheeled chairs will be necessary for the step that requires them to push on each other to observe the action-reaction phenomenon.
- When the student groups are using the spring scales to observe the equal and opposite forces, caution them not to pull so hard as to exceed the maximum reading on the scale. In addition, the students should not twist the scale in their hand while the scale is attached to the second student's hand because this will also twist the second student's hand.
- Have stacks of books or other support material available for the students to use to support the meter sticks for *Part B*.

Safety Requirements

- Using a wheeled chair will most likely be a safer alternative than using the skateboards; however, student interest and engagement is better with the skateboard, with little additional risk. If skateboards are used when the two students push off on each other, choose two students who are not too dissimilar in mass to prevent one student from accelerating too much, and possibly losing balance. All students on the skateboards should wear appropriate safety equipment, including helmets and knee and elbow pads. Make certain that the area the students will use while on the skateboards is clear of all obstructions.

NOTES

Meeting the Needs of All Students

Differentiated Instruction: Augmentation and Accommodations

Learning Issue	Reference	Augmentation and Accommodations
Describing motion	***Investigate*** Part A	**Augmentation** • Students with language issues struggle to describe qualitative features in a way that is meaningful. In small groups, ask students to brainstorm a list of words or phrases that describe motion. Compile the list and then ask students to complete *Investigate, Part A*. • Encourage students to use diagrams or drawings to support their descriptions of motion.
Vocabulary Scientific measurements	***Investigate*** Part B ***Physics to Go*** Question 2	**Augmentation** • Students are asked to measure the deflection of the meter stick and record the values for comparison. • Explain the meaning of deflection using the diagram in *Investigate, Part B, Step 4*. • Model an appropriate way to measure the deflection. Put a taut string across the distance between the stack of books to mark the original position of the meter stick before any weights are placed on it. This will give students a consistent reference point to measure from. • Measuring from the surface of the table or floor to the lowest position of the meter stick may be easier for some students. Then they can calculate the deflection by subtracting the deflected position of the meter stick from its original position.
Reading comprehension Understanding key concepts	***Physics Talk***	**Augmentation** • Students often do not complete longer reading assignments and may miss important information in this section if they do not read the entire section. Break the reading into smaller chunks using some of the following ideas. • Ask students to draw diagrams or sketches to represent each of the four bulleted examples near the beginning of this section. • Provide direct instruction on how to draw free-body diagrams. Ask students to write a procedure, in their own words, that could be used to teach new students how to draw free-body diagrams. • Provide opportunities for guided practice to draw free-body diagrams and identify equal and opposite forces. • Ask students to summarize Newton's third law and most of the content in *Physics Talk* in a bulleted format.

Strategies for Students with Limited English-Language Proficiency

Learning Issue	Reference	Augmentation
Following complex procedures	***Investigate***	Break down the *Investigate* into smaller chunks that allow students to comprehend each portion of the activity before moving on to the next one. This approach will allow students to get comfortable following the procedures outlined within each step, and to internalize new concepts and any new vocabulary that is introduced. Lead a brief class discussion after each step to allow students the opportunity to demonstrate acquired knowledge and understanding of vocabulary, instructions, and concepts.
Comprehension	***Investigate*** Part A Steps 3.a) and 3.b)	Once students have finished answering the questions in their *Active Physics* log, look through their work and discuss it with them to check for accurate representations and understanding of the forces involved in walking or running. Or you may wish to draw and label your own diagrams and project them on the overhead. Hold a class discussion about the diagrams and allow students time to make corrections to the drawings in their logs if necessary.
Vocabulary comprehension	***Investigate*** Part B Steps 1 and 4.a)	ELL students may not be familiar with the word "inanimate." Allow them sufficient time to read the paragraph and infer the meaning from context. Explain that "animate" means "alive" or "full of life." It may help to think of animals, which generally can control their movement, as opposed to plant life, which generally cannot. Have students classify objects in the room as animate or inanimate. Check their understanding before moving on. The word "deflection" will likely be unfamiliar to ELL students. The illustration will help put the term in context. However, before moving on, be sure students know that deflection means the amount an object has moved away from its normal, or resting, position.
Comprehension Cooperative learning	***Active Physics Plus***	Once students have finished drawing their free-body diagrams and answering the questions, have them work in pairs to interpret each other's diagrams. The phrase "force of tension" may be challenging for ELL students. Explain that "tension" comes from a Latin root word that means "to stretch." Have students brainstorm common usages of the words "tense" and "tension," for example tensing up their muscles, or tightening a violin string. Clarify that tension is a force that can vary—the force of tension will increase as more force is applied to the string, up to the point at which the string can stretch no more and breaks.

SECTION 6

Teaching Suggestions and Sample Answers

What Do You See?

This illustration captures the viewer's attention on many levels. Draw students' attention to the changing facial expressions of the two persons in the illustration. Guide their intuitive interest from a general observation to the more specific purpose of the topic by asking questions related to the person pushing on the wall. Why does the boy appear to be so disgruntled in one visual but so cheerful in the other? Why is the girl so disinterested before but so happy later? How did the boy push back? These are some questions among others that should lead to an engaging discussion. You might want to bring in earlier discussions on force and motion.

Chapter 2 Physics in Action

Section 6 Newton's Third Law: Run and Jump

What Do You See?

Learning Outcomes

In this section, you will

- **Provide** evidence that forces come in pairs, with each force acting on a different object.
- **Use** Newton's third law to analyze physical situations.
- **Describe** how Newton's third law explains much of the motion in your everyday life.

What Do You Think?

The high-jump record is 2.45 m (about 8 ft) for men and 2.09 m (about 6 ft) for women.

- **Pretend that you have just met somebody who has never jumped before. What instructions could you provide to get the person to jump up (that is, which way do you apply the force when you push with your feet)?**

Record your ideas about this question in your *Active Physics* log. Be prepared to discuss your responses with your small group and the class.

Investigate

In *Part A* of this *Investigate*, you will observe what happens when an object pushes or pulls on another object. In *Part B*, you will observe how a meter stick applies an upward force on a mass.

Part A: Push, Push Back and Pull, Pull Back

1. Carefully stand or sit on a skateboard or sit on a wheeled chair near a wall. (Your teacher may have one person demonstrate this part of the activity for safety reasons.) By touching only the wall, not the floor, cause yourself to move away from the wall to "coast" across the floor.

198

Students' Prior Conceptions

This section provides evidence that forces come in pairs and that each force acts on a different object. It introduces Newton's third law, the action-reaction law, leading to student understanding of the motions of everyday life. The stage for the major conservation law, the conservation of momentum, builds.

1. **Students may continue to assign forces to animate objects only.** You could use this section to spiral learning and create ladders built upon previous activities in this chapter. This technique also works when encouraging students to align their thinking about subsequent preconceptions.

2. **Students may believe that a force implies motion. If there is a force there must be motion.** Teachers should emphasize that forces always occur in equal and opposite pairs; these equal and opposite pairs always act on different objects; inanimate objects can push back.

You are likely to get a few humorous responses from your students. This *What Do You See?* illustration prompts students to begin thinking about action reaction forces. Your task is to encourage them and at the same time draw responses that set the stage for Newton's third law. Use those responses to steer students toward the purpose of this section. Encourage them to notice the smaller images and suggest that other images also contribute to the overall effect of the artist's intent. Remind students that they will have a chance to revisit their ideas and recognize how their knowledge has progressed.

What Do You Think?

Students will have a difficult time coming up with good instructions. You may try "acting out" some instructions as a way of demonstrating that the instructions must be very precise. Accept all explanations and encourage students to write down their answers in their *Active Physics* log. Reassure them that at a later stage they will get the time to edit their responses, and that they should treat this activity as a means of improving what they might already know or are about to learn.

What Do You Think?

A Physicist's Response

To jump, I push downward on the floor. When I do, the floor pushes upward on me. My push on the floor bends or deflects it a little bit and this deflection springs back in the form of the upward force. If the floor were too weak to generate the 10-cm jump, then I could not jump that high because the floor would break instead. When I tried to push hard enough to jump the 10 cm, the floor would break if it were that weak.

Investigate

This investigation is best done as a teacher-led demonstration for safety reasons. Students will not easily accept that the wall exerts a force on them. At the right point in the section, you might ask a second student to "be the wall" by allowing the first student to push off his or her hands. The second student should quickly realize that if he or she does not push the first student, the first student does not accelerate.

NOTES

Part A

1.a)

Your motion is accelerated when you push away from the wall. The acceleration lasts for a short distance when you speed up from zero to your top speed, after which you move at a constant velocity, and then slow down. The direction of acceleration is away from the wall.

1.b)

Motion is at a constant speed just after the initial acceleration, when the force is no longer acting on you. If you neglected friction, you would keep moving until another force acted on you to slow you down or stop you. During this part of the section, encourage students to rely on their motion to determine the forces acting on them. This will help to put preconceived notions out of their minds as they think through the situation.

1.c)

The force is supplied by the wall. The force must be acting in the direction of acceleration, away from the wall.

1.d)

You push on the wall, in a direction toward the wall.

Use words and diagrams to record answers to the following questions in your log:

a) When is your motion accelerated? (Recall that acceleration is the change in velocity over time.) For what distance does the accelerated motion last? In what direction do you accelerate?

b) When is your motion at constant speed? If you ignore the effects of friction, how far should you travel? (Remember Galileo's principle of inertia and Newton's first law when answering this question.)

c) Newton's second law, $F = ma$, says that a force must be acting when acceleration occurs. What is the source of the force, the push or pull, that causes you to accelerate in this case? Identify the object that pushes on your mass (body plus skateboard) to cause the acceleration. Also identify the direction of the push that causes you to accelerate.

d) Obviously, you do some pushing, too. On what object do you push? In what direction?

e) How do you think, on the basis of both amount and direction, the following two forces compare?

- The force exerted by you on the wall.
- The force exerted by the wall on you.

2. Once again, as a class demonstration, two students can stand on skateboards. With extreme caution, the students should push on each other's palms.

a) Describe the motion of student A.

b) What force caused the motion of student A?

c) Describe the motion of student B.

d) What force caused the motion of student B?

3. Do a "thought experiment" about the forces involved when you are running or walking on a horizontal surface. Use words and sketches to answer the following questions in your log:

a) Since you move forward, not backward, there must be a force in the forward direction that causes you to accelerate. Identify where the forward force comes from, and compare its amount and direction to the backward force exerted by your shoe with each step.

b) Would it be possible to start walking or running on an extremely slippery surface (like an ice-skating rink) when wearing ordinary shoes? Discuss why or why not in terms of forces.

4. You and a member of your group will now see if you can apply unequal forces on each other. Clip two spring scales together. Each of you will pull on one scale. Try to pull so that one of you pulls with twice the force of the other. Do not pull on the scales so hard that they read a measurement above the highest value. You will have applied unequal forces if you can make one scale read twice the value of the other scale.

a) Record your results in your log.

b) In a diagram, draw the force exerted by you on your partner and the force exerted by your partner on you.

1.e)

Although the two forces are equal but opposite in direction, the students may not come to this conclusion at this point.

2.a)

Student A accelerates from student B while the push is taking place.

2.b)

The force of student B on student A made student A accelerate.

2.c)

Student B accelerates away from student A while the push is taking place.

2.d)

The force of student A on student B

Teaching Tip

If two large spring scales or bathroom scales are available, two students can push on the bathroom scales back-to-back so the reading on each can be seen by the class. The students will be able to clearly see that the two readings are equal.

3.a)

You push your foot on the ground backward from yourself. How strongly you can push your foot parallel to the surface of the sidewalk depends on how much frictional force exists between the sole of the shoe and the sidewalk's surface. If the shoe does not slip on the surface, then the sidewalk surface reacts with an equal and opposite force that causes your body to accelerate forward.

3.b)

On the slippery ice surface, the force of friction is greatly reduced. You are unable to apply much of a backward force on the ice surface because your shoe will slide on the surface. In turn, the opposite force of the sidewalk will also be minimal with the result that you accelerate by such a small amount that it is not noticeable. There is no significant force to push you forward.

4.a)

The students will find that the readings on the two scales are always equal.

4.b)

Students drawing should appear as below with equal length arrows.

Force of A on B　　Force of B on A

◄——————■——————►

CHAPTER 2

Part B

The weights specified in this section work with a meter stick of a certain stiffness. Be prepared to adjust the weights used by the students in order for them to work well with the meter sticks being used.

1.

Students read about free-body diagrams.

Teaching Tip

If the students are having trouble identifying the forces that act on a body (the free-body diagram), have them draw a circle around the object for which they are trying to identify the forces. Anything that touches the circle is a source of force. The only other forces that can act on the object are forces like gravity that do not need to touch objects to exert a force. This will allow the students to identify more forces more easily.

2.

Students set up the meter stick with a coin or washer in the center.

3.a)

Nothing happens, or else the meter stick bends very slightly to produce a force equal to the weight of the object.

4.a)

The more weight that is added to the meter stick, the greater the deflection of the stick.

4.b)

As the meter stick is deflected, the restoring forces in the wood build up until they exert an upward force equal to the downward force of the weight. A graph of the deflection vs. the weight added would look something like the one shown on the next page. Ask the students to consider many situations in which an object deflects and produces a force equal and opposite to the one causing the deflection. A chair does this when they sit on it, a floor does this when they are standing, the layer of ice does this when they are skating on a pond, etc. Encourage them to consider slight variations of this, but with the same basic principles. When they hang on a rope, it stretches in order to produce an upward force equal and opposite to their weight. Bungee cords are no different from ropes; they just have to stretch a much larger distance to generate the same force as a rope.

Chapter 2 Physics in Action

Part B: Observing a Meter Stick Push Back

1. When you hold up a book, you apply an upward force and gravity applies a downward force of equal strength. As a result, the book has no acceleration. A *free-body diagram* (a diagram showing the forces acting on an object) to illustrate this is shown below. When a book sits on a table, gravity applies a downward force and the table supplies an upward force of equal strength. The free-body diagram illustrating the force of gravity on a book lying on a table is similar to the one of the hand holding up a book.

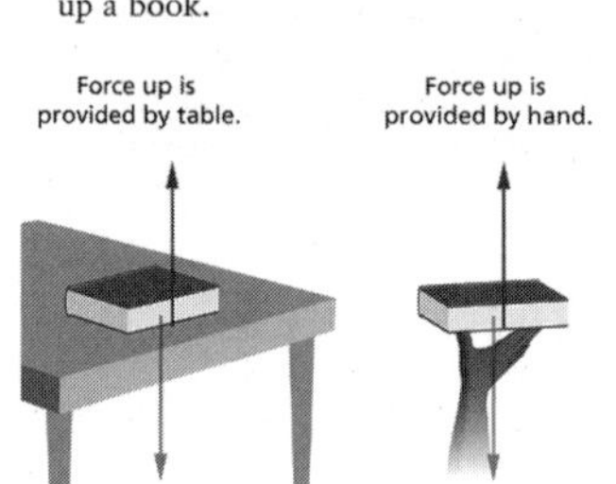

Walls, tables, and floors are extremely stiff, making it difficult to understand how they can produce forces. In this part of the *Investigate*, you will use something much less stiff, like a meter stick, to uncover how inanimate objects produce forces.

2. Set up a meter stick with a few books for support as shown.

3. Place a washer or coin in the center of the meter stick.

a) In your log, record what happens.

4. Remove the washer and replace it with a 100-g mass (weight of 100-g mass = 1.0 N). Continue to place a few more 1.0 N weights on the center of the meter stick. Note what happens as you place each weight on the stick.

a) Measure the deflection of the meter stick for each 1.0 N of weight and record the values for these deflections.

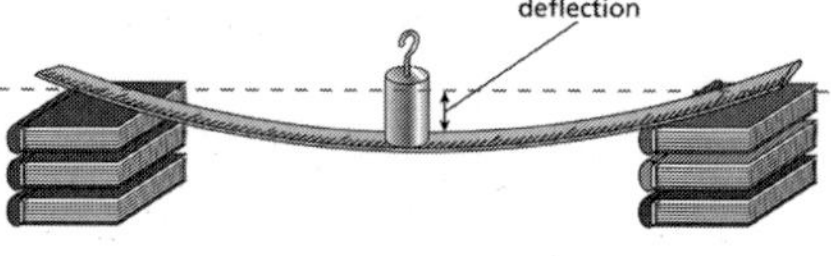

b) How does the deflection of the meter stick compare to the weight it is supporting? In your log, sketch a graph to show this relationship.

c) Remove the weights one at a time, noticing the change in deflection. Once all the weights have been removed, place the washer or coin back in the center of the meter stick. Do you think that the meter stick is deflecting? Write a concluding statement concerning the washer and the deflection of the meter stick.

d) Draw the forces acting on the 100-g mass when it is at rest on the meter stick. (This is a free-body diagram.)

Active Physics 200

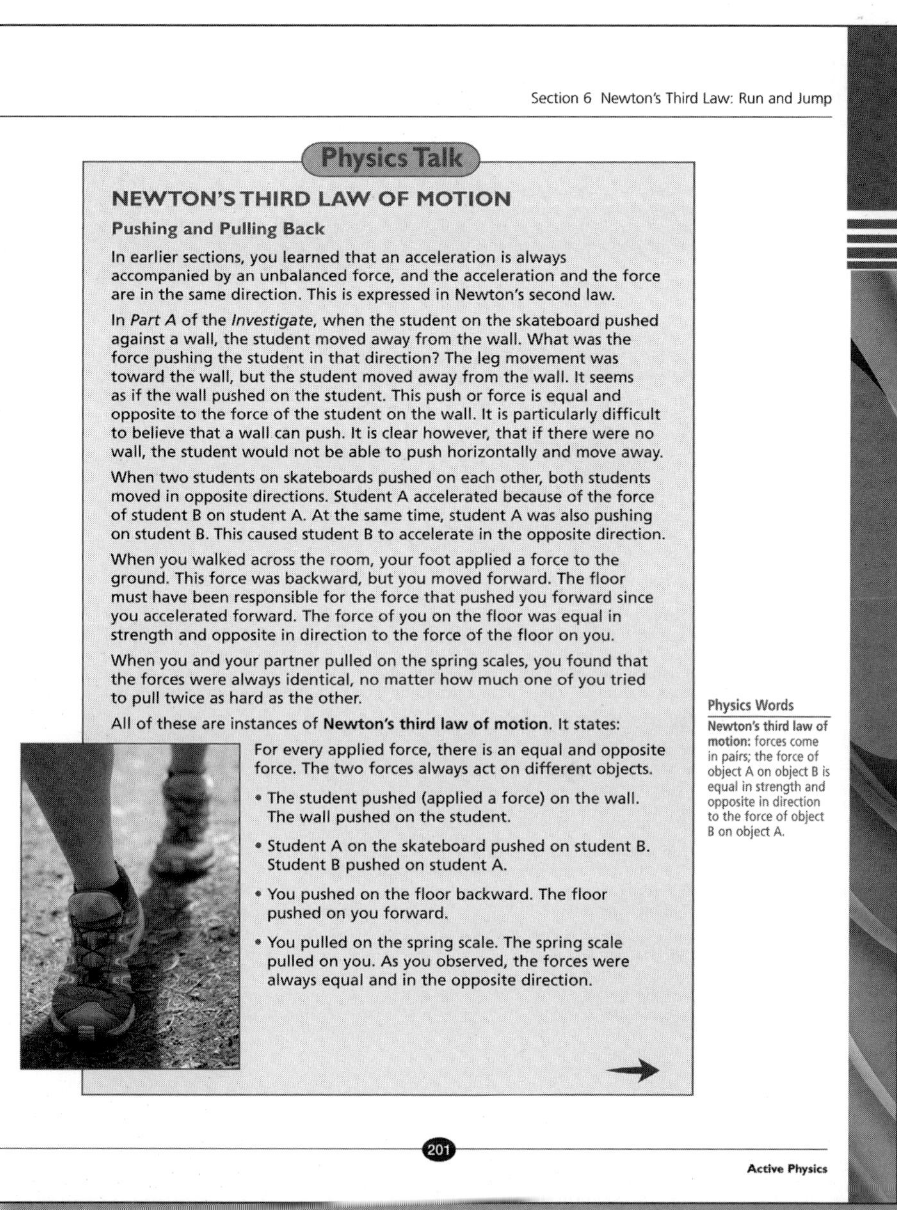

Section 6 Newton's Third Law: Run and Jump

Physics Talk

NEWTON'S THIRD LAW OF MOTION

Pushing and Pulling Back

In earlier sections, you learned that an acceleration is always accompanied by an unbalanced force, and the acceleration and the force are in the same direction. This is expressed in Newton's second law.

In *Part A* of the *Investigate*, when the student on the skateboard pushed against a wall, the student moved away from the wall. What was the force pushing the student in that direction? The leg movement was toward the wall, but the student moved away from the wall. It seems as if the wall pushed on the student. This push or force is equal and opposite to the force of the student on the wall. It is particularly difficult to believe that a wall can push. It is clear however, that if there were no wall, the student would not be able to push horizontally and move away.

When two students on skateboards pushed on each other, both students moved in opposite directions. Student A accelerated because of the force of student B on student A. At the same time, student A was also pushing on student B. This caused student B to accelerate in the opposite direction.

When you walked across the room, your foot applied a force to the ground. This force was backward, but you moved forward. The floor must have been responsible for the force that pushed you forward since you accelerated forward. The force of you on the floor was equal in strength and opposite in direction to the force of the floor on you.

When you and your partner pulled on the spring scales, you found that the forces were always identical, no matter how much one of you tried to pull twice as hard as the other.

All of these are instances of **Newton's third law of motion**. It states:

For every applied force, there is an equal and opposite force. The two forces always act on different objects.

- The student pushed (applied a force) on the wall. The wall pushed on the student.
- Student A on the skateboard pushed on student B. Student B pushed on student A.
- You pushed on the floor backward. The floor pushed on you forward.
- You pulled on the spring scale. The spring scale pulled on you. As you observed, the forces were always equal and in the opposite direction.

Physics Words

Newton's third law of motion: forces come in pairs; the force of object A on object B is equal in strength and opposite in direction to the force of object B on object A.

201

Active Physics

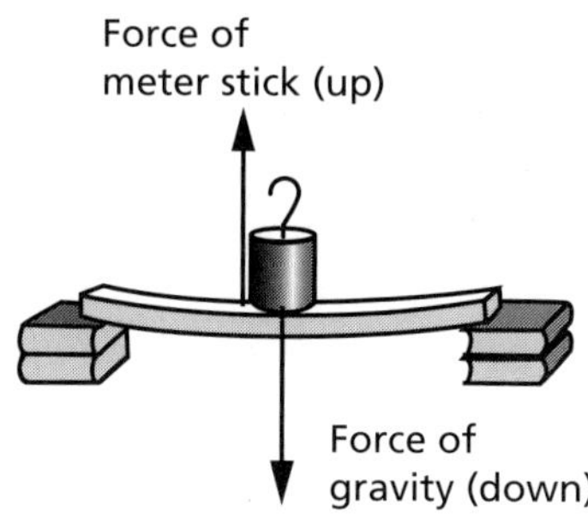

Physics Talk

Ask the students to read the *Physics Talk*, and then write down in their *Active Physics* logs what they have understood so far. Encourage students to draw free-body diagrams that explain Newton's third law. The idea that inanimate objects can push back might be a useful starting point for a discussion. Emphasize that forces always act in pairs, including why a mass cannot exert a force in one direction without being acted upon by an equal force in the opposite direction. Providing different opportunities for students to identify the different forces acting on an object will give them a visual reference, and enhance their understanding. Try to come up with examples of action-reaction pairs that are in different contexts to what is discussed in the *Physics Talk* to see if students revert to their prior conceptions. For example, in contrast to a person being accelerated by the floor while walking across, ask what force accelerates a car forward when the driver steps on the gas. Another possibility is to ask a student how the forces compare when a bug strikes the windshield of a car.

At this stage, you can ask students to explain each aspect of

bend of meter stick

number of weights added

4.c)

The students should conclude that the meter stick deflects with the coin on it, but the deflection is too small to measure.

4.d)

The forces are its weight (0.98 N directed downward) and the force of the meter stick on the mass (also equal to 0.98 N) directed upward. The total or net force on the mass is zero.

CHAPTER 2

Newton's third law and compare it with the previous two laws of motion. Have them illustrate their explanation through diagrams. Students may need some practice drawing free-body diagrams. Ask students to point out the steps of the *Investigate* that demonstrate Newton's third law. An important question to ponder would be why a bend in an inanimate object is produced when a force is applied to it. Review the concept of the normal force being provided by a surface. Point out that the normal force is always a reaction force, that it does not appear by itself, but only in response to another force being applied. Students should also understand the significance of the **center of mass.** As you bring your discussion of the *Physics Talk* to an end, discuss the term "reaction" from Newton's third law and its usage in the context of the law.

Newton's third law states that forces always come in equal and opposite pairs. If you push on the wall, the wall pushes on you with the same force. If you press your finger against the table, the table presses against your finger with the same force. You cannot touch someone without someone touching you back. The equal and opposite pairs of forces in Newton's third law always act on different objects. When you and your partner pulled on the spring scales, the forces were equal and you could not do anything to make the forces unequal. One spring scale pulled on your partner's finger and the other spring scale pulled on your finger. The two equal forces acted on different objects. The two forces that were applied can be shown in a diagram.

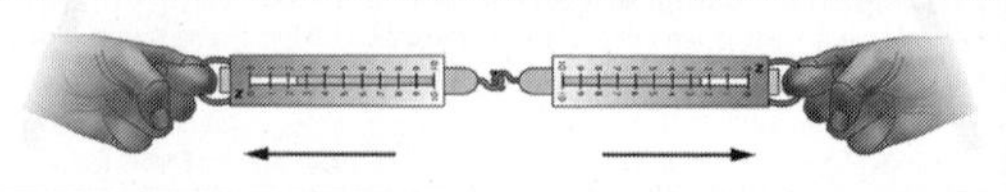

Inanimate Objects Can Push Back

The belief that a wall or a floor can apply a force is troublesome. How does a floor push, and how does it push with different amounts of force on different objects? In *Part B* of the *Investigate*, the masses on the meter stick provided evidence of how an inanimate object can apply a force. When a large mass was placed on the meter stick, you noticed a bend in the meter stick. This bend provides a force. The force of the meter stick on the mass in the upward direction was exactly equal to the force of gravity on the mass in the downward direction. The mass was therefore able to stay at rest. A smaller mass on the meter stick required a smaller force and the meter stick bent a bit less. The washer required a very small force, and the meter stick bent such a small amount that it may not have been observable.

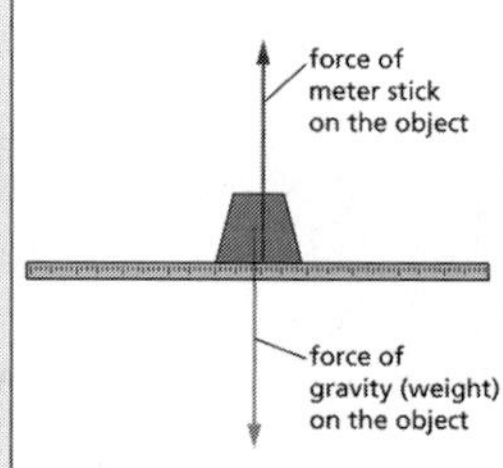

A force diagram of the forces on the mass can be drawn. Recall that a diagram that shows the forces acting on an object is also called a **free-body diagram**.

When you stand on the floor, your mass is pulling you down. You would fall if the floor were not applying an equal force up on you. The floor provides that force by bending just a bit. If you stand in the center of a trampoline, the bend is quite noticeable; however, floors made of wood or concrete provide less of a bend.

Physics Words

free-body diagram: a diagram showing the forces acting on an object.

Section 6 Newton's Third Law: Run and Jump

How to Draw a Free-Body Diagram

A free-body diagram is a diagram used to show the relative strength and the direction of all the forces acting on an object in a given situation. In a free-body diagram, each force is represented by an arrow. The direction of the arrow is the direction of the force. The size of the arrow is the strength of the force. Each arrow is labeled to show the type of force acting. Often, the actual object is drawn as a box. The weight of the object can be represented by an arrow emerging from the **center of mass.** This is the point at which all the mass of an object is considered to be concentrated. The other forces can be represented by arrows emerging from the contact point (such as the table on the book).

Physics Words

center of mass: the point at which all the mass of an object is considered to be concentrated.

Identifying the Opposite and Equal Forces of Newton's Third Law

Do not confuse *Part A* with *Part B* of the *Investigate*. In *Part A*, you found evidence for Newton's third law, which states that forces always come in pairs. These forces always act on different objects. You pushed on the wall and the wall pushed on you. Your weight applied a force to the floor and the floor pushed up on you. You pulled your partner's finger with the spring scale and your partner pulled your finger with the spring scale.

In *Part B*, you found evidence that an inanimate object can apply a force by bending. When the 100-g mass did not move as it rested on the meter stick, you drew two forces acting on the mass. The force of gravity pulled down on the mass. The meter stick pushed up with an equal force on the mass. These two forces are not the equal and opposite forces of Newton's third law. They are two forces on the same object, not equal and opposite forces acting on different objects.

You can combine your knowledge from *Parts A* and *B* of the *Investigate* to get the full picture. When the mass sits on the meter stick, there are two pairs of forces.

First pair of forces of Newton's third law: The meter stick pushes up on the mass and the mass pushes down on the meter stick. (Do not worry about the force on the meter stick here.)

Second pair of forces of Newton's third law: Earth pulls down on the mass with a force of gravity and the mass pulls up on Earth with an equal force of gravity. (Do not worry about the force on Earth here.)

In this situation, the focus is only on the mass. The two forces on the mass are the force of gravity pulling downward, and the force of the meter stick pushing upward. Each of these has an equal and opposite force on a different object and that is not a concern because attention is restricted only to the mass.

CHAPTER 2

Drawing Free-Body Diagrams

When you drew the force diagram (free-body diagram) of the forces on the mass, you only showed the forces on the mass and you did not show the force on the meter stick or the force on Earth.

When you drew the forces on the 100-g mass resting on the meter stick, you drew them from a point located in the center of the 100-g mass. The force of gravity acts on every little part that makes up the 100-g mass, and if you add all these forces on different parts together, it equals the weight of the 100-g mass (Weight = mg = (100 g)(10 m/s^2) = 1.0 N). So instead of many forces on different parts of the 100-g mass, you can consider a single force equal to the weight acting on the 100-g mass. But on what part of the 100-g mass should this single force act? The proper point is called the center of mass of the 100-g mass.

How Newton Described the Third Law of Motion

Newton's third law of motion can be stated in three equivalent ways:

- For every force applied to object A by another object B, there is an equal and opposite force applied to object B by object A.
- If you push or pull on something, that something pushes or pulls back on you with an equal amount of force in the opposite direction. This is an inescapable fact — it happens every time.
- Forces always come in pairs.

In one of the most important science books of all time, *Principia*, Isaac Newton wrote his third law of motion. It is interesting both historically and in terms of understanding physics to read Newton's third law in his own words:

"To every action there is always opposed an equal reaction: or, the mutual actions of two bodies upon each other are always equal, and directed to contrary parts."

In reading a passage written so long ago, you should be aware that slightly different meanings may be associated with some words. For example, your definition of the word "reaction" is probably something that happens after whatever causes it. Newton did not mean this.

The equal and opposite reaction happens instantaneously with the action that causes it. There is no delay. It is not a reaction, in today's use of the word, but an equal and opposite force that occurs at exactly the same time.

Challenging Newton's Third Law

After learning Newton's third law, students are often traditionally challenged to explain how a horse can pull a cart or how a person can pull a chair across the room. The argument goes like this: "If I pull on the chair then the chair pulls on me with an equal force. Therefore, the two forces cancel and nothing should move. Newton's third law must be wrong."

This would actually be true if the person and chair were on very slippery ice and there was no traction. However, on the ground, there are additional forces as shown in the diagram below.

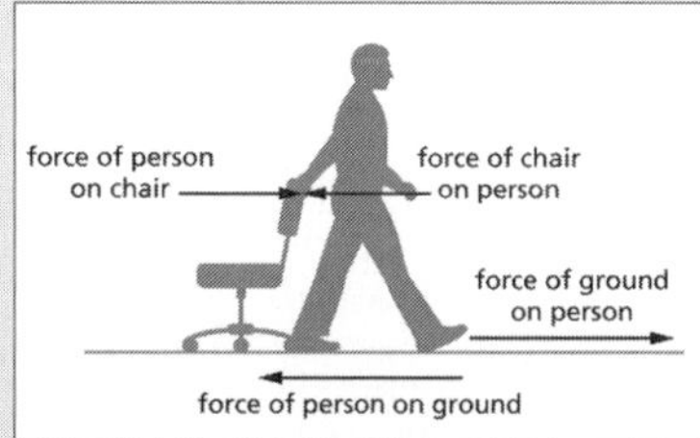

Assume that the chair is on wheels, as in this *Investigate*. The person pulling the chair applies a force to the ground. By Newton's third law, the ground pulls on the person. These two forces are equal and opposite. That force moves the person forward as the person pulls on the chair.

The person pulls on the chair with a small force. The chair pulls back on the person with an equal small force. Because there is only one force on the chair, the chair moves forward. There are two forces on the person — the force of the ground and the force of the chair. These are not equal. The force on the ground is larger than the force on the chair, thus, moving the person forward.

Checking Up

1. Describe Newton's third law of motion.
2. Earth pulls down a mass with a force of gravity. What is the equal and opposite force acting in this situation?
3. What does a free-body diagram illustrate?

Checking Up

1.

Newton's third law of motion says that every action has an equal and opposite reaction. If you push or pull on something you will experience the same push or pull in the opposite direction.

2.

The mass pulls up on Earth with a force that equals the gravitational pull.

3.

A free-body diagram illustrates the direction of all the forces acting on an object in a given situation.

2-6a Blackline Master

CHAPTER 2

Active Physics Plus

This section gives students the chance to explore Newton's third law in more depth, and also provides an opportunity for the teacher to have students explain problems by drawing diagrams. Developing the *Active Physics Plus* as an interactive tool by having students work in groups will help the entire class to understand how forces work.

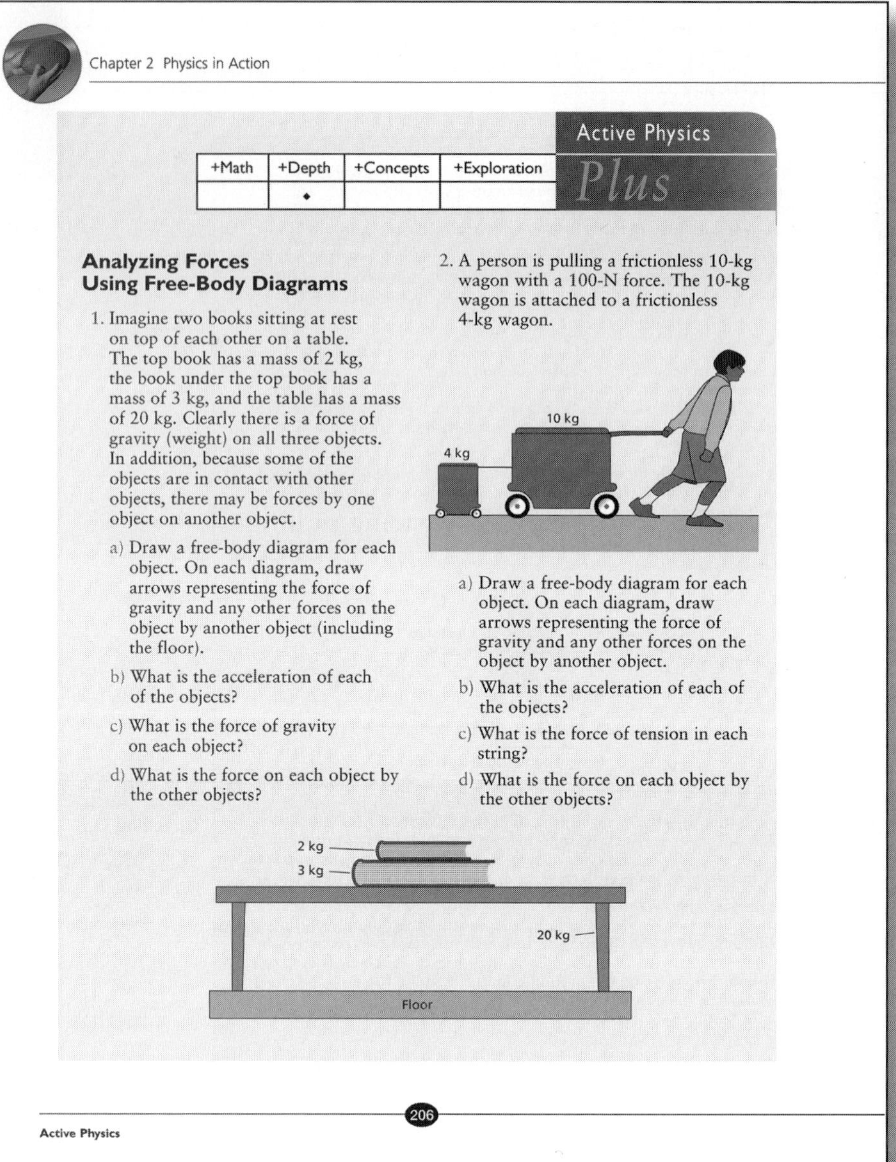

1.a)

The 2-kg book has two forces on it: its weight downward and the force from the 3-kg book upward. The 3-kg book has three forces on it: its weight downward, the force from the 2-kg book downward, and the force from the 20-kg table upward. The table has three forces on it: its weight downward, the force from the 3-kg book downward, and the force from the floor upward.

1.b)

All three objects are stationary, so the acceleration of each is zero.

1.c)

The force of gravity on the 2-kg book is 19.6 N, on the 3-kg book is 29.4 N, and on the 20-kg table is 196 N.

1.d)

Because the total or net force on each object is zero (since its acceleration is zero), the force from the 3-kg book on the 2-kg mass must be 19.6 N upward. By Newton's third law, the 2-kg book must exert an equal and opposite force on the 3-kg book. Therefore the combination of its weight and the downward force from the 2-kg book is 29.4 N plus 19.6 N, or 49.0 N. Since the total or net force on the 3-kg book is zero, the upward force from the table on the 3-kg book must be 49.0 N. Again, the third law indicates that the 3-kg book exerts a downward force on the table of 49 N. This combined with its weight of 196 N means the floor must exert an upward force on the table of 49 N plus 196 N, or 245 N. Notice that it is easy to check this last answer. The total weight of the table and books is 25 kg multiplied by 9.8 m/s^2, or 245 N.

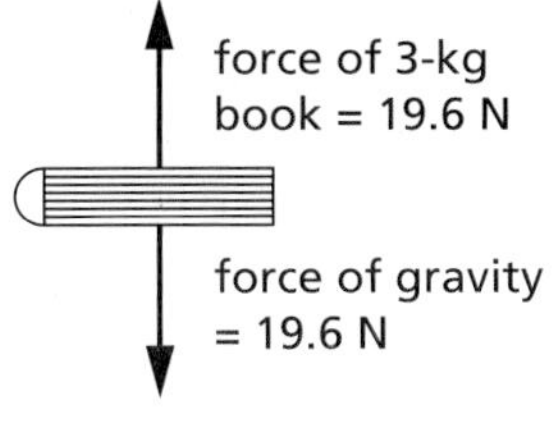

forces on
2-kg book

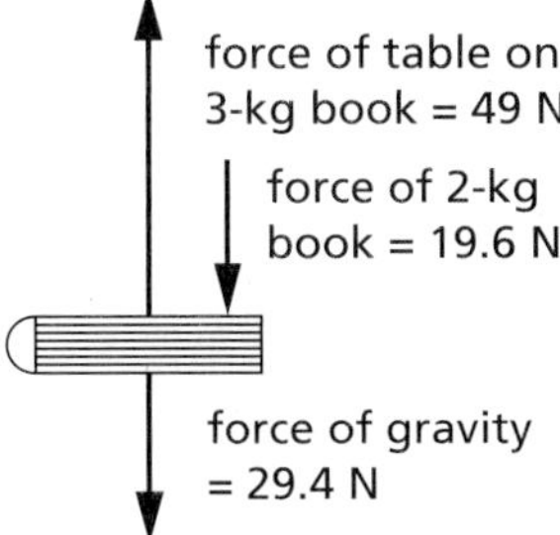

forces on
3-kg book

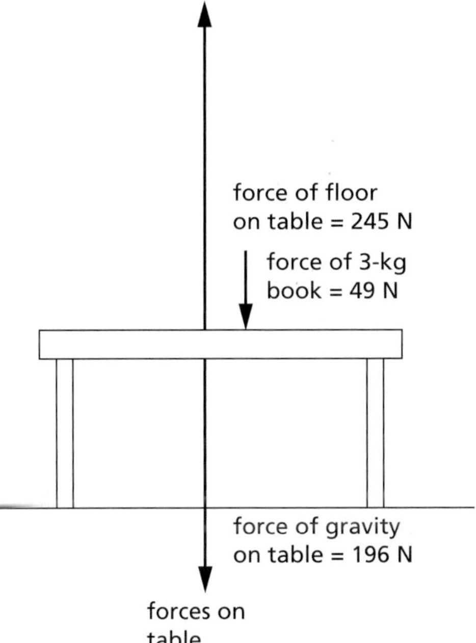

forces on
table

2.b)

Using Newton's second law,

$\Sigma F = ma$

$100 \text{ N} = (14 \text{ kg})a$

$a = 7.1 \text{ m/s}^2$

All the objects will accelerate together at the same rate.

2.c)

To find the force on the 4-kg wagon, let the tension in the string pulling the 4-kg wagon be T_1.

$\Sigma F = ma$

$T_1 = (4 \text{ kg})(7.1 \text{ m/s}^2) = 28.4 \text{ N}$

The tension in the string pulling the 10-kg wagon is 100 N.

2.d)

The force on the 4-kg cart by the 10-kg cart is 28.4 N, and the force on the 10-kg cart by the 4-kg cart is also 28.4 N (equal and opposite forces). The force on the 10-kg cart by the person pulling is 100 N, and the 10-kg cart pulls back on the person with a force of 100 N, making the net force on the 10-kg cart 71.6 N (so it may accelerate at 7.1 m/s²). The force on the person by the ground will depend upon the person's mass, but will be greater than 100 N (so the person can also accelerate at 7.1 m/s²).

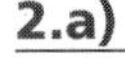

2.a)

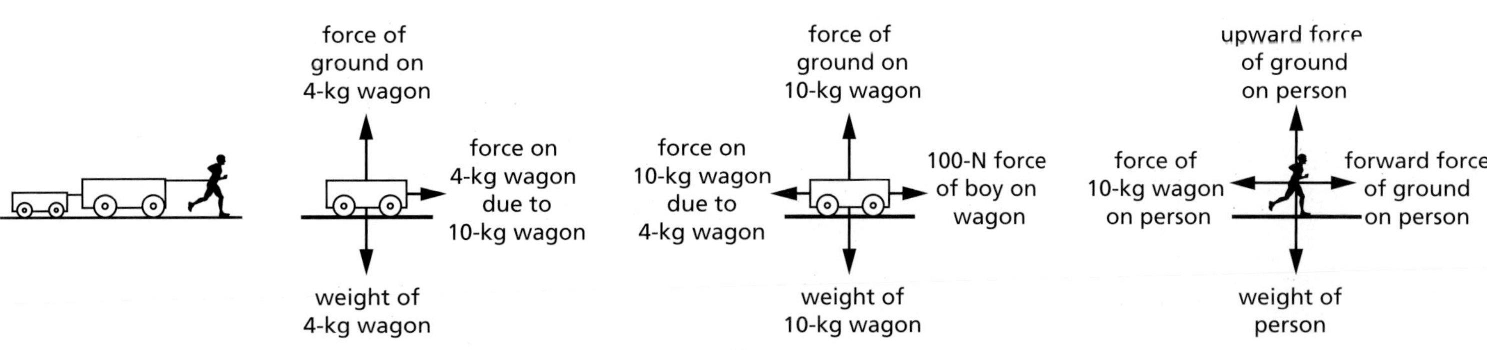

NOTES

Section 6 Newton's Third Law: Run and Jump

What Do You Think Now?

At the beginning of this section, you were asked the question

- **Pretend that you have just met somebody who has never jumped before. What instructions could you provide to get the person to jump up (that is, which way do you apply the force when you push with your feet)?**

When you jump, in what direction do you push on the floor? When you push on the floor, what direction does the floor push on you? What happens to the floor that results in an upward force on you? Use your investigations of Newton's third law to justify your answers.

Physics Essential Questions

What does it mean?

In a video clip, a player catches a football. In another video clip, a soccer ball is caught by the goalie. Newton's third law states that there are equal and opposite forces in each video. Identify the forces of Newton's third law for each of these situations.

How do you know?

Two children pull on spring scales and neither child moves. What evidence do you have that the forces of the students are equal and opposite?

Why do you believe?

Connects with Other Physics Content	Fits with Big Ideas in Science	Meets Physics Requirements
Force and motion	Models	* Experimental evidence is consistent with models and theories

* In relying on observation to come up with explanations, physics often includes ideas that do not seem plausible to someone who has not thought about it. Why do you believe that a table bends when you put a plate on it?

Why should you care?

In sports, forces are exerted by both animate and inanimate objects. In your sports voice-over, you will need to comment on one or both of these situations. Give an example from the sport you have selected in which a force is exerted by an inanimate object.

Active Physics

What Do You Think Now?

Now that students have investigated Newton's third law, they should be able come up with useful instructions. You might want to point out the role of acceleration and mass in determining the strength of a person's force and the direction in which it will act. Sharing the answer provided in *A Physicist's Response* would give you an opportunity to discuss the *What Do You Think Now?* question in more detail.

CHAPTER 2

Physics Essential Questions

What does it mean?

There is a force of the player on the football and a force of the football on the player. There is a force of the goalie on the soccer ball and a force of the soccer ball on the goalie.

How do you know?

The spring scales read the same value regardless of how each is pulled.

Why do you believe?

Because the gravitational force is pulling on the plate, there must be a force pushing upward on the plate. The table is the only thing in contact with the plate. The assumption is that the table bent a small amount to provide that force.

Why should you care?

When a basketball is dribbled, the ball hits the floor and bounces up. The floor must be providing a force to change the direction of the ball.

Reflecting on the Section and the Challenge

Reflecting on Newton's third law will challenge students to think of how they can connect their understanding of equal and opposite forces to the *Chapter Challenge*. You could point out to them how this section, in gradual steps, should have given them more confidence in describing a player's technique on a sports field. Have them read the examples in the *Reflection on the Section and Challenge* and draw a visual of one of these examples so they can review what they have learned, and think of the different ways they encounter Newton's third law.

Chapter 2 Physics in Action

Reflecting on the Section and the Challenge

According to Newton's third law, each time an athlete acts to exert a force on something, an equal and opposite force acting on the athlete happens in return. There will probably be countless examples of this in your video production. When you kick a soccer ball, the soccer ball exerts a force on your foot. When you push backward on the ground, the ground pushes forward on you (and you accelerate). When a boxer's fist exerts a force on the other boxer's body, the body of the other boxer exerts an equal force on the first boxer's fist. You can now use the same sports video sequence of a sport to describe how it illustrates all three of Newton's laws of motion.

Physics to Go

1. When an athlete is preparing to throw a shot put, does the ball exert a force on the athlete's hand equal and opposite to the force the hand exerts on the ball? Explain your answer.
2. When you sit on a chair, the seat of the chair pushes up on your body with a force equal and opposite to your weight. How does the chair "know" exactly how hard to push up on you—are chairs intelligent? Is there any "deflection" going on?
3. You have weighed yourself by stepping on a scale many times. How do you think a simple bathroom scale works?
4. For a hit in baseball, compare the force exerted by the bat on the ball to the force exerted by the ball on the bat. Why do bats sometimes break?
5. Compare the amount of force experienced by each football player when a big linebacker tackles a small running back.
6. Identify the forces active when a hockey player "hits the boards" at the side of the rink at high speed.
7. Newton's second law, $F = ma$, suggests that when catching a baseball in your hand, a great amount of force is required to stop a high-speed baseball in a very short time interval. The great amount of force is needed to provide the great amount of acceleration required (in this case negative acceleration). Use Newton's third law to explain why baseball players prefer to wear gloves for catching high-speed baseballs. Use a pair of forces in your explanation.
8. ***Preparing for the Chapter Challenge***
   a) Write a sentence or two explaining the physics of an imaginary sports clip using Newton's third law. How can you make this description more exciting so that it can be used as part of your sports voice-over?
   b) Describe how deflection of the ground can produce a force. What would make this description more exciting and therefore a valuable part of your sports voice-over?

Physics to Go

1.

Yes, the forces are equal and opposite. The force of the shot put on the hand keeps the hand from accelerating much faster under the applied force.

2.

The restoring forces within the material making up the chair build up until the upward force exerted by the chair equals the downward force caused by your weight. If someone sits in your lap, the chair "bends" (or is otherwise deformed) more, resulting in a higher reaction force, equal and opposite to the combined weight. No intelligence is required by the chair, only the people who designed it.

3.

A spring system in the scale bends to produce the opposite force. A needle or a sensor is attached to that spring to indicate the amount of force needed to oppose the weight.

4.

The forces on the ball and the bat are equal and opposite; sometimes the force exerted by the ball on the bat is enough to break the wood of the bat.

5.

The forces on the players are equal and opposite, but the smaller player experiences a greater acceleration, which can have more harmful effects on the human body than a lesser acceleration.

6.

The forces are equal and opposite; the hockey player is more likely than the boards to complain about the pain involved.

7.

Gloves having padding which compresses and/or webbing which deforms when the ball hits the glove. The "softness" of a glove reduces the force which the glove exerts on the ball. This force is lower than the force that a stationary hand would need to exert to stop the ball. The lesser force causes the ball to decelerate at a lower rate, also reducing the force which the ball exerts on the glove during stopping. A sure way to reduce the forces during a collision is to increase the amount of time that the objects exert forces on each other during the collision; that is one reason why air bags reduce injuries in automobile collisions.

Preparing for the Chapter Challenge

8.a)

Student answers will vary, but each one should indicate that the force exerted on one object is equal and opposite to the force exerted on the other. In addition, the students should point out that although the forces are equal and opposite, they never act on the same body, and thus do not "cancel" each other.

8.b)

Student answers again will vary, but each should indicate that the deflection of the ground is similar to a compressing spring, which is how the force is provided.

NOTES

CHAPTER 2

NOTES

Section 6 Newton's Third Law: Run and Jump

Inquiring Further

Forces acting on you in an elevator

Ask the manager of a building that has an elevator for permission to use the elevator for a physics experiment. Your teacher may be able to help you make the necessary arrangements.

Stand on a bathroom scale in the elevator and record the force indicated by the scale while the elevator is:

- At rest.
- Beginning to move upward (upward acceleration).
- Appearing to move upward at constant speed.
- Beginning to stop while moving upward (downward acceleration).
- Beginning to move downward (downward acceleration).
- Appearing to move downward at constant speed.
- Beginning to stop while moving downward (upward acceleration).

For each of the above conditions of the elevator's motion, Earth's downward force of gravity is the same. If you are accelerating upward, the floor must be pushing up on you with a force larger than the force due to gravity.

a) Make free-body force diagrams that show the vertical forces acting on your body when standing on a scale in the elevator.

b) Use Newton's laws of motion to explain how the forces acting on your body are responsible for the kind of motion (at rest, constant speed, positive and negative acceleration) that your body experiences.

209

Active Physics

Inquiring Further

The force reading on the scale is equal to the student's weight.

- The force is greater than the student's weight.
- The force is equal to the student's weight.
- The force is less than the student's weight.
- The force reading is less than the student's weight.
- The force is equal to the student's weight.
- The reading is more than the student's weight.

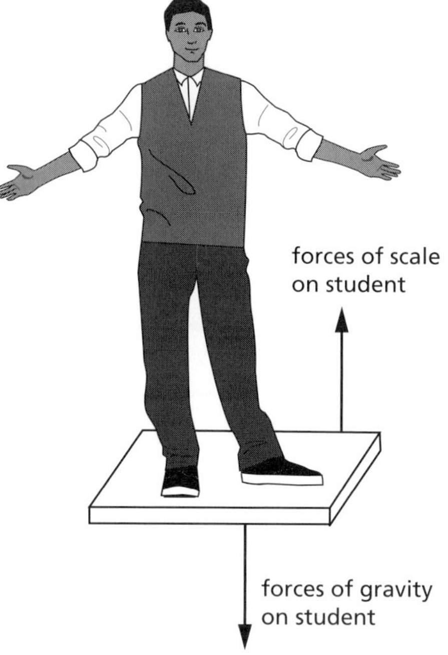

When the force of the scale on the student is equal to the force of gravity on the student (the student's weight), the forces are equal and opposite, and according to Newton's first law the student may either be at rest or traveling with constant velocity (either up or down).

When the force of the scale is greater than the student's weight, the student will be accelerating upward, either by having an increasing upward velocity or a decreasing downward velocity.

When the force of gravity is less than the student's weight, the student will be accelerating downward, either by increasing the downward velocity or decreasing the upward velocity.

CHAPTER 2

SECTION 6 QUIZ

2-6b Blackline Master

1. Which statement explains why a book resting on a table is in equilibrium?
   a) There is a net force acting downward on the book.
   b) The weight of the book equals the weight of the table.
   c) The acceleration due to gravity is 9.8 m/s^2 for both the book and the table.
   d) The weight of the book and the table's upward force on the book are equal in magnitude, but opposite in direction.

2. A book that is resting on a shelf 1.5 m above the floor has a weight of 12 N. The force the shelf exerts upon the book must be
   a) 8 N
   b) 12 N
   c) 18 N
   d) 24 N

3. While closing a door of mass 10 kg, a student pushes on it with a force of 25 N. As the door is accelerating to close, what force does the door exert on the student?
   a) 5 N
   b) 25 N
   c) 250 N
   d) 35 N

4. A man is attempting to push a chair across the floor with constant speed by exerting a force of 50 N. When will the chair exert a force of 50 N on the man?
   a) only if the man and the chair are at rest
   b) only if the man and the chair are accelerating
   c) only if the man and the chair are moving with constant velocity
   d) in all of the above cases

5. A man standing on a chair weighs 800 N. In order for the chair to support the man, the chair must exert an upward force of
   a) less than 800 N.
   b) 800 N.
   c) more than 800 N.

SECTION 6 QUIZ ANSWERS

1. d) The book is in equilibrium so there must be no net force acting on the book. The downward force of the book on the table due to its weight must be counterbalanced by the upward force of the table to result in zero net force on the book.

2. b) By Newton's third law, the force of the book on the shelf must be equal and opposite to the shelf's force on the book.

3. b) By Newton's third law, the forces must be equal and opposite. The fact that the door is accelerating only means that the force the person is exerting on the door will be difficult to maintain as the door moves away.

4. d) Newton's third law holds in all cases, regardless of motion or other factors.

5. b) The force must be exactly equal and opposite to the man's weight. If the force were greater than 800 N, the man would experience a net upward force and would accelerate upward.

NOTES

SECTION 7

Frictional Forces: The Mu of the Shoe

Section Overview

In this section, students investigate the effect of different surfaces and different weights on the force of friction. By pulling a shoe across a surface, the students examine the factors that affect the force of friction between the shoe and the surface. They measure and record the amount of force needed to keep the shoe sliding on the surface at a slow, constant speed when the mass of the shoe is varied. The force required to slide an object on a surface at constant speed determines the force of friction. They also measure the weight of the shoe and determine that the perpendicular (normal) force exerted by the surface on the object is equal in magnitude to the weight but opposite in direction. The students then discover that the ratio of the force of friction to the normal force is a constant known as the *coefficient of friction*. The coefficient of friction is found to be independent of the force holding the surfaces together, and depends only upon the nature of the surfaces in contact. Finally, the students investigate various surfaces to determine their coefficients of friction.

Background Information

To accelerate an object (change its speed, its direction, or both), a net force must act on the object. If a person is standing still on a sidewalk and wants to get moving, he or she must somehow cause a force to be exerted on the body's mass, which is accomplished by the application of a force in the backward direction, parallel to the surface of the sidewalk. How much the person can push his or her feet depends on how much frictional force can be sustained by the interaction of the sole of the shoe and the sidewalk's surface. If the shoe does not slip on the surface, the sidewalk surface's equal and opposite reaction to the rearward force of friction causes the body to accelerate forward; if there is ice on the sidewalk, the available force will be reduced, and shoes may just slide on the surface with the result that one goes nowhere.

The maximum frictional force, F_{max}, that can be generated between the surfaces of two materials in contact but not sliding relative to each other is expressed by the equation $F_{max} = \mu_s F_N$. F_N is the "normal force" (perpendicularly pushing the two surfaces together) and μ_s is the "coefficient of static or starting friction" (for the pair of materials from which the surfaces are made). The word "normal" in this context means "perpendicular." Expect students to need some time to get comfortable with the use of this word. In the above example, the normal force, F_N, would be equal to your weight.

The value of μ_s depends on the quality of the two materials in contact. If the two surfaces are sliding relative to each other, the frictional force, F, is given by the equation $F = \mu F_N$, where μ is the "coefficient of sliding friction" (sometimes called the coefficient of "kinetic"—meaning "moving"—friction). The coefficient of static friction is larger than the coefficient of sliding friction. The coefficient of sliding friction applies when the surfaces are moving with respect to each other in a sliding mode and the formula $F = \mu F_N$ gives the value of the frictional force. It is this frictional force that is measured in this section because it is a bit easier conceptually.

When objects are not sliding relative to each other, the frictional force can be anything between zero and $\mu_s F_N$. When objects are sliding relative to each other, the frictional force is simply μF_N. The frictional force generated between the shoe and the sidewalk due to pushing the foot rearward, as gravity and the upward restoring force of the sidewalk squeeze the sole of the shoe and the surface of the sidewalk together, is answered by a corresponding equal forward push by the sidewalk on the foot. The latter force, the forward push by the sidewalk, is the push you "feel" and which causes you to accelerate forward. It is not always the case that the normal force, F_N, is equal to the weight of the object. In the case of an object on a sloped surface, the normal force is less than the weight, equaling the component or effectiveness, of the object's weight in the direction perpendicular to the sloped surface. Other examples of cases where F_N is not equal to the weight of an object bearing on a surface would include the frictional force between belts riding on pulleys in machines; in such cases, tensioned springs usually are used to force the surfaces together to provide sufficient F_N to prevent sliding, or, intentionally as when stopping a machine, to reduce tension to cause F_N to be reduced to an amount where a belt will slide on a pulley. Another example of an increased normal force would be when an athlete's sudden thrust downward creates a much larger force than the weight of the athlete. This is a temporary condition because the athlete's body will soon be accelerated upward, lifting that foot away from the surface. Similar conditions may occur when the athlete descends from this leap for the next stride.

Crucial Physics

- For any two surfaces sliding past each other, the ratio of the force necessary to keep the surfaces sliding and the force between the surfaces perpendicular to the surfaces is called coefficient of sliding friction.
- The coefficient of sliding friction depends on the nature of the two surfaces, not on the force between the surfaces perpendicular to the surfaces.

Learning Outcomes	Location in the Section	Evidence of Understanding
Apply the definition of the coefficient of sliding friction, μ.	***Investigate*** Steps 2.a) and 3	Students measure and record the weight of a shoe and the amount of force in newtons needed to keep the shoe moving at a slow constant speed. They then use an equation to find the coefficient of friction.
Measure the coefficient of sliding friction between soles of athletic shoes and a variety of surfaces.	***Investigate*** Steps 3-5	Students measure and record the amount of force needed to pull different shoes on a variety of surfaces and then by recording the weight, as well, calculate the coefficient of friction.
Calculate the effects of frictional forces on the motion of objects.	***Investigate*** Steps 4 and 5	Students use different surfaces and weights to measure the force required to slide an object on each surface.

CHAPTER 2

Section 7 Materials, Preparation, and Safety

Materials and Equipment

PLAN A		
Materials and Equipment	**Group (4 students)**	**Class**
Calculator, basic	1 per group	
Weight, slotted, 100 g	10 per group	
Scale, spring, 0-20 N	2 per group	
Surface, felt, rough, 6 in. x 36 in.	1 per group	
Box, friction, 5 3/4 in. (length) x 3 3/4 in. (width) x 3 in. (depth) (with hook)	1 per group	
Access to a flat surface (such as a table, floor or other open space)*	1 per group	
Shoe, athletic*	1 per group	

*Additional items needed not supplied

PLAN B		
Materials and Equipment	**Group (4 students)**	**Class**
Calculator, basic	1 per group	
Weight, slotted, 100 g	10 per class	
Scale, spring, 0-20 N	2 per group	
Surface, felt, rough, 6 in. x 36 in.	1 per group	
Box, friction, 5 3/4 in. (length) x 3 3/4 in. (width) x 3 in. (depth) (with hook)	1 per group	
Access to a flat surface (such as a table, floor or other open space)*	1 per group	
Shoe, athletic*	1 per group	

*Additional items needed not supplied

Note: Time, Preparation, and Safety requirements are based on Plan A, if using Plan B, please adjust accordingly.

Time Requirements

This *Investigate* should take one class period or 40 minutes.

Teacher Preparation

- Prior to the day of the *Investigate*, tell your students they will be measuring the friction of their athletic shoes. Ask the students to wear a pair of athletic shoes or bring one to class on that day. If you choose not to use athletic shoes, the same activity may be done by pulling material such as a wooden block across the surface, however students are much more engaged when they are using their shoes.

- In addition, you will need several samples of different surfaces to measure the frictional force. Sandpaper and/or a rough cloth taped to a lab table work quite well as a surface different from the lab table. Have masses available to place inside the shoes to increase the shoe's weight.

Safety Requirements

No particular safety precautions are required for this investigation.

NOTES

Meeting the Needs of All Students

Differentiated Instruction: Augmentation and Accommodations

Learning Issue	Reference	Augmentation and Accommodations
Designing an experiment	***Investigate*** Step 2	**Augmentation** • Check to make sure groups choose variables that can be safely tested using the classroom equipment. • Prior to beginning the experiment, check for data tables that will allow students to record useful data. • Carefully plan groups by mixing students of different learning styles and strengths. A student who does poorly on tests may be great at designing an experiment. **Accommodation** • Provide a graphic organizer to guide student planning. The organizer could have sections for a purpose, hypothesis, diagram of the lab setup, procedure, data table, analysis, and conclusions. • Use the experiment that has already been designed in *Steps 3* and *4*.
Following directions Using a spring scale Reading comprehension	***Investigate*** Steps 2.b), 3, and 4	**Augmentation** • Model how to properly use the spring scale. Students may also need to learn how to "read" the spring scale, especially those with visual motor and executive function issues. • Students have learned about the forces that act on objects earlier in this chapter. For the *Investigate*, Step 2.b), ask students what forces they think might be acting on the shoe before they read this step. The students may be able to identify all four forces without having to read this task. Activating prior knowledge may empower students who struggle with reading comprehension. • Assist students with setting up a data table that will allow them to record all of the required data including the type of surface, the shoe description, the weight of the shoe, the force required to slide the shoe, and μ. **Accommodation** • Pair students intentionally to include someone with strong reading skills and someone who struggles with reading. • Provide a step-by-step task checklist and time limits for students who really struggle to maintain focus/attention.
Understanding the concept of a ratio	***Physics Talk***	**Augmentation** • Provide direct instruction to teach the concept of a ratio because many students may not remember the math vocabulary word, *coefficient of sliding friction* (μ).
Solving word problems	***Physics to Go*** Step 5	**Augmentation** • Students are asked to rearrange the coefficient of friction formula to solve for a "horizontal force." • Remind students that the helpful circle can assist them in rearranging the formula if they struggle with algebra. • Students may have no idea that they are being asked to find the force of friction in this question.

Learning Issue	Reference	Augmentation and Accommodations
Choosing a formula and solving word problems	***Investigate*** Step 6	**Augmentation** • Students with executive-function, reading, and math-learning issues struggle to extract information from a word problem and use the numbers to find a solution. • Review key phrases such as "what is," "find the value of," and "calculate the." These phrases give students clues about what the problem is asking them to find. • Require that students show their work, especially for multi-step problems that expect students to use answers they have calculated as given values. • Remind students to use their formula sheet. Also, tell students that they may need to use formulas from other sections or chapters. **Accommodation** • Provide students with a page of blank problem-solving boxes. • Limit choice for the formulas. Show students the four formulas they need to solve these problems, and then ask them to choose the appropriate formula for each step.

Strategies for Students with Limited English-Language Proficiency

Learning Issue	Reference	Augmentation
Vocabulary comprehension	***Investigate***	The concept of variables is crucial to scientific investigation. Explain to students that the root of the word "variable" is "vary," meaning "to change." A variable is a quantity that can change. When scientists conduct an experiment, they change only one variable at a time, which ensures that any change in the results must be caused by the change in the variable.
Demonstrating understanding	***Investigate*** Step 3.b)	To give students a chance to demonstrate their understanding of the forces acting on the shoe, have them draw a free-body diagram that shows the direction of the four forces, and label the forces as well. Then have students indicate the two forces necessary for figuring out the value of μ. Encourage students to swap diagrams and check one another's work.
Vocabulary comprehension	***Physics Talk,*** Coefficient of Sliding Friction, μ	To help ELL students understand "material," "surface texture," "moisture," and "lubrication" of surfaces and their effect on friction, bring in some samples of different materials, such as smooth plastic, wood, brick, and sandpaper, for students to touch. Allow students to push a coin, pencil eraser, or other small object over the different surfaces. Then dampen part of each material with water and part with mineral oil and allow students to do the same things again. Hold a class discussion about the friction associated with each dry material and what change wetting it makes. Encourage students to use the terms in context during the discussion.
Vocabulary comprehension Understanding concepts	***Active Physics Plus***	Thus far, students have been working only with sliding friction. Here, they encounter static friction. They likely will be able to infer the meaning of "static"—not moving—from context. Reinforce their inference by telling them that static in science refers to bodies at rest and to balanced forces. Once students have grasped the meaning of "static," allow them time to write in their *Active Physics* log the description of how static friction works. Encourage them to write in their own words. Collect the logs and check descriptions for accuracy. Correct any misunderstandings of science and improper English before proceeding.
Vocabulary comprehension	***Physics to Go*** Step 9	ELL students are likely familiar with the word "copy" as a synonym for "duplicate" (a photocopy, for example). But they may need help understanding the term to mean "the words that make up an advertisement."

SECTION 7

Teaching Suggestions and Sample Answers

What Do You See?

The person slipping on ice captures this section's theme. This is an opportunity to steer students toward the concept of the coefficient of friction. You may want to ask them to pay close attention to each object in the picture and query whether one particular object stands out in relation to the title of the section. As this is a time to stimulate interest, remind students that they should primarily focus on giving answers and revisit this visual at a later stage.

Chapter 2 Physics in Action

Section 7 Frictional Forces: The Mu of the Shoe

What Do You See?

Learning Outcomes

In this section, you will

- **Apply** the definition of the coefficient of sliding friction, μ.
- **Measure** the coefficient of sliding friction between the soles of athletic shoes and a variety of surfaces.
- **Calculate** the effects of frictional forces on the motion of objects.

What Do You Think?

A shoe store may sell as many as 100 different kinds of sport shoes.

- **Why do some sports require special shoes?**
- **Why would different features of a shoe be useful for different sports?**

Record your ideas about these questions in your *Active Physics* log. Be prepared to discuss your responses with your small group and the class.

Investigate

In this *Investigate*, you will examine how difficult it is to pull a shoe across a surface.

1. Take an athletic shoe. Use a spring scale to measure the weight of the shoe, in newtons.

    a) Record a description of the shoe (such as its brand) and the shoe's weight, in your log.

   b) List some things (which scientists call "variables") that may affect the force required to pull the shoe.

210

Students' Prior Conceptions

This section introduces the idea that friction occurs between the surfaces of objects and that the nature of the attraction of the surface particles to one another gives rise to static friction and subsequently to kinetic or sliding friction when one object slides or moves over another.

1. **Students may confuse friction with inertia and inertial mass.** This preconception may persist even after inertia was introduced in *Sections 1* and *2*.
2. **Students believe that the force needed to accelerate an object at rest works against friction; inertia and inertial mass are not relevant.** Reviewing concepts studied in earlier activities is a good instructional technique to apply here.
3. **Frictional forces only are due to irregularities in surfaces moving past each other.** Significant changes in the nature of either contact surface (molecular attraction of surface molecules, type of material, surface texture, moisture, or lubrication) affects the value of μ, the coefficient of friction.

What Do You Think?

As students answer these questions, ask them to think of the shoes they wear daily and how they are different from the sports shoes they wear on the field. You might want to remind them that this is the time for them to write their responses without any hesitation. It is important that at this stage students discuss answers among themselves. Facilitate their discussion and make sure that your students are engaged in the process of preparing for the concepts they are about to learn.

What Do You Think?

A Physicist's Response

A shoe for a specific surface should have the best traction possible without causing too much stickiness, which might delay the athlete. A second consideration is the safety of athletes in contact sports where having a shoe that sticks too strongly to the ground when the player collides with another player might result in a broken leg. A shoe for many surfaces would have to reflect some compromise as to what works best on average.

Certain shoes require additional features to improve the athlete's performance. Golf shoes, and baseball and football cleats use spikes that allow the athletes to get extra traction on grass surfaces. Basketball shoes have higher tops to protect players' ankles from injury. The boots used by skiers have additional support to withstand the strains of the long skis.

Investigate

1.a)

Answers will vary depending on the brand of shoe used and the size of the shoe. If you wish, you may use blocks of wood; some instructors prefer the increased precision. Students really like using shoes though.

Teaching Tip

Although using a shoe may not appear to be the most sanitary method of performing this activity, the students' enthusiasm for it more than makes up for this aspect. Keeping a spray can of deodorant/disinfectant handy is usually a good idea. If the students do not list the weight of the shoe as a variable, ask them if a heavy crate is easier or harder to push across a floor.

1.b)

Some variables the students may come up with include shoe size, tread style, shoe brand, weight of shoe, whether the shoe is high or low cut, etc.

The students should quickly realize that if they are to use just one shoe, factors such as size, brand, etc., would not be variables for that shoe. They may also think that the "weight" (mass) is not a variable, but point out that objects can easily be slipped inside the shoe to increase its weight. The students' design should show how a variable might affect the shoe's frictional force. Check it for safety and feasibility using available equipment.

NOTES

2. Design an experiment that would allow you to determine how one of the variables you listed affects the force required to pull the shoe. Include in your design:
   - What you will be able to conclude as a result of your experiment.
   - What data you will record.
   - What tools you will use to measure your data.
   - How you will analyze your data.

   a) Record your procedure in your log.

Your teacher may ask you to continue with your experimental design or, because of equipment or safety concerns, ask you to proceed with the experiment as described below.

3. Your teacher may ask you to use blocks of wood instead of shoes for greater precision in your results. However, wooden shoes are not recommended for sports. Place the shoe on a horizontal surface (either rough or smooth) designated by your teacher. Attach the spring scale to the shoe as shown below (low down at the toe or heel or hooked onto the front lace loop) so that the spring scale can be used to slide the shoe across the surface while, at the same time, the scale can be read. Be sure to keep the spring scale parallel to the surface.

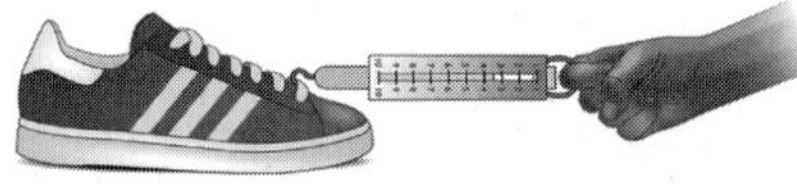

   a) Record a description of the surface in your log.

   b) Measure and record the amount of force, in newtons, needed to keep the shoe sliding on the surface at a slow, constant speed. Be careful to pull horizontally so that you do not tend to lift the shoe or pull downward on the shoe. Do not measure the force needed to start the shoe moving. Measure the force needed to keep it sliding at a slow, constant speed. How will you determine if the shoe is moving with a constant speed?

When the shoe is being pulled across the surface, there are four forces acting on it. The first is the horizontal force you apply and is measured by the spring scale. But since the shoe moves at a slow, constant speed, it is not accelerating horizontally. There must be a second force on the shoe of equal strength and in the opposite direction to the force you apply. This second horizontal force is the force due to *friction* between the shoe and the surface. The third force is the downward force of gravity on the shoe, which is equal to the weight of the shoe. But since the shoe is not accelerating downward, there must be a fourth force on the shoe. This is the force of the table on the shoe; this force is equal in strength and in the opposite direction to the shoe's weight. Since this force is directed perpendicularly to the surface, it is often called the *normal force*, since the word "normal" sometimes means "perpendicular to."

The *coefficient of sliding friction*, symbolized by the Greek letter μ (mu), is calculated using the following equation:

$$\mu = \frac{\text{force of friction}}{\text{perpendicular force exerted by the surface on the object}}$$

2.a)

Students should record a data table in their logs that includes mass, force of spring scale, surface type, etc. Students should also realize that graphing the data will be a good analysis tool.

3.a)

Students should record the surface type. Surfaces may be varied by taping sandpaper or rubber mats to the lab tables, or the students may use the floor. Using a carpeted area may prove problematic, since the loops of the carpet will provide a very uneven frictional force as they bend under the applied force.

3.b)

Encourage students to strive for consistent results. Hooking the scale to a low point on the shoe helps make it slide rather than tilt. Ask them if they notice whether the force necessary to start the shoe moving is more or less than the force necessary to keep it moving at a slow, constant speed. Students should look for a consistent average reading on the spring scale.

3.c)

μ results will vary. Generally, the coefficient of friction should be less than 1, and in the region of 0.3-0.8.

4.a)

Any masses that will approximately double the weight of the shoe or block will be fine. Adding the other shoe is an easy way to double the weight. If wooden blocks are used, be certain the students use the same side of the block for this part of the activity as for *Step 3*.

4.b)

Students should find that μ remains approximately the same as mass is added to the shoe.

Teaching Tip

To demonstrate that the frictional force between two surfaces is roughly independent of the contact area, you may want to do the following demonstration. Pull a shoe with a spring scale and measure the force required to overcome friction. Now double the mass of the shoe by adding masses equal to the mass of the shoe. The frictional force, and the force required to pull it with the spring scale, should double. Now remove the masses from the first shoe, and tie a second identical shoe side-by-side with the first so they can both be pulled at the same time. The mass and normal force is doubled as before, now with twice the contact area. The force required to pull the two-shoe combination will be essentially the same as the one shoe with masses added to double it.

Chapter 2 Physics in Action

Example:

Brand X athletic shoe has a weight of 5.0 N. If 1.5 N of applied horizontal force is required to cause the shoe to slide with constant speed on a smooth concrete floor, what is the coefficient of sliding friction?

The force of friction is equal to the applied horizontal force because there was no acceleration. The perpendicular force exerted by the surface on the shoe must be equal to the force exerted by the shoe on the surface (which is the weight of the shoe). In this case, that would be the same as the weight of the shoe.

$$\mu \text{ on concrete} = \frac{1.5 \not{N}}{5.0 \not{N}} = 0.30$$

c) Use the data you have gathered to calculate μ, the coefficient of sliding friction for this particular kind of shoe on the particular kind of surface used. Show your calculations in your log.

4. Add something to the shoe to approximately double its weight. Pull the shoe with a spring scale.

a) Record the force to pull the heavier shoe.

b) Calculate μ for the heavier shoe, showing your work in your log.

c) Taking into account possible errors of measurement, does the weight of the shoe seem to affect μ? Use data to answer the question in your log.

d) How do you think the weight of an athlete wearing the shoe would affect μ? Why?

5. Place the shoe on the second surface designated by your teacher. Repeat the procedure.

a) Make a sketch (free-body diagram) to show the forces acting on the shoe.

b) Calculate μ for this new surface and the shoe.

c) How does the value of μ for this surface compare to the value of μ for the first surface used? Suggest reasons for any difference in μ.

d) Would it make any difference if you used the empty shoe or the weighted shoe to calculate μ in this step? Explain your answer.

Physics Talk

FRICTION

Analyzing the Forces Acting on the Shoe

In this *Investigate*, you pulled the shoe at a constant velocity. Newton's second law informs you that motion with a constant velocity happens only when there is no net force on the shoe. So all the forces on the shoe must add up to zero.

You applied a horizontal pulling force and measured the value of the force with the spring scale. The shoe moved at a slow, constant speed. It was not accelerating horizontally. Therefore, there must be a second force on the shoe of equal strength and in the opposite direction to the force you applied. This second force was the force due to **friction** between the shoe and the surface. The pulling force you applied was equal to the frictional force, and since the two forces were in opposite directions, the net or

Physics Words

friction: a force that resists relative motion between two bodies in contact.

212

Active Physics

4.c)

Taking into account possible errors in measurement, students should find that the value of μ is not affected by the weight of the shoe. Ask students how the horizontal force and weight changed from trial to trial, hinting that μ stays the same because it is the ratio of these two forces.

4.d)

Students should understand that the mass of the athlete would not matter, since μ was independent of mass.

5.a)

Student sketch (see example below).

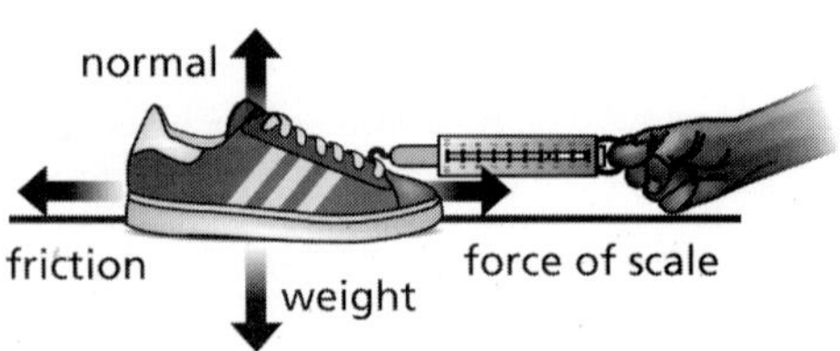

5.b)

Students calculate the new value of μ from their data.

5.c)

Students should recognize that the value of μ will be different for different surfaces. The "rougher" the surface, the greater the coefficient of friction. For example, the coefficient of friction for rubber on dry concrete is 140 times greater than that for rubber on ice.

5.d)

The previous step in this *Investigate* indicates that the weight of the shoe should not make a difference in the coefficient of friction. If time permits, urge the students to change the weight and recalculate μ, just to show that the first result was general. Whether the students explore this more or not, make sure you emphasize that the value of μ does not depend on the weight of the object.

Physics Talk

The *Physics Talk* will help the students in understanding that the different forces may act on an object moving with constant velocity. Find out if students understand why the shoe moved at a slow constant speed and did not accelerate horizontally. Be sure to ask them why the shoe did not accelerate even though there was a constant force exerted by the spring scale. Students should also be able to apply the same concept to explaining why the shoe did not move in the vertical direction. Ask them to illustrate the normal force and how its value is determined. Students should draw free-body diagrams to show all the forces acting on the shoe as it slides across a level surface that has friction. The students should be able to identify the direction of the frictional force as always being opposed to the direction of motion.

Students should also realize that when the shoe is at rest, there is no frictional force acting, since a horizontal force of friction would be unbalanced and cause the shoe to accelerate in the direction of friction. The dependence of the frictional force and the coefficient of friction on different surfaces should be emphasized, as well as how forces cancel each other when an object is at rest. Have students explain why the coefficient of friction does not have units, although the force of friction is expressed in newtons. Have the students write down the definition of the coefficient of friction in their *Active Physics* log. Emphasize that the value of μ is different for different surfaces. Finally, point out that although the coefficient of friction between two surfaces will not change, the frictional force between these surfaces will be different if the surface is tilted due to the normal force being different.

NOTES

NOTES

2-7a Blackline Master

total force due to them was zero. Note that you actually measured the pulling force but used its value as the value for the frictional force. This is perfectly fine, since the two forces are equal in strength.

In the *Investigate*, the shoe did not move in the vertical direction. Newton's second law informs you that the vertical forces on the shoe must add up to zero. The downward force of gravity on the shoe (weight) must be equal to the upward force applied to the shoe by the surface. Since this force is directed perpendicularly to the surface, it is often called the **normal force**, since the word "normal" sometimes means "perpendicular to." This force is equal in strength and in the opposite direction to the shoe's weight. Note that you measured the weight of the shoe, but used its value as the value for the normal force. Again, this is perfectly fine, since the two forces are equal in strength.

A free-body diagram can help you see the relationships among the four forces when the shoe moves with a constant speed.

Coefficient of Sliding Friction, μ

The **coefficient of sliding friction**, symbolized by μ, is defined as the ratio of two forces:

$$\mu = \frac{\text{force of friction}}{\text{perpendicular force exerted by the surface on the object (normal force)}} = \frac{F_f}{F_N}$$

The force of friction is equal to the force required to slide the object on the surface with a constant speed.

Physics Words

normal force: the force acting perpendicularly or at right angles to a surface.

coefficient of sliding friction: a dimensionless quantity symbolized by the Greek letter μ; its value depends on the properties of the two surfaces in contact and is used to calculate the force of friction.

CHAPTER 2

Checking Up

1.

According to Newton's first law, an object in motion will continue with constant velocity unless there is an unbalanced force. Because the shoe travels with constant speed, the forces of the spring scale must be exactly balanced by the force of friction.

2.

The coefficient of friction is defined as the force of friction divided by the normal force. Because this is a force divided by a force, the units cancel.

3.

The coefficient of friction only depends upon the two surfaces in contact.

Chapter 2 Physics in Action

Note the following about the coefficient of sliding friction:

- μ does not have any units because it is a force divided by a force; it has no unit of measurement.
- μ usually is expressed in decimal form, such as 0.85 for rubber on dry concrete (0.60 on wet concrete).
- μ is valid only for the pair of surfaces in contact when the value is measured; any significant change in either of the surfaces (such as the kind of material, surface texture, moisture, or lubrication on a surface, etc.) may cause the value of μ to change.
- The situation in this section was chosen deliberately so that the "perpendicular force exerted by the surface on the object" was exactly equal to the weight. If the surface were tilted, or if the pulling force were angled upward or downward, then the force exerted by the surface on the object would be more difficult to determine.

The Greek Alphabet

There are not enough letters in the English alphabet to provide the number of symbols needed in physics, so letters from another alphabet, the Greek alphabet, are also used as symbols. The Greek alphabet has been used for centuries to write the Greek language. It is one of the oldest alphabets in use today.

The letters of the Greek alphabet are often used in physics and mathematics. The letter μ (pronounced like "mew" and rhymes with "you") traditionally is used in physics as the symbol for the "coefficient of sliding friction." There are other Greek letters you may use in *Active Physics* or other physics courses that are shown below. For example, you have already used the Greek letter "Δ" to represent "a change in."

Checking Up

1. Why can you say that the force of friction is equal to the force reading on the spring scale when pulling an athletic shoe across a surface with a constant speed?
2. Why does the coefficient of friction have no units?
3. What determines the coefficient of friction?

214

Active Physics

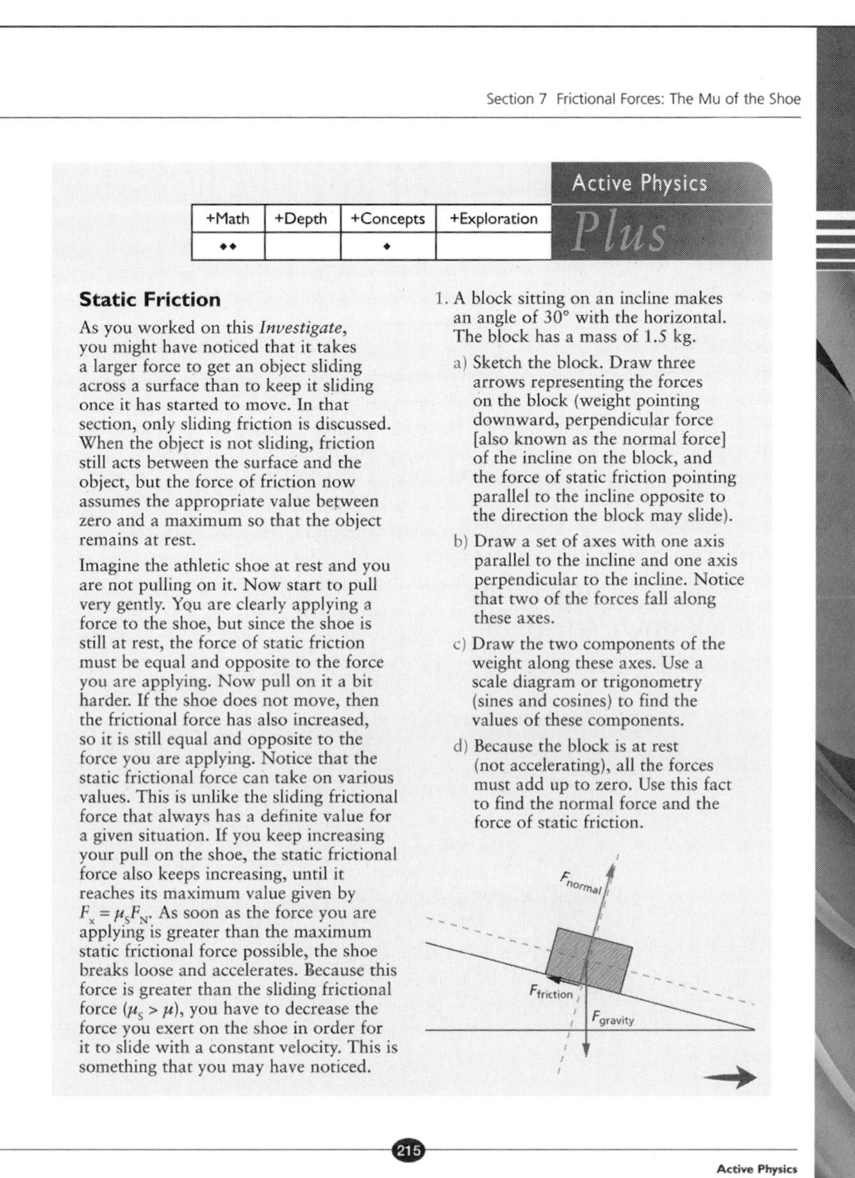

Section 7 Frictional Forces: The Mu of the Shoe

Active Physics *Plus*

+Math	+Depth	+Concepts	+Exploration
••		•	

Static Friction

As you worked on this *Investigate*, you might have noticed that it takes a larger force to get an object sliding across a surface than to keep it sliding once it has started to move. In that section, only sliding friction is discussed. When the object is not sliding, friction still acts between the surface and the object, but the force of friction now assumes the appropriate value between zero and a maximum so that the object remains at rest.

Imagine the athletic shoe at rest and you are not pulling on it. Now start to pull very gently. You are clearly applying a force to the shoe, but since the shoe is still at rest, the force of static friction must be equal and opposite to the force you are applying. Now pull on it a bit harder. If the shoe does not move, then the frictional force has also increased, so it is still equal and opposite to the force you are applying. Notice that the static frictional force can take on various values. This is unlike the sliding frictional force that always has a definite value for a given situation. If you keep increasing your pull on the shoe, the static frictional force also keeps increasing, until it reaches its maximum value given by $F_x = \mu_s F_N$. As soon as the force you are applying is greater than the maximum static frictional force possible, the shoe breaks loose and accelerates. Because this force is greater than the sliding frictional force ($\mu_s > \mu$), you have to decrease the force you exert on the shoe in order for it to slide with a constant velocity. This is something that you may have noticed.

1. A block sitting on an incline makes an angle of 30° with the horizontal. The block has a mass of 1.5 kg.
   a) Sketch the block. Draw three arrows representing the forces on the block (weight pointing downward, perpendicular force [also known as the normal force] of the incline on the block, and the force of static friction pointing parallel to the incline opposite to the direction the block may slide).
   b) Draw a set of axes with one axis parallel to the incline and one axis perpendicular to the incline. Notice that two of the forces fall along these axes.
   c) Draw the two components of the weight along these axes. Use a scale diagram or trigonometry (sines and cosines) to find the values of these components.
   d) Because the block is at rest (not accelerating), all the forces must add up to zero. Use this fact to find the normal force and the force of static friction.

215 Active Physics

Active Physics Plus

This section highlights the difference between static and kinetic friction. Static friction is a reaction force that only appears in response to a force that is trying to accelerate an object. The result of static friction is a force just large enough to cancel the applied force and leave the object at rest. Once the force being applied exceeds the value that static friction can apply, the object starts to move, and kinetic friction takes over. The coefficient of kinetic friction is always less than the coefficient of static friction. Point out to students that this section explains static friction. Ask them to write down the equation for maximum frictional value. You might want to mix students in groups and have them explore this concept in more detail. Ask students to explain to each other how a shoe is made to slide with constant velocity.

The formula for static friction on an inclined plane can be determined as follows. For an object at rest on an inclined plane, the components of the weight perpendicular and parallel to the plane are

$F_\perp = mg\cos\theta$, which is also equal to the normal force (F_N)

$F_\parallel = mg\sin\theta$, which is equal to the component of the weight acting down the plane

Using the equation for static friction when the shoe is at rest on the plane tells us that the force of friction up the plane must balance the component of weight acting down the plane, or

$mg\sin\theta = F_f = \mu_s F_N = \mu_s(mg\cos\theta)$

Solving this equation for the coefficient of static friction (μ_s) gives

$\mu_s = \sin\theta/\cos\theta = \tan\theta$

This provides an alternative method for finding the coefficient of friction. Simply raise an inclined plane slowly until the object starts to slide, and the tangent of the angle where the slide begins is the coefficient of static friction. While some students might be able to solve the problems given in this section, some may not. Ask students who are able to get answers to share their solutions with other students in their group.

CHAPTER 2

1.a)

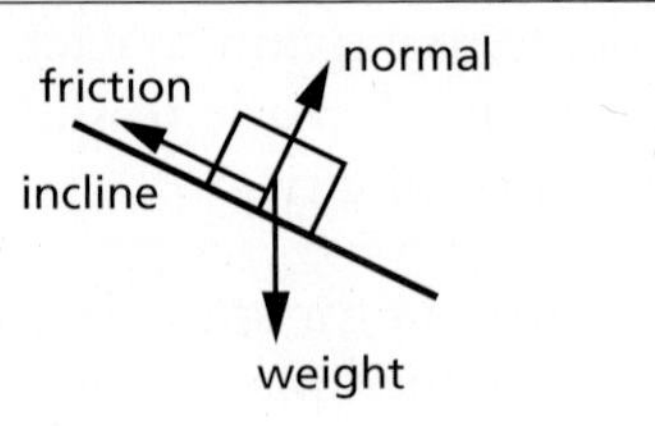

1.b)

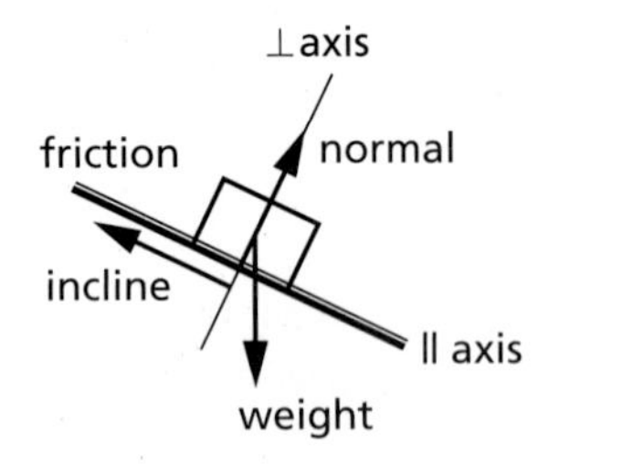

1.c)

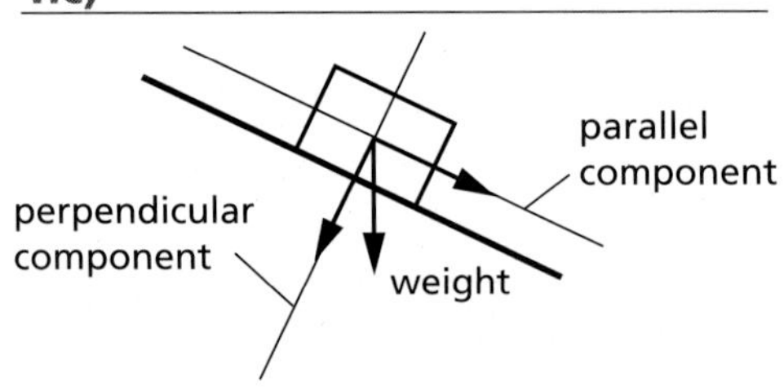

Component of weight down the incline = $mg \sin\theta$ =

$(1.5 \text{ kg})(9.8 \text{ m/s}^2)\sin 30^\circ = 7.4 \text{ N}$

Component of weight perpendicular to the incline = $mg \cos\theta$ =

$(1.5 \text{ kg})(9.8 \text{ m/s}^2)\cos 30^\circ = 13 \text{ N}$

1.d)

Because the block is at rest, the normal force must equal the component of the weight pressing against the plane, and the component of the weight tending to pull the block down the plane must be balanced by the force of friction.

Therefore, the normal force = 13 N.

Component of weight down the incline = $mg \sin\theta$

Component of weight perpendicular to the incline = $mg \cos\theta$

$F_f = \mu F_N$

$mg \sin\theta = \mu mg \cos\theta$

$\mu = \dfrac{mg \sin\theta}{mg \cos\theta} = \tan\theta$

$F = F_N \tan\theta = (13 \text{ N}) \tan(30°) =$

$(13 \text{ N})(0.58) = 7.5 \text{ N}$

NOTES

NOTES

2.

The coefficient of static friction is

$$\mu = \frac{mg\sin\theta}{mg\cos\theta} = \tan\theta.$$

2.a)

The coefficient of friction found by pulling the object horizontally across a table is

$$F_f = \mu mg$$

$$\mu = \frac{F_f}{mg}.$$

2.b)

The coefficient of friction from tilting is $\mu = \dfrac{mg\sin\theta}{mg\cos\theta} = \tan\theta$.

2.c)

The two values for the coefficient of friction should agree to within 10%.

What Do You Think Now?

Encourage your students to revise their earlier responses in their logs and come up with more definite responses that indicate how much they have understood the new concepts, and to what extent they can now apply them to their answers in the *What Do You Think Now?* section.

Chapter 2 Physics in Action

2. Imagine that you can vary the angle of the incline in the previous problem. As you increase the angle, at first, the block just stays where it is. However, at some angle θ, the block begins to slide down the incline. Let the mass of the block be m so you can work only with symbols. The coefficient of friction is equal to the tangent of this angle.

$$\mu = \tan\theta$$

(Tangent is a trigonometry function that you can find on many calculators.)

a) Find the frictional force by pulling a block at a constant speed across a horizontal table. Calculate μ.

b) Find the μ by tilting the table so that once started, the block can move with a constant speed. Measure the angle of the tilt, and calculate $\tan\theta$.

c) Compare the values in *a)* and *b)*.

What Do You Think Now?

At the beginning of this section, you were asked the following:

- **Why do some sports require special shoes?**
- **Why would different features of a shoe be useful for different sports?**

Record your ideas about these questions now. Use the concept of sliding friction to answer the questions this time.

Section 7 Frictional Forces: The Mu of the Shoe

Physics

Essential Questions

What does it mean?

An athlete may complain that the field is slippery. How can you describe the same situation using the terms friction, coefficient of sliding friction, forces, and the symbol μ?

How do you know?

A running shoe is pulled along the ground at a constant speed. How do you know that a frictional force was equal to the pulling force?

Why do you believe?

Connects with Other Physics Content	Fits with Big Ideas in Science	Meets Physics Requirements
Force and motion	Change and constancy	* Experimental evidence is consistent with models and theories

* Physicists often believe in invisible forces. Friction is invisible; it happens without you seeing it. A sliding object slows down and stops. If you did not believe in friction, could Newton's first and second laws explain this motion?

Why should you care?

Friction, or the lack of enough friction, is critical to sports. Describe two parts of your sport where friction is critical and what you are going to say about it in your voice-over.

Reflecting on the Section and the Challenge

Many athletes seem more concerned about their shoes than other items of equipment, and for good reason. Small differences in the shoes (or skates or skis) can affect performance. Athletic shoes have become a major industry because people in all "walks" of life have discovered that athletic shoes are great just about anywhere. Now that you have studied friction, you know about a major aspect of what makes shoes function well. You are prepared to do physics commentary on athletic footwear and other effects of friction in sports when the need exists to be "sure-footed." Your sports commentary may discuss the μ of the shoe, the change in friction when a playing surface gets wet, and the need for friction when running. You may also wish to discuss the use of cleats on certain surfaces or the friction of tires on the road in stock car races. No matter which sport you choose, friction will play an important role.

Active Physics

Reflecting on the Section and the Challenge

Have students summarize the *Reflecting on the Section and the Challenge* from their textbooks. Discuss how they could apply what they know about friction to the sports commentary in their *Chapter Challenge*. Ask them to reflect on athletic footgear, and why it is so crucial to the outcome of a sport. It is important for students to realize the role of different surfaces in determining the effects of friction. Encourage students to explain the connections between the *Investigate* and *Physics Talk*, including the *Checking Up* questions. This strategy will help them see how their understanding has developed and how they can benefit from it in *Preparing for the Chapter Challenge*.

CHAPTER 2

Physics Essential Questions

What does it mean?

There is a low coefficient of friction, μ, causing a small frictional force.

How do you know?

Because the shoe moved at a constant speed, the net force on it must be zero. Therefore, the frictional force must be equal to the pulling force.

Why do you believe?

If you did not believe in friction, then this behavior would violate Newton's first and second laws.

Why should you care?

When you run, you need friction between your shoes and the ground. In auto racing, the friction of the wheels on the ground provides the force for acceleration and for turning.

Physics to Go

Chapter 2 Physics in Action

Physics to Go

1. Think of a sport and changing weather conditions that would cause an athlete to want to increase friction to have better footing. Name the sport, describe the change in conditions, and explain what the athlete might do to increase friction between the shoes and the surface of the ground.
2. Think of a sport in which athletes desire to have frictional forces as small as possible and describe what the athletes do to reduce friction.
3. If a basketball player's shoes provide an amount of friction that is "just right" when she plays on her home court, can she be sure the same shoes will provide the same amount of friction when playing on another court? What details about the other court would she need to know to answer this question?
4. Tennis is played on clay, grass and hard surfaces. Please explain why you think tennis players have or don't have different shoes for each surface.
5. A cross-country skier who weighs 600 N has chosen ski wax that provides $\mu = 0.03$. What is the minimum amount of horizontal force, perhaps from a tailwind, that would keep the skier coasting at constant speed across level snow?
6. A vehicle having a mass of 1000 kg had an accident on a wet, but level, concrete road under foggy conditions. The tires were measured to have $\mu = 0.55$ on wet concrete. The driver locked the brakes, skidded for 6 seconds, and then hit the guardrail causing a very small dent because the vehicle stopped just as it touched the guardrail. The driver claimed to be driving 65 miles per hour (29 m/s). You have been hired as an investigator to determine if the driver is telling the truth.
   a) What is the weight of the vehicle?
   b) The frictional force produces the negative acceleration (often called deceleration) that reduces the velocity of the vehicle from its initial unknown speed to zero. Find the value of the frictional force.
   c) Use the frictional force to calculate the acceleration (remember that it is a negative number).
   d) Using the acceleration and the time over which acceleration occurred, calculate the change in speed that the acceleration would produce.
   e) Use the change in speed to find the original speed of the vehicle when the brakes were applied. Write a statement of your findings, including your opinion of the driver's claim.

Active Physics 218

1.

In football, for example, players change the length of shoe cleats or sometimes wear shoes without cleats to improve footing in bad weather.

2.

Downhill skiers use wax to reduce friction; ice skaters "sharpen" their skates, which makes sure they are as smooth as possible (and have the sharp edges needed for ice skating).

3.

A common misconception is that the coefficient of friction—and, therefore, the force of friction—depends only on the shoe. In fact, it depends as much on the nature of the surface beneath the shoe. The athlete cannot be assured that the same amount of frictional force will be present when the same shoe is used on a court having a different surface. The athlete would need to know what the surface is made of and what preparation is used on the surface before a game.

4.

The way a tennis player responds to a shot varies greatly with the court type. Clay courts allow the player to slide across the court easily, while "hard courts" bring the player to a sudden stop. A shoe with a smooth bottom on clay allows the player to glide into a shot, while on a "hard court" the player would want a shoe that grips firmly to allow a quicker change in direction.

5.

The minimum amount of horizontal force, F, that would keep the skier coasting at constant speed would be $F = (0.03)(600\text{ N}) = 18\text{ N}$.

6.a)

Weight of vehicle $= mg =$

$(1{,}000\text{ kg})(10\text{ m/s}^2) = 10{,}000\text{ N}$

6.b)

Frictional (stopping) force $=$

$\mu F_N = (0.55)(10{,}000\text{ N}) = 5{,}500\text{ N}$

6.c)

Acceleration $= F/m =$

$(-5{,}500\text{ N})/(1{,}000\text{ kg}) = -5.5\text{ m/s}^2$

6.d)

Initial speed $= at =$

$(-5.5\text{ m/s})(6\text{ s}) =$

$-33\text{ m/s} = -75\text{ mi/h}$

6.e)

Since the final speed was zero, the initial speed was 33 m/s or

Section 7 Frictional Forces: The Mu of the Shoe

7. In some sports, the air or water have limiting effects on motion similar to sliding friction. Do you think that the forces of "air resistance" and "water resistance" remain constant or do they change when speeds change? Use examples from your own experience with these forms of resistance as a basis for your answer.
8. If there is a maximum frictional force between your shoe and the track, does that set a limit on how fast you can start (accelerate) in a sprint? Does that mean you cannot have more than a certain acceleration even if you have incredibly strong leg muscles? What is done to solve this problem?
9. How might an athletic shoe company use the results of your experiment to "sell" a shoe? Write copy for such an advertisement.
10. Explain why friction is important to running. Why are cleats used in football, soccer, and other sports?
11. *Preparing for the Chapter Challenge*
    Choose a sport and describe an event in which friction with the ground or the air plays a significant role. Create a voice-over or script that uses physics to explain the action.

Inquiring Further

Frictional force and the weight of a player

You found that for a given coefficient of sliding friction, the frictional force increased with the weight of the shoe. Of course, if a person is wearing the shoe, it is the weight of the person that determines the frictional force (assuming the person has only one foot on the surface). This means that the heavier the athlete, the greater the frictional force. Does this imply that the heavier a person is, the greater his or her acceleration can be?

a) Calculate the force of sliding friction for a 50-kg person using a shoe with a μ of 0.6.
b) Calculate the acceleration of the 50-kg person due to the force of sliding friction.
c) Calculate the force of sliding friction for a 90-kg person using a shoe with a μ of 0.6.
d) Calculate the acceleration of the 90-kg person due to the force of sliding friction.
e) Can the heavier person achieve a greater acceleration using the same shoes?

219

Active Physics

75 mi/h. The driver has a problem because the laws of physics will prevail in court.

7.

Any sports that involve objects moving through air or water will involve fluid resistance. For objects moving through these media, the resistance depends upon the speed. Any student who has held a hand outside a car window while the car is in motion has felt the force of the air increase with speed. Air resistance affects all sports that use objects traveling through the air, but for sports such as sky diving it is critical. Without the limiting effects of air resistance, sky diving would quickly die out as a sport.

8.

Yes, the maximum frictional force between your shoe and the track does place a limit on acceleration. Sprinters use blocks when beginning a race, thereby providing a surface that is not horizontal, and does not depend on the weight of the runner.

9.

See the **Assessment Rubric** following this chapter in the *Teacher's Edition.*

10.

Without friction it would be impossible to walk or run, since it is the force of friction between the shoe and the surface that provides the forward force to propel the athlete. Cleats increase the friction between the shoe and the ground by increasing the amount of surface in contact, and also by changing the situation from plain friction into a gear-like intermeshing with the top layer of soil.

11.

Preparing for the Chapter Challenge

The voice-over should include why friction is important to athletes, how the normal force affects the force of friction, and how athletes choice of athletic shoes is important.

Inquiring Further

a)

$F = (0.6)(50\text{ kg})(10\text{ m/s}^2) = 300\text{ N}$

b)

$a = (300\text{ N})/(50\text{ kg}) = 6\text{ m/s}^2$

c)

$F = (0.6)(90\text{ kg})(10\text{ m/s}^2) = 540\text{ N}$

d)

$a = (540\text{ N})/(90\text{ kg}) = 6\text{ m/s}^2$

e)

No

CHAPTER 2

SECTION 7 QUIZ

2-7b Blackline Master

1. A runner whose weight is 800 N wishes to accelerate at 5 m/s^2. What minimum coefficient of friction is required for her to accelerate without slipping?

   a) 0.25 b) 0.50

   c) 0.30 d) 0.40

2. A box decelerates as it moves to the right along a horizontal surface as shown in the diagram.

   V

   Which vector best represents the direction of the force of friction on the box?

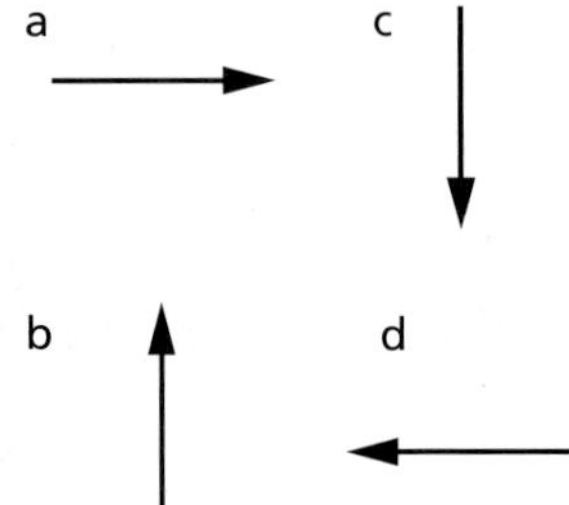

3. Sand is often placed on an icy road because the sand

   a) decreases the coefficient of friction between the car and the road.

   b) increases the coefficient of friction between the car and the road.

   c) increases the normal force of the car on the road.

   d) decreases the normal force of the car on the road.

4. A different force is applied to each of the four 1-kg blocks to slide them across a uniform level surface at constant speed. In which diagram is the coefficient of friction the smallest?

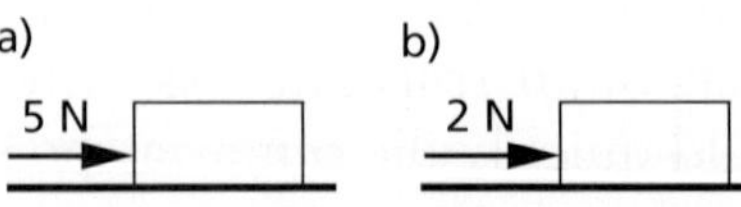

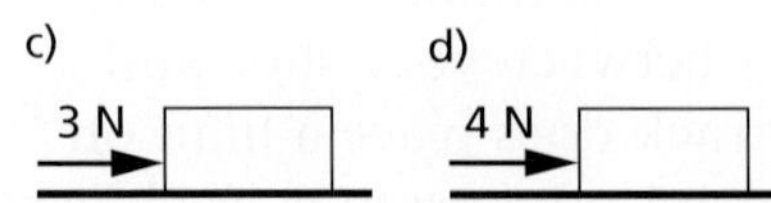

5. Jill is pulling a 200-N sled through the snow at constant speed with a constant horizontal force of 10 N. What is the coefficient of friction between the sled and the snow?

a) 0.02 b) 0.05

c) 0.20 d) 20

SECTION 7 QUIZ ANSWERS

1 a) The force of friction accelerates the runner. To accelerate at 5 m/s^2 the frictional force $\mu F_N = ma$, where the normal force equals the weight, mg. Thus $\mu mg = ma$ or, $\mu = a/g = (5\text{ m/s}^2)/(10\text{ m/s}^2) = 0.5$.

2 d) The force of friction will oppose the motion.

3 b) Rough surfaces generally have a higher coefficient of friction than smooth surfaces, so sand will increase the coefficient.

4 b) $\mu = F_f/F_N$ and here all the objects will have the same normal force, since they are all 1-kg blocks. Hence, the lowest friction force would correspond to the lowest μ.

5 b) $\mu = F_f/F_N = (10\text{ N})/(200\text{ N}) = 0.05$

CHAPTER 2

SECTION 8

Potential and Kinetic Energy: Energy in the Pole Vault

Section Overview

Students design experiments to investigate how energy is converted from one form to another. They blast a penny in the air from the end of a ruler to measure the maximum height to which the penny flies upward. They roll balls across a ramp, which collide with a ruler attached to a pencil, at three different speeds. In each case, students record the amount the ruler deflects as it moves back to its original position. The experiments simulate factors that determine the amount of energy stored in a pole-vaulter's bar that helps to launch the vaulter up in the air. Students identify different forms of energy and solve problems to see how the total energy of a system remains constant. This section also details the concept of work and how energy relates to work.

Background Information

Concepts involving energy, introduced in this activity, include:

- energy in three forms: kinetic energy, gravitational potential energy, and spring potential energy (the energy stored in a spring),
- transformations among and conservation of the above-listed forms of energy, and
- the equivalence of *work* and *energy*.

Students are introduced first to the concept of kinetic energy, and then to how this form of energy may be transferred to other forms of energy. The kinetic energy of the runner in a pole vault is converted into the spring potential energy of the bent pole, which is then lost and converted into the gravitational potential energy of the rising pole-vaulter. These forms of energy are then related back to the basic concept of work, and it is shown that the work done by the runner accelerating is the source of the kinetic energy that undergoes all of these transformations. It is then noted that whenever work is done, the energy of the system changes. For a closed system such as the pole-vault example, energy may be transferred from one form to another without any net loss. If work is done, then the energy of the system may increase (as in the example of the runner doing work to accelerate) or it may decrease (for example, as a driver applies the brakes to a car).

The *Physics Talk* in this section confines itself to discussing the forms of energy and how to calculate the energy stored in kinetic, gravitational potential, and spring potential energy. A more formal treatment of conservation of energy is given in *Section 9*.

The treatment of the joule as the unit of energy in this activity is "soft" because groundwork for defining the joule in a mechanical context has not yet been established. At this stage without benefit of a definition of the joule, dimensional analysis can be used to satisfy yourself (and, only if it seems necessary, your students) that all of the equations given in the *Investigate* for forms of energy at least have the same unit:

- Kinetic energy $= \frac{1}{2}mv^2$, Unit:
  $(\mathrm{kg})(\mathrm{m/s})^2 = (\mathrm{kg})(\mathrm{m}^2)/(\mathrm{s}^2)$
- Gravitational potential energy $= mgh$, Unit:
  $(\mathrm{kg})\left[\mathrm{m}/(\mathrm{s})^2\right]\mathrm{m} = (\mathrm{kg})(\mathrm{m}^2)/(\mathrm{s}^2)$
- Potential energy stored in a spring $= \frac{1}{2}kx^2$ (the spring constant, k, has a unit N/m which is equivalent to $\mathrm{kg/s^2}$), Unit:
  $(\mathrm{kg/s^2})\,\mathrm{m}^2 = (\mathrm{kg})(\mathrm{m}^2)/(\mathrm{s}^2)$

As shown on the previous page, all of the forms of energy defined in the activity share the same unit, (kg)(m^2)/(s^2). Because 1 joule is the amount of work done when 1 newton of force is active through one meter of distance $(1 \text{ joule} = 1 \text{ newton·meter})$, and because one newton is the force that will cause 1 kg to accelerate at 1 m/s^2, one joule also can be expressed as $\left(1 \text{ kg m/s}^2 \cdot 1\text{m}\right) = \left(\text{kg}\right)\left(\text{m}^2\right)/\left(\text{s}^2\right)$. Therefore, the equations are legitimate because the unit (kg)(m^2)/(s^2) is equivalent to the standard energy unit, the joule. The information above will prepare you to address questions about units, if they arise.

Crucial Physics

- Kinetic energy is energy due to motion and is given by half the mass times the square of the velocity.
- Gravitational potential energy is energy due to an object's position in Earth's gravitational field. It equals the mass of the object times *g* times the vertical height from a reference point where the gravitational potential energy has been set to zero.
- Elastic potential energy is energy due to a spring being compressed or stretched. It is equal to half the spring constant times the square of the compression or stretching distance.
- Energy can be transformed from one kind to another, and if there is no external force on a system, the total energy is conserved (does not change).
- Work occurs any time an object moves with a force parallel to the motion. Work is equal to the force parallel to the motion times the distance moved.
- Energy is "stored work." Work done on an object raises its energy. That energy later on can be converted back to work done by the object on another object.

Learning Outcomes	Location in the Section	Evidence of Understanding
Apply equations for kinetic energy, gravitational potential energy, and elastic potential energy.	***Physics Talk*** Sample Problems 1-3 ***Physics to Go*** Questions 3 and 6-12	Students apply equation of kinetic and elastic and gravitational potential energy to solve problems.
Recognize that restoring forces are active when objects are deformed.	***Investigate*** Steps 1-3 and 5-9	Students recognize that when the ruler deflects from its original position energy is stored.
Apply the equation for the force necessary to compress or stretch a spring.	***Physics Talk*** Sample Problem 2 ***Physics to Go*** Questions 9-11 and 12.d)	Students solve problems by applying equation for the force necessary to compress the spring and the energy stored in compressed spring.
Measure the transformations among the different forms of energy.	***Investigate*** Steps 1-3 and 5-9	Students recognize that the deformation of the ruler is caused by the kinetic energy of the rolling ball, and that the gravitational potential energy gained by the penny comes from the deformation of the ruler.
Conduct simulations of the transformation of energy involved in the pole vault.	***Investigate*** Steps 1-9	Students design experiments to simulate how energy is transferred from one form to another.

CHAPTER 2

Section 8 Materials, Preparation, and Safety

Materials and Equipment

PLAN A

Materials and Equipment	Group (4 students)	Class
Ball, steel, 1 in. (diameter)	1 per group	
Ruler, metric, 30 cm	2 per group	
Meter stick, wood	1 per class	
C-clamp, steel, 3 in.	1 per group	
Ramp, starting, 20-degree angle	1 per group	
Pen, marking, felt tip	1 per group	
Tape, masking, 3/4 in. x 60 yds		6 per class
Cards, index, unlined, 3 in. x 5 in., pkg. of 100		1 per class
Coins (pennies, nickels)*	5 per group	

*Additional items needed not supplied

PLAN B

Materials and Equipment	Group (4 students)	Class
Ball, steel, 1 in. (diameter)	1 per group	
Ruler, metric, 30 cm	2 per group	
Meter stick, wood	1 per class	
C-clamp, steel, 3 in.	1 per group	
Ramp, starting, 20-degree angle	1 per group	
Pen, marking, felt tip	1 per group	
Tape, masking, 3/4 in. x 60 yds		6 per class
Cards, index, unlined, 3 in. x 5 in., pkg. of 100		1 per class
Coins (pennies, nickels)*	5 per group	

*Additional items needed not supplied

Note: Time, Preparation, and Safety requirements are based on Plan A, if using Plan B, please adjust accordingly.

Time Requirement

This *Investigate* should take approximately one class period or 40 min.

Teacher Preparation

- Assemble the required material for the *Investigate*, including the coins or washers to be flipped upward by the rulers. Try out the methods the students will use to clamp the rulers to the lab tables they will use. You may need additional "filler" blocks to allow the rulers to be clamped in the vertical orientation. The rulers or plastic strips the students will use should be very flexible, and not prone to break when bent down.

Safety Requirements

- Students should be cautioned to make certain their safety goggles are in place all during this *Investigate*. Do not use stiff plastic rulers to flip the coins upward. They may shatter into sharp pieces. Make certain the students are not above the area where the coins are being flipped as they measure the height achieved by the coin.
- Student enthusiasm for this investigation runs very high. Caution the students that this is not a contest to see who can embed coins in the ceiling!

NOTES

CHAPTER 2

Meeting the Needs of All Students

Differentiated Instruction: Augmentation and Accommodations

Learning Issue	Reference	Augmentation and Accommodations
Understanding vocabulary	***Learning Outcomes*** ***Physics to Go*** Questions 1, 2, and 13-15	**Augmentation** • Students are expected to explain the transformations of energy. Define transformation and provide examples of different kinds of transformations. Students should also be able to provide some good examples.
Designing an experiment	***Investigate*** Steps 1 and 2	**Augmentation** • Students with attention and behavior concerns often focus better on tasks that they are motivated to complete because the task is high-interest or their idea. These students may be more successful in designing their own experiment rather than following the directions of the previously designed experiment. • When allowing students to design their own experiment, make sure that they have a sound experimental design and data-collection method before they begin.
Following directions	***Investigate*** Steps 3-9	**Augmentation** • Provide a physical model of each experiment setup. • Set time limits for each step of the *Investigate* and use a timer to keep students on task. • Make a list, presented orally and visually, of the data that should be recorded at the end of the experiment. **Accommodation** • Provide a checklist that includes each step of the experiment so that students can mark off each task as it is completed.
Understanding vocabulary	***Physics Talk*** ***Physics Essential Questions***	**Augmentation** • Provide a definition and examples for the word "conserve." Students could generate a class list of what people conserve and why.
Understanding energy transformation	***Physics Talk***	**Augmentation** • Energy transformation is difficult for students to understand because they cannot visibly see the energy "moving," instead they see the effects of energy transformation. • Provide a drawing or graphic of a ball being thrown into the air that has the types of energy along the labeled trajectory. • Ask students to think of a new situation in which energy is transferred in a similar way, sketch the situation, and label the energy transformations. • Students might find it difficult to grasp that for work to be done, the object must move over a distance. Provide examples such as pushing or pulling as hard as one can on an object without it moving. Ask students if any work has been done.
Learning many formulas at one time	***Physics Talk***	**Accommodation** • Provide a table that includes the variables, symbols, and units for each of the formulas. This graphic organizer will help students have more confidence in problem-solving. Students can identify the given values more easily when they become comfortable with the units.

Learning Issue	Reference	Augmentation and Accommodations
Finding key phrases in word problems	***Physics Talk*** ***Physics to Go*** Steps 3, 8-11	**Augmentation** • Solving word problems is a difficult skill for most students, especially ninth graders. • Ask students to highlight or underline the key phrases in word problems to emphasize what the problem is asking them to find. • Generate a class list of key phrases for problem solving, including "how much," "what is," "calculate the," "solve for the," etc. • Mastering the above skill makes it easier for students to choose the correct formula needed to solve a problem. **Accommodation** • Highlight the key phrases that students are expected to find in word problems and fade this accommodation away as they become more proficient in problem-solving.

Strategies for Students with Limited English-Language Proficiency

Point out new vocabulary words in context and practice using the words as much as possible throughout the section. As you work through the section, have students write the terms in their *Active Physics* log and add the definitions in their own words.

catapult	plausible
elastic potential energy	potential energy
gravitational potential energy	release
joule	simulate
kinetic energy	work
law of conservation of energy	

Consider giving students a cloze activity when you reach the end of the section. Cloze activities are useful tools for summarizing material and for giving English-language learners opportunities to practice writing complete sentences using science vocabulary. Cloze activities are most effective when used frequently, to build students' abilities with more complex sentences.

Ask students to give you sentences describing what they did in the section, and telling what important lessons they learned. Their comments should include the vocabulary words listed above. You may wish to offer a first sentence as an example. For instance, "We investigated how much energy is stored in a pole vaulter's pole." Write simple sentences, and work them into paragraphs. Model using a topic sentence, supporting statements, and a closing sentence. Model the process of editing, in which students make corrections that improve the sentences.

There is a lot of information in this section, so you will likely end up with a few paragraphs. Once the paragraphs are complete and students agree that they accurately summarize what they did and learned, have them copy down the paragraphs. Explain that there will be a brief quiz on the paragraphs tomorrow at the beginning of the class. The quiz will be on the same paragraphs that they wrote down, but with several blanks where some terms were. The students will need to fill in the blanks with the terms that are missing. Tell students how the quiz will be graded. Prepare this quiz by keying in the paragraphs and then going back and removing every vocabulary term and replacing it with a blank. Choose a variety of words to leave out—nouns, verbs, adjectives, etc. They can be science-content words, but they do not have to be. Score the quiz by allotting two points for every blank, one point for the correct term or word (or perhaps another word with the correct meaning), and a second point for the correct spelling of that term or word.

SECTION 8

Teaching Suggestions and Sample Answers

What Do You See?

This illustration is full of images that students can ponder and discuss with their peers. Interrelated concepts are represented by humorous sketches that stimulate curiosity. Remind students that the main purpose of the *What Do You See?* is to get them to think about motion and relate it to the title of this section. As students feel more comfortable in expressing their idea about the topic, keep them engaged by asking open-ended question that draw attention to key concepts of this section.

Chapter 2 Physics in Action

Section 8 **Potential and Kinetic Energy: Energy in the Pole Vault**

What Do You See?

Learning Outcomes

In this section, you will

- **Apply** equations for kinetic energy, gravitational potential energy, and elastic potential energy.
- **Recognize** that restoring forces are active when objects are deformed.
- **Apply** the equation for the force necessary to compress or stretch a spring.
- **Measure** the transformations among the different forms of energy.
- **Conduct** simulations of the transformations of energy involved in the pole vault.

What Do You Think?

You would need a fence more than 6.0 m (about 20 ft) high to keep the world champion pole vaulter out of your yard.

- **If champion pole vaulters can clear a 6.0-m high bar with a 5.5-m long pole, why can't they vault over a 12.0-m high bar with a pole 11.0 m long?**
- **What factors (variables) do you think limit the height a pole vaulter has been able to attain?**

Record your ideas about these questions in your *Active Physics* log. Be prepared to discuss your responses with your small group and the class.

Investigate

Pole vaulters rely on the energy stored in their flexible poles to soar to remarkable heights. In this section, you will design an experiment that simulates the factors that determine the amount of energy stored in a pole vaulter's pole by launching a penny with a flexible ruler.

220

Students' Prior Conceptions

This section culminates in students' understanding the relationships among forces, the concept of work, and transformations of gravitational and potential energy.

1. **In general, students confuse the concepts of force, work, and energy; they may use them interchangeably.**
2. **Students associate energy only with animate objects; energy is linked with force and movement.**
3. **Preconceptions 1 and 2 may lead students to think of energy as a causal agent that can be stored in some objects as an ingredient or catalyst similar to a fuel that is used up.**
4. **If a fuel is used up, then it cannot be transferred from one form of energy to another.**

The nature of these student preconceptions makes it paramount for teachers to interview students while they are doing the investigation to encourage them to see through observations, measurements, and mathematical modeling that kinetic energy, gravitational potential energy, and spring potential energy all involve forces acting over distances to transform the different forms of energy. In the absence of friction, energy is conserved in these transformations. In the presence of friction, energy is converted to heat which is no longer available to energy transformations within the system.

What Do You Think?

Ask students to answer the question without any hesitation, even if they are not sure about their responses. Encourage them to discuss their answers and record them in their *Active Physics* log book. Reassure your students that while they may not have the right answers at this stage, they will gradually learn more as they progress in the section. At this stage, students' answers don't have to be correct. What matters most is that students think and reflect with ease. Facilitate discussions in small groups as well the whole class.

What Do You Think?

A Physicist's Response

The speed at which the athlete can run and the runner's kinetic energy determine the height a pole vaulter may attain in the pole vault. Although the arm strength of the pole vaulter may add a small amount to the height, it is the transformation of the athlete's kinetic energy to gravitational potential energy that determines the height attained, not the pole length. Because humans have a limiting speed at which they can run, they also have a limit as to the height they may reach with the pole vault. Human limitations on the kinetic energy and the amount of input work that the vaulter is able to provide limit the height much more than the length of the pole. A vaulter using an 8-m pole would not be able to reach a height any higher than if he or she used a 5.5-m pole.

In addition to the kinetic energy of the vaulter, the other factors that determine the height attained are the efficiency of the pole in converting kinetic energy into gravitational potential energy, and the work the vaulter is able to do with his arms at the top of the rise to gain a little bit of additional height.

NOTES

NOTES

Investigate

1.a)

The students record that the more the ruler is deflected, the higher the penny travels.

1.b)

Students will record that if the ruler is clamped toward an end that allows most of the ruler to flex, the penny will not go as high for an equal deflection since the spring constant will be lower for a "longer" ruler.

2.

How the ruler is clamped to the table, how far it is deflected, and the mass of the coin determine how high the penny travels. The students will form conclusions about how high the coin goes as a function of one of the factors described above. The students should list the data that will be recorded in a table and list which variables are being kept constant to present their data in the form of a graph for quick analysis.

Section 8 Potential and Kinetic Energy: Energy in the Pole Vault

Wear safety goggles during this *Investigate*.

1. Hold one end of a ruler on the table and press down on the other end. Try to get a penny (or some other small mass) to travel close to the height of the ceiling without hitting the ceiling.
   a) Record your technique for blasting the penny high in the air.
   b) What factors about the ruler and how it is positioned determine the height the penny achieves?
2. Design an experiment to test one of the variables and its effect on the height of the penny.
   Include in your design:
   - What you will be able to conclude as a result of your experiment.
   - What data you will record.
   - What tools you will use to make your measurements.
   - How you will analyze your data.

   Your teacher may ask you to continue with your experimental design or, because of equipment or safety concerns, may ask you to proceed with the experiment as described below.
3. Carefully clamp a ruler in a vertical position so that the clamp is near the bottom end of the ruler. The top end should extend a few centimeters above the edge of a tabletop.

   Tape a pencil or pen to the surface of the ruler near the top end of the ruler so that the writing end of the pen extends to one side of the top end of the ruler. If the top end of the ruler moves as it is bent, the pencil moves with it.

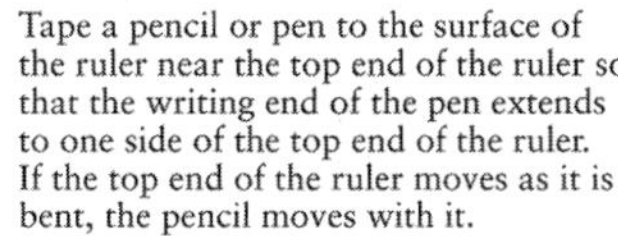

4. Set up a ramp as shown in the bottom left diagram. Three different starting points on a ramp will be used to roll a ball across the tabletop at three different speeds. Each time the ball rolls, it will strike the ruler near the top end, causing the ruler to bend. A marking surface held in contact with the tip of the pencil or pen will be used to measure the deflection.
5. Roll the ball from three different starting points on the ramp to achieve a low, medium, and high speed. In each case, measure the amount of deflection of the end of the ruler as indicated by the length of the pencil mark.
   a) Record the amount of deflection in each case.
   b) If the rolling ball represents the running vaulter and the ruler represents the pole in this model of the pole vault, how does the amount of bend in the pole depend on the vaulter's running speed? Record your data and response in your log.
6. Carefully clamp a ruler flat-side down to a tabletop so that two-thirds of the ruler's length extends over the edge of the table as shown below.

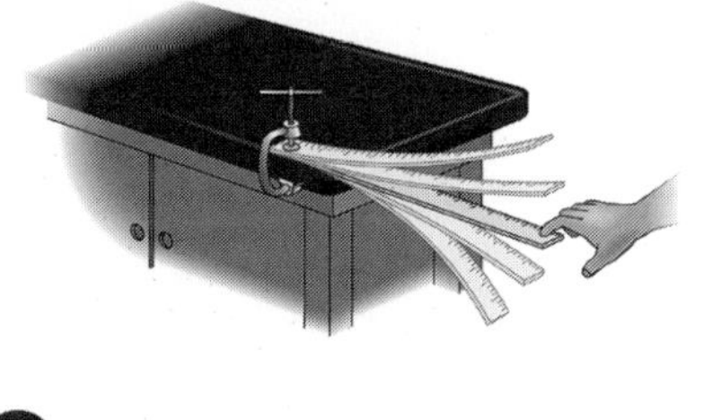

A more careful analysis shows that if the distance up the ramp is doubled, the speed of the ball increases by the square root of 2. To double the speed of the ball at the bottom of the ramp, the ball must be placed four times further up the ramp. Doubling the speed should approximately double the ruler deflection.

5.a)

The students will record their data for ruler deflection depending upon starting position on the ramp.

5.b)

The students will start with a simple conception that the higher the speed, the greater the deflection of the pole. Do not discourage this view at this point.

6.

If students clamp less than the recommended amount of the ruler, the measurement of small deflection differences will become more difficult.

CHAPTER 2

3.

The student should keep all variables constant except one, and then record the height as that variable is changed. There are many variations on how the next activity can be done. Any way different speeds of the ball can be achieved is fine. Likewise, any method to determine how far the ruler deflects is also fine.

Teaching Tip

A track of some sort may be necessary to make certain the rolling ball hits the ruler.

4.

Prior to doing *Step 5*, the students may expect, as they linearly increase the distance up the ramp, to see a linear increase in the speed of the ball coming down the ramp and striking the ruler.

7.

Any coin will do but uniformity of coins for each group allows a better comparison of data between groups.

Teaching Tip

If the penny falls off the ruler easily, the students can tape a cut-down paper cup onto the ruler to hold the penny in place until it is launched.

8.

Placing a second reference ruler or straightedge next to the deflecting ruler may assist the deflection-measurement process.

8.a)

Students record the height the penny rises for the 2-cm deflection of the ruler. It might be expected that as the deflection of the ruler is doubled, the energy in the ruler increases by a factor of 4 due to the equation $SPE = \frac{1}{2}kx^2$, and thus the penny's height should increase by a factor of 4. The situation in practice is more complex, however, so the students should be expected only to see an increase of more than double.

9.a)

Encourage the students to take several trials for each ruler deflection to ensure uniform results.

9.b)

The larger the amount of deflection, the higher the coin travels. The relationship should not be expected to be linear; that is, twice the deflection may cause the coin to travel more than twice as high.

Chapter 2 Physics in Action

7. Place a penny on the top surface of the ruler at the outside end.

8. Bend the clamped ruler downward. Use a second ruler to measure a 2-cm downward deflection from the unbent position of the ruler. Prepare to measure the maximum height to which the penny flies upward. Use the position of the penny when the ruler is relaxed as the "zero" vertical position of the penny. Release the ruler, launching the penny.

a) Record in your log the height that the penny travels.

9. Repeat *Step 8* for ruler deflections of 4 cm and 6 cm.

a) In each case, record the maximum height of the "vaulted" penny.

b) How is the height that the penny reaches related to the amount of deflection of the ruler?

c) If the ruler represents the pole in this model of the pole vault, and the penny is the pole vaulter, how does the amount of bend in the pole affect the height that the pole vaulter can attain? Record your response in your log.

Physics Talk

LAW OF CONSERVATION OF ENERGY

When a force acts on an object, the speed and position of the object may change. In many cases, the speed and position of the object change in a way that makes it possible for the speed and position to change back to their original values. Throwing a ball vertically into the air is a good example of this. A force acting on the ball gives it an upward speed. That speed then decreases as the ball travels upward and is acted upon by a gravitational force. When the ball reaches the very top of its trajectory, its vertical position has increased and its vertical speed has decreased to zero. But you know that as time continues, the ball will fall, returning to its original position and increasing its speed to its original value right after the force was applied. This idea that a force can change the position and speed of an object in a way that allows the position and speed to change back prompted scientists studying motion to wonder: was there some quantity that was not changed in these situations?

To identify what was not changed in these situations, scientists came up with the concept of energy. Energy comes in various forms. Two very important forms of energy are **kinetic energy** (energy associated with motion) and **gravitational potential energy** (energy associated with position). When forces act on objects, energy changes from one form

Physics Words

kinetic energy: energy associated with motion.

gravitational potential energy: the energy an object possesses because of its vertical position from Earth.

222

Active Physics

9.c)

The more bend in the pole, the higher in the air the pole vaulter can go.

Physics Talk

As students read about the law of conservation of energy, have them focus on the example of a ball that travels upward and then is acted upon by a gravitational force. Emphasize that the velocity decreasing to zero means that the kinetic energy is also zero but the ball now has maximum potential energy due to gravity. Point out that when the transformation of energy takes place the energy is not being lost, but transformed to a different form. Consider drawing two columns for the gravitational potential energy and kinetic energy on the board to show what happens at different stages of a ball's trajectory. Calculations may be

Section 8 Potential and Kinetic Energy: Energy in the Pole Vault

to another, but the sum of the kinetic and **potential energy** (the total energy) remains constant. That is why it is often possible for the objects to reverse the transformation of one form of energy to the other and return to their past positions and speeds. The concept that the total energy remains constant is referred to as the **law of conservation of energy**.

Energy and Work

While a ball is rising or falling, the sum of the gravitational potential energy and the kinetic energy remains constant. For the ball to start to rise, a force had to be applied to the ball over a distance. In the case of throwing a ball up in the air, the force acting on the ball is the force of your hand acting over a distance in an upward direction. For the ball to stop, another force had to be applied to the ball over another distance. In this case, it is gravitational force acting in a downward direction as the ball rises over a distance into the air. Whenever a force is applied to an object over a distance (in the same direction or opposite direction of the force), **work** is done. Work is a precisely defined physics quantity that equals the force multiplied by the distance. Whenever work is done, the energy of an object changes. Therefore, one very appropriate way to think about energy is to consider it "stored work."

Conservation of Energy in the Pole Vault

The coin in the investigation is a good example of work and conservation of energy. You applied a force to the ruler to bend it a certain distance. This was the work done on the ruler to add energy. After that, the ruler had **elastic potential energy**. When you released the ruler, that energy was transferred to the coin as the ruler applied a force to the coin over a certain distance. The coin now had kinetic energy. It traveled up in the air and the kinetic energy became gravitational potential energy as the coin rose. At its peak, the coin stopped momentarily. It now had no kinetic energy, but did have gravitational potential energy. As the coin began to fall, it gained kinetic energy as it lost gravitational potential energy. At all points during the rise and fall, the sum of the kinetic energy and gravitational energy of the coin was constant.

The pole vault is another wonderful example of the law of conservation of energy. The forms of energy are changed, or transformed, from one to another during a vault, but, in principle, the total amount of energy in the system of the vaulter and the pole remains constant.

Physics Words

potential energy: energy associated with position.

law of conservation of energy: energy cannot be created or destroyed; it can be transformed from one form to another, but the total amount of energy remains constant.

work: the product of the displacement and the force in the direction of the displacement.

elastic potential energy (also called spring potential energy): the energy of a spring due to its compression or stretch.

223

Active Physics

used to show that as the ball's gravitational energy is increasing, its kinetic energy is decreasing. Encourage students to think about the concept of energy remaining constant, regardless of its transformations. You could reemphasize the question, "Was there some quantity that was not changed in these situations?"

Have students distinguish between the different forces that propel the ball, either upward or downward.

Discuss the definition of work and ask students to write down the definition of work in their *Active Physics* log. To help relate the concept of work to energy, explain to the students that the work done by the thrower's hand while it was in contact with the ball is equal to the ball's increase in kinetic energy as it starts to rise. As students read about the conservation of energy, ask them to provide examples of situations where work is stored as elastic potential energy and then changes into kinetic energy. Discuss examples of different forms of energy transformations that an object goes through and have students take a short quiz to gauge their understanding. Concentrate on the general idea of energy being conserved as it passes through the various transformations between work and the different forms of energy. The *Physics Talk* provides several examples of calculating different forms of energy. It does not specifically provide numerical examples of using the conservation of energy principle to determine the energy at various stages when transformations occur. This aspect of conservation of energy will be dealt with more explicitly in the following section.

Chapter 2 Physics in Action

Food energy provides muscular energy for the vaulter to run, gaining an amount of kinetic energy. Some of the vaulter's kinetic energy is used to catapult the vaulter with an initial speed upward and the remaining kinetic energy is converted into an amount of elastic potential energy as the vaulter does work on the pole as it bends. As the bent pole straightens, its elastic potential energy is transferred to the vaulter to increase the vaulter's gravitational potential energy as the vaulter's height increases.

Richard Feynman's Explanation of the Conservation of Energy

In making measurements of the ruler's deflection and the height of the coin, you were investigating conservation of energy — one of the most important principles of science. Richard Feynman, an American physics giant of the twentieth century, provides a story that may help you to understand energy conservation.

In his story, a child plays with 28 blocks. Every day the child's mother counts the blocks and always finds the total to be 28. On one occasion, she only finds 27 blocks, but then realizes that one block is hidden in a box. On another day, she finds only 25 blocks, but can see that the water in a pail is higher than expected. By measuring the height difference, and knowing something about the original height of the water and the volume of a block, she determines that 3 blocks are below the surface of the water. Feynman equates counting the blocks with measuring the total energy. There were 28 blocks and there will always be 28 blocks. If there are 28 units of energy, then there will always be 28 units of energy.

Is There an Equation?

Physicists always ask if there is an equation that can help them understand and explain the model of observed events.

Work:

You can calculate the work done using the following equation.

$$W = F \cdot d$$

where F is the force applied in newtons (N) and
d is the distance in meters (m) over which the force was applied.

The following equations can be used to calculate the different forms of energy.

Elastic (spring) potential energy:

$$EPE = \frac{1}{2}kx^2$$

where k is the spring constant in newtons per meter (N/m) and
x is the amount of bending in meters (m).

Gravitational potential energy:

$$GPE = mgh$$

where m is the mass of the object in kilograms (kg),
g is the acceleration due to gravity in meters per second squared (m/s^2), and
h is the height in meters (m) through which the object is lifted.

Kinetic energy:

$$KE = \frac{1}{2}mv^2$$

where m is the mass in kilograms (kg) of the moving object and
v is the speed in meters per second (m/s) of the object.

SI Units of Work or Energy

The unit of work or energy is called the joule (J). From the formula for work, you can see that $1\ J = 1\ N \cdot m$. From the formulas for gravitational potential energy and kinetic energy you can see that $1\ J = 1\ kg \cdot m^2/s^2$, which makes sense since $1\ N = 1\ kg \cdot m/s^2$. Work and energy are scalar quantities. They have no direction.

When solving problems involving work and energy units, it is important to remember that the following are all the same unit.

$$J = 1\ N \cdot m = 1\ \frac{kg \cdot m^2}{s^2}$$

You may sometimes see $1\ kg \cdot m^2/s^2$ written as $1\ kg \cdot m^2 \cdot s^{-2}$. They are the same unit.

CHAPTER 2

Sample Problem 1

A weightlifter uses a force of 325 N to lift a set of weights 2.00 m off the ground. How much work did the weightlifter do?

Strategy: You can use the following equation for calculating work:

$$W = F \cdot d$$

Given:

$F = 325\text{ N}$

$d = 2.00\text{ m}$

Solution:

$$\begin{aligned} W &= F \cdot d \\ &= 325\text{ N} \times 2.00\text{ m} \\ &= 650\text{ N} \cdot \text{m or } 650\text{ J} \end{aligned}$$

Work done by the weightlifter is 650 J.

Sample Problem 2

How much energy is stored in a pole with a spring constant of 15 N/m if it is deflected 1.6 m?

Strategy: You can use the following equation for calculating elastic potential energy:

$$EPE = \frac{1}{2}kx^2$$

Given:

$k = 15\text{ N/m}$

$x = 1.6$

Solution:

$$\begin{aligned} EPE &= \frac{1}{2}kx^2 \\ &= \frac{1}{2} \cdot 15\ \frac{\text{N}}{\cancel{\text{m}}} \cdot (1.6\text{ m})^{\cancel{2}} \\ &= 19.2\text{ N} \cdot \text{m or } 19\text{ J} \end{aligned}$$

Elastic potential energy in the pole is 19 J.

Sample Problem 3

One of the highest pop flies ever recorded in baseball was about 172 m. What is the gravitational potential energy of a baseball with a mass of 145 g that is hit that high into the air? Use the value of 9.8 m/s² for the acceleration due to gravity.

Strategy: You can use the following equation for calculating gravitational potential energy:

$$GPE = mgh$$

Given:

$$m = 145 \text{ g or } 0.145 \text{ kg}$$
$$h = 172 \text{ m}$$
$$g = 9.8 \text{ m/s}$$

Solution:

$$\begin{aligned} GPE &= mgh \\ &= 0.145 \text{ kg} \times 9.8 \text{ m/s}^2 \times 172 \text{ m} \\ &= 244 \text{ kg} \bullet \text{m}^2/\text{s}^2 \text{ or } 244 \text{ J} \end{aligned}$$

The gravitational potential energy of the baseball is 244 J.

Sample Problem 4

A football player has a mass of 100.0 kg and runs at a speed of 6.0 m/s. What is his kinetic energy?

Strategy: You can use the equation for calculating kinetic energy.

$$KE = \frac{1}{2}mv^2$$

Given:

$$m = 100.0 \text{ kg}$$
$$v = 6.0 \text{ m/s}$$

Solution:

$$\begin{aligned} KE &= \frac{1}{2}mv^2 \\ &= \frac{1}{2} \times 100 \text{ kg} \times \left(6\frac{\text{m}}{\text{s}}\right)^2 \\ &= \frac{1}{2} \times 100 \text{ kg} \times 6\frac{\text{m}}{\text{s}} \times 6\frac{\text{m}}{\text{s}} \\ &= 1800 \text{ kg} \bullet \text{m}^2/\text{s}^2 \text{ or } 1800 \text{ J} \end{aligned}$$

The kinetic energy of the football player is 1800 J.

Checking Up

1. What is required for the energy of an object to change?
2. From where does the penny that is launched into the air get its energy?
3. From where does the pole vaulter get the energy needed to bend the pole and then rise over the bar?
4. What are the units for work, kinetic energy, gravitational potential energy, and spring potential energy?

Checking Up

1.

Work must be done on an object to change its total energy.

2.

The penny gets kinetic energy from the elastic potential energy of the ruler.

3.

The energy to bend the pole comes from the kinetic energy of the running pole vaulter, and the energy to go over the bar comes from the stored energy in the pole and the work done by the pole vaulter's muscles.

4.

The unit for all kinds of energy and work is joules, which means they are all equivalent forms.

Active Physics Plus

Students explore the change in energy due to work done by learning the strategies of solving sample problems. As students solve each problem, ask them to focus on how one type of energy is equated to another to solve a variable comprising the energy equation. Draw their attention to the statement "energy is stored work," and have them explain how this concept is shown by sample problems.

Active Physics *Plus*

+Math	+Depth	+Concepts	+Exploration
♦ ♦			

Energy Is "Stored Work"

The energy equations you have been using can be related to the work that increases the energy of an object. Work has a very special meaning in physics. Work is done on an object when a force is applied over a certain distance. The distance must be in the same direction as the force. This can be written as:

$$W = F \cdot d$$

Work done on a spring gives the spring elastic or spring potential energy. When the spring is released, that spring potential energy can move an object.

Imagine applying a force to stretch a spring. Some springs are easy to stretch and others require a large force to stretch. The difficulty of stretching a spring is defined by a number for each spring, called the spring constant k. The force required to stretch a spring with spring constant k, a distance x, is given by the equation $F = kx$. A larger stretch requires a larger force.

The average force will be halfway between the zero force (to start the stretch) and the final force for the last bit of stretch. The final force is kx. The initial force is 0. The average force is:

$$F_{avg} = \frac{kx+0}{2} = \frac{1}{2}kx$$

The total stretch of the spring is x. The work done is:

$$W = F \cdot d = \left(\frac{1}{2}kx\right)x = \frac{1}{2}kx^2$$

which is the expression for the elastic potential energy that was given to you earlier. Now you know where it comes from!

You can also calculate the work done to lift an object of mass m up through a distance h. In order for you to move the object vertically, you must apply an upward force that is just equal to its weight. You don't want to apply a force greater than this, because then there will be an upward unbalanced force that will cause acceleration, increase the speed of the mass, and give it kinetic energy.

$$W = F \cdot d = mgh$$

This is the expression for the gravitational potential energy given earlier.

You can also calculate the work done in accelerating an object from an initial velocity v_i to a final velocity v_f with a constant force. All you need to remember is that the acceleration, a, is equal to the change in velocity, $v_f - v_i$, divided by the time t, and that the distance traveled d is equal to the average velocity, $(v_f + v_i)/2$, times the time t.

$$W = F \cdot d$$

Since $F = ma$, then

$$W = mad$$

$$= m\left[\left(\frac{v_f - v_i}{t}\right)\right]\left[\left(\frac{v_f + v_i}{2}\right)\right]t$$

$$= \frac{1}{2}mv_f^2 - \frac{1}{2}mv_i^2$$

which is the expression for the change in kinetic energy you saw earlier.

Section 8 Potential and Kinetic Energy: Energy in the Pole Vault

Although the energy equations were stated first, notice that they result from the idea that whenever work is done, energies change. The energy equations simply reflect the amount of work done. This is where the statement "energy is stored work" originates.

Sample Problem I

Your teacher gives you a pop-up toy. When you push down on it, it sticks to the desk for a moment and then pops into the air.

a) If the toy has a mass of 100.0 g and leaps 1.20 m off the table, how much potential energy does it have at its point of maximum height? (Use $g = 9.80\ \text{m/s}^2$.)

Strategy:

The toy at its peak has a type of energy that depends on its position in Earth's gravitational field. So you need to use the formula for gravitational potential energy.

Given:

$$m = 100.0\ \text{g or } 0.100\ \text{kg}$$
$$h = 1.20\ \text{m}$$

Solution:

$$GPE = mgh$$
$$= (0.100\ \text{kg})(9.8\ \frac{\text{m}}{\text{s}^2})(1.20\text{m})$$
$$= 1.18\ \text{kg}\cdot\text{m}^2/\text{s}^2 \text{ or } 1.18\ \text{J}$$

b) When the toy jumps off the desk, with what speed does it leave?

Strategy: At the point where it jumps off the desk, the toy has its maximum amount of kinetic energy. This is what becomes the potential energy at the peak of its trajectory. Because energy is conserved, these two values will be equal—kinetic energy at the bottom equals the potential energy at the peak.

Given:

$$GPE = 1.18\ \text{J}$$

Solution:

$$GPE = KE$$
$$KE = \frac{1}{2}mv^2$$

Since you know that the KE must be equal to 1.18 J and you know the mass is 0.100 kg, you can use your calculator to find a value for v such that $\frac{1}{2}mv^2 = 1.18\ \text{J}$.

Alternatively, you can practice your algebra skills and find the value directly.

You can use algebra to rearrange the equation to solve for v.

$$v = \sqrt{\frac{KE}{\frac{1}{2}\text{m}}}$$
$$= \sqrt{\frac{1.18\text{J}}{\frac{1}{2}(0.100\ \text{kg})}}$$
$$= 4.90\ \text{m/s}$$

c) If you push the toy down 2.0 cm to make it stick to the desk, what is the spring constant of the spring in the toy?

Strategy: The kinetic energy to make the toy leap off the desk came from doing work on the spring and storing it as elastic potential energy. Using the conservation of energy, this energy was then transformed into kinetic energy.

CHAPTER 2

Given:

$x = 2.0\text{ cm} = 0.020\text{ m}$

Solution:

$EPE = 1.18\text{ J}$

You can use your calculator to find a value for k, such that:

$\frac{1}{2}kx^2 = 1.18\text{ J}$

or you can use algebra again to rearrange the equation to solve for k.

$$k = \frac{1.18\text{ N}\cdot\cancel{\text{m}}}{\frac{1}{2}(0.020\ \cancel{\text{m}})(0.020\text{ m})}\ (1\text{ J} = 1\text{ N}\cdot\text{m})$$

$$= 5900\ \frac{\text{N}}{\text{m}}$$

d) What force was needed to compress the spring the 2.00 cm?

Strategy: Now that you know the compression and the spring constant, it is possible to find the amount of force required to press down on the spring.

Given:

$k = 5900\text{ N/m}$

$x = 0.0200\text{ m}$

Solution:

$$F = kx$$

$$= \left(5900\ \frac{\text{N}}{\cancel{\text{m}}}\right)(0.0200\ \cancel{\text{m}})$$

$$= 118\text{ N or }120\text{ N}$$

Sample Problem 2

At what height, above the ground, could a tennis ball ($m = 57$ g) be dropped to give it the same kinetic energy it has when traveling at 45 m/s? (Neglect air resistance.)

Strategy: Assume that you are looking for the vertical position that will yield a speed of 45 m/s the instant before the ball touches the ground. The problem can be solved in one step using conservation of energy.

Given:

$m = 57\text{ g} = 0.057\text{ kg}$

$v = 45\text{ m/s}$

$g = 9.8\text{ m/s}^2$

Solution:

$$GPE = KE$$

$$mgh = \frac{1}{2}mv^2$$

Notice that you do not have to take into account the mass of the ball.

$$h = \frac{v^2}{2g}$$

$$= \frac{\left(45\ \frac{\text{m}}{\text{s}}\right)^2}{2\left(9.8\ \frac{\text{m}}{\text{s}^2}\right)}$$

$$= \frac{2025\ \frac{\text{m}^2}{\text{s}^2}}{19.6\ \frac{\text{m}}{\text{s}^2}}$$

$$= 103.3\text{ m or }100\text{ m}$$

When the ball is traveling at 45 m/s, it has a kinetic energy of 58 J. You can calculate this:

$$KE = \frac{1}{2}mv^2 = \frac{1}{2}(0.57\text{ kg})\ (45\text{ m/s})^2 = 58\text{ J}$$

If the tennis ball were positioned at a location 103 m above Earth, the gravitational potential energy of the ball would also equal 58 J.

$$GPE = mgh = (0.57\text{ kg})(9.8\text{ m/s}^2)(103\text{ m})$$

$$= 58\text{ J}$$

What Do You Think Now?

At the beginning of this section, you were asked the following

- **If champion pole vaulters can clear a 6.0-m high bar with a 5.5-m long pole, why can't they vault over a 12.0-m high bar with a pole 11.0 m long?**
- **What factors (variables) do you think limit the height a pole vaulter has been able to attain?**

Use energy conservation to explain how to determine why there is a limit to the height pole vaulters have been able to attain.

Physics Essential Questions

What does it mean?

In attempting to understand the physical world, physics often discovers quantities that remain unchanged while other quantities change. What does it mean when you say energy is conserved during the pole-vault event?

How do you know?

Conservation of energy is an important concept in physics because it is observed to be the case over and over again in the physical world. What did you observe in this activity that made the concept of conservation of energy plausible?

Why do you believe?

Connects with Other Physics Content	Fits with Big Ideas in Science	Meets Physics Requirements
Force and motion	* Conservation laws	Good, clear, explanation, no more complex than necessary

* In physics, organizing principles like the conservation of energy are used to explain a wide range of phenomena. Although you may never have seen a rugby match, why do you believe that you can use conservation of energy to describe the event?

Why should you care?

Conservation of energy is such an important concept of physics because it is important in so many situations. It is going to be important in your sports voice-over. Give an example in which your commentary is going to discuss conservation of energy.

What Do You Think Now?

Have students review their answers in light of what they have learned so far in this section. Ask them questions that lead students to answer the *What Do You Think Now?* questions. Share the *Physicist's Response* with them. Invite students to discuss their answers. Remind them that at this stage they should review and revise their original responses. While students are revising their responses, emphasize how they can use their knowledge of energy conservation to determine the maximum height pole vaulters can attain.

Physics Essential Questions

What does it mean?

Although the height and the speed of the vaulter change, the sum of the gravitational potential energy, *GPE*, and the kinetic energy, *KE*, remains constant.

How do you know?

The spring toy had kinetic energy at one point and then had an equal amount of gravitational potential energy at another point.

Why do you believe?

Conservation of energy holds for all events, including all sports.

Why should you care?

When a pop fly is hit in baseball, the ball first has kinetic energy. There is a loss in kinetic energy as the ball rises and an equal gain in gravitational potential energy.

Reflecting on the Section and the Challenge

This is the time for students to reflect on what their commentary could cover if they were sportscasting a pole-vault event. They should be able to describe the law of conservation of energy in relation to other sports and include it in their voice-over narration. Have the students write a short summary of their sportscast, so that they can discuss it with their peers and incorporate physics concepts that apply to the sport they want to include in their *Chapter Challenge*. Point out that their work will be evaluated on the basis of how clearly they explain the transformation of energy in a system after force is applied. Use the terms *force*, *work*, and *energy* frequently in your discussion so that students can relate to each term in context.

Chapter 2 Physics in Action

Reflecting on the Section and the Challenge

In this section, you were told that throughout the event of pole vaulting, energy changes from one form to another, but the total amount of energy in the system at all instants remains the same. (A small amount of energy may be transformed into internal energy by making a dent in the end of the pit that stops the pole or by raising the temperature of the pole as it bends.) Therefore, a sportscaster covering the pole vault event has many opportunities to explain what is happening in terms of the law of conservation of energy.

As a sportscaster, you can also describe the law of conservation of energy as it applies to a baseball rising in the air. In soccer, when you kick the ball, you do work on the ball and compress it. This elastic potential energy becomes kinetic energy. If the ball rises, some of that kinetic energy becomes gravitational potential energy. From a physics perspective, the behavior of a golf ball or a tennis ball is identical. High-speed photographs can show the compression of the balls. Since conservation of energy is one of the organizing principles of all science, you may want to include this in your voice-over.

Physics to Go

1. Describe the energy transformations in the shot put.
2. Describe the energy transformations in golf.
3. Assume that a vaulter is able to carry a vaulting pole while running as fast as Carl Lewis in his world record 100-m dash (around 12 m/s). Also assume that all of the vaulter's kinetic energy is transformed into gravitational potential energy. What vaulting height could that person attain? (Hint: Use the equation $\frac{1}{2}mv^2 = mgh$.)
4. Why does the length of the pole alone not determine the limit of vaulting height?
5. The temperature of some poles increases slightly as they flex. Use the law of conservation of energy to explain how this would affect performance.
6. The women's pole vault world record as of spring 1997 was 4.55 m, set by Emma George. What do you estimate was Emma's speed prior to planting the pole? Use conservation of energy for your prediction.
7. Sergei Bubka held the world record for the pole vault as of spring 1997 at 6.14 m. How did Sergei's speed compare with Emma George's speed? (See *Question 6*.)
8. A 2.0-kg rock is dropped off a 100-m high cliff.

232 Active Physics

Physics to Go

1.

The shot putter imparts a launching speed to the projectile, which is kinetic energy. By giving the shot two speeds in the same direction, the speeds add together. One of the motions is provided by the spinning motion of the shot putter before release, and the other is provided by the thrusting action of the shot putter's arm. In both cases, the athlete does work that is transformed into the kinetic energy that the ball has upon release. The horizontal component of the shot's speed is maintained while the projectile is in the air. The vertical component of the speed can be used to calculate the part of the kinetic energy, which will be transformed into gravitational potential energy, allowing prediction of the height to which the shot will rise at the peak of its flight. Students are more likely to write that the athlete uses muscles to transfer kinetic energy to the shot. Some of this energy makes the shot rise, converting kinetic energy into potential energy. Then the shot falls as potential energy is converted back to kinetic energy. When the shot hits the ground, the kinetic energy transforms into internal energy as the temperature of the shot and ground increases slightly.

2.

In golf, the golfer does work on the golf club to give the end of the club a great deal of kinetic energy. Upon impact, the end of the golf club does work on the golf ball, giving the golf ball kinetic energy. As with the shot after release, the golf ball has both horizontal and vertical speed components. The energy transformations are therefore the same as with the shot.

3.

Lewis' maximum speed in the dash = 12 m/s. Solving for h in the equation given in the problem: $h = v^2/2g = 7.3$ m.

4.

The vaulter's kinetic energy is determined by the running speed and the amount of work he or she does using their arms to lift their body. This energy, which includes the work done, determines the maximum height the pole vaulter can reach, regardless of the length of the vaulting pole. The pole could, however, have an effect if it is either too short or too long.

5.

The amount of internal energy (due to the temperature rise) generated during the bending of the pole would "rob" part of the potential energy that can be stored in the pole by causing it to bend less than it would if no internal energy were generated and by causing the pole to straighten with less "straightening" speed than if no internal energy were generated; overall, heating of the pole would reduce the vaulter's height.

6.

$v = \sqrt{2gh} = \sqrt{2\left(10 \text{ m/s}^2\right)\left(4.6 \text{ m}\right)} = \sqrt{92 \text{ m}^2/\text{s}^2} = 9.6$ m/s. Notice that Emma's height was rounded to two significant figures in the above substitution.

7.

$v = \sqrt{2gh} = \sqrt{2\left(10 \text{ m/s}^2\right)\left(6.1 \text{ m}\right)} = \sqrt{120 \text{ m}^2/\text{s}^2} = 11$ m/s. Therefore, Sergei's speed was greater than Emma's. Notice that while Sergei's speed was only about 15% greater than Emma's, his vault height (6.1 m compared to Emma's 4.6 m) was about 32% higher. This is due to the squared effect of speed on kinetic energy; squaring the ratio of speeds verifies that the ratio of heights should be 1.32:1.00, as shown below:

$$\left(\frac{6.1 \text{ m/s}}{4.6 \text{ m/s}}\right)^2 = (1.33)^2 = 1.76$$

NOTES

CHAPTER 2

NOTES

8.a)

Using the law of conservation of energy, the gravitational potential energy (GPE) at the top of the cliff is equal to the kinetic energy (KE) at the bottom.

$$GPE_i + KE_i = GPE_f + KE_f$$

$$mgh + 0 = 0 + \tfrac{1}{2}mv^2$$

$$v = \sqrt{2gh}$$

$$v = \sqrt{2\left(9.8 \text{ m/s}^2\right)\left(100 \text{ m}\right)} =$$

44 m/s

8.b)

Yes, the speed is independent of the mass. All objects fall at the same rate as long as air resistance is small enough to ignore. In the energy equations, you can see that the mass drops out.

9.a)

$$W = \tfrac{1}{2}kx^2 =$$

$$\tfrac{1}{2}(1500 \text{ N/m})(0.25 \text{ m})^2 = 47 \text{ J}$$

9.b)

The work becomes spring potential energy (SPE). This SPE becomes kinetic energy (KE) of the arrow.

$$KE = \tfrac{1}{2}mv^2$$

$$47 \text{ J} = \tfrac{1}{2}(0.1 \text{ kg})v^2$$

$$v = \sqrt{\frac{2(47 \text{ J})}{0.1 \text{ kg}}} = 30.7 \text{ m/s}$$

10.a)

$$W = \tfrac{1}{2}kx^2 =$$

$$\tfrac{1}{2}(315 \text{ N/m})(0.30 \text{ m})^2 = 14.2 \text{ J}$$

10.b)

$$F = kx = (315 \text{ N/m})(0.30 \text{ m}) = 95 \text{ N}$$

11.

The gravitational potential energy (GPE) of the car will become kinetic energy (KE) of the car. This KE will then do work on the spring and compress it.

The compression of the spring will store this energy as spring potential energy (SPE). You can therefore compare the GPE to the SPE and not concern yourself with the KE.

$$mgh = \tfrac{1}{2}kx^2$$

$$x = \sqrt{\frac{2mgh}{k}}$$

$$x = \sqrt{\frac{2\left(0.04 \text{ kg}\right)\left(9.8 \text{ m/s}^2\right)\left(1 \text{ m}\right)}{\left(18 \text{ N/m}\right)}} =$$

0.21 m

12.a)

$$F(\text{in newtons}) = ma =$$

$$\left(1 \text{ kg}\right)\left(1 \text{ m/s}^2\right) = 1 \text{ N} = 1\left(\text{kg}\cdot\text{m}\right)/\text{s}^2$$

12.b)

$$GPE = mgh = \left(1 \text{ kg}\right)\left(1 \text{ m/s}^2\right)\left(1 \text{ m}\right) =$$

$$\left(1 \text{ N}\right)\left(1 \text{ m}\right) = 1 \text{ J}$$

SECTION 8 QUIZ ANSWERS

1. c) $KE = \frac{1}{2}mv^2 \quad v = \sqrt{\frac{2KE}{m}} = \sqrt{\frac{2(300,000\text{ J})}{(1500\text{ kg})}} = 20\text{ m/s}$

2. b) $KE = \frac{1}{2}mv^2 \quad m = \frac{2KE}{v^2} = \frac{2(16\text{ J})}{(4\text{ m/s})^2} = 2.0\text{ kg}$

3. d) $GPE = mgh = (5\text{ kg})(9.8\text{ m/s}^2)(2\text{ m}) = 100\text{ J}$

4. a) $SPE = \frac{1}{2}kx^2 = \frac{1}{2}(120\text{ N/m})(0.20\text{ m})^2 = 2.4\text{ J}$

5. b) $KE = GPE$ gives $\frac{1}{2}mv^2 = mgh$

   Solving for h we have

   $$h = \frac{v^2}{2g} = \frac{(35\text{ m/s})^2}{2(10\text{ m/s}^2)} = 61\text{ m}$$

NOTES

CHAPTER 2

SECTION 9

Conservation of Energy: Defy Gravity

Section Overview

By calculating an athlete's time in air, called "hang time," students find out if the skater or basketball player at any point is able to counter the effect of gravity by hanging in the air. They watch a video of the skater doing a triple-axle jump, both in slow motion and at normal speed, and multiply the number of frames (pictures completed every 1/30 s) by the time the skater remains in air. Students then locate their center of mass to make a better estimate of how high they are able to jump. They jump as high as they can to calculate and record the vertical height through which their center of mass shifts during the jump. They use the concepts of work, spring potential energy (*SPE*), gravitational potential energy (*GPE*), and kinetic energy (*KE*) to help them analyze the energy transformations that take place during the jump. The students measure the position of their center of mass at the ready, launch, and peak positions and use these values along with their mass and the acceleration of gravity to calculate the energy forms at the various positions. Using the principle of conservation of energy, the students create a chart to compares value of the energies at the three different positions, as well as the sum of the *GPE*, *SPE*, and *KE* at these positions.

Background Information

Read the *Physics Talk* in the student text for this section before proceeding. Energy is simply stored work, and work is transformed into kinetic energy and gravitational potential energy in a vertical jump. Research has shown that the location of the center of mass within the jumper's body varies only slightly for the body positions assumed during the process of the vertical jump. The force that lifts and accelerates the body's center of mass during a vertical jump is provided by muscles of the leg, ankles, and feet. The method of analysis used for this activity assumes that the muscular force is constant as the body rises from "ready" to "launch" positions. This is not entirely accurate—in a real jump, the force varies—but is a reasonable approximation of reality.

The center of mass of an object is the only idea introduced in this section. The center of mass is the point at which the entire mass of an object may be thought of as being concentrated for purposes of analyzing the translational (along a path) or rotational (spinning) motion of the object. For practical purposes, the location of the center of mass of an object having only one significant dimension—such as a straight stick, loaded teeter-totter, twirler's baton, screwdriver, or wrench—corresponds to the object's balance point. For a two-dimensional object—such as a sheet of plywood cut into any shape—the location of the center of mass corresponds to the balance point located on either of the two large, flat surfaces of the object. To the extent that a two-dimensional object—such as a triangle cut from a sheet of plywood—may have significant thickness and, therefore, actually be three-dimensional, the center of mass would be located within the object, "in line" with the balance point, at the center of the thick dimension. For objects with simple three-dimensional shapes—such as homogeneous or symmetrically layered spheres such as bowling balls, basketballs, cubes, rectangular solids and cylinders—the center of mass is located within the center. An alternative to balancing an object to locate the center of mass is to suspend the object from any point that is not the center of mass. When suspended, gravity serves to orient the object so that its center of mass is located directly below the point of suspension. Earth "views" an object near it as a "point mass" (located at the object's center of mass) and pulls the point mass as close to Earth as possible. A line extended straight downward from the point of suspension passes through the object's center of mass. The intersection of two such lines, corresponding to two points of suspension, locates the object's center of mass.

Crucial Physics

- The center of mass of an object is the point that moves due to forces on the object as if all the mass of the object were located at that point.
- Energy is "stored work" and energy comes in different forms.
- Energy is conserved when there are no outside forces acting on the objects under consideration.

Learning Outcomes	Location in the Section	Evidence of Understanding
Measure changes in height of the body's center of mass during a vertical jump.	***Investigate*** Steps 4-6	Students record the height in meters of their body's distance from its center of mass to the floor at three different positions of their vertical jump.
Calculate changes in the gravitational potential energy of the body's center of mass during a vertical jump.	***Investigate*** Step 6	Students record the distance from the floor to their center of mass at the "peak position." Then they subtract the height at the peak of the jump height from the launch height.
Apply the definition of work.	***Investigate*** Steps 3 and 5 ***Physics Talk***	Students' work, which is required to lift themselves from the ready to the launch position, is their weight times the change in height.
Recognize how work is related to energy.	***Investigate*** Steps 5 and 6 ***Physics Talk***	Students do work with their leg muscles to provide the energy of the jump that is converted into kinetic energy.
Apply the joule as a unit of work and conservation of energy to the analysis of a vertical jump, including weight, force, height, and time of flight.	***Investigate*** Step 7 ***Physics Talk***	Students use the same units for all measures of energy and work in their calculations of the energy transformations between work, *SPE*, *GPE*, and *KE* during the jump.
Describe the concepts of work and conservation of energy to the analysis of a vertical jump, including weight, force, height, and time of flight	***Physics Talk***	Students read the *Physics Talk* and learn about conservation of energy based on tables providing a breakdown of total energy.

Section 9 Materials, Preparation, and Safety

Materials and Equipment

PLAN A		
Materials and Equipment	**Group (4 students)**	**Class**
Meter stick, wood	2 per group	
Calculator, basic	1 per group	
Tape, masking, 3/4 in. x 60 yds		6 per class
MBL or CBL Technology (to record probeware activity*		1 per class
Probeware, motion detector*		1 per class
Scale, bathroom*		1 per class
TV with VCR*		1 per class

*Additional items needed not supplied

PLAN B		
Materials and Equipment	**Group (4 students)**	**Class**
Meter stick, wood		2 per class
Calculator, basic	1 per group	
Tape, masking, 3/4 in. x 60 yds		6 per class
MBL or CBL Technology (to record probeware activity*		1 per class
Probeware, motion detector*		1 per class
Scale, bathroom*		1 per class
TV with VCR*		1 per class

*Additional items needed not supplied

Note: Time, Preparation, and Safety requirements are based on Plan A, if using Plan B, please adjust accordingly.

Time Requirement

This *Investigate* should take at least one class period or 40 minutes.

Teacher Preparation

- Obtain a TV and VCR (or DVD player) that has single-frame advance capability to show the students stop-frame video of an athlete jumping. Have a common bathroom scale available for students to determine their mass, but place it in a location in the classroom where students can obtain their mass privately.
- Set up a motion detector, computer, and interface if available. Attach the motion detector to the classroom ceiling so the students can jump underneath it to measure their jump height.

Safety Requirements

Students who do not wish to jump should not be forced to do so. Check with the school nurse to see if any students might have physical limitations that would preclude their jumping in class.

NOTES

NOTES

Meeting the Needs of All Students

Differentiated Instruction: Augmentation and Accommodations

Learning Issue	Reference	Augmentation and Accommodations
Understanding vocabulary	***What Do You Think?*** ***Investigate*** Steps 1.c) and 2 ***Physics to Go*** Step 2	**Augmentation** • Explain the meaning of "defy" and provide examples. **Accommodation** • Show a video clip of a bobsled team in action before students are asked to answer this question because they may have no idea what bobsledding means.
Visual perception and attention	***Investigate*** Steps 1.a) and 2.a)	**Augmentation** • Ask students to count the individual frames in the video for the figure skater and basketball player. • Students with vision issues could be assigned to record the class data on the board and help do the hang time calculations.
Scientific measurement	***Investigate*** Step 4	**Augmentation** • Students struggle with metric conversions. Tell them to find the vertical distance in centimeters. Then show them how to do the conversion. • Students often leave out decimal points and do not understand the importance of decimals in determining the value of numbers. Teach this concept in the context of money ($0.50 versus $50.00).
Measuring "peak position"	***Investigate*** Step 6	**Augmentation** • Teach strategies for marking "peak position" and then measuring it. For example, make one group member responsible for being the height-marker of the "peak position." Then have a pair of students measure the distance from the floor to the peak position. • Students could also jump near the blackboard and mark the peak position with chalk before measuring.
Recording data for calculations	***Investigate*** Steps 3-8	**Augmentation** • To help students record data accurately, model how to set up a table or record sheet before beginning the *Investigate*. **Accommodation** • Provide a blank copy of a data table for students to tape into their logs.
Creating a chart to understand an essential concept	***Investigate*** Step 7.d)	**Augmentation** • Model and ask students to turn their paper to landscape format to create any tables/charts that are larger than four columns. This will accommodate students with large handwriting and fine-motor issues. • Ask students to first draw the template of a table based on the teacher's sample. The table should fill most of the page. • If students are using lined paper, they can use the vertical lines on the paper to space out their columns (about five lines per column). Encourage them to quickly draw the appropriate number of rows. • Make sure students include a column for total energy. **Accommodation** • Provide a blank chart for students to tape into their logs and record the values.

Learning Issue	Reference	Augmentation and Accommodations
Understanding the conservation of energy principle	***Physics Talk***	**Augmentation** • Students struggle to conceptualize energy transfer and conservation because they cannot see the energy changing from one form to another. Also, the most common way for students to understand this principle is to compare the total energy of a system before and after the transfer of energy. Show students that the total energy of a system is always conserved. • Ask pairs of students to make drawings to represent the energy transfers involved when a person jumps. Students could use the values presented in the tables in this section to mathematically support their sketches. **Accommodation** • Provide a teacher-made worksheet or sketch of a person jumping on which students can add "types of energy" labels or "number of joules" labels.
Solving multi-step problems and using more than one formula in an assignment	***Physics Talk*** ***Physics to Go***	**Accommodation** • Provide opportunities to practice individual formulas for mastery before combining concepts into one multi-step problem that requires the use of a few different formulas and concepts.
Solving for velocity using the *KE* formula	***Physics Talk*** Sample Problem d)	**Augmentation** • Students struggle to understand the concept of square root and square root functions. Provide direct instruction in how to solve for velocity using the *KE* formula. Make sure that students understand that v^2 means v multiplied by v. **Accommodation** • Some students may not be developmentally or academically prepared at this time to solve for velocity using the *KE* formula.

Strategies for Students with Limited English-Language Proficiency

Learning Issue	Reference	Augmentation
Vocabulary comprehension	***What Do You Think?***	Given the "Defy Gravity" focus of this section, it is essential that all students understand the concept of hang time. Hang time is how long an athlete stays in the air after jumping. Some athletes stay in the air so long they appear to defy, or overcome, gravity.
Understanding scientific concepts	***Physics Essential Questions***	Students likely think of a theory as a guess or an assumption based on little understanding or information. But the word "theory" has a different meaning in science. A theory, in science, is an organized body of knowledge known to accurately predict and explain a specific set of phenomena in the natural world. The law of conservation of energy is a scientific theory that has stood the test of time. Remind students of the definition of "model" from *Section 5:* A model is anything that accurately represents what we know about how the natural world behaves. A theory is just as much a model as a mathematical model or a physical model.

Two important aspects of learning a new language are speaking and writing in that language. Some ELL students will be self-conscious and shy about speaking in front of their peers, while others will be less reluctant to try. Be sure to encourage all ELL students to speak in class, and give them opportunities to write on the board from time to time. Experience will broaden their comfort level. Over time, the shy students will get increasingly less self-conscious about speaking in front of their classmates.

CHAPTER 2

SECTION 9

Teaching Suggestions and Sample Answers

What Do You See?

There are many points of interest that your students are most likely to bring up when they are asked to discuss the illustration. Each image provides a focus that relates to the main topic. Consider various aspects of students' responses and encourage them to articulate their thoughts without hesitation. The *What Do You See?* stage provides the centerpiece on which you can build the rest of your students' understanding.

Chapter 2 Physics in Action

Section 9 Conservation of Energy: Defy Gravity

What Do You See?

Learning Outcomes

In this section, you will

- **Measure** changes in height of the body's center of mass during a vertical jump.
- **Calculate** changes in the gravitational potential energy of the body's center of mass during a vertical jump.
- **Apply** the definition of work.
- **Recognize** how work is related to energy.
- **Apply** the joule as a unit of work and energy using equivalent forms of the joule.
- **Describe** the concepts of work and conservation of energy to the analysis of a vertical jump, including weight, force, height, and time of flight.

What Do You Think?

No athlete can escape the pull of gravity.

- **Does the "hang time" of some athletes defy the pull of gravity?**
- **Does a world-class figure skater defy gravity to remain in the air long enough to do a triple axel?**

Record your ideas about these questions in your *Active Physics* log. Be prepared to discuss your responses with your small group and the class.

Investigate

In this *Investigate*, you will trace the energy conversions that take place as you jump vertically.

1. Your teacher will show you a slow-motion video of a world-class figure skater doing a triple axel jump. The image of the skater will appear to "jerk," because a video camera completes one "frame," or one complete picture every $\frac{1}{30}$ s. When the video is played at normal speed, you perceive the action as continuous. Played at slow motion, the individual frames can be detected and counted. The time interval between frames is $\frac{1}{30}$ s.

234

Students' Prior Conceptions

The concepts of force, motion of the center of mass of an object, free fall, weight, work, energy, and conservation of energy culminate to perplex students as they master the concepts building throughout the activities in this chapter. It is essential for students to explain the concepts in their own words and for teachers to listen to and probe student understanding so that the preconceptions listed in *Sections 1-8* are not overlooked. Explaining work, identifying the forces involved, examining weight and the center of mass, and considering the height of the vertical jump along with the time of the flight all act to perplex students in their analysis of the conservation of energy during a vertical jump. It is vital for the teacher to recognize students' preconceptions that continue to hinder understanding in order to help compare prior conceptions with observable behavior in order to prevent many misconceptions.

What Do You Think?

While recording their ideas, some students might ask you what hang time means. Instead of being locked down in a question and answer session, gently remind them that their task for the *What Do You Think?* questions is to record their answers in their *Active Physics* logs. Ask them to ponder what they already know and how they can respond to these questions. You might want to emphasize that the main purpose of this task is to stimulate their scientific thinking to prepare them for the inquiry-driven approach followed in *Active Physics*.

What Do You Think?

A Physicist's Response

No, the hang time does not defy the pull of gravity. Hang time is mere illusion. The pull of gravity continues to act, but this cannot be detected because the people's eyes cannot detect the rapid movement that happens in the flash of a second.

The world-class figure skater cannot defy gravity long enough to do a triple-axle. Each turn brings him very slightly toward the ground.

NOTES

CHAPTER 2

NOTES

Investigate

1.a)

The skater is in the air for 15 frames.

1.b)

Time in air (s) = Number of Frames

$(15)(1/30\text{ s}) = \frac{1}{2}$ s

1.c)

During the time frame viewed on the video, the skater's position is constantly changing. Although there is no hang time, the vertical change in position at the peak is at its slowest, giving the impression of a hang time. If students continue to say they observe hang time, ask them for a suitable definition of hang time. If they come up with a useable definition, they should not be able to observe it in the video.

2.a)

The basketball player is in the air for 31 frames.

a) As a class, count and record in your log the number of frames during which the skater is in the air.

b) Calculate the skater's time in the air or "hang time." (Show your calculation in your log.)
Time in air (s) = (No. of frames) ($\frac{1}{30}$ s)

c) Did the skater "hang" in the air during any part of the jump, appearing to "defy gravity"? If necessary, view the slow-motion sequence again to make the observations necessary to answer this question in your log. If your observations indicate that hanging did occur, be sure to indicate the exact frames during which it happened.

2. Your teacher will show you a similar slow-motion video of a basketball player whose hang time is believed by many fans to defy gravity.

a) Using the same method as above for the skater, show in your log the data and calculations used to determine the player's hang time during the "slam dunk."

b) Did the player hang? Cite evidence from the video in your answer.

3. How much force and energy do you use to do a vertical jump? You use body muscles to "launch" your body into the air, and, it is primarily your leg muscles that provide the force. First, analyze only the part of jumping that happens before your feet leave the ground. Find your body mass, in kilograms, and your body weight, in newtons, for later calculations. Remember, a mass of 1 kg has a weight of about 10 N. If you do not wish to use data for your own body, you may use the data for another person or one of your favorite athletes. If you know your body weight in pounds, you can find your mass in kilograms:

- First convert your weight in pounds to newtons.
Weight (N) = Weight (lb) (4.38 N/lb)
- Use your body weight, in newtons, and the equation Weight = mg to calculate your body mass (m), in kilograms.

4. Every object has a point called the center of mass. This point is special because when a force is exerted on the object, the center of mass moves as if all the mass of the object were located there. Your center of mass is in the middle of your body near your waist. Place a patch of tape on either the right or left side of your clothing (above one hip) at waist level. Crouch as if you are ready to make a vertical jump. While crouched, have an assistant measure the vertical distance, in meters, from the floor to the level of your body's center of mass (C of M).

center of mass

a) In your log, record the distance, in meters, from the floor to your C of M in the "ready position."

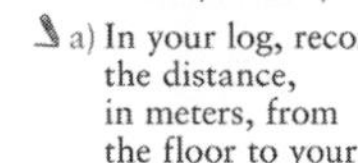

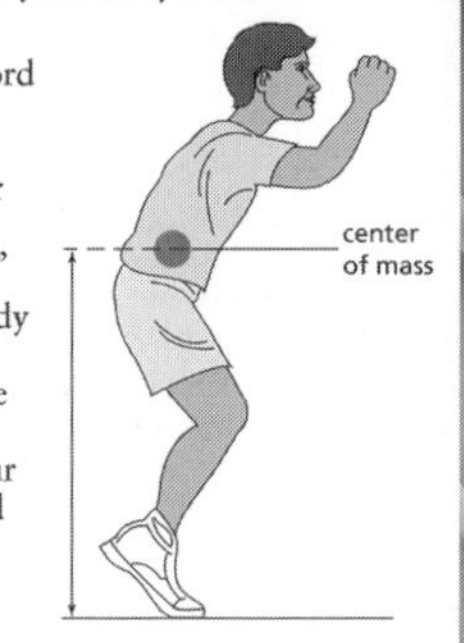

5. Straighten your body and rise to your tiptoes as if you are ready to leave the floor to launch your body into a vertical jump, but don't jump yet.

Time in air (s) = Number of Frames

$(31)(1/30\ \text{s}) = 31/30\ \text{s} =$

$1\frac{1}{30}\ \text{s} = 1.03\ \text{s}$

2.b)

During the time frame viewed on the video, the basketball player's position is constantly changing. There is no hang time but due to the slow vertical change of position it may appear that way. (As the ball is moving upward before the player leaves the ground, on the way down his arms are extended and lift the ball into the net, giving the illusion of hanging in the air.)

3.a)

Students' answers will vary according to their weight in pounds. Be aware that some students may be sensitive about their weight and be reluctant to write it down. To avoid this problem, you may choose to only have one student in the group, or class, do the jumping and everyone in the class use this weight and height in their calculations. Expect to help students when applying their own data to replicate the calculations presented as an example in *Physics Talk*. Ask the students about the forces acting on them between the "ready" and "launch" positions and between the "launch" and "peak" positions.

4.a)

Students record the distance from the floor to their center of mass in the crouched or "ready" position.

Teaching Tip

If the student is going to jump near the wall, the group can just place pieces of masking tape on the wall to correspond to the "ready" and "launch" positions, and then measure the distance between them.

5.a)

Students record the distance from the floor to their center of mass in the launch position.

5.b)

The students calculate the difference between the ready and launch heights.

6.a)

Students measure the height of the center of mass from the floor at the peak of the jump. One student should indicate the position with their finger, and hold that position while another measures the height after the student has landed.

6.b)

The students calculate vertical jump height by subtracting the height at the peak of the jump height from the launch height.

Teaching Tip

A motion detector can easily be attached to the ceiling of the classroom for the students to check the height of their jump. This quickly becomes a competitive exercise for many students, so you may want to limit its use to only serve as an additional check on the height reached by the jumpers in each group. To determine the height of the jump, have the students stand underneath the motion detector. Start the detector, and then have the students crouch down and jump directly upward toward the detector. The height of their jump is the difference between their distance from the detector when standing, and the minimum distance from the detector reached while jumping.

Hold this launch position while an assistant measures the vertical distance from the floor to the level of your center of mass.

a) In your log, record the distance, in meters, from the floor to your C of M in the "launch position."

b) By subtraction, calculate and record the vertical height through which you use your leg muscles to provide the force to lift your center of mass from the "ready position" to the "launch position."

6. Now it's time to jump! Have a group member ready to observe and measure the vertical height from the floor to the level of your center of mass at the peak of your jump. When your group member is ready to observe, jump straight up as high as you can. (Can you hang at the peak of your jump for a while to make it easier for your group member to observe the position of your center of mass? Try it, and see if your group member thinks you are successful.)

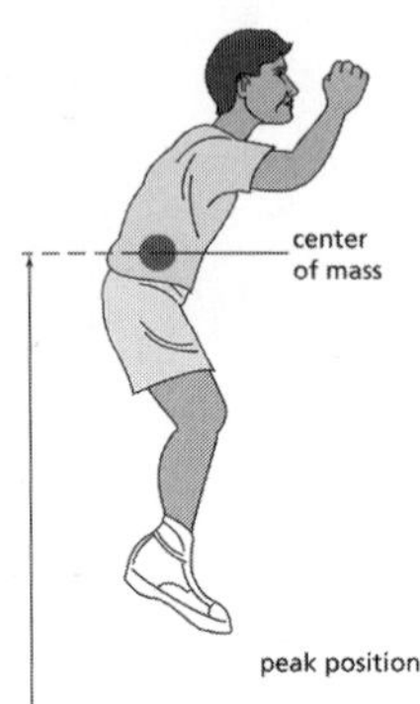

a) In your log, record the distance from the floor to your C of M at the "peak position."

b) By subtraction, calculate and record the vertical height through which your center of mass moved during the jump from the launch position to the peak position.

7. The jump can be analyzed at the three positions. In the ready position, you have only elastic potential energy (*EPE*). In the peak position, you have only gravitational potential energy (*GPE*). In the launch position, you have some *GPE* and some kinetic energy (*KE*).

a) The best place to start to analyze your jump is the peak position. At this point in your jump, all the elastic potential energy you started with in the ready position has been transformed to gravitational potential energy. Use the equation $GPE = mgh$ where h is the distance between your peak position and your ready position. Calculate your gravitational potential energy at the peak position (in joules) and write it in your log.

b) The next step is to realize that when you are in the ready position, all your energy is in the form of elastic potential energy. Therefore, the law of conservation of energy tells you that the amount of elastic potential energy in the ready position must equal the amount of gravitational potential energy in the peak position. Write this amount of energy in your log as your elastic potential energy (*EPE*).

c) When you are in the launch position, the elastic potential energy you had in the ready position has been transformed into both gravitational potential energy and kinetic energy. Use the equation, $GPE = mgh$, where h is the distance between the ready position and the launch position.

Calculate your gravitational potential energy in the launch position. The law of conservation of energy tells you that the rest of the elastic potential energy in the ready position must be kinetic energy, so subtract the gravitational potential energy at the launch position from the elastic potential energy at the ready position to find the kinetic energy at the launch position.

7.a)

Answers will vary. The *GPE* will equal mgh where h is the difference in height measure in *Step 6.b)*.

7.b)

The students copy their answer from *Step 7.a)* here.

7.c)

The students calculate their potential energy in the launch position by mgh where h is the height measure in *Step 5.b)*. The *KE* at the launch is then the difference between the values of *Step 7.a)* and the energy they measured in the first part of this step.

Write both numbers in the proper places in your log. Compute the total energy at the launch position.

d) The conservation of energy states that the total energy in the three positions should be equal. Create a chart that compares the sum of *GPE*, and *EPE*, and *KE* at the ready position, the launch position, and the peak position.

8. An ultrasonic ranging device coupled to a computer or graphing calculator, which can be used to monitor position, speed, acceleration, and time for moving objects, may be available at your school. If so, it can be used to monitor a person doing a vertical jump. This would provide interesting information to compare to the data and analysis that you have already done.

Physics Talk

CONSERVATION OF ENERGY

In this *Investigate*, you jumped and measured your vertical leap. You went through a chain of energy conversions where the total energy remained the same, in the absence of air resistance. You began by lifting your body from the crouched "ready position" to the "launch position." The work that you did was equal to the product of the applied force and the distance. The work done must have lifted you from the ready position to the launch position (an increase in gravitational potential energy) and also provided you with the speed to continue moving up (an increase in kinetic energy). After you left the ground, your body's gravitational potential energy continued to increase, and the kinetic energy decreased. Finally, you reached the "peak position" of your jump, where all of the energy became gravitational potential energy. On the way down, that gravitational potential energy began to decrease and the kinetic energy began to increase.

When you are in the ready position, you have elastic potential energy. If you were a spring, the elastic potential energy would be present due to compression of the material making up the spring. In your case, the potential energy you are going to use is present due to chemical reactions waiting to happen in your muscles. As you move toward the launch position, you have exchanged your elastic potential energy for an increase in gravitational potential energy and an increase in kinetic energy. As you rise in the air, you lose the kinetic energy and gain more gravitational potential energy. You can show this in a table.

Energy→ Position↓	Elastic potential energy	Gravitational potential energy = mgh	Kinetic energy $= \frac{1}{2}mv^2$
ready position	maximum	0	0
launch position	0	some	maximum
peak position	0	maximum	0

The energy of the three positions must be equal. In this first table, the sum of the energies in each row must be equal.

7.d)

The students create a chart that shows the $SPE = KE + GPE$ for the top of the jump.

8.

The students may try to jump under a motion detector. The height of the jump and the height during the ready position may be easily measured with this device.

2-9a Blackline Master

Physics Talk

As students read the *Physics Talk*, have them revise their definitions of gravitational potential energy, spring potential energy, and kinetic energy. Ask them to write a brief explanation of how energy conversions took place during the *Investigate*, how work done brought about a change in energy, and what happened to the total energy of a system. Once students have written their explanation, divide them into groups and have them share their responses, then invite them to discuss their responses with the whole class. Suggest to students that they can illustrate their explanation with diagrams showing an athlete in different positions. Have them label each diagram. A useful visual would be the use of bar graphs to indicate the amount of energy in the form of *GPE*, *SPE*, and *KE* in the various positions, as well as the total energy of the system. The students should realize that the total energy remains constant for the system, and should have a total energy graph that indicates this in all positions.

Choose the results of one member of the class (or yourself if you did the jump) as an example of how to calculate the energy in the various positions, as it is shown in the *Physics Talk*. Have them work out the values of gravitational potential energy and kinetic energy at their launch position. The height of jump of a second class member may be chosen, or perhaps that of an Olympic athlete as a comparison for additional practice. A person jumping on a trampoline provides another opportunity for students to examine energy conversion. A discussion of why jumping on a trampoline yields ever-increasing heights (up to a point), while successive jumps on a hard floor do not, should convince the students that the *SPE* stored in the deformed trampoline is restored to the jumper, while the hard floor does not respond in a similar manner. They should know how jumping from a hard surface produces a difference in the kinetic energy of rebound as compared to jumping from a soft surface like a trampoline.

When students are introduced to other forms of energy like light energy, chemical energy, heat energy, and sound energy, reemphasize that each type of energy can be measured and the total energy of a closed system is always conserved when this energy is transformed from one form into another. Write one of the sample problems in the *Physics Talk* on the board and discuss each step of the strategy. Be sure to emphasize the connection between work and energy, and how doing work

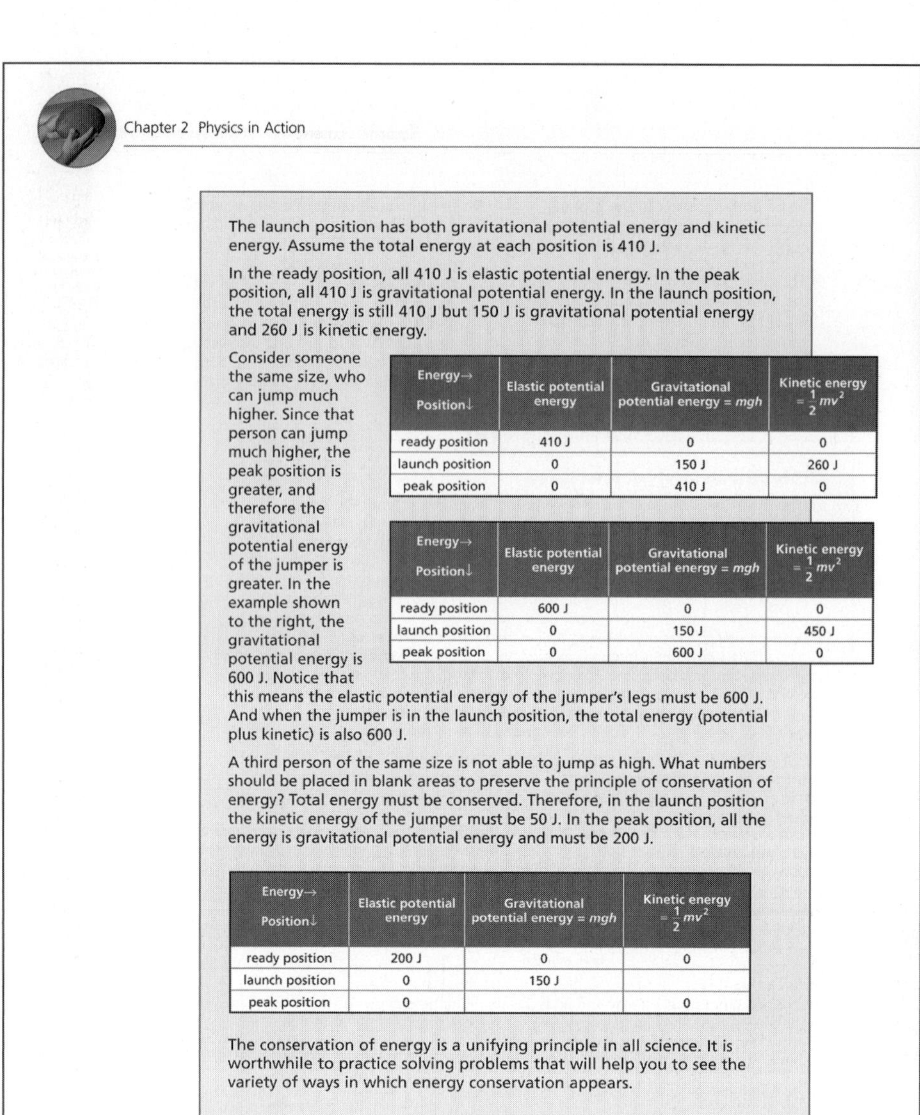

Chapter 2 Physics in Action

The launch position has both gravitational potential energy and kinetic energy. Assume the total energy at each position is 410 J.

In the ready position, all 410 J is elastic potential energy. In the peak position, all 410 J is gravitational potential energy. In the launch position, the total energy is still 410 J but 150 J is gravitational potential energy and 260 J is kinetic energy.

Energy→ Position↓	Elastic potential energy	Gravitational potential energy = mgh	Kinetic energy $= \frac{1}{2}mv^2$
ready position	410 J	0	0
launch position	0	150 J	260 J
peak position	0	410 J	0

Consider someone the same size, who can jump much higher. Since that person can jump much higher, the peak position is greater, and therefore the gravitational potential energy of the jumper is greater. In the example shown to the right, the gravitational potential energy is 600 J. Notice that this means the elastic potential energy of the jumper's legs must be 600 J. And when the jumper is in the launch position, the total energy (potential plus kinetic) is also 600 J.

Energy→ Position↓	Elastic potential energy	Gravitational potential energy = mgh	Kinetic energy $= \frac{1}{2}mv^2$
ready position	600 J	0	0
launch position	0	150 J	450 J
peak position	0	600 J	0

A third person of the same size is not able to jump as high. What numbers should be placed in blank areas to preserve the principle of conservation of energy? Total energy must be conserved. Therefore, in the launch position the kinetic energy of the jumper must be 50 J. In the peak position, all the energy is gravitational potential energy and must be 200 J.

Energy→ Position↓	Elastic potential energy	Gravitational potential energy = mgh	Kinetic energy $= \frac{1}{2}mv^2$
ready position	200 J	0	0
launch position	0	150 J	
peak position	0		0

The conservation of energy is a unifying principle in all science. It is worthwhile to practice solving problems that will help you to see the variety of ways in which energy conservation appears.

Active Physics 238

on a system will change the system's total energy, while energy transformations within the system do not.

Section 9 Conservation of Energy: Defy Gravity

A similar example to jumping from a hard floor into the air is jumping on a trampoline (or your bed, when you were younger). If you were to jump on the trampoline, the potential energy from the height you are jumping would provide kinetic energy when you landed on the trampoline. As you continued down, you would continue to have kinetic energy because you would still be losing gravitational potential energy. However, the trampoline bends and/or the springs holding the trampoline stretch. Either way, the trampoline or springs gain elastic potential energy at the expense of your kinetic energy and changes into gravitational potential energy.

Energy→ Position↓	Elastic potential energy	Gravitational potential energy = mgh	Kinetic energy $= \frac{1}{2}mv^2$
High in the air position	0	2300 J	0
Landing on the trampoline position	0	500 J	1800 J
Lowest point on the trampoline position	2300 J	0	0

The conservation of energy is one of the great discoveries of science. You can describe the type of energy in words (elastic potential energy, gravitational potential energy, and kinetic energy). There is also sound energy, light energy, chemical energy, electrical energy, nuclear energy, and the internal energy that reveals itself through temperature. These words, however, do not give the complete picture. Each type of energy can be measured and calculated. In a system not exchanging energy with objects external to it, the total of all the energies at any one time must equal the total of all the energies at any other time. That is what is meant by the conservation of energy.

If you choose to look at one object in the system, that one object can gain energy. For example, in the collision between a player's foot and a soccer ball, the soccer ball can gain kinetic energy and move faster. Whatever energy the ball gained, you can be sure that the foot lost an equal amount of energy. The ball gained energy, the foot lost energy, and the "ball and foot" total energy remained the same. The ball gained energy because work (force on the ball over a distance in the same direction) was done on it. The foot lost energy because negative work (force on the foot over a distance in opposite directions) was done on it. The total system of "ball and foot" neither gained nor lost energy.

Physics provides you with the means to calculate energies. You may wish to practice some of these calculations now. Never lose sight of the fact that you can calculate the energies because the sum of all of the energies remains the same.

→

CHAPTER 2

Chapter 2 Physics in Action

Sample Problem

A trainer lifts a 5.0-kg equipment bag from the floor to the shelf of a locker. The locker shelf is 1.6 m off the floor.

a) How much force will be required to lift the bag off the floor?

b) How much work will be done in lifting the bag to the shelf?

c) How much potential energy does the bag have as it sits on the shelf?

d) If the bag falls off the shelf, how fast will it be going when it hits the floor?

Strategy: This problem has several parts. It may look complicated, but if you follow it step-by-step, it should not be difficult to solve.

Given:

$$m = 5.0 \text{ kg}$$
$$h = 1.6 \text{ m}$$
$$a = 9.8 \text{ m/s}^2$$

Strategy:

a) Why does it take a force to lift the bag? It takes a force because the trainer must act against the pull of the gravitational field of Earth. This force is called weight, and you can solve for it using Newton's second law.

Solution:

$$\begin{aligned} F &= ma = w \\ &= (5.0 \text{ kg})(9.8 \text{ m/s}^2) \\ &= 49 \text{ kg} \cdot \text{m/s}^2 \text{ or } 49 \text{ N} \end{aligned}$$

A force of 49 N is required to lift the bag.

Strategy:

b) The information you need to find the work done on an object is the force exerted on it and the distance it travels. The distance was given and you calculated the force needed. Use the equation for work.

Solution:

$$\begin{aligned} W &= F \cdot d \\ &= (49 \text{ N})(1.6 \text{ m}) \\ &= 78.4 \text{ N} \cdot \text{m or } 78 \text{ J} \end{aligned}$$

The work done lifting the bag is 78 J.

Strategy:

c) The amount of potential energy depends on the mass of the object, the acceleration due to gravity, and the height of the object above what is designated as zero height (in this case, the floor). You have all the needed pieces of information, so you can apply the equation for potential energy.

Section 9 Conservation of Energy: Defy Gravity

Solution:

$$GPE = mgh$$
$$= (5.0 \text{ kg})(9.8 \text{ m/s}^2)(1.6 \text{ m})$$
$$= 78.4 \text{ kg}\cdot\text{m}^2/\text{s}^2 \text{ or } 78 \text{ J}$$

Should you be surprised that this is the same answer as *Part b)*? No, because you are familiar with energy conservation. You know that the work is what gave the bag the potential energy it has. So, in the absence of work that may be converted to internal energy because of friction, which you did not have in this case, the work equals the potential energy.

Strategy:

d) The bag has some potential energy. When it falls off the shelf, the potential energy becomes kinetic energy as it falls. Just before it strikes the ground in its fall, it has zero potential energy and all kinetic energy. You calculated the potential energy. Conservation of energy tells you that the kinetic energy will be equal to the potential energy. You know the mass of the bag so you can calculate the velocity with the kinetic energy formula.

Solution:

$$KE = \frac{1}{2}mv^2$$

You can use your calculator to find a value for v, such that:

$$78 \text{ J} = \frac{1}{2}mv^2$$

or you can practice your algebra and solve for v

$$v^2 = \frac{KE}{\frac{1}{2}m}$$
$$= \frac{78 \text{ J}}{\frac{1}{2}(5.0 \text{ kg})}$$
$$= 31 \text{ m}^2/\text{s}^2$$
$$v = \sqrt{31 \text{ m}^2/\text{s}^2}$$
$$= 5.6 \text{ m/s}$$

The bag will be traveling 5.6 m/s when it hits the ground.

Checking Up

1. Where does the energy come from that allows the jumper to move from the ready position to the launch position?
2. In the launch position, what types of energy will the student have? What types of energy will the student have at the peak of the jump?
3. What are three other types of energy beside potential and kinetic?

Active Physics

Checking Up

1.

The energy is provided by the student's muscles.

2.

In the launch position, the students will have gravitational potential energy and kinetic energy. At the peak of the jump, the students will have gravitational potential energy.

3.

Other forms of energy include sound, electrical, chemical, nuclear, and internal energy or heat. However, all these energies may be classified as either kinetic or potential when observed on an atomic level.

Active Physics Plus

While students are solving problems, ask them how they could explain that kinetic energy is a scalar quantity. One justification for kinetic energy being a scalar can be seen from the formula

$$KE = \frac{1}{2}mv^2$$

Because the energy is proportional to velocity squared, it would make no difference if the velocity were in the positive or negative direction. The square of either would be positive and the previous direction would be lost. On a simpler level, there simply is no direction associated with kinetic energy because the work necessary to cancel the energy would be independent of the direction in which the object moves.

Point out that each answer should highlight the transformation of energy and how the law of conservation of energy is helpful in finding the value of a quantity. Ask students to share their answers with their classmates, then have a whole-class discussion on each problem.

Chapter 2 Physics in Action

Active Physics Plus

+Math	+Depth	+Concepts	+Exploration
•	•		

Kinetic Energy as a Scalar Quantity

One of the fascinating aspects of kinetic energy is that it is a scalar quantity. It makes no difference what direction an object is going. All objects with the same mass and speed have the same amount of kinetic energy, regardless of the directions of their motions. Objects moving on frictionless tracks frequently change directions. In computing the kinetic energy, this makes no difference. Just use speed to find kinetic energy and don't worry about direction.

1. A roller coaster is poised at the top of a hill 50 m high.
   a) How fast will it be going when it goes over the top of the next hill on the track that is only 30 m high?
   b) From a practical point of view, why is it advantageous that this ride is mass independent?
2. A water balloon (m = 300 g) is launched horizontally from a platform 2 m above the ground with a slingshot. The slingshot (k = 60 N/m) is stretched 40 cm before launch. How far from the platform will the balloon strike the ground?
3. In a motorcycle jumping exhibition, a rider zooms down an incline starting 25 m off the ground at rest. At the bottom of the incline the track slopes upward and ends 5 m above the ground, at which point the motorcycle is airborne. The mass of the motorcycle and rider is 200 kg. While on the track, the motorcycle receives 200,000 J of energy from the engine and loses 50,000 J to friction. To what height above the ground does the motorcycle ascend when airborne if its horizontal velocity at the highest point while airborne is 40 m/s?

What Do You Think Now?

At the beginning of this section, you were asked

- **Does the "hang time" of some athletes defy the pull of gravity?**
- **Does a world-class figure skater defy gravity to remain in the air long enough to do a triple axel?**

How would you answer these questions now that you have closely analyzed a jump? Do world-class athletes defy gravity in any way during a slam dunk or a triple axel?

242

Active Physics

1.a)

$\frac{1}{2}mv^2 = mg\Delta h$. Solving for v we have $v = \sqrt{2g\Delta h} = 20$ m/s.

1.b)

Velocity is the same regardless of the number of people in the roller coaster. If the ride were people dependent, numerous problems might arise. If the cart went faster with more people, increasing the number of people would increase the chances of a collision between carts. Also, the "thrill" of the ride would depend upon how many people were in the car. Finally, the stresses created on the roller-coaster structure would depend upon the cart's speed, which would vary greatly if the speed was mass dependent.

2.

The spring potential energy is

$$\frac{1}{2}kx^2 = \frac{1}{2}(60 \text{ N/m})(0.4 \text{ m})^2 = 4.8 \text{ J}$$

The kinetic energy of the launched balloon equals the spring potential energy. Setting these two equal and solving for v we have

$$\frac{1}{2}mv^2 = 4.8 \text{ J}$$

$$v = \sqrt{\frac{2(4.8 \text{ J})}{m}} = 5.7 \text{ m/s}$$

The vertical fall distance determines the time in the air for an object launched horizontally or

$$d = \frac{1}{2}at^2$$

$$t = \sqrt{\frac{2d}{a}} = \sqrt{\frac{2(2\text{ m})}{(10\text{ m/s}^2)}} = 0.63\text{ s}$$

The horizontal distance traveled is the horizontal speed multiplied by the time in the air or

$d = vt$, which gives $d =$

$(5.7\text{ m/s})(0.63\text{ s}) = 3.6\text{ m}$

3.

At the start, the energy of the motorcycle is all gravitational potential energy:

$PE = mgh =$

$(200\text{ kg})(9.8\text{ m/s}^2)(20\text{ m}) =$

$39{,}000\text{ J}$.

The motorcycle gains 200,000 J of energy and loses 50,000 J of energy while on the ramp, so its total energy when airborne equals the net energy gained, which is

$200{,}000\text{ J (from the motor)} +$

$39{,}000\text{ J(from } PE) -$

$50{,}000\text{ J(due to friction)} =$

$189{,}000\text{ J}$.

At the highest point while airborne, the motorcycle has a kinetic energy of $KE = \frac{1}{2}mv^2 =$

$\frac{1}{2}(200\text{ kg})(40\text{ m/s})^2 = 160{,}000\text{ J}$.

Therefore, the rest of the energy, $189{,}000\text{ J} - 160{,}000\text{ J} = 29{,}000\text{ J}$ must be gravitational potential energy. So, the height can be found using

$PE = mgh$ or $h = \frac{PE}{mg} =$

$$\frac{29{,}000\text{ J}}{(200\text{ kg})(9.8\text{ m/s}^2)} = 15\text{ m}.$$

What Do You Think Now?

The *What Do You Think Now?* questions give students the opportunity to bring up any questions that they might have, revisit their responses, and modify them where necessary. You might want to return to the *What Do You See?* illustration and ask students if they now understand which concept the artist was trying to capture. Encourage them to discuss their ideas. Consider sharing *A Physicist's Response* with them. At this stage, students should show confidence in their answers.

CHAPTER 2

NOTES

NOTES

Physics

Essential Questions

What does it mean?

How can a jump be described as an example of the conservation of energy?

How do you know?

Conservation of energy is not merely a description of *GPE*, *EPE*, and *KE*. Each of these energies can be calculated. How did you calculate the total energy of your jump?

Why do you believe?

Connects with Other Physics Content	Fits with Big Ideas in Science	Meets Physics Requirements
Force and motion	 Conservation laws	Good, clear, explanation, no more complex than necessary

* The conservation of energy is a major organizing principle of all science. This theory is one of the great achievements of science. When someone throws a baseball straight up, the ball gains kinetic energy, *KE*, which gets converted into gravitational potential energy, *GPE*. How can you say that energy is conserved if the ball began with no energy and then gained *KE* and *GPE*?

Why should you care?

Physics says that objects on the surface of Earth cannot "defy gravity." Therefore, this must be true for all sports events. Give an example from your sport in which it is clear that the motion of a person or object is exactly what physics says it has to be.

Reflecting on the Section and the Challenge

Work, the force applied by an athlete to cause an object to move (the athlete's own body can be the object in some cases), multiplied by the distance the object moves while the athlete is applying the force explains many things in sports. For example, the vertical speed of any jumper's takeoff (which determines height and "hang time") is determined by the amount of work done against gravity by the jumper's muscles before takeoff. You will be able to find many other examples of work in action in sports videos, and now you will be able to explain them. In creating a description of a sporting event, you may decide to describe the work done and then move to a description of the energy transformations — how kinetic energy may become gravitational potential energy.

Reflecting on the Section and the Challenge

After students have read *Reflecting on the Section and the Challenge* in their *Active Physics* textbooks, focus attention on the important aspects of this section by highlighting key physics concepts. Discuss how work is related to energy transformations. Ask students to reflect on a specific event in a sports video that illustrates the work done by an athlete, as well as the energy transformations that took place during the sporting event. Emphasize that concepts learned in this section could be applied to their *Chapter Challenge*.

Physics Essential Questions

What does it mean?

The energy of the muscles provides the body with kinetic energy which propels the body in the air where it gains gravitational potential energy. The sum of *EPE*, *KE*, and *GPE* is constant for any point in the jump.

How do you know?

$KE = \frac{1}{2}mv^2$

$GPE = mgh$

$EPE = \frac{1}{2}kx^2$

Why do you believe?

The energy came from the work done by your arm. Chemical energy in your muscles provided the kinetic energy of the ball.

Why should you care?

The conservation of energy can explain how a high jumper uses the muscles in her legs to give her body kinetic energy. This kinetic energy and more chemical energy from the legs propels her into the air, where she gains gravitational potential energy and clears the bar.

Physics to Go

1.

Work = $Fd = (mg)d =$

$(50 \text{ kg})(10 \text{ m/s}^2)(1 \text{ m}) = 500 \text{ J}.$

2.

Before jumping on the sled, team members do work while running and pushing the sled to give the sled and their bodies kinetic energy; therefore, $Fd = \frac{1}{2}mv^2$. After the team has jumped on the sled, the total energy of the team and sled is equal to the kinetic energy gained during the pushing phase plus the gravitational potential energy = mgh, where h is the vertical distance from the top to the bottom of the hill. At the bottom of the hill, the kinetic energy of the sled should be equal to the kinetic energy gained during the pushing phase plus the loss in potential energy due to coming down the hill, $\frac{1}{2}mv^2 + mgh$. The brake must do enough work to cause the sled to lose all of its kinetic energy by exerting a force in the direction opposite the sled's motion.

Chapter 2 Physics in Action

Physics to Go

1. Calculate the work a male figure skater does when lifting a 50-kg female skating partner's body a vertical distance of 1 m in a pairs competition, if she does nothing to propel herself upward and just lets him lift her.
2. Describe the energy transformations during a bobsled run, beginning with team members pushing to start the sled and ending when the brake is applied to stop the sled after crossing the finish line. Include both work and energy in your answer and ignore friction.
3. Suppose that a person who saw the video of the basketball player used in the *Investigate* said, "He really can hang in the air. I've seen him do it. Maybe he was just having a 'bad hang day' when the video was taken, or maybe the speed of the recording or playback was not accurate." How might you and the person go about seeing if the person's statements are correct?
4. If someone claims that a law of physics can be defied or violated, should they be required to provide observable evidence, or should someone else need to prove that the claim is not true? Who do you think should have the burden of proof? Discuss this issue within your group and write your own personal opinion in your log.
5. Identify and discuss two ways in which an athlete can increase his or her maximum vertical jump height.
6. Calculate the amount of work, in joules, done when a:
   a) 1.0-N weight is lifted a vertical distance of 1.0 m.
   b) 1.0-N weight is lifted a vertical distance of 10 m.
   c) 10-N weight is lifted a vertical distance of 1.0 m.
   d) 0.10-N weight is lifted a vertical distance of 100 m.
   e) 100-N weight is lifted a distance of 0.10 m.
7. List how much gravitational potential energy, in joules, each of the weights in *Question 6* above would have after being lifted.
8. List how much kinetic energy, in joules, each of the weights in *Questions 6* and 7 would have at the instant before striking the ground if dropped.
9. How much work is done on a go-cart if you push it with a force of 50.0 N parallel to its path and move it a distance of 43 m, ignoring any friction that may exist?
10. What is the kinetic energy of a 62-kg cyclist if she is moving on her bicycle at 8.2 m/s?
11. A net force of 30.00 N acts on a 5.00-kg wagon that is initially at rest.
    a) What is the acceleration of the wagon?
    b) If the wagon travels 18.75 m, what is the work done on the wagon?

Active Physics 244

3.

It is apparent that the person wants to believe that the player can defy gravity and is attempting to justify that belief by rejecting scientific evidence. It could be said that the person is not reflecting open-mindedness, a desirable attribute in scientific pursuits. To address the situation, you may offer to video tape in slow motion the player, as many times as necessary, to ascertain if in any of the cases the player is able to "hang." After several attempts, even this person may agree he was mistaken.

4.

The burden of proof rests with the person making the claim.

5.

One way is to increase the force the athlete is able to exert using muscles, and the other way is to lose weight without decreasing muscular force.

6.a)

$1.0 \text{ N} \times 1.0 \text{ m} = 1 \text{ J}$

6.b)

$1.0 \text{ N} \times 10 \text{ m} = 10 \text{ J}$

6.c)

$10 \text{ N} \times 1.0 \text{ m} = 10 \text{ J}$

6.d)

$0.10 \text{ N} \times 100 \text{ m} = 10 \text{ J}$

6.e)

$100 \text{ N} \times 0.10 \text{ m} = 10 \text{ J}$

7.

All answers are the same as for *Question* 6 on the previous page.

8.

All answers are the same as for *Question* 6 on the previous page.

9.

$W = Fd = (50.0 \text{ N})(43 \text{ m}) =$

2150 J (2200 J)

10.

$KE = \frac{1}{2}mv^2 = \frac{1}{2}(62 \text{ kg})(8.2 \text{ m/s})^2$

$= 2084$ J (2100 J)

11.a)

$F = ma \quad a = F/m =$

$30.0 \text{ N}/5.0 \text{ kg} = 6 \text{ m/s}^2$

11.b)

$W = Fd = (30.0 \text{ N})(18.75 \text{ m}) =$

563 J

NOTES

CHAPTER 2

NOTES

12.a)

$W = Fd \quad d = W/F =$

$(40{,}000\ \text{J})/(3200\ \text{N}) =$

12.5 m (12 m)

12.b)

$F = ma \quad a = F/m =$

$(3200\ \text{N})/(1200\ \text{kg}) = 2.7\ \text{m/s}^2$

13.

The work done is equal to the change in *KE*. The final *KE* is 0.

The initial *KE* can be found.

$KE = \frac{1}{2}mv^2 =$

$\frac{1}{2}(0.150\ \text{kg})(40\ \text{m/s})^2 = 120\ \text{J}$

14.

The change in *KE* is equal to the work done. Calculate the change in *KE* and then calculate the distance.

$KE = \frac{1}{2}mv^2 =$

$\frac{1}{2}(64.0\ \text{kg})(15.0\ \text{m/s})^2 = 7200\ \text{J}$

$W = Fd$

$d = W/F =$

$(7200\ \text{J})/(417\ \text{N}) = 17\ \text{m}$

Section 9 Conservation of Energy: Defy Gravity

12. Assume you do 40,000 J of work by applying a force of 3200 N to a 1200-kg car (ignore friction).
    a) How far does the car move during the time you are doing work on it?
    b) What is the acceleration of the car?
13. A baseball (m = 150.0 g) is traveling at 40.0 m/s. How much work must be done to stop the ball?
14. A boat exerts a force of 417 N pulling a water skier (m = 64.0 kg) from rest. The skier's speed becomes 15.0 m/s. Over what distance was this force exerted?
15. Create a chart showing the *GPE*, *EPE*, and *KE* and their sum at different positions for a pole vault (running, full bend of the pole while on the ground, peak height, landing, and collapsing on the cushion).
16. Create a chart showing the *GPE*, *EPE*, and *KE* and their sum at different positions for a person on a trampoline (at peak height, upon landing on the trampoline, and at lowest point of the trampoline).
17. Create a chart showing the *GPE*, *EPE*, and *KE* and their sum at different positions for a skier at the top, middle, and bottom of a slope.
18. ***Preparing for the Chapter Challenge***
    Use the law of conservation of energy to prepare an exciting voice-over for one part of the action in the video you have selected to use.

15.

A sample table might look like the one below.

	KE	*GPE*	*EPE*	Sum of energies
Running	1000	0	0	1000
Pole bent	100	0	900	1000
At peak	150	850	0	1000
Landing	1000	0	0	1000
On cushion	0	0	0	0

16.

	KE	*GPE*	*EPE*	Sum of energies
Peak height	0	1000	0	1000
Landing	900	100	0	1000
Lowest point	0	0	1000	1000

17.

	KE	*GPE*	*SPE*	Sum of energies
Top of Mt	0	1000	0	1000
Middle	500	500	0	1000
Bottom	1000	0	0	1000

18.

Preparing for the Chapter Challenge

The students should choose a sports event that includes elastic potential energy and a thorough description of what is happening during the energy transformations. Many sports that include rackets, bats, or other objects striking a ball, puck, etc., will have stored energy. A baseball transition might go as follows: work is done by the pitcher to give the ball *KE*, the loss of *KE* into spring *PE* when the ball strikes the bat, the change from elastic *PE* to *KE* as the ball leaves the bat, the transformation of some of the *KE* to Gravitational *PE* as the ball rises, then returning to *KE* as the ball falls, and finally the work done by the fielder who catches and stops the ball. Additional discussion might include the energy lost to heat in the bat, and the air due to frictional forces.

SECTION 9 QUIZ

2-9b Blackline Master

1. A baseball bat strikes a baseball of mass that is at rest, and as a result the baseball flies away from the bat. The kinetic energy gained by the baseball as it leaves the bat is due to the

   a) loss of potential energy of the ball as it falls back to the ground.

   b) work done by the bat on the ball.

   c) gain in potential energy as the ball rises to the peak of its trajectory.

   d) loss of mass after the ball is hit.

2. A catapult with a spring constant of 10,000 N/m is used to launch a target from the deck of a ship. The spring is compressed a distance of 0.5 m before the 1.56-kg target launched. What is the target's velocity as it leaves the spring?

   a) 10 m/s　　b) 20 m/s

   c) 30 m/s　　d) 40 m/s

3. A ball is thrown vertically upward. As the ball rises, the ball's total energy

   a) decreases.

   b) increases.

   c) remains the same.

4. A 10-kg mass at rest on a horizontal frictionless table is accelerated by a force of 20 N. How much work is needed to accelerate the ball to a velocity of 10 m/s?

   a) 5 J　　b) 200 J

   c) 500 J　　d) 2000 J

5. An object is lifted upward at a constant speed. The gain in potential energy of the object is due to the

   a) work done on the object.

   b) total force applied to the object.

   c) change in kinetic energy of the object.

   d) weight of the object.

SECTION 9 QUIZ ANSWERS

1. b) work done by the bat on the ball. This work will not be equal to the loss of *GPE* when the ball falls down, since it will still be moving forward when it lands, and will maintain some *KE*.

2. d) $SPE = KE$ or $\frac{1}{2}kx^2 = \frac{1}{2}mv^2$. Canceling the $\frac{1}{2}$'s on each side gives $(10{,}000\ \text{N/m})(0.5\ \text{m})^2 = (1.56\ \text{kg})v^2$, and $v = 40$ m/s.

3. c) by conservation of energy

4. c) The work needed is equal to the increase in $KE = \frac{1}{2}mv^2 = 500$ J.

5. a) Because the object is lifted with constant speed, there is no change in *KE*, so all the work goes into increasing *GPE*.

NOTES

Chapter Assessment

Physics You Learned

This section is a useful resource that helps students to revise and recall prior learning. Important concepts that students investigated in *Physics in Action* are briefly summarized and related equations are listed across the summary. Students could use the *Physics You Learned* table throughout their reading of the chapter for a quick reference to an essential concept that they investigated earlier in a previous section.

To help students review all the equations and concepts, once they have reached the end of the chapter, consider posting a visual display of important equations on the walls of your classroom. Divide students into groups and ask them to take turns to make up a list of five questions relating to the concepts they have learned and have each team member answer those questions. This strategy is useful in having students evaluate their knowledge of the physics concepts they have learned in the chapter. A key function of this section is that it gives students an opportunity to review for their *Physics Practice Test*.

Physics You Learned

Physics Concepts	Is There an Equation?
When an object is moving, it will continue to move at constant speed in a straight line unless there is an unbalanced force to change its motion. If the object is at rest, it stays at rest unless there is an unbalanced force. This is known as **Newton's first law.**	
The tendency of an object to resist changing its motion is called **inertia**. Inertia is measured in the same units as mass.	
A **frame of reference** is the specific point of view from which a particular measurement is made. Different frames of reference yield different measurements.	
The acceleration is defined as the change in velocity with respect to time.	$a = \frac{\Delta v}{\Delta t}$
A **force** is measured in the SI unit newtons.	
The acceleration of an object (a) is directly proportional to the net force applied (F_{net}), and inversely proportional to the object's mass (m). This is known as **Newton's second law.**	$a = \frac{F_{net}}{m}$
The weight (F_g) of an object is equal to an object's mass (m) multiplied by the strength of Earth's gravitational field (g). Weight is the force of Earth's gravity acting on an object.	$F_g = mg$
Using significant figures ensures that any calculations made do not indicate a level of precision greater than the measurements.	
Active Physics Plus: The net force (F_{net}) on an object in equilibrium is zero. When an object is in equilibrium (either at rest or traveling with constant velocity) the vector sum of all the forces acting on the object equals zero.	$F_{net} = 0$
Active Physics Plus: When forces act at right angles on the same body, the net force is determined by using the **Pythagorean theorem.**	$F_{net} = \sqrt{(F_1^2 + F_2^2)}$
The shape of a projectile's path is a **parabola** if there is no air resistance.	
Active Physics Plus: The vertical velocity and the horizontal velocity of an object are independent of one another, and can be used separately to determine aspects of a projectile's flight. The total velocity can be calculated from the horizontal and vertical components using the Pythagorean theorem.	$v = \sqrt{(v_y^2 + v_x^2)}$
The horizontal distance traveled by a projectile (d_{horiz}) equals the projectile's horizontal speed (v_{horiz}) multiplied by the time of flight. The vertical distance covered by a projectile (d_{vert}) depends upon the acceleration due to gravity (a_g) and the time the object is in flight (t) squared. The horizontal and vertical motions of a projectile are independent of each other.	$d_{horiz} = (v_{horiz})t$ $d_{vert} = \frac{1}{2} a_g t^2$

CHAPTER 2

The maximum range of a projectile returning to the same height as the launch point occurs when it is launched at 45° degrees to the horizontal.	
Active Physics Plus — When an object is projected at an angle to the horizontal, the motion may be analyzed after the velocity is broken into vertical and horizontal components.	
Forces come in pairs. Whenever a force is exerted on a mass b $(F_{a \text{ on } b})$, the mass b exerts an equal force in the opposite direction on the mass a $(-F_{b \text{ on } a})$. This is known as **Newton's third law.**	$(F_{a \text{ on } b} = -F_{b \text{ on } a})$
The normal force is a force that acts perpendicular to a surface.	
A **free-body diagram** is a sketch of all the forces acting on an object.	
The force of friction (F_f) equals the coefficient of friction (μ) multiplied by the normal force (F_N). Friction is a force acting between two bodies in contact that resists the relative motion of those bodies. It always acts parallel to the surfaces in contact.	$F_f = \mu F_N$
The coefficient of friction (μ) is a dimensionless constant.	
Active Physics Plus — The coefficient of static friction (μ_s) on an inclined plane equals the tangent of the angle the plane makes with the horizontal $(\tan\theta)$.	$\mu_s = \tan\theta$
An object's **kinetic energy** (KE) is proportional to the object's mass (m) multiplied by its velocity squared (v^2). Kinetic energy is an object's energy of motion.	$KE = \frac{1}{2}mv^2$
Gravitational potential energy (GPE) is proportional to an object's mass (m) multiplied by its vertical height above Earth (Δh) and the acceleration due to gravity (g). Gravitational potential energy is energy due to an object's vertical position above Earth's surface.	$GPE = mg\Delta h$
Elastic potential energy (EPE) is proportional to the spring constant of the material (k) multiplied by the material's change in length (x) squared. Elastic potential energy is energy stored in a material due to its compression or stretch.	$EPE = \frac{1}{2}kx^2$
Active Physics Plus — Work (W) done on an object can increase its kinetic energy (ΔKE). When work is done on an object moving on a horizontal surface, the kinetic energy of the object increases.	$W = \Delta KE = \left(\frac{1}{2}mv_f^2 - \frac{1}{2}mv_i^2\right)$
Work (W) is the product of the force exerted on an object (F), and the displacement in the direction of the force (d). Work done on an object increases its energy and may change an object's kinetic or potential energy.	$W = Fd$
The **law of conservation of energy** states that energy may change its forms, but not its amount. The total amount of energy remains the same during any changes in form.	$Energy_{before} = Energy_{after}$

Physics Chapter Challenge

A second cycle of the *Engineering Design Cycle* reinforces students' understanding of previous concepts. Students go through the same series of steps to determine how they can make their final presentation more effective. Review the *Chapter Challenge* in a class discussion so students understand each aspect of the *Goal*. Ask a student to summarize the *Goal* and if it meets your expectations write it on the board. Suggest to students that they should also keep referring to the rubric that they have developed in class to have the criteria and constraints before them as they go through the various phases of the *Engineering Design Cycle*.

The *Inputs* section is specifically designed for students to analyze what they have learned in each section so that they can apply important physics concepts to complete their *Goal*. Encourage students to reflect on the summary of each section that is provided in the *Student Edition*. Ask them to jot down important ideas that can be incorporated into their *Chapter Challenge* to improve the quality of their voice-over for an exciting sports event. You might want them to list additional *Inputs* they now have, since their presentation of the *Mini-Challenge*.

To ensure that students participate in a substantive reflection of each section's summary, have them share their *Inputs* with their classmates, in pairs or in groups. During this phase of the *Engineering Design Cycle*, ask them to connect their personal experiences in sports to Newton's laws of motion and Galileo's law of inertia to evaluate how these laws affect the outcome of a sports event. A sharing of individual sports experiences may generate more ideas that leads to additional *Inputs*.

Physics Chapter Challenge

You will now be completing a second cycle of the *Engineering Design Cycle* as you prepare for the *Chapter Challenge*. The goals and criteria remain unchanged. However, your list of *Inputs* has grown.

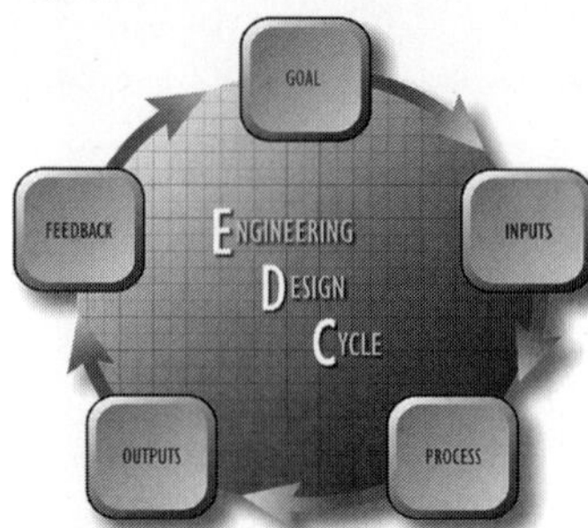

Goal

Your challenge for this chapter is to create a voice-over for an exciting sports event that will educate an audience on the physics behind the action. Review the *Goal* as a class to make sure you are familiar with all the criteria and constraints.

Inputs

You now have additional physics information to help you identify and analyze the various physics concepts that apply to sports activities. You have completed all the sections of this chapter and learned the physics content you will need to complete your challenge. This is part of the *Input* phase of the *Engineering Design Cycle*. Your group needs to apply these physics concepts to put together your presentation.

You also have the additional *Input* of your own personal experience with sports as well as the feedback you received following your *Mini-Challenge* presentation.

Section 1 You investigated Galileo's law of inertia and learned how it relates to the mass of an object. You read about Newton's first law and learned about reference frames for measuring the speed of an object.

Section 2 You measured speed by making speed vs. time graphs using a ticker timer for objects with constant speeds and changing speeds. You also explored the concept of acceleration, or the rate at which speed increases or decreases.

Section 3 You investigated the relationship between forces on an object and the acceleration and change in velocity that they produce. You also read about Newton's second law, which helps to calculate the unbalanced force, mass, or acceleration of an object.

Section 4 You used models to learn about the horizontal and vertical motion of a projectile. You also explored how the horizontal speed and total height of a projectile affects the horizontal distance that it travels.

Section 5 You measured constant acceleration due to gravity and discovered how it causes all objects to speed up as they fall toward Earth. You also used calculations and models to describe the trajectory, or path, of a projectile.

Section 6 You studied examples of force pairs and considered Newton's third law as an explanation for the forces caused by inanimate objects. You also learned to use force diagrams to clearly represent forces on objects.

Section 7 You measured the force of sliding friction between a sports shoe and various surfaces and calculated the coefficient of sliding friction for the different combinations. You also studied the impact of friction on the movement of objects.

Section 8 You explored the idea of conservation of energy and tracked energy through a system as it changed from potential to kinetic, back to potential, and so on. You also learned that energy could be stored by stretching or bending objects and that restorative forces could transform that stored energy back into motion.

Section 9 You calculated work and gravitational potential energy changes for objects that are lifted and learned that gravity applies a constant force on objects moving near the surface of Earth.

Process

In the *Process* phase, you must decide what information you have that you will use to meet the *Goal*. Deciding what physics topics to include is the first step. Select sports footage that is exciting in terms of game play or competition between rivals. Three minutes of exciting sports action from almost any sport will have examples of most of the physics topics you have studied in this chapter. You may want to watch your film several times and simply list the examples that you see. Once you have a list, try to pick about five that are equally spaced throughout the clip. If you use a computer-based video, you may be able to use slow motion and replay certain features to highlight them.

Gather data that will allow you to calculate the magnitude, or size, of forces, masses, and accelerations that you plan to focus on in your sports action. Make sure you know the SI units for each quantity and include a sample calculation for any estimates you make. For example, if you can find the mass of a baseball and time how long it takes to travel a distance on the field, you can estimate its speed and the size of the force required for the pitcher or the batter to accelerate it to get it to that speed in a limited amount of time.

Once you have selected the five examples, have a person in your group write a short script for each one. Then arrange the scripts into one narrative and see what it sounds like. Refine the script each time you practice. You may consider inserting some humor or dramatic narration for emphasis and entertainment. After all, people are not usually watching sports for its educational content. You will not get the job if you cannot keep the viewers in their seats. Manage your time to make sure your group has an opportunity to rehearse before you present. If you are going to record the narration, you may need extra time to edit the final product. Even the experts make mistakes during live narrations!

Outputs

Presenting your information to the class is your design-cycle *Output*. You will provide an auditory sample for the class either through a live reading of the script or a replay of your recorded narration. Some ad-libbing will be fine for entertainment value, but a purely unscripted narration will score very poorly for this challenge.

Feedback

Your classmates will give you *Feedback* on the overall appeal and the accuracy of your presentation, based on the criteria of the design challenge. This feedback will likely become part of your grade but could also be useful for additional design iterations. Remember that you will be viewing other design solutions for the same challenge. The different design solutions may represent feedback in the form of alternative ways you might have solved the problem. No design is perfect; there is always room for optimization or improvement. From your experience with the *Mini-Challenge*, you can see how the design cycle is structured to continuously refine almost any idea.

Active Physics

The *Process* phase should start with a careful selection of physics topics that students will include in their final presentation. Through a process of weighing what most effectively accomplishes their *Goal*, students should carefully study the data they have. Encourage them to replay sports video actions several times and pick examples that are relevant to their *Chapter Challenge*. While students are viewing the sports video, they should gather data that allows them to estimate the magnitude of the forces, masses and accelerations they are describing. The data that students collect should be useful for the sports action for which they intend to provide a voice-over narration. Instead of directly moving on to the commentary of a sports action, a group member should put together a written narrative with the help of the sports footage that is selected. Point out to students that they should practice and refine their written script to make it more lively and entertaining, while at the same time delivering the essential physics content.

During the *Output* phase when students present a live reading of their written script or replay their recorded narration, ask your class to pay close attention to the reading. The class environment should be relaxed and purpose driven. A clear focus on the objective will pave the way for a constructive *Feedback*. For this phase of the design cycle, students should have a copy of the rubric before them. Once they receive comments on their performance, ask them to focus on how their experience with design iterations helped improve the quality of their work. Also, point out that their will always be room for further refinement of any design they have chosen to present for their *Chapter Challenge*.

Physics Connections to Other Sciences

This section shows how intrinsic the laws of motion are to understanding natural phenomena that are studied under different scientific disciplines. Body mass becomes an impediment to animals with large bodies, massive molecules diffuse more slowly, and the combined mass of mountain ranges have so much inertia that it causes them to rise at the point of intersection when tectonic plates collide. These examples, among others, point to the significance of the connections physics makes with other sciences. To have a thorough understanding of any scientific process, it is essential to approach it from multiple-scientific perspectives. Students should, therefore, look for instances of a scientific concepts in other branches of science with which they are familiar. Encourage them to follow an interdisciplinary approach to learning.

Physics Connections to Other Sciences

Here are some examples of how the concepts you studied in this chapter relate to other sciences.

Newton's First Law – Inertia

Biology Animals with large body mass are generally unable to change direction quickly when in motion. Smaller animals can often elude their larger predators by making sharp turns while moving quickly.

Chemistry Massive molecules diffuse more slowly than less massive ones, allowing chemists to separate molecules and atoms by mass.

Earth Science When tectonic plates collide, the inertia of their combined mass can cause mountain ranges to rise at the point of intersection.

Newton's Second Law

Biology A flea is able to exert tremendous force for its size, allowing it to accelerate its body into huge jumps, up to 13 in. (33 cm), or 200 times the length of their bodies.

Chemistry The electric force of attraction between water molecules causes them to accelerate and join, forming water droplets.

Earth Science The force of gravity causes water to accelerate as it passes over a waterfall, increasing the erosive power of the water at the bottom of the fall.

Newton's Third Law

Biology An octopus propels itself through water by shooting out a forceful stream of water similar to jet exhaust. The reaction force of the ejected water on the octopus may cause an acceleration of up to 30 m/s^2.

Chemistry When an atom of gas in a balloon strikes the inner surface of the balloon, the force the balloon wall exerts on the atom to change its direction is equal to the force the atom exerts on the balloon. It is this combined force of countless atoms that keeps the balloon inflated.

Earth Science When a hurricane strikes land, the wind exerts tremendous force on topographical features and objects on the land. The reaction force of these objects on the moving air causes it to slow down, which is why hurricanes eventually dissipate as they move over land.

Projectile Motion

Biology Seagulls will often drop clamshells onto rocks to break them open, demonstrating a natural understanding of projectiles.

Chemistry The force of gravity causes settling of insoluble particles in a mixture with water. This is enhanced by spinning of a centrifuge.

Earth Science Volcanic eruptions often blast large rocks into the air. These rocks behave as projectiles, and their landing place may be accurately predicted.

Friction

Biology Air friction limits the speed at which a bird can fly. Friction, exerted as drag, pushes against the outstretched skin of a flying squirrel, allowing it to land safely after jumping from a high tree.

Chemistry Frictional forces provide the activation energy required to begin many chemical processes, such as the lighting of a match.

Earth Science The force of friction between tectonic plates holds them in place as they attempt to slide past each other. When this friction is eventually overcome, earthquakes often result from the sudden, rapid movement of the plates.

Conservation of Energy

Biology The chemical energy in the food a frog eats is converted into potential energy in its leg muscles. When the frog jumps, this energy is transformed into kinetic energy.

Chemistry When two atoms bond together, their electric potential energy decreases and their kinetic energy increases by the same amount.

Earth Science Earth continually receives energy from the Sun and radiates energy back out to the universe at the same rate. But the usefulness, or quality, of energy received from the Sun and that radiated by Earth; is greater. This is important for the continuance of events on Earth, which are characterized by energy conversions to forms that are less useful, or of lower quality. The higher quality energy received from the Sun compensates for the loss of energy quality in energy conversions on Earth.

Ask students to read the information provided in their student text on the connections of physics to biology, chemistry, and Earth science. Consider dividing students into groups of four and assign each group biology, chemistry, or Earth science connections. Ask each group to read the information given in the text, discuss, and paraphrase their understanding of each connection. For example, you could assign a group two biology connections, two chemistry, or two Earth science connections. Then ask student volunteers from each group to present their information in a whole-class discussion.

Physics
At Work

A. Dean Bell

Writer/Director; New York, NY

A. Dean Bell is an award-winning filmmaker, television writer, director, and producer. He wrote and directed the highly acclaimed show *SportsFigures* that aired on ESPN for 12 years.

SportsFigures is an educational television series designed to teach the principles of physics and mathematics through sports. Bell, having never even taken physics in high school while growing up in Rochester, New York, said he was not worried that he lacked a physics background when it came to writing and directing the show. "I was learning physics from the show's advisors and I felt that my discovery process could be translated into the show," he said.

SportsFigures won four Clarion Awards for best children's television program, and a number of Parents' Choice Awards. Bell knew that when *SportsFigures* was awarded these crowning achievements, his aim to combine education and entertainment had been achieved.

SportsFigures may have taped its last season, but it is still shown in reruns on *Cable In The Classroom*. "It is also used and available in school libraries across the country," said Bell.

Bell is also an assistant professor at SUNY Purchase, New York, his alma mater, where he has been teaching directing and screenwriting since 1995. "I tell my college students that as writers, don't hesitate to take that science course. You just might need to do something like write and direct a television series that teaches physics someday."

Rick Angelo

Producer, ESPN; Fairfield, CT

Rick Angelo began his career in sports television in 1995. Today, Angelo produces games for all college sports for ESPN.

Angelo believes physics plays a phenomenal role in his job. "Everything with sports has something to do with physics," he said. Producers use graphics and animation of the players to show a viewer the athleticism of the athlete. "We will use graphics to show the speed of a ball and what makes it the perfect pitch, to show the viewer what phenomenal athletes they are watching," said Angelo. "Physics enhances our stories about the athletes."

Sandra Giddins

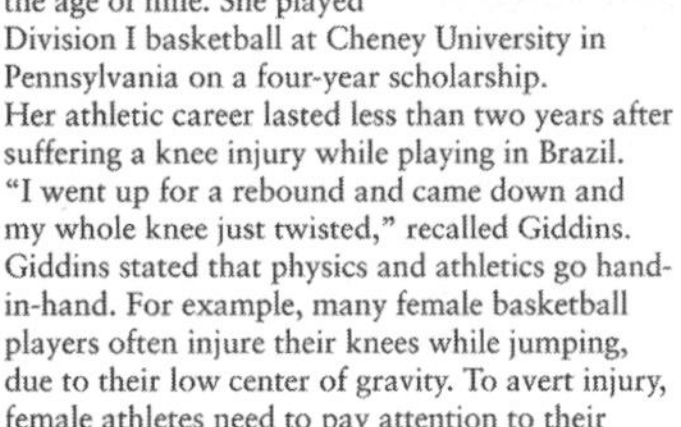

Community Center Director and former Professional Athlete & Coach; Queens, NY

Sandra Giddins grew up in Yonkers, New York and started playing basketball at the age of nine. She played Division I basketball at Cheney University in Pennsylvania on a four-year scholarship. Her athletic career lasted less than two years after suffering a knee injury while playing in Brazil. "I went up for a rebound and came down and my whole knee just twisted," recalled Giddins. Giddins stated that physics and athletics go hand-in-hand. For example, many female basketball players often injure their knees while jumping, due to their low center of gravity. To avert injury, female athletes need to pay attention to their body's center of mass.

251

Active Physics

Physics At Work

Physics At Work provides profiles of three accomplished professionals who employ physics to understand and improve the quality of their work. All three professionals have one aspect in common: an appreciation of physics. Have students read and discuss each profile.

A close reading of A. Dean Bell's career illustrates how his highly acclaimed show *SportsFigures* was designed to teach the principles of physics and mathematics. Bell felt that his understanding of physics that came from the show's advisors could be translated into the show. Interestingly, Bell had never taken physics in high school but was able to relate to the subject so deeply that he produced one of the finest television shows in this genre that won him many awards. You might want to ask students what valuable lesson they get when they read A. Dean Bell's Profile. Also ask them if they can relate their *Chapter Challenge* to Bell's discovery process of physics.

Rick Angelo's profile demonstrates the importance of physics in the world of sports. Angelo's belief that physics plays a phenomenal role in his job, illustrates the role of physics in sports television shows. The production techniques that Angelo uses to show the viewer how the expert skills of athletes are based in technology rooted in physics concepts. Ask students if they can describe in detail how physics is used to enhance Rick Angelo's stories and what laws of physics he might be using to produce his television shows.

Sandra Giddins, a former professional athlete and coach, who started playing basketball at an early age recounts her success on the field in an impressive profile. Her story though has a tragic note. She suffered a knee injury that could have been prevented if she had paid attention to her body's center of mass while jumping—a prime example of how the laws physics affect the safety of an athlete on the field. In a discussion of Sandra Giddins's profile, encourage students to reflect on their sports experiences and how Newton's laws of physics determine their safety on the field.

CHAPTER 2

Physics Practice Test

The *Physics Practice Test* is provided as a Blackline Master in your *Teacher Resources CD.*

2c Blackline Master

Content Review

1. b

2. d

The inertia of an object is measured in terms of its mass, so the object with the largest mass has the largest inertia. Inertia is the resistance of a mass to a change in its state of motion. Inertia is often confused with momentum.

3. d

4. a

5. c

6. c

7. d

Physics Practice Test

Before you try the Physics Practice Test, *you may want to review sections 1-7, where you will find* 29 Checking Up *questions,* 7 What Do You Think Now? *questions,* 28 Physics Essential Questions, 77 Physics to Go *questions, and* 11 Inquiring Further *questions.*

Content Review

1. A cart is rolling along a frictionless, horizontal surface. Which of the following describes the motion of the cart as it continues to roll along the surface?
   a) The cart will slow down as it runs out of the forward force.
   b) The cart will continue to roll with constant speed.
   c) The cart will continue to roll with constant speed only if it is rolling downhill.
   d) The cart will slow down as it uses up its speed.

2. Which object has the most inertia?
   a) a 0.001-kg bumblebee traveling at 2 m/s
   b) a 0.1-kg baseball traveling at 20 m/s
   c) a 5-kg bowling ball traveling at 3 m/s
   d) a 10-kg tricycle at rest

3. An athlete walks with a piece of ticker tape attached to herself with the tape timer running, and produces the tape shown below.

   beginning

   According to the tape, she was traveling with
   a) constant velocity.
   b) positive acceleration.
   c) negative acceleration.
   d) constant velocity, then negative acceleration.

4. A track coach with a meter stick and a stopwatch is trying to determine if a student is walking with constant speed. He should
   a) measure the walker's speed at regular intervals to see if it is always the same.
   b) measure the total distance the student travels and the total time to get the average speed.
   c) measure the beginning and ending speeds only to see if they are the same.
   d) use the meter stick to measure the student's stride length and time how long it takes to take one step.

5. If a cart is traveling with uniform negative acceleration, what conclusions can be drawn about the forces acting on the cart?
   a) The cart must be frictionless.
   b) The cart must be rolling downhill.
   c) The cart must have a net unbalanced force acting on it.
   d) No force is needed; the cart will naturally slow down.

6. A student wants to set up an experiment to determine the effect of a net force on an object's acceleration. To do this, she should
   a) vary the force acting on the object and the mass of the object at the same time.
   b) vary the mass of the object, but not the force acting on the object.
   c) vary the force acting on the object, but not the object's mass.
   d) keep both the force acting on the mass and the mass of the object constant as it rolls along a horizontal surface.

7. A 2-kg block is dropped from the roof of a tall building at the same time a 6-kg ball is thrown horizontally from the same height. Which statement best describes the motion of the block and the motion of the ball? (Disregard air resistance.)
   a) The 2-kg block hits the ground first because it has no horizontal velocity.
   b) The 6-kg ball hits the ground first because it has more mass.
   c) The 6-kg ball hits the ground first because it is round.
   d) The block and the ball hit the ground at the same time because they have the same vertical acceleration.

8. A pitching machine launches a baseball horizontally with no spin. Which of the following statements correctly describes the ball's motion in the air as the launch speed is increased?
a) The ball's acceleration increases, and the distance it falls in one second decreases.
b) The ball's acceleration remains the same, and the distance the ball falls in one second decreases.
c) The ball's acceleration remains the same, and the distance the ball falls in one second increases.
d) The ball's acceleration remains the same, and the distance the ball falls in one second remains the same.

9. A punter on a football team can kick the ball at an angle of either 30° or 80°. If he wants to maximize both the amount of time the ball spends in the air and the distance the ball travels, at which angle should he kick the ball?
a) the 30° angle because the ball goes further
b) the 80° angle because the ball goes further
c) the 30° angle because the ball spends more time in the air
d) the 80° angle because the ball spends more time in the air

10. A student is holding a book that has a weight of 20 N in his hand while sitting in a chair. The man claims that the book must be attracting Earth with a force of 20 N. His claim must be
a) false because books do not attract objects.
b) false because Earth is much larger than the book.
c) true because the book has more inertia than Earth.
d) true due to Newton's third law of action-reaction.

11. Which diagram of a 5-kg mass resting on a table correctly represents the force of the table on the mass?

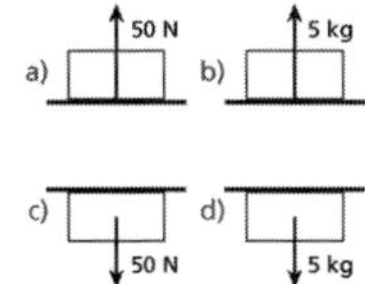

12. Two students have a "tug-of-war" on a smooth gym floor. One student has a mass of 70 kg and is wearing socks, but no athletic shoes. The other student has a mass of 60 kg and is wearing athletic shoes. The student most likely to win will be
a) the 60-kg student because he can pull harder on the 70-kg student.
b) the 70-kg student because he can pull harder on the 60-kg student.
c) the 60-kg student because he experiences a greater frictional force with the floor.
d) the 70-kg student because he experiences a greater frictional force with the floor.

13. Automobiles with front-wheel drive that have the engine located over the drive wheels have better traction in snow than automobiles with rear-wheel drive. This is most likely because
a) the tires on front-wheel drive automobiles have a higher coefficient of friction than rear-wheel drive automobiles.
b) the front tires encounter the snow first.
c) the front tires have a higher normal force than the rear-wheel tires because the engine is heavier than the rear of the automobile.
d) the front wheels are used for steering.

14. A student whose mass is 60 kg and a bicycle with a mass of 20 kg are at rest on a horizontal road. The student exerts a force of 120 N to accelerate the bike over a distance of 48 meters. What is the velocity of the bicycle and rider at the end of the 48 meters?
a) 3 m/s
b) 6 m/s
c) 8 m/s
d) 12 m/s

15. A basketball player is able to jump to a vertical height of 1.25 m. A student calculates that the player must have left the floor with a velocity of 5 m/s. The student can prove this claim by using
a) conservation of energy.
b) the principle of friction.
c) Newton's third law of motion.
d) the principle of inertia.

8. d

9. d

10. d

11. a

12. c

13. c

14. d

Here it is assumed that the student is riding the bicycle, so that the force the student exerts is to accelerate both the student and the bicycle together (total mass = 80 kg). Using $\Sigma F = ma$ gives $120 \text{ N} = (80 \text{ kg})(a)$ or $a = 1.5 \text{ m/s}^2$. To find the velocity, $v^2 = 2a\Delta d$ or $v^2 = 2(1.5 \text{ m/s}^2)(48 \text{ m}) = 144 \text{ m}^2/\text{s}^2$; therefore, $v = 12$ m/s.

15. a

CHAPTER 2

Critical Thinking

16.a)

The coefficient of friction between the steel block and the table surface could be determined by pulling the block along the surface with constant velocity by a horizontal pull parallel to the table's surface, using a spring scale. The materials you would need include a spring scale, a balance to measure the weight of the steel block if the spring scale did not work, and a method to attach the spring scale to the steel block, such as a string.

16.b)

The measurements you would need to take would include the weight of the steel block, and the reading on the spring scale when the block is sliding across the table with constant velocity.

16.c)

The coefficient of friction could be determined using the formula μ = frictional force/normal force.

17.a)

The upward force supplied by the park bench is exactly equal to your weight.

17.b)

When a force is applied to the bench, the bench flexes like a spring until it reaches a point where the spring-like force it is exerting upward just balances the force of your weight downward. The bench then stops bending and stays in that position until the force is removed, and it returns to its original, un-flexed position.

Practice Test *(continued)*

Critical Thinking

16. Design an experiment to measure the coefficient of friction between a steel block and the surface of your classroom lab table.
    a) What measuring tools will you need?
    b) What measurements will you take to determine the coefficient of friction?
    c) Show how you will use this data to calculate the coefficient of friction.

17. When you sit on a park bench, the bench exerts an upward force on you.
    a) Compare the force exerted by the park bench on you to your weight.
    b) Explain how the bench is able to provide the force required.

18. During an activity to measure how high a student can jump, the following measurements were made by the student's lab partners:
    - Mass = 65 kg
    - Increase in height of the student's center of mass during jump from the crouched down (ready) position = 0.60 m
    - Change in height from the ready position to the exact point where the student's feet leave the ground = 0.35 m

    a) How much gravitational potential energy did the student have at the peak of the jump?
    b) How much spring potential energy did the student's legs have as he was crouched in the ready position?
    c) Explain why the kinetic energy the student had as he left the ground was less than the spring potential energy when in the crouched down, ready position.

19. A ball is kicked horizontally off a tall building as shown.
    a) Draw a sketch of the ball's positions at 0.1 s intervals for the first 0.4 s as the ball falls to the ground.
    b) Draw arrows to represent the ball's horizontal velocity at positions described in *a)*.
    c) Draw arrows to represent the ball's acceleration for the positions described in *a)*.
    d) Draw arrows to represent the ball's vertical velocity in the positions described in *a)*.

20. Before leaving Earth, the mass of an astronaut is measured to be 60 kg. The astronaut lands on the Moon and measures the acceleration of gravity to be 1.6 m/s^2.
    a) What would the astronaut's weight be on Earth?
    b) What would the astronaut's weight be on the Moon?
    c) What would the astronaut's mass be on the Moon?
    d) Explain your answers to *a)* and *b)* using Newton's second law.

21. Four forces act on a 10-kg mass as shown in the diagram. What would the acceleration of the mass be?

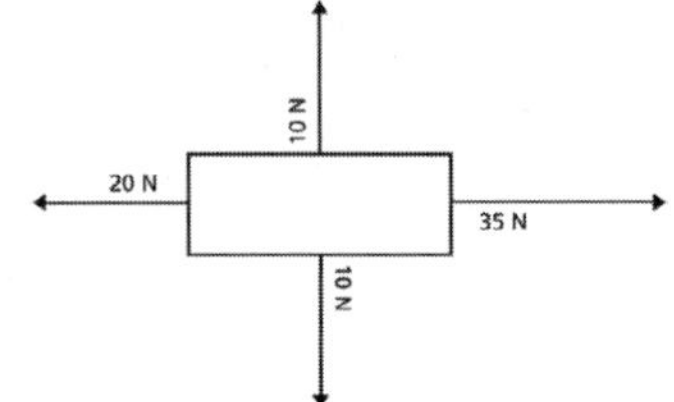

22. A soccer ball is kicked so that at the peak of its trajectory it has a horizontal speed of 15 m/s, and is 5 m above the ground. How far away from the kicker does the soccer ball land?

23. A motorcycle rider starts out on top of a ramp 10 m high, and then rides down and jumps the motorcycle as shown. The rider is at the peak of his jump at 5 m. How fast is the motorcycle going horizontally at this point?

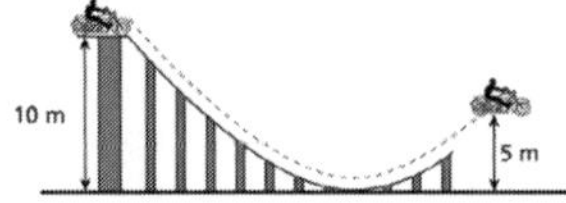

18.a)

$$GPE = mg\Delta h = (65\ \text{kg})(10\ \text{m/s}^2) \times (0.6\ \text{m}) = 390\ \text{J}$$

18.b)

$$SPE = GPE_{\text{peak}} - GPE_{\text{ready}} = mgh = (65\ \text{kg})(10\ \text{m/s}^2)(0.6\ \text{m}) = 390\ \text{J}$$

18.c)

The spring potential energy in the legs is greater than the student's kinetic energy when he leaves the ground because some of the spring potential energy has gone into the energy needed to raise the student's center of mass to the lift-off position.

19.a)

Diagram should be similar to what is shown below:

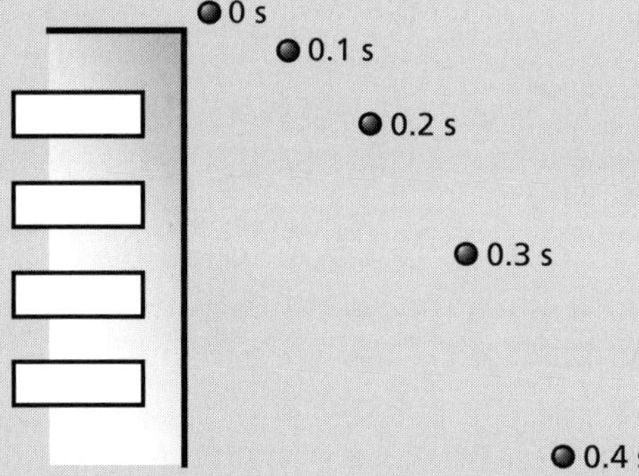

19.b)

Diagram should look similar to the one shown below, showing all equal length, horizontal arrows.

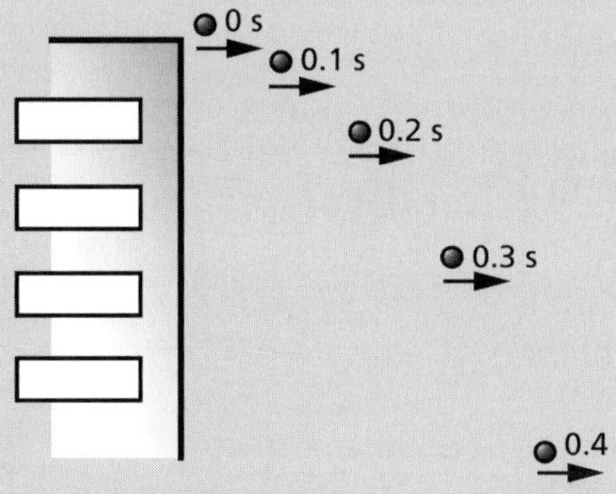

19.c)

Diagram should look similar to the one at right with all equal length vertical arrows.

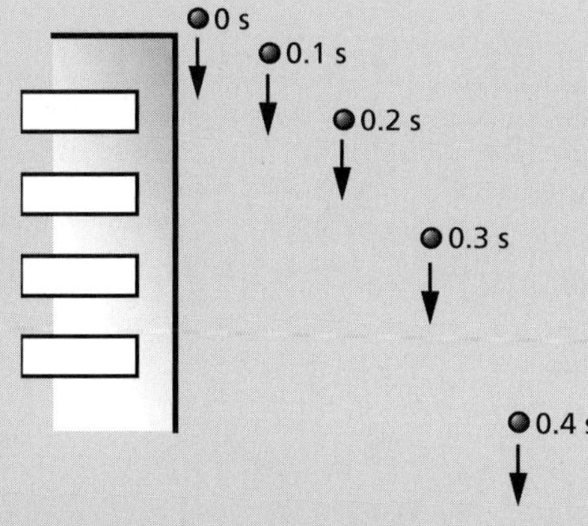

19.d)

Diagram should look similar to the one at right with arrows increasing uniformly in length.

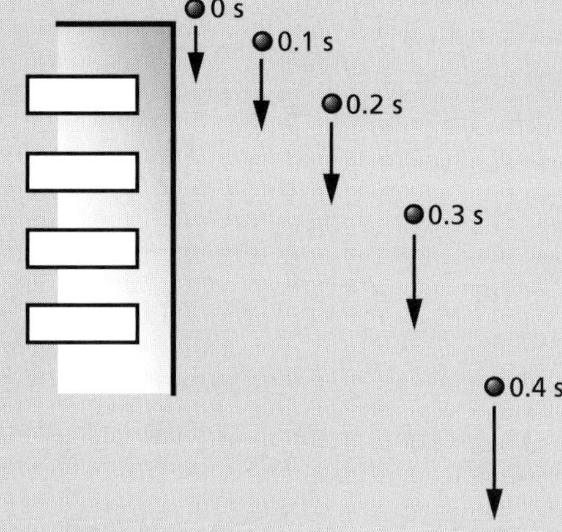

20.a)

Astronaut's weight on Earth,
$W = mg_{\text{earth}} = (60\text{ kg})(9.8\text{ m/s}^2) =$
590 N

20.b)

Astronaut's weight on the Moon,
$W = mg_{\text{moon}} = (60\text{ kg})(1.6\text{ m/s}^2) =$
96 N

20.c)

Astronaut's mass on the Moon is still 60 kg.

20.d)

Newton's second law says the acceleration of an object is equal to the net force applied divided by the object's mass. To make a falling object accelerate at the acceleration of gravity, the force the planet attracts the mass with must be equal to the weight.

21. Active Physics *Plus*

The two forces acting vertically (toward the top or bottom of the page) are each 10 N and cancel each other. The two forces acting in horizontally yield a net force of 15 N. Using the formula $\Sigma F = ma$ and assuming that acceleration to the right is positive gives $35\text{ N} - 20\text{ N} = (10\text{ kg})(a)$ or $a = 1.5\text{ m/s}^2$.

22. Active Physics *Plus*

The time for the ball to fall back to the ground is the same time as required to reach the peak. Because horizontal velocity has no effect on time of fall, use $d = \frac{1}{2}at^2$ to find out how long it takes to fall from the peak height of 5 m. Therefore, $5\text{ m} = \frac{1}{2}(10\text{ m/s}^2)(t^2)$ or $t = 1$ s. The total time of flight is 2 s. The horizontal distance traveled, $d_x = v_x t = (15\text{ m/s})(2\text{ s}) = 30\text{ m}$.

23. Active Physics *Plus*

Using conservation of energy, the *GPE* at the top of the ramp must be equal to the sum of the *GPE* and the *KE*. So, $GPE_{\text{peak}} = GPE_{\text{jump}} + KE_{\text{jump}}$. Solving for the velocity gives $mgh_{\text{peak}} = mgh_{\text{jump}} + \frac{1}{2}mv^2_{\text{jump}}$. After canceling the mass, the equation becomes $gh_{\text{peak}} = gh_{\text{jump}} + \frac{1}{2}v^2_{\text{jump}}$. Substituting in the values and solving for v, gives
$(10\text{ m/s}^2)(10\text{ m}) =$
$(10\text{ m/s}^2)(5\text{ m}) + \frac{1}{2}(v^2)$ or
$\frac{1}{2}v^2 = 50\text{ m}^2/\text{s}^2$ or $v = 10\text{ m/s}$.

Sample Assessment Rubric

Introduction

Development of rubrics as a template for identifying performance criteria has been shown to increase student achievement. A typical rubric clearly denotes each category to be evaluated and provides specific, required criteria for defining excellence, proficiency, and below-proficiency levels of performance. The sample rubrics for each chapter are intended to serve as guidelines. It should be understood that assessment is more effective when students and teachers tailor it to fit their needs. You are encouraged to work with your colleagues and especially with students to customize the rubric and the criteria. Decisions should be made together with respect to the curricular goals of the project within the particular context. For example, a class may choose to add one requirement in lieu of another, or to change the relative weighting of categories. It is helpful to remember the following recommendations:

1. Assessment should directly address the goals of the *Chapter Challenge*.

   Attention has been paid to the suggested rubrics in addressing the goals of the chapter, and the *Physics You Learned* section should serve as a guide for students and teachers working with the challenge. You may choose to make changes to the rubric in order to emphasize goals important to their context.

2. Students should participate in the assessment of their own performance.

   Students submit their rubric along with the grade they have given themselves. This not only encourages students to take ownership of the project, but it becomes a useful assessment tool for the teacher. If a student earns a "C" and gives himself or herself a "C," the conversation is very different than if a student were to earn a "C" and give himself or herself an "A." The question you might have for the first student is, "Why didn't you choose to do more?" While you might need to review the criteria with the second student and help understand what it takes to get an "A." After the teacher has graded the assignment, students have an opportunity to revise their work, and resubmit it for the "revision" grade. Emphasis should be placed not only on the finished project, but on progress with the rubric during revision.

3. Assessment should begin from a foundation of a proficient level of performance, providing ladders for students to achieve higher orders of thinking.

Finally, the rubric is built from a foundation of proficiency (meets standards). An analogy for this level is that in the real world their is a minimum acceptable standard for performance. A CD must play without skipping, and a shirt must have all of its buttons. Anything less, is substandard. This rubric works the same way. To get a "C" or better, the work must meet all the standards, and fall into the "proficient" category. Work not meeting all standards must be revised and resubmitted. Beyond proficiency, students can do work which shows mastery and may therefore earn a "good" or "excellent" rating (a "B" or an "A"). The scoring column of the rubric includes suggested point ranges for each level of mastery.

For further discussion, see: "Assessment of Laboratory Investigations," Eisenkraft, Arthur and Anthes-Washburn, Matthew. Assessment: Research and Practical Approaches, eds. Coffey, Douglas and Stearns. NSTA Press 2008.

Guide to the Sample Assessment Rubric

Assessment via this rubric will assign students to one of three major groups:

Excellent: Work meets all standards and demonstrates extensive evidence of mastery.

Good: Work meets all standards and demonstrates moderate evidence of mastery.

Proficient: Work meets standards without further evidence of mastery.

Please note that these groups are written at the top of the rubric page as a reminder to students.

In the table, there are three main groups of criteria—Mastery, Meets Standards, and Interventions. A student or team of students should achieve all of the criteria in order to satisfactorily complete the project. Anything less than this will require that the student make another attempt using the Interventions listed in the last column. This is the foundation, or floor of expectations. As teachers, we have to beware that our floor of expectations does not become a ceiling for some students.

1. In the first column, there are suggestions for demonstrating mastery. Completing one or more of these may raise a student or team from Proficient, to Good or Excellent.
2. In the second column, the criteria to meet the standards for the assignment are listed.
3. Some students may have trouble meeting the standards in the Meeting Standards column. The last column provides Interventions, or suggestions for how a student might meet the requirements of the project.
4. In the Scoring column, students submit their own grade, and you respond with a grade and feedback. Students receive a final grade after a revision is submitted. The range of scores for Excellent, Good, and Proficient allow you to assign points that match a student or team's demonstrated mastery. Thus, a student who barely meets standards can receive a different score than one who shows a higher level of mastery.

Implementing the Sample Assessment Rubric

- Modify the rubric with discussions from students.
- Hand out the rubric.
- Review the rubric so that you are confident that students understand each component.
- Have students complete the Scoring column for their work in the chapter by placing checks in each of the boxes. Have students assign themselves a point value for each component.
- Have students total their score for the rubric.
- Collect the student self-appraisal of their work.
- Use the rubric to grade the student work.
- Grade students' work after you and the student agree on the grade. Encourage the student to improve their work for the next chapter. If you and a student disagree, have an appropriate conversation with the student about his or her work and how it could be improved.

The *Sample Assessment Rubric* on the following page is provided as a *Blackline Master* on your *Teacher Resources CD*.

2c **Blackline Master**

Sample Assessment Rubric

Mastery (Students may show mastery through these or other ideas provided by students and teachers.)	**Meets Standards**	**Scoring** (To be discussed by students and teacher)	**Interventions** (Guiding questions and instructions for students falling short of the Standards)
• Support the explanation of physics principles with a "chalk talk." Make quantitative analysis of the sport using reasonable values and formulas learned in the chapter. • Create a "mashup" of your video, adding commentary and analysis of physics with voiceover, graphics, and notes. Use video editing software or social media site, such as voicethread.com. (Teacher approval required.)	**1. Physics Principles** • Explain at least four physics principles from "Physics You Learned," in the context of your sport. • Use scientific vocabulary consistently and precisely. • Use appropriate scientific symbols for units of measurement.	**Maximum:** 50 Points ***Excellent:** 45–50 **Good:** 40–44 **Proficient:** 35–39 *Student Self Grade:* *Teacher Grade:* *Revision:*	• Why do objects keep moving without being pushed or pulled? • How can we measure all kinds of motion? • How do objects speed up, slow down or change direction? • Why can't we see forces? How do we know when they are acting? • Where does energy come from and where does it go? • Do frictional forces help the player, hinder her or both?
• Present your voice-over "live" within the action of the sport – no pausing the video! • Produce a "demo tape" of your work that might be presented in an audition for a sports announcer position.	**2. Quality of the Presentation** • Connect explanation of physics principles to the action of the sport. • Prepare and practice your presentation with the team. • Cooperate with your team to ensure that all members participate. • Keep your presentation within the agreed-upon time limit. (Recommended—5 minutes) • Answer questions presented by the audience.	**Maximum:** 25 Points ***Excellent**: 23–25 **Good:** 20–22 **Proficient:** 18–19 *Student Self Grade:* *Teacher Grade:* *Revision:*	• First, write out your script of the voiceover. • Then, practice with your team.
• Prepare the presentation for sharing via a poster, Web page, or other medium.	**3. Quality of Written Script** • Organize the report so it is easy to follow and understand. • Use correct sentence structure. • Use correct spelling, punctuation, and grammar. • Use the correct number of pages (determined by class and teacher). Suggested–2-3 pages double spaced.	**Maximum:** 25 Points ***Excellent:** 23–25 **Good:** 20–22 **Proficient:** 18–19 *Student Self Grade:* *Teacher Grade:* *Revision:*	• Follow the suggestions of your teacher and submit a revised script.
		TOTAL: ***Excellent:** 90–100 **Good:** 80–89 **Proficient:** 70-79	

* **Excellent:** Work meets all standards and demonstrates rich evidence of mastery.
Good: Work meets all standards and demonstrates moderate evidence of mastery.
Proficient: Work meets standards without further evidence of mastery.

Chapter 3

SAFETY

CHAPTER 3

Safety

Chapter Overview

Chapter Challenge

The *Chapter Challenge* provides motivation for students to demonstrate their understanding of the physics content presented in this chapter. Students are asked to design and build a prototype safety system that protects an egg in a moving cart during a collision. They are required to provide an oral presentation and a written report in which they describe the physics concepts involved in the safety system that protects automobile, motorcycle, bicycle, or train passengers during a collision. Students present and test their prototypes in class. The class determines the type of collision an egg-carrying cart will undergo after students build their prototypes. For example, they may have the cart crash into another cart or a stationary object.

Students may find the challenge difficult at first, but by the end of the chapter students should have enough understanding of the concepts of force, the work-energy theorem, impulse, momentum, and conservation of momentum to be successful in completing the challenge. Toward the end of each section, students are asked to reflect on the section and consider its meaning for the challenge. At the end of most *Physics to Go* questions is a question designed to guide students in applying the content of that section to the challenge. The *Inquiring Further* is also geared toward helping students meet the challenge.

As you review the assignments, reassure students that while they may feel unprepared now, by the end of the chapter, they should have the necessary knowledge, skills, and vocabulary to respond adequately. To facilitate cooperative work in groups, have each student take individual responsibility for different tasks that make up the challenge.

The criteria for the challenge and a rubric for assessing student performance should be determined with the class based on the *Criteria for Success* in the student text. In your grading criteria, include factors such as the physics principles applied and discussed correctly, the proper use of physics terms, clarity of expression, possible improvements to their design, consumer acceptance and market potential, and credibility.

Chapter Summary

The students investigate and apply ideas involving force, energy, work, momentum, and conservation of momentum as they design and build a prototype of a safety feature to protect an egg in a colliding cart. Students

- identify, describe, and compare features of crash safety designs.
- describe, explain, and apply Newton's laws of motion.
- describe and apply the relationship between force, pressure, and area.
- describe and apply the work-energy theorem to determine force, energy, initial speeds, and stopping distances during collisions.
- describe the collisions that occur during a car crash and how they can affect the human body.
- define momentum, elastic collisions, and impulse.
- analyze collisions in one dimension using the law of conservation of momentum.
- apply Newton's laws, the work-energy theorem, the relationship between impulse and momentum, and conservation of momentum to design and build a "crumple zone" that absorbs some of the energy of a vehicle during a collision.
- apply Newton's laws, the work-energy theorem, the relationship between impulse and momentum, and conservation of momentum to design and build a safety device to directly reduce the net force acting on a person in a vehicle during a collision.

Key Physics Concepts

Section Summaries	Physics Principles
Section 1 Accidents Students identify and evaluate safety features in automobiles. Students then consider what safety features they could use for various vehicles and for their design of a safety system.	**Identifying criteria for building a safety feature**
Section 2 Newton's First Law of Motion: Life and Death before and after Seat Belts Students explain what occurs to passengers during a collision using Newton's first law. They read about the concept of pressure and apply this concept while designing and testing a seat belt to safely secure a clay passenger in a cart undergoing a collision.	**Newton's first law** **Pressure**
Section 3 Energy and Work: Why Air Bags? Students investigate and observe how spreading the force of an impact over a greater distance reduces the amount of damage done to an egg during a collision. They describe and explain their observations using the work-energy theorem.	**Average velocity** **Newton's second law** **Work** **Kinetic energy** **Work-energy theorem**
Section 4 Newton's Second Law of Motion: The Rear-End Collision Students explore the effects of rear-end collisions on passengers, focusing on whiplash. They use Newton's laws to describe how whiplash occurs. They also describe, analyze, and explain situations involving collisions using Newton's first and second laws.	**Newton's first law** **Newton's second law**
Section 5 Momentum: Concentrating on Collisions After observing various collisions, students are introduced to the concept of momentum. Through measurements taken during various collisions, they determine the mass of a cart. Students then calculate and consider the momentum of various objects.	**Linear motion** **Momentum**
Section 6 Conservation of Momentum Students investigate the law of conservation of momentum by measuring the masses and velocities of objects before and after collisions. Students then analyze various collisions by applying the law of conservation of momentum.	**Newton's second law** **Newton's third law** **Momentum** **Law of conservation of momentum**
Section 7 Impulse and Changes in Momentum: Crumple Zone Students design a device on the outside of a cart to absorb energy during a collision to assist in reducing the net force acting on passengers inside the vehicle. Students use probes to measure the velocity of the vehicle and the force acting on the vehicle during impact, and then describe the relationship between impulse ($F\Delta t$) and change in momentum ($m\Delta v$).	**Newton's second law** **Impulse** **Momentum** **Work-energy theorem**

Chapter Concept Map

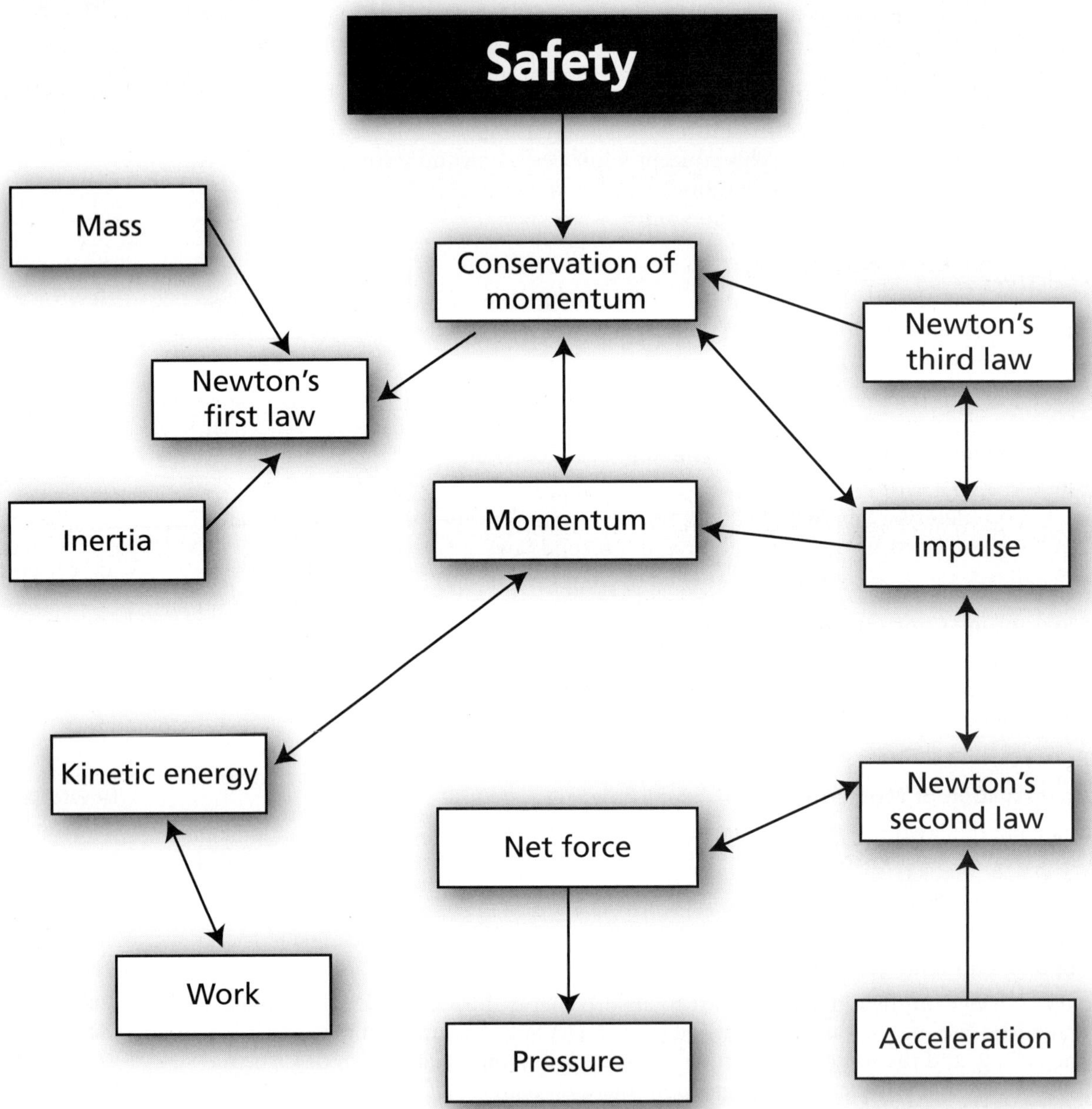

Understanding by Design*

The *Understanding by Design* template focuses on the three stages of backward design:

- **Identify desired results**
- **Determine acceptable evidence**
- **Plan learning experiences**

What overarching (enduring) understandings are desired?

Not all accidents can be prevented but you can ensure that an accident is less severe for a driver, a passenger and a pedestrian.

- Physical laws can describe and predict what will occur during an accident.
- Safety measures based on physics principles can protect you during an accident.
- Seat belts, air bags, headrests and other technologies are based on physics principles.
- Accidents can be analyzed by using physics principles.

What will students understand as a result of this chapter?

- Many safety features have been added to automobiles such as seat belts, head restraints, air bags, crumple zones, etc.
- Newton's first law states that an object in motion will stay in motion unless acted upon by a force. If your vehicle stops suddenly, you continue to travel forward.
- Pressure is defined as the force per unit area. A seat belt made from a wire could slice through your body, while a seat belt of wide cloth will stop you without much pressure.
- A moving object has kinetic energy $KE = \frac{1}{2}mv^2$. To change that KE, there must be work (a force applied over a distance, $W = Fd$). To stop someone during an accident requires a certain amount of work. The person can be stopped with a large force over a very short distance or a small force over a larger distance. Cushioned dashboards and air bags are used to lengthen the distance over which the smaller force can bring you to a halt.
- A driver hit from behind will experience whiplash. This is where the head moves back and then snaps forward. Whiplash can be explained using Newton's first law and Newton's second laws ($F = ma$).
- When two moving vehicles collide, the change in motion depends upon the relative masses of the vehicles. This is described in terms of the momentum of the vehicles, $p = mv$.
- In all collisions, the total momentum before the collision is equal to the total momentum after the collision. This law of the conservation of momentum is true of collisions from planets to microscopic particles.
- The total momentum in a collision is conserved but the momentum of each vehicle can change. The change in momentum is equal to the force applied over a given time $\Delta(mv) = F\Delta t$.

What are the overarching "essential" questions?

- How do Newton's laws relate to automobile safety?
- How do seat belts, air bags, headrests, collapsible steering columns and other safety devices protect you?
- How can people reconstruct an accident after it has taken place to determine liability?
- What is the law of conservation of momentum?
- What physics principles would you take into account if you were designing a safety device for a bicycle or motorcycle?

What "essential" questions will focus this chapter?

- What safety features exist in automobiles?
- How do safety features in vehicles today improve upon those in cars from 30 years ago?
- How do seat belts protect us during a car accident?
- How does the thickness of the seat belt affect its utility?
- Why does an egg not break when it lands on a padded surface but does break when it lands on concrete?
- How is work related to the energy during a collision?
- What is whiplash? How can the events during whiplash be explained using Newton's first and second laws?
- How does the size of a vehicle impact its damage during a collision?
- What laws of physics are used by traffic-accident investigators to reconstruct an accident?
- How can "cushioning" decrease the severity of an accident?

* Grant Wiggins and Jay McTighe, *Understanding by Design* (Merril/Prentice Hall, 1998), 181.

Pacing Guide

The *Pacing Guide* below is designed so that you have the option to complete the first eight chapters of *Active Physics* during the school year. The *Plan A Pacing Guide* allows the students to complete all the *Investigates*. If you are a new teacher, or unfamiliar with the program, you may have difficulty adhering to *Pacing Guide A*. *Pacing Guide B* suggests places where either time or equipment may be saved if it becomes necessary to complete the chapter in the allotted time. To reach this goal, many of the investigations are whole-class *Investigates* rather than small-group *Investigates*. This will save time and require less equipment than the optimal inquiry-based instruction that the curriculum is intended to provide. In order to choose which plan is best for you, please consult the *Implementation Chart* following this guide.

Note: Each "day" assumes a 45-minute class period, or one half of a 90-minute block.

Day	Plan A (small-group *Investigates*)	Homework (for Plan A and Plan B)	Day	Plan B (combination of whole-class and small-group *Investigates*)	Plan B Equipment Reduction
1	*Scenario, Your Challenge, Criteria for Success,* Scoring Rubric. ***Section 1*** Answer *What Do You See? What Do You Think?*	Look over local papers from the previous week and find any articles available of automobile crashes or safety to report to the class.	1	**See Plan A.**	
2	Carry out the *Investigate* with a class discussion of what constitutes a safety device. Discuss *Physics Talk* and how any of the safety devices listed might have helped in the accidents the students discovered in their homework. Answer *What Do You Think Now?* and *Reflecting on the Section and the Challenge.*	Read the *Physics Talk* and answer the *Checking Up* questions. Answer *Physics to Go* Questions 1-5.	2	**See Plan A.**	
3	Review *Checking Up* questions and *Physics to Go.* ***Section 2*** Answer *What Do You See? What Do You Think?* Students should conduct *Investigate* Parts A and B.	Read the *Physics Talk* and answer the *Checking Up* questions.	3	**See Plan A.**	
4	Discuss *Section 2 Physics Talk* and review *Checking Up* questions. Complete *What Do You Think Now?* and *Reflecting on the Section and the Challenge.* If possible, show a portion of the PBS video from the Nova series "Escape" on car crashes, or some of the safety and car video online to illustrate the problems associated with car crashes.	Complete *Physics to Go* Questions 1-4, 6-8 and 10.	4	**See Plan A.**	

Day	Plan A (small-group *Investigates*)	Homework (for Plan A and Plan B)	Day	Plan B (combination of whole-class and small-group *Investigates*)	Plan B Equipment Reduction
5	Discuss *Physics to Go*. **Section 3** Answer *What Do You See?* and *What Do You Think?* Students conduct *Investigate* Steps 1-7. Complete *Investigate* Step 8 as a demonstration using the students to throw the egg.	Read the *Physics Talk* sections on Energy and Work, Kinetic Energy and Work and Change in Kinetic Energy. Answer *Checking Up* Questions 1 and 2 and *Physics to Go* Questions 1-4.	5	**See Plan A.**	
6	Review results of previous day's *Investigate*. Review the *Checking Up* and *Physics to Go* questions. Discuss the *Physics Talk* on kinetic energy and work, and speed, kinetic energy and stopping distance. Review the numerical examples in the *Physics Talk*, and show why energy and work are expressed in the same units from the equations.	Read the remainder of the *Physics Talk* and answer *Checking Up* Questions 3 and 4 and *Physics to Go* Questions 5-8 and 10.	6	**See Plan A.**	
7	Review *Physics to Go* homework. Students complete the *Reflecting on the Section and the Challenge* and record their answers in their logs. **Section 4** Answer *What Do You See?* and *What Do You Think?* Students complete *Investigate* Steps 1-7.	Read and summarize the *Physics Talk* in your logs. Answer *Checking Up* Questions 1-3.	7	Review *Physics to Go* homework. Students complete *Reflecting on the Section and the Challenge* and note their answers in their logs. **Section 4** Answer *What Do You See?* and *What Do You Think?* Teacher does *Investigate* Steps 1-9 as a demonstration.	Requires two dynamics carts, one ramp, one pound of modeling clay, ringstand, crossarm, right angle clamp, scissors and file folders, masking tape, meterstick, weights and wire.
8	Review results of *Section 4 Investigate* Steps 1-7. Students complete *Investigate* Sections 8-9. Review the *Physics Talk* discussing how Newton's first and second laws influence whiplash. Answer *What Do You Think Now?* and *Reflecting on the Section and the Challenge*.	Answer *Physics to Go* Questions 1-7.	8	Review the *Physics Talk* discussing how Newton's first and second laws influence whiplash. Answer *What Do You Think Now?* and *Reflecting on the Section and the Challenge*. Discuss *Physics to Go* problems. Students begin working on the *Mini-Challenge* in their groups. Students will need to decide on the safety device they wish to construct and the type of vehicle that will have the safety device.	

Pacing Guide *(continued)*

Day	Plan A (small-group *Investigates*)	Homework (for Plan A and Plan B)	Day	Plan B (combination of whole-class and small-group *Investigates*)	Plan B Equipment Reduction
9	Discuss *Physics to Go* questions. As a review, have the students do the Whiplash quiz (*Physics to Go* Question 8). Students start work on the *Mini-Challenge* in their groups. Students will need to decide on the safety device they wish to construct and the type of vehicle that will have the safety device.	Conduct an Internet or other media search for any information on a similar safety device to yours already in use.	8	**See Plan A.**	
10	Students work in groups and compare notes on safety device search. Students discuss what is needed for a consensus first draft. Review the first draft, and aspects of the oral presentation the students will make.	Prepare a first draft of the written report to share with your group.	9	**See Plan A.**	
11	Student presentations of the *Mini-Challenge*. **Section 5** Answer *What Do You See?* and *What Do You Think?*	Read the *Physics Talk* and write the answers to the *Checking Up* questions in your log books.	10	**See Plan A.**	
12	Students conduct the Investigate. Discuss the *Physics Talk* and review the *Checking Up* questions. Ask students *What Do You Think Now?* Students discuss with their groups *Reflecting on the Section and the Challenge* and record their answers in their logs.	Answer *Physics to Go* Questions 1–6 and 7 (*Preparing for the Chapter Challenge*).	11	**See Plan A.**	
13	Review the *Physics to Go* questions from the previous day. **Section 6** Answer *What Do You See?* and *What Do You Think?* Students complete the *Investigate* Steps 1-3.	Read the *Physics Talk* and summarize in your logs, then record your answers to the *Checking Up* questions.	12	Review the *Physics to Go* questions from the previous day. **Section 6** Answer *What Do You See?* and *What Do You Think?* Teacher conducts *Investigate* all steps as a class demonstration. Review the *Physics Talk* and discuss the numerical examples. Discuss the *Checking Up* questions.	Requires only two dynamics carts, 1 mass set, 1 velocity meter setup (stopwatch, ruler, tickertape timer, velocimeter or computer and motion detector), scale, masking tape and modeling clay.
14	Review the results of previous day's *Investigate*. Students conduct *Investigate* for collisions 4–6. Review the *Physics Talk* and discuss the numerical examples. Discuss the *Checking Up* questions.	Answer *Physics to Go* Questions 1–8.			

Day	Plan A (small-group *Investigates*)	Homework (for Plan A and Plan B)	Day	Plan B (combination of whole-class and small-group *Investigates*)	Plan B Equipment Reduction
15	Discuss previous day's *Physics to Go*. Discuss *Physics to Go* Question 11 with the class. Students should discuss *Preparing for the Chapter Challenge* with their groups. Answer *What Do You Think Now?* and *Reflecting on the Section and the Challenge.*	Answer *Physics to Go* Questions 9, 10, 13, 14.	**13**	**See Plan A.**	
16	Review the *Physics to Go* homework. ***Section 7*** Answer *What Do You See?* and *What Do You Think?* Students conduct Part A of the *Investigate*.	Read *Physics Talk* sections up to "Work and Energy" or "Impulse and Momentum" and summarize in your logs, and answer the *Checking Up* questions.	**14**	Review the *Physics to Go* homework. ***Section 7*** Answer *What Do You See?* and *What Do You Think?* Students conduct Part A of the *Investigate*, while you conduct Part B as a class demonstration. Start a discussion of the *Physics Talk*. Review the *Checking Up* questions.	Only requires one dynamics cart, computer interface and probes to measure the force and velocity, three sets of cushioning materials, ramp, ringstand, crossarm, right angle clamp, masking tape and index cards.
17	Review the results of Part A. Students conduct Part B of the *Investigate*. Start a discussion of the *Physics Talk*. Review the *Checking Up* questions.	Read the *Physics Talk* from "Work and Energy or Impulse and Momentum" to end. Students complete *Physics to Go* Questions 1–5.			
18	Finish discussion of the *Physics Talk*. Review the *Physics to Go* questions.	Answer *Physics to Go* Questions 6–10 and *Preparing for the Chapter Challenge* (Question 11).	**15**	**See Plan A.**	
19	Review the previous night's *Physics to Go*, and *Reflecting on the Section and the Challenge*. Show the students the materials available for their design of the safety system for the challenge. Students start work on the *Chapter Challenge*, designing and building the safety system.	Work on the *Chapter Challenge* written and oral report.	**16**	**See Plan A.**	
20	Students continue to work on the *Chapter Challenge*, comparing written reports for a consensus report, and completing construction and testing of safety device.	Prepare for class presentations of the *Chapter Challenge*.	**17**	**See Plan A.**	
21	*Chapter Challenge* presentations.	Study for *Physics Practice Test.*	**18**	**See Plan A.**	
22	*Physics Practice Test.*		**19**	**See Plan A.**	

Implementation Chart

Hopefully, as you become more experienced and comfortable with the curriculum, you will shift to small-group *Investigates*. Accordingly, below is an *Implementation Chart* that suggests a three-year timetable to expand the student's role in the chapter by having them do more of the *Investigates*. Although this will require a slightly greater expenditure of time and more equipment, the benefits to the student will be manifest. Eventually, your goal should be to have the students complete almost all the investigations rather than you having to provide the maximum opportunity for inquiry.

	Section 1 Investigate	Section 2 Investigate	Section 3 Investigate	Section 4 Investigate	Section 5 Investigate	Section 6 Investigate	Section 7 Investigate
Year 1	Small-group	Small-group	Small-group	Whole-class	Small-group	Whole-class	Whole-class
Year 2	Small-group	Small-group	Small-group	Small-group	Small-group	Small-group	Whole-class
Year 3	Small-group	Small-group	Small-group	Small-group	Small-group	Small-group	Small-group

Chapter 3 Materials and Equipment

The following tables contain lists of materials and equipment needed to complete all the experiments. The tables are organized as follows:

- Multimedia Needed
- Durables
- Consumables
- Additional Items Needed Not Supplied

Multimedia contains all the software and content videos that are necessary to complete the sections. All the multimedia items needed will be included in the Multimedia DVD/CD Set. In addition to the Multimedia DVD/CD Set, other multimedia items are available such as the *ExamView*™ Test Generator and the Constructing Physics Understanding simulations.

Durables are items which are not consumed during an experiment. They can be used several times. **Consumables** are items that are used up during each class and must be resupplied for future classes. Both the durables and consumables are broken down by group and by class. A group consists of four students. While the group size will be determined by the teacher based upon logistics and availability of equipment, the information is based upon recommended group size of four students. Items listed per class are based on a class size of 40 students.

Additional Items Needed Not Supplied consists of items that are needed to complete the sections, but are not supplied by **It's About Time**®.

The first column gives the section number(s) in which each item will be used. The second column provides information needed for ordering the particular item. The third column gives the item description. The fourth and fifth columns will indicate quantity, based on either group or class needs. The last column provides a space for you to fill in your total item quantity. You will notice certain rows are shaded and have an asterisk in the last column. This is to indicate a duplicate item listing, based on the needs of different sections. You will not need to figure the items in the shaded rows into your total items needed, as they have already been accounted for in the corresponding non-shaded row.

The *Materials and Equipment* lists are broken down into either Plan A or Plan B corresponding to the *Pacing Guide*.

PLAN A

Section	ISBN	Multimedia	Group (4 Students)	Class	Total Items Needed
2	978-1-60720-041-3	Multimedia DVD/CD Set		1 per classroom	

Section	Item No.	Durables	Group (4 Students)	Class	Total Items Needed
2	RS-6719	Ribbon, 12 in. (length), varying widths	3 per group		
2	WS-6731	Wire, #22, bare copper, ft	1 per group		
2, 4, 5, 6, 7	GC-0001	Dynamics cart	2 per group		
2, 4, 6	MC-5418	Clay, modeling, lb	2 per group		
2, 4, 7	SR-6737	Inclined plane for lab cart	1 per group		
2, 4, 7	SH-7212	Large ring stand	1 per group		
2, 4, 7	ST-0002	Rod, aluminum, 12 in. (length) x 3/8 in. (diameter) (to act as cross arm)	1 per group		
2, 4, 7	HO-0002	Holder, right angle, cast iron	1 per group		
2, 4, 7	SS-1281	Scissors	1 per group		
3	PE-5600	Petri dish, sterile, disposable	1 per group		
3, 6, 7	RS-2826	Ruler, metric, in/cm	1 per group		
4, 7	SM-7616	Meter stick	1 per group		
4, 5, 6	WS-0376	Weight, slotted, 100 g	6 per group		
6	SS-7779	Stopwatch	1 per group		
7	TT-6100	AC Ticker tape timer	1 per group		
7	WS-6732	Wood piece, 1 in. x 2 in. x 2 in.	1 per group		
7	VE-0012	Self-adhesive Velcro dot (part of variety of cushioning material set)	1 per group		
7	CO-0012	Cork/cardboard piece, 1/2 in. x 1 in. (part of variety of cushioning material set)	6 per group		
7	TA-0016	Double-sided tape (to adhere pieces, part of variety of cushioning material set)	6 per group		
5, 6	EL-4181	Scale, electronic, 0-1500 g, 0.01 g readability		1 per class	
3	PA-7614	Paper clips, pkg 100		1 per class	
Section	**Item No.**	**Consumables**	**Group (4 Students)**	**Class**	**Total Items Needed**
2, 4, 7	FF-7704	File folders	3 per group		
2, 4, 5, 6, 7	TS-2662	Tape, masking		6 per group	
3	WS-7546	Plastic wrap, roll		1 per class	
3	FL-0007	Flour, 5 lb		1 per class	
4	MW-2460	Wire, magnet, #24, 60 ft roll		1 per class	
7	SS-7722	Ball of string		1 per class	
7	RB-6108	Rubber bands, #64, pack		1 per class	

**Please note duplicate item.*

CHAPTER 3

Section	Item No.	Durables	Group (4 Students)	Class	Total Items Needed
7	CS-5011	Unlined index card, 3 in. x 5 in., pkg 100		1 per class	
7	GS-7706	Graph paper, pkg of 50		1 per class	
Mini-Challenge	EP-1100	Easel pad		1 per class	
Mini-Challenge	CM-0008	Markers		6 per class	
Section	**Item No.**	**Additional Items Needed Not Supplied**	**Group (4 Students)**	**Class**	**Total Items Needed**
2		Concrete block or similar barrier (wall)		1 per class	
3		Raw egg	2 per group		
3		Large bed sheet		1 per class	
3, 5, 6, 7		Access to a smooth level surface	1 per group		
6		Photogate probeware		1 per class	
6, 7		Motion detection probeware		1 per class	
6, 7		MBL or CBL technology to record probeware activity		1 per class	
7		Stack of books	1 per group		
7		Force probeware		1 per class	
5		Spring or loop of thin metal for dynamics cart	1 per group		
7		Wood block, 4 cm x 4 cm x 3 cm	1 per group		

**Please note duplicate item.*

NOTES

PLAN B					
Section	**ISBN**	**Multimedia**	**Group (4 Students)**	**Class**	**Total Items Needed**
2	978-1-60720-041-3	Multimedia DVS/CD set		1 per classroom	

Section	Item No.	Durables	Group (4 Students)	Class	Total Items Needed
2	RS-6719	Ribbon, 12 in. (length), varying widths	3 per group		
2	WS-6731	Wire, #22, bare copper, ft	1 per group		
2	GC-0001	Dynamics cart	1 per group		*
5	GC-0001	Dynamics cart	2 per group		
4, 6	GC-0001	Dynamics cart		2 per class	*
7	GC-0001	Dynamics cart		1 per class	*
2	MC-5418	Clay, modeling, lb	2 per group		
4, 6	MC-5418	Clay, modeling, lb		2 per class	*
2	SR-6737	Inclined plane for lab cart	1 per group		
4, 7	SR-6737	Inclined plane for lab cart		1 per class	*
2	SH-7212	Large ring stand	1 per group		
4, 7	SH-7212	Large ring stand		1 per class	*
2	HO-0002	Holder, right angle, cast iron	1 per group		
4, 7	HO-0002	Holder, right angle, cast iron		1 per class	*
2	ST-0002	Rod, aluminum, 12 in. (length) x 3/8 in. (diameter) (to act as cross arm)	1 per group		
4, 7	ST-0002	Rod, aluminum, 12 in. (length) x 3/8 in. (diameter) (to act as cross arm)		1 per class	*
2, 7	SS-1281	Scissors	1 per group		
4,	SS-1281	Scissors		1 per class	*
3	PE-5600	Petri dish, sterile, disposable	1 per group		
3, 7	RS-2826	Ruler, metric, in/cm	1 per group		
6	RS-2826	Ruler, metric, in/cm		1 per class	*
4	SM-7616	Meter stick	1 per group		
7	SM-7616	Meter stick		1 per class	*
5	WS-0376	Weight, slotted, 100 g	6 per group		
4, 6	WS-0376	Weight, slotted, 100 g		6 per class	*
6	SS-7779	Stopwatch		1 per class	
6	TT-6100	AC ticker tape timer		1 per class	
7	WS-6732	Wood piece, 1 in. x 2 in. x 2 in.	1 per group		
7	VE-0012	Self-adhesive Velcro dot (part of variety of cushioning material set)		6 per class	
7	CO-0012	Cork/cardboard piece, 1/2 in. x 1 in. (part of variety of cushioning material set)		6 per class	

**Please note duplicate item.*

CHAPTER 3

Section	Item No.	Durables	Group (4 Students)	Class	Total Items Needed
7	TA-0016	Double-sided tape (to adhere pieces, part of variety of cushioning material set)		6 per class	
5, 6	EL-4181	Scale, electronic, 0-1500 g, 0.01 g readability		1 per class	
3	PA-7614	Paper clips, pkg 100		1 per class	
Section	**Item No.**	**Consumables**	**Group (4 Students)**	**Class**	**Total Items Needed**
2, 7	FF-7704	File folders	3 per group		
4	FF-7704	File folders		3 per class	
2, 4, 5, 6, 7	TS-2662	Tape, masking		6 per group	
3	WS-7546	Plastic wrap, roll		1 per class	
3	FL-0007	Flour, 5 lb		1 per class	
4	MW-2460	Wire, magnet, #24, 60 ft roll		1 per class	
7	SS-7722	Ball of string		1 per class	
7	RB-6108	Rubber bands, #64, pack		1 per class	
7	CS-5011	Unlined index card, 3 in. x 5 in., pkg 100		1 per class	
7	GS-7706	Graph paper, pkg of 50		1 per class	
Mini-Challenge	EP-1100	Easel pad		1 per class	
Mini-Challenge	CM-0008	Markers		6 per class	
Section	**Item No.**	**Additional Items Needed Not Supplied**	**Group (4 Students)**	**Class**	**Total Items Needed**
2		Concrete block or similar barrier (wall)		1 per class	
3		Raw egg	2 per group		
3		Large bed sheet		1 per class	
3, 5, 6, 7		Access to a smooth level surface	1 per group		
6		Photogate probeware		1 per class	
6, 7		Motion detection probeware		1 per class	
6, 7		MBL or CBL technology to record probeware activity		1 per class	
7		Stack of books	1 per group		
7		Force probeware		1 per class	
5		Spring or loop of thin metal for dynamics cart	1 per group		
7		Wood block, 4 cm x 4 cm x 3 cm	1 per group		

**Please note duplicate item.*

NOTES

Teacher Resources

Blackline Masters

Available in *Teacher Resources* and on *Color Overheads* and *BLMs* CD.

Chapter Supports

Title	Point of Use	Blackline Masters
Standard for Excellence	*Chapter Challenge* Introduction	3-a

Section Quizzes

Title	Point of Use	Blackline Masters
Section 1 Quiz	Section 1	3-1b
Section 2 Quiz	Section 2	3-2a
Section 3 Quiz	Section 3	3-3a
Section 4 Quiz	Section 4	3-4d
Section 5 Quiz	Section 5	3-5a
Section 6 Quiz	Section 6	3-6b
Section 7 Quiz	Section 7	3-7c

Section Supports

Title	Point of Use	Blackline Masters
Data Table Step 3a	Section 1 – *Investigate*	3-1a
Illustration of Clay Figure	Section 4 – *Investigate*	3-4a
Illustration of Carts	Section 4 – *Investigate*	3-4b
Templates	Section 4 – *Investigate*	3-4c
Collision Types 1-6	Section 6 – *Investigate*	3-6a
Example Graphs and Area Under Curve	Section 7 – *Investigate*	3-7a
Equation Box	Section 7 – *Physics Talk*	3-7b

Chapter Assessment

Title	Point of Use	Blackline Masters
Physics Practice Test	End-of-Chapter	3b
Sample Assessment Rubric	End-of-Chapter	3c

Color Overheads

Available on *Color Overheads* and *BLMs* CD.

Title	Point of Use
Physics Corner	*Chapter Challenge* Introduction
What Do You See?	Section 1
What Do You See?	Section 2
What Do You See?	Section 3
Stopping Distance	Section 3 – *Physics Talk*
What Do You See?	Section 4
What Do You See?	Section 5
What Do You See?	Section 6
Sample Problem 1	Section 6 – *Physics Talk*
What Do You See?	Section 7
Illustration of setup	Section 7 – *Investigate Step 2*

NOTES

Chapter Challenge

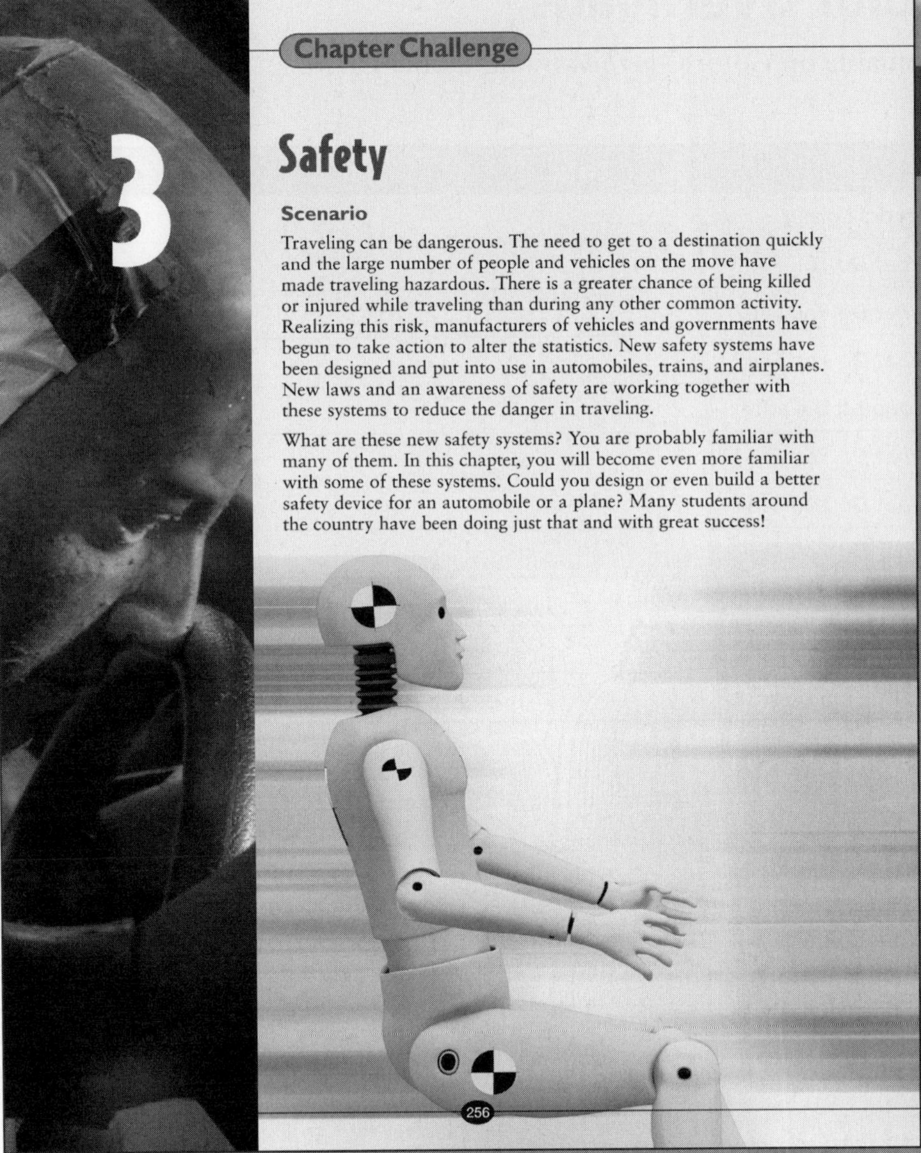

Scenario

Consider having students read the *Scenario* aloud. Ask them about the safety devices used in various vehicles. Have the class discuss ideas while categorizing these devices either on the board or on an overhead. Ask students how their knowledge of physics will help them build a safety system that will reduce the risk of injury or death in an accident. Encourage students to share their stories and experiences that relate to the topic.

Keep the *Scenario* focused on the main topic by prompting students to support their responses. You may want to invite a professional who can discuss the safety features in vehicles. Emphasize the importance of how the knowledge of physics will help improve the design of safety features.

Review the title of each section in the *Table of Contents,* and remind students that the content of each section corresponds to physics concepts they will need to apply to complete the challenge. Reassure the class that while they may not feel prepared now, by the end of the chapter they will have the skills and vocabulary to respond adequately. Suggest strategies to students to keep track of the information presented in each section, such as making a concept map during the chapter, outlining information, or keeping a record of concepts and examples they deem important in their *Active Physics* logs. Avoid providing students with too many examples of what could be done to complete the challenge, as this may limit their creativity and confidence in seeking their own solutions.

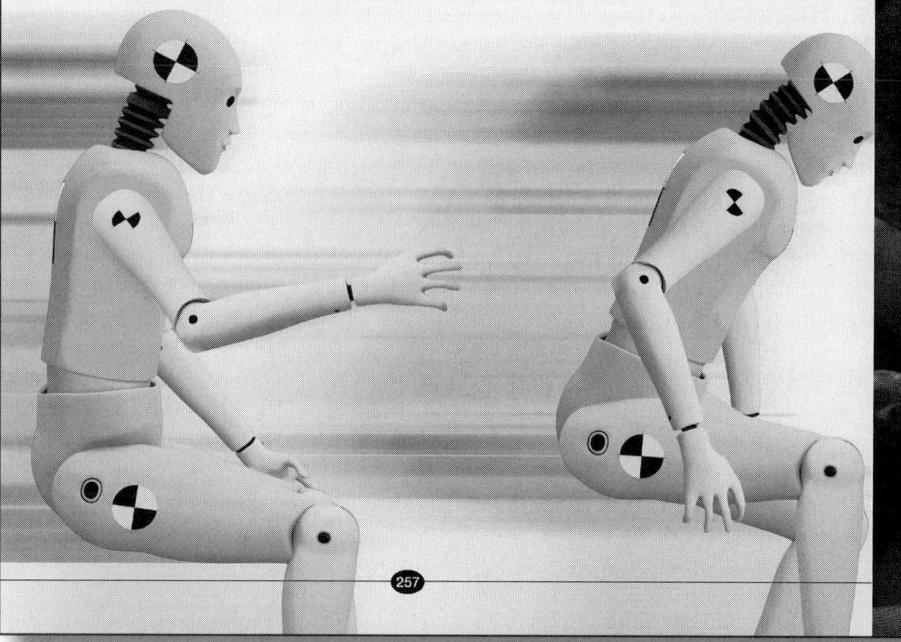

Your Challenge

Your design team will develop a safety system for protecting automobile, airplane, bicycle, motorcycle, or train passengers during a collision. To illustrate this safety system, you will design and build a prototype safety system to protect an egg in a moving cart that undergoes a collision.

When the design teams bring their final products to class, all teams will display their safety systems around the room. Each design team will give a five-minute oral report. Each design team will also be asked to submit a written and/or multimedia report.

Each safety system will also be tested in a collision. Your class as a whole will determine what kind of collision the egg-carrying cart will undergo. It could be a collision with another cart, or it could be a collision with a stationary object. Your class will also decide how fast the carts will be moving prior to the collision. As in real life, you will not be certain about the details of the collision that you are going to have to protect against when you design your system.

257

Your Challenge

Have a class discussion about the challenge and the expectations. Let the class know that each design team will build a prototype that will be tested in class, present an oral presentation to the class, and turn in a written report. Emphasize that all students should be able to explain the main physics concepts during their presentation.

Let the class know that each prototype will carry an egg as a passenger, and undergo a collision. The type of collision will be decided upon by the class, but not until after they have completed designing and building their prototypes. Emphasize that this means that students will be learning about collisions. If students have completed other *Active Physics* chapters, ask them what physics concepts from these chapters they think they can apply to the challenge.

Assure students that as the class progresses through the chapter, they will become more familiar with the content and will be able to connect the physics principles to their *Chapter Challenge*. Remind students that each new concept builds upon the others, providing tools to strengthen their presentation.

Criteria for Success

Have a class discussion about the criteria for evaluating the safety system, oral presentation, and written report. Develop a rubric with the class to use during this chapter. Consider starting the discussion by recording a list of important criteria for the *Chapter Challenge*. Make frequent references to the physics principles that are likely to be considered. The criteria should include accurate and clear explanations with original ideas. It is important to recognize each student's contribution during this discussion. This will facilitate student understanding on the assessment criteria. The criteria listed in the *Sample Criteria for Excellence* should be included, as well as any state standards and benchmarks. Consider using the Blackline Master of the *Sample Criteria for Excellence* while guiding the class discussion.

Chapter Challenge

Read aloud the criteria that the class has decided upon. This may bring up other interesting points that could be used to reinforce or modify the criteria. After the class has agreed upon each criterion, develop a point system with the class for grading the challenge. Consider asking students what details of each portion of the challenge are necessary for an A, B, C, D, and so on, or have the class decide how many points each part is worth. For example, you may want to have students assign fewer points to the aesthetic qualities of the report or presentation, and more points for explanations using physics principles. The assessment rubric should place the greatest emphasis on showing understanding of physics concepts, and should include the criteria in the student text. Make sure that students understand the criteria and the rubric before they begin their work.

Criteria for Success

You and your classmates will work with your teacher to define the criteria for evaluating your safety system, your presentation, and your written report. After discussion about what features should be included in the grading of each part of the project, you should then read some of the suggestions below. Then you will determine the relative importance of the assessment criteria. Next, assign point values to each. The following are suggestions of point values for each part.

Standard for Excellence	
1. The quality of your safety-system model or prototype, and the ability of the prototype to protect the egg during the collision	40 points
2. The quality of a five-minute oral report • the need for the system • the method used to develop the working model • the demonstration of the working model • the discussion of the physics concepts involved • the description of the next-generation version of the system • the answers to questions posed by the class	30 points
3. The quality of a written and/or multimedia report • the information from the oral report • the documentation of the sources of expert information • the discussion of consumer acceptance and market potential • the discussion of the physics concepts applied in the design of the safety system	30 points

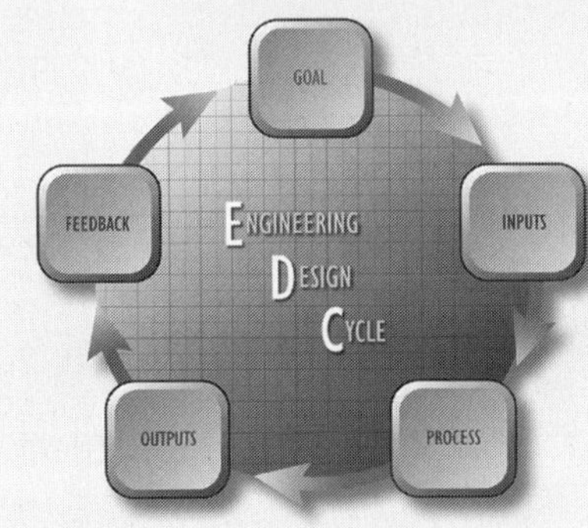

Engineering Design Cycle

You have now heard about the *Chapter Challenge* to design a new safety system for a vehicle of your choice. You will use a simplified *Engineering Design Cycle* to help your group create the best safety system you can with the materials that are available to you. Defining the *Goal* is the first step in the *Engineering Design Cycle*. Since you have already read the criteria for the challenge and considered some of the constraints, you have already begun.

You probably don't have a complete vision of how you will create your safety system yet, but the chapter sections will help you. As you experience each one of the sections you will be gaining *Inputs* to use in the design cycle. These *Inputs* will include new physics concepts, vocabulary, and even calculations that will help you to create a successful design.

258

Standard for Excellence

It is important for students to decide on the criteria for grading the *Chapter Challenge* themselves. Deciding the criteria will give them a voice in determining how their projects will be judged. You can help them arrive at a list of criteria by outlining those aspects of the *Chapter Challenge* that should be rated. Some criteria are listed in the *Standard for Excellence* table with suggested point values. An example of further describing and/or adding criteria and how the points could be distributed is provided in the student text. Remember the primary purpose of establishing the criteria necessary to earn an 'A' is to motivate students to succeed and to keep them focused.

As the *Chapter Challenge* approaches, you will want to develop a more comprehensive assessment rubric which will determine grading and expectations that meet your own evaluation system. A complete *Sample Assessment Rubric* for *Chapter 3* is located at the end of the chapter in this *Teacher's Edition*.

3-a Blackline Master

Engineering Design Cycle

Discuss with the class the *Engineering Design Cycle*. Consider using a projection of

During the *Process* step you will combine *Inputs* with your ideas, consider design criteria, compare and contrast potential solutions, and most importantly, make safety-system design decisions.

The first of your *Outputs* in your design cycle will be the safety-system concepts that your group presents to the class as part of the *Mini-Challenge*, including any models, diagrams, and charts you may use to clarify the information you present.

Finally, you will receive *Feedback* from your classmates and your teacher about what parts of your presentation are accurate and which parts need to be refined. You will repeat the design cycle during the second half of the chapter when you gain more *Inputs*, refine your safety system, and make your final safety system presentation.

Physics Corner

Physics in *Safety*

- Acceleration
- Change in momentum
- Collisions
- Effect of forces on motion
- Energy and work
- Force, pressure, and area
- Inertia
- Impulse
- Kinetic energy = $\frac{1}{2}mv^2$
- Law of conservation of momentum
- Mass
- Momentum = $m \times v$
- Newton as a unit of force
- Newton's laws of motion
- Physical properties of matter
- Velocity
- Work = $F \cdot d$

259

the *Engineering Design Cycle* to focus the discussion. Let the class know that they will be going through the entire cycle during the first half of the chapter. Then discuss each part of the cycle as needed. During the discussion, ask students to provide specific examples of each step. Emphasize that students will be using this simplified *Engineering Design Cycle* when designing their safety systems. Describe how students should first set their *Goal* using the criteria and then consider the *Inputs* they receive after each section. Emphasize that these *Inputs* will be important physics concepts that they will need to apply to complete their safety system and the rest of the challenge. Describe how in the *Mini-Challenge,* students are expected to complete the *Process* step that involves combining the *Inputs* with their ideas and the criteria of the challenge, along with comparing and contrasting potential solutions. Discuss the *Outputs* as the concepts that student groups present to the class in the *Mini-Challenge* and the *Challenge.* Emphasize that the entire class will be providing *Feedback* on the *Outputs*. It is important for students to realize that the *Feedback* from the *Mini-Challenge* should be used to refine their design and written report before they complete the *Chapter Challenge.*

Physics Corner

The *Physics Corner* illustrates the physics concepts presented in the chapter, which students are expected to apply to meet the *Chapter Challenge.* Ask your students if they are familiar with any of the concepts, and encourage them to provide definitions for the terms they know. Consider having students describe where in the illustration each concept is being displayed. Note any misconceptions students might have and discuss these when the concept is addressed in the chapter. Consider using the overhead of the *Physics Corner* while guiding the class discussion.

You should find that the students are motivated while they are actively engaging in the learning process. As the *Chapter Challenge* approaches, review the *Physics Corner* to help them keep track of the physics concepts they have investigated.

SECTION 1

Accidents

Section Overview

The section begins by having students assess their knowledge of risks involved with vehicles. Students answer questions designed to deal with common misconceptions about automobile safety. A list of common safety features found in automobiles today is provided. Students use this list to consider how safety features have changed over time. They apply their knowledge of safety features by identifying safety features in various modes of transportation, describing what type of collision they protect passengers from, and how they protect passengers. Students then record new ideas in their logs for designing their prototype of safety features to protect passengers during a collision.

Background Information

Accidents occur whenever a vehicle collides with another object. The major factors that influence accidents are slowed reaction times and road hazards, such as snow, potholes, or animals. The physics involved in accidents is the same. The fundamental concepts of conservation of energy and conservation of momentum determine how severe the accident is or how much damage is done to the objects and passengers involved in the collision. The more energy and momentum transferred, the greater the damage. Safety features are designed to avoid collisions and to assist in absorbing and dispersing energy and momentum during a collision.

There are two main categories of safety features: active and passive. Active safety features use information from the vehicle's external environment to change and improve the response of the vehicle before and during a collision. The ultimate goal of an active safety feature is to avoid a collision. Examples of active safety features include the following:

- Intelligent speed adaptation—uses global positioning satellites and an electronic throttle to control the speed of a vehicle, not allowing the vehicle to go over the speed limit.
- Turn signals and brake lights—allow other drivers to determine how the vehicle is changing its motion.
- Dynamic steering response and variable-assist power steering—change how much force is needed to turn the steering wheel, requiring less force when driving at slow speeds and more force when driving at higher speeds. They reduce overcompensation when drivers try to avoid collisions at high speeds.
- Traction control—prevents wheels from spinning and operates the brakes or the throttle to restore traction.
- Four-wheel drive—allows each wheel to have the power to rotate. This feature reduces sliding and over and under-compensation of turning.
- Reverse backup sensors—alert drivers to items behind their vehicle when driving in reverse.
- Electronic stability control—monitors an automobile's stability; can decrease power, apply brakes, and adjust steering compensation if the automobile seems to be going out of control.
- Lateral support—alerts drivers when they leave their lane.
- Mirrors—provide visibility.
- Low center-of-gravity—provides better handling.
- Anti-lock braking assist, brake-assist systems, dynamic brake control, etc. —braking systems that reduce skidding and sliding.
- Driver-state sensor—monitors a driver's eyelid movement.

Passive safety features are designed to help minimize injury during a collision. Some examples of passive safety features include the following:

- Seat belts—absorb some of the energy and limit the forward motion of an occupant, keeping the occupant from being ejected from the vehicle. Shoulder harnesses restrain the upper body, absorb energy, and prevent a secondary collision with items in the vehicle.
- Air bags—restrain motion and absorb some of the energy during a collision.
- Crumple zone—absorbs the energy of a collision, diverting it from the occupant compartment.
- Side-impact bars—reduce deformation of the side of a vehicle during impact.
- Collapsible steering column—reduces impact between driver and steering wheel.
- Fuel pump shut-off—turns off the fuel pump during a collision.
- Car seats—reduce energy of impact for children.
- There are also features that protect pedestrians involved in accidents:
    - bumpers are soft;
    - no sharp edges on the exterior of the vehicle;
    - and the shape of vehicles prevent the hit pedestrian from hitting other parts of the vehicle.

Crucial Physics

- Identify and describe factors that affect safety while driving.
- Compare and contrast safety features designed to protect passengers during collisions.

Learning Outcomes	Location in the Section	Evidence of Understanding
Evaluate your understanding of safety.	***Investigate*** Steps 1-3	Students take a test about safety.
Identify and evaluate safety features in selected automobiles.	***Investigate*** Step 3 ***Physics Talk*** ***Physics Essential Questions*** ***Physics to Go*** Questions 1, 6 ***Inquiring Further*** Question 1	Students consider a list of common safety features and describe the conditions under which they protect passengers and how they protect passengers. Students identify and/or evaluate safety features of new vehicles.
Compare and contrast the safety features in selected automobiles.	***Investigate*** Step 3 ***Physics Talk*** ***Inquiring Further*** Question 2	Students compare and contrast common safety features in automobiles.
Identify safety features required for other modes of transportation (in-line skates, skateboards, bicycles).	***Physics to Go*** Questions 2-4 ***Inquiring Further*** Question 1	Students identify safety features that would reduce the risk of injury during a collision for other modes of transportation, such as in-line skates, skateboards, and bicycles.

NOTES

Section 1 Materials, Preparation, and Safety

Materials and Equipment

No materials or equipment are needed for this section.

Time Requirement

Approximately 30 minutes are required for this investigation.

Teacher Preparation

- This section is centered primarily on reading and answering the questions in the *Investigate*. After the students have answered the questions, there should be a class discussion to investigate the understanding students now have about accidents, and to help the students open their minds to the seriousness of accidents. As the majority of students are entering a very dangerous time of their lives, in terms of learning to drive and statistically being prone to accidents, the discussion should be serious. Allow more time if a film and a discussion are added. A review of early automobile safety problems would provide additional background. An Internet search for Ralph Nader and vehicle safety should provide much information.

Safety Requirements

- No safety precautions are necessary for this section. A review of early automobile safety problems would provide background information.

NOTES

Meeting the Needs of All Students

Differentiated Instruction: Augmentation and Accommodations

Learning Issue	Reference	Augmentation and Accommodations
Understanding the purpose of a section	***Learning Outcomes***	**Augmentation** • Students often complete sections as discrete items that are not connected to a common goal. This makes it much more difficult to analyze results, draw conclusions, and establish meaning from new learning. Explicitly review the *Learning Outcomes* at the beginning of each section to set a purpose and make students aware of the big picture.
Creating a table from a model	***Investigate*** Step 3	**Augmentation** • Copying tables is a tedious and time-consuming task for students with visual-motor, graphomotor, and focus difficulties. Ask students to hold their papers in portrait format. Then students should divide their papers into four columns—three skinny columns and a wider one—to record an explanation. Next, students can mark rows using single lines on the paper, mark rows for every two lines, or mark the rows as they complete the explanations. Ask students to label the columns and rows using the table in the book. • Point out the photographs as a reference tool for filling in the table. **Accommodation** • Provide students with a blank chart to fill in and tape into their logs.
Reading comprehension	***Physics Talk*** ***Checking Up***	**Augmentation** • Many students are able to read and decode words, but they really struggle to understand the meaning of a text. Instruct students to read the *Checking Up* questions to provide a purpose for the reading and to guide understanding.

Strategies for Students with Limited English-Language Proficiency

Point out new vocabulary words in context and practice using the words as much as possible throughout the section.

anticipation
attributed to
beneficial
casualty
chrome
restraint
dashboard
steering column
designation
submerged

manufacturing
minimizing
overcompensate
pedestrian
hazard
collision
compile
fatality
statistics
transportation

After students have finished the *Inquiring Further* in which they review advertising brochures and research auto safety, have them write and illustrate a one-page brochure of their own. (Not all students will have collected brochures during the *Inquiring Further*, therefore, you may need to help students understand that a brochure is a small publication giving product information or advertising a product. Showing them a few sample brochures may be helpful.) Let them choose one of the following two scenarios:

- You sell automobiles. Write and illustrate a one-page brochure advertising a vehicle you sell. Be sure to tell possible buyers why it is the safest vehicle they can buy. Use at least five vocabulary words from the list in your brochure.

- You work for the National Highway Traffic Safety Administration (NHTSA). Write and illustrate a one-page brochure that tells people why they should always wear their seat belts. Use at least five vocabulary words from the list in your brochure.

Instruct students to write in complete sentences. You may wish to have students work in pairs. Review and correct the students' work. Rapid feedback about students' sentences is essential, because the sentences and errors will be fresh in their minds. A quick and powerful method for providing this feedback is to prepare a list of examples of sentences containing errors from the students' work. Divide examples into the following categories: incorrect science, incorrect usage of vocabulary, incorrect sentence structure, and incorrect grammar. Choose several examples from the collected work to use in each category and edit the sentences until they contain only one or two obvious errors. At the beginning of class the next day, provide each student with a page containing a double-spaced, typed list of the incorrect sentences, with headings for the four categories. Allow students 10 minutes to silently make corrections to the sentences. Then place a copy of the list on the overhead projector and ask students to suggest ideas for how to correct the sentences. During the exercise, guide students toward correct science and correct English usage.

NOTES

CHAPTER 3

SECTION 1

Teaching Suggestions and Sample Answers

What Do You See?

The *What Do You See?* illustration provides an interesting way to introduce students to the physics concepts being developed. Have a class discussion of the illustration. Consider using an overhead of the illustration as a focal point for the discussion. Elicit students' initial impressions of what they see, what dangerous features of a collision are portrayed, what safety features are shown, and why it is significant in the context of this section. You are likely to get many different responses. Focus on the responses that provide an opportunity for you to get the students engaged in the physics concepts presented in this section. For example, if students mention some of the safety features illustrated, you might ask how they provide protection against

Chapter 3 Safety

Section 1 Accidents

What Do You See?

Learning Outcomes

In this section, you will

- **Evaluate** your understanding of safety.
- **Identify** and evaluate safety features in selected automobiles.
- **Compare** and contrast the safety features in selected automobiles.
- **Identify** safety features required for other modes of transportation (in-line skates, skateboards, bicycles).

What Do You Think?

Chances are that you will be in an accident one day involving some means of transportation, such as an automobile, in-line skates, or bicycle.

- **How can you protect yourself from serious injury should an accident occur?**

Record your ideas about this question in your *Active Physics* log. Be prepared to discuss your response with your small group and the class.

Investigate

In this section, you will test your knowledge of the risks involved in vehicle collisions. You will also investigate some of the safety features available in vehicles built after 1960.

1. Many people think that they know the risks involved with day-to-day transportation. The "test" on the next page will check your knowledge of these risks.

   The statements are organized in a true and false format. Record a T in your log for each statement you believe is true and an F if you believe the statement is false. Your teacher will supply the correct answers, based on statistics, at the end of the section.

260

Students' Prior Conceptions

This section gives students the opportunity to articulate their understanding of what actually happens during a vehicle collision. They identify and evaluate prior knowledge continuously, as they progress through the phases of their evaluation of safety features.

1. **Students believe that collision severity is the same for both automobiles in a head-on collision.** Giving students opportunities to investigate what happens to passengers during collisions with and without seat belts provides them with cognitive and mathematical tools needed to explore the impact of forces involved in collisions. Modeling what happens to a passenger and to the automobiles during the same impact will show students that the severity of a collision can be high to the automobile but at the same time minimal to the passenger wearing a seat belt. The mechanical devices employed for safety reduce the impact of the collision to the passenger and/or the automobile. Modeling what happens between colliding automobiles should lead students to explore the nature of the mass and the velocity of automobiles involved in accidents. Examining what happens during each of the three collisions in every accident should reinforce correct student thinking.

injury or what physics concepts they think are involved with the safety features they described.

What Do You Think?

Ask students to use the illustration and their own experiences to help them answer the *What Do You Think?* question. Emphasize to students that there is no correct or incorrect response, but rather, opinions and ideas are being elicited. Remind students to record their ideas in their log. Have a class discussion, eliciting and recording students' ideas. Encourage students to ask classmates questions during the discussion. Students might discuss front air bags, seat belts, and child seats. Ask them what they think these features do to help reduce injury, and have them describe their ideas as best they can in terms of physics concepts. Reassure students that as they go through the chapter, they will investigate the physics involved with collisions and safety features.

What Do You Think?

A Physicist's Response

Accidents occur when two objects collide. The severity of the accident depends on how much energy is transferred to the objects, and the structural design of the objects. The greater the energy exchanged and the more localized it is during a collision, the greater the damage. Protecting yourself from injury during an accident requires the use of safety features. Some safety features are designed to aid in the prevention of a collision (such as mirrors and directional signals), and some are designed to reduce the energy transferred to a passenger or a localized area of a vehicle during a collision (such as seat belts and air bags). Objects with more mass or moving faster tend to cause greater damage during a collision.

Investigate

Decide how you wish students to take the safety quiz. The quiz can be given individually, or students could work in groups of two to three and discuss each question.

After all students have completed the quiz, create a bar graph on the board, listing the questions by letter with some key words. Have each student or team bring up a small sticky note for each question to create a bar graph that represents the class's responses.

NOTES

a) More people die of cancer than in automobile accidents.

b) Your chances of surviving a collision improve if you are thrown from the automobile.

c) The fatality rate in motorcycle accidents is less than in automobiles.

d) A large number of people who wear seat belts are killed in a burning or submerged automobile.

e) If you do not have a child restraint seat, you should place the child in your seat belt with you.

f) You can react fast enough during an accident to brace yourself against the impact of the collision.

g) Most people die in traffic accidents during long trips.

h) A person not wearing a seat belt in your vehicle poses a hazard to you.

i) Traffic accidents occur most often on Monday mornings.

j) Male drivers between the ages of 16 and 19 are most likely to be involved in traffic accidents.

k) Automobile accidents resulting in casualties are most frequent during the winter months due to snow and ice.

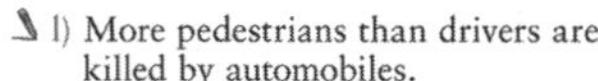

l) More pedestrians than drivers are killed by automobiles.

m) The greatest number of roadway fatalities can be attributed to poor driving conditions.

n) The greatest number of females involved in traffic accidents are between the ages of 16 and 20.

o) Unrestrained occupant casualties are more likely to be young adults between the ages of 16 and 19.

2. Calculate your score. Give yourself one point for a correct answer. You might want to match your score against the descriptors given below.

14-15 points: Expert Analyst

11-13 points: Assistant Analyst

8-10 points: Novice Analyst

7 points and below: Myth Believer

a) Record your score in your log. Were you surprised about the extent of your knowledge? Some of the reasons behind these facts will be better understood as you continue through this chapter.

3. Look at the photographs of two automobiles. One was built prior to 1960 and the other was built after 2000.

Teaching Tip

Students may get into a lively discussion of the relative safety of various ages of drivers. Acknowledge that young drivers are often very safe drivers, but point out that statistically younger drivers are involved in more accidents than older drivers.

1.a)

True. According to the Statistical Abstract of the United States, in 2005, there were less than one and a half times as many deaths from respiratory cancer as from motor-vehicle accidents. When all forms of cancer are considered, this figure is nearly five times higher than for motor-vehicle accidents.

Deaths in 2005 (in units of deaths/100,000 population) were because of the following:

Motor vehicle accidents—39.7 (leading cause of death by accidents)

Major cardiovascular disease—288.8 (leading cause of death listed)

Respiratory cancer—53.7

All forms of cancer—188.7

1.b)

False. Thirty percent of the people involved in automobile accidents die from being thrown from the vehicle during the collision. One report indicates that at least 50% of these fatalities need not have occurred. In 2005 just over 4% of reported motor-vehicle accidents resulted in a death, according to the Statistical Abstract of the United States.

1.c)

False. Motorcycles are less common, but they are much more dangerous.

1.d)

False. Few people are killed because of burning or drowning. One study indicates that it is less than 0.1%.

1.e)

False. The child can be sandwiched by your mass. The abdomens of small children are particularly vulnerable to impact.

1.f)

False. You do not have enough time to react even at 50 km/h (about 31 mi/h).

1.g)

False. Most people die within 40 kilometers of their home (about 25 miles).

CHAPTER 3

Safety features you may find in an automobile are listed in the first column of the following table. Explain why each safety feature may protect the driver, a passenger, or a pedestrian during an accident. You will record this in the second column in a table in your log. In the third column, state whether you think that the safety feature was present in most pre-1960 automobiles (yes/no). In the fourth column, state whether the safety features are in all new automobiles (1), in some new automobiles (2), or in very few new automobiles, (3).

a) Copy and complete the table in your log.

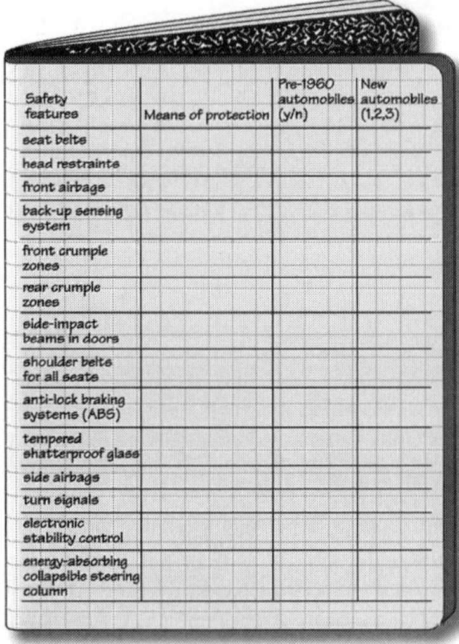

Safety features	Means of protection	Pre-1960 automobiles (y/n)	New automobiles (1,2,3)
seat belts			
head restraints			
front airbags			
back-up sensing system			
front crumple zones			
rear crumple zones			
side-impact beams in doors			
shoulder belts for all seats			
anti-lock braking systems (ABS)			
tempered shatterproof glass			
side airbags			
turn signals			
electronic stability control			
energy-absorbing collapsible steering column			

1.h)

True. The person becomes a projectile once the car stops.

1.i)

False. Most traffic accidents occur during Friday afternoon or evening traffic.

1.j)

True. That accounts for the higher insurance rates.

1.k)

False. The months of May, June, and August have the greatest number of casualty collisions.

1.l)

False. Pedestrians only account for 6.1% of traffic fatalities.

1.m)

False. More accidents take place when road conditions are dry.

1.n)

True. Young female drivers, like their male counterparts, are responsible for the greatest number of accidents. Inexperience may be the major contributing factor.

1.o)

True.

2.a)

Students should calculate their scores after the class has made a graph of their responses. Using the bar graph created with sticky notes, discuss the class's overall responses. If a question generally had incorrect responses, ask them how surprised they were to find out the answer. A short discussion should follow with the expectation that the students will be able to better understand the physics of the accidents after the next few sections. Then have them complete *Step 3*.

3.a)

Students should record the table and their responses in their *Active Physics* logs. Compare students' initial ideas with the correct responses on the following page.

Teaching Tip

Students may not be familiar with safety features that were available in automobiles prior to 1960. You may want to update this to prior to 1990 or even 2000. If you choose 2000, point out that passenger-side air bags did not become common until around 2000, side-curtain airbags until after 2000, and electronic stability control and backup-sensing systems were not available in most automobiles until 2007.

3-1a Blackline Master

Safety feature	Means of protection	Pre-1960 automobiles (y/n)	New automobiles (1, 2, 3)
seat belts	keeps driver and passengers inside of car	n	1, all
head restraints	prevents whiplash	n	1, all
front air bags	cushions during a collision	n	1, all (driver's side)
back-up sensing system	allows to see in blind spots while backing up	n	3, few
front crumple zones	increase collision distance reducing impact	n	1, 2, all, some
rear crumple zones	increase collision distance reducing impact	n	2, some
side impact beams in doors	resists side penetration	n	2, some
shoulder belts for all seats	keeps passengers in seats during collision	n	1, all
anti-lock braking systems (ABS)	helps maintain control/ prevents skids	n	2, some
tempered shatterproof glass	helps prevent cuts	y	1, all
side air bags	protects head/torso in side collisions	n	2, some
turn signals	warns other drivers of actions	y	1, all
electronic stability control	helps resist rollovers	n	2, 3, some, few
energy absorbing collapsible steering column	prevents chest trauma	n	1, all

NOTES

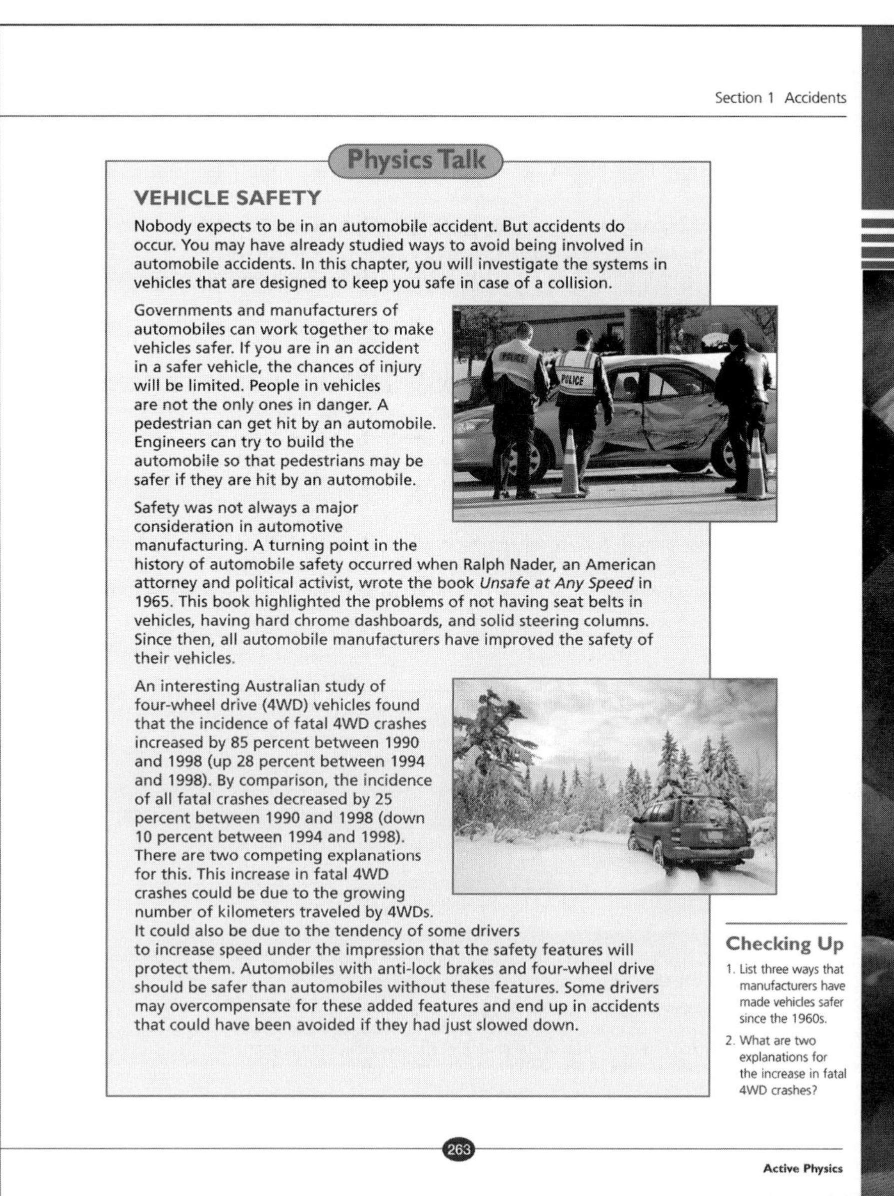

Section 1 Accidents

Physics Talk

VEHICLE SAFETY

Nobody expects to be in an automobile accident. But accidents do occur. You may have already studied ways to avoid being involved in automobile accidents. In this chapter, you will investigate the systems in vehicles that are designed to keep you safe in case of a collision.

Governments and manufacturers of automobiles can work together to make vehicles safer. If you are in an accident in a safer vehicle, the chances of injury will be limited. People in vehicles are not the only ones in danger. A pedestrian can get hit by an automobile. Engineers can try to build the automobile so that pedestrians may be safer if they are hit by an automobile.

Safety was not always a major consideration in automotive manufacturing. A turning point in the history of automobile safety occurred when Ralph Nader, an American attorney and political activist, wrote the book *Unsafe at Any Speed* in 1965. This book highlighted the problems of not having seat belts in vehicles, having hard chrome dashboards, and solid steering columns. Since then, all automobile manufacturers have improved the safety of their vehicles.

An interesting Australian study of four-wheel drive (4WD) vehicles found that the incidence of fatal 4WD crashes increased by 85 percent between 1990 and 1998 (up 28 percent between 1994 and 1998). By comparison, the incidence of all fatal crashes decreased by 25 percent between 1990 and 1998 (down 10 percent between 1994 and 1998). There are two competing explanations for this. This increase in fatal 4WD crashes could be due to the growing number of kilometers traveled by 4WDs. It could also be due to the tendency of some drivers to increase speed under the impression that the safety features will protect them. Automobiles with anti-lock brakes and four-wheel drive should be safer than automobiles without these features. Some drivers may overcompensate for these added features and end up in accidents that could have been avoided if they had just slowed down.

Checking Up

1. List three ways that manufacturers have made vehicles safer since the 1960s.
2. What are two explanations for the increase in fatal 4WD crashes?

263

Active Physics

Physics Talk

The *Physics Talk* discusses some of the issues and history of vehicle safety. It also presents a study conducted on four-wheel drive vehicles that provides an example of competing theories. This provides a good opportunity to discuss how scientists may have competing theories and the need for evidence to ascertain which theory, if any, is correct.

Oftentimes when there are competing theories, part of each theory may be correct.

Consider asking students what factors or variables might affect the results provided and record their responses. Then ask students what ideas they have to reduce the number of accidents that occur with four-wheel drive vehicles and how they could minimize passenger injuries.

Checking Up

1.

Some possible safety features are seat belts, air bags, padded dashboards, four-wheel drive, anti-lock brakes, energy-absorbing collapsible steering columns, and so on.

2.

Fatal 4WD crashes may be due to the fact that drivers of 4WD vehicles have a greater tendency to increase speed and may drive less cautiously because they think they have safety features that will protect them from accidents and minimize injury. Another factor that may have affected the number of fatalities involved with 4WD vehicles is that the 4WD vehicles in the study were driving greater distances.

What Do You Think Now?

Have students revisit the *What Do You See?* illustration and ask students if they can identify any other safety features they did not already list that are shown in the illustration. Discuss how safety features can reduce the number of injuries and revisit the *What Do You Think?* question, asking students what other ideas they have. Consider sharing the information in *A Physicist's Response* and elicit students' opinions on it. Students should apply ideas from the *Investigate*.

What Do You Think Now?

At the beginning of this section, you were asked

- **How can you protect yourself from serious injury should an accident occur?**

In light of all the safety features you have investigated so far, how would you protect yourself in the event of an accident? What safety device do you think is most effective and why? What actions will not protect you in an accident?

Physics Essential Questions

What does it mean?

Automobiles today have improved safety devices over older models. Describe three safety features of an automobile and explain how each feature provides passenger safety.

How do you know?

How do you know that safety has become a concern for automobile manufacturers?

Why do you believe?

Connects with Other Physics Content	Fits with Big Ideas in Science	Meets Physics Requirements
Forces and motion	Conservation laws	 Good clear explanation, no more complicated than necessary

* The laws of physics do not change from day to day. Auto manufacturers add safety devices to automobiles in anticipation of accidents occurring. Compare bicycle helmet laws with laws of physics.

Why should you care?

Safer automobiles can reduce injuries to drivers, passengers, and pedestrians in the event of an accident. How is minimizing injuries in transportation accidents beneficial to society?

Reflecting on the Section and the Challenge

Automobiles accidents can cause serious injuries in a number of different ways. If there are no restraints or safety devices in a vehicle, or if the vehicle is not constructed to absorb any of the energy of the collision, even a minor collision can cause serious injury. Until the early 1960s, automobile design and construction did not even consider passenger safety.

Physics Essential Questions

What does it mean?

Seat belts prevent you from being ejected from your seat. Air bags protect you in an accident. A padded dashboard protects your head if it collides into it.

How do you know?

There are many safety features, such as seat belts, that are mandated for vehicles.

Why do you believe?

A bicycle helmet law is passed by legislature. A law of physics is not changed by a vote of the public.

Why should you care?

If there are fewer accidents, there will be fewer hospitalizations, which will save money.

The general belief was that a heavy automobile was a safe automobile. While there is some truth to that statement, today's lighter automobiles may be safer than some of the large, heavy automobiles of the past.

In completing the *Chapter Challenge*, you will want to discuss which safety concerns you are addressing in your improved safety device, and the physics behind each improvement.

Physics to Go

1. Review and list 10 safety features found in today's new automobiles. As you compile your list, write next to each safety feature one or more of the following designations:

   F–effective in a front-end collision.

   R–effective in a rear-end collision.

   S–effective in a collision where the automobile is struck on the side.

   T–effective when the automobile rolls over or turns over onto its roof.
2. Make a list of safety features that could be used for bicycling.
3. Make a list of safety features that could be used for in-line skating.
4. Make a list of safety features that could be used for skateboarding.
5. What safety features do you think should be in every automobile used today?
6. Ask family members or friends if you may evaluate the safety of their automobile. Discuss and explain your evaluation to the automobile owners. Record your evaluation and their response in your log.
7. ***Preparing for the Chapter Challenge***

   The safety survey may have provided you ideas for constructing a prototype of a safety system used for transportation. In your log, record ideas that have been generated from this section.

Inquiring Further

1. **Safety and sales**

   Interview a salesperson of automobiles or bicycles, or collect brochures from various automobile and bicycle manufacturers. What new safety features are presented by the salesperson or in the brochures? How much of the advertising is devoted to safety?
2. **Vehicle safety ratings**

   Do an Internet search for automobile safety features and ratings. You may wish to visit the National Highway Traffic Safety Administration Web Site. Compare vehicles from different categories, such as vans, sports cars, or pickup trucks.

Reflecting on the Section and the Challenge

Read as a class the three paragraphs in *Reflecting on the Section and the Challenge*, or have students read them to themselves. Have a discussion based on the student text. Ask students why it might be true that a heavier automobile is a safer automobile. Then ask why they think new automobiles are safer now, even though they are much lighter. Ask students what types of safety features they are considering for their prototype and why. Then emphasize that when they complete the *Chapter Challenge* they will need to clearly state the safety issue they are addressing, the improvements they have made, and the physics concepts behind each improvement.

Physics to Go

1.

Students should describe at least ten safety features and the type of collision that each safety feature focuses on. Students may use the following chart in the *Investigate*.

Safety Feature	Collision Type
side-impact beams in door	S
shoulder belts	F, R, S, T
anti-lock brakes	F, R, S
side air bags	S
collapsible steering column	F, R
rear crumple zones	R
front crumple zones	F
head restraints	F, R, S
back-up sensing systems	R
mirrors	R, S
side air bags	S
front air bags	F, R, S
knee air bags	F, R, S
padded front seats	F, R
padded roof frame	T
head rests	F, R
padded console	F, R, S, T
padded sun visor	F, S, T
padded doors and arm rests	S
padded steering wheel	F
padded gear level	F, S
padded door pillars	S, T

2.

The students' lists should describe how each safety feature reduces collisions and/or injuries. For example, students might describe mirrors on helmets or handlebars that help avoid collisions, or padded handlebars that absorb energy during a collision, reflectors and/or lights for visibility, better braking systems, knee pads, elbow pads, and helmets.

3.

Students' responses should describe how each safety feature reduces collisions and/or injuries. For example, students might describe mirrors on helmets that help drivers avoid collisions; knee pads and elbow pads that absorb energy during a collision; reflectors or lighting on clothing to increase visibility; or helmets.

4.

Students' lists should describe how each safety feature reduces collisions and/or injuries. For example, students might describe mirrors on helmets that help drivers avoid collisions; or helmets, knee pads, and elbow pads that absorb energy during a collision. Lighting or reflectors on clothing increase visibility.

5.

Student suggestions will vary, and will probably include seat belts, anti-lock brakes, various types of air bags and other items from the list preceeding the *Physics Talk*. Students may not realize that items such as a dashboard and ceiling padding would be safety items so they may not include them. Newer features such as traction control may also be suggested. Many students may also suggest four-wheel drive. If this is suggested it may be worthwhile to probe the students as to why they believe this is a safety item. Many students are under the false impression that four-wheel drive allows them to drive safely faster in the snow. Point out that more will be learned about this later in the chapter.

6.

Students should record their responses in their log, as well as the responses of the vehicle owner. Students should describe the safety features, the types of collisions it is constructed for, and how it helps drivers avoid a collision or reduce injury.

7.

Preparing for the Chapter Challenge

Students should record ideas in their logs for safety features of their prototype that were generated from this section.

Inquiring Further

1. Safety and sales

Have students describe the vehicles they are writing about, the safety features in the vehicles, and how much advertising is devoted to safety in the selling of an automobile. Consider asking students what the amount of safety advertising implies.

2. Vehicle safety ratings

Students should compare and contrast safety features and ratings of several new vehicles.

NOTES

SECTION 1 QUIZ

3-1b Blackline Master

1. Which of the following would not be considered a safety device in an automobile?

   a) seat belt
   b) speedometer
   c) turn signals
   d) tempered glass

2. Which of the following safety devices will make it less likely for an automobile to get into a crash with another vehicle?

   a) turn signals
   b) seat belts
   c) tempered glass
   d) head restraints

3. The age group most likely to be involved in an automobile accident is

   a) 16-20.
   b) 20-30.
   c) 40-50.
   d) 50-60.

4. Which of the following safety devices will protect the driver of an automobile during a collision?

   a) electronic stability control
   b) back-up sensing system
   c) anti-lock brake system
   d) shoulder belts

5. Four-wheel drive vehicles should be safer than two-wheel drive vehicles on snowy roads. Which of the following is a possible cause of why the driver of a four-wheel drive vehicle might be more likely to have a crash than the driver of a two-wheel drive vehicle in the snow?

   a) Four-wheel drive vehicles are larger than two-wheel drive vehicles.
   b) Drivers of four-wheel drive vehicles are older than drivers of two-wheel drive vehicles.
   c) The drivers of four-wheel drive vehicles are less cautious on snowy roads than are the drivers of two-wheel drive vehicles.
   d) Four-wheel drive vehicles have fewer safety features than two-wheel drive vehicles.

SECTION 1 QUIZ ANSWERS

1. b) speedometer
2. a) turn signals
3. a) 16-20.
4. d) shoulder belts
5. c) The drivers of four-wheel drive vehicles are less cautious on snowy roads than are the drivers of two-wheel drive vehicles.

NOTES

SECTION 2

Newton's First Law of Motion: Life and Death before and after Seat Belts

Section Overview

Using a model consisting of a cart and a driver made of clay, students investigate what happens to the driver, with and without a seat belt, during collisions with various impact speeds. They also investigate the effect of different types of seat belts on the driver during a collision. Newton's first law is then discussed using student observations as evidence. Students read about three main collisions involving vehicles: the collision between a vehicle and another object, the collision between a person in a vehicle and an object in the vehicle that brings the person to rest, and the collision between organs inside the body and the body walls that bring the organs to rest. Connections are made between students' observations of different types of seat belts and the concept of pressure (force per unit area). Students apply this concept to solve various problems and engage in a discussion of how Newton's first law and the concept of pressure affect the design of safety features.

Background Information

Scientists see events that occur as interactions between two or more objects. To analyze and explain interactions, scientists use two main approaches: force and/or energy considerations. Force considerations involve vector quantities, and energy considerations involve scalar quantities. In this section, students use one-dimensional force considerations and focus on Newton's first law.

Newton's first law is helpful in describing and explaining why there are so many subsequent collisions that occur during a collision involving a vehicle and a passenger and why safety is so important. Collisions between a person and the object that stops him or her within a vehicle, as well as collisions between internal organs and the walls of the body, can cause severe injuries—including brain damage, aneurysms, broken bones, and damaged organs—or death.

Newton's first law is often called the law of inertia. It states that the motion of an object (whether moving or at rest) will not change unless there is a net external force exerted on the object.

Students are also introduced to the concept of pressure. Pressure, or force per unit area, is an important factor during a collision. When a force is spread out over a greater area (lower pressure), the overall force on any part is less. For example, think about someone standing on snow with snowshoes versus someone standing on snow in stilts. For equal masses, the stilts will sink through the snow whereas the snowshoes will stay on top because they decrease the force per unit area of pressure exerted on the snow.

Crucial Physics

- Newton's first law—an object remains at rest or in motion with constant velocity unless a net external force acts on it.
- The inertia of an object is proportional to its mass.
- Pressure is the amount of force acting perpendicular to a surface on an object divided by the cross-sectional area that it acts on.

Learning Outcomes	Location in the Section	Evidence of Understanding
Explain Newton's first law of motion.	***Physics Talk*** ***Physics Essential Questions*** ***Physics to Go*** Questions 1-4	Students discuss Newton's first law and describe the three parts of it. Students use Newton's first law to explain their observations in the *Investigate* and why a passenger keeps moving when a vehicle suddenly stops.
Describe the role of seat belts.	***Investigate*** Part B Step 3.b) ***Physics Talk*** ***Reflecting on the Section and the Challenge***	Students observe how seat belts affect the outcome of a clay passenger during a collision. Using their observations, students describe the role of seat belts during collisions.
Identify the three collisions in every accident.	***Physics Talk*** ***Physics to Go*** Questions 4, 8	Students identify and describe the three types of collisions that occur during an accident, and identify possible consequences of these collisions.
Compare the effectiveness of various wide and narrow seat belts.	***Investigate*** Part B: Step 3 ***Reflecting on the Section and the Challenge*** ***Physics to Go*** Question 5	Students observe and describe how the effectiveness of seat belts varies with the width of the seat belt. Students' observations are later used to support the concept of pressure.
Express the relationship between pressure, force, and area.	***Physics Talk*** ***What Do You Think Now?*** ***Physics to Go*** Questions 5, 9	Students describe the relationship between pressure, force, and area and provide supporting examples based on their observations. Students apply the relationship between force, area, and pressure to solve problems.

Section 2 Materials, Preparation, and Safety

Materials and Equipment

PLAN A

Materials and Equipment	Group (4 students)	Class
Multimedia DVD/CD Set		1 per class
Ribbon, 12 in. (length), varying widths	3 per group	
Wire, #22, bare copper, ft	1 per group	
Dynamics cart	2 per group	
Clay, modeling, lb	2 per group	
Inclined plane for lab cart	1 per group	
Large ring stand	1 per group	
Rod, aluminum, 12 in. (length) x 3/8 in. (diameter) (to act as cross arm)	1 per group	
Holder, right angle, cast iron	1 per group	
Scissors	1 per group	
File folders	3 per group	
Tape, masking		6 per group
Concrete block or similar barrier (wall)*		1 per class

*Additional items needed not supplied

PLAN B

Materials and Equipment	Group (4 students)	Class
Multimedia DVD/CD Set		1 per class
Ribbon, 12 in. (length), varying widths	3 per group	
Wire, #22, bare copper, ft	1 per group	
Dynamics cart	1 per group	
Clay, modeling, lb	2 per group	
Inclined plane for lab cart	1 per group	
Large ring stand	1 per group	
Holder, right angle, cast iron	1 per group	
Rod, aluminum, 12 in. (length) x 3/8 in. (diameter) (to act as cross arm)	1 per group	
Scissors	1 per group	
File folders	3 per group	
Tape, masking		6 per group
Concrete block or similar barrier (wall)*		1 per class

*Additional items needed not supplied

Note: Time, Preparation, and Safety requirements are based on Plan A, if using Plan B, please adjust accordingly.

Time Requirement

This *Investigate* should take at least one class period (40–50 min). Allow extra time for variations of the seat belts on the molded clay figures.

Teacher Preparation

- This investigation requires students to crash a loaded cart into a wall, or other suitable barrier. Test out different barriers (walls, desks, bricks, homemade structures) prior to laboratory day.
- If the students will be making measurements on the floor, inform the students the day prior to the investigation so they may wear appropriate clothing.
- Students may either perform this investigation on large lab tables or on the floor. Doing the investigation on the floor is preferable to prevent the dynamics carts from rolling off the tables. In addition, walls may be used as stopping devices for the dynamics carts.
- When the dynamics carts leave the ramp and travel onto the horizontal surface, a transition surface such as a manila file folder taped to the ramp and the floor or table will smooth the transition.

Safety Requirements

- Plastic dynamics carts may be damaged during the collision and are not recommended for this investigation.
- During the collisions, students should wear goggles, particularly if plastic dynamics carts are used.
- If the investigation is done on lab tables, special care should be taken so that the heavy barrier does not fall off the table.

Meeting the Needs of All Students

Differentiated Instruction: Augmentation and Accommodations

Learning Issue	Reference	Augmentation and Accommodations
Describing the effects of a collision on a passenger	***Investigate*** Part A: Steps 2-3 ***Investigate*** Part B: Step 3	**Augmentation** • Students will probably enjoy crashing their automobiles and passengers into the wall, but they may struggle to describe the effects of these collisions. Giving two minutes, ask students to individually write a list of possible descriptors for injuries sustained in a collision. Then ask students to share their ideas with the whole group and create a comprehensive list of descriptors. **Accommodation** • Students who struggle with spelling often do not write down their thoughts for fear of appearing ignorant. Provide these students with a typed list of descriptors that they can use to complete this section.
Determining significant injury	***Investigate*** Part B: Step 2	**Augmentation** • Some students are very literal learners and may not be able to decide what a "significant injury" means. After the class compiles the list in the above augmentation, ask students to decide which injuries are severe. Mark those injuries with a star or another symbol of your choice.
Reading comprehension Understanding essential concepts	***Physics Talk***	**Augmentation** • After this section of reading, students need to understand Newton's first law, the three collisions in an accident, and force per unit area. Students may struggle to extract all of this information from reading the *Physics Talk* section. • Provide direct instruction to teach these concepts. • Use the jigsaw strategy to allow students to teach these concepts to the class. Divide the class into three expert groups (A, B, and C). Give these groups a time limit (such as 20 minutes) to read the section on their topic and gather supplemental information from other provided resources (textbooks, the Internet, etc.). Then students should form triads or be assigned to triads (A, B, and C) to teach their classmates the information that they became experts on. Students could teach the class using notes, examples, a poster, and so on. **Accommodation** • Provide guided notes for students to complete using the *Physics Talk* section.
Vocabulary comprehension	***Physics Talk*** Three Collisions in One Accident!	The word "conveyance" is used often among people who work with modes of transportation. In addition to seeing it used here, your students are likely to run into it in their research later on. Be sure they know that it means "a machine that moves people or objects from one place to another," and in context here it means "automobile." Some students may have encountered the similar term "conveyor" (also spelled "conveyer") as a belt used to move bales of hay on a farm, or as a moving sidewalk in an airport.
Vocabulary comprehension	***Reflecting on the Section and the Challenge***	Students know from their experiments in Part B of the *Investigate* that a seat belt can alter the shape of their clay passenger. Therefore, they may be able to deduce the meaning of the word "distortion" in context. Be sure they understand that distortion means the way in which an object has been deformed from its original shape.
Vocabulary comprehension	***Inquiring Further*** Step 2	Be sure students begin their investigations of hydraulic brake systems by looking up the meaning of the word "hydraulic." They should know that it means operated by a fluid, usually water, under pressure.

Strategies for Students with Limited English-Language Proficiency

ELL students benefit greatly from writing practice. Have students describe in writing the three collisions involved in any automobile accident. Then ask pairs of students to exchange papers and comment on each other's sentences and descriptions.

Comparing and contrasting is an important science skill that can be difficult for students. Breaking down the work into simpler steps helps students master this skill. Ask students to compare (say what is similar about) and contrast (say what is different about) driver safety without a seat belt and driver safety with two kinds of seat belts by filling in a chart like the one below. A few boxes have been filled in to get them started. Direct students to think over all the information they have learned during this section and then have them use the circumstances from their experiments during the *Investigate* to complete the chart. Finally, ask students to use the chart to write a paragraph comparing and contrasting driver safety with seat belts and driver safety without seat belts. The paragraph should have a topic sentence, a body with supporting sentences, and a concluding sentence.

Evaluation of Seat Belt Choices

	Prevented or Lessened Collision 1 (✓ = yes, × = no)	Prevented or Lessened Collision 2 (✓ = yes, × = no)	Prevented or Lessened Collision 3 (✓ = yes, × = no)	Caused Visible Injury (1 = severe, 2 = moderate 3 = little to none)	Likely Caused Internal Injury (1 = severe, 2 = moderate, 3 = little to none)
No Seat Belt	×			1	
Wire Seat Belt					
Wide Seat Belt		✓			

NOTES

CHAPTER 3

SECTION 2

Teaching Suggestions and Sample Answers

What Do You See?

Have a class discussion on students' descriptions of the illustration. Consider using an overhead of the illustration as a focal point for the discussion. Elicit students' initial impressions of what they see in the illustration. Ask students questions such as: What dangerous features of a collision are portrayed? What safety features are shown? Why are these significant in the context of this section?

What Do You Think?

Discuss the description of a collision as compared to being struck by bowling balls and ask students to use the illustration and their own experiences to

Chapter 3 Safety

Section 2

Newton's First Law of Motion: Life and Death before and after Seat Belts

What Do You See?

Learning Outcomes

In this section, you will

- **Explain** Newton's first law of motion.
- **Describe** the role of seat belts.
- **Identify** the three collisions in every accident.
- **Compare** the effectiveness of various wide and narrow seat belts.
- **Express** the relationship between pressure, force, and area.

What Do You Think?

In a collision, you cannot brace yourself and prevent injuries. Instead of thinking about bracing yourself against a collision when an automobile is going 50 km/h (about 30 mph), think about 10 bowling balls, a mass of 45 kg (a weight of about 100 lb), all hurtling toward you at 50 km/h. You could not use your arms and legs to stop these fast-moving bowling balls. The two situations are equivalent.

- **Suppose you had to design a seat belt for a race car that can go 300 km/h (about 200 mph). How would it be different from one available on a passenger automobile?**

Record your ideas about this question in your *Active Physics* log. Be prepared to discuss your responses with your small group and the class.

Investigate

In this section, you will be investigating what happens to a passenger involved in an automobile accident without and with a seat belt.

266

Students' Prior Conceptions

Identifying types of collisions and the role of safety restraints offer students the opportunity to apply Newton's laws to real-world situations, emphasizing the essential question, "Why should you care?"

1. **A force is needed to keep an object moving with a constant speed.** During the investigations in *Chapters 1* and *2*, students measure distance and time to see what happens when unbalanced forces act on objects either at rest or in motion. This leads students to believe that an object in motion will remain in constant motion in the absence of an unbalanced force. Coupled with this understanding, students recognize how forces affect changes in velocity and cause acceleration. As your class begins *Chapter 3*, consider asking questions during discussions, particularly when students are offering explanations, to ensure that students' language reflects accepted principles. This will enable you to mediate and guide student cognition when their perception of reality differs from the physics of motion.

2. **Students confound inertia with friction believing that objects resist acceleration from a state of rest due to friction.** Friction is a concept that students view as something that always impedes motion. They need to experience the difficulty of changing the state of motion of a heavy mass and a lighter mass that have similar shapes and the same coefficient of friction on the surface upon which the mass moves in order to mentally alter their prior knowledge.

help them answer the question. Remind students to record their ideas in their log. Have a class discussion, eliciting and recording students' responses to the question. Encourage students to ask questions of each other during the discussion. Ask what the main difference is between collisions that involve an average driver and those involving a race-car driver, and how this might affect the design of a seat belt. Remind students that they will revisit this question at the end of the section. Record misconceptions and address them at appropriate times during the section. Focus on the responses that provide an opportunity for you to get the students engaged in the physics concepts. Consider emphasizing that there are no "right" answers and that all answers are acceptable. These questions elicit students' prior knowledge. Let students know they should refer to their answers while they are being introduced to new physics concepts. Point out that when they are aware of what they think, they will be better able to add to what they know.

What Do You Think?

A Physicist's Response

When designing a seat belt for a race-car driver, it is important to keep in mind the higher speeds at which race cars travel. The faster a car is going, the more energy it has. Considering the work-energy theorem, the greater the energy and the greater the force needed or the greater the distance needed to stop the motion. Because the distance between a driver and the dashboard is small, it is important for the seat belt to supply an ample force to stop the driver in this short distance. However, based on observations made in the previous *Investigate*, it is important to spread this force out over as great an area as possible to avoid injuries. Therefore, a seat belt designed to restrain a race-car driver should cover a greater amount of the driver's body, be cushioned to allow it to act over a greater distance, and be easy to get out of in case the vehicle catches fire during a collision.

Investigate

Let students know that they will be making observations of various crashes with and without seat belts. Emphasize that students should record their observations clearly in their *Active Physics* logs. Discuss and demonstrate the equipment they will be using. Show them how to increase the speed of the cart by increasing the height of the ramp, and how to attach the seat belts (tape is sufficient). Emphasize to students that the seat belts have to fit snugly around the driver.

3. **Students often think that a force is needed to sustain motion.** Students' experiences and observations involve friction acting on moving objects, and therefore, they may not realize that a net force is not needed to keep an object in motion. The concept is illustrated through the "second collision" in which the driver's body continues to move after the automobile stops. Consider demonstrating how objects continue their motion as friction is minimized. Then have them imagine what would happen if there were no friction.
4. **Most students cannot differentiate force from pressure.** To illustrate the difference between force and pressure, have students stand on two feet, on one foot, and then on their toes. Then ask them to correlate the pressure in each of the three cases based on the area of contact. Point out that pressure exerted by a force increases with decreasing area. This simple experiment empowers students to restructure their prior conception.

NOTES

Section 2 Newton's First Law of Motion: Life and Death before and after Seat Belts

Part A: Accidents Without Seat Belts

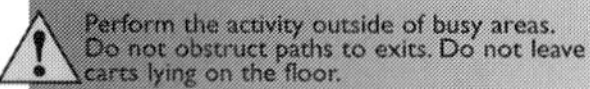
Perform the activity outside of busy areas. Do not obstruct paths to exits. Do not leave carts lying on the floor.

1. In this section, you will investigate automobile crashes where the driver or passenger does not wear seat belts. Your model automobile is a laboratory cart. Your model passenger is molded from a lump of soft clay.

   Obtain a lump of soft clay. Mold the clay to represent a human figure.

2. With the "passenger" in place, send the "automobile" at a low speed into a wall.
   a) Describe, in your log, what happens to the "passenger."
3. Repeat the collision at a higher speed.
   a) Compare and contrast this collision with the previous one. "Compare and contrast" requires you to find and record at least one similarity and one difference. A better response includes more similarities and differences.

Part B: Accidents With Seat Belts

1. You will test the suitability of different materials for use as seat belts. Your model automobile is, once again, a laboratory cart. Your model passenger is molded from a lump of soft clay.

   Give your passenger a seat belt by stretching a thin piece of wire across the front of the passenger. Attach the wire to the cart.

2. Make a collision by sending the cart down a ramp. Start with small angles of incline. Increase the height of the ramp until you see significant injury to the clay passenger.

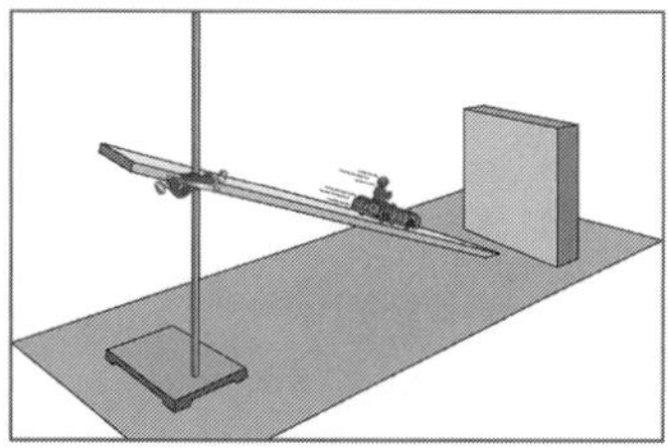

   a) In your log, note the height of the ramp at which significant injury occurs.
3. Use at least two other kinds of seat belts (ribbons, cloth, and so on). Use the same angle of ramp and release height as in *Step 2*.
   a) In your log, compare the injury that occurs to the "passenger" using the other kinds of seat-belt material.

267

Active Physics

Part A: Accidents Without Seat Belts

Teaching Tip

Using non-toxic, flour-based modeling compound or a similar material may be preferable to using modeling clay, which is oil based to prevent drying. Students with younger siblings often have non-toxic, flour-based modeling compound at home and are willing to bring it to class.

Teaching Tip

If you are using plastic laboratory carts, you may want to attach a piece of wood or some other fairly rigid material to the front of the cart where it will strike the wall to protect the surface. Students' enthusiasm for this investigation occasionally overcomes their sense of judgment, and equipment may get damaged.

1.

Students mold the clay to represent a human figure.

2.a)

Make sure students realize that the higher the ramp, the faster the speed of the cart down the ramp, and the faster and harder the figure will crash into the wall. Students should observe that for both cases the passengers continued in motion when the cart stopped (similarity) but the degree of damage to the passenger and how far the passenger moves depends upon the cart's speed (difference).

3.a)

Students should list similarities and differences between the two collisions. Some similarities might include both cars were moving and came to a stop very quickly; the "passenger" went flying off the cart in both scenarios; and the clay passenger was dented in both cases, indicating some damage occurred. Differences might include the cart in the second trial was moving much faster; the passenger in the second trial flew farther off the cart; and the passenger's "injuries" were worse in the second case.

Part B: Accidents With Seat Belts

Teaching Tip

When attaching the "seat belts" to the carts with tape, it is often wise to double the belt back over the tape, and then tape again to ensure the belt does not slide under the tape.

1.

Have students prepare their materials if needed. Students should attach the seat belt to the cart. Tape is often sufficient, and the seat belt should be snug against the clay passenger.

Teaching Tip

Tape a manila folder near the bottom of the ramp and also to the surface when sending carts down the ramp. This will allow the cart to make a smooth transition from the ramp to the tabletop, so it will maintain its speed.

2.a)

Students should record the starting height where the seat belt seems to cut into the clay driver.

3.a)

Students should note that when a wider seat belt is used (such as one made of ribbon), it is unlikely that a cut will be observed in the clay driver.

3.b)

Have students support their response with their reasoning. As will be discussed in the *Physics Talk*, the wider seat belt distributes the force required over a larger area, making it less likely to cut into the clay driver.

4.a)

Students should observe that just as in a "real-world" collision, the crash dummy moves forward when the car stops until it is brought to rest by something, such as an air bag, a steering wheel, or the windshield.

4.b)

Check that students support their response with their observations.

b) What accounts for the difference in injury?

4. Crash dummies cost thousands of dollars! Watch the video presentation of a vehicle in a collision, with a crash dummy in the driver's seat. You may have to observe the video more than once to answer the following questions.

a) In the collision, the automobile stops abruptly. What happens to the crash-dummy driver?

b) What parts of the crash-dummy's body are in the greatest danger? Explain what you saw.

Physics Talk

SEAT BELTS AND NEWTON'S FIRST LAW OF MOTION

The Three Parts of Newton's First Law

Newton's first law of motion (also called the law of **inertia**) is one of the foundations of physics. You probably have already encountered Newton's first law. It states:

> An object at rest stays at rest, and an object in motion stays in motion in a straight line with constant speed unless acted upon by a net, external force.

There are three distinct parts to Newton's first law.

Part 1 says that objects at rest stay at rest. This hardly seems surprising.

Part 2 says that objects in motion stay in motion in a straight line with constant speed. This may seem strange indeed. After looking at the collisions in this section, this should seem clearer. The automobile and the clay passenger were moving at constant speed. Even though the automobile stopped, the clay passenger continued moving at the same constant speed until it hit the barrier.

Part 3 says that Parts 1 and 2 are only true when the net force, or total of all forces, on the object is zero. An object may have forces acting on it and still have no change in its motion.

Physics Words

Newton's first law of motion: an object at rest stays at rest, and an object in motion stays in motion in a straight line with constant speed unless acted upon by a net, external force.

inertia: the natural tendency of an object to remain at rest or to remain moving with constant speed in a straight line.

The head and torso of the crash dummy are in the most danger, because the lower portions of the body are more likely to be constrained by the lap portion of the seat belt.

Consider having a class discussion on students' observations.

Physics Talk

This *Physics Talk* describes Newton's first law and how it applies to the simulated automobile accident students observed in the *Investigate*. It introduces three types of collisions experienced by people involved in collisions. Students are also introduced to the concept of pressure (force per unit area) and this concept is used to explain their observations of different seat belts.

Begin a class discussion by asking students what they know about Newton's first law.

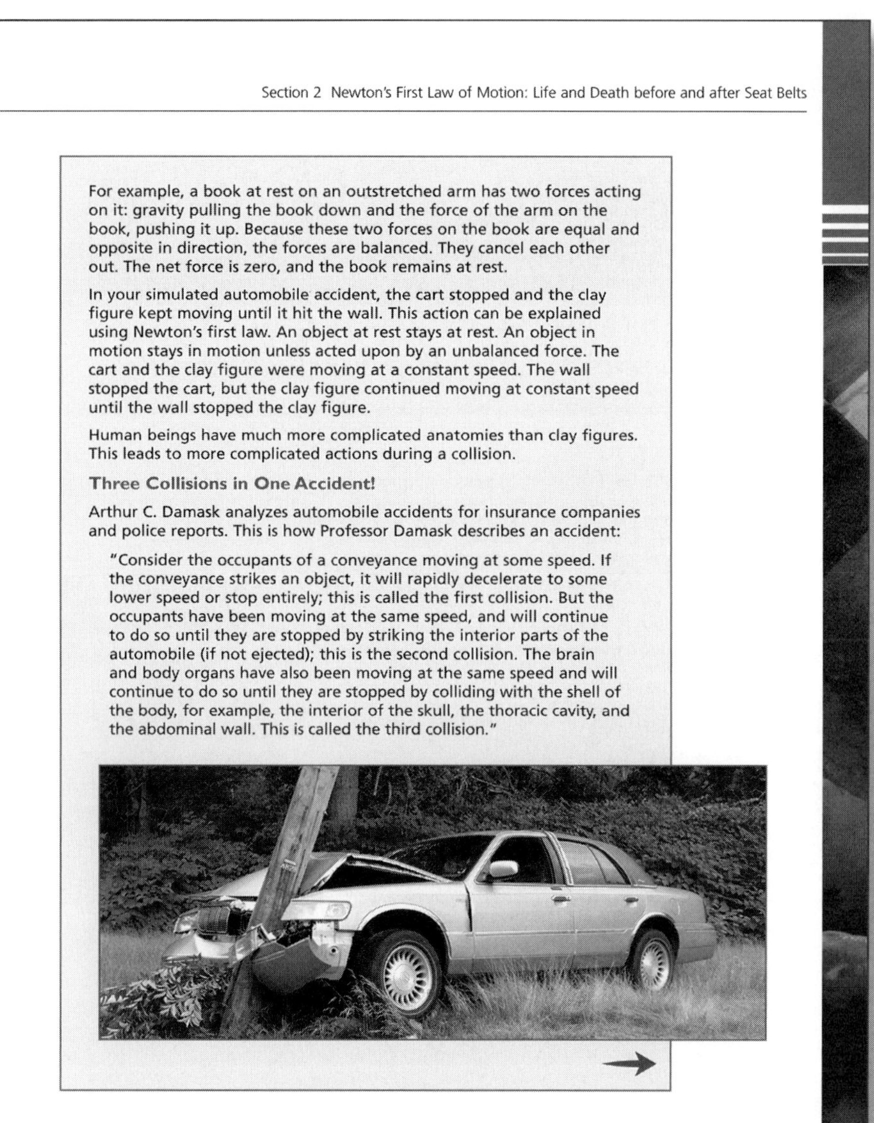

Section 2 Newton's First Law of Motion: Life and Death before and after Seat Belts

For example, a book at rest on an outstretched arm has two forces acting on it: gravity pulling the book down and the force of the arm on the book, pushing it up. Because these two forces on the book are equal and opposite in direction, the forces are balanced. They cancel each other out. The net force is zero, and the book remains at rest.

In your simulated automobile accident, the cart stopped and the clay figure kept moving until it hit the wall. This action can be explained using Newton's first law. An object at rest stays at rest. An object in motion stays in motion unless acted upon by an unbalanced force. The cart and the clay figure were moving at a constant speed. The wall stopped the cart, but the clay figure continued moving at constant speed until the wall stopped the clay figure.

Human beings have much more complicated anatomies than clay figures. This leads to more complicated actions during a collision.

Three Collisions in One Accident!

Arthur C. Damask analyzes automobile accidents for insurance companies and police reports. This is how Professor Damask describes an accident:

> "Consider the occupants of a conveyance moving at some speed. If the conveyance strikes an object, it will rapidly decelerate to some lower speed or stop entirely; this is called the first collision. But the occupants have been moving at the same speed, and will continue to do so until they are stopped by striking the interior parts of the automobile (if not ejected); this is the second collision. The brain and body organs have also been moving at the same speed and will continue to do so until they are stopped by colliding with the shell of the body, for example, the interior of the skull, the thoracic cavity, and the abdominal wall. This is called the third collision."

269

Active Physics

Students should support their ideas. If students have completed *Physics in Action*, have them support their ideas with their observations from that chapter. Discuss the three parts of Newton's first law using the information in the student text.

Encourage students to make connections to each part with the observations they made during the *Investigate*. Consider asking students what they observed as they increased the speed before the collision and how this fits in with Newton's first law.

Continue the discussion, focusing on what occurs to humans (and animals) during a collision using the description by Arthur C. Damask in the student text. Emphasize that a person in a vehicular collision continues to move, even after the vehicle has stopped. The person moves until he or she collides with an object (dashboard, seat belt, and windshield) that stops him or her. Describe how the organs inside the person's body continue moving until they are stopped by the person's body walls. Consider eliciting students' ideas about why all three collisions are of concern. Discuss the concerns using the information in the student text.

Transition the discussion by asking students how seat belts helped their clay figure and what observations they made. Make connections between their observations and the width or area of the belt they used. Describe how the force each seat belt exerted to stop the passenger was the same, but the results were very different. Then introduce the concept of pressure (force per area where the force is perpendicular to the surface). Describe that when the force is distributed over a larger area it is not as great at any given point. Provide examples for students, such as what happens when they try to push a book through a piece of cardboard, or a nail through a piece of cardboard with the same force. Ask students to provide examples.

Discuss the SI units used for pressure, the pascal (Pa). $1 \text{ Pa} = 1 \text{ N/m}^2$. Consider demonstrating the example at the end of the *Physics Talk* by pulling a clay figure with the same force using just a spring scale, and a spring scale attached to a wide belt.

Checking Up

1.

An object does not change its motion (at rest or moving with constant speed in a straight line) unless an unbalanced external force acts on it. Ask students to provide examples of Newton's first law.

2.

The driver in a collision continues his or her motion until a net force acts on him or her, following Newton's first law. Students should realize that just before the collision, the driver and automobile are moving with the same speed. They may mention that what stops the person from moving is the second of the three collisions described in the student text.

3.

Students should describe the third collision using Newton's first law. A person's organs continue to move, undisturbed, until a net external force acts on them. The net force that stops the internal organs is the force applied to the organs from the inner walls of the body as the organs collide into it. Students may not realize that although organs are connected to parts of the body, they are not rigidly attached to any part of the body, so they have the ability to move slightly within the body cavity.

4.

Inertia is the tendency of an object to remain at rest or in motion and is dependent on the object's mass. Consider asking students for examples.

Newton's first law of motion can explain these three collisions when an automobile strikes a pole:

- First collision—The automobile strikes the pole. The pole exerts the force that brings the automobile to rest.
- Second collision—When the automobile stops, the body keeps moving. The structure of the automobile exerts the force that brings the body to rest.
- Third collision—The body stops, but the heart, the brain, and other organs keep moving. The body wall exerts the force that brings the organs to rest.

Even with all the safety features in automobiles, some deaths due to accidents cannot be prevented. In one recorded accident, only a single automobile was involved, with only the driver inside. The automobile failed to follow the road around a turn, and it struck a telephone pole. The seat belt and the air bag prevented any serious injuries apart from a few bruises, but the driver died. An autopsy showed that the driver's aorta (a large blood vessel) had burst, at the point where it leaves the heart. The man's organs were damaged during the "third" collision, when the heart collided with the skeleton.

Force per Unit Area: Designing a Safer Seat Belt

In *Part B* of the *Investigate*, you used a seat belt to stop the clay passenger. Newton's first law states that an object at rest will remain at rest and an object in motion will remain in motion unless acted upon by an unbalanced **force**. The cart stopped, but the passenger continued to move forward until a force acted upon it. In this part, the force stopping the passenger was the force exerted by the seat belt.

Some of the seat belts you used did not work as well as others. Each time you repeated the investigation, the stopping force that the belt exerted on the clay was the same. The force was the same because you released the cart from the same height each time. Yet different materials had different effects on the clay passenger. For example, the wire cut far more deeply into the clay than a broader material did.

The stopping force that each of the seat belts exerted on the clay was approximately the same. When a thin wire was used, all the force was concentrated onto a small area. By replacing the wire with a broader strip of material, you spread the force out over a much larger area of contact.

Force that is spread out over a given area is called **pressure**.

Physics Words

force: an interaction between two objects that can result in an acceleration of either or both objects.

pressure: force per area where the force is normal (perpendicular) to the surface; measured in N/m^2 (newtons per meter squared) or Pa (pascals).

5.

Descriptions should include pressure being force per area and how when the force is spread out over a greater area, it prevents injury and helps to stop a greater area of the person's motion. Consider asking students what they think happens to a person's head during a collision. This will be discussed in *Section 4*.

Pressure is defined as force per unit area. The pressure is much smaller with a ribbon, for example, than with a wire. It is the pressure, not the force, that determines how much damage the seat belt does to the body. A force applied to a single rib might be enough to break a rib. If the same force is spread out across many ribs, the force on each rib can become too small to do any damage. While the total force does not change, the pressure on each rib becomes much smaller.

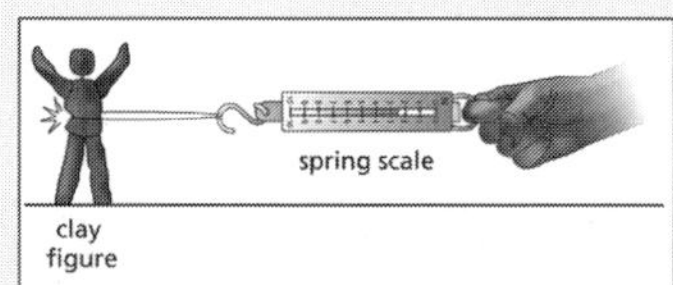

Force can be measured using a spring scale.

Checking Up

1. Explain Newton's first law of motion.
2. Why does the driver in an automobile collision remain in a state of motion when the automobile suddenly stops moving?
3. Use Newton's first law of motion to describe the three collisions.
4. Describe inertia.
5. Why did a broad band of material work better as a seat belt than a narrow wire?

Active Physics *Plus*

+Math	+Depth	+Concepts	+Exploration
◆	◆		

Calculating Pressure

Physicists are rarely satisfied just to know that the pressure decreases as the width of the seat belt increases. In physics, once a relation between two things is known, everyone wants to know, "Is there an equation?" or "Can you describe this mathematically?"

Pressure is the force per unit area:

$$P = \frac{F}{A}$$

where F is force in newtons (N)
A is area in meters squared (m^2) and
P is pressure in newtons per meter squared (N/m^2) which is also called a pascal.

Sample Problem

Two students have the same mass and apply a constant force on the ground of 450 N while standing in the snow.

Student X is wearing snowshoes that have a base area of 2.0 m^2. Student Y, without snowshoes, has a base area of 0.1 m^2.

Active Physics Plus

This *Active Physics Plus* provides an opportunity for students to increase the depth of their understanding of pressure, and describe Newton's first law and the concept of pressure using an algebraic representation. Introduce the mathematical representation to the class and discuss the sample problems provided in the student text. As a class, discuss students' responses to the questions.

1.a)

$P = (10\ \text{N})/(1.0\ \text{m}^2) = 10\ \text{N/m}^2$

1.b)

$P = (10\ \text{N})/(0.2\ \text{m}^2) = 50\ \text{N/m}^2$

1.c)

$P = (10\ \text{N})/(15\ \text{m}^2) = 0.7\ \text{N/m}^2$

1.d)

$P = (10\ \text{N})/(0.04\ \text{m}^2) = 250\ \text{N/m}^2$

2.a)

$P = (700\ \text{N})/(0.04\ \text{m}^2) =$

$17{,}500\ \text{N/m}^2$

2.b)

$P = (700\ \text{N})/(0.02\ \text{m}^2) =$

$35{,}000\ \text{N/m}^2$ (the pressure is twice as much since the area is half as much)

What Do You Think Now?

Have students review their previous answers to the question. Conduct a survey on how many students would change their answer, and ask how they would answer this question now. Point out that scientists often change their ideas as they gather more information. Consider discussing the information in *A Physicist's Response*.

Why does the student without snowshoes sink into the snow?

Strategy: This problem involves the pressure that is exerted on the snow surface by each student. You can use the equation that relates force and area to compare the pressure exerted by each student.

Given:

$F = 450\ \text{N}$

$A_x = 2.0\ \text{m}^2$

$A_y = 0.1\ \text{m}^2$

Solution:

Student Y

$$P = \frac{F}{A} = \frac{450\ \text{N}}{0.1\ \text{m}^2} = 4500\ \text{N/m}^2$$

Student X

$$P = \frac{F}{A} = \frac{450\ \text{N}}{2.0\ \text{m}^2} = 225\ \text{N/m}^2$$

Student Y sinks into the snow because the pressure that Student Y exerts on the snow is much greater than the pressure exerted by Student X.

1. What is the pressure exerted when a force of 10 N is applied to an object with each of the following areas?
   a) $1.0\ \text{m}^2$
   b) $0.2\ \text{m}^2$
   c) $15\ \text{m}^2$
   d) $400\ \text{cm}^2$
2. A person who weighs 155 lb exerts approximately 700 N of force on the ground while standing. If the person's shoes cover a total area of $400\ \text{cm}^2$ ($0.04\ \text{m}^2$), calculate the following:
   a) the average pressure the person's shoes exert on the ground
   b) the pressure the person would exert by standing on one foot

What Do You Think Now?

At the beginning of this section, you were asked the following:

- **Suppose you had to design a seat belt for a race car that can go 300 km/h (about 200 mph). How would it be different from one available on a passenger automobile?**

Using Newton's first law of motion, explain why a seat belt is an important safety feature in a vehicle. Now that you have also investigated the relationship between force and area, what would you need to consider when designing a seat belt for a race car? How do your ideas now compare to the ideas you previously recorded in your log?

Section 2 Newton's First Law of Motion: Life and Death before and after Seat Belts

Physics
Essential Questions

What does it mean?

Newton's first law is a very important part of physics because it describes how objects move in the absence of forces. Use Newton's first law of motion to explain why a passenger keeps moving when a vehicle suddenly stops.

How do you know?

What evidence do you have from your experiment that collisions at higher speeds will have a greater effect on the passenger?

Why do you believe?

Connects with Other Physics Content	Fits with Big Ideas in Science	Meets Physics Requirements
Forces and motion	Systems	* Good clear explanation, no more complicated than necessary

* Laws in physics can be applied in a wide range of situations. Describe what happens to the passengers when a bus stops quickly. How is this an example of Newton's first law?

Why should you care?

How does what you learned about Newton's first law of motion in this section help you design a safety device for a collision even though you do not know the exact circumstances of the collision?

Reflecting on the Section and the Challenge

In this section, you discovered that an object in motion continues in motion until an unbalanced force stops it. An automobile will stop when it hits a pole. However, the passenger will keep moving until something else stops the passenger, such as the interior of the automobile. The greater the speed of the automobile and passenger prior to the collision, the more damage the passenger will suffer.

Have you ever heard someone say that they can prevent an injury by bracing themselves against a collision in an automobile? This is not true. Even if your muscles were strong enough (which they are not), your bones would break in a serious accident. Restraining devices help to stop the movement of the body. With a restraining system, the force of impact is absorbed by the interior surfaces of the automobile.

In this section, you also gathered data to provide evidence on the effectiveness of seat belts as restraint systems. The seat belt was effective in applying a force to stop the clay passenger. The material used for the seat belt and the width of the restraint affected the distortion of the clay passenger. By applying the force over a greater area, the pressure exerted by the seat belt during the collision can be reduced.

Reflecting on the Section and the Challenge

Using the information in the student text, review Newton's first law and how it applies to the three types of collisions that occur during an accident, and how pressure is an important concept to keep in mind when designing safety features such as seat belts. Emphasize that these concepts are important for the *Chapter Challenge* and that students should consider these concepts as they design their prototype and form their explanations.

Physics Essential Questions

What does it mean?

Newton's first law states that an object in motion stays in motion. When the vehicle stops, the passenger keeps moving until something stops him or her.

How do you know?

At higher speeds, the clay figure was more damaged during the crash.

Why do you believe?

When the bus stops, the passengers lean or fall forward. The passengers were moving and continue to move even though the bus has stopped.

Why should you care?

You know that the passenger will move forward if the vehicle is suddenly stopped. Therefore, you have to protect the passenger while he or she continues to move forward.

Physics to Go

1.a)

Students' responses should contain the following:

You and the automobile are moving forward. The brakes apply a force and the automobile stops. Newton's first law states that an object in motion will remain in motion unless a force acts upon it. In this case, the force stops the car from moving forward. You stop moving because the interior of the automobile—seat, seat belts, dashboard, and floor—apply a force to stop you.

1.b)

You and the car are stopped. The engine provides a force that causes the car to move forward. Following Newton's first law, inertia will keep you at rest until a force acts upon you. The force that pushes you forward is exerted on you by the seat-back, which pushes you forward at the same rate as the car. To provide this force, the back of the seat must compress like a spring, and when sufficiently compressed, applies a forward force to accelerate you at the same rate as the car. The passenger is likely to attribute this compression to being "pushed back" in the seat, but actually the seat is pushing him or her forward.

1.c)

The object was moving with the same velocity as the automobile. The force, which caused the automobile to stop, was not acting on the object, only a small amount of friction from the surface of the object was resting on what was acting on the object. The inertia of the object kept it moving in a straight path (until it collided with something in the automobile to change its motion, for example, the driver or the windshield).

1.d)

During a collision, the seat belt helps provide the force necessary to stop the passenger when the vehicle stops. The seat belt typically does this over a larger area than other parts of the vehicle, such as the steering wheel or windshield, so there is less injury to the driver and passengers.

2.

Students' are to give two examples that should describe clearly how Newton's first law applies. Some students might include the sensation of being pushed forward or backward when accelerated,

It is important to note that not every safety restraint system will be a seat belt or harness, but that all restraints attempt to reduce the pressure exerted on an object by increasing the area over which a force is applied.

Physics to Go

1. Describe how Newton's first law applies to the following situations:
   a) You step on the brakes to bring your vehicle to a safe stop.
   (Sample answer: You and the vehicle are moving forward. The brakes apply a force and the vehicle stops. Newton's first law states that an object in motion will remain in motion unless an unbalanced force acts upon it. In this case, the force stops the vehicle from moving forward. You stop moving because the interior of the automobile–seat, seat belts, dashboard, floor—apply a force to stop you.)
   b) You step on the accelerator to get going.
   c) You step on the brakes, and an object in the back of the automobile comes flying forward.
   d) A vehicle is involved in a collision and a passenger is wearing a seat belt.
2. Give two more examples of how Newton's first law applies to vehicles or people in motion.
3. The skateboard, shown in the diagram at the right, strikes the curb. Draw a diagram indicating the direction in which the person on the skateboard moves after the impact. Use Newton's first law to explain the direction of movement.
4. Explain, in your own words, the three collisions during a single automobile accident as described by Professor Damask in the *Physics Talk*.

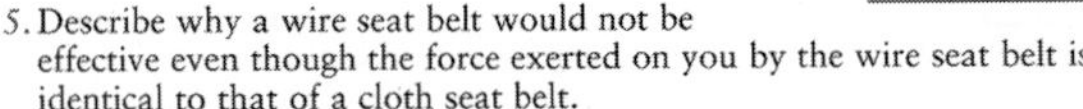

5. Describe why a wire seat belt would not be effective even though the force exerted on you by the wire seat belt is identical to that of a cloth seat belt.
6. Do you think laws making seat belts mandatory are fair? In answering this question, consider how using seat belts affect the society as a whole.
7. Suppose one of your friends wanted a ride in your automobile, but refused to wear a seat belt.
   a) Give two arguments that he or she might make against wearing a seat belt.
   b) How can you challenge these arguments using what you have learned about Newton's first law of motion?

for example, being slammed into the seat of a bus, as it accelerates from the stop. This occurs because the passenger continues his or her motion until something in the vehicle acts on the passenger to change his or her motion.

3.

A force from the curb has stopped the skateboard, however, according to Newton's first law, the person on the skateboard will continue to move in the same direction as before the collision, because there is nothing to stop him from changing that motion. This is why there is the need for a helmet, knee pads, and gloves!

4.

Students' responses should include the following: The first collision is the vehicle hitting a tree. The second collision is the driver or passenger hitting something in the vehicle. The third collision occurs when the internal organs in the driver or passenger hit the internal walls of the body, or the brain hits the internal surface of the skull.

5.

Students' responses should indicate an understanding that if the restraining belt causes an acting pressure on the passenger that that is too high, it will injure the passenger. A wire seat belt would be ineffective (and dangerous) because the force it exerts on the body to stop it during a collision occurs over too small of an area. In a severe collision, the pressure could be sufficient to cut deeply into the wearer's body, possibly causing more injury than not using a seat belt.

6.

Students should support their response. For example, people who don't use seat belts may end up having greater injuries, thereby increasing the cost of health care, or may lose control of their vehicle more easily, thereby causing more damage to their own vehicle or others. Have students explore the concept of social responsibility. What constitutes the need for any law? How does the use or lack of use of safety restraints affect us as a society? How do economics affect the passing of legislation in this area? Forty-nine states and the District of Columbia have mandatory seat-belt laws. In some states, seat-belt laws date back to 1985.

7.a)

Two of the more common reasons for not wanting to wear seat belts are the myths that if a vehicle catches fire, or if it goes into water, you will be able to exit more easily if you are not wearing a seat belt. Other reasons suggested may be that the seat belt is too constraining or uncomfortable, or that the passenger will brace him or herself in time for the collision.

7.b)

The force required to make a person stop in time to prevent injury is too big to be provided by bracing. The people who wear seat belts during a collision are far more likely to be conscious and not badly hurt, making it easier to exit a vehicle that catches fire or goes into the water. If the seat belt locks, a device can be stored in the vehicle to cut through the seat belt.

8.

The boxing glove hitting a boxer's head corresponds to the "second collision" in the Damask model. This is when the body first makes contact with an external object. The "third collision" occurs when the internal parts of the body collide. In this case, the skull moves backward, and the brain collides with the interior of the skull.

9. Active Physics *Plus*

Students should show reasoning with their response. Five hundred N divided by 1000 nails would only be 0.5 N per nail, which is not sufficient to puncture the skin. For a rough estimate of spacing, assume the 1.6-m-tall person is 0.5 m wide on average. This would give an area of $A = l \times w = 0.8 \text{ m}^2$, or 8000 cm^2. Dividing by 1000 nails means each nail should occupy 8 cm^2. If the area is a square, then each side should be the square root of 8 cm^2, or about 2.8 cm on a side. The nails in a square array should be approximately 2.8 cm from one another. Some students might note that the difficult task is getting on and off the bed of nails rather than lying on the bed of nails. The easiest way to do so would be to have the person lying flat, lowered onto and then lifted off of the bed of nails.

NOTES

Section 2 Newton's First Law of Motion: Life and Death before and after Seat Belts

8. Use the diagrams below to compare the second and third collisions described by Professor Damask with the impact of a punch during a boxing match.

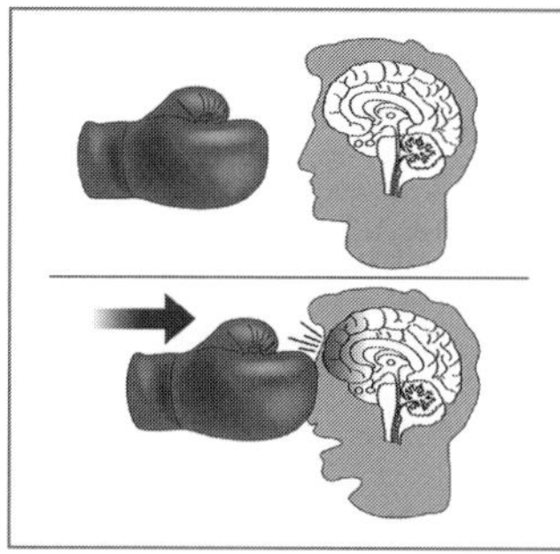

9.  A famous demonstration around the world has a person lying on a bed of nails. This looks both painful and frightening for those who do not understand physics. Your skin is able to tolerate a certain amount of pressure without tearing. You can hold heavy barbells in your hands, but the pressure is very low because the area of contact is considerably large. In contrast, you can puncture your skin with a single nail because the total area of one nail is very small.

Assume that you can push on the tip of a large, dull nail with a force of 1 N without experiencing pain or puncturing the skin. If you weigh 500 N and you had 1000 nails, then the force would only be 0.5 N per nail. How far apart should the nails be for a person who is 1.6 m tall and 0.5 m wide?

10. *Preparing for the Chapter Challenge*

Describe modifications to a seat belt that you would make if the seat belt were to be used in the following situations:

a) in a plane when it experiences an air pocket and suddenly drops down

b) in a train

c) in a bus

275

Active Physics

10.

Preparing for the Chapter Challenge

Students should describe why they have suggested the modifications using the concepts of pressure and Newton's first law. Suggestions are provided for the scenarios listed in the student text.

10.a)

When a plane suddenly drops down, the passenger lifts out of the seat due to Newton's first law. A possible seat belt for this situation would be one that goes over one or both shoulders such as a three-point buckling system as found on infant seats, or a cushioned bar as found on roller coasters.

10.b)

A seat belt similar to an automobile's would suffice for a train.

10.c)

A seat belt similar to an automobile's would suffice for a bus. School buses could take into account the smaller passenger size, allowing for adjustable shoulder straps.

Inquiring Further

1. Opinions about wearing seat belts

Students should have a minimum of five surveys for each age group: Group A = 15 to 24 years, Group B = 25 to 59 years, and Group C = 60 years and older. Check that students have the same number of surveys in each group. Students should compile, analyze, and synthesize their data. Consider having students present their data to the class.

2. Brakes in an automobile

Students should research brake systems of cars and hydraulic systems. They should relate their findings to pressure. Information on hydraulic brakes and how they work can be obtained by doing an Internet search on "hydraulic brakes." Some information your students should include for common braking systems follows. When you push against a brake, it actually pushes against a plunger in what is called the master cylinder. This forces a fluid (brake fluid) through the braking unit at each wheel. The fluid does not compress by any significant amount, so as it goes through the system it maintains the same pressure. No air is in the system because that would compress and not be as effective. The fluid usually presses on a piston that pushes the brake pads against a disk that is attached to a wheel, or the fluid is forced into a cylinder, which then pushes brake shoes out until they push against a drum attached to the wheel. In both cases, friction is used to slow the wheel.

Inquiring Further

1. Opinions about wearing seat belts

Determine what opinions people in your community hold about wearing seat belts. Survey at least five people in each of theses age groups: Group A = 15 to 24 years, Group B = 25 to 59 years, and Group C = 60 years and older. Survey the same number of individuals in each age group. Ask each individual to fill out a questionnaire. Compare the opinions of the different groups.

A sample questionnaire is provided below. Eliminate any question that you feel is not relevant. Develop questions of your own that help you understand what attitudes people in your community hold about wearing seat belts. The answers have been divided into three categories: 1 = agree; 2 = will accept, but do not hold a strong opinion; and 3 = disagree. Try to keep your survey to between five and ten questions.

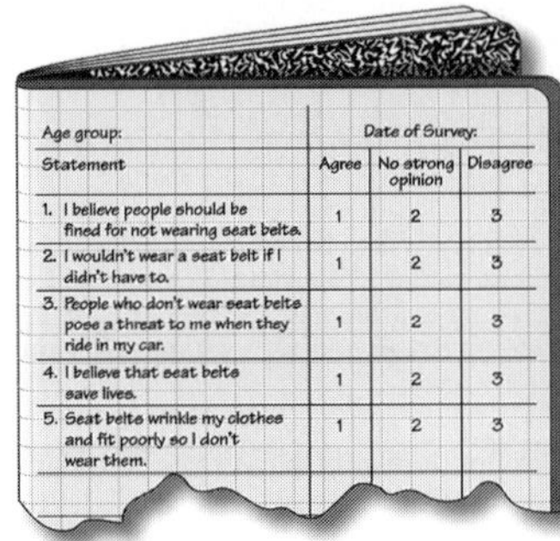

Age group:	Date of Survey:		
Statement	Agree	No strong opinion	Disagree
1. I believe people should be fined for not wearing seat belts.	1	2	3
2. I wouldn't wear a seat belt if I didn't have to.	1	2	3
3. People who don't wear seat belts pose a threat to me when they ride in my car.	1	2	3
4. I believe that seat belts save lives.	1	2	3
5. Seat belts wrinkle my clothes and fit poorly so I don't wear them.	1	2	3

2. Brakes in an automobile

How is your foot able to stop an automobile? How can the small force of your foot on the brake create a large enough force on the brakes of the automobile to stop the automobile? Investigate how the hydraulic systems in automobile brakes work and relate this to your study of pressure in this section.

Students could also discuss how a machine such as brakes can change the input force and direction.

SECTION 2 QUIZ

3-2a Blackline Master

1. Which person has the greatest inertia?

   a) A 110-kg wrestler resting on a mat.

   b) A 90-kg man walking at 2 m/s.

   c) A 70-kg man running at 5 m/s.

   d) A 50-kg girl sprinting at 8 m/s.

2. Two equal-mass safety dummies are being used to test seat belts. Dummy A has a wide seat belt and Dummy B has a narrow seat belt. Both dummies are crashed into a barrier while in the same car. Which statement below best describes the forces needed to stop each dummy?

   a) Dummy A's force is bigger than Dummy B's.

   b) Dummy B's force is bigger than Dummy A's.

   c) The forces on the two dummies are equal.

   d) The forces cannot be determined unless you know the size of the seat belts.

3. In the collision in *Question 2*, which statement below correctly describes the pressure exerted by the seat belt on the dummies during the stopping process?

   a) Dummy A's pressure is greater than Dummy B's.

   b) Dummy B's pressure is greater than Dummy A's.

   c) The pressures of the two seat belts are equal because the dummies have equal mass.

   d) The pressures of the two seat belts are equal because both dummies stop in the same time.

4. When a rapidly descending elevator is quickly stopped, blood tends to drain away from the head of a rider in the elevator. The principle that best describes this phenomenon is

   a) objects at rest tend to remain at rest.

   b) pressure varies with area and the applied force.

   c) for every action there is an equal and opposite reaction.

   d) objects in motion tend to continue in motion with constant speed.

CHAPTER 3

5. A 50-kg mass is moving at a constant speed of 10 m/s in a straight line. When the mass begins to slow down, which of the following must be occurring?

   a) The object must be losing some of its inertia.

   b) The object must be losing some of its mass.

   c) A net force must be acting opposite the object's motion.

   d) No force is needed because objects slow down naturally.

SECTION 2 QUIZ ANSWERS

1. a) A 110-kg wrestler resting on a mat.
2. c) The forces on the two dummies are equal.
3. b) Dummy B's pressure is greater than Dummy A's.
4. d) objects in motion tend to continue in motion with constant speed.
5. c) A net force must be acting opposite the object's motion.

NOTES

CHAPTER 3

SECTION 3

Energy and Work: Why Air Bags?

Section Overview

Students consider how an air bag protects a passenger during an automobile accident by using an egg to simulate the head of a passenger during a collision. They investigate what occurs during a collision between an egg and a hard surface, and what occurs when the surface is covered with about 2 cm of a softer substance such as flour for various dropping heights. Then students observe a demonstration of an egg being thrown at a large sheet. Their observations provide evidence for the conclusion that when the force is spread out over a larger distance and/or area the impact on the egg is reduced. Students use the work-energy theorem to explain their observations. They then explain how an air bag protects a passenger during a collision.

Background Information

Models are very important to scientists, particularly when the phenomenon they are studying cannot be easily observed because it is too small or big, too fast or slow, or too dangerous. Automobile accidents are dangerous, happen quickly, and involve many different interactions between objects. The first modeled automobile accidents used real people. Eventually crash dummies were developed and used. Today, crash dummies with special sensors are used. These new crash dummies simulate human movement during a crash much better than previous models, and their sensors measure the amount of velocity, acceleration, force, and torque acting on them during a collision.

The forces encountered when an air bag inflates to stop a passenger can be discussed using either the work-energy theorem or impulse and momentum. In both cases, a force acts on the automobile passenger to decrease both the kinetic energy and the momentum. Using the work-energy theorem, when work is done on an object in the direction opposite to its motion, the work causes a decrease in the object's kinetic energy. If the work is done over a larger distance, a smaller force is required to achieve the same decrease in kinetic energy.

An inflated air bag will stop an occupant of the car in a distance of 0.20 m or less, depending upon impact speed. This should be compared to the passenger being stopped by the dashboard or vehicle window in a distance of less than 0.02 m or a decrease by a factor of 10 for the force required. This lowered force is the reason injuries are greatly reduced when a passenger strikes an air bag rather than an interior surface of the vehicle. In addition, air bags often spread the force over a larger area, decreasing the pressure on a given part of the passenger.

The work-energy theorem was introduced in *Chapter 2*. The work-energy theorem states that the work done by a net force acting on a rigid object to move it through a distance d is equal to the change in kinetic energy, or

$$\vec{F} \cdot \vec{d} = \Delta KE$$

Only the component of force in the direction of the displacement does work. This is written mathematically as a scalar product between the force and the displacement. In component form this equation can be expressed as

$$F_x x + F_y y + F_z z = \Delta KE$$

A more general expression of the work-energy theorem is rewritten in terms of the sum of forces acting on the object. These forces are then placed into two groups, the conservative and nonconservative forces acting on the object. A conservative force is one that changes the potential energy of the system and is path independent; for example, the gravitational force. A nonconservative force does not change the potential energy of the system and is path dependent; for example, friction. Based on this, the work-energy theorem may be written as

$$\vec{F}_c \cdot \vec{d} + \vec{F}_{nc} \cdot \vec{d} = \Delta KE$$

$$-\Delta U + \vec{F}_{nc} \cdot \vec{d} = \Delta KE$$

$$\vec{F}_{nc} \cdot \vec{d} = \Delta KE + \Delta U$$

This states that all the work done by non-conservative forces acting on the object is equal to the change in kinetic energy plus the change in potential energy.

Crucial Physics

- The net work done on an object is equal to the change in the object's kinetic energy.
- The net work done on an object moving in a straight line is equal to the net force applied in or opposite to the direction of the displacement, multiplied by the displacement.
- The force required to stop a moving object depends upon the available distance to stop the object. Stopping in a short distance requires a large force, while stopping over a longer distance requires a smaller force.

Learning Outcomes	Location in the Section	Evidence of Understanding
Model an automobile air bag.	***Investigate*** Steps 5, 8	Students model collisions, with and without air bags, using a raw egg colliding with a countertop and on a surface covered 2-cm deep with flour, sand, or rice. Students record their observations, noting that the egg does not break when it lands on the flour. Students also observe that an egg thrown at a sheet does not break. Through these models, students obtain evidence that during a collision if the force is applied over a greater distance (and area) it does not do as much damage.
Relate the energy of a moving object to the work required to stop the object.	***Physics Talk*** ***What Do You Think Now?*** ***Physics Essential Questions*** ***Physics to Go*** Questions 1, 4, 6	Students describe the relationship between the change of energy of an object and the work needed to stop it, using their observations. Students describe and apply the work-energy theorem to solve problems.
Demonstrate an understanding about the relationship between the force of an impact and the stopping distance.	***What Do You Think Now?*** ***Physics Essential Questions*** ***Physics to Go*** Questions 1, 3, 4, 6, 7 ***Inquiring Further*** Question 2	Students describe how less force is needed to stop an object if it is applied over greater distances using the work-energy theorem. Students should apply the work-energy theorem to solve problems.

Section 3 Materials, Preparation, and Safety

Materials and Equipment

PLAN A		
Materials and Equipment	**Group (4 students)**	**Class**
Petri dish, sterile, disposable	1 per group	
Ruler, metric, in/cm	1 per group	
Paper clips, pkg 100		1 per class
Plastic wrap, roll		1 per class
Flour, 5 lb		1 per class
Raw egg*	2 per group	
Large bed sheet*		1 per class
Access to a smooth level surface*	1 per group	

*Additional items needed not supplied

PLAN B		
Materials and Equipment	**Group (4 students)**	**Class**
Petri dish, sterile, disposable	1 per group	
Ruler, metric, in/cm	1 per group	
Paper clips, pkg 100		1 per class
Plastic wrap, roll		1 per class
Flour, 5 lb		1 per class
Raw egg*	2 per group	
Large bed sheet*		1 per class
Access to a smooth level surface*	1 per group	

*Additional items needed not supplied

Note: Time, Preparation, and Safety requirements are based on Plan A, if using Plan B, please adjust accordingly.

Time Requirement

Allow one period for the students to design and run the tests. If the investigation is run over two days, students can bring in their own materials to test.

Teacher Preparation

- The raw eggs used in this section should be placed in plastic wrap to prevent spillage.
- Fasten a paper clip around the plastic wrap to keep it taut around the egg so cracks will be easily visible.
- Have students drop the egg so that it falls in the same position each time (in other words—don't let it land on its "side" once and then on one of its "ends" next).
- Conduct the investigation before class to determine what difficulties your students may have and to get an idea of the smash height for the egg colliding into the hard surface, and the depth of indentation when it is dropped onto the softer material.
- Consider ways to make cleanup efficient and easy in your classroom. You may want to have a box lid with a place cut out for the hard surface and for the dish of soft material. This way, if anything spills, it will land in the lid.
- For the demonstration, it is best to get a twin-size flat or fitted sheet. The sheet will be folded in half and two students should hold the ends of the sheet. The goal is to catch the egg in the sheet. For a fitted sheet, the egg rolling downward after hitting the sheet could get caught by the fitted part of the sheet.
- Mark a location on the sheet to indicate where the thrower should aim. In case of an accident, place the egg in a sealed bag and consider having some paper on the floor under the location where the egg should hit. If possible, try this out before class. There are videos of this demonstration on the Internet. Try the keywords "egg throw, sheet."
- Consider obtaining a video on safety to show students. Local automotive dealers have safety videos about air bags and ABS brakes. Many dealers will lend, or may even give a copy to you. You may also ask the local American Automobile Association affiliate in your area to provide safety videos; some will even send instructors to give safety talks. Some local driving companies (taxi, trucking, courier services) may also have safety supervisors who would be able to come to the classroom to talk about safety.

- The *Active Physics Transportation* content video has excellent footage showing air bags inflating. The Insurance Institute for Highway Safety has produced a wonderful video on the physics of car crashes called *Understanding Car Crashes–It's Basic Physics!* It comes with a well-written teacher's guide and is available from their Web site, which you can find by doing an Internet search for The Insurance Institute for Highway Safety.

Safety Requirements

- Check to see if any students have egg allergies or flour allergies. If so, let students know that the eggs will be contained in plastic. If an egg allergy is severe, you may want to find an alternative activity for the student to do. If students have a flour allergy, fine dry sand can be used as a substitute.
- Students holding the sheet during the egg throw should wear goggles.
- If an egg is broken and leaks out of the plastic, clean the residue immediately and wash the area with soap and water.
- When the students are throwing the egg into the bed sheet, have all the students except those holding the sheet stand behind the thrower. Make certain the area behind the sheet is clear of all obstructions for several meters. If the investigation is done in a hall, make certain no one is in the hall on the side of the sheet opposite the thrower.
- If the egg bounces out of the sheet and falls on the floor, use the cleanup procedures described previously.

NOTES

Meeting the Needs of All Students

Differentiated Instruction: Augmentation and Accommodations

Learning Issue	Reference	Augmentation and Accommodations
Measuring the indentation	***Investigate*** Step 6	**Augmentation** • As noted in *Step 6*, measuring the indentation left in the landing material is a challenging task for most students. This step provides an opportunity for differentiation. • Some students will be able to independently measure the indentation. • Some students will need the teacher to quickly model a method or two for measuring the indentation. • A few students may need a small group or one-on-one assistance to measure the indentation. **Accommodation** • Use hand-over-hand techniques (physical guidance) to assist students in measuring the indentation.
Conceptualizing an equation	***Physics Talk*** Kinetic Energy ***Physics to Go*** Question 3	**Augmentation** • The equation for kinetic energy is more complex than most of the equations students have learned so far and may require more instruction than previous equations. Students often understand how to square a number but are confused by what v^2 represents. Also, they do not know how to input $\frac{1}{2}$ into their calculators to perform the required computation. • Show students that $KE = 0.5 \cdot m \cdot v \cdot v$ is the same equation. This can lead into lessons about fractions and decimals, using the calculator for computation, and squaring a number. • The conceptual understanding of this equation is explained further in *Physics Talk, Speed and Kinetic Energy* with an example that includes values for comparison.
Understanding dimensional analysis	***Physics Talk*** Science Skills	**Augmentation** • Some students are not developmentally ready to understand this abstract concept. If students are just beginning to learn algebra and their number sense is weak, they will have a difficult time understanding what the letters for units represent. If a student struggles to understand what a unit represents, he or she will really struggle to derive units. • For these students, it may be better to have them memorize the units associated with different variables for the purpose of problem-solving and to do more work on dimensional analysis when their skills are more developed.
Solving problems	***Sample Problems*** ***Physics to Go*** Questions 4, 6, 8	**Augmentation** • Students need opportunities to practice problem-solving with new equations and receive timely feedback to make sure they learn correct procedures. • Give pairs of students the same problem to solve on a whiteboard or using a whole sheet of paper. Ask them to solve the problem individually and then compare their results with their partner to check their answers. • Remind students to draw problem-solving boxes as introduced in *Chapter 1* or to carefully show all of the steps used to solve the problem. • Provide many opportunities for students to practice identifying which variable the problems are asking students to solve for and which equation to use. **Accommodation** • Provide students with a blank sheet of problem-solving boxes. • Highlight the important information in a word problem to assist students. Withdraw this accommodation as students become more independent.

Learning Issue	Reference	Augmentation and Accommodations
Solving problems	***Sample Problems*** ***Physics to Go*** Questions 4, 6, 8	**Augmentation** • Students need opportunities to practice problem-solving with new equations and receive timely feedback to make sure they learn correct procedures. • Give pairs of students the same problem to solve on a whiteboard or using a whole sheet of paper. Ask them to solve the problem individually and then compare their results with their partner to check their answers. • Remind students to draw problem-solving boxes as introduced in *Chapter 1* or to carefully show all of the steps used to solve the problem. • Provide many opportunities for students to practice identifying which variable the problems are asking students to solve for and which equation to use. **Accommodation** • Provide students with a blank sheet of problem-solving boxes. • Highlight the important information in a word problem to assist students. Withdraw this accommodation as students become more independent.
Sketching a graph	***Active Physics Plus*** Steps 1-3	**Augmentation** • These graphs may support students' understanding of the relationships between work, kinetic energy, speed, and mass. However, some students will struggle to sketch these graphs independently or to understand the derivation. • Ask for volunteers who were able to create accurate graphs to share their graphs with the class and explain what they have learned from the graphs. • Design teacher-made graphs and ask students to use the *Physics Talk* section and apply their understanding of the new equations to explain the concepts that the graphs represent.
Understanding essential concepts	***Physics Essential Questions*** ***Physics to Go*** Questions 1, 7	**Augmentation** • Students have been asked to read, solve problems, and interpret graphs related to work and change in kinetic energy. Now they are being asked to synthesize the essential concepts based on the work they have completed. • Allow students to Think-Pair-Share to answer the *Physics Essential Questions*, and then ask them to answer the *Physics to Go* questions independently. • Think-Pair-Share means that students think for a couple of minutes about a question. Then students talk to a partner about the question and formulate an answer. Lastly, the pairs share their answers with the whole class. This strategy allows students to discuss their understandings and rethink their misconceptions. **Accommodation** • Provide direct instruction to teach these essential concepts.

Strategies for Students with Limited English-Language Proficiency

Learning Issue	Reference	Augmentation
Understanding concepts Vocabulary comprehension	***What Do You Think?***	The concept of energy management is crucial to automotive engineers, and it will be crucial to your students when they design their prototype safety system. Hold a class discussion about what energy management means in the context of automobile accidents. Ask why engineers manage the energy instead of getting rid of some of the energy. Students may need some guidance to remember the law of conservation of energy. Help students understand that energy management here means designing automobiles to control or direct the flow of energy during a collision.
Vocabulary comprehension	***Investigate*** Step 7	Help students infer the meaning of "dissipated" in context. Guide them to think about what happens when a force is applied over a distance. They should be able to discern that the energy, like the force, is also spread out, or dissipated, over a distance. Some students may have encountered the term in the context of a crowd dissipating after a ball game, or fumes spreading out after a toxic spill.
Understanding concepts Vocabulary comprehension	***Dimensional Analysis***	"Dimensional analysis" is a big term for a simple concept: making sure that you are comparing apples to apples. To help students understand, share with them the following anecdote: The *Climate Observer*, a spacecraft sent to explore Mars, was lost during a 1999 mission. NASA and other teams of scientists had planned the mission together. It was determined that one team used standard units (feet, inches) to make its calculations and another team used metric units (centimeters, meters). As a result, the *Observer's* on-board computer received incorrect numerical data when establishing its course. It went slightly off course, but the error added up to 100 km by the time the probe reached Mars. The mission was ruined in part because of inconsistent use of units.
Understanding concepts Vocabulary comprehension	***Active Physics Plus***	In dimensional analysis, checking that each term has the same dimension may require deriving, or determining, equations and the fundamental quantities (mass, length, time) from other equations. The math can be quite involved, but the concept is basic. When you work through *Deriving the Equation for the Relationship Between Work and Change in Kinetic Energy*, be sure students are able to follow the derivation in the example. Pay close attention to their graphs, as they are another tool with which you can check student understanding.
Understanding concepts Vocabulary comprehension	***Inquiring Further*** Step 2	When students design their landing pad, they need to understand the constraints, or limits, under which they have to work. Hold a class discussion to elicit ideas students have about the constraints for building a landing pad. Point out to students that they will have to recognize the constraints when they design their prototype safety system as well.

Section 3 Energy and Work: Why Air Bags?

Section 3 Energy and Work: Why Air Bags?

What Do You See?

Learning Outcomes

In this section, you will

- **Model** an automobile air bag.
- **Relate** the energy of a moving object to the work required to stop the object.
- **Demonstrate** an understanding about the relationship between the force of an impact and the stopping distance.

What Do You Think?

Automotive engineers often think in terms of energy management when they design safety systems. A good example of energy management is the use of an air bag to protect passengers during an accident.

- **How does an air bag protect you during an accident?**

Record your ideas about this question in your *Active Physics* log. Be prepared to discuss your responses with your small group and the class.

Investigate

In this section, you will use an egg to simulate the head of a passenger traveling in an automobile. An egg is a fairly good model of a human head. It has a fairly thin outer shell with a fragile interior.

1. Wrap an egg in plastic wrap or bag to minimize cleanup, as shown in the diagram on the next page.
2. Drop egg # 1 onto a hard surface (such as a counter) from a very low height. Start with a drop of 2 cm. Check the egg for any cracks. Try to drop the egg so that it always lands the same way, for example, on its side.

SECTION 3

Teaching Suggestions and Sample Answers

What Do You See?

Have students consider the illustration and the title of the section. Ask them what they think the illustration depicts and record their responses. When students discuss the air bag, ask them to compare the drivers in the illustration, and what difference the air bag may have made. This elicitation of students' initial ideas provides a focus for the science content. Encourage students to comment on how they perceive the image in context of the title of this section. Remind them that they will get a chance to return to this illustration later during this section.

CHAPTER 3

Students' Prior Conceptions

Students' understanding of the relationship between the force of the impact and the distance through which this force acts is crucial to the connections they make among the conservation of momentum, the energy involved in collisions due to mass and velocity, and the work involved in transferring energy to and from the system. This sets the stage for subsequent investigations within this chapter.

1. **Students believe that traveling at low velocities does not cause extensive or severe impact damage during collisions.** Students tend to overlook the kinetic energies involved in collisions. They recognize the transfer of energy from one object to another and realize that an object must absorb energy upon impact, but mathematical modeling forms the basis for them to alter their perception that driving at a slow speed does not cause damage during a collision. Analysis gives students the confidence to reorganize their ideas. As they calculate the energies involved in stopping a large mass moving at a slow velocity and then compare the work needed to bring a small mass moving with a large velocity to rest, they change their disregard of the physical impact of collisions with low speeds.

What Do You Think?

Discuss how automotive engineers use energy considerations in designing automobiles and their safety features. Then ask students to consider the *What Do You Think?* question. Record students' responses and encourage them to ask questions to clarify and support their ideas. Consider asking students what they think energy has to do with the air bag. Then emphasize to students that the physics they learn in this section will help them to answer this question and to solve the challenge.

Investigate

Ask students why models are used. During the discussion point out the importance of models in studying phenomena that are too big, small, fast, slow, or dangerous. Describe the model students will be using and how the egg represents a person's head. Discuss the procedure for the *Investigate* and the cleanup. Emphasize to the students that the eggs should be completely covered in plastic wrap, and that they should drop their eggs so that they land the same way every time.

Teaching Tip

Surfaces upon which it is easy to measure indentation are often ones that are quite messy. Flour seems to work well, but it scatters over a large area when the egg strikes. Place the flour or sand in a container such as a Petri dish or a tuna can. The sides of the container help to contain the flour or sand.

What Do You Think?

A Physicist's Response

An air bag can protect passengers in many ways in an accident, but it can also harm a child or small adult. When an automobile undergoes a collision, it has a large deceleration, however, the driver and passengers still move at about the speed of the vehicle until they collide with something. Colliding with a rigid object, such as a dashboard, can cause severe injury because it stops the passenger in a short distance and time, which requires great force. If a passenger can be slowed down over a greater distance and time, the force acting on the passenger to stop him or her is reduced. Reducing the force acting on a passenger reduces the severity of the passenger's injury during a collision.

As an automobile undergoes a collision, a sensor in the vehicle releases the air bags when it senses the vehicle decelerating above a certain amount. A chemical reaction transpires that releases nitrogen gas (commonly found in air) into the air bag. The gas fills the bag at about 200 mi/h (322 km/h). All of this happens in about 1/25 of a second. The back of the air bag has vents or holes through which the gas dissipates after the crash, so when a passenger collides with the air bag, he or she does not hit a rigid object but rather hits this bag that releases the nitrogen gas out of the vents in the back of the bag. This slows down the persons involved in a collision to a halt over a distance, and considerably absorbs their energy of motion, so the collision does not cause them to have such a rapid deceleration (or a great force acting on them).

Air bags can also cause injury. One problem that occurs with air bags is when the passenger is too close to the air bag as it deploys, for instance, two to three inches. A safe distance is about ten inches from the air bag.

Air bags can be fatal for children. Different types of air bags can cause different problems. It is best to read the manufacturers' warnings about the air bags in a vehicle and how they pertain to children. In general, children under the age of 13 years should ride buckled up in a properly installed, age-appropriate seat in the back seat of the vehicle.

1.

Students should make sure the eggs are covered in plastic wrap and sealed with a paper clip.

2.

Eggs should be dropped so that they always land the same way, for example, on their side.

Teaching Tip

If the students are having difficulty measuring the depression made by the egg, several methods may help. A few are listed below:

- If using flour in a tuna can, fill the can to the rim. The students can then place a ruler across the top of the rim, and measure down to the bottom of the depression.
- An alternative method would be to mark the edge of the egg that is even with the top of the surface with a marking pen. Removing the egg and placing it on a flat surface should allow the students to get a fairly accurate depression measurement.

NOTES

CHAPTER 3

3.a)

Students should observe that as the distance increases, the likelihood of the shell cracking increases. A distance of 5 cm is usually more than sufficient to crack the egg.

4.a)

After the first crack appears, subsequent cracks will be created more easily and increasing the drop height by only a few additional centimeters should cause the contents of the egg to leak.

5.a)

Students should observe that the egg leaves an indentation in the surface.

6.

One method to measure the indentation is to draw a line using a felt-tipped pen around the surface of the egg that is just at the level of the sand. Removing the egg from the sand and placing it on a hard surface such as a table will then allow the students to measure the distance between the bottom of the egg and the position of the line. This is the depth of the indentation made in the sand.

7.a)

Students should observe that the damage to egg # 1 is much greater because the stopping distance (indentation) is much smaller for this egg. Egg # 2 stops over a greater distance, requiring a smaller force to bring it to rest, and therefore, less damage occurs.

Chapter 3 Safety

For best results, hold the egg with your thumb and index finger as shown in the diagram below.

3. Gradually increase the height of the drop in increments of 2 cm until you get a crack in the shell.
   a) Record this as crack height in your log.
4. Continue dropping from greater heights until you get a full break in the shell and the yolk spills out.
   a) Record this as smash height # 1.
5. Now create a softer surface for a second egg to fall on. Try a bed of flour, sand, or rice about 2 cm thick. Drop egg # 2 from smash height # 1. Try to drop the egg so that it lands in the same way as your first egg.
   a) Record your observations.
6. Measure the depth of the indentation left in the landing material for the drop at smash height # 1. This can be challenging. Try measuring how much of the egg is still sticking out above the original level of the landing material. Then take the difference between the amount above the surface and the total height of the egg. This should be what remains below the surface, or the indentation depth.
   a) Record your measurement.
7. Compare the damage of egg # 1 and egg # 2 when dropped from the smash height # 1. When dropped from the same height, egg # 1 and egg # 2 have the same speed just before hitting the landing material. The material must supply a force over a distance (that is, the indentation of the material) for the energy to be dissipated.
   a) Compare the force (damage) and distance (indentation) of egg # 1 and egg # 2 when dropped from smash height # 1.
8. The next part of the *Investigate* is best done as a class demonstration. Take a large bed sheet to an area with a clear throwing area. Choose a volunteer that has a good throwing arm, such as a pitcher from the softball or baseball team. Have two other students be the "catchers." They should design a target for the egg-thrower by stretching the sheet out, holding the top two corners of the bed sheet over their heads and the lower two corners a little lower than their waists. The goal for the catchers is to catch the egg in the sheet by creating a cup or scoop at the bottom, which will prevent the egg from rolling off the sheet. Have the pitcher throw the egg as hard as she or he can at the center of the sheet. The pitcher should try to break the egg when it hits the sheet. Everyone else should observe the motion of the sheet when the egg hits it. Have the egg in a plastic bag. It will be a bit harder to throw, but much easier to clean up.
   a) Explain why the sheet cannot exert a force large enough to break the egg.

Active Physics 278

8.a)

It is likely that no matter how hard the thrower throws the egg, it will not break because the sheet "gives" or cushions the stopping motion of the egg. In comparison with dropping the egg on a hard surface, this stopping distance is very large.

Have a class discussion on students' observations and record them to refer to later during the chapter.

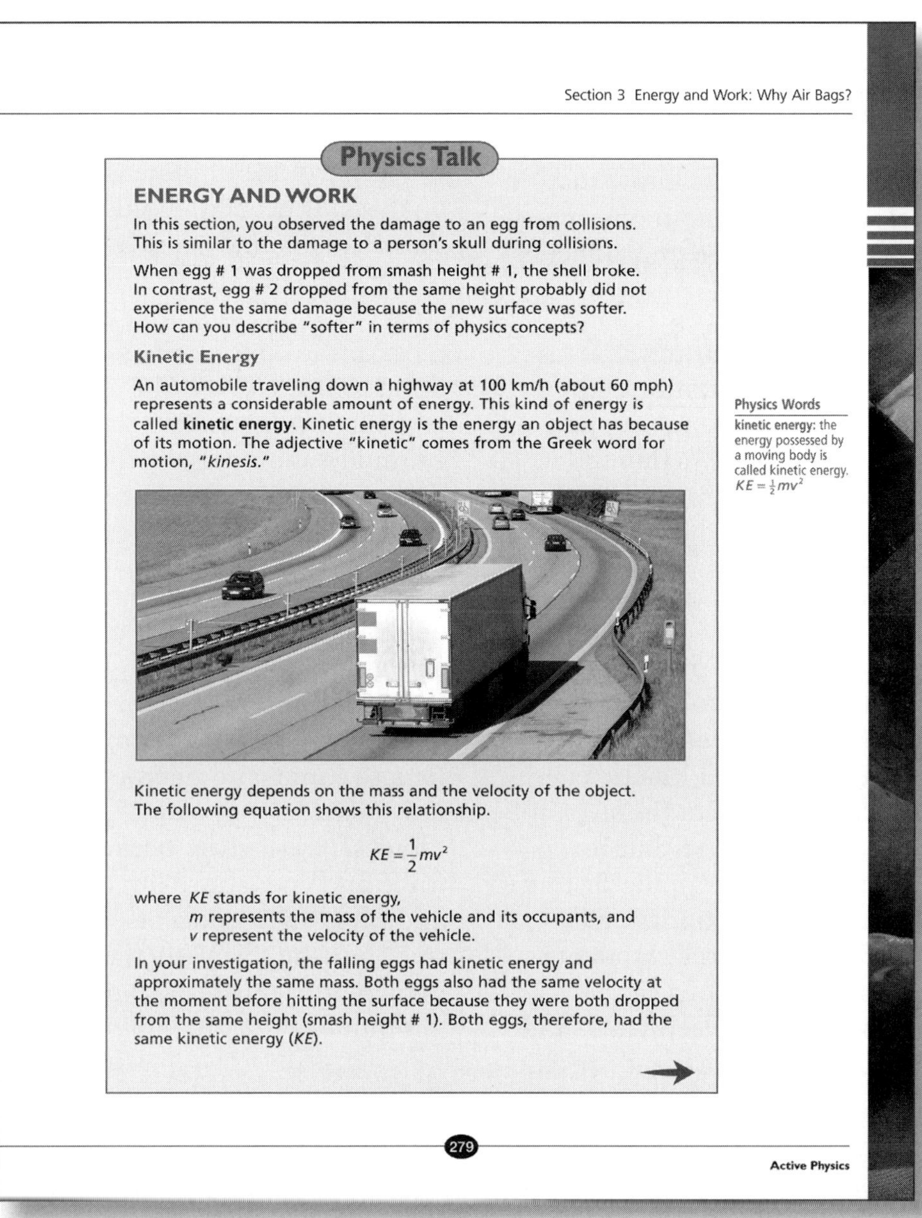

Section 3 Energy and Work: Why Air Bags?

Physics Talk

ENERGY AND WORK

In this section, you observed the damage to an egg from collisions. This is similar to the damage to a person's skull during collisions.

When egg # 1 was dropped from smash height # 1, the shell broke. In contrast, egg # 2 dropped from the same height probably did not experience the same damage because the new surface was softer. How can you describe "softer" in terms of physics concepts?

Kinetic Energy

An automobile traveling down a highway at 100 km/h (about 60 mph) represents a considerable amount of energy. This kind of energy is called **kinetic energy**. Kinetic energy is the energy an object has because of its motion. The adjective "kinetic" comes from the Greek word for motion, "*kinesis*."

Physics Words

kinetic energy: the energy possessed by a moving body is called kinetic energy. $KE = \frac{1}{2}mv^2$

Kinetic energy depends on the mass and the velocity of the object. The following equation shows this relationship.

$$KE = \frac{1}{2}mv^2$$

where KE stands for kinetic energy,
m represents the mass of the vehicle and its occupants, and
v represent the velocity of the vehicle.

In your investigation, the falling eggs had kinetic energy and approximately the same mass. Both eggs also had the same velocity at the moment before hitting the surface because they were both dropped from the same height (smash height # 1). Both eggs, therefore, had the same kinetic energy (KE).

279

Active Physics

Physics Talk

Students are introduced to kinetic energy, work, and the work-energy theorem. These concepts are needed to understand collisions and to design safety features. Relate these concepts to the *Investigate* and students' observations. Consider asking students how they would describe their observations in physics terms. Record students' ideas and revisit them during and after the presentation of physics concepts.

Teaching Tip

When having the student throw the egg at the bed sheet, adhere to the following precautions:

- A fitted sheet may prevent excess mess. When a student throws the egg, the egg will hit the sheet, and then slide down the sheet, getting caught by the edge of the fitted sheet.
- Mark a location on the sheet where the thrower should aim. In case of an accident, place the egg in a sealed bag and consider having some paper on the floor, under the location where the egg should hit.
- Have students hold the sheet tilted at a slight angle away from the thrower to make it easier to catch the egg at the bottom of the sheet.
- Show students holding the sheet how to hold it. It should be held with one hand at the top and one hand at the bottom, with one student on each side. All the students who are observing the demonstration should be behind the thrower.
- Warn the students who are holding the sheet to hold it firmly! A loosely held sheet will be pulled out of their hands by the egg's impact.
- Most importantly, the student who throws the egg should be no more than 3 ft away from the sheet. Do not be swayed by a student who claims to be a star pitcher on the baseball team, and can easily hit the sheet from 30 ft away. They have been known to miss with disastrous consequences.

CHAPTER 3

Initiate a discussion on kinetic energy. Remind students that this is the energy an object has because of its motion. Discuss the relationship between kinetic energy, mass, and speed. Ask students to describe how the kinetic energy of the egg changed when it collided with the hard and soft surfaces. Emphasize that the kinetic energies of the eggs were the same, but the kinetic energy changed much more quickly during the collision with the hard surface than the soft surface.

Remind students that work is done whenever a net force acts on an object to displace it. Emphasize that work is done only if the net force acting on the object is in the same or opposite direction to the motion of the object. Ask students if a net force was acting on the egg before it crashed or/and when it crashed and have them explain their answer. They should realize that while the egg's motion was changing, a net force acted on it. Describe how the net force exerted on the egg during the collision did work on the egg to change its motion and bring it to rest. Let students know that the total work done on an object is equal to the change in its kinetic energy.

Discuss how important it is to stop an occupant's motion in a vehicle that is in an accident, how to do it safely using the work-energy theorem from the examples provided in the student text, and how this relates to their observations with the egg.

Highlight the information presented in the student text concerning the relationship between speed and kinetic energy, and how this affects the stopping distance. Emphasize that for a given force, the stopping distance will be directly proportional to the speed squared. Students will construct graphs of the relationships between work and kinetic energy, work and speed, and work and mass later in this section.

Remind students of SI units and the importance of dimensional analysis. If needed, discuss how dimensional analysis is a way of making sure the equation is correct by checking that each term of the equation has the same dimensions. Review the examples provided, emphasizing that these examples should assist them in designing a safety device for the challenge.

Discuss the units and dimensional analysis of force, work, and energy. Consider pointing out that these units are all based on just a few fundamental units (length, time, and mass). New units, such as "newtons" and "joules" were given because force and energy are important and often-used concepts, and to honor the scientists they are named after (Sir Isaac Newton and James Prescott Joule).

NOTES

NOTES

To stop the egg, a surface had to apply a force over some distance. This distance can be seen as the indentation of the surface. For a hard surface, you cannot see the indentation. For the soft surface, you were able to measure the indentation.

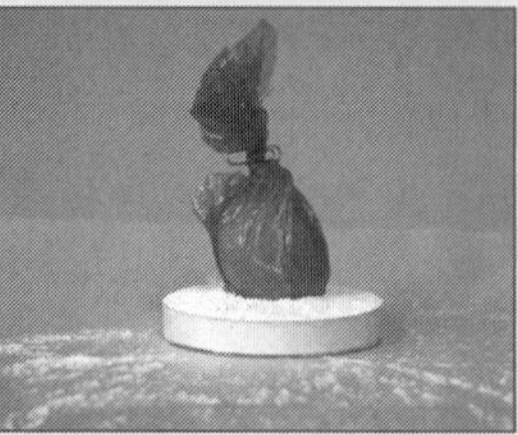

Work and Change in Kinetic Energy

In order to stop an automobile safely, the braking system must decrease the speed of the vehicle and effectively eliminate the kinetic energy by applying a great deal of force over a large distance. When force is applied over a distance, *work* is being done. **Work** is calculated using the following equation:

$$\text{Work} = \text{force} \times \text{distance}$$

$$W = F \cdot d$$

Physics Words

work: the amount of force applied on an object over a certain distance; $W = F \bullet d$

Work is equal to the change in kinetic energy. Work can either increase the kinetic energy or decrease the kinetic energy depending on the direction of the applied force and the distance (displacement) that the object moves.

You can write this as a new equation using the symbol Δ to represent "change in."

$$W = \Delta KE$$

Getting rid of the kinetic energy that an automobile and its occupants have before a collision can be done in different ways. It can be done either safely, or dangerously (causing injury to the passengers). This is what automotive safety engineers call energy management. They need to create a system that transfers the energy of the automobile safely. The work that is needed to stop the automobile traveling at a given speed is a fixed quantity, so it could be done with a small force and a large distance, or a large force and a small distance.

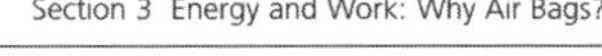
Section 3 Energy and Work: Why Air Bags?

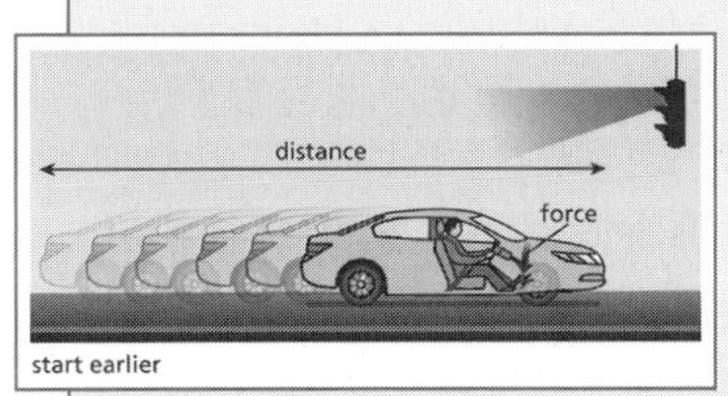

start earlier

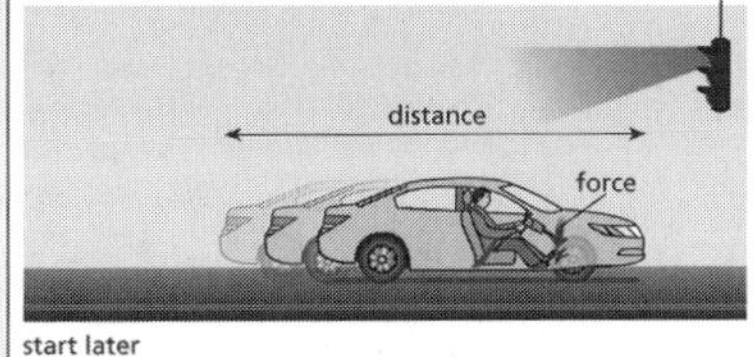

start later

Work = force • **distance** (safely) Work = **force** • distance (dangerously)

If the change in kinetic energy (*KE*) equals 5000 J (joules), different forces and distances can all provide this change.

Kinetic energy	Force	Distance	Work = $F \cdot d$
5000 J	50,000 N	0.10 m	5000 J
5000 J	10,000 N	0.50 m	5000 J
5000 J	5000 N	1.00 m	5000 J

You should see that the smaller the stopping distance, the larger the force. You noticed this in the *Investigate* in which egg # 1 hitting the hard surface had more damage than egg # 2 hitting the soft surface which has a larger indentation or distance. This is what happens in a collision, where the automobile and its passengers come to a stop in a very small distance. If a passenger strikes the interior of the automobile, such as an unpadded steering wheel or windshield, the stopping distance will be very small. That would require an extremely large force. Such a force would certainly cause serious injury. That is why air bags are used. An air bag increases the distance over which the stopping force is applied. The force is reduced considerably.

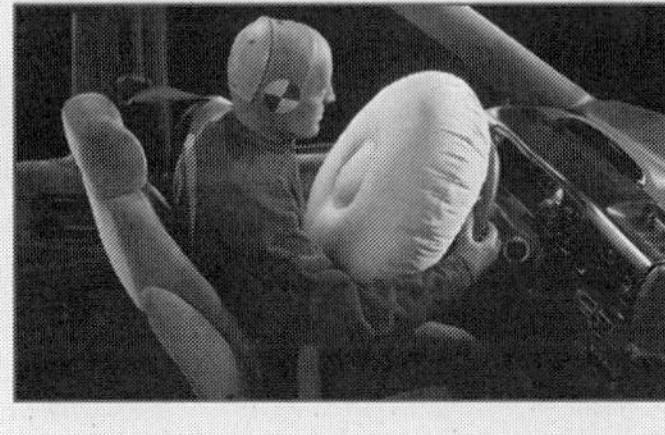

CHAPTER 3

Speed and Kinetic Energy

You noticed that the work, W, equals the change in kinetic energy, KE. The fact that $W = \Delta KE$, also helps to explain why slowing down is always a good safety move.

To stop an automobile moving at 9 m/s (20 mi/h) requires the force of the brakes to be applied over a fairly large distance. An automobile traveling at 9 m/s requires about 6 m to stop safely. Imagine that an automobile going three times as fast would require three times the distance to stop, but this is not the case. If you carefully examine the formula for kinetic energy ($KE = \frac{1}{2}mv^2$), you will notice that the energy is proportional to the square of the velocity. That is, if you triple the speed, the KE is not three times greater, but it is nine times greater (3^2 or 3×3). It would take an automobile traveling at 27 m/s nine times the distance to stop than at 9 m/s, assuming that the brakes apply the same force. That means that it would require 54 m (9×6 m) to stop safely. The fact that kinetic energy is proportional to the square of the velocity also explains why high speed greatly increases the damage done during a collision. The automobile at three times the speed has nine times as much kinetic energy requiring nine times the distance to stop safely.

If you assume that the mass of a car is 1000 kg, you can see the effect of speed on kinetic energy and stopping distance for a given braking force. In the table below, you should notice the following:

- The speed has tripled from 9 m/s in the first row to 27 m/s in the second row.
- The kinetic energy has increased nine times as the speed tripled.
- The work required to stop the car is equal to the kinetic energy at all speeds.
- The braking force of the car is constant irrespective of the speed.
- The stopping distance increased by a factor of nine when the speed tripled.

Speed (meters per second)	$KE = \frac{1}{2} mv^2$ (joules)	Work to stop car $W = \Delta KE$ (joules)	Braking force (newtons)	Stopping distance (meters)
9	40,500	40,500	6740	6
27	364,500	364,500	6740	54

Section 3 Energy and Work: Why Air Bags?

On dry roads, you can safely stop a car traveling at 9 m/s in 6 m. Drivers are used to this speed-to-distance relationship after many years of driving. When the road is wet or icy, there is less force between the tires and the road. With less force you need a larger distance to stop, since the work is still the same for a given amount of kinetic energy. Alternatively, you can decrease the speed of your car so that the kinetic energy decreases and the required work and stopping distance will be smaller. The key to safety when road conditions change is to slow down. Slowing down will permit a safer stop and will also significantly decrease the kinetic energy (and damage) if there is an accident.

For example, imagine a 70.0-kg passenger traveling at a speed of 13 m/s. That person's kinetic energy is

$$KE = \frac{1}{2}mv^2$$
$$= \frac{(70.0\text{ kg}) \times (13\text{ m/s})^2}{2}$$
$$= 5915\text{ J or } 5900\text{ J (rounded off to two significant figures)}$$

Dimensional Analysis

Notice how the numbers and units were handled in the solution. The numbers were multiplied together to get the numerical answer:

$$\frac{1}{2} \times 70 \times 13^2 = \frac{70 \times 13 \times 13}{2}$$
$$= 5915$$

CHAPTER 3

The units were also multiplied together in the same way:

$$kg \times \left(\frac{m}{s}\right)^2 = kg \times \left(\frac{m}{s}\right) \times \left(\frac{m}{s}\right)$$
$$= kg \cdot \frac{m^2}{s^2}$$

This derived SI unit is given a special name. The unit for energy is called a joule (J).

$$1\ J = kg \cdot \frac{m^2}{s^2} \text{ or } kg \cdot m^2/s^2$$

Paying attention to units is an important problem-solving skill and tool. It is called dimensional analysis.

To stop the person, something has to do the 5900 J of work to get rid of that energy. In an accident, it could be the windshield. What would happen if the person strikes the windshield? Since the windshield is fairly rigid, it might only give 3.0 cm (0.030 m) in stopping the person. So the work done by the windshield is

$$\text{Work} = \text{force} \times \text{distance}$$
$$5900\ J = \text{force} \times 0.3\ m$$
$$\text{Force} = \frac{5900\ J}{0.030\ m}$$
$$= 196{,}667\ N \text{ or } 197{,}000\ N \text{ when rounded off}$$

That is a lot of force exerted on the skull. What happens if the passenger strikes a fully inflated air bag instead of the windshield? Suppose that the air bag creates a stopping distance of 30.0 cm (10x greater stopping distance). The amount of work to be done is still the same, 5900 J. But this time, it is applied over a greater distance than the 3 cm of the windshield.

$$\text{Work} = \text{force} \times \text{distance}$$
$$5900\ J = \text{force} \times 0.3\ m$$
$$\text{Force} = \frac{5900\ J}{0.3\ m}$$
$$= 19{,}667\ N \text{ or } 19{,}700\ N$$

This is still a lot of force, but it is much less than before (10x smaller force). Air bags are not the only system in the automobile that is designed to absorb energy. Seat belts and crumple zones in the frame of the automobile also help a lot. You will learn more about crumple zones in a later section.

The work done by the air bag decreases the kinetic energy of the person. However, energy in the entire system must remain the same. In this case, the kinetic energy of the person decreases while the energy of the air bag increases. An air bag with increased energy may become a bit hotter as all the molecules in the air bag gain some kinetic energy. Some of the energy during the collision may have produced some sound energy as well.

SI Units of Force, Work, and Energy

Notice that dimensional analysis was used, once again, in calculating the force and the work.

Newton's second law states that force is equal to mass multiplied by acceleration.

$$F = ma$$

The units of mass (kg) multiplied by the units of acceleration (m/s^2) provide the units for force ($\frac{kg \cdot m}{s^2}$ or $kg \cdot m/s^2$).

Since force is such an important concept in physics, this unit is given a special name, the newton (N).

$$1\ N = 1\frac{kg \cdot m}{s^2} \text{ or } 1\ kg \cdot m/s^2$$

Work is equal to force multiplied by distance. $W = F \cdot d$

Referring only to the units

$$\begin{aligned} W &= N \cdot m \\ &= \frac{kg \cdot m}{s^2} \cdot m \\ &= \frac{kg \cdot m^2}{s^2} \text{ or } kg \cdot m^2 \cdot s^{-2} \\ &= J \end{aligned}$$

Work and kinetic energy are equivalent, and therefore both are expressed in joules.

Also, notice the unit in the following calculation:

$$F = \frac{W}{d}$$
$$= \frac{J}{m}$$
$$= \frac{kg \cdot m^2 \cdot s^{-2}}{m}$$
$$= \frac{kg \cdot m}{s^2} \text{ or } kg \cdot m/s^2$$
$$= N$$

Sample Problem

A total of 12,000 J of work is required to stop a 45-kg cart.

a) What speed would the cart be traveling before it was brought to a stop?

Strategy: This problem involves work required to stop a moving object. Work and kinetic energy are equivalent. That is, 12,000 J of work are necessary to change 12,000 J of *KE* of the cart to 0 J. You are given the mass of the cart. You can use the equation that relates kinetic energy and mass to calculate the speed of the cart.

Given:

$KE = 12,000$ J

$m = 45$ kg

Solution:

$$KE = \frac{1}{2}mv^2$$
$$v^2 = \frac{2KE}{m}$$
$$v^2 = \frac{2(12,000)\,J}{45\,kg}$$
$$v^2 = 533.3\,\frac{J}{kg}$$
$$v = \sqrt{533.3\,\frac{\cancel{kg} \cdot m^2 \cdot s^{-2}}{\cancel{kg}}}$$
$$= 23\,m/s$$

The speed of the cart was 23 m/s.

b) If the cart stopped in a distance of 3 m, what force was needed to stop the cart?

Strategy: This problem involves the kinetic energy and distance it takes to stop. You can use the equation that relates work and distance to calculate the force.

Given:	*Solution:*
$W = 12{,}000$ J	$W = F \cdot d$
$d = 3$ m	$F = \frac{W}{d}$
	$F = \frac{12{,}000 \text{ J}}{3 \text{ m}}$
	$F = 4000$ J/m
	$= 4000$ N

Checking Up

1. What factors determine a body's kinetic energy?
2. When work is done on an object, what is the effect on its kinetic energy?
3. How does the force needed to stop a moving object depend upon the distance the force acts?
4. What is the unit of kinetic energy? What is the unit for work?

Active Physics *Plus*

+Math	+Depth	+Concepts	+Exploration

Deriving the Equation for the Relationship between Work and Change in Kinetic Energy

In the *Physics Talk*, the equation for work and the relationship that work equals the change in kinetic energy were stated. By using algebra, this relationship can be derived from the definitions of work, Newton's second law and a motion equation that emerges from the definition of velocity and acceleration. (The derivation of the motion equation also uses algebra that is not shown here.)

Derive the equation for

W = change in KE

$W = Fd$

Since $F = ma$

$W = mad$

Using the definitions of $a = \Delta v / \Delta t$ and average $v = \Delta d / \Delta t$, you can derive one of the motion equations:

$$v_f^2 = 2ad + v_i^2$$

$$W = \frac{m(v_f^2 - v_i^2)}{2}$$

$$= \frac{1}{2}mv_f^2 - \frac{1}{2}mv_i^2$$

Work is required to change the KE of an automobile so that it stops.

$W = \Delta\left(\frac{1}{2}mv^2\right)$, where Δ is a symbol for "change."

In order to demonstrate your understanding of this equation:

1. Sketch a graph of the work necessary to bring an automobile to a stop as a function of the automobile's kinetic energy.
2. Sketch a graph of the work required to bring an automobile to a stop as a function of the automobile's speed.
3. Sketch a graph of the work required to bring an automobile to a stop as a function of the automobile's mass.

Checking Up

1.

The factors that determine a body's kinetic energy are mass and speed.

$KE = \frac{1}{2}mv^2$

2.

An object's kinetic energy changes when work is done on an object. The total work done on the object is equal to the change in its kinetic energy. For constant mass this is written as

$W_{net} = \Delta KE = \frac{1}{2}mv_{final}^2 - \frac{1}{2}mv_{initial}^2$

3.

The force needed to bring a moving object to rest can be found using the work-energy theorem. This shows that as the distance the net force acts on the object increases, the amount of force needed decreases. This only applies to the net force in or opposite to the direction of motion.

$W_{net} = F_{net \parallel d}\, d = \Delta KE =$

$\frac{1}{2}mv_{final}^2 - \frac{1}{2}mv_{initial}^2$

$$F_{net \parallel d} = \frac{\Delta KE}{d}$$

4.

Kinetic energy is measured in joules, as is work.

Active Physics Plus

This *Active Physics Plus* derives the equation for the relationship between work and energy based on the equation for force and the kinematic equations. This provides students with more mathematical depth. They are also asked to graph and analyze the net work done on an automobile as a function of its kinetic energy, mass, and initial speed to show their understanding of the factors in the work-energy theorem.

Discuss how the work-energy theorem can be derived from the definition of work, following the information in the student text. Have students demonstrate their understanding of the work-energy theorem by graphing relationships involving work.

1.

Students' graphs should include axes labels and units, and should look similar to the following linear graph, indicating a linear relationship between work and change in kinetic energy to bring a vehicle to rest. The change in kinetic energy is negative because the vehicle comes to rest, and hence, the work done is negative. This is because the force doing the work opposes the direction of motion.

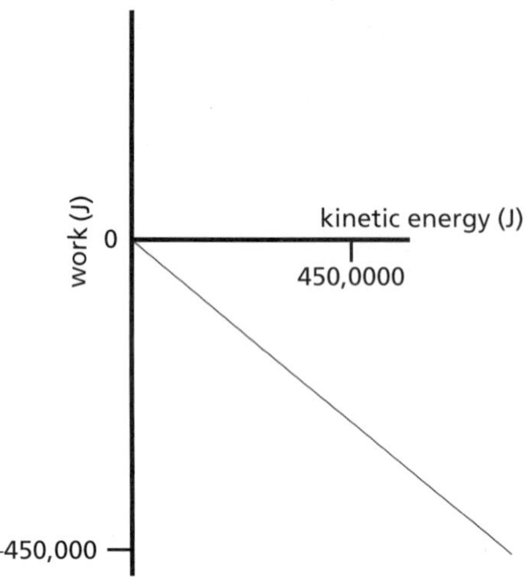

2.

Students' graphs should look similar to the graph shown below, indicating that the amount of negative work (work done by a force opposing the direction of motion) increases as the square of the initial speed. The work done is negative because the force doing the work is opposite to the direction of motion.

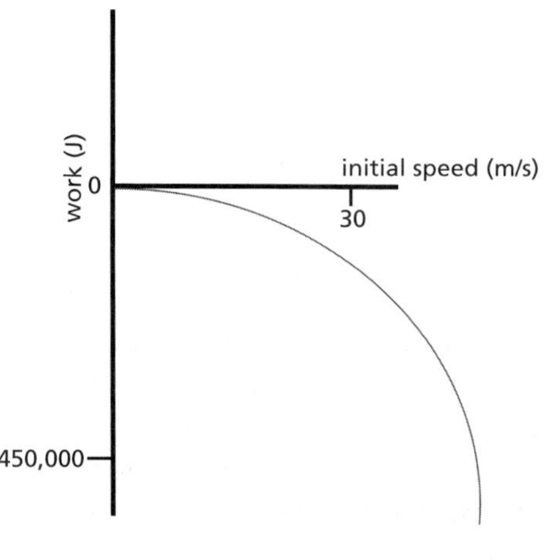

What Do You Think Now?

At the beginning of the section, you were asked

- **How does an air bag protect you during an accident?**

Explain how an air bag can protect you. Be sure to include the ideas of work and change in kinetic energy in your explanation.

Physics

Essential Questions

What does it mean?

During a crash, you will eventually stop. It can occur when you hit the hard dashboard or the softer air bag. Physics can be used to determine how much force will be exerted on you. Explain how two different forces can cause the same change in the kinetic energy of an object. Include the definition of work in your explanation.

How do you know?

How did you test whether a material is good for cushioning an egg during a collision?

Why do you believe?

Connects with Other Physics Content	Fits with Big Ideas in Science	Meets Physics Requirements
Forces and motion	* Conservation laws	Experimental evidence is consistent with models and theories

* Energy is an organizing principle of all science. It states that the total energy of a system remains the same. It also allows for a force to be applied by an external force to change the energy. Explain how you can change the energy of an egg or an automobile during a collision without violating the conservation of energy.

Why should you care?

How will you use the physics concept of $Work = F \cdot d = \Delta KE$ to help design your safety system?

3.

Students' graphs should look similar to the graph shown to the right.

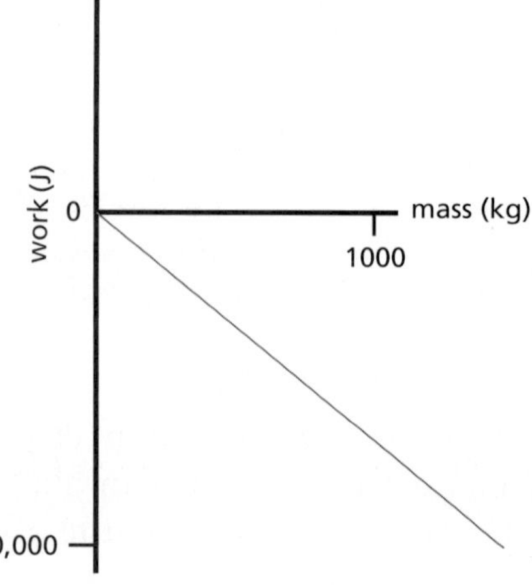

Reflecting on the Section and Challenge

In this section, you found that softer surfaces were better able to protect an egg during a collision. Similarly, softer surfaces such as air bags are able to protect you by extending the distance it takes to stop you in an automobile accident. Without the air bag, you will hit something else that will stop you in a shorter distance.

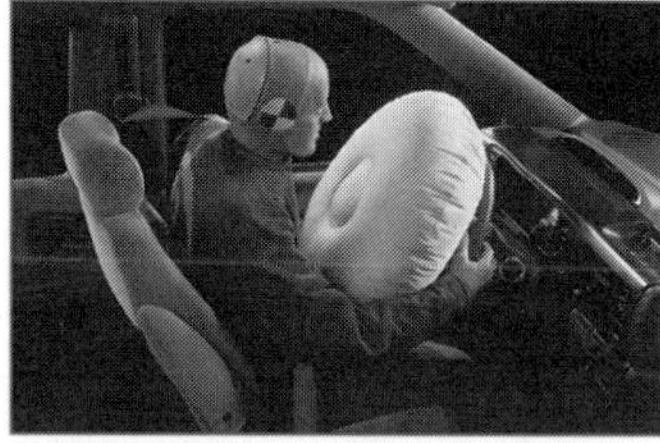

Work produces a change in the kinetic energy. A large force over a short distance or a small force over a large distance are two ways in which work can produce the same change in kinetic energy. The large force can injure you. With an air bag, the stopping distance is larger and therefore, the force required to stop you is smaller.

Energy and work must be considered in designing your safety system. Stopping an object over a large distance reduces the damage. The harder a surface is, the shorter the stopping distance, and the greater the damage. In part, this provides you with a clue to the use of padded dashboards and visors in newer vehicles. Understanding energy and work allows designers to reduce damage both to vehicles and passengers.

What Do You Think Now?

Students should reflect on their earlier answers to the *What Do You Think?* question and revise them based on their current understanding of the work-energy theorem. You might want to provide students with *A Physicist's Response* to give them a better understanding of how air bags work and the physics behind how they can reduce the severity of injury during a collision.

Reflecting on the Section and the Challenge

Emphasize to students that the main physics concept in this section is the work-energy theorem, which can explain why a smaller force can be used to stop a moving object (or person) in a collision if it can be applied over a greater distance. Remind students that they should use this information when they are designing and building a prototype and preparing an explanation for the *Chapter Challenge*.

Physics Essential Questions

What does it mean?

The change in kinetic energy is due to a force applied over a distance (defined as work). A large force over a short distance can do the same work as a small force over a large distance.

How do you know?

The egg was dropped on different surfaces. The damage to the egg was less for certain "soft" materials.

Why do you believe?

An egg or an automobile can do work on an object by applying a force over a distance and changing the kinetic energy of the object. The object receives energy that was transferred to it from the egg or the automobile. The amount of energy the object receives is equal to the reduction in energy of the egg or automobile. The energy was transferred by the egg or automobile applying a force over a distance on the object. The total energy of the system (egg or automobile and the object) stays the same.

Physics to Go

1.a)

The padding of the mitt increases the stopping distance of the ball, as does the catcher's hand when it recoils backward as the ball strikes the glove. Because the stopping distance is increased, less force is required to bring the ball to rest.

1.b)

When a person jumps onto the ground from a large height, he or she bends his or her knees, which flex as the person's feet strike the ground. This increases the stopping distance and thus, less force is needed by the legs to stop the person.

1.c)

When a person bungee jumps, kinetic energy is gained as he or she falls. To stop the person, the bungee rope provides a small force over a very long distance as the rope stretches. This small force reduces the jumper's speed (and his or her kinetic energy) to zero, allowing the jumper to be safely stopped without injury.

1.d)

The mat used in professional wrestling is generally soft and has a certain amount of spring to it that allows the wrestlers to fall onto the mat without being hurt. The mat's padding and spring allow the wrestlers to be stopped over a longer distance, requiring less force to bring them to rest. The mat is also constructed to make a pronounced sound when a wrestler falls on it, adding to the drama.

Chapter 3 Safety

Physics to Go

1. There are many situations in which the force of an impact is reduced by increasing the stopping distance. Explain how each of the following actions reduces the force by increasing the stopping distance. Use the terms force, energy, and stopping distance in your answers.
   a) Catching a hard ball with a catcher's mitt
   b) Jumping to the ground from a large height and bending your knees
   c) Bungee jumping
   d) A wrestler's mat
2. If you triple the speed from 20 km/h to 60 km/h, by what factor does the kinetic energy increase?
3. Two eggs are thrown at a blanket. One egg is thrown at twice the speed of the other. If you assume that in both cases the force exerted by the blanket on the egg is the same, how far will the faster egg travel once it hits the blanket as compared to how far the slower egg travels when it hits the blanket?
4. If the work to stop an object is 60 J, list in your log three force and distance combinations that can stop the object.
5. Copy and complete the following table. The first row has been completed for you. In all the other rows, one of the values is missing.

Kinetic energy	Mass	Speed
500 J	1000 kg	1 m/s
	1000 kg	20 m/s
100,000 J		20 m/s
50,000 J	500 kg	
	1000 kg	30 m/s

6. A person with a mass of 60.0 kg is traveling at 18 m/s.
   a) How much kinetic energy does this person have?
   b) How much work is required to stop the person?
   c) Calculate the force required to stop the person if the stopping distance is 50.0 m.

290

Active Physics

Physics Essential Questions

Why should you care?

The work-energy theorem is helpful in the design of a safety system because from it, you realize that the kinetic energy of the passenger must be reduced by applying a small force over a long distance. If the kinetic energy of the passenger is reduced over a short distance, the force on the passenger will be too great and will increase the chance of injury to the passenger.

For *Questions 7 and 8*, choose the best answer from those provided.

7. An egg dropped on a pillow is less likely to break than an egg dropped on concrete because:
   a) The pillow provides a larger force over a larger stopping distance.
   b) The pillow spreads the force out over a longer stopping distance.
   c) The egg loses less energy when it is stopped by the pillow as compared to when it is stopped by the concrete.
   d) The concrete contains sharp bits of sand and rock that break the shell.
8. A 60.0-kg runner has 1920 J of kinetic energy. At what speed is she running? Show your work.
   a) 5.66 m/s
   b) 32.0 m/s
   c) 8.00 m/s
   d) 64.0 m/s
9. Active Physics Plus In *Active Physics Plus*, you used the equation $v_f^2 = 2ad + v_i^2$. Demonstrate your algebra skills, and use the definitions of $a = \Delta v/\Delta t$ and average $v = \Delta d/\Delta t$, to derive the equation.
10. ***Preparing for the Chapter Challenge***

    What safety devices do you know that decrease the force experienced by passengers by increasing the distance over which the collision does work on the passenger? What are the characteristics of those devices that allow them to manage energy better? What can you put in your safety design to manage the energy of a collision?

Inquiring Further

1. **Are air bags always safe?**

   Conduct an Internet search on air bag safety. Do air bags sometimes cause injuries? What are the dangers? What is the physics behind those dangers? How can people reduce the chance of injury due to air bags?
2. **Design a landing pad**

   Design your own landing pad with whatever materials your teacher approves. A pillow might be a good landing pad, but it has one drawback that would make it unsuitable for an automobile. It takes up too much space. Automobile engineers are also limited by other constraints in the design of safety devices. For this problem, you are limited to a landing pad height of 5 cm.
   a) What is the maximum height that you can drop your egg onto your landing pad without getting a crack in the shell?
   b) Do a survey of what other students found and check out their landing pads.
   c) What are the characteristics or properties of a good landing pad?

2.

Kinetic energy is proportional to the square of the speed. If the speed increases by a factor of three, the kinetic energy increases by a factor of three squared, or nine times the energy.

3.

Students should use the work-energy theorem. When two eggs are thrown at a blanket, the egg thrown at twice the speed of the slower egg will have four times the kinetic energy. If the blanket applies the same amount of force to stop both eggs, the one thrown at twice the speed will need four times the distance to be stopped by the blanket. This is why speed is a determining factor in the seriousness of automobile accidents.

4.

Students' responses should include any force and distance combinations that when multiplied together equal 60 N · m. They should be aware that the force points opposite to the direction of motion. Some possible force and distance combinations to stop an object with a kinetic energy of 60 J are as follows:

Force	Distance
60 N	1 m
30 N	2 m
20 N	3 m
10 N	6 m

5.

Kinetic energy	Mass	Speed
500 J	1000 kg	1 m/s
200,000 J	1000 kg	20 m/s
100,000 J	500 kg	20 m/s
50,000 J	500 kg	14.1 m/s
450,000 J	1000 kg	30 m/s

6.a)

$KE = \frac{1}{2}mv^2 =$

$\frac{1}{2}(60.0\ \text{kg})(18\ \text{m/s})^2 = 9720\ \text{J}.$

6.b)

Students should note that the work required to stop the person is equal to the change in kinetic energy, or –9720 J. The negative sign arises because the force doing the work must oppose the direction of motion.

6.c)

194 N. Students should use the relationship that the force times the distance is equal to the work done.

7.b)

The correct choice is *b)*. The pillow spreads out the force over a larger stopping distance.

CHAPTER 3

8.c)

$KE = \frac{1}{2}mv^2$

$1920\text{ J} = \frac{1}{2}(60\text{ kg})v^2$

$$v = \sqrt{\frac{2(1920\text{ J})}{60\text{ kg}}} = 8\text{ m/s}$$

Active Physics Plus

9.

Students should note that for constant acceleration

$$a = \frac{\Delta v}{\Delta t} = \frac{(v_{\text{final}} - v_{\text{initial}})}{t}$$

and that for constant acceleration the average velocity is given by

$$v_{\text{average}} = \frac{\Delta d}{\Delta t} = \frac{d}{t} \text{ and,}$$

$$v_{\text{average}} = \frac{(v_{\text{final}} + v_{\text{initial}})}{2}.$$

Combining these two equations and solving for t you have

$$\frac{d}{t} = \frac{(v_{\text{final}} + v_{\text{initial}})}{2}$$

$$t = \frac{2d}{(v_{\text{final}} + v_{\text{initial}})}.$$

Substituting this value of t in the equation for the acceleration you have

$$at = (v_{\text{final}} - v_{\text{initial}})$$

$$a\frac{2d}{(v_{\text{final}} + v_{\text{initial}})} = (v_{\text{final}} - v_{\text{initial}})$$

$$2ad = (v_{\text{final}} - v_{\text{initial}})(v_{\text{final}} + v_{\text{initial}}) = v_{\text{final}}^2 - v_{\text{initial}}^2.$$

Rearranging the terms, you have

$$2ad = v_{\text{final}}^2 - v_{\text{initial}}^2$$

$$v_{\text{final}}^2 = v_{\text{initial}}^2 + 2ad.$$

10.

Preparing for the Chapter Challenge

Students should describe more than one safety device and how the devices increase the distance over which the collision does work on the passenger. Students should note that for each device, the increase in distance over which the collision does work reduces the amount of force needed to stop the vehicle, which reduces the amount of force needed to stop the passenger. In this case, the energy transferred to the passenger is reduced. The total energy is still conserved; however, some of that energy may go into deforming the vehicle, producing a sound, heating, or another form of energy. Some examples are provided below.

- The rear bumper might be mounted on a piston or a spring, which compresses when the bumper strikes an object. Even without a spring or piston, the bumper can dent before the rigid frame of the car strikes the object. This increases the distance of the collision, reducing the amount of force that acts, and decreases the amount of energy transferred to the passenger during the collision.
- The crush or crumple zones allow the vehicle to compress like an accordion. Previously, the sheet-metal parts of the vehicle were welded continuously along the seams, conveying the impact from object to occupants. After the front of the vehicle collides, the crumpling of the vehicle increases the distance over which the force of the collision acts, until the rest of the vehicle stops. This greatly reduces the force of the collision and decreases the amount of energy transferred to the passenger.
- The collapsible steering wheel crushes on impact. Rather than impaling the driver's body on the rigid shaft, the steering column telescopes inward, increasing the stopping distance. Increasing the stopping distance of the passenger decreases the force acting on the passenger.
- Soft padding on all surfaces increases the stopping distance and reduces force.

Inquiring Further

1. Are air bags always safe?

Students should research this question and their responses should contain the following information.

Air bags are not always safe, and can cause injury in certain situations. Air bags deploy at high speeds. If a driver or passenger is too close to the air bag, he or she could suffer injuries such as broken bones (rib cage) from air bag deployment. Small passengers, pregnant women, and small children (under the age of 12) can be severely injured or killed by air bags. For a child with little mass and a much shorter stature, air bags can be fatal. If an infant in a rear-facing child seat were to be placed in the front seat, his or her head would receive the greatest impact from the air bag.

Because the air bag must fill in a fraction of a second, the gas filling the air bag does so at extremely high speeds (around 200 mi/h) and pressure. The motion of this gas is toward the person and the person is moving toward the air bag. If the two collide while the air bag is filling, the force of impact is great over a short distance and time. In this case, the force acting on the moving passenger is enough to break bones. Once the air bag is filled, the gas inside quickly comes to equilibrium and starts to slowly evacuate the vents in the back of the air bag. As a person collides with a gas-filled air bag, the gas is pushed out of these vents slowly. The person's speed is decreased to zero over a distance, reducing the force exerted on the person to stop him or her.

To avoid the dangers associated with air bags, children less than 12 years of age should sit in the back seat of the vehicle. Passengers and drivers should have the air bag at least 10 in. from their breastbone, pointed toward their breastbone.

2. Design a landing pad.

Let students know what materials they are allowed to use or have them get your approval of the materials they wish to use. Emphasize that the landing pad can be no more than 5 cm high.

2.a)

Students should list the maximum height from which they dropped their egg without its shell cracking.

2.b)

Students should record what other students' maximum heights were and describe their landing pads. Consider asking students what physics concepts are shown, for example, landing pads that allowed for greater heights before cracking the egg must have dissipated the force over a larger area and/or allowed for the force to act over a greater distance.

2.c)

Students should note that a good landing pad is one that acts over a greater distance, requiring less force to stop the object, and dissipates the force over a greater area.

NOTES

CHAPTER 3

SECTION 3 QUIZ

3-3a Blackline Master

1. In a demonstration, a student throws a raw egg at a bed sheet held by two other students, and the egg does not break. The best reason to explain why the egg does not break is that

   a) The sheet is made of soft cotton.

   b) The student threw the egg at an angle so it struck the sheet at a glancing blow.

   c) The sheet stops the egg over a large distance, leading to a small force.

   d) The egg landed on its end, which is harder to break.

2. A 50-kg girl is running along a track at 4 m/s. What is the girl's kinetic energy?

   a) 200 J b) 400 J

   c) 800 J d) 650 J

3. A 1.0-kg cart traveling at a constant 2.0 m/s across a level lab table 0.80 m above the floor is stopped in a distance of 0.20 m. The amount of work required to stop the cart is

   a) 1 J. b) 2 J.

   c) 2.5 J. d) 4 J.

4. In *Question 3*, if the cart stopped in half the distance, the force required to stop the cart would be

   a) half as great.

   b) the same.

   c) twice as great.

   d) four times as great.

5. In *Question 3*, if the cart stopped in half the distance, the amount of work required to stop it would be

   a) half as great.

   b) the same.

   c) twice as great.

   d) four times as great.

SECTION 3 QUIZ ANSWERS

1. c) The sheet stops the egg over a large distance, leading to a small force.
2. b) 400 J
3. b) 2 J.
4. c) twice as great.
5. b) the same.

NOTES

CHAPTER 3

SECTION 4

Newton's Second Law of Motion: The Rear-End Collision

Section Overview

Students explore the effects of rear-end collisions on passengers, focusing on whiplash. They model a driver by constructing a clay figure with its head attached to its body using a clay neck and a piece of wire. They investigate what happens to the clay figures during rear-end collisions between a moving vehicle and a stationary vehicle, and observe the differences between collisions involving drivers using a seat with a headrest and without a headrest. Students review Newton's second law and read about how the head whips back and then forward during a collision, which causes whiplash. Newton's first and second laws are applied to explain how whiplash occurs, and students use their observations as evidence to support the explanation. Students then describe, analyze, and explain situations involving collisions using Newton's first and second laws.

Background Information

Newton's second law states that if a net external force acts on an object, then the object's motion will change. If the mass of the object is constant then Newton's second law is expressed as

$$\vec{F} = m\vec{a}$$

When the net external force acts in the direction of motion the object speeds up. When the net external force acts opposite to the direction of motion, the object slows down. When the net external force acts in a direction different from the direction of motion, then the object begins to move in that direction.

Whiplash is a term commonly used to describe injury to the muscles, ligaments, tendons, and/or bones in the neck due to hyperextension and hyperflexion of the neck. Whiplash can occur in any situation in which the neck gets snapped back and forth.

Crucial Physics

- Newton's first law—an object at rest will remain at rest and an object in motion will continue its motion with constant velocity (speed and direction) unless a net external force acts on the object.
- Newton's second law—when an object is acted upon by a net external force its motion changes such that the object accelerates in the direction of the force. The acceleration of the object is directly proportional to the net external force, and inversely proportional to the object's mass.

Learning Outcomes	Location in the Section	Evidence of Understanding
Evaluate, from simulated collisions, the effect of rear-end collisions on the neck muscles.	***Investigate*** Steps 7-9 ***What Do You Think Now?***	Students record their observations of what occurs to a clay driver, in particular, its head and neck, during a rear-end collision simulation involving two types of carts with seats that collide with each other in the presence and absence of headrests. Students explain their observations using Newton's first law. After reading about whiplash, students describe their observations using Newton's first and second laws.
Describe the causes of whiplash injuries.	***Physics Talk*** ***What Do You Think Now?*** ***Physics to Go*** Questions 1, 8	Students discuss and describe what whiplash is and what causes it.
Provide examples of Newton's first and second laws of motion in automobile crashes.	***Physics Talk*** ***What Do You Think Now?*** ***Physics Essential Questions*** ***Physics to Go*** Questions 1, 4, 5	Students describe various situations involving automobile collisions using Newton's first and second laws.
Analyze the role of safety devices in preventing whiplash injury.	***Investigate*** Steps 7-9 ***Physics Essential Questions*** ***Reflecting on the Section and the Challenge*** ***Physics to Go*** Questions 5-7	Students use their observations to support their ideas and analysis on how headrests help prevent whiplash injury.

Section 4 Materials, Preparation, and Safety

Materials and Equipment

PLAN A

Materials and Equipment	Group (4 students)	Class
Dynamics cart	2 per group	
Clay, modeling, lb	2 per group	
Inclined plane for lab cart	1 per group	
Large ring stand	1 per group	
Rod, aluminum, 12 in. (length) x 3/8 in. (diameter) (to act as cross arm)	1 per group	
Holder, Right Angle, Cast Iron	1 per group	
Scissors	1 per group	
Meter sticks	1 per group	
Weight, slotted, 100 g	6 per group	
File folders	3 per group	
Tape, masking		6 per group
Wire, magnet, #24, 60 ft roll		1 per class

*Additional items needed not supplied

PLAN B

Materials and Equipment	Group (4 students)	Class
Dynamics cart		2 per class
Clay, modeling, lb		2 per class
Inclined plane for lab cart		1 per class
Large ring stand		1 per class
Holder, Right Angle, Cast Iron		1 per class
Rod, aluminum, 12 in. (length) x 3/8 in. (diameter) (to act as cross arm)		1 per class
Scissors		1 per class
Meter sticks	1 per group	
Weight, slotted, 100 g		6 per class
File folders		3 per class
Tape, masking		6 per group
Wire, magnet, #24, 60 ft roll		1 per class

*Additional items needed not supplied

Note: Time, Preparation, and Safety requirements are based on Plan A, if using Plan B, please adjust accordingly.

Time Requirement

Allow one period for the students to design and run the tests. If the investigation is run over two days, students can try different configurations for the seats to protect the passengers.

Teacher Preparation

- The two templates for the driver's seat will need to be prepared in advance. Based on the width of the cart you will be using, the templates may have to be enlarged or reduced on a photocopy machine so that they fit securely on the cart.
- Build a model of the driver and seats to demonstrate to students how they should look when they are done constructing the models. Conduct the investigation in advance to determine what your students may have difficulties with.
- When increasing the mass of the cart, added masses should be taped in place so they do not shift during the collision.

Safety Requirements

- In all collision experiments, goggles and closed-toe shoes are of extra importance.
- Plastic dynamics carts may be used for this investigation if a spring bumper is used.
- See the caution for smoothing the transition from ramp to floor in *Section 1*.
- If the investigation is performed on a lab table, caution the students to stand back so any falling masses do not land on their feet.

Meeting the Needs of All Students

Differentiated Instruction: Augmentation and Accommodations

Learning Issue	Reference	Augmentation and Accommodations
Creating clay passengers	***Investigate*** Steps 2 and 3	**Augmentation** • Students who struggle with sensory integration and fine motor skills may have a difficult time making a clay passenger, either because they do not want to touch the clay or because they cannot form a shape similar to a human in a reasonable amount of time. Intentionally assign partners to compensate for these concerns. • Give students a time limit for creating their clay passengers. Emphasize that the directions require a torso, neck, and head that can sit up in a cart, but the details on the clay passenger are not important. **Accommodation** • Prepare clay passengers in advance and provide students with a clay passenger at the start of the *Investigate*.
Making scientific measurements Following a list of directions to build something	***Investigate*** Steps 5-7	**Augmentation** • Ask groups of students to read *Steps 5-7* and decide how they are going to make all of the required measurements to setup the ramp. Refer students to the diagram of the setup as a resource. • If students feel confident that they can set up the ramp independently and complete the measurements, allow them to begin. If students need help, instruct them to ask specific questions to clarify their confusion. (For example: How can I measure 2.7 m?) **Accommodation** • Provide one-on-one or small group assistance to teach students measurement skills. • Provide a checklist of steps necessary to complete the setup. 1) Set up a ramp that is 40 cm high. 2) Use a piece of stiff paper at the bottom of the ramp. 3) Mark off a distance of 2.7 m or 270 cm from the bottom of the ramp. 4) Etc.
Reading comprehension Imagining a scenario	***Physics Talk*** Newton's First and Second Laws of Motion and Whiplash ***Checking Up*** Questions 1-3 ***Physics Essential Questions***	**Augmentation** • The descriptions in this section clearly explains what happens during a rear-end collision, but students who struggle with reading will skim over information and miss important concepts. Ask students to act out a rear-end collision as a group. As they are acting out the collision, tell them to try to annotate the scenario using Newton's laws. • Some students will be able to write or explain what happens in a rear-end collision in their own words after they read this section. If students can explain the actions in their own words, they will be able to apply the concept more easily later. • Ask students to draw a comic strip of the events that happen in a rear-end collision and include physics vocabulary in their comics. • Allow students to experience their head's motion being different from their body's motion by asking them to sit in a desk chair with wheels and push the chair quickly from behind. This experience will be a good frame-of-reference for students to refer to when they are answering questions. • Show an Internet video clip of a person getting whiplash using video available on the Internet. **Accommodation** • Read the *Physics Talk* section aloud and have a group discussion to apply Newton's laws to the whiplash scenario.

Strategies for Students with Limited English-Language Proficiency

Learning Issue	Reference	Augmentation
Understanding concepts	***Physics Talk***	Once students have completed the *Investigate* and have read through the description of how a rear-end collision causes whiplash, hold a class discussion. Try and let students figure out how a headrest can reduce the extent of injury from whiplash. Encourage ELL students to participate actively in the discussion and practice their speaking skills. You may wish to use a notebook or file folder to serve as a headrest, and demonstrate how it prevents the head from fully hyperextending, thus reducing injury to the neck bones and vertebral column.
Cooperative learning	***Active Physics Plus***	Have students work in pairs or groups. Assign each group one of the given speeds (3 m/s, 5 m/s, 10 m/s, or 15 m/s) and have them calculate the force on the driver's neck muscles for the conditions given. Have each group draw the corresponding graph. Ask an ELL student to copy his or her group's graph on the board and explain it to the class. Repeat the process with a different student for the shorter time interval given in *Question 3*.
Vocabulary comprehension	***Reflecting on the Section and the Challenge***	Students may have difficulty discerning the meaning of "susceptible," which means "likely to be affected with or by."
Vocabulary comprehension	***Physics to Go Whiplash Quiz*** Question 5	The meaning of "threshold" may be difficult to determine from context. Help students understand that it is the point separating conditions that will produce an effect (in this case, cause injury) from conditions that will not produce an effect (not cause injury). Students may benefit from another example in a context they have likely encountered: Lengthy exposure to noise levels above 85 dB will damage hearing, while lengthy exposure to noise levels below 85 dB will not. Eighty-five dB is the threshold for hearing damage.
Research skills	***Inquiring Further***	ELL students may need some guidance in finding library and Internet sources of relevant information. To help them practice their language skills, encourage them to choose a presentation style that requires at least some time speaking to the class.

Consider finishing this section with a cloze activity. Cloze activities are useful tools for summarizing material and for giving English-language learners an opportunity to practice using their science vocabulary words in context. Write the following paragraph on the board, replacing the underlined words with write-on-lines. Encourage volunteers to fill in the blanks.

An automobile is hit from behind and begins to move. Newton's second law says the automobile accelerates because of the unbalanced force acting on the back of the automobile. The driver's torso moves with the automobile, but the driver's unsupported head does not. Then the neck muscles whip the head forward. The head moves until it is ahead of the torso. Newton's first law tells why an object in motion stays in motion. Then the neck muscles pull the head back to its usual position. The driver suffers a whiplash injury.

NOTES

SECTION 4

Teaching Suggestions and Sample Answers

What Do You See?

Have students consider the illustration and the *Learning Outcomes* of this section. Use an overhead of the illustration to help focus the discussion. Ask students what they think the illustration depicts and record their responses. This elicitation helps to get their initial ideas and provides a focus for the science content. Consider asking students how the illustration captures Newton's first and second laws.

What Do You Think?

These preliminary questions are designed to elicit students' prior knowledge and to focus students on the issue. After students record their responses in their log, have them discuss their ideas with

Chapter 3 Safety

Section 4 Newton's Second Law of Motion: The Rear-End Collision

What Do You See?

Learning Outcomes

In this section, you will

- **Evaluate,** from simulated collisions, the effect of rear-end collisions on the neck muscles.
- **Describe** the causes of whiplash injuries.
- **Provide** examples of Newton's first and second laws of motion in automobile crashes.
- **Analyze** the role of safety devices in preventing whiplash injury.

What Do You Think?

The whiplash effect is a serious injury that is caused by a rear-end collision. It is the focus of many lawsuits, the inability to work, and discomfort.

- **What is whiplash?**
- **Why is it more prominent in rear-end collisions?**

Record your ideas about these questions in your *Active Physics* log. Be prepared to discuss your responses with your small group and the class.

Investigate

Whiplash injury can occur in automobile collisions at surprisingly low speeds—as low as 2.7 m/s (6 mi/h). In this section, you will simulate a rear-end collision between two carts—a "bullet" cart and a "target" cart. The bullet cart will be moving at a speed of approximately 2.7 m/s (6 mi/h) and will strike a stationary target cart.

292

Students' Prior Conceptions

The examination of rear-end collisions and the subsequent analysis of the transfer of energies and the work involved in these collisions provides real-life examples that lead students to alter their prior conceptions.

1. **Collision severity is less for the automobile that hits a second automobile in a rear-end collision than for the automobile being hit.** This misconception arises due to a misunderstanding of action-reaction forces and a misconstrued application of the laws of motion. Hands-on inquiry, measurement, and analysis require the student to examine the laws of motion and to reconstruct a model for what happens to each vehicle during the collision.

2. **In a collision, the extent of injury to passengers is related to the speed of the automobile in which they are traveling.** Students generally believe that traveling at low velocities does not cause extensive or severe impact damage during collisions. To remove this misconception, refer to *Investigate, Steps 7* and *8,* and the subsequent analysis in the *Physics Talk* of whiplash during a rear-end collision, based on Newton's laws of motion. This is also a good time to review the velocity squared rule developed in *Chapter 1, Section 5.*

their group members to come up with their best group response. Then have a class discussion and record the class's ideas. Remind them that the correctness of their response is not the issue; what matters is their ability to construct and to convey their ideas in response to the questions.

Investigate

Discuss the investigation steps with the class. Show students the completed driver's seats and make these available for students to view while they are assembling their driver's seats. Describe for students how their clay figures should be constructed and emphasize the importance of using the wire to model the spinal column and neck, and that the seated clay figure should have its head just above the seat without the headrest.

What Do You Think?

A Physicist's Response

In whiplash, neck muscles, ligaments, and/or tendons become damaged. Whiplash occurs when the head moves far back, and then far forward, causing the neck muscles and vertebrae to hyperextend and then hyperflex. The injury caused is even more severe if the head is turned to the side rather than forward. To understand why whiplash is more prominent in rear-end collisions, assume a driver is at rest within the automobile. Applying Newton's first law, the driver will remain at rest unless a net external force acts on the driver. If the vehicle is rear-ended, there is a net external force acting on the vehicle during the collision. Applying Newton's second law, the vehicle accelerates forward. The back of the seat pushes the driver forward and the driver's torso moves forward with the automobile. The driver's head, however, is not pushed forward by the seat, and obeys Newton's first law, so it stays back until the neck muscles holding the head to the body pull the head forward. At this point, the hyperextended muscles whip the head forward, accelerating it according to Newton's second law. The head then continues its forward motion following Newton's first law, until it gets ahead of the body, and the neck muscles attached to the body are hyperflexed and pull it back to its usual position. If the seat has a padded headrest it can reduce injury because it can support the head so that it does not hyperextend. A padded headrest will also absorb some of the energy so that the head does not hyperflex.

In a front-end collision, the vehicle accelerates backward because of the unbalanced force exerted on it. According to Newton's first law, the passenger remains in place until an unbalanced force is exerted on the passenger. The seat belt exerts a force on the passenger causing the body to be pushed back, and the head, according to Newton's first law, would remain where it was until the neck muscles pulled the head backward. The head would then continue moving backward until it was pulled forward to its normal location. Whiplash can still occur in this case, but usually the muscles in the neck are not as hyperflexed and hyperextended during a front-end collision. The head, which naturally has a greater forward than backward mobility, is not affected as severely as during a rear-end collision. Also, the body bends forward with the head so the neck muscles are not flexed as much initially before the head moves back.

1.

Two templates are provided for a driver's seat – one that has a headrest and one that does not. Have students fold along the solid lines and cut along the dotted lines to create the two seats.

2. - 3.

Once the seats are built, have each group of students create two identical clay dummies to fit the seats, so that the heads of the dummies will be just above the back of the driver's seat that is without a headrest. The head of each dummy needs to be created as a separate part with a small piece of clay that represents the neck. Do not use the clay neck to attach the head of the dummy to the torso. Use a 1-inch piece of 24-gauge wire to secure the head through the neck to the torso, as shown in the image. Attach the seats to the carts with tape to keep them in place to have a realistic collision. Students should place the clay figure in the seat with the headrest with its head against the headrest. The clay figure in the seat without a headrest should have its shoulders level with the top of the seat.

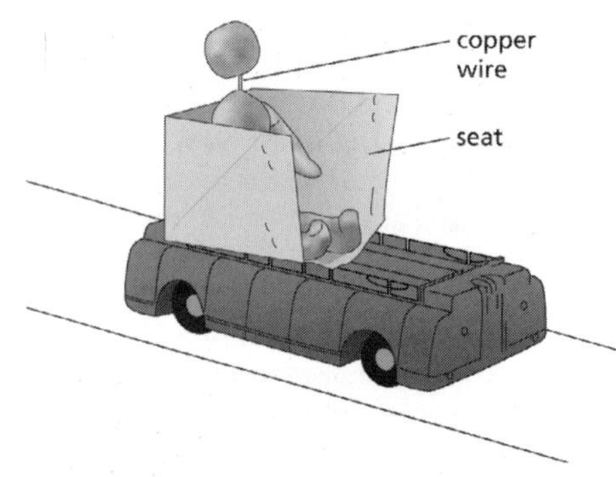

3-4a Blackline Master

4.

Students should create seat belts using masking tape, or ribbon if preferred.

5.

To set up the ramp so that a speed of about 2.7 m/s (about 6 mi/h) will occur as the car traverses the level portion of the track, almost the entire height of 40 cm will be needed. Check that students have placed a stiff piece of paper at the bottom of the ramp to smooth out the junction between the ramp and table. Check that students have marked the position on the ramp with, for example, a piece of masking tape.

NOTES

Section 4 Newton's Second Law of Motion: The Rear-End Collision

You will need to create two clay passengers and two different driver's seats—one with a headrest and one without a headrest.

1. Your teacher will provide you with a set of two templates to create driver's seats. Use Template A to create a driver's seat with no headrest. The clay passenger should fit into this seat so that the shoulders of the passenger are level with the top of the seat. (There is no built-in support for the head.)
2. Create the clay passenger in two sections. In the first section, create the torso and legs out of clay.

In the second section, create the head and connect it to the torso with a small piece of clay rolled in the shape of a neck so that the head is not sitting directly on the shoulders. Use a 2.5 cm piece of # 26 wire to fasten the head-neck-torso combination. The wire represents the spinal column in the neck. Do not press the head onto the neck, but rather allow it to rest on it held in place by the wire. The passenger and seat will go in the target cart.

3. Create a similar second clay passenger to go in the bullet cart. Use Template B to create a driver's seat with a headrest.
4. Use masking tape to create seat belts for both dummies.
5. Set up a ramp about 40 cm high, as shown in the diagram on the following page. Use a piece of stiff paper (like card stock) at the bottom of the ramp to smooth out the bump when the cart comes off the ramp.

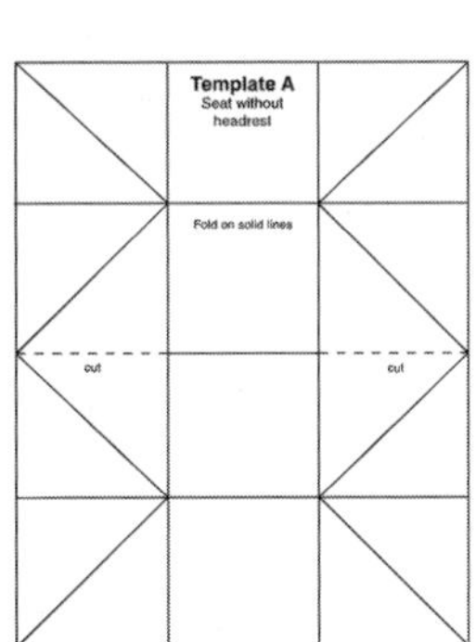

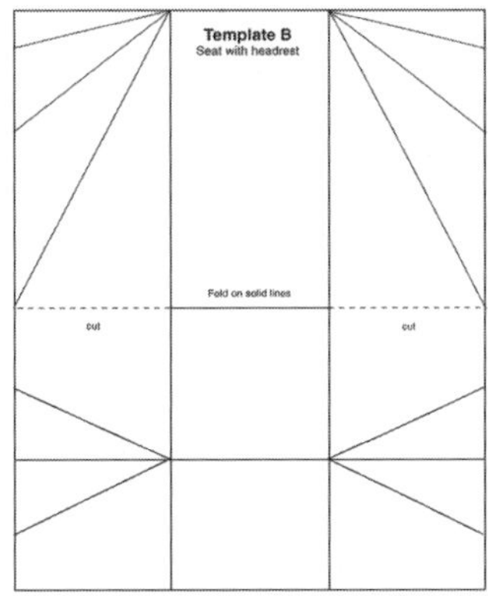

Active Physics

6.

Students should place the target cart about 50 cm from the end of the ramp. (For the first run, *Steps 1-7*, the target cart has the seat with no headrest.)

7.a)

Students should notice that the unsupported head of the passenger in the target cart is bent backward after the collision.

7.b)

Newton's first law states that an object at rest (the clay model's head) remains at rest until acted upon by an unbalanced force, so the head tends to remain at rest while the rest of the "body" is pushed rapidly forward. The head begins to move forward when the body begins to pull it forward.

8.

With the target cart having a seat with a headrest, the students should notice that the head is not bent backward, as in the previous step. Newton's first law states that an object at rest remains at rest until acted upon by an unbalanced force. The head and body move forward together because they have the same unbalanced force acting on them from the seat and headrest.

Teaching Tip

To double and triple the mass of the bullet cart, carts may be stacked one on top of the other, or masses may be added. If masses are added, tape them in place to ensure that they do not slide during the collision.

9.

Students should observe that as the mass of the bullet cart is increased, the amount that the head bends back in the target cart without the headrest increases. They should notice no change if the target cart has the seat with the headrest.

3-4b Blackline Master

3-4c Blackline Master

CHAPTER 3

Physics Talk

This *Physics Talk* discusses Newton's second law, whiplash, and how Newton's first and second laws can be used to explain the cause of whiplash. Consider eliciting students' initial ideas about Newton's first and second laws. Ask them what they remember about each of these laws and to provide an example of each law. Elicit their ideas on how Newton's first and second laws can be used to explain whiplash. Then have a class discussion on Newton's first and second laws of motion and whiplash.

Emphasize to students that Newton's second law states that if an unbalanced force acts on an object of unchanging mass, its motion will change, which means it will accelerate. An accelerating object may change its speed, its direction, or both, as was discussed in *Chapters 1* and *2*. Discuss whiplash, describing what happens within the body. Use the information in the student text to guide the discussion. Point out to students that if the head is turned during this motion of the neck and head, the injury is increased due to greater forces acting on the soft tissues in the neck. Emphasize that whiplash is a term used when any of the soft tissues in the neck (muscles, ligaments, and tendons) are damaged due to the sudden motion of the head forward and/ or backward.

Describe how each interaction during a rear-end collision can be explained using Newton's first and second laws of motion, and how they can explain the entire motion of the head that causes whiplash to occur. Discuss each bulleted point, describing how Newton's first and second laws help to explain what is happening to the head during a collision. Check student understanding by asking them to explain their observations during the *Investigate* using Newton's laws.

Summarize Newton's second law and emphasize what net force, force, acceleration, and mass are. Discuss the examples in the student text. Check students' understanding by asking questions about these concepts or having them provide an example similar to the one in the student text. Ask students to provide examples of these concepts.

Chapter 3 Safety

Release the bullet cart from the position on the ramp. This produces an average speed of approximately 2.7 m/s (6 mi/h). This may take several trials.

6. Now place the target cart about 50 cm from the end of the ramp.

7. Release the bullet cart from the same position on the ramp to produce a 2.7 m/s (6 mi/h) collision with the target cart.

a) What happens to the clay passenger's head in the target cart?

b) Use Newton's first law of motion to explain your observations.

8. Repeat this experiment (*Steps 1-7*) by exchanging the carts so that the cart with the headrest is the target cart.

9. Repeat this experiment using a bullet cart with a mass two to three times the mass of the target cart. You may be able to tape two or three carts together or one on top of the other to get a cart with about two to three times the mass. You may also increase the mass of the bullet car by adding masses to the cart.

bullet cart
ramp
40 cm
target cart
card stock
75 cm
50 cm

Physics Talk

NEWTON'S SECOND LAW OF MOTION

Newton's first law informs us what happens to objects if no net force acts upon them. Knowing that objects at rest have a tendency to remain at rest and that objects in motion will continue in motion does not provide enough information to analyze collisions. **Newton's second law of motion** allows you to make predictions about what happens when an unbalanced external force is applied to an object. If you were to place a collision cart on a level surface, it would not move. However, if you begin to push the cart, it will begin to move.

Newton's second law of motion states:

> If a body is acted on by an unbalanced force, it will accelerate in the direction of the unbalanced force. The acceleration will be larger for smaller masses. The acceleration can be an increase in speed, a decrease in speed, or a change in direction.

Newton's second law of motion indicates that the change in motion is determined by the net force acting on the object, and the mass of the object itself. Physicists are never satisfied with a verbal explanation and always ask, "Is there an equation that can describe this precisely?"

Physics Words

Newton's second law of motion: if a body is acted on by an unbalanced force, it will accelerate in the direction of the unbalanced force. The acceleration will be larger for smaller masses. The acceleration can be an increase in speed, a decrease in speed, or a change in direction.

294

Active Physics

Section 4 Newton's Second Law of Motion: The Rear-End Collision

Newton's second law does have such an equation that can be written as:

$$F = ma \text{ or } a = \frac{F}{m}$$

From this equation, one can see that "if a body is acted upon by an unbalanced force, it will accelerate in the direction of the unbalanced force." One can also see that the "acceleration will be larger for smaller masses."

What is Whiplash?

In the collision in the investigation you completed, where the target cart did not have a headrest, the clay head swung back during the collision. **Whiplash** is a serious injury that can be caused by a rear-end collision. The back of the automobile seat pushes forward on the torso of the driver and the passengers and their bodies lunge forward. The head remains still for a very short time. The body moving forward and the head remaining still causes the head to snap backward. The neck muscles and bones of the vertebral column (spine) become damaged. The same muscles must then snap the head back to its place atop the shoulders.

Physics Words

whiplash: the common name for a type of neck injury to muscles of the neck.

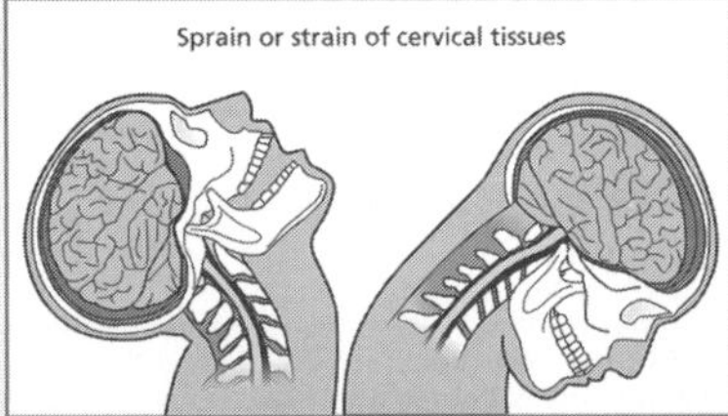

A headrest can prevent whiplash injury. The headrest must be adjusted for the height of the passenger.

Newton's First and Second Laws of Motion and Whiplash

The activity in this section demonstrated the effects of a rear-end collision. Newton's first law and Newton's second law can help explain the "whiplash" injury that passengers suffer during this kind of collision.

Imagine looking at the rear-end collision in slow motion. Think about all that happens.

- An automobile is stopped at a red light. This is the automobile in which the driver is going to receive a whiplash injury. It was the target cart in your investigation. The driver is at rest within the automobile.

CHAPTER 3

- The stopped automobile gets hit from the rear.
- The automobile begins to move. The back of the seat pushes the driver forward and the driver's torso moves with the automobile. The driver's head is not supported and tends to stay back where it is.
- The neck muscles hold the head to the torso as the body moves forward. The muscles then "whip" the head forward. The head keeps moving until it gets ahead of the torso. The neck muscles stop the head, and pull it back to its usual position. Ouch!

Let's repeat the description of the collision and insert all of the places where Newton's first law and Newton's second law apply (in color).

- An automobile is stopped at a red light. This is the automobile in which the driver is going to receive a whiplash injury. The driver is at rest within the automobile. Newton's first law: An object at rest stays at rest, and an object in motion stays in motion unless acted upon by an unbalanced, outside force.
- The stopped automobile gets hit from the rear.
- The automobile begins to move. Newton's second law: The automobile accelerates because of the unbalanced, outside force from the rear: $F = ma$. The back of the seat pushes the driver forward and the driver's torso moves with the automobile. Newton's second law: The torso accelerates because of the unbalanced, outside force from the back of the seat: $F = ma$. The driver's head is not supported and stays back where it is. Newton's first law: an object (the driver's head) at rest stays at rest.
- The neck muscles hold the head to the body as the body moves forward. The muscles then "whip" the head forward. Newton's second law: The head accelerates because of the unbalanced force of the muscles: $F = ma$. The head keeps moving until it gets ahead of the torso. Newton's first law: An object (the head) in motion stays in motion. The head is stopped by the neck muscles. The muscles pull the head back to its usual position. Newton's second law: The head accelerates (slows down) because of the unbalanced force from the neck muscles: $F = ma$. Ouch!

Newton's second law informs you that all accelerations are caused by *unbalanced, outside* forces. It does not say that all forces cause accelerations, only those that are unbalanced.

Section 4 Newton's Second Law of Motion: The Rear-End Collision

An object at rest may have many forces acting upon it. When you hold a book in your hand, the book is at rest. There is a force of gravity pulling the book down. There is a force of your hand pushing the book up. These forces are equal and opposite. The "net" force on the book is zero because the two forces balance each other. There is no acceleration because there is no "net" force.

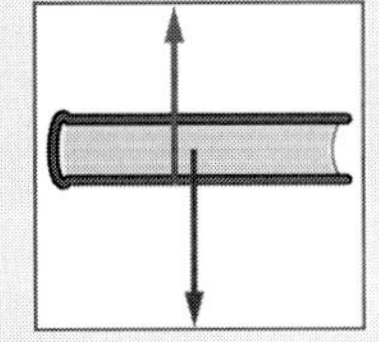

As an automobile moves down the highway at a constant speed, there are forces acting on the automobile but there is no acceleration. No acceleration indicates that the net force must be zero. The force of the engine on the tires and road moving the automobile forward must be equal in magnitude and opposite in direction to the force of the air pushing backward on the automobile. These forces balance each other in this case, where the speed is not changing. There is no net force and there is no acceleration. The automobile stays in motion at a constant speed.

A similar situation occurs when you push a book across a table at constant speed. The push is to the right and the friction is to the left, causing opposing motion. If the forces are equal in size, there is no net force on the book. The book does not accelerate—it moves with a constant speed.

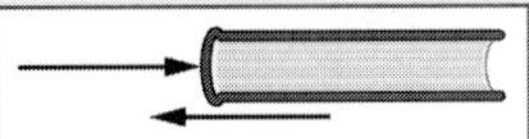

Checking Up

1. What type of safety devices can prevent a whiplash injury?
2. Describe how a whiplash injury occurs in a vehicle collision.
3. Use Newton's first and second laws of motion to analyze how a whiplash injury occurs during a rear-end collision.

Active Physics *Plus*

+Math	+Depth	+Concepts	+Exploration
♦			

Using Equations to Analyze a Whiplash Injury

1. Using the model of whiplash and assuming that the driver's head has a mass of 5 kg, calculate the force on the driver's neck muscles during the collision when the target vehicle gets hit and moves at 3 m/s, 5 m/s, 10 m/s, and 15 m/s. (To give a better sense of these speeds, 27 m/s = 60 mi/h.) Assume this change in motion occurs in 0.2 s (seconds).
2. Draw a graph showing the relationship between force on the neck muscles versus speed of the automobile coming from behind.
3. Repeat the analysis assuming that the time is only 0.1 s (seconds).

Checking Up

1.

A headrest can reduce the delay between the motion of the body and the motion of the head by not allowing the head to go back very far and by absorbing some of the kinetic energy of the head. This helps to reduce and/or prevent whiplash injury.

2.

Whiplash can be caused by a rear-end collision. The back of the automobile seat pushes forward on the body of the driver. The body lunges forward, but the head remains still for a very short time. This causes the head to snap backward and then forward. The neck muscles and bones can become damaged as the motion of the head pushes and then pulls the muscles, tendons, ligaments, and vertebrae one way and then the other. The same muscles then snap the head back into place.

3.

Students should explain how whiplash occurs using Newton's first and second laws. Their responses should contain the information in the *Physics Talk*, but should be in their own words. Consider asking students to describe another situation in which whiplash could occur, for example, on a roller coaster while skateboarding, and so on.

Active Physics Plus

This *Active Physics Plus* provides students with the opportunity to apply the physics concepts of Newton's first and second laws to problem-solve situations algebraically.

1.

The mass, the initial speed, the final speed, and the time the change in speed occurs are given. To calculate the force ($F = ma$), the acceleration is needed. To find the average acceleration, use the equation:

$$a = \frac{v_{\text{final}} - v_{\text{initial}}}{t}.$$

For $v_{\text{initial}} = 3$ m/s, $a = -15$ m/s^2, and $F = -75$ N.

For $v_{\text{initial}} = 5$ m/s, $a = -25$ m/s^2, and $F = -125$ N.

For $v_{\text{initial}} = 10$ m/s, $a = -50$ m/s^2, and $F = -250$ N.

For $v_{\text{initial}} = 15$ m/s, $a = -75$ m/s^2, and $F = -375$ N.

The negative sign indicates that the acceleration and the force point opposite to the direction of the initial velocity.

2.

Students' graphs should look similar to the one below and at right.

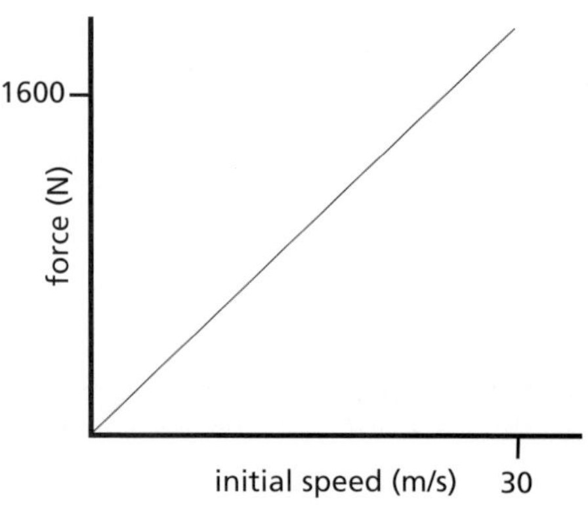

What Do You Think Now?

At the beginning of the section, you were asked

- **What is whiplash?**
- **Why is it more prominent in rear-end collisions?**

Revisit your initial ideas about whiplash and rear-end collisions. Based on your investigation of Newton's first and second laws, how would you answer these questions now?

What does it mean?

Use Newton's second law of motion to explain what happens to a passenger in a rear-end collision.

How do you know?

How do you know that headrests can improve passenger safety during a rear-end collision?

Why do you believe?

Connects with Other Physics Content	Fits with Big Ideas in Science	Meets Physics Requirements
Forces and motion	Systems	* Optimal prediction and explanation

* In physics, there is often more than one correct way to describe an event. Describe a rear-end collision using Newton's first law and compare it with your explanation using Newton's second law.

Why should you care?

How can the possibility of a rear-end collision be factored into the design of your safety system?

Reflecting on the Section and the Challenge

Whiplash is a serious injury that can occur during rear-end collisions. The bones that attach the spinal column to the skull are called attachment bones. They are supported by the least amount of muscle. Unfortunately, these smaller bones, with less muscle support, make this area particularly susceptible to injury. The brainstem is very susceptible to damage following whiplash. The brainstem is vital because it regulates blood pressure and breathing movements. By restraining the movement of the head and neck muscles, you can protect against the most severe aspects of whiplash.

3.

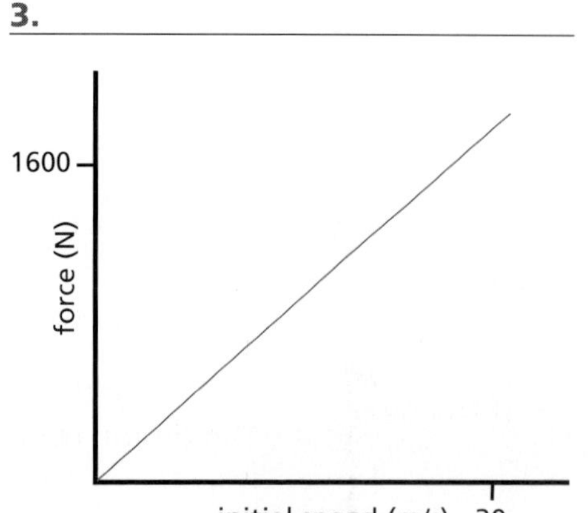

Physics to Go

1. Why are neck injuries common during rear-end collisions?
2. Explain why packages in the back of a truck move forward if it comes to a quick stop.
3. As a bus accelerates, the passengers on the bus are jolted toward the back of the bus. Explain what causes the passengers to be apparently pushed backward.
4. Why would the rear-end collision demonstrated in the *Investigate* be more dangerous for someone driving a motorcycle than driving an automobile?
5. Explain in which type of collision headrests serve the greater benefit: during a head-on collision or a rear-end collision?
6. What additional devices have been placed in automobiles to help reduce the impact of rear-end collisions?
7. Consider how your safety device will help prevent whiplash injuries following a collision. What part of the restraining device prevents the movement of the head?
8. As a way to help you learn more about whiplash, rate your whiplash knowledge by taking this whiplash quiz and then check your knowledge against the answers given at the end.

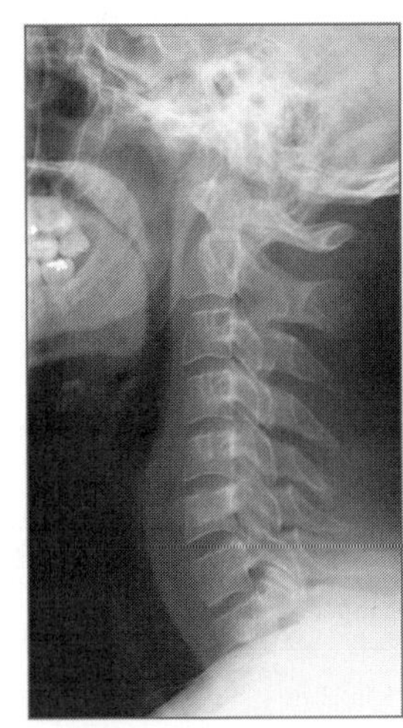

Whiplash Quiz

1. The range of collision speed in which most (nearly 80 percent) rear-impact whiplash injuries occur is:
   a) 10-25 mi/h (4.5 – 11 m/s or about 16-40 km/h)
   b) 15-30 mi/h (7 – 13 m/s or about 25-47 km/h)
   c) 1-5 mi/h (0.5 – 2.3 m/s or about 2-8 km/h)
   d) 6-12 mi/h (2.7 – 5.4 m/s or about 10-19 km/h)

What Do You Think Now?

Revisit the *What Do You Think?* questions, and review students' initial ideas. Ask students how they would now answer these questions using Newton's first and second laws. After discussing how students would revise their answers, consider discussing *A Physicist's Response* with the class to give them a better understanding of what whiplash is and why it occurs in rear-end collisions.

Reflecting on the Section and the Challenge

Have a discussion on why the neck area is so susceptible to serious injury, and the possible severity of whiplash, using the information in the student text. Then emphasize to the class that by restraining the movement of the head and neck muscles, drivers and passengers can protect against the possible severe effects of whiplash. Ask students what evidence they have of this from their investigation and how

Physics Essential Questions

What does it mean?

A vehicle is hit from behind and accelerates forward. The force of the back of the seat of the vehicle pushes on the passenger and he or she accelerates forward.

How do you know?

Without a headrest, as the passenger accelerates forward, the passenger's head would remain where it was until the neck muscles pulled it forward. This would cause the body to move in front of the head. A headrest applies a force to the head to keep it moving with the body.

Why do you believe?

A vehicle is at rest and stays at rest until hit with a force from behind (Newton's first law).

A vehicle begins to accelerate forward due to the force (Newton's second law).

A passenger remains at rest while the vehicle accelerates (Newton's first law).

The back of the seat accelerates the passenger forward (Newton's second law).

Why should you care?

The design could include something like headrests to protect the head if the body abruptly moves forward.

CHAPTER 3

Newton's first and second laws explain this. Discuss how they could use this information for the *Chapter Challenge*.

Physics to Go

1.

Responses should include a description of whiplash and how it is explained using Newton's first and second laws. For example, during a rear-end collision, the automobile gets hit from the rear. The automobile begins to move because an unbalanced force is acting on it during the collision (Newton's second law). The back of the seat pushes the driver forward and the driver's torso moves forward with the automobile because an unbalanced force is acting on the driver (Newton's second law). The driver's head is not supported and stays back where it is because at this moment, no unbalanced forces are acting on it, obeying Newton's first law. The neck muscles hold the head to the body as the body moves forward. When the muscles have been extended as far as possible, they apply an unbalanced force on the head and pull the head forward, accelerating the head (Newton's second law). The head keeps moving until it gets ahead of the torso because an object in motion stays in motion until an unbalanced force acts on it (Newton's first law). When the muscles on the other side of the neck have reached their maximum stretch, they pull the head back, accelerating it, and moving it to its usual position (Newton's second law).

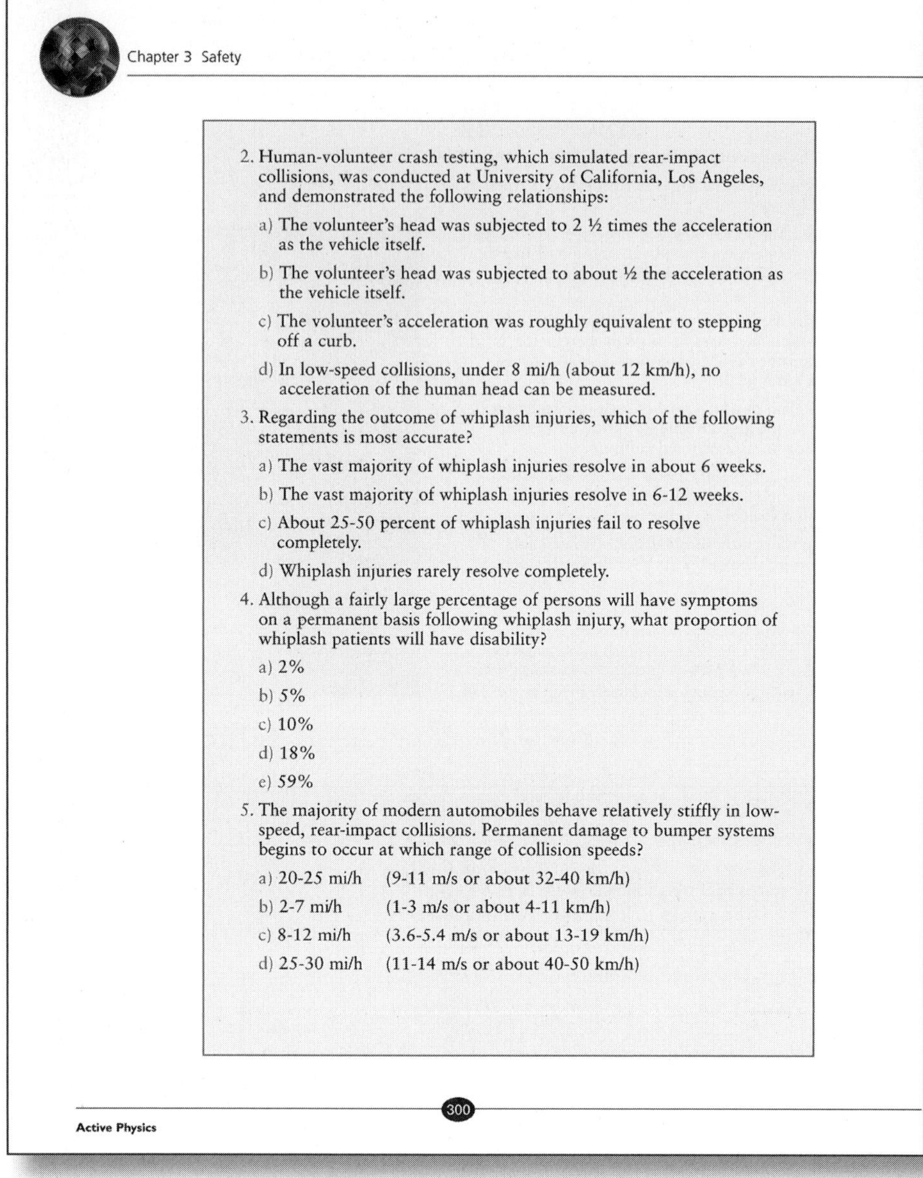

Chapter 3 Safety

2. Human-volunteer crash testing, which simulated rear-impact collisions, was conducted at University of California, Los Angeles, and demonstrated the following relationships:
   a) The volunteer's head was subjected to 2 ½ times the acceleration as the vehicle itself.
   b) The volunteer's head was subjected to about ½ the acceleration as the vehicle itself.
   c) The volunteer's acceleration was roughly equivalent to stepping off a curb.
   d) In low-speed collisions, under 8 mi/h (about 12 km/h), no acceleration of the human head can be measured.
3. Regarding the outcome of whiplash injuries, which of the following statements is most accurate?
   a) The vast majority of whiplash injuries resolve in about 6 weeks.
   b) The vast majority of whiplash injuries resolve in 6-12 weeks.
   c) About 25-50 percent of whiplash injuries fail to resolve completely.
   d) Whiplash injuries rarely resolve completely.
4. Although a fairly large percentage of persons will have symptoms on a permanent basis following whiplash injury, what proportion of whiplash patients will have disability?
   a) 2%
   b) 5%
   c) 10%
   d) 18%
   e) 59%
5. The majority of modern automobiles behave relatively stiffly in low-speed, rear-impact collisions. Permanent damage to bumper systems begins to occur at which range of collision speeds?
   a) 20-25 mi/h (9-11 m/s or about 32-40 km/h)
   b) 2-7 mi/h (1-3 m/s or about 4-11 km/h)
   c) 8-12 mi/h (3.6-5.4 m/s or about 13-19 km/h)
   d) 25-30 mi/h (11-14 m/s or about 40-50 km/h)

Active Physics 300

2.

Explanations should use Newton's first law. For example, packages in the back of a truck move forward if the truck comes to a quick stop because an object in motion continues its motion unless an unbalanced force acts on it. The packages were moving at the same speed as the truck before the truck's brakes were applied. After the truck applied it brakes, the packages continued to move forward because the friction between the packages and the surfaces they were in contact with was not great enough to hold them in place. Remind students that this is similar to passengers moving forward when a vehicle stops abruptly, and the importance of seat belts restraining passengers so that they do not hit the vehicle parts in front of them.

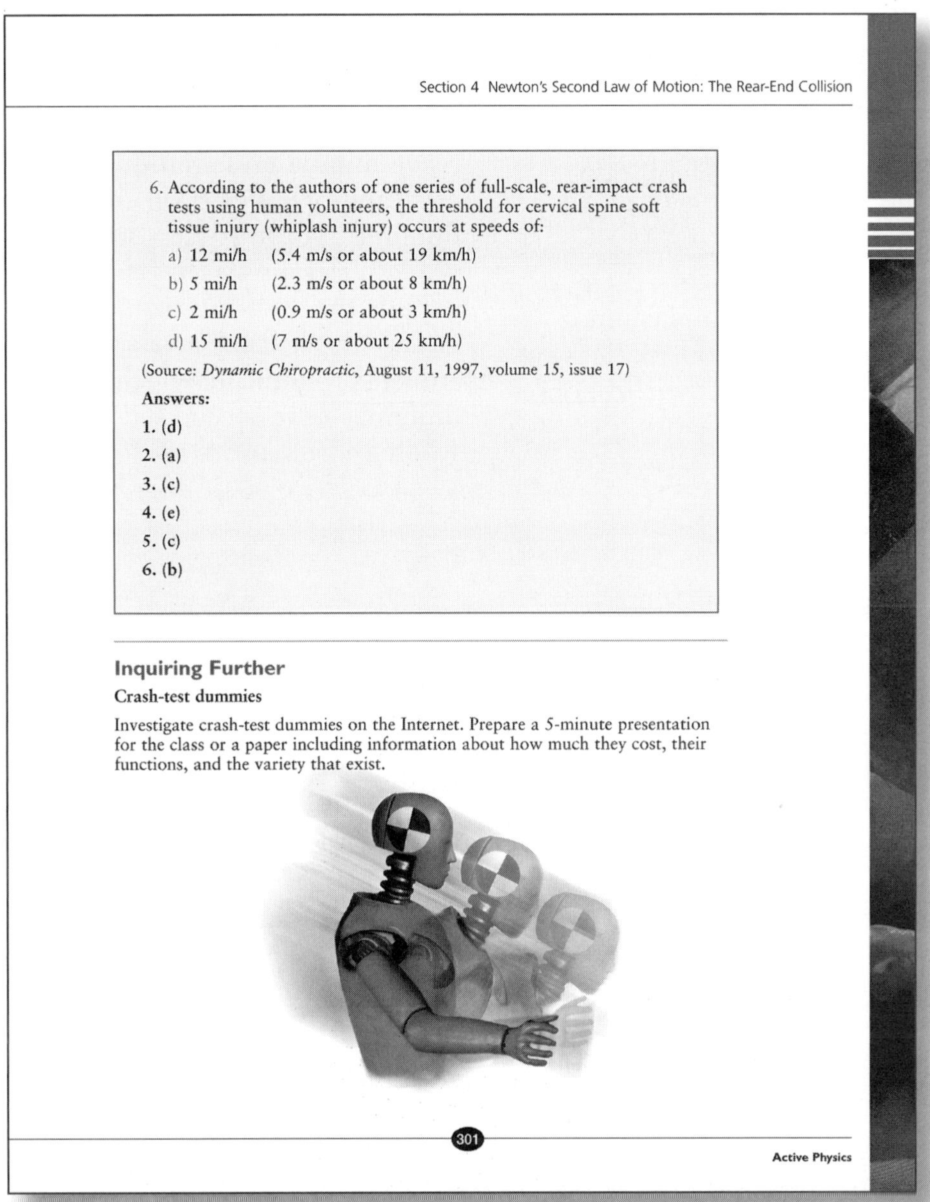

6. According to the authors of one series of full-scale, rear-impact crash tests using human volunteers, the threshold for cervical spine soft tissue injury (whiplash injury) occurs at speeds of:
   a) 12 mi/h (5.4 m/s or about 19 km/h)
   b) 5 mi/h (2.3 m/s or about 8 km/h)
   c) 2 mi/h (0.9 m/s or about 3 km/h)
   d) 15 mi/h (7 m/s or about 25 km/h)

(Source: *Dynamic Chiropractic*, August 11, 1997, volume 15, issue 17)

Answers:
1. (d)
2. (a)
3. (c)
4. (e)
5. (c)
6. (b)

Inquiring Further

Crash-test dummies

Investigate crash-test dummies on the Internet. Prepare a 5-minute presentation for the class or a paper including information about how much they cost, their functions, and the variety that exist.

3.

Students should use Newton's first and second laws in their explanations. When the bus accelerates, the passengers are at rest and remain at rest until an unbalanced force acts on them from the seat bottoms pushing the passengers forward, following Newton's second law. Meanwhile, the passengers' torsos are still at rest (obeying Newton's first law) until they come in contact with the back of the seat, and the back of the seat pushes them forward (Newton's second law), or until their muscles pull them forward. So, the passengers are not really "pushed backward," rather they try to remain in the same place until they are pushed forward.

4.

Students' responses should refer to their observations of an increased mass hitting the target cart, and their graphs from the previous section in which they showed that there is a greater amount of work done to stop an object as the mass increases. This can also be explained by conservation of energy. The energy being transferred to a vehicle and a motorcycle are the same. Because the motorcycle has less mass, it will have a greater speed after the collision than if a vehicle were involved in the same collision. Because of this, it will take more work or more energy to stop the motorcycle and its driver after the collision. This also results in the damage and severity of injury being greater for the motorcycle.

5.

Students' responses should indicate that headrests are important for both front-end and rear-end collisions, however, the headrest is most important for rear-end collisions in which the vehicle moves forward as the person's head tries to remain in place. In this case, the headrest helps to absorb some of the energy from the head as it pushes into the head, and prevents the neck muscles from hyperextending because it does not allow the head to be in a position behind the back. This also reduces the amount the head whips forward, reducing the hyperflexing of the neck muscles. For a front-end collision, the passenger's body moves backward and the head stays in place. The neck muscles are hyperflexed. Usually the neck is not as hyperflexed as in a rear-end collision because the body also bends forward, reducing

CHAPTER 3

the amount of flexion in the neck muscles. The neck muscles then whip the head back into the headrest, which absorbs some of the energy and prevents the neck muscles from being hyperextended.

6.

Students should list devices besides headrests that help reduce the impact of a rear-end collision such as seat belts, air bags, and padding in seats.

7.

Students' responses should include the features they plan to use in their design that will restrain head movement during a collision.

8.

Students should record their answers for the whiplash quiz and their grade. Consider having students reflect on their responses and showing their work for any questions they answered incorrectly.

Inquiring Further

Consider having students work in groups, or individually, to prepare a five-minute presentation to the class and/or a report on crash-test dummies. As you assess students' work, check to see that students include the cost, function, and variety that exist. Encourage students to indicate which crash dummies have special sensors and where those sensors are located.

NOTES

SECTION 4 QUIZ

3-4d Blackline Master

1. In an experiment, a student rests a figure made of modeling clay on a lab cart at rest on a table. The student then collides a second lab cart of equal mass with the first into the stationary cart, and notices the modeling-clay figure topple over backward. The physical principle that explains why the figure tumbled over is

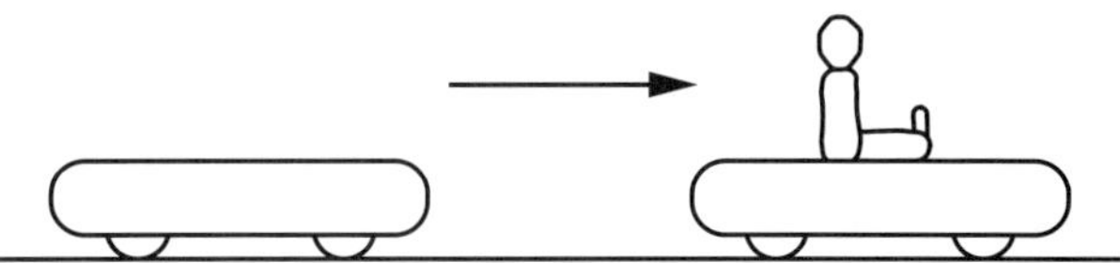

   a) Newton's first law.
   b) Newton's second law.
   c) the law of action-reaction.
   d) whiplash.

2. In the collision in *Question 1*, the modeling-clay passenger would be prevented from toppling over backward by

   a) using a seat belt.
   b) having an air bag in front.
   c) sitting in a chair with a head restraint.
   d) Any of the above.

3. In which of the diagrams below does the block have a net external force acting on it?

   a) A
   b) B
   c) C
   d) D

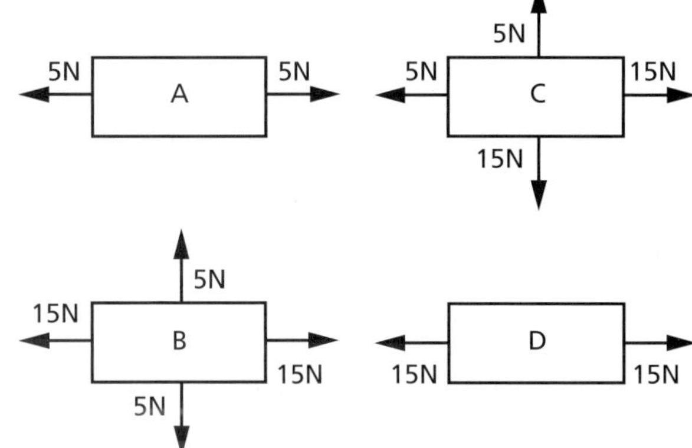

4. The diagram at right shows all the forces acting on block A. As a result, block A cannot be

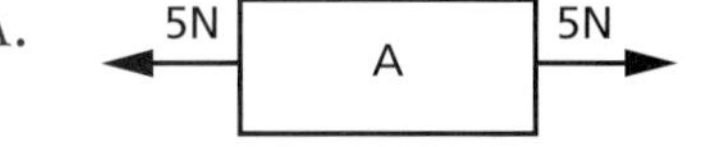

   a) at rest.
   b) moving with constant speed to the right.
   c) moving at constant speed upward.
   d) accelerating.

5. A constant net force of 20 N acts on an object of mass 5 kg. As the mass of the object is increased, the acceleration of the object will

   a) decrease.
   b) increase.
   c) remain the same.
   d) increase until the mass equals the force.

SECTION 4 QUIZ ANSWERS

1. a) Newton's first law.
2. c) sitting in a chair with a head restraint.
3. c) C.
4. d) accelerating.
5. a) decrease.

NOTES

NOTES

CHAPTER 3

Chapter Mini-Challenge

The *Chapter Mini-Challenge* marks the midpoint of *Chapter 3*. Briefly review the *Engineering Design Cycle*, and have students review their goals for the *Chapter Challenge*.

Remind students that the *Inputs* have been the information introduced in each section, and in each section, they have reflected on how the physics content can be applied to their *Chapter Challenge*. Review the physics content that the class has been introduced to so far by asking students about each section and/or reviewing the list in the student text.

Let students know that the *Mini-Challenge* is part of the *Process* in the *Engineering Design Cycle*. Describe how they should create drafts of their five-minute oral report and their written report or multimedia report for the *Mini-Challenge*. Students should draw detailed plans of the model safety system they intend to build. Emphasize that they should select the vehicle that they will use to design a safety system, write a description of how the physics concepts introduced so far will be used to create a safety system, and a presentation method of this information to the class.

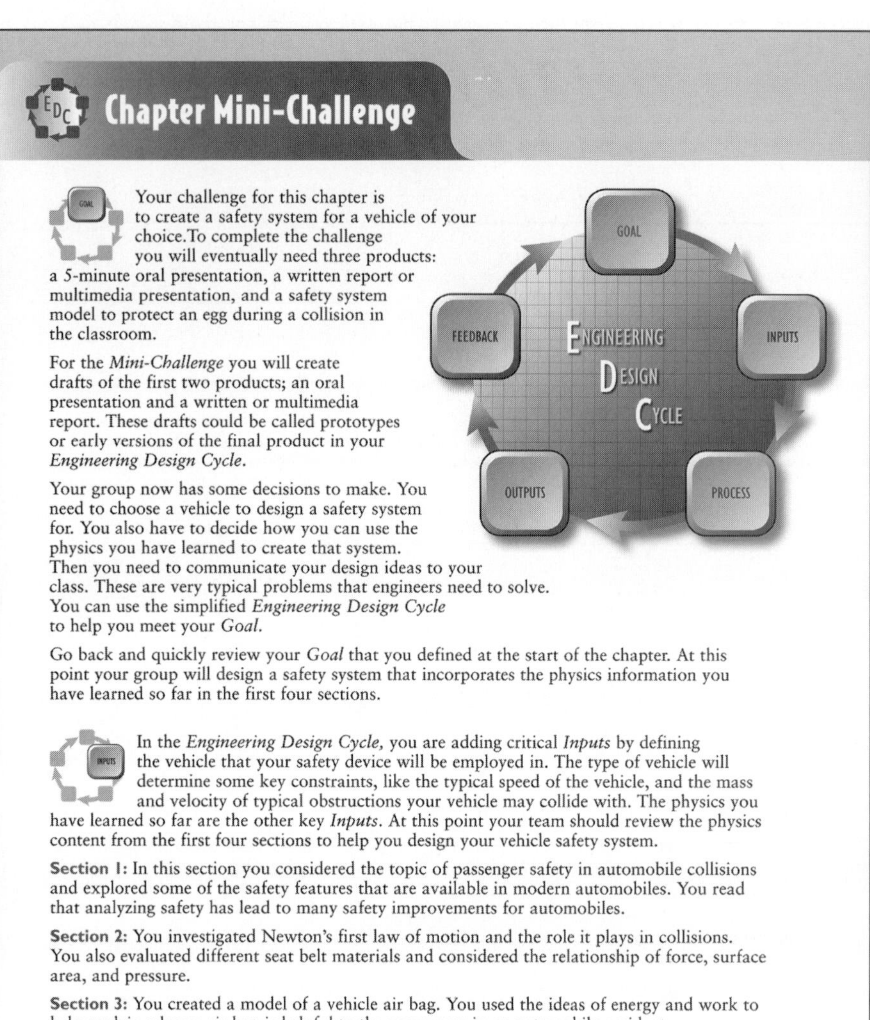

Chapter Mini-Challenge

Your challenge for this chapter is to create a safety system for a vehicle of your choice.To complete the challenge you will eventually need three products: a 5-minute oral presentation, a written report or multimedia presentation, and a safety system model to protect an egg during a collision in the classroom.

For the *Mini-Challenge* you will create drafts of the first two products; an oral presentation and a written or multimedia report. These drafts could be called prototypes or early versions of the final product in your *Engineering Design Cycle*.

Your group now has some decisions to make. You need to choose a vehicle to design a safety system for. You also have to decide how you can use the physics you have learned to create that system. Then you need to communicate your design ideas to your class. These are very typical problems that engineers need to solve. You can use the simplified *Engineering Design Cycle* to help you meet your *Goal*.

Go back and quickly review your *Goal* that you defined at the start of the chapter. At this point your group will design a safety system that incorporates the physics information you have learned so far in the first four sections.

In the *Engineering Design Cycle,* you are adding critical *Inputs* by defining the vehicle that your safety device will be employed in. The type of vehicle will determine some key constraints, like the typical speed of the vehicle, and the mass and velocity of typical obstructions your vehicle may collide with. The physics you have learned so far are the other key *Inputs*. At this point your team should review the physics content from the first four sections to help you design your vehicle safety system.

Section 1: In this section you considered the topic of passenger safety in automobile collisions and explored some of the safety features that are available in modern automobiles. You read that analyzing safety has lead to many safety improvements for automobiles.

Section 2: You investigated Newton's first law of motion and the role it plays in collisions. You also evaluated different seat belt materials and considered the relationship of force, surface area, and pressure.

Section 3: You created a model of a vehicle air bag. You used the ideas of energy and work to help explain why an air bag is helpful to the passengers in an automobile accident.

Section 4: You learned about Newton's second law that explains how objects respond when forces are applied to them by exploring rear-end collisions. Newton's second law describes the relationship between the force applied to a mass and the acceleration it will experience.

302

Time will be a key constraint for the work you present for the *Mini-Challenge*. You will want to organize your group members so that everyone knows what to do. Assigning roles and responsibilities may be the fastest way to get started. You may want to have a "scriptwriter" to record group thoughts, an "artist" to quickly sketch ideas, a "researcher" who checks chapter sections and returns facts, definitions, and formulas, a "math wizard" who performs calculations and checks the units for all numbers, and maybe even a "multimedia designer" who creates digital files, finds images, or creates a slide presentation of your ideas. The specific roles are not as important as making sure that everyone has a job and knows how to contribute to the group's success.

Once you are organized, you may want to use a short brainstorm to come up with different safety-system ideas. You can use physics principles to help focus your ideas. For instance, ask the group "How can we use a change in force to make a passenger safer?" As soon as you have a list of potential ideas, examine them to see if you want to combine any ideas. For instance, a seat belt and an air bag are both useful, but they work even better together. Finally, vote on an idea that your group will work on.

The time restraints you have to work may prevent you from creating a complete design, so you should focus on describing ideas that you will use based on the physics principles you have learned. You may have an idea for a new seat belt, but instead of detailing what color it will be, focus on explaining why you think a seat belt is necessary and how its specific features will keep the passenger safe. If you have extra time, you can always go back and add more detailed and stylish design elements.

Presenting your information to the class are your design cycle *Outputs*. You will not have to build an egg safety system model yet, but you will want to present some information about how you expect it to work. Even if you change your design before the *Chapter Challenge*, the more ideas you explore now, the more prepared you will be for the *Chapter Challenge*.

Your classmates will give you *Feedback* on the accuracy and the overall appeal of your presentation based on the criteria of the design challenge. This feedback will become an *Input* for your final design in the *Chapter Challenge*.

Remember to correct any parts of your design that were not complete or accurate. Finally, store all of your information in a safe place so that it will be ready to use in the *Chapter Challenge*.

As you complete the remaining sections, look for additional information that will help you improve your safety system. You may learn information that will help you explain why your system is effective by adding details, or even numerical calculations.

303

Discuss the time constraints with your class—about one class period to prepare and another to present. Emphasize that this will require groups to be organized. Use the information in the student text to suggest tasks group members might tackle to help them complete the *Mini-Challenge*, and eventually the *Chapter Challenge*. While students are preparing, assess their understanding and how they are working within their groups.

During the presentations, determine if students need extra help and address misconceptions that may have arisen. Students should refer to their list of criteria to see how closely they are meeting each criterion. Have each student in the audience provide written *Feedback* to the presenting group. This allows the presenting group to refer back to, reflect upon, and build on the comments and suggestions they receive. To promote constructive criticism, have students include at least two positive comments for each safety system presented.

Remind students that as they complete the remaining sections of this chapter, they should look for ways to improve their safety system, and for information that will assist them in explaining their system. Encourage the students to lead the discussion by having them ask and answer questions, as well as discuss ideas for improvement they would like to include in their own safety system.

SECTION 5

Momentum: Concentrating on Collisions

Section Overview

Students investigate head-on, nearly elastic collisions between two carts in one dimension. In these collisions, one cart is moving and the other is stationary. Students record their observations for carts colliding with equal masses, different masses, and varying initial speeds of the moving cart. Students then apply their observations and analysis to determine information about a stationary cart involved in a collision. From their investigations, students are introduced to the concept of momentum. They then calculate and consider the momentum of various objects.

Background Information

Momentum is a vector quantity describing the motion of an object. Momentum is equal to the mass of an object multiplied by its velocity or $\vec{p} = m\vec{v}$.

The momentum of an object does not change unless a net external force acts on the object. Newton's first law or the law of inertia states that an object at rest will remain at rest, and an object in linear motion will continue its motion, unless acted on by a net force. Another way to state Newton's first law is that the momentum of an object does not change unless a net external force acts on the object, or

If $\vec{F}_{\text{net}} = \vec{0}$, then $\frac{d\vec{p}}{dt} = \vec{0}$.

One reason that momentum is such an important quantity in physics is that the total momentum of any group of objects (a system) remains the same unless external forces act on the objects (the system). This is known as the law of conservation of momentum, which is further discussed in the next section.

A net force changes an object's momentum. Sometimes, Newton's second law is written as

$$\vec{F}_{\text{net}} = \frac{d\vec{p}}{dt} = m\frac{d\vec{v}}{dt} + \frac{d\vec{m}}{dt}v = m\vec{a} + \frac{d\vec{m}}{dt}v.$$

This form is general enough to also account for changes in mass, such as with fuel-burning vehicles. Students will not consider changes in mass during this curriculum.

Photons (quanta of light energy) are massless but have momentum. A photon's momentum is calculated from the relativistic energy expression

$$E = mc^2 = \sqrt{p^2c^2 + m_0^2c^4},$$

and setting the rest mass to zero. This leads to Planck's relationship $p = \frac{E}{c} = \frac{hf}{c} = \frac{h}{\lambda}$, where h is Planck's constant, f is the frequency, c is the speed of light, and λ is the wavelength. Planck's relationship is discussed in another chapter.

Momentum may be referred to as a measure of an object's inertia of motion, or its resistance to change its motion, but as was stated in *Chapter 2, Physics in Action*, "Newton explained that an object's **mass** is a measure of its inertia, or tendency to resist a change in motion. Given different masses moving at the same speed, the one with the greatest mass has the greatest inertia." Momentum is dependent on the inertial reference frame it is in, but conservation of momentum is not. Any object moving with a constant velocity appears to be at rest in a reference frame moving with the same velocity. Hence, a ball held in one's hand, while in a vehicle moving at constant speed, is in an inertial reference frame. To the person in the vehicle, the ball appears to be at rest and appears to have a value of zero for its momentum. To an observer on the side of the road, the ball has a momentum equal to the mass of the ball multiplied by its velocity, which would be the same as the velocity of the vehicle. If the passenger in the vehicle holding the ball were in a boxcar with no windows moving at a constant speed, that person would have no idea that he or she was in motion relative to someone on the ground. One of the important rules of physics is that all of the laws of physics must be obeyed in any inertial reference frame. This means that all the concepts of momentum, force, energy, and so on, have to hold true whether the person applying them is in the boxcar moving with a constant velocity, or whether he or she is in a laboratory cart that is not moving relative to the ground below.

In the *Investigate,* students explore collisions that are nearly elastic (no energy is transferred to the surroundings). For an elastic collision both the kinetic energy and the momentum of the system are conserved quantities. Because these laws must hold true in any inertial reference frame, it is easiest to move into a reference frame where one object is initially at rest. According to the law of conservation of momentum, for elastic collisions, you have the following equation:

$$m_1\vec{v}_{1\text{initial}} + m_2\vec{v}_{2\text{initial}} = m_1\vec{v}_{1\text{final}} + m_2\vec{v}_{2\text{final}}.$$

From the conservation of kinetic energy, you have the following equation:

$$\tfrac{1}{2}m_1v^2_{1\text{initial}} + \tfrac{1}{2}m_2v^2_{2\text{initial}} = \tfrac{1}{2}m_1v^2_{1\text{final}} + \tfrac{1}{2}m_2v^2_{2\text{final}}.$$

In the *Investigate,* the second cart has zero initial speed. Using this condition, you can find the following mathematical relationships for a head-on, elastic collision:

$$v_{1\text{final}} = \frac{m_1 - m_2}{m_1 + m_2}v_{1\text{initial}} \quad \text{and} \quad v_{2\text{final}} = \frac{2m_1}{m_1 + m_2}v_{1\text{initial}}.$$

From these it is easy to see the following:

- if $m_1 = m_2$, then $v_{1\text{final}} = 0$, and $v_{2\text{final}} = v_{1\text{initial}}$.
- if m_1 is much greater than m_2, then $v_{1\text{initial}} \approx v_{1\text{final}}$, and $v_{2\text{final}} \approx 2v_{1\text{initial}}$.
- if m_2 is much greater than m_1, then $v_{1\text{initial}} \approx -v_{1\text{final}}$, and $v_{2\text{final}} \approx 0$.

Crucial Physics

- Momentum is mass multiplied by velocity.
- During a collision, momentum is transferred between colliding objects. The momentum is transferred from an object that initially has a greater magnitude of momentum to an object that initially has a smaller magnitude of momentum. Sometimes this transfer is imperceptible. If the two objects have the same magnitude of momentum, a transfer may occur, resulting in a change of direction or speed of one or both objects.

Learning Outcomes	Location in the Section	Evidence of Understanding
Apply the definition of momentum.	***Physics Talk*** ***Physics Essential Questions*** ***Physics to Go*** Questions 1-7	Students compare and calculate the momentum of different objects.
Conduct analyses of the momentum of pairs of objects in one-dimensional collisions.	***Investigate*** Steps 5 and 6 ***Physics Essential Questions***	Students conclude from analyses of their observations that during collisions between two carts (one at rest), the cart at rest increases its motion when the colliding cart has increased speed or mass, and its motion does not change as much as its mass increases. Students consider how two sets of observations can determine the masses of colliding carts.

NOTES

Section 5 Materials, Preparation, and Safety

Materials and Equipment

PLAN A		
Materials and Equipment	**Group (4 students)**	**Class**
Dynamics cart	2 per group	
Weight, slotted, 100 g	6 per group	
Scale, electronic, 0-1500 g, 0.01 g readability		1 per class
Tape, masking		6 per group
Access to a smooth level surface*	1 per group	
Spring or loop of thin metal for dynamics cart*	1 per group	

*Additional items needed not supplied

PLAN B		
Materials and Equipment	**Group (4 students)**	**Class**
Dynamics cart	2 per group	
Weight, slotted, 100 g	6 per group	
Scale, electronic, 0-1500 g, 0.01 g readability		1 per class
Tape, masking		6 per group
Access to a smooth level surface*	1 per group	
Spring or loop of thin metal for dynamics cart*	1 per group	

*Additional items needed not supplied

Note: Time, Preparation, and Safety requirements are based on Plan A, if using Plan B, please adjust accordingly.

Time Requirement

Allow one period for the students to investigate the collisions.

Teacher Preparation

- Obtain dynamics carts with a spring mounted to one end to allow elastic collisions. If carts with springs are unavailable, a spring can be made from "strap steel" from a lumberyard. This is scrap metal that is used to hold bundles of wood together. When attached firmly to the sides of a cart and allowed to extend over the front, it makes an excellent spring.
- Caution the students to keep the collision speeds for the carts low and that care must be taken so that the carts are lined up properly to have a direct hit on the spring between the carts.
- During the collision between the small-mass cart and the large-mass cart, the small-mass cart may travel much faster than the large-mass cart. The students should be prepared to catch the small-mass cart quickly so that it doesn't fall on the floor.
- Decide where students should conduct the investigation. A long-level area clear of obstructions is needed for each group. To prevent carts from falling, consider using the floor.
- Conduct the investigation before class to determine where your students may have difficulties. Decide if you wish to stop students at any step to have a class discussion.

Safety Requirements

- In all collision experiments, goggles are of extra importance. Closed-toe shoes are also advised.

Meeting the Needs of All Students

Differentiated Instruction: Augmentation and Accommodations

Learning Issue	Reference	Augmentation and Accommodations
Describing collisions	***Investigate*** Steps 2-7	**Augmentation** • Students often have difficulty writing subjective observations with enough detail to be used for meaningful comparisons. They write phrases such as "It moved." Explicitly teach students how to describe motion in detail. • Provide structured activities to practice describing motion. Jump into the air and ask students to describe the motion. Show video clips of animals and/or athletes jumping and ask students to describe the motion. Students will understand that saying, "It jumped," is an ineffective way to describe this motion. Then students can make a list of words they have used to effectively describe motion. **Accommodation** • Provide students with a list of words that describe motion including forward, backward, fast, slow, moving cart, target cart, and so on.
Understanding essential concepts	***Physics Talk***	**Augmentation** • Many students can conceptualize that a large moving object can cause damage because of its momentum, but they really struggle to conceptualize the idea that very small objects that are moving very fast can also have a lot of momentum and cause damage. Show an Internet video clip of a sandblaster at work.
Understanding vectors	***Physics Words***	**Augmentation** • Without frequent review, students often forget the difference between scalar and vector quantities. Review the meaning of *scalar* and *vector*. Add *momentum* to the scalar/vector list created during *Chapter 1, Section 4*. • If students created a list of words to describe motion earlier in this section, explicitly make the connection that describing motion usually involves using a direction (vector).

Strategies for Students with Limited English-Language Proficiency

Point out new vocabulary words in context and practice using the words as much as possible throughout the section. As you work through the section, have students write the terms in their *Active Physics* logs and add the definitions in their own words.

absolute masses	obstructions
circumstances	pedestrian
disastrous	relative masses
elastic collision	stationary
enormous	transfer of momentum

Consider giving students a cloze activity when you reach the end of the section. Cloze activities are useful tools for summarizing material and for giving English-language learners opportunities to practice writing complete sentences using science vocabulary. Cloze activities are most effective when used frequently, to build students' abilities with more complex sentences.

Ask students to give you sentences describing what they did in the section, telling what important lessons they learned. Their comments should include the vocabulary words listed above. You may wish to offer a first sentence as an example. For instance, "You investigated momentum in collisions between two carts of different masses and different velocities." Write simple sentences, and work them into paragraphs. Some sentences should compare and contrast the transfer of momentum in the different collisions studied. Model using a topic sentence, supporting statements, and a closing sentence. Model the process of editing, in which students make corrections that improve the sentences.

There are many types of collisions covered in this section, so you will likely end up with a few paragraphs. Once the paragraphs are complete and students agree that these paragraphs accurately summarize what they did and learned, have the students copy them down. Explain that there will be a brief quiz on the paragraphs the next day at the beginning of class. The quiz will consist of the same paragraphs they wrote down, but with blanks where several terms were. The students will need to fill in the blanks with the terms that are missing. Tell students how the quiz will be graded. Prepare this quiz by keying in the paragraphs and then going back and removing every vocabulary term and replacing it with a blank. Choose a variety of words to leave out—nouns, verbs, adjectives, and so on. These terms can be science content words, but they do not have to be. Score the quiz by allotting two points for every blank: one point for the correct term or word (or perhaps another word with the correct meaning), and a second point for the correct spelling of that term or word.

SECTION 5

Teaching Suggestions and Sample Answers

What Do You See?

Have students consider the illustration and the title of the section. Use an overhead of the illustration to help focus the discussion. Ask students to describe the illustration and what they think is happening. Ask them what physics ideas they think pertain to the situation and how they might be useful. This elicitation helps to get students' initial ideas and provides a focus for the science content.

Chapter 3 Safety

Section 5 Momentum: Concentrating on Collisions

What Do You See?

Learning Outcomes

In this section, you will

- **Apply** the definition of momentum.
- **Conduct** analyses of the momentum of pairs of objects involved in one-dimensional collisions.

What Do You Think?

Automobile collisions are a leading cause of injury and death among teenagers.

- **A small sports automobile hits a heavy truck in a collision. What factors determine the outcome for the passengers of the two vehicles?**
- **Which driver will sustain worse injuries? Why?**

Record your ideas about these questions in your *Active Physics* log. Be prepared to discuss your responses with your small group and the class.

Investigate

1. You will stage a head-on collision between two collision carts of equal mass. One of the carts will have a spring or loop of thin metal attached to it so that the carts will collide with a "bounce" rather than with a "thud." A collision of this sort is called an elastic collision. This will serve as a model for the collision of vehicles. Find a level area clear of obstructions, such as the classroom floor, where the one cart can slide into a second cart at rest.

304

Students' Prior Conceptions

Being able to identify the momentum for each object involved in a collision both before and after the impact is an overriding strategy and an analytical tool that must be developed by the students during this section. The simple dictum "momentum before is equal to the momentum after" followed by verbal and/or mathematical statements identifying the mass and the velocity of each object should be included when students analyze what happens during collisions. Remind them that velocity is a vector and the vector nature of momentum is important in describing collisions. Exploring the transfer of energy during these interactions serves as a ladder for student cognition.

1. **Students have difficulty appreciating that all collisions involve equal forces acting on the separate bodies in opposite directions.** Provide opportunities for students to investigate and analyze collisions and ask them questions to ascertain their understanding through all of the sections embedded in *Chapter 3*.

2. **Higher impact speed causes more severe distortions than collisions with lower impact velocities.** This interpretation of prior student knowledge appears throughout this chapter. In this section, you may find it useful to evaluate the student engineering designs that mitigate the effects of a high speed collision to check that a shift in

What Do You Think?

Ask students to record their responses in their *Active Physics* logs, then have them discuss it with their group members to come up with their best group response. Have a class discussion on each group's ideas and record these ideas. Refer to these initial ideas during the section.

The concept of momentum is introduced later in this section. It is not expected that students will apply this concept in their supporting reasons; however, students can try to use reasoning based on the work-energy theorem as they did in the previous section to describe their ideas. They should incorporate the idea that as energy transferred to a passenger increases the likelihood of sustaining an injury increases.

What Do You Think?

A Physicist's Response

The factors that determine the outcome for the passengers of the two vehicles are the mass of each vehicle, the initial velocities of the two vehicles, and the safety features of each vehicle. The faster the vehicles are moving toward each other, and the more massive they are, the more energy they each have to transfer to the other vehicle. This results in more energy being transferred to the passengers in the other vehicle when they collide and increases risk of injury.

Trucks have more mass than small sports cars, so if they are traveling at the same speed toward each other, or even if the truck is traveling at a slightly lower speed than the sports car, the truck will have more energy than the sports car and will transfer more energy to the sports car during the crash. Because more energy will be transferred to the sports car than the truck, the passengers in the sports car are more likely to be injured than the passengers in the truck. Since the sports car does not have a lot of mass, it doesn't take much force to change its motion. Not as much work is required to move the sports car as is needed to move the truck, allowing more energy to be transferred to the passengers. For the very massive truck, most of the energy would go into stopping the truck or changing its motion, and not as much energy would be transferred to moving the passengers. However, serious injuries could still occur for the passengers in the truck.

In general, more massive vehicles are safer for passengers, but in practice, more massive vehicles may not be safer for passengers depending on their structure, the safety features that have been added, and the type of collision. Scientists use a physical quantity known as momentum to analyze collisions. The momentum is given by the mass multiplied by the velocity. Scientists consider how momentum changes from before a collision to after the collision. As the change in momentum increases, the chance of sustaining injury increases, as does the severity of the injury.

student understanding is occurring. Ensure that mass of colliding objects is also considered while students explain momentum and impact velocities.

3. **Students believe that backward momentum or recoil is less than forward momentum in a collision.** Applying the theory of the conservation of momentum, and measuring the mass and velocity of two objects that separate from each other after a collision in which the objects move in opposite directions facilitates students' assessment of forward momentum. They are able recognize the equality of recoil in an action-reaction situation.

Investigate

If you decided to have a discussion during the *Investigate* to guide the students through some of the steps, then let them know now at which step they should stop. Otherwise, have a discussion on the *Investigate* steps and let students begin and complete it. Let students know that the collisions they will be investigating are almost elastic collisions. Perfectly elastic collisions only occur on the molecular level. All other collisions will lose some energy to things such as sound and heat. Emphasize to students that in an elastic collision, the kinetic energy of the colliding objects before the collision is equal to the kinetic energy of the colliding objects after the collision. This means that no energy is transferred to the surroundings and no energy is used to permanently deform either object.

Teaching Tip

Using carts with springs actually reduces the force of the carts on the occupants because the springs lengthen the time that the carts interact. This has no effect on the momentum principles being studied.

Teaching Tip

Students should use carts that are in good condition with very low friction. If much friction is present, the carts will slow down immediately after the collision. Have students make their observations for just before, during, and just after the collision.

1.

Student groups should set up the materials and practice a head-on elastic collision between a moving cart and a stationary cart.

2.a)

Students should observe that when the two carts of equal mass have a head-on, elastic collision, the first cart should stop after the collision (or nearly so), and the second cart should leave with a speed equal to that of the moving cart before the collision. Students should record their observations in their log and construct a diagram similar to the following.

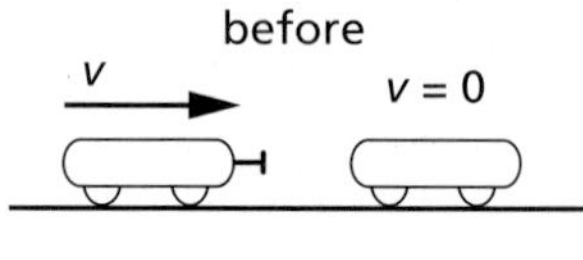

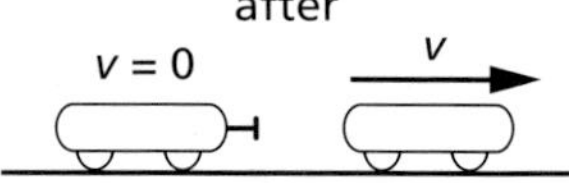

NOTES

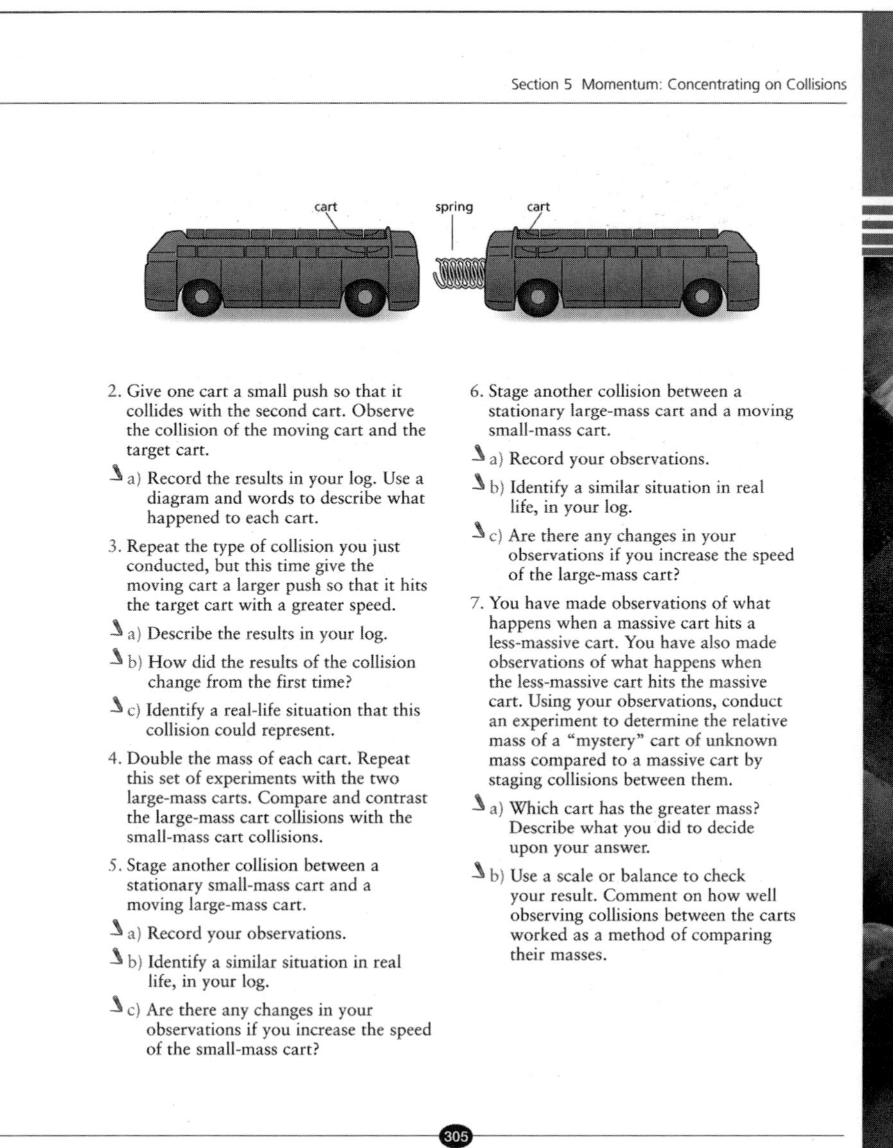
Section 5 Momentum: Concentrating on Collisions

2. Give one cart a small push so that it collides with the second cart. Observe the collision of the moving cart and the target cart.
   a) Record the results in your log. Use a diagram and words to describe what happened to each cart.
3. Repeat the type of collision you just conducted, but this time give the moving cart a larger push so that it hits the target cart with a greater speed.
   a) Describe the results in your log.
   b) How did the results of the collision change from the first time?
   c) Identify a real-life situation that this collision could represent.
4. Double the mass of each cart. Repeat this set of experiments with the two large-mass carts. Compare and contrast the large-mass cart collisions with the small-mass cart collisions.
5. Stage another collision between a stationary small-mass cart and a moving large-mass cart.
   a) Record your observations.
   b) Identify a similar situation in real life, in your log.
   c) Are there any changes in your observations if you increase the speed of the small-mass cart?
6. Stage another collision between a stationary large-mass cart and a moving small-mass cart.
   a) Record your observations.
   b) Identify a similar situation in real life, in your log.
   c) Are there any changes in your observations if you increase the speed of the large-mass cart?
7. You have made observations of what happens when a massive cart hits a less-massive cart. You have also made observations of what happens when the less-massive cart hits the massive cart. Using your observations, conduct an experiment to determine the relative mass of a "mystery" cart of unknown mass compared to a massive cart by staging collisions between them.
   a) Which cart has the greater mass? Describe what you did to decide upon your answer.
   b) Use a scale or balance to check your result. Comment on how well observing collisions between the carts worked as a method of comparing their masses.

305
Active Physics

3.a)

Students' descriptions and diagrams should indicate that the results are similar to those in *Step 2*, except with a larger incoming and outgoing velocity.

3.b)

Students' responses should note that the outgoing velocity of the target cart would be higher.

3.c)

Students might identify this collision as similar to a rear-end collision between two automobiles of approximately equal mass.

4.

Students observe results like they observed in *Step 3*. The incoming mass should stop, and the target mass moves off with a velocity equal to that of the incoming cart. Similarities in the two collisions would be that in both cases, the moving cart came to rest (or almost to rest) while the target cart moved off with a speed almost equal to that of the moving cart before the collision. Also, in both cases, the target cart continued forward in the same direction as the moving cart was traveling before the collision.

Differences could include the difference in speed between the carts in the two different collisions before and after. The target cart in the second collision will tend to roll further than the target cart in the first collision before coming to a stop.

5.a)

Students should observe that when the moving large-mass cart strikes the stationary small-mass cart, the small-mass cart moves off at a higher velocity than the moving cart had before the collision. The moving cart does not come to a complete stop immediately after the collision, but continues forward at a reduced speed.

5.b)

Students may describe a situation where a heavy truck hits a sports vehicle, or a pickup truck striking a compact car from the rear.

5.c)

As the speed of the large-mass cart increases, the departing speed of both the small-mass and the large-mass carts increases after the collision.

6.a)

Students should observe that when the moving small-mass cart strikes the stationary large-mass cart, the small-mass cart bounces backward, and the large-mass cart moves off with a low speed in the same direction as the small mass had initially.

6.b)

Students might state that this is analogous to a compact vehicle or sports car striking a truck in a rear-end collision.

6.c)

Increasing the speed of the moving small-mass cart before the collision will increase the speed of rebound, and also the speed at which the large-mass cart moves off after the collision.

7.a)

Students should describe the procedures they took to determine the relative mass of a mystery cart. The students should find the cart that has the larger mass using the principles they observed in *Steps 5* and *6*. If the moving cart rebounds after the collision it will have the smaller mass. If it continues moving forward after the collision, it will have the larger mass. If the carts have equal mass, then the incoming cart will come to rest after the collision and the second cart will move away.

7.b)

Using a balance should confirm the results stated above. This result is only for a relative mass measurement.

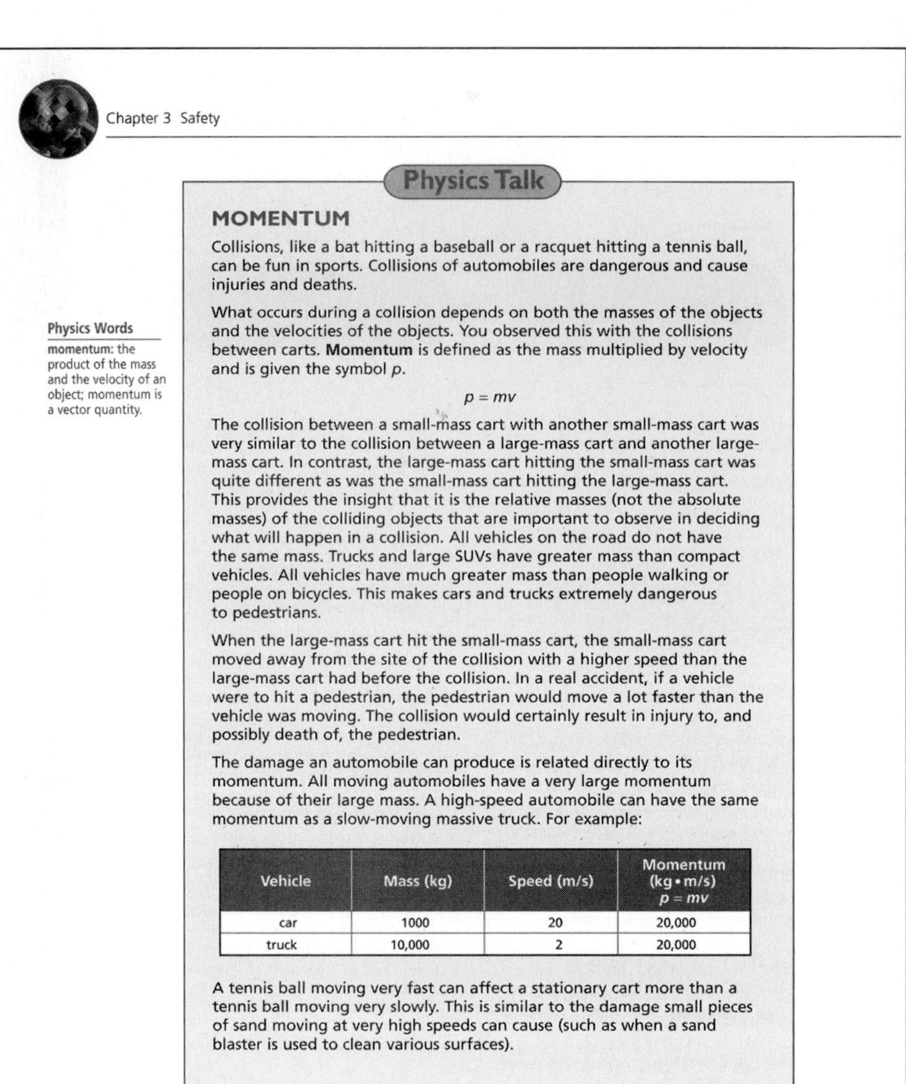

Chapter 3 Safety

Physics Talk

MOMENTUM

Collisions, like a bat hitting a baseball or a racquet hitting a tennis ball, can be fun in sports. Collisions of automobiles are dangerous and cause injuries and deaths.

What occurs during a collision depends on both the masses of the objects and the velocities of the objects. You observed this with the collisions between carts. **Momentum** is defined as the mass multiplied by velocity and is given the symbol p.

$$p = mv$$

Physics Words
momentum: the product of the mass and the velocity of an object; momentum is a vector quantity.

The collision between a small-mass cart with another small-mass cart was very similar to the collision between a large-mass cart and another large-mass cart. In contrast, the large-mass cart hitting the small-mass cart was quite different as was the small-mass cart hitting the large-mass cart. This provides the insight that it is the relative masses (not the absolute masses) of the colliding objects that are important to observe in deciding what will happen in a collision. All vehicles on the road do not have the same mass. Trucks and large SUVs have greater mass than compact vehicles. All vehicles have much greater mass than people walking or people on bicycles. This makes cars and trucks extremely dangerous to pedestrians.

When the large-mass cart hit the small-mass cart, the small-mass cart moved away from the site of the collision with a higher speed than the large-mass cart had before the collision. In a real accident, if a vehicle were to hit a pedestrian, the pedestrian would move a lot faster than the vehicle was moving. The collision would certainly result in injury to, and possibly death of, the pedestrian.

The damage an automobile can produce is related directly to its momentum. All moving automobiles have a very large momentum because of their large mass. A high-speed automobile can have the same momentum as a slow-moving massive truck. For example:

Vehicle	Mass (kg)	Speed (m/s)	Momentum (kg • m/s) $p = mv$
car	1000	20	20,000
truck	10,000	2	20,000

A tennis ball moving very fast can affect a stationary cart more than a tennis ball moving very slowly. This is similar to the damage small pieces of sand moving at very high speeds can cause (such as when a sand blaster is used to clean various surfaces).

Active Physics 306

Physics Talk

This *Physics Talk* introduces momentum and the transfer of momentum, and discusses the importance of momentum during collisions.

Have a class discussion on momentum using the information in the student text. Introduce what momentum is (mass multiplied by velocity), and how it is written mathematically. Emphasize that it is a vector quantity. Students are not introduced to the law of conservation of momentum until the next section. However, students should be able to analyze various elastic collisions by considering the relative masses and initial speeds of each of the objects. Encourage students to make connections between their observations and the examples provided in the student text.

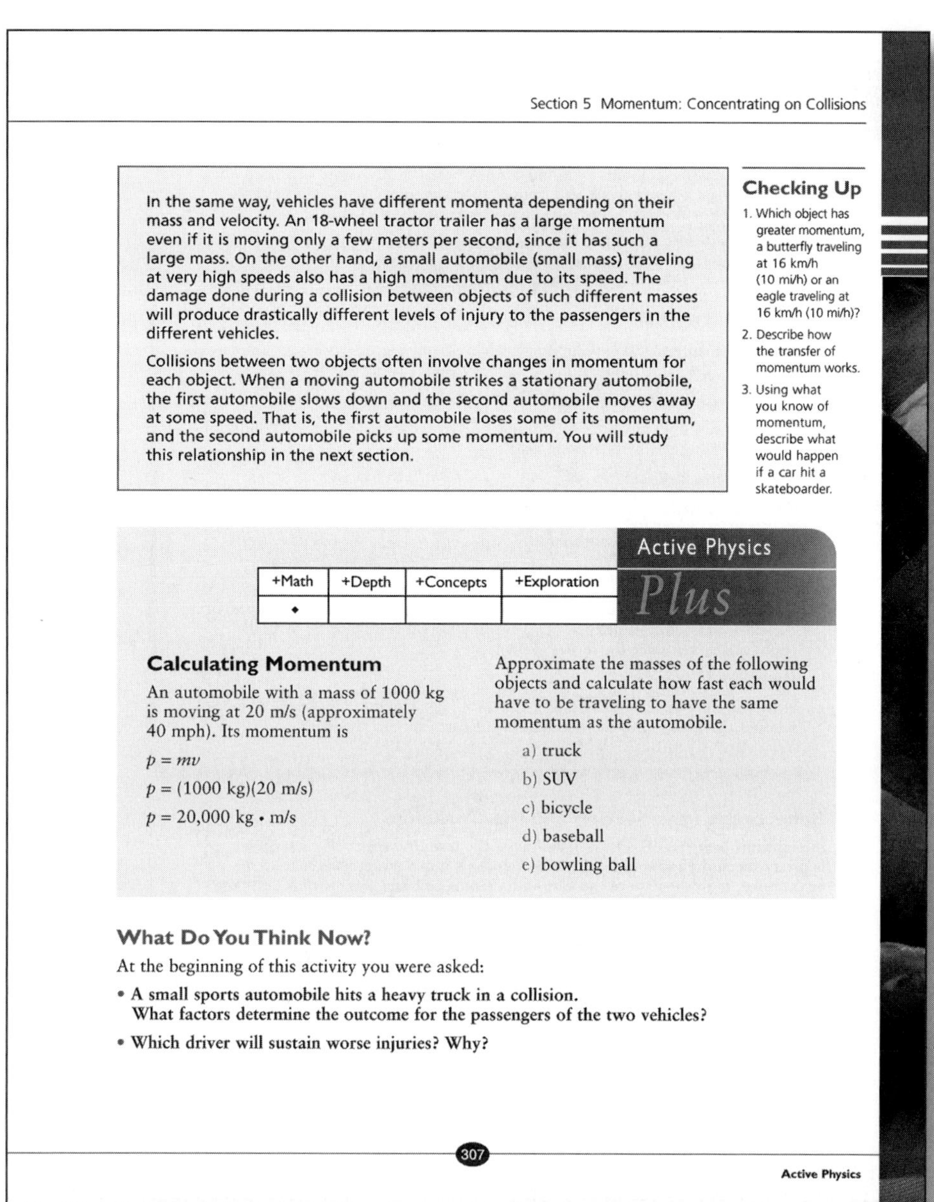

Section 5 Momentum: Concentrating on Collisions

In the same way, vehicles have different momenta depending on their mass and velocity. An 18-wheel tractor trailer has a large momentum even if it is moving only a few meters per second, since it has such a large mass. On the other hand, a small automobile (small mass) traveling at very high speeds also has a high momentum due to its speed. The damage done during a collision between objects of such different masses will produce drastically different levels of injury to the passengers in the different vehicles.

Collisions between two objects often involve changes in momentum for each object. When a moving automobile strikes a stationary automobile, the first automobile slows down and the second automobile moves away at some speed. That is, the first automobile loses some of its momentum, and the second automobile picks up some momentum. You will study this relationship in the next section.

Checking Up

1. Which object has greater momentum, a butterfly traveling at 16 km/h (10 mi/h) or an eagle traveling at 16 km/h (10 mi/h)?
2. Describe how the transfer of momentum works.
3. Using what you know of momentum, describe what would happen if a car hit a skateboarder.

Active Physics *Plus*

+Math	+Depth	+Concepts	+Exploration
◆			

Calculating Momentum

An automobile with a mass of 1000 kg is moving at 20 m/s (approximately 40 mph). Its momentum is

$p = mv$

$p = (1000\text{ kg})(20\text{ m/s})$

$p = 20{,}000\text{ kg} \cdot \text{m/s}$

Approximate the masses of the following objects and calculate how fast each would have to be traveling to have the same momentum as the automobile.

a) truck

b) SUV

c) bicycle

d) baseball

e) bowling ball

What Do You Think Now?

At the beginning of this activity you were asked:

- **A small sports automobile hits a heavy truck in a collision. What factors determine the outcome for the passengers of the two vehicles?**
- **Which driver will sustain worse injuries? Why?**

307 Active Physics

Checking Up

1.

If two objects have the same speed, then the object with the greater mass has the greater momentum. In this case, the eagle has greater momentum than the butterfly.

2.

Students' descriptions should include that a change in motion indicates a transfer of momentum. Encourage students to provide examples such as those from their *Investigate* observations. They may say, "When two objects of equal mass collide elastically, where initially one object is at rest, after the collision the object that was at rest moves away and the object that was initially moving stops moving. This shows a transfer of momentum from one object to the other."

3.

Students' responses should indicate an understanding of how mass plays an important role in collisions and the transfer of momentum. Encourage students to support their responses using their observations from the *Investigate*. When a vehicle strikes a skateboarder, it transfers momentum to the skateboarder. The skateboarder's speed will increase in the direction that the vehicle was moving in, and the skateboarder is likely to sustain severe injuries.

Active Physics Plus

This *Active Physics Plus* is geared toward increasing students' ability to apply the concept of momentum to solve problems algebraically. Review the example of calculating the momentum of an automobile. Have a discussion of students' responses for *Steps a)* through *e)*. Students may have different estimates of the mass of each object however they should not be very far off.

a)

The approximate mass of a truck is 10,000 kg (about 22,000 lbs). For a truck to have a momentum of $p = 20{,}000\text{ kg} \cdot \text{m/s}$, it must have a speed of

$$v = \frac{p}{m} = \frac{20{,}000\text{ kg} \cdot \text{m/s}}{10{,}000\text{ kg}} =$$

2 m/s (≈ 4 mi/h), about fifteen times the speed of a garden snail.

CHAPTER 3

Based on the relative amounts of momentum, what is the outcome of a head-on collision between a heavy truck and a small sports automobile if both have the same speed? How do your ideas now compare to your initial ideas?

Physics Essential Questions

What does it mean?

Define momentum and explain under what circumstances a compact automobile could have the same momentum as a more massive sport utility vehicle.

How do you know?

Explain, using two sets of observations, how you can determine the relative masses of two carts by observing collisions.

Why do you believe?

Connects with Other Physics Content	Fits with Big Ideas in Science	Meets Physics Requirements
Forces and motion	* Conservation laws	Good clear explanation, no more complicated than necessary

* Physicists define new quantities because they are useful in describing real events. Mass and velocity are two easily observable quantities. Why did physicists introduce the term momentum?

Why should you care?

How will the design of your safety system for the *Chapter Challenge* take into account the speed, thus momentum, of the cart carrying the egg?

Reflecting on the Section and the Challenge

In collisions between objects of equal mass, the resulting effect on each object is pretty much the same. But when one vehicle is much more massive than the second one, as in the case of the cart and a tennis ball or a heavy truck colliding head-on with a sports automobile, the results for the smaller vehicle is often disastrous. The object with the low mass suffers the most drastic damage.

The key to understanding collisions is to calculate the momentum of each colliding object. Momentum is mass multiplied by velocity. This is also written as $p = mv$. Objects with a small mass and high speed can have the same momentum as massive objects moving at slow speeds. An automobile with a mass of 1000 kg moving at 5 m/s (10 mph) has an enormous momentum and would severely injure any pedestrian upon contact.

b)

The approximate mass of an SUV is 4000 kg (about 8800 lbs). For an SUV to have a momentum of $p = 20{,}000 \text{ kg}\cdot\text{m/s}$, it must have a speed of

$$v = \frac{p}{m} = \frac{20{,}000 \text{ kg}\cdot\text{m/s}}{4000 \text{ kg}} =$$

5 m/s $(\approx 11 \text{ mi/h})$, about half the average walking speed.

c)

The approximate mass of a bicycle is 14 kg (about 30 lbs). For a bicycle to have a momentum of $p = 20{,}000 \text{ kg}\cdot\text{m/s}$, it must have a speed of

$$v = \frac{p}{m} = \frac{20{,}000 \text{ kg}\cdot\text{m/s}}{14 \text{ kg}} =$$

1400 m/s $(\approx 3100 \text{ mi/h})$, about the wind speed of a tornado.

d)

The approximate mass of a baseball is 0.15 kg (about 0.33 lbs). For a baseball to have a momentum of $p = 20{,}000 \text{ kg}\cdot\text{m/s}$, it must have a speed of

$$v = \frac{p}{m} = \frac{20{,}000 \text{ kg}\cdot\text{m/s}}{0.15 \text{ kg}} =$$

130,000 m/s $(\approx 290{,}000 \text{ mi/h})$,

almost 400 times the speed of sound in air at sea level and 0°C, and more than the escape velocity needed to leave Earth's atmosphere.

e)

The approximate mass of a bowling ball is 5 kg (about 12 lbs). For a bowling ball to have a momentum of $p = 20{,}000 \text{ kg}\cdot\text{m/s}$, it must have a speed of

$$v = \frac{p}{m} = \frac{20{,}000 \text{ kg}\cdot\text{m/s}}{5 \text{ kg}} =$$

4000 m/s $(\approx 9000 \text{ mi/h})$, ten times greater than the speed of sound in air.

What Do You Think Now?

Revisit the *What Do You Think?* questions, and review students' initial ideas. Then ask students how they would answer these questions now using what they know about momentum. After discussing how students would revise their answers, consider discussing with the class *A Physicist's Response* to give them a better understanding of what whiplash is and why it occurs in rear-end collisions.

Section 5 Momentum: Concentrating on Collisions

Physics to Go

1. Suppose an automobile collides with another automobile that is stopped. If both automobiles have the same mass, what do you expect to happen in the resulting collision?
2. Describe the collision between two vehicles of equal mass moving toward each other at equal speeds.
3. Describe the collision between two vehicles of very different masses moving toward each other at equal speeds.
4. Why do football teams prefer offensive and defensive linemen who weigh about 140 kg (about 300 lb)?
5. What determines who will get knocked backward when a big vehicle collides with a smaller vehicle in a head-on collision?
6. A 1000-kg automobile is moving at 10.0 m/s. At what speed would a 10,000-kg truck need to travel in the same direction so that the momentum of the two would be equal?
7. ***Preparing for the Chapter Challenge***

   Use the words mass, velocity, and momentum to write a paragraph that gives a detailed "before and after" description of what happens when a moving vehicle hits a stationary vehicle of equal mass in a direct collision.

309 Active Physics

Reflecting on the Section and the Challenge

Have a discussion on how collisions involving objects of similar and different masses affect the amount of damage and possible injury. Then emphasize that momentum is a key concept in understanding collisions. Remind students that momentum is the mass multiplied by the velocity, and that the greater the momentum of an object, the greater the damage it can do when it collides with another object.

Physics Essential Questions

What does it mean?

Momentum is the product of mass and velocity $p = mv$. A massive sport utility vehicle with a small velocity can have the same momentum as a compact vehicle with a large velocity.

How do you know?

If a moving vehicle collides elastically with a stationary vehicle and both vehicles move forward after the collision, then the moving vehicle is more massive. If the moving vehicle moves backward after the collision, it is less massive.

Why do you believe?

Momentum is defined because in a collision it is the momentum, not the mass nor the velocity alone that informs you about the collision.

Why should you care?

The more momentum the egg has, the more protection it will need.

CHAPTER 3

Physics to Go

1.

Students should refer to the observations from the *Investigate* to support their responses. If the collision is elastic, then the momentum of the moving automobile is transferred to the automobile initially at rest. The automobile initially moving will come to rest, while the automobile initially at rest will move off with a speed equal to that of the other automobile's initial speed.

2.

Since students have not yet been introduced to the law of conservation of momentum, they should reason this problem out using evidence from the *Investigate*. Students should assume the collision is elastic, and consider what would happen if the first vehicle was moving and the second vehicle was not moving, but at rest. This would result in the same answer as in *Question 1*, resulting in all the momentum of the first vehicle being transferred to the second vehicle. Similarly, if the second vehicle was moving and the first vehicle was at rest, all of the second vehicle's momentum would be transferred to the first vehicle. Since the vehicles are the same mass, after the collision both vehicles will move away from each other with the same speed.

3.

If two vehicles, one much more massive than the other, traveling at equal speeds collide head-on, the more massive vehicle will continue moving in its original direction at a reduced speed. The less massive vehicle will reverse its direction. This is one of the reasons why heavier vehicles are considered safer. Encourage students to use their observations from the *Investigate* to support their responses.

4.

It is more difficult to change an object's momentum as its mass increases hence, it is more difficult to change the momentum of a massive person. The more mass means the more inertia, which means the more massive the linemen, the more difficult it is to get them moving if they are stopped, or stop them if they are moving.

5.

For an elastic collision, it is the relative masses of the vehicles that determine which vehicle gets knocked backward if one of the vehicles is initially at rest.

Note: Because one can always move into an inertial reference frame where one of the objects is initially at rest, the description is as general as both objects initially moving toward each other.

6.

The momentum of the automobile is $p_{car} = (1000\text{ kg})(10\text{ m/s}) =$

$10{,}000\text{ kg}\cdot\text{m/s}$.

The momentum of the truck is the same as that of the automobile, so its speed can be calculated using

$$v = \frac{p}{m} = \frac{(10{,}000\text{ kg}\cdot\text{m/s})}{10{,}000\text{ kg}} = 1\text{ m/s}.$$

7.

Preparing for the Chapter Challenge

Consider having students support their claims with the evidence from the *Investigate*. Students should have the following information in their paragraph:

When vehicles of equal mass collide elastically, and one vehicle is initially stationary, the vehicles trade conditions following the collision. The vehicle that was originally moving is stationary after the collision, and the vehicle that was originally stationary is moving at about the same speed as the initial speed of the other vehicle.

SECTION 5 QUIZ

3-5a Blackline Master

1. The diagram at right shows a collision between two carts of equal mass. Cart A has a spring on the front and is moving forward, while cart B is at rest. After cart A hits cart B, which of the following will occur?

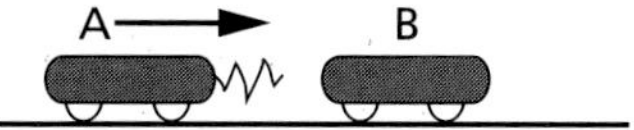

   a) Cart A stops and cart B moves forward.

   b) Cart A continues forward with the same speed as cart B.

   c) Cart A bounces backward and cart B stays at rest.

   d) Cart A bounces backward and cart B moves forward.

2. In *Question 1*, cart B now has a mass double that of cart A when the collision occurs. After cart A hits cart B, which of the following will occur?

   a) Cart A stops and cart B moves forward.

   b) Cart A continues forward with the same speed as cart B.

   c) Cart A bounces backward and cart B stays at rest.

   d) Cart A bounces backward and cart B moves forward.

3. Football players A and B both have the same momentum when running downfield. Football player A has a mass of 100 kg, while player B has a mass of 120 kg. Which statement below must be true?

   a) Player A is running faster than player B.

   b) Player B is running faster than player A.

   c) Both players must be running with the same speed.

   d) Both players must have the same kinetic energy.

4. A student does an experiment to check the mass of a cart. The student sends a 1.0-kg cart with a spring attached at the front end into a collision with a cart of unknown mass. After the collision, the student notes that the 1.0-kg cart moves forward with reduced speed, and the unknown cart moves forward at a faster speed than the 1.0-kg cart. What does this experiment show about the mass of the unknown cart?

   a) The unknown cart is more than 1.0 kg.

   b) The unknown cart is 1.0 kg.

   c) The unknown cart is less than 1.0 kg.

   d) No information about the mass of the unknown cart can be obtained from this experiment.

5. The product of an object's mass and velocity is

a) force.

b) kinetic energy.

c) weight.

d) momentum.

SECTION 5 QUIZ ANSWERS

1. a) Cart A stops and cart B moves forward.
2. d) Cart A bounces backward and cart B moves forward.
3. a) Player A is running faster than Player B.
4. c) The unknown cart is less than 1.0 kg.
5. d) momentum.

NOTES

SECTION 6

Conservation of Momentum

Section Overview

Students begin by considering how traffic investigators use physics to analyze accidents. They investigate completely inelastic collisions in which the two colliding carts stick together after the collision, and how the factors of initial speed and mass affect inelastic collisions. Using their measurements, students calculate the momentum before and after the collision, and observe that momentum is conserved. They then describe and analyze both elastic and inelastic collisions by applying the law of conservation of momentum. A class discussion on the law of conservation of momentum and the different types of collisions follows. Connections are made between Newton's third law and the conservation of momentum. Students then discuss the physics concepts that traffic investigators use to analyze accidents and how they pertain to the *Chapter Challenge*. Students conclude by applying what they have learned to analyze collision situations.

Background Information

The law of conservation of momentum is fundamental in physics. It states that the momentum of a system with no external forces acting on it before a collision is equal to the momentum of the system after the collision. This can be derived from Newton's laws, and is shown in the student text in the *Active Physics Plus*, for systems composed of objects with constant nonzero mass. More information is provided about momentum in the previous section's *Background Information.*

Crucial Physics

- Momentum is a conserved vector quantity. If no external forces are acting on a system, then the momentum of a system before a collision is equal to the momentum of the system after the collision.

Learning Outcomes	Location in the Section	Evidence of Understanding
Understand and apply the law of conservation of momentum to collisions.	***Investigate*** Steps 3 and 4 ***Physics Talk*** ***What Do You Think Now?*** ***Physics Essential Questions*** ***Reflecting on the Section and the Challenge*** ***Physics to Go*** Questions 1-17	Students conclude from their measurements and analysis that momentum is conserved. Students describe and apply the law of conservation of momentum.
Measure the momentum before and after a moving mass strikes a stationary mass in a head-on collision.	***Investigate*** Steps 1-3 ***Physics Essential Questions***	Students measure the masses and speeds of two objects before and after a collision. Using their measurements, students calculate the momentum for each object. Through analysis, they conclude that momentum is conserved. Students describe how their measurements could be improved.

Section 6 Materials, Preparation, and Safety

Materials and Equipment

PLAN A		
Materials and Equipment	**Group (4 students)**	**Class**
Dynamics cart	2 per group	
Clay, modeling, lb	2 per group	
Ruler, metric, in/cm	1 per group	
Weight, slotted, 100 g	6 per group	
Stopwatch	1 per group	
Scale, electronic, 0-1500 g, 0.01 g readability		1 per class
Tape, masking		6 per group
Access to a smooth level surface*	1 per group	
Photogate probeware*		1 per class
Motion detection probeware*		1 per class
MBL or CBL technology to record probeware activity*		1 per class

*Additional items needed not supplied

PLAN B		
Materials and Equipment	**Group (4 students)**	**Class**
Dynamics cart		2 per class
Clay, modeling, lb		2 per class
Ruler, metric, in/cm		1 per class
Weight, slotted, 100 g		6 per class
Stopwatch		1 per class
AC Ticker tape timer		1 per class
Scale, electronic, 0-1500 g, 0.01 g readability		1 per class
Tape, masking		6 per group
Access to a smooth level surface*	1 per group	
Photogate probeware*		1 per class
Motion detection probeware*		1 per class
MBL or CBL technology to record probeware activity*		1 per class

*Additional items needed not supplied

Note: Time, Preparation, and Safety requirements are based on Plan A, if using Plan B, please adjust accordingly.

Time Requirement

Allow two class periods for the students to investigate the collisions.

Teacher Preparation

- Decide where students should conduct the investigation. A long level area clear of obstructions is needed for each group. To avoid carts falling, consider using the floor.
- Conduct the investigation before class to determine where your students may have difficulties. Decide if you wish to stop students at any step to have a class discussion.
- Masses of 1 and 2 kg are recommended, but certainly not required. Instead, one laboratory cart could collide with another identical laboratory cart to provide a 1:1 mass ratio, and a 2:1 (and 1:2) ratio could be obtained by loading one cart on top of another to double its mass, or any other variation.
- The colliding objects need to stick together upon colliding for the "sticky collision" to move as a single object after the collision. Clay magnets or fabric hook-and-loop fasteners are very convenient for this purpose. Placing lumps of clay on the colliding surfaces will not make the carts stick together completely, but they will almost do so.
- For an inelastic collision, the spring side of a dynamics cart should not be used.
- Tape any added masses to the dynamics carts in place to ensure they stay in the cart.
- When colliding two moving carts, caution students to carefully align the carts so an on-center collision will occur.
- Determine how your class will measure the velocity of the carts. Velocity before and after the collisions may be measured in several ways. The least-accurate measurement would be with stopwatches over a known distance due to timing error, and the carts slowing down. Tape timers attached to each cart would be more accurate, but are extremely difficult to coordinate. The best method would be to have a motion detector and associated equipment to measure the velocities

immediately before and after the collision. To help the motion detector lock onto the cart, attach a piece of cardboard about the size of an index card to each cart to improve the reflections.
Be sure the area around the collision point is clear of extraneous material to prevent confusion in the motion detector.

Safety Requirements

- Students should wear closed-toed shoes.
- In all collision experiments, goggles are of extra importance.
- If this investigation is done on a lab table, take precautions so that no masses fall off the table onto a student's foot.
- Caution students to keep the cart velocities low to prevent damage to the carts or other problems.

NOTES

CHAPTER 3

Meeting the Needs of All Students

Differentiated Instruction: Augmentation and Accommodations

Learning Issue	Reference	Augmentation and Accommodations
Copying data tables	***Investigate*** Steps 1.a) and 3.a)	**Augmentation** • Students who have difficulty with fine-motor tasks, paying attention to details, and copying from one place to another will spend a lot of time copying these data tables. Verbal direction and helpful hints could be given to quicken the copying; however, in this case, it may be most effective to provide a copy of the tables for students to tape into their logs. This allows more time for data collection and analysis.
Recalling how to measure velocity	***Investigate*** Step 1.b)	**Augmentation** • If the class has not measured velocity in a week or longer, students who struggle with short- and long-term memory tasks may forget how to measure velocity. Accurate measurements are important in this section to support and reinforce students' understanding of the conservation of momentum. If the data does not show that momentum is conserved, it is very confusing for students. • Review the class's method for measuring velocity. Give students an opportunity to ask clarifying questions about measuring velocity before they begin collecting data.
Applying concepts	***Investigate*** Step 4	**Augmentation** • Organizing the 12 collision sketches on one page will make it easier for students to compare the momentums in different types of collisions. Direct students to make a chart that takes up the whole page. The chart should have three columns and seven rows. The first column would be the "Type of Collision," the second column the "Before" sketch, and the third column the "After" sketch. Provide a visual model of the table when the instructions are given. • Provide students with ideas for representing the objects and their motion. For example, a rectangle may be a sufficient representation of an object, and arrows may be the best way to represent motion. • Some students may not be able to visualize the collisions without seeing real carts crash together. Provide carts for students to collide as they are making their sketches. **Accommodation** • Provide a blank chart for students to record their sketches and tape into their logs. • Demonstrate one collision at a time for the whole class, and ask students to create their sketches after each demonstration.

Strategies for Students with Limited English-Language Proficiency

Point out new vocabulary words in context, and practice using the words as much as possible throughout the section.

analysis	reconstruct
conservation of momentum	sticky collision
crucial	tread marks
depart	tremendous
hallmark	uncertainties
optical illusion	victim

An important skill that students should learn in science is how to organize data into tables. This section presents an opportunity to practice this skill. Students should be able to distinguish between rows and columns (they often confuse the two). They should give appropriate information labels to rows, and headers to columns. Tables should be numbered and have headings. Point out examples of tables in books. Ask students how they use the information contained in the organization of a table to learn about what is presented in the table. Consider giving each group examples of one table that is well organized and one that is not, and then ask students to describe what information is missing from the poorly organized table.

Two important aspects of learning a new language are speaking and writing in that language. Some ELL students will be self-conscious about speaking in front of their peers, while others will be less reluctant to try. Be sure to encourage all ELL students to speak in class, and give them opportunities to write on the board from time to time. Experience will broaden their comfort level. Over time, the shy students will become increasingly less self-conscious about speaking in front of their classmates.

With that in mind, hold a class discussion to review *Section 6*. Call on ELL students to answer or address the bulleted items below.

- In your own words, what is the law of conservation of momentum? (The total momentum before a collision equals the total momentum after a collision, if no net external forces act on the system.)
- Is momentum a scalar quantity or a vector quantity? Explain. (The direction of momentum is important; it is the same direction as velocity.)
- What does the following equation mean, in words? $m_1 v_{1\text{initial}} + m_2 v_{2\text{initial}} = (m_1 + m_2) v_{\text{final}}$ (For an inelastic collision in which the objects stick together after they collide, the momentum before a collision equals the momentum after a collision.)
- Solve the equation above for v_f using these values: $m_1 = 750$ kg, $v_{1\text{initial}} = 4$ m/s, $m_2 = 250$ kg, $v_{2\text{initial}} = 8$ m/s. [$v_{\text{final}} = 5$ m/s]
- **Critical Thinking:** Do these values substituted in the equation above represent a collision between two moving vehicles or a collision between one moving vehicle and one stationary vehicle? (Two moving vehicles.) How do you know? (Neither starting momentum is zero.)

SECTION 6

Teaching Suggestions and Sample Answers

What Do You See?

Review the illustration with the students. Consider using the overhead of the illustration as a focal point for the discussion. Elicit initial impressions of what students see in the illustration, focusing on their description of the collision. Ask students what they think is meant by conservation of momentum. Discuss how students think the illustration connects with the conservation of momentum.

What Do You Think?

Ask students to think about what traffic-accident investigators can determine by analyzing tire marks on the road and damage to the vehicles. Record students' ideas and ask students to consider what physics principles these investigators use to "reconstruct" the accident. Have students record their ideas in their log and discuss their answers with their group. Then have a class discussion on students' ideas. Record misconceptions and make sure they are addressed at appropriate times during the section. Focus on the responses that provide an opportunity for you to get the students engaged in the physics concepts presented in this chapter.

Chapter 3 Safety

Section 6 Conservation of Momentum

What Do You See?

Learning Outcomes

In this section, you will

- **Understand** and apply the law of conservation of momentum to collisions.
- **Measure** the momentum before and after a moving mass strikes a stationary mass in a head-on collision.

What Do You Think?

Traffic-accident investigators can determine what happened during an automobile accident by analyzing tire marks and the damage to the automobiles.

- **What physics principles do the traffic-accident investigators use to "reconstruct" the accident?**

Record your ideas about this question in your *Active Physics* log. Be prepared to discuss your responses with your small group and the class.

Investigate

In this *Investigate*, you will use two collision carts of equal masses that will stick together after a collision. Before the collision, one cart will be moving and the other cart will be at rest. After the collision, the two carts should stick together and move as a single object. You may use clay, magnets, or fabric hook-and-loop fasteners to stick the carts together, depending on what is available.

1. Stage a "sticky" collision between the two carts with equal masses. Measure the velocity, in meters per second, of the moving mass before the collision and the velocity of the combined masses after the collision.

310

Consider emphasizing that there are no "right" answers and that all answers are acceptable. The purpose of these questions is to elicit students' prior knowledge and to get students to think about the physics involved. Ask them to refer to their answers while they are being introduced to new physics concepts. Point out that when they are aware of what they think, they will be better able to add to what they know.

Section 6 Conservation of Momentum

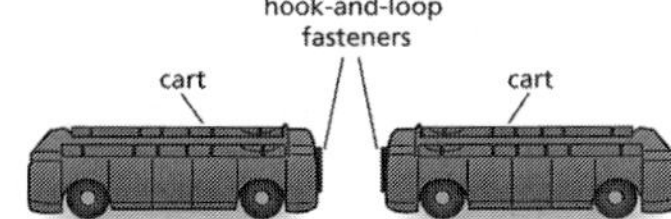

You can measure the velocities with a ruler and stopwatch, with a ticker-tape timer, with a velocimeter or a computer and motion detector.

a) Prepare a data table in your log similar to the one shown below. Provide enough horizontal rows in the table to enter data for at least four collisions.

Sticky Head-on Collisions:
One Object Moving before Collision

Mass of Object 1 (kg)	Mass of Object 2 (kg)	Velocity of Object 1 before Collision (m/s)	Velocity of Object 2 before Collision (m/s)	Mass of Combined Objects after Collision (kg)	Velocity of Combined Objects after Collision (m/s)
1.0	1.0		0.0	2.0	
2.0	1.0		0.0	3.0	
1.0	2.0		0.0	3.0	
			0.0		

b) Record the measured values of the velocities in the first row of the data table.

2. Stage other sticky collisions using the masses listed in the second and third rows of the data table. Then stage one or more additional collisions using other masses. Measure the velocities before and after each collision.

a) Enter the measured values in the data table.

3. Organize a table for recording the momentum of each object before and after each of the above collisions.

a) Prepare a table similar to the following example in your log.

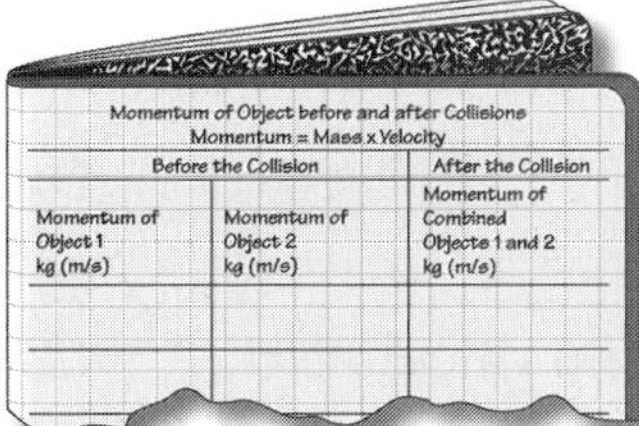

Momentum of Object before and after Collisions
Momentum = Mass x Velocity

Before the Collision		After the Collision
Momentum of Object 1 kg (m/s)	Momentum of Object 2 kg (m/s)	Momentum of Combined Objects 1 and 2 kg (m/s)

b) Calculate the momentum of each object before and after each of the above collisions and enter each momentum value in the table.

c) Calculate and compare the total momentum before each collision to the total momentum after each collision.

d) Allowing for minor variations due to uncertainties of measurement, write in your log a general conclusion about how the momentum before a collision compares to the momentum after a collision.

4. There are a variety of collisions involving two objects. In each collision, momentum is conserved and the same equation is used. The equation gets simpler when one of the objects is at rest and has zero momentum.

311 Active Physics

What Do You Think?

A Physicist's Response

Scientists and traffic-accident analyzers can determine many things from skid marks on the street and damage done to vehicles involved in a collision. This information is related to the direction vehicles were moving in before and after the crash and the speed of the vehicles. The physics principles used to "reconstruct" an accident are many. The most important concept used in collision analysis is the law of conservation of momentum, which states that the sum of momentums of the objects involved in a collision before the collision is equal to the sum of momentums of the objects after the collision. Conservation of energy is also used, as well as Newton's laws of motion.

Students' Prior Conceptions

The accuracy of mass and speed measurements before the moving mass strikes the stationary mass and the subsequent measurements of the two moving masses, after the collision, enables students to examine the concept of conservation of momentum. Calculating the momentum of the moving mass before it strikes a stationary mass and the subsequent momentum of the masses after the collision makes it possible for students to investigate the law of conservation of momentum. In addition to measurements of mass and velocity, students must determine the physical conditions of each mass before and after the collision to ascertain if any deformation has occurred. Knowledge about kinetic energy changing to sound, light, and heat also explains to students how energy is conserved when all parts of a system are considered.

1. **Momentum is a vector.** Students do not recognize the importance of momentum when a collision involves a moving object and a stationary object. Sources of error in measurement, external unbalanced forces, and the accuracy of the model used for investigating collisions are important discussion points. You should review the concept of relative motion, encouraging students to establish the directions for the vectors with respect to a start position. It is useful to pose scaffolding questions to support inquiry-based learning.

CHAPTER 3

Investigate

If you decided to have a discussion to guide the students through some of the steps in the *Investigate*, then let them know now at which step they should stop. Otherwise, have a discussion on the *Investigate* steps and let students begin and complete it.

1.

Have each group set up the materials and practice a completely sticky (inelastic) collision between a moving cart and a stationary cart.

1.a)

Students create a table in their logs to record their data.

1.b)

Students measure and record the incoming and outgoing velocities of the carts and record these values in their data table.

Teaching Tip

If masses are added to the carts to reach the 1 and 2 kg masses suggested, tape the masses to the carts to prevent sliding. Smaller masses may be used as long as the ratios are kept the same as those in the charts.

2.a)

Students repeat *Step 1* using the masses listed, and record them in their data table. Make sure students secure the masses to the carts by taping them to the carts.

3.a)

Students should create another table in their logs for the momentum of the objects before and after the collision.

3.b)

Students should calculate the momentum for each cart before the collision, and the momentum of the combined "stuck-together" carts after the collision. Students realize that momentum is mass multiplied by velocity.

3.c)

Students compare the total momentum of the incoming carts (the sum of the initial momentums of the carts) to the momentum of the combined carts after the collision.

3.d)

Students should conclude that within the uncertainties of the measurements the total momentum before the collision is equal to the total momentum after the collision.

4.a)

Have students draw sketches describing the objects before and after the collisions and record the momenta they know for the six types of collisions listed. For the examples provided, the class can try setting conditions such as equal masses, or one mass being twice the mass of the other. The sign for the direction of the velocity should be included if doing sample calculations. For Collision Type 5, the carts could be moving toward each other from opposite directions, or toward each other while traveling in the same direction. Afterward, they may be moving in the opposite directions if the collision was head-on or in the same directions if the collision was from the rear.

NOTES

3-6a Blackline Master

Collision Type 1

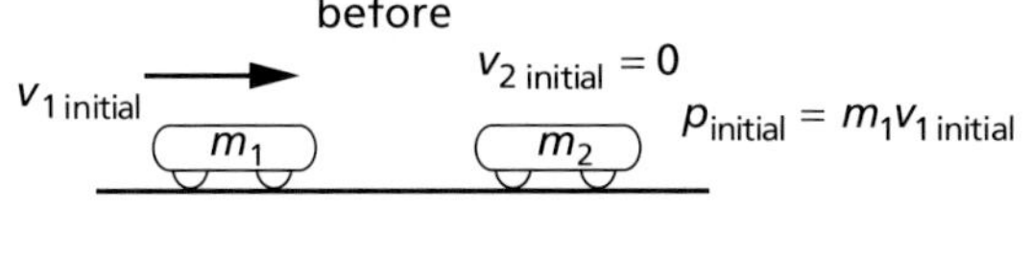

after

$v_{1\,\text{final}}$

$$v_{\text{final}} = \frac{m_1}{(m_1 + m_2)} v_{1\,\text{initial}}$$

$$p_{\text{final}} = (m_1 + m_2) v_{\text{final}}$$

Have students determine the final velocity if the two masses are equal and the initial velocity of the moving cart is 1 m/s to the right. In this collision, the final velocity of the carts stuck together after the collision is 0.5 m/s to the right.

Collision Type 2

before

$v_{1\,\text{initial}} = 0$ $v_{2\,\text{initial}} = 0$

$$p_{\text{initial}} = 0$$

after

$v_{1\,\text{final}}$ $v_{2\,\text{final}}$

$$v_{2\,\text{final}} = -\frac{m_1}{m_2} v_{1\,\text{final}}$$

$$p_{\text{final}} = m_1 v_{1\,\text{final}} + m_2 v_{2\,\text{final}} = 0$$

If the two masses are equal, and the final velocity of the right side cart is 1 m/s to the right, have students determine the final velocity of the left-side cart. In this explosion, the final velocity of the cart on the left will be 1 m/s to the left.

Collision Type 3

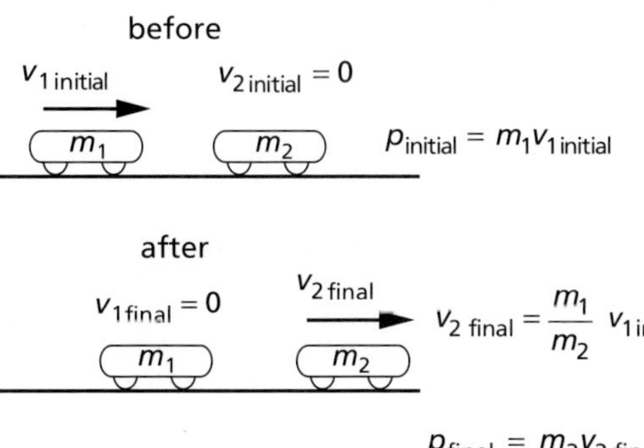

If the two masses are equal, and the initial velocity of the cart on the left is 1 m/s to the right, consider having students determine the final velocity of the cart that was initially stationary. Using the conservation of momentum, the final velocity of the cart on the right will be 1 m/s to the right, and the velocity of the cart on the left will be zero.

Collision Type 4

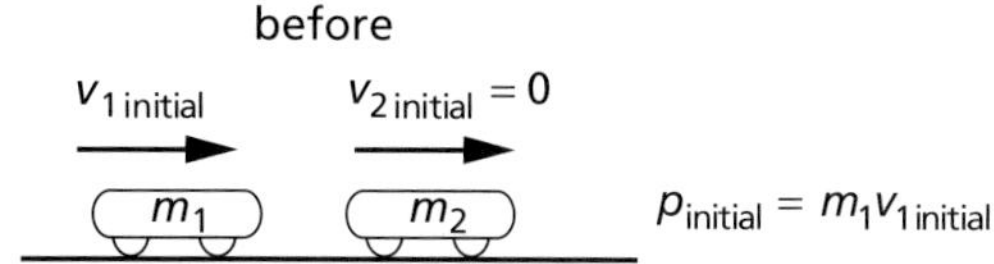

after

$v_{1\,\text{final}}$ $v_{2\,\text{final}}$

$$v_{2\,\text{final}} = \frac{m_1}{m_2}(v_{1\,\text{initial}} - v_{1\,\text{final}})$$

$$p_{\text{final}} = m_1 v_{1\,\text{final}} + m_2 v_{2\,\text{final}}$$

If the cart on the left has twice the mass of the cart on the right, and the initial velocity of the cart on the left is 1 m/s to the right, its final velocity after the collision is 0.5 m/s to the right. Have students determine the final velocity of the cart that was initially stationary. The final velocity of the cart on the right will be 1 m/s to the right.

Collision Type 5

Emphasize that for this collision, the two objects could be moving toward each other from opposite directions (a head-on collision), or they could be going in the same direction where one cart is faster than the other and hits it from behind.

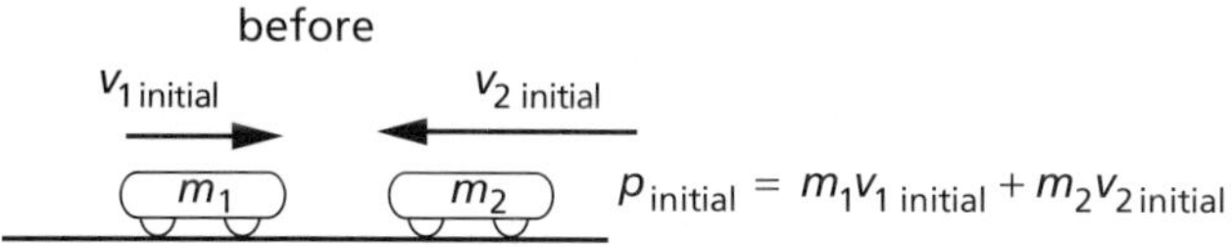

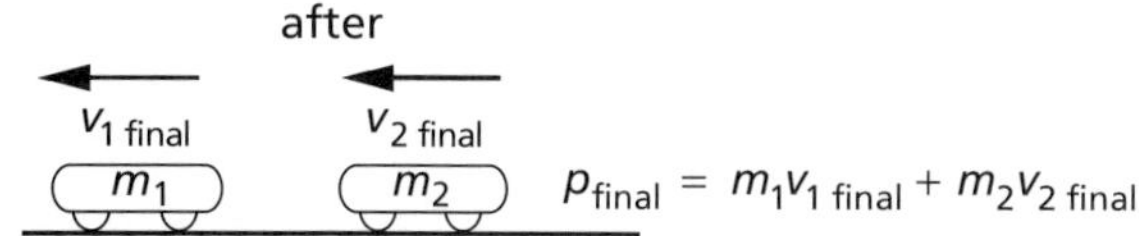

Consider having students determine the final velocity of the cart on the right for the following conditions. The cart on the left is twice the mass of the cart on the right. The initial velocity of the cart on the left is 0.5 m/s to the right. The initial velocity of the cart on the right is 2 m/s to the left, and its final velocity is 0.5 m/s to the left. In this case, the final velocity of the cart on the right is 0.25 m/s to the left.

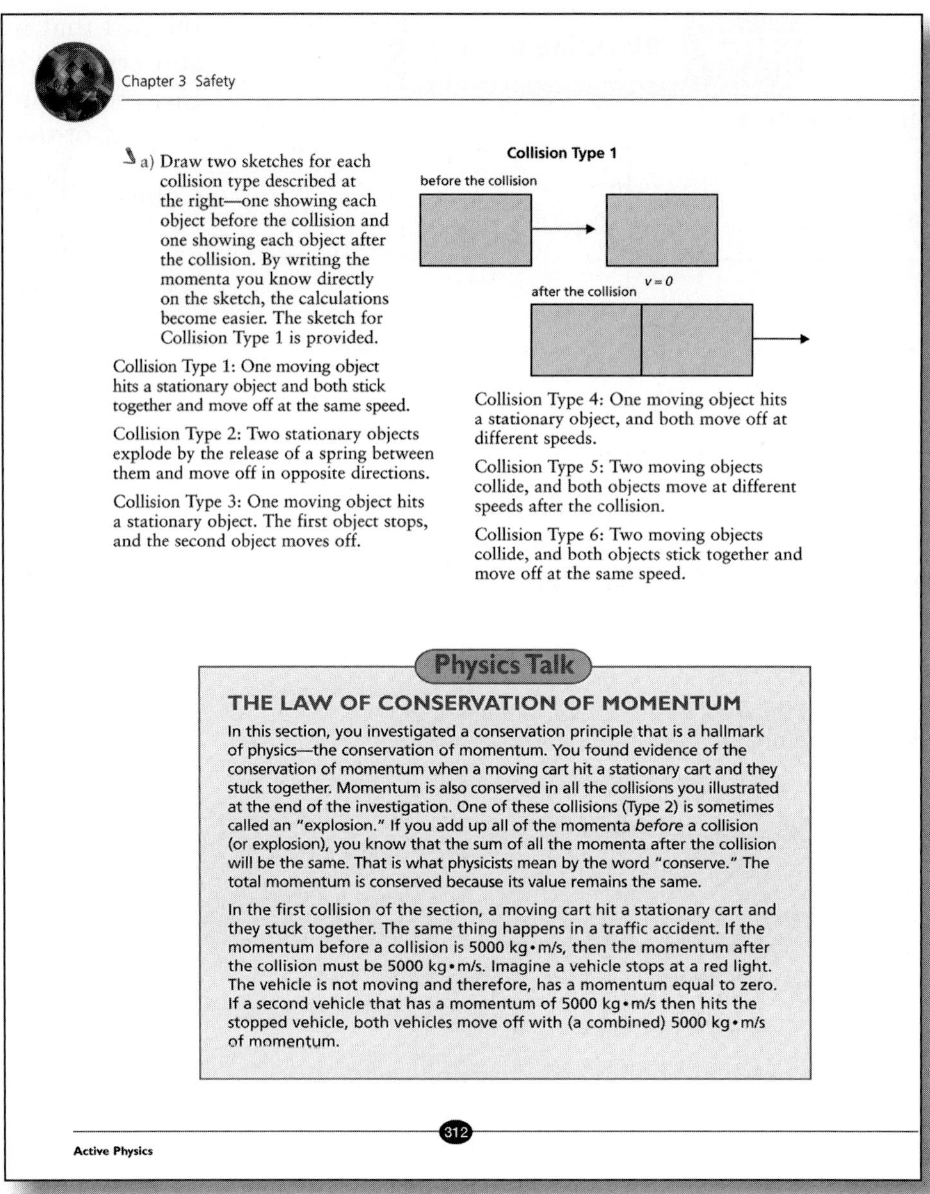

a) Draw two sketches for each collision type described at the right—one showing each object before the collision and one showing each object after the collision. By writing the momenta you know directly on the sketch, the calculations become easier. The sketch for Collision Type 1 is provided.

Collision Type 1: One moving object hits a stationary object and both stick together and move off at the same speed.

Collision Type 2: Two stationary objects explode by the release of a spring between them and move off in opposite directions.

Collision Type 3: One moving object hits a stationary object. The first object stops, and the second object moves off.

Collision Type 4: One moving object hits a stationary object, and both move off at different speeds.

Collision Type 5: Two moving objects collide, and both objects move at different speeds after the collision.

Collision Type 6: Two moving objects collide, and both objects stick together and move off at the same speed.

Physics Talk

THE LAW OF CONSERVATION OF MOMENTUM

In this section, you investigated a conservation principle that is a hallmark of physics—the conservation of momentum. You found evidence of the conservation of momentum when a moving cart hit a stationary cart and they stuck together. Momentum is also conserved in all the collisions you illustrated at the end of the investigation. One of these collisions (Type 2) is sometimes called an "explosion." If you add up all of the momenta *before* a collision (or explosion), you know that the sum of all the momenta after the collision will be the same. That is what physicists mean by the word "conserve." The total momentum is conserved because its value remains the same.

In the first collision of the section, a moving cart hit a stationary cart and they stuck together. The same thing happens in a traffic accident. If the momentum before a collision is 5000 kg•m/s, then the momentum after the collision must be 5000 kg•m/s. Imagine a vehicle stops at a red light. The vehicle is not moving and therefore, has a momentum equal to zero. If a second vehicle that has a momentum of 5000 kg•m/s then hits the stopped vehicle, both vehicles move off with (a combined) 5000 kg•m/s of momentum.

Collision Type 6

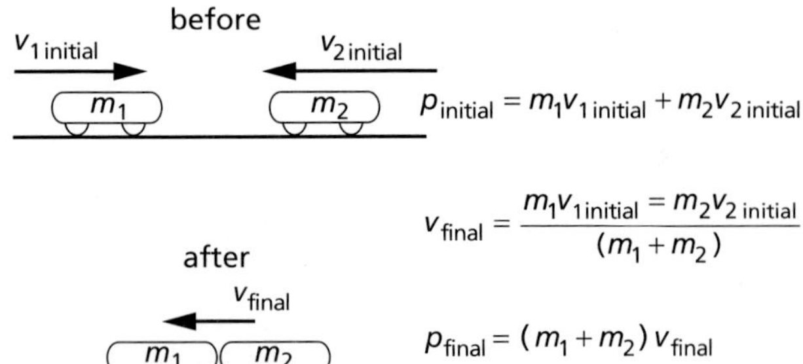

If the cart on the left is twice the mass of the cart on the right, the initial velocity of the cart on the left 1 m/s to the right, and the initial velocity of the cart on the right is 4 m/s to the left, have students determine the final velocity of the stuck-together carts after the collision. In this case, the final velocity of the stuck-together carts is 0.67 m/s to the left.

Total momentum stays the same or, in other words, total momentum is conserved.

You can write this as an equation and substitute values for the masses and velocities of the vehicles.

$$\text{Momentum BEFORE collision} = \text{Momentum AFTER collision}$$
$$m_1v_1 + m_2v_2 = (m_1 + m_2)v_f$$
$$(1000\text{ kg})(5\text{ m/s}) + (1000\text{ kg})(0\text{ m/s}) = (1000\text{ kg} + 1000\text{ kg})(2.5\text{ m/s})$$
$$5000\text{ kg}\cdot\text{m/s} = 5000\text{ kg}\cdot\text{m/s}$$

You add the masses "after the collision" because they stuck together and travel with the same speed. Conservation of momentum is an experimental fact. Physicists have compared momentum before and after collisions between pairs of objects ranging from railroad cars slamming together to subatomic particles impacting one another at near the speed of light. Never have any exceptions been found to the statement, "The total momentum before a collision is equal to the total momentum after the collision if no external forces act on the system." This statement is known as the **law of conservation of momentum**. In all collisions between vehicles and trucks, between protons and protons, between planets and meteors, the momentum before the collision equals the momentum after the collision.

Physics Words

law of conservation of momentum: the total momentum before a collision is equal to the total momentum after the collision if no external forces act on the system.

A single cue ball hits a rack of 15 billiard balls and they all scatter. It would seem as if everything has changed. Physicists have discovered that in this collision, as in all collisions and explosions, nature does keep at least one thing from changing—the total momentum. The sum of the momentum of all of the billiard balls immediately after the collision is equal to the momentum of the original cue ball. Nature conserves momentum. Irrespective of the changes you can see, the total momentum undergoes no change whatsoever. The objects may move in new directions and with new speeds, but the momentum stays the same.

Physics Talk

This *Physics Talk* introduces the law of conservation of momentum. Conservation of momentum is a fundamental physics concept. Emphasize the importance of this law and check students' comprehension of it.

Have a class discussion on the law of conservation of momentum using the information in the student text. Go through the examples with students and make connections with students' observations from the *Investigate*.

Describe how the momentum of a system is always conserved if no net external force is acting on the system. This is an extremely important physics concept that has never been negated. Describe how this law holds true for everything that scientists have studied, from stars to the inner makings of the atom. Discuss the special conditions for some situations. For example, for a perfectly elastic collision, kinetic energy is conserved. For all other collisions, the kinetic energy is not conserved. Discuss both sample problems in the *Student Edition* as a class.

Chapter 3 Safety

Conservation of momentum is crucial physics for analyzing any collision between objects. Traffic-accident investigators use tread marks, the positions of the automobiles, and the damage to the automobiles to understand what happened at the time of the accident. They also use the physics principle of momentum conservation to help them in their analysis.

If you know the masses and velocities of two objects before a collision, you can accurately predict the velocities after the collision. Physics allows you to predict the future!

Solving conservation of momentum problems is easy. Using the definition of momentum, $p = mv$, calculate each object's momentum before the collision. Calculate each object's momentum after the collision. The total after the collision must equal the total before the collision.

Sample Problem I

A boy and a girl are riding in bumper cars at an amusement park. A 75-kg boy and car are moving to the east at 3.00 m/s toward a 50-kg girl and car who are moving toward him (west) at 1.80 m/s. If they catch up, stick together, and then move away together, what is their final velocity?

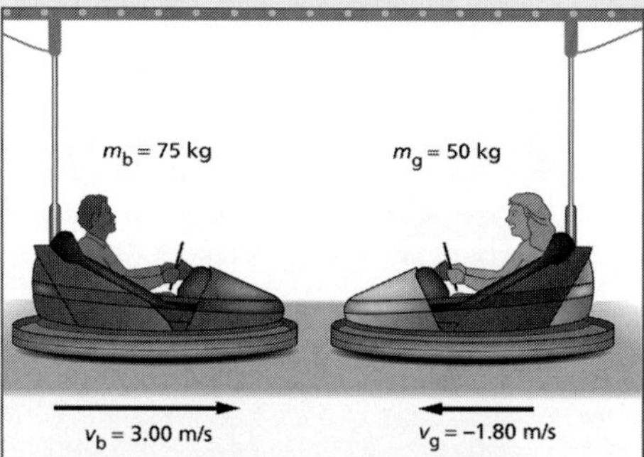

Strategy: This is a problem involving the law of conservation of momentum. The momentum of an isolated system before an interaction is equal to the momentum of the system after the interaction. As you are working through this problem, remember that the *v* in this expression is velocity and that it has direction as well as magnitude. Make east the positive direction, and then west will be negative.

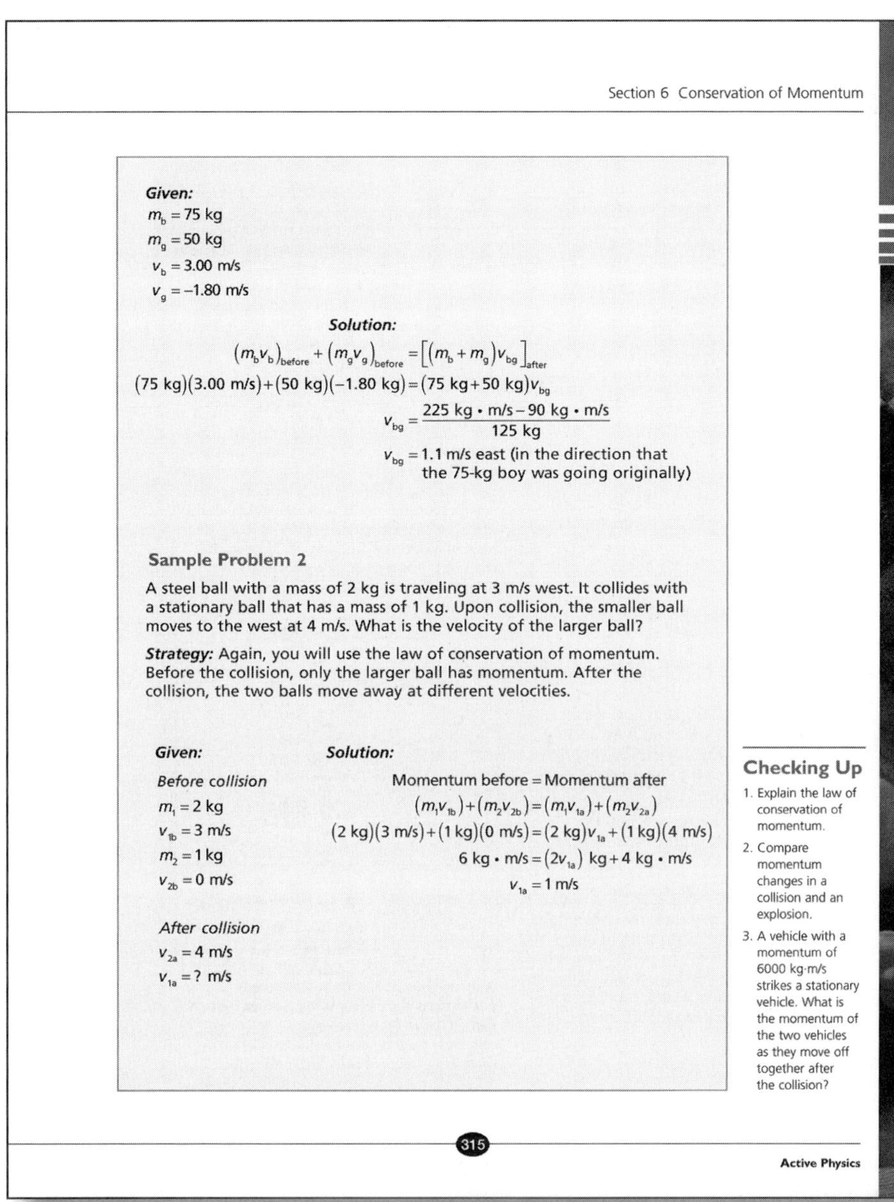

Given:

$m_b = 75$ kg

$m_g = 50$ kg

$v_b = 3.00$ m/s

$v_g = -1.80$ m/s

Solution:

$$(m_b v_b)_{before} + (m_g v_g)_{before} = [(m_b + m_g)v_{bg}]_{after}$$

$$(75 \text{ kg})(3.00 \text{ m/s}) + (50 \text{ kg})(-1.80 \text{ kg}) = (75 \text{ kg} + 50 \text{ kg})v_{bg}$$

$$v_{bg} = \frac{225 \text{ kg} \cdot \text{m/s} - 90 \text{ kg} \cdot \text{m/s}}{125 \text{ kg}}$$

$v_{bg} = 1.1$ m/s east (in the direction that the 75-kg boy was going originally)

Sample Problem 2

A steel ball with a mass of 2 kg is traveling at 3 m/s west. It collides with a stationary ball that has a mass of 1 kg. Upon collision, the smaller ball moves to the west at 4 m/s. What is the velocity of the larger ball?

Strategy: Again, you will use the law of conservation of momentum. Before the collision, only the larger ball has momentum. After the collision, the two balls move away at different velocities.

Given:

Before collision

$m_1 = 2$ kg

$v_{1b} = 3$ m/s

$m_2 = 1$ kg

$v_{2b} = 0$ m/s

After collision

$v_{2a} = 4$ m/s

$v_{1a} = ?$ m/s

Solution:

Momentum before = Momentum after

$$(m_1 v_{1b}) + (m_2 v_{2b}) = (m_1 v_{1a}) + (m_2 v_{2a})$$

$$(2 \text{ kg})(3 \text{ m/s}) + (1 \text{ kg})(0 \text{ m/s}) = (2 \text{ kg})v_{1a} + (1 \text{ kg})(4 \text{ m/s})$$

$$6 \text{ kg} \cdot \text{m/s} = (2v_{1a}) \text{ kg} + 4 \text{ kg} \cdot \text{m/s}$$

$$v_{1a} = 1 \text{ m/s}$$

Checking Up

1. Explain the law of conservation of momentum.
2. Compare momentum changes in a collision and an explosion.
3. A vehicle with a momentum of 6000 kg·m/s strikes a stationary vehicle. What is the momentum of the two vehicles as they move off together after the collision?

Checking Up

1.

The law of conservation of momentum states that the total momentum of a system just before a collision is equal to the total momentum of the system just after a collision, provided that there are no net external forces acting on the system.

2.

For both a collision and an explosion, the total momentum before the event is equal to the total momentum after the event. In a collision, the total momentum of both objects just before the collision is equal to the total momentum of both objects just after the collision. Before the collision, one or both objects may be moving, having a nonzero momentum. After the collision, one or both objects may be moving, having a nonzero momentum. During an explosion, the momentum of the object before the explosion is equal to the sum of the momentums of the particles of the object after the explosion. If before the explosion the object is not moving, then the initial momentum of the system is zero. In this case, the sum of momentums of all the particles after the explosion is equal to zero.

3.

Students should apply the law of conservation of momentum to solve the problem.

$p_{initial} = p_{final}$. Because the initial momentum of the system is 6000 kg · m/s, the final momentum is also 6000 kg · m/s. Ask students what they think the final speed is compared to the initial speed. Students should realize that the final speed is less than the initial speed because the mass increased.

$$p_{initial} = p_{final}$$

$$m_1 v_{1initial} = (m_1 + m_2) v_{final}$$

CHAPTER 3

Active Physics Plus

This *Active Physics Plus* provides an opportunity for students to increase the depth of their understanding of conservation of momentum and how it ties in with Newton's laws. It also provides a more mathematical approach and briefly discusses two-dimensional collisions. When discussing two-dimensional collisions of equal mass, consider drawing velocity vectors for the momentum equation and show students how they add tip-to-tail to get the resultant vector equal to the initial velocity. This also clearly shows the right angle that is formed.

Chapter 3 Safety

Active Physics *Plus*

+Math	+Depth	+Concepts	+Exploration
◆◆	◆◆		

Conservation of Momentum and Newton's Laws

Conservation of momentum can be shown to emerge from Newton's laws. Newton's third law states that if object A and object B collide, the force of object A on B must be equal and opposite to the force of object B on A.

$$F_{\text{A on B}} = -F_{\text{B on A}}$$

The negative sign shows mathematically that the equally sized forces are in opposite directions. Since $F = ma$ by Newton's second law:

$$m_B a_B = -m_A a_A$$

$$\frac{m_B \Delta v_B}{\Delta t} = \frac{-m_A \Delta v_A}{\Delta t}$$

$$m_B\left(\frac{v_f - v_i}{\Delta t}\right)_B = -m_A\left(\frac{v_f - v_i}{\Delta t}\right)_A$$

Since the change in time must be the same for both objects (A acts on B for as long as B acts on A), then Δt can be eliminated from both sides of the equation.

Combining the initial velocities (v_i) on one side of the equation and the final velocities (v_f) on the other side of the equation:

$$m_A v_{iA} + m_B v_{iB} = m_A v_{fA} + m_B v_{fB}$$

Newton's laws have yielded the conservation of momentum. The momentum of object A before the collision plus the momentum of object B before the collision equals the momentum of object A after the collision plus the momentum of object B after the collision.

This equation works in one-dimensional collisions, and also in the extraordinarily complex two-dimensional collisions of multi-vehicle collisions. Momentum is a vector. When analyzing a two-dimensional collision, you must add the momenta using vector diagrams or vector mathematics.

Consider a game of billiards where a cue ball (the white ball) hits a stationary object ball (the red ball). The two balls have the same mass. If the cue ball moves up and to the right (as shown), then the object ball must move down and to the right so that the total momentum is conserved.

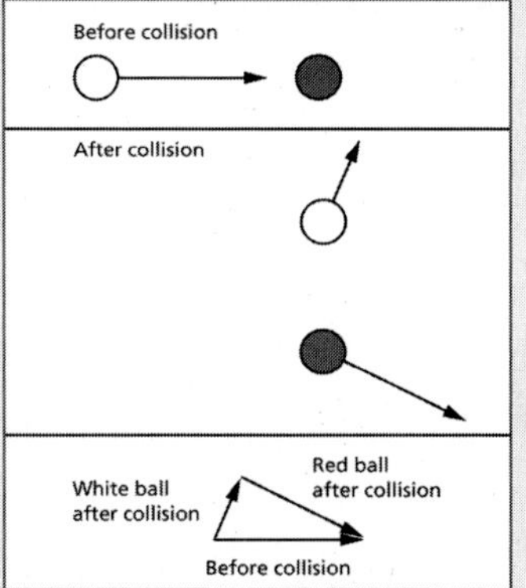

You can see in the vector diagram above that the momentum vector before the collision (the horizontal vector) is identical to the sum of the two other momenta vectors after the collision.

316

Active Physics

1.

With the assumptions that momentum is conserved and 50% or half of the kinetic energy is conserved, students should calculate the following:

From the law of conservation of momentum

$p_{\text{initial}} = p_{\text{final}}$

$(1000 \text{ kg})(10 \text{ m/s}) =$

$(1000 \text{ kg})v_{\text{car final}} + (70 \text{ kg})v_{\text{person final}}$

$v_{\text{car final}} = 10 \text{ m/s} - 0.07v_{\text{person final}}$

Using the condition that half of the kinetic energy is conserved

$\frac{1}{2}KE_{\text{initial}} = KE_{\text{final}}$

$\frac{1}{2}(1000 \text{ kg})(10 \text{ m/s})^2 =$

$\frac{1}{4}(1000 \text{ kg})v^2_{\text{car final}} +$

$\frac{1}{4}(70 \text{ kg})v^2_{\text{person final}}$

Substituting

$v_{\text{car final}} = 10 \text{ m/s} - 0.07v_{\text{person final}}$, found from the law of conservation of momentum you have

$$200{,}000 \text{ J} = 1000 \text{ kg} \times \left(100 \frac{\text{m}^2}{\text{s}^2} - 1.4\frac{\text{m}}{\text{s}} v_{\text{person final}} + 0.0049 v^2_{\text{person final}}\right) + (70 \text{ kg})v^2_{\text{person final}}$$

$$200{,}000 \text{ kg}\frac{\text{m}^2}{\text{s}^2} = 100{,}000 \text{ kg}\frac{\text{m}^2}{\text{s}^2} - 1400 \text{ kg}\frac{\text{m}}{\text{s}} v_{\text{person final}} + (4.9 \text{ kg})v^2_{\text{person final}} + (70 \text{ kg})v^2_{\text{person final}}$$

$$(74.9)v^2_{\text{person final}} - \left(1400 \frac{\text{m}}{\text{s}}\right)v_{\text{person final}} - 100{,}000\frac{\text{m}^2}{\text{s}^2} = 0.$$

A collision where objects bounce off of each other is an elastic collision. In a perfectly elastic collision, kinetic energy is conserved: the total kinetic energy before the collision is equal to the total kinetic energy after the collision.

When the colliding objects are equal in mass, you can use the conservation of momentum and the conservation of kinetic energy to prove that the angle between the objects after the collision is 90°. In the vector diagram for momentum conservation, you can see that the three vectors create a triangle. Look at the following equation for the conservation of kinetic energy.

$$KE_{\text{Before}} = KE_{\text{After}}$$

$$\tfrac{1}{2}mv_{iA}^2 = \tfrac{1}{2}mv_{fA}^2 + \tfrac{1}{2}mv_{fB}^2$$

Since all the masses are equal:

$$v_i^2 = v_{fA}^2 + v_{fB}^2$$

You may recognize this as a form of the Pythagorean theorem for a right triangle $a^2 + b^2 = c^2$. The vector triangle for momentum must therefore be a right triangle. The cue ball and the object ball must depart at right angles. You can now use your physics knowledge when you play billiards!

1. The mathematics of collisions becomes a bit difficult since the kinetic energy depends on the square of the velocity. Try to solve for the speed of a pedestrian when hit by an automobile. Assume that momentum is conserved. Also assume that 50 percent of the kinetic energy is conserved. Assume a 1000-kg automobile moving at 10 m/s (only 20 mph) collides with a 70-kg pedestrian at rest. The calculation will give you a sense of how dangerous an automobile is because it has such tremendous momentum.

What Do You Think Now?

At the beginning of the activity you were asked:

- **What physics principles do the traffic-accident investigators use to "reconstruct" the accident?**

Revisit your initial responses. Now that you investigated how the law of conservation of momentum works, how would you answer this question?

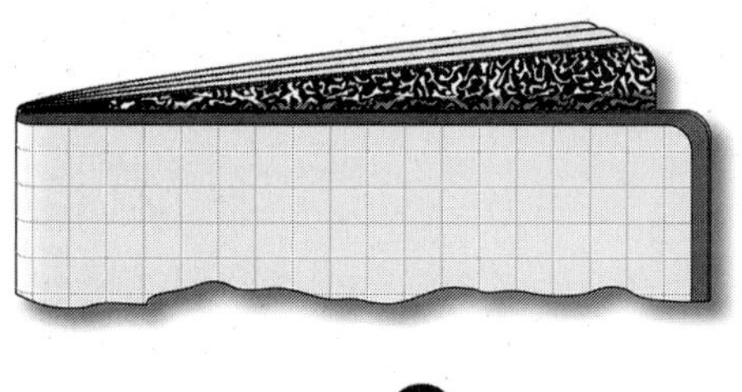

What Do You Think Now?

Ask students to review their previous responses to the *What Do You Think?* question. Ask students how they would answer this question now and survey the class for how many students changed their ideas. Point out that scientists often change their ideas as they gather more information. Consider discussing the information in *A Physicist's Response*.

Using the quadratic equation:

If $ax^2 + bx + c = 0$ then $x = \dfrac{-b \pm \sqrt{b^2 - 4ac}}{2a}$

$v_{\text{person final}} =$

$$\frac{(1400\ \text{m/s}) \pm \sqrt{(-1400\ \text{m/s})^2 - 4(74.9)(-100{,}000\ \text{m}^2/\text{s}^2)}}{2(74.9)}$$

$v_{\text{person final}} = -28\ \text{m/s}$ or $47\ \text{m/s}$.

There are two possible values. From this, students should select the value that can be explained by physics. The negative value would indicate that the person went flying in the opposite direction. This does not make sense. The positive value indicates the same direction as the vehicle was originally moving in, which does make sense in physics. Therefore,

$v_{\text{person final}} = 47\ \text{m/s} = 169\ \text{km/h} = 105\ \text{mi/h}$.

Reflecting on the Section and the Challenge

Using the information in the student text, review the law of conservation of momentum and its importance. Emphasize that it always holds true for all collision types. Use the examples provided in the student text. Remind students that this concept is important to keep in mind for the *Chapter Challenge* and that they should consider these concepts as they design their prototype and in their explanations.

Chapter 3 Safety

Physics
Essential Questions

What does it mean?

What does it mean to "conserve" momentum in a collision?

How do you know?

All experiments have measurement uncertainties. How accurately were you able to measure the conservation of momentum in your activity? How could the measurements be improved?

Why do you believe?

Connects with Other Physics Content	Fits with Big Ideas in Science	Meets Physics Requirements
Forces and motion	* Conservation laws	Good clear explanation, no more complicated than necessary

* In physics, concepts are introduced to help make complex situations appear simpler to understand. Describe in words all that happens when a cue ball hits the 15 balls at the beginning of a billiards match. Describe the same event using the concepts of conservation of momentum and conservation of energy.

Why do you care?

How would the principle of conservation of momentum influence the design of a safety system that must protect against collisions with a much more massive object?

Reflecting on the Section and the Challenge

The law of conservation of momentum is a very powerful tool for explaining collisions in traffic accidents. The law works even when one of the objects involved in a collision "bounces back," reversing the direction of its velocity and therefore, its momentum. Whether describing a collision between two people, or between a moving automobile and a tree, you can describe how the total momentum is conserved.

Active Physics 318

Physics Essential Questions

What does it mean?

To conserve momentum means that the sum of the momentums of all objects just before a collision is equal to the sum of the momentums of all objects just after the collision.

How do you know?

(Answer depends on student results.) The results can be improved by minimizing the friction in the carts.

Why do you believe?

The cue ball has a certain amount of energy and momentum when it strikes the 15 stationary balls. After the collision, the sum of the momentums of the 16 balls is equal to the original momentum of the cue ball. Also, the sum of the energies of the 16 balls is equal to the energy the cue ball had just before the collision if the collision is elastic. (The collision is not perfectly elastic; some energy is transferred to the surroundings as sound and heat.)

Why should you care?

A massive object, if moving, will have an enormous amount of momentum and energy that could be transferred during a collision. Designing a safety system to protect against this may require lots of padding so that when hit, the passengers experience a smaller force over a longer distance.

Physics to Go

For the following problems, show your work, as always, and also show a diagram of the momentum "before" and "after."

1. One cart hits and sticks to a second cart. Make a diagram showing the carts before and after the collision. How does the speed of the carts after the collision compare with the initial speed of the moving cart?
2. Two 1-kg carts are each moving toward each other at 2 m/s. They collide and each reverses directions, moving in the opposite directions at 2 m/s. Draw a diagram showing the carts before and after the collision.
   a) Calculate the momentum of each cart before the collision. (Hint: Since they are moving in opposite directions, one momentum will be positive and one will be negative.)
   b) Calculate the total momentum before the collision.
   c) Calculate the total momentum after the collision.
3. In an automobile crash, a vehicle that was stopped at a red light is rear-ended by another vehicle. The vehicles have the same mass. If the tire marks show that the two vehicles moved after the collision at 4 m/s, what was the speed of the vehicle before the collision?
4. Given that the total momentum before a collision must equal the total momentum after the collision, how can one of the cars gain momentum?
5. Vehicle A and vehicle B collide and vehicle A loses 4000 kg•m/s of momentum. What is the change in momentum of vehicle B? What is the total change in momentum due to the collision?
6. A railroad car with a mass of 2000 kg coasting at 3.0 m/s overtakes and locks together with an identical car coasting on the same track in the same direction at 2.0 m/s. What is the speed of the cars after they lock together?
7. In a hockey game, an 80.0-kg player skating at 10.0 m/s overtakes and bumps from behind a 100.0-kg player who is moving in the same direction at 8.00 m/s. As a result of being bumped from behind, the 100.0-kg player's speed increases to 9.78 m/s. What is the 80.0-kg player's velocity (speed and direction) after the bump?
8. A 3-kg hard steel ball collides head-on with a 1-kg hard steel ball. The balls are moving at 2 m/s in opposite directions before they collide. Upon colliding, the 3-kg ball stops. What is the velocity of the 1-kg object after the collision? (Hint: Assign velocities in one direction as positive; then any velocities in the opposite direction are negative.)

Physics to Go

1.

Students' diagrams should look similar to the following diagram. Students should note that the final velocity of the carts after they collide and stick together is in the same direction as the initial velocity, and the final speed is less than the initial speed. Encourage students to record the momentums for the before and after situations on their diagram and to solve for the final velocity of the carts.

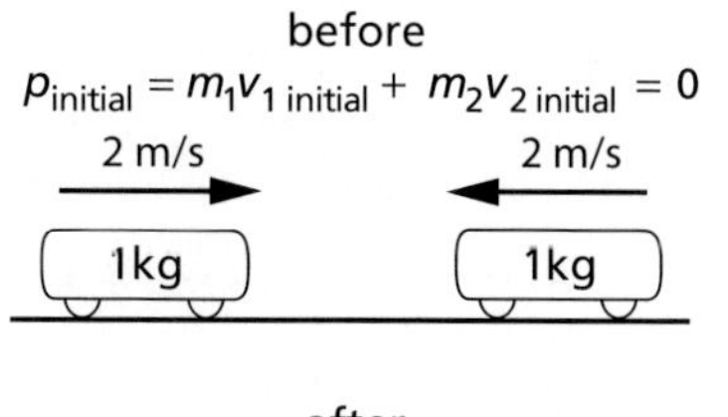

2.a)

Students' diagrams should look similar to the one below.

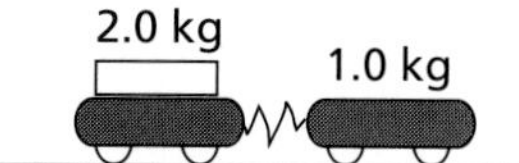

Velocity and momentum are vector quantities. This is indicated by opposite directions of the velocities having opposite signs. If the direction pointing to the right is assumed to be positive then students should calculate the following momenta:

$p_{\text{right cart initial}} = (1\text{ kg})(-2\text{ m/s}) =$ $-2\text{ kg}\cdot\text{m/s}$, where the negative sign indicates it moves to the left.

$p_{\text{left cart initial}} = (1\text{ kg})(+2\text{ m/s}) =$ $+2\text{ kg}\cdot\text{m/s}$, where the positive sign indicates it moves to the right.

2.b)

Adding the momentum of each cart before the collision to

$p_{\text{total initial}} = 2\text{ kg}\cdot\text{m/s} - 2\text{ kg}\cdot\text{m/s} = 0$

2.c)

The total momentum after the collision is also zero due to the conservation of momentum, and because the carts have just reversed direction.

3.

Students apply the law of conservation of momentum.

$p_{\text{total before}} = p_{\text{total after}}$

$mv_{\text{initial}} + 0 = 2m(4\text{ m/s})$

$v_{\text{initial}} = 8\text{ m/s}$

4.

Students should realize that even though the total momentum before and after a collision does not change, momentum can be transferred between objects during a collision. The magnitude (how much) of each object's individual momentum can increase or decrease, and/or change direction, but the magnitude and direction of the total momentum cannot change.

5.

Students should apply the law of conservation of momentum. It is important that students realize that the change in momentum of one object must be equal in size and opposite in direction to the change in momentum of the other object.

$p_{\text{total before}} = p_{\text{total after}}$

$m_A v_{A\text{ initial}} + m_B v_{B\text{ initial}} =$

$m_A v_{A\text{ final}} + m_B v_{B\text{ final}}$

$m_A v_{A\text{ initial}} - m_A v_{A\text{ final}} =$

$m_B v_{B\text{ final}} - m_B v_{B\text{ initial}}$

$-\left(m_A v_{A\text{ final}} - m_A v_{A\text{ initial}}\right) =$

$m_B v_{B\text{ final}} - m_B v_{B\text{ initial}}$

This says that the change in momentum of vehicle B is equal and opposite to the change in momentum of vehicle A. Because vehicle A changes its momentum by 4000 kg · m/s in the "negative" direction, vehicle B must change its momentum by 4000 kg · m/s in the direction opposite to vehicle A. Students should recognize that even though the vehicles exchange momentum during the collision, the total momentum of the system consisting of both vehicles does not change. That is, the total change of momentum is zero.

9. A 45-kg female figure skater and her 75-kg male skating partner begin their ice-dancing performance standing at rest in face-to-face position with the palms of their hands touching. When their dance-music starts, both skaters "push off" with their hands to move backward. If the female skater moves at 2.0 m/s relative to the ice, what is the velocity of the male skater? (Hint: The momentum before the skaters push off is zero.)
10. A 0.35-kg tennis racquet moving to the right at 20.0 m/s hits a 0.060-kg tennis ball that is moving to the left at 30.0 m/s. The racquet continues moving to the right after the collision, but at a reduced speed of 10.0 m/s. What is the velocity of the tennis ball after it is hit by the racquet?
11. A stationary 3-kg hard steel ball is hit head-on by a 1-kg hard steel ball moving to the right at 4 m/s. After the collision, the 3-kg ball moves to the right at 2 m/s. What is the velocity (speed and direction) of the 1-kg ball after the collision? (Hint: Direction is important.)
12. A 90.00-kg hockey goalie, at rest in front of the goal, stops a puck (m = 0.16 kg) that is traveling at 30.00 m/s. At what speed do the goalie and puck travel after the save?
13. A 45.00-kg girl jumps from the side of a pool into a raft (m = 0.08 kg) floating on the surface of the water. She leaves the side at a speed of 1.10 m/s and lands on the raft. At what speed will the girl and the raft begin to travel across the pool?
14. Two cars collide head on. Initially, automobile A (m = 1700.0 kg) is traveling at 10.00 m/s north and automobile B is traveling at 25.00 m/s south. After the collision, automobile A reverses its direction and travels at 5.00 m/s while automobile B continues in its initial direction at a speed of 3.75 m/s. What is the mass of automobile B?
15. A proton ($m = 1.67 \times 10^{-27}$ kg) traveling at 2.50×10^5 m/s collides with an unknown particle initially at rest. After the collision, the proton reverses direction and travels at 1.10×10^5 m/s. Determine the change in momentum of the unknown particle.
16. A 0.04-kg bullet moving at 200.0 m/s is shot into a 20.00-kg block initially at rest on an icy pond. What is the velocity of the bullet-block combination? The coefficient of friction between the block and the ice is 0.15. How far would the block slide before coming to rest?
17. ***Preparing for the Chapter Challenge***

    In your description of your design for a safety device, you will need to include an explanation of how it works to reduce injuries. In the past, you have used the ideas "energy management" and pressure to describe what happens in a crash. You can also think of a collision as a transfer of momentum. How does momentum explain what happens when a truck with great mass and an automobile with small mass collide?

6.

This is a completely inelastic collision or a "sticky" collision. Students should apply the conservation of momentum to calculate the unknown quantity.

$p_{\text{total before}} = p_{\text{total after}}$

$m_1 v_{1\text{initial}} + m_2 v_{2\text{initial}} = \left(m_1 + m_2\right) v_{\text{final}}$

$(2000\text{ kg})(3\text{ m/s}) +$

$(2000\text{ kg})(2.0\text{ m/s}) =$

$(2000\text{ kg} + 2000\text{ kg}) v_{\text{final}}$

$10{,}000\text{ kg}\cdot\text{m/s} = (4000\text{ kg}) v_{\text{final}}$

$v_{\text{final}} = 2.5\text{ m/s}$

7.

This is a completely inelastic collision. Apply the conservation of momentum to calculate the unknown quantity.

$p_{\text{total before}} = p_{\text{total after}}$

$m_1 v_{1\text{initial}} + m_2 v_{2\text{initial}} =$

$m_1 v_{1\text{final}} + m_2 v_{2\text{final}}$

$(80.0\text{ kg})(10.0\text{ m/s}) +$

$(100.0\text{ kg})(8.00\text{ m/s}) =$

$(80.0\text{ kg}) v_{1\text{final}} +$

$(100.0\text{ kg})(9.78\text{ m/s})$

$1600\ \text{kg}\cdot\text{m/s} =$

$(80.0\ \text{kg})v_{1\text{final}} + (978\ \text{kg}\cdot\text{m/s})$

$622\ \text{kg}\cdot\text{m/s} = (80.0\ \text{kg})v_{1\text{final}}$

$v_{1\text{final}} = 7.78\ \text{m/s}$, in the same direction.

8.

This is a completely inelastic collision. Apply conservation of momentum to calculate the unknown quantity. The direction of travel of the 3-kg ball before the collision is assigned as positive:

$p_{\text{total before}} = p_{\text{total after}}$

$m_1v_{1\text{initial}} + m_2v_{2\text{initial}} =$

$m_1v_{1\text{final}} + m_2v_{2\text{final}}$

$(3\ \text{kg})(2\ \text{m/s}) + (1\ \text{kg})(-2\ \text{m/s}) =$

$(3\ \text{kg})(0\ \text{m/s}) + (1\ \text{kg})v_{2\text{final}}$

$4\ \text{kg}\cdot\text{m/s} = (1\ \text{kg})v_{2\text{final}}$

$4\ \text{m/s} = v_{2\text{final}}$

The 1-kg ball bounces back after the collision, moving in the opposite direction at twice the speed it had coming into the collision.

9.

Students should assign a positive direction. In the solution below, the direction of the female skater after push off is assigned as positive:

$p_{\text{total before}} = p_{\text{total after}}$

$m_1v_{1\text{initial}} + m_2v_{2\text{initial}} =$

$m_1v_{1\text{final}} + m_2v_{2\text{final}}$

$0 = (45\ \text{kg})(2\ \text{m/s}) + (75\ \text{kg})v_{2\text{final}}$

$-90\ \text{kg}\cdot\text{m/s} = (75\ \text{kg})v_{2\text{final}}$

$-1.2\ \text{m/s} = v_{2\text{final}}$

The male skater moves at 1.2 m/s in the direction opposite the female skater.

10.

In the solution below, the right direction is assigned as positive:

$p_{\text{total before}} = p_{\text{total after}}$

$m_1v_{1\text{initial}} + m_2v_{2\text{initial}} =$

$m_1v_{1\text{final}} + m_2v_{2\text{final}}$

$(0.35\ \text{kg})(20\ \text{m/s}) +$

$(0.060\ \text{kg})(-30\ \text{m/s}) =$

$(0.35\ \text{kg})(10\ \text{m/s}) + (0.060\ \text{kg})v_{2\text{final}}$

$5.2\ \text{kg}\cdot\text{m/s} =$

$3.5\ \text{kg}\cdot\text{m/s} + (0.060\ \text{kg})v_{2\text{final}}$

$28.33\ \text{m/s} = v_{2\text{final}}$

The ball moves to the right after the collision with a speed of 28.33 m/s.

11.

In the solution below, the right direction is assigned as positive:

$p_{\text{total before}} = p_{\text{total after}}$

$m_1v_{1\text{initial}} + m_2v_{2\text{initial}} =$

$m_1v_{1\text{final}} + m_2v_{2\text{final}}$

$(3\ \text{kg})(0) + (1\ \text{kg})(4\ \text{m/s}) =$

$(3\ \text{kg})(2\ \text{m/s}) + (1\ \text{kg})v_{2\text{final}}$

$4\ \text{kg}\cdot\text{m/s} = 6\ \text{kg}\cdot\text{m/s} + (1\ \text{kg})v_{2\text{final}}$

$-2\ \text{m/s} = v_{2\text{final}}$

The 1.0-kg ball moves to the left after the collision with a speed of 2 m/s.

12.

The collision described is a completely inelastic collision.

$p_{\text{total before}} = p_{\text{total after}}$

$m_1v_{1\text{initial}} + m_2v_{2\text{initial}} = (m_1 + m_2)v_{\text{final}}$

$(90\ \text{kg})(0) + (0.16\ \text{kg})(30.00\ \text{m/s}) =$

$(90.16\ \text{kg})v_{\text{final}}$

$0.05\ \text{m/s} = v_{\text{final}}$

The goalie and puck move in the direction of the incoming puck with a speed of 0.05 m/s.

13.

This is a completely inelastic collision.

$p_{\text{total before}} = p_{\text{total after}}$

$m_1v_{1\text{initial}} + m_2v_{2\text{initial}} = (m_1 + m_2)v_{\text{final}}$

$(45.00\ \text{kg})(1.10\ \text{m/s}) +$

$(0.08\ \text{kg})(0\ \text{m/s}) = (45.08\ \text{kg})v_{\text{final}}$

$1.1\ \text{m/s} = v_{\text{final}}$

The girl and raft move in the direction of the girl's jump with a speed of just less than 1.1 m/s. Because the value is rounded off, it appears from this answer that the speed is the same as that of the girl's initial jump. This is because the mass of the raft is much less than the girl's mass, so the decrease in speed is minimal.

14.

Students should indicate a positive and a negative direction. In the answer below, north is taken as positive.

$p_{\text{total before}} = p_{\text{total after}}$

$m_Av_{A\text{initial}} + m_Bv_{B\text{initial}} =$

$m_Av_{A\text{final}} + m_Bv_{B\text{final}}$

$(1700.0\ \text{kg})(10.00\ \text{m/s}) +$

$(m_B)(-25.00\ \text{m/s}) =$

$(1700.0\ \text{kg})(-5.00\ \text{m/s}) +$

$(m_B)(-3.75\ \text{m/s})$

$17{,}000\ \text{kg}\cdot\text{m/s} - m_B(25.00\ \text{m/s}) =$

$-8500\ \text{kg}\cdot\text{m/s} - m_B(3.75\ \text{m/s})$

$25{,}500\ \text{kg}\cdot\text{m/s} = m_B(21.25\ \text{m/s})$

$m_B = 1200\ \text{kg}$

15.

Use the conservation of total momentum to solve the change in momentum of the unknown particle.

$p_{\text{total before}} = p_{\text{total after}}$

$m_1 v_{1\text{initial}} + m_2 v_{2\text{initial}} =$

$m_1 v_{1\text{final}} + m_2 v_{2\text{final}}$

$m_1 v_{1\text{initial}} - m_1 v_{1\text{final}} =$

$m_2 v_{2\text{final}} - m_2 v_{2\text{initial}} =$

the change in momentum of particle 2.

$(1.67 \times 10^{-27} \text{ kg})(2.5 \times 10^5 \text{ m/s}) -$
$(1.67 \times 10^{-27} \text{ kg})(-1.1 \times 10^5 \text{ m/s}) =$

$m_2 v_{2\text{final}} - m_2 v_{2\text{initial}}$

$6.01 \times 10^{-22} \text{ kg} \cdot \text{m/s} =$

$m_2 v_{2\text{final}} - m_2 v_{2\text{initial}}$

The change in momentum of the proton is equal and opposite to the change in momentum of the unknown particle.

16.

This is a completely inelastic collision. Hence,

$p_{\text{total before}} = p_{\text{total after}}$

$m_1 v_{1\text{initial}} + m_2 v_{2\text{initial}} = (m_1 + m_2) v_{\text{final}}$

$(0.04 \text{ kg})(200 \text{ m/s}) +$
$(20.00 \text{ kg})(0\text{m/s}) = (20.04 \text{ kg}) v_{\text{final}}$

$0.4 \text{ m/s} = v_{\text{final}}$

After the bullet embeds itself in the block, the block-bullet system moves with a speed of 0.4 m/s in the direction in which the bullet was traveling.

Students should calculate the frictional force acting on the block and the distance traveled by using the work-energy theorem. The only force acting on the block-bullet system after the collision and in the direction of motion is the force of friction from the ice surface acting on the block, which is given by the coefficient of friction multiplied by the mass and the acceleration due to gravity. This force does work on the block to change its initial kinetic energy to zero.

Using the work-energy theorem,

$\Delta KE = Fd$

$\frac{1}{2} m v_{\text{final}}^2 - \frac{1}{2} m v_{\text{initial}}^2 = -\mu_k mgd$

Students should realize that friction does negative work on the block-bullet system because it acts in a direction opposite to the direction of motion.

$0 - \frac{1}{2}(20.04 \text{ kg})(0.4 \text{ m/s})^2 =$
$-(0.15)(20.04 \text{ kg})(9.8 \text{ m/s}^2) d$

$d = 0.05 \text{ m}$

17.

Preparing for the Chapter Challenge

Student answers will vary. The descriptions should include how a collision between the truck and the automobile will involve a transfer of momentum to the automobile, but the total momentum of the truck-automobile system remains the same.

NOTES

SECTION 6 QUIZ

3-6b **Blackline Master**

1. A 1200-kg automobile is traveling at a speed of 20 m/s when it collides with a truck at rest with a mass of 10,000 kg. What is the momentum of the vehicle before the collision?

   a) 60 kg · m/s
   b) 24,000 kg · m/s
   c) 34,000 kg · m/s
   d) 200,000 kg · m/s

2. In *Question 1*, if the automobile and the truck stick together after the collision, what is the momentum of the combined mass?

   a) 60 kg · m/s
   b) 24,000 kg · m/s
   c) 34,000 kg · m/s
   d) 200,000 kg · m/s

3. A 2.0-kg cart moving with a speed of 3 m/s collides with a 1.0-kg cart that is at rest. After the collision, the two carts stick together. What is the velocity of the combined mass of the two carts after the collision?

   a) 1.0 m/s
   b) 2.0 m/s
   c) 3.0 m/s
   d) 1.5 m/s

4. A 2.0-kg cart and a 1.0-kg cart are at rest on a level surface with a compressed spring between them as shown. When the spring is released, if the 1.0-kg cart moves off with a speed of 3.0 m/s, what is the speed of the 2.0-kg cart?

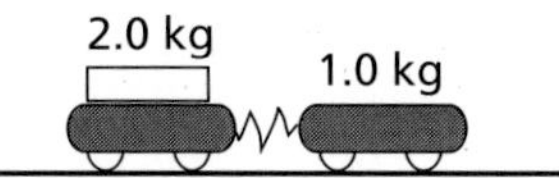

   a) 1 m/s
   b) 2 m/s
   c) 3 m/s
   d) 1.5 m/s

5. In a circus act, a 100-kg "human cannonball" is fired from a 400-kg cannon on wheels. After the cannon is fired, which equation below describes the system?

   a) speed of the cannon + the speed of the human = 0
   b) mass of the cannon + the mass of the human = 0
   c) momentum of the cannon + the momentum of the human = 0
   d) velocity of the cannon + the velocity of the human = 0

CHAPTER 3

SECTION 6 QUIZ ANSWERS

1. b) 24,000 kg · m/s
2. b) 24,000 kg · m/s
3. b) 2.0 m/s
4. d) 1.5 m/s
5. c) momentum of the cannon + the momentum of the human = 0

NOTES

NOTES

SECTION 7

Impulse and Changes in Momentum: Crumple Zone

Section Overview

Students consider the ideas of conservation of energy and reduction of force on passengers in designing and testing a crumple zone. Using probes, they measure the force and velocity over time during a collision involving a cart fitted with different types of cushions. By analyzing the data, students find the relationship between the force exerted, the time the force acts on the object, and the change in momentum. A class discussion follows on the relationship between impulse and changes in momentum, and how work and energy or impulse and momentum can be used to analyze a collision. Examples using both approaches are presented. Students then apply the concepts of impulse and change in momentum as they create graphs to show how factors involved in collisions, such as force, change for various situations. Students apply the concepts of impulse, conservation of momentum, and the work-energy theorem to solve various problems.

Background Information

A change in momentum only occurs if a net external force acts on an object. During a collision, each object involved in the collision experiences a force from the other object. The amount of force acting on the object at any instant is given by

$$F = \frac{dp}{dt}.$$

The average force acting on the object is given by

$$F = \frac{\Delta p}{\Delta t} = m\frac{\Delta v}{\Delta t},$$

for constant mass. From this equation, one can see that the force needed to change an object's momentum increases as the time involved decreases. For example, if you wish to bring a moving object to rest, it requires more force if you bring it to rest in half a second than if you bring it to rest in one second. This equation is often rearranged and written as $F\Delta t = m\Delta v$.

The quantity $F\Delta t$ is called the impulse. When a vehicle stops and as the stopping time increases, the required force to stop the vehicle decreases. If the stopping time is quite short, as in a collision with a fixed object, the stopping force can be extremely large. For a given velocity, the change in the momentum of the vehicle is the same, regardless of whether the vehicle is stopped over a long time or a short time.

This is very important in designing safety features to protect passengers during a collision, because the goal is to reduce the force acting on the person. Previously, this was discussed using the concepts of the work-energy theorem during the secondary and tertiary collisions, when air bags were reviewed. This section focuses on interpreting the situation using the impulse and change in momentum approach, and by reducing the forces exerted on the passenger by increasing the time required to stop the vehicle, and thus reducing the energy transferred to the passenger. To do this, the vehicle is designed to crumple in order to absorb some of the energy during the collision by increasing the distance the force acts and the time of the collision. These crumple zones allow for deformation of the vehicle during the collision. From the energy approach, some of the initial energy of the vehicle is transferred to deforming the vehicle, leaving less energy to be transferred to the passenger.

Crucial Physics

- Impulse is the change in momentum of an object. An impulse is equal to the net external force acting on an object multiplied by the time that the force acts on the object. A net external force acting on an object causes a change in the object's momentum.
- Stopping an object requires changing its velocity from an initial value to zero. By increasing the amount of time an external force acts on the object, and/or the distance over which the external force acts on the object, the amount of force required to stop the object is reduced.
- For a force vs. time graph, the impulse is equal to the area under the graph, and equal to the object's change in momentum.

Learning Outcomes	Location in the Section	Evidence of Understanding
Design a device that is able to absorb the energy of a collision and reduce the net force on an object in an automobile.	***Investigate*** Part A: Steps 1-5	Students construct "crumple zones," a design feature that reduces the amount of energy and force acting on a cart so that an object in the cart does not fall out during a collision. Students measure the force and velocity of the cart over time during the collision.
Describe collisions and crumple zones in terms of momentum, impulse, and force.	***Investigate*** Part A: Step 5, Part B: Step 7 ***Physics Talk*** ***Physics Essential Questions*** ***Physics to Go*** Questions 1-3, 11	Students analyze data from the cart collisions and find a relationship between impulse and change in momentum. Students read about and participate in a class discussion on crumple zones in terms of change in momentum and impulse, and are able to distinguish this approach from the work-energy approach.
Apply the concept of impulse in the analysis of collisions.	***Physics to Go*** Questions 8-9, 11	Students should apply the concepts discussed in the *Physics Talk* to analyze automobile collisions.
Use a computer's motion probe (sonic ranger) to determine the velocity of moving vehicles.	***Investigate*** Part B: Steps 2-5	Students use a sonic ranger and a computer to measure and graph the velocity of a cart during a collision.
Use a computer's force probe to determine the force exerted during a collision.	***Investigate*** Part B: Steps 2-5	Students use a force probe and a computer to measure and graph the force of a cart during a collision
Compare the change of momentum of a model vehicle before a collision with the impulse applied during a collision.	***Investigate*** Part B: Step 7	Students analyze data and find the relationship between impulse and momentum.
Explore ways of using cushions to increase the time that a force acts during a primary collision.	***Investigate*** Part B: Steps 5-7	Students measure how using different cushions attached to the outside of a cart affect the force and velocity of a cart during a collision.

Section 7 Materials, Preparation, and Safety

Materials and Equipment

PLAN A		
Materials and Equipment	**Group (4 students)**	**Class**
Dynamics cart	2 per group	
Inclined plane for lab cart	1 per group	
Large ring stand	1 per group	
Rod, aluminum, 12 in. (length) x 3/8 in. (diameter) (to act as cross arm)	1 per group	
Holder, right angle, cast iron	1 per group	
Scissors	1 per group	
Ruler, metric, in/cm	1 per group	
Meter sticks	1 per group	
AC Ticker tape timer	1 per group	
Wood piece, 1 in. x 2 in. x 2 in.	1 per group	
Self-adhesive Velcro dot (part of variety of cushioning material set)	1 per group	
Cork/cardboard piece, 1/2 in. x 1 in. (part of variety of cushioning material set)	6 per group	
Double-sided tape (to adhere pieces, part of variety of cushioning material set)		
File folders	3 per group	
Tape, masking		6 per group
Ball of string		1 per class
Rubber bands, #64, pack		1 per class
Unlined index card, 3 in. x 5 in., pkg 100		1 per class
Graph paper, pkg of 50		1 per class
Access to a smooth level surface*	1 per group	
Motion detection probeware*		1 per class
MBL or CBL technology to record probeware activity*		1 per class
Stack of books*	1 per group	
Force probeware*		1 per class
Wood block, 4 cm x 4 cm x 3 cm*	1 per group	

*Additional items needed not supplied

PLAN B		
Materials and Equipment	**Group (4 students)**	**Class**
Dynamics cart		1 per class
Inclined plane for lab cart		1 per class
Large ring stand		1 per class
Holder, right angle, cast iron		1 per class
Rod, aluminum, 12 in. (length) x 3/8 in. (diameter) (to act as cross arm)		1 per class
Scissors	1 per group	
Ruler, metric, in/cm	1 per group	
Meter sticks		1 per class
Wood piece, 1 in. x 2 in. x 2 in.	1 per group	
Self-adhesive Velcro dot (part of variety of cushioning material set)		6 per class
Cork/cardboard piece, 1/2 in. x 1 in. (part of variety of cushioning material set)		6 per class
Double-sided tape (to adhere pieces, part of variety of cushioning material set)		6 per class
File folders	3 per group	
Tape, masking		6 per group
Ball of string		1 per class
Rubber bands, #64, pack		1 per class
Unlined index card, 3 in. x 5 in., pkg 100		1 per class
Graph paper, pkg of 50		1 per class
Access to a smooth level surface*	1 per group	
Motion detection probeware*		1 per class
MBL or CBL technology to record probeware activity*		1 per class
Stack of books*	1 per group	
Force probeware*		1 per class
Wood block, 4 cm x 4 cm x 3 cm*	1 per group	

*Additional items needed not supplied

Note: Time, Preparation, and Safety requirements are based on Plan A, if using Plan B, please adjust accordingly.

Time Requirement

At least two periods are required to give students time to design and test their crumple-zone designs.

Teacher Preparation

- Consider having students bring in materials to use as cushions for the second part of *Investigate*. Make sure cushion materials that you have tested to compare results are also supplied. Conduct the investigation before class to determine where your students may have difficulties. Decide if you wish to stop students at any step to have a class discussion on the procedure.
- If your materials are limited, use a large-screen monitor or LCD projector to display the data to the class. Prepare a kit for each team consisting of a toy cart and a different cushioning material. Each team can prepare one cart and run it as a demonstration station.
- Try a few sample runs in advance to make sure that the results fall within limits of the available probes. Adjust the height of the ramp accordingly. Demonstrate the operation of the sensors and the software to the class if necessary. Providing materials produces a certain regularity and predictability to the investigation. You may want to present this investigation as a challenge a day in advance. The students are sure to bring some unusual cushioning materials from home.
- When the dynamics carts leave the ramp and travel onto the horizontal surface, a transition surface such as a manila file folder taped to the ramp and the floor or table will smooth the transition.
- A ramp that has a lower incline works better when placing the motion detector at the top of the ramp.
- Place the force probe near the bottom of the ramp so the motion detector will be able to read the velocity up to the point of collision.
- Clear the area being used of all extraneous equipment so the motion detector has a clear image of the moving cart.

Safety Requirements

- Caution the students to keep the collision speeds low to protect the equipment.
- In all collision experiments, goggles are of extra importance.
- Students should wear closed-toe shoes.
- Caution students to keep the cart velocities low to prevent damage to the carts or the force probe.

NOTES

Meeting the Needs of All Students

Differentiated Instruction: Augmentation and Accommodations

Learning Issue	Reference	Augmentation and Accommodations
Copying graphs accurately	***Investigate*** Part B: Steps 4-6	**Augmentation** • It is important that students copy the graphs accurately for comparison. Since students are copying both velocity vs. time and force vs. time graphs, remind students to label the axes carefully. • Model an example of the level of accuracy required to create a graph well enough to use for comparison. • Provide graph paper for students to draw their graphs and then tape into their logs. **Accommodation** • Allow students to print the graphs created by the computer probe programs and tape them into their logs.
Comparing graphs	***Investigate*** Part B: Step 7 ***Physics to Go*** Question 10	**Augmentation** • Some students need help comparing the force vs. time graphs because they do not notice the details that differentiate graphs. Show students a few different graphs and ask them to brainstorm a list of the important details shown by the graphs. The list should include general shape of the graph, slope, axis scales, units, area under the graph, and so on. Ask students to highlight differences on the graph. **Accommodation** • Provide copies of the graphs with differences already highlighted.
Summarizing information in a chart	***Investigate*** Part B: Step 7.d)	**Augmentation** • Summarizing information is a challenge for students who struggle to differentiate the importance of details. • Ask students what information should be included in the summarizing chart. After a list is compiled, ask students if the organization or order of the information is important. For example, since mass and velocity are used to calculate momentum, it would make sense for those values to be included next to each other. It may also be helpful for impulse and change in momentum to be next to each other for comparison.
Using equations and calculations to understand a concept	***Physics Talk***	**Augmentation** • Some students struggle to conceptualize the values that numbers represent and may not be able to independently make sense of these series of calculations. Use this opportunity to review how to calculate changes in velocity, acceleration, force, and momentum. • Model each calculation step-by-step or ask pairs of students to explain the calculations to each other. • Ask students why some of the calculations yield negative values. Explicitly make the connection that vector quantities can be represented with negative numbers because of direction. • Make analogies or provide examples so students can understand what the calculated values represent. How much force is –3000 N? How fast is –15 m/s?

Learning Issue	Reference	Augmentation and Accommodations
Being too literal	***Physics Talk***	**Augmentation** • Students who are very literal thinkers may misunderstand the "small force exerted over a long time" statement. A long time may be several minutes or even hours for some students. Ask students how much time it took for their carts to come to a stop in the *Investigate*. Then ask students how long "a long time" is when comparing crumple zones. **Accommodation** • Some students may need a very concrete example to conceptualize lengths of time. Play music for one second, 10 seconds, 30 seconds, a minute, etc. Then ask students which trial seemed most reasonable in relation to their carts coming to a stop.
Learning essential concepts using equations	***Physics Talk***	**Augmentation** • Direct and inverse relationships help students understand concepts, but the struggle with manipulating fractions often gets in the way of student understanding. For the *Newton's second law explanation* and the *Momentum/impulse explanation*, substitute values in place of the variables to make the relationship more concrete. For example, show students what happens to acceleration if Δt is 10 s, 20 s, or 100 s.

NOTES

CHAPTER 3

Strategies for Students with Limited English-Language Proficiency

Learning Issue	Reference	Augmentation
Vocabulary	***What Do You Think?***	To help students understand the word "crumple," crumple a piece of paper in your hands.
Following complex procedures	***Investigate***	Break down the *Investigate* into smaller chunks that allow students to comprehend each portion of the investigation before moving on to the next one. This approach will allow students to get acquainted with the apparatus, get comfortable following the procedures outlined within each step, and to internalize new concepts that are introduced. Lead a brief class discussion after each step to allow students the opportunity to demonstrate acquired knowledge and understanding.
Following directions Vocabulary comprehension	***Investigate*** Part A: Step 2 Part B: Steps 1.a), 2, 4.b), 5	To properly follow each direction, ELL students may need help with some of the vocabulary, such as "disturbing," "detector," "barrier," "transparency," and "coasting."
Understanding concepts	***Investigate*** Part B: Step 3	Hold a class discussion to determine what your students know about sound waves and reflection. Give them a brief tutorial on how a sonic ranger works, geared toward their level of understanding.
Understanding concepts Vocabulary comprehension	***Physics Talk***	When students encounter the term "impulse," be sure they understand that it means a change in momentum. Also, help them grasp that the two factors that affect impulse—force and time—are inversely proportional. For the same collision, a small force exerted over a long time produces the same change in momentum (the same impulse) as a large force exerted over a short time.
Comprehension	***Physics Talk*** ***Active Physics Plus***	Students may benefit from an oral representation of tables and graphs. When students look at the graphs, ask them which graph represents the collision that likely would have caused less injury. Make sure students correctly interpret the labels on the graph axes before proceeding. Be sure they correctly identify the graph showing less force over a longer time, and allow students time to discuss reasons for their choices.
Vocabulary comprehension	***Reflecting on the Section and the Challenge***	Students need to know that they must give a rationale for the design decisions they have made when they have completed and are demonstrating their safety devices. That is, they need to provide reasons for going with the design they chose, and the reasons must be based on sound physics principles.
Applying concepts	***Physics to Go*** Step 4	Hold a class discussion to address this question. Students should be able to deduce that bending your knees when landing from a jump increases the time during which the landing force is applied, thereby reducing the amount of force on your legs.

NOTES

SECTION 7

Teaching Suggestions and Sample Answers

What Do You See?

Have students describe the illustration in the student text. Discuss how students think the illustration connects to the title of the section and the *Learning Outcomes*. Ask them what in the illustration might be a device capable of absorbing the energy of a collision or reducing the net force of the collision, or where such a device could be. Consider using an overhead of the illustration as a focal point for the discussion.

What Do You Think?

Describe to the class what a crumple zone is using the information in the student text. Ask what part of the vehicle in the image they think is the crumple zone and why. Ask students to record in their logs their responses to the *What Do You Think?* question. Have a class discussion on students' ideas. Record misconceptions and make sure they are addressed at appropriate times during the section. Focus on the responses that provide a chance for you to get the class engaged in a discussion of reducing the energy transferred to the passengers, and reducing the force acting on the passengers.

Investigate

If you decide to guide students through some of the steps during the *Investigate*, let them know at which step they should stop. Otherwise, have a discussion on the *Investigate* steps, and let them begin and complete the *Investigate*. Let them know what guidelines they should follow for time, space, and materials. Demonstrate how to use the force and velocity probes before Part B.

What Do You Think?

A Physicist's Response

Crumple zones are very important because they reduce the amount of energy transferred to passengers in a vehicle during a collision, and the amount of force acting on the passengers during the collision. It is important to realize that crumple zones must follow conservation of momentum and conservation of energy. The momentum change of the vehicle for a given speed is the same, as is the change in kinetic energy. By increasing the time of the collision and the distance over which the collision occurs, one can reduce the amount of force needed to stop the vehicle. This also reduces the amount of force that acts on the passenger. From the energy approach, energy is needed to deform an object, and if the object is permanently deformed, that energy cannot be retrieved and transferred to the passenger. Energy is also transformed into heat energy when deformation occurs. Because of this, it is important to design crumple zones that will permanently deform and will absorb the energy of the collision. Very rigid objects and very elastic objects do not do this well.

Crumple zones are built not only on vehicles but also around areas where collisions are likely to occur or where collisions could cause other danger. Examples include canisters of sand near roadwork or bridge supports, and railings along bridges.

Students' Prior Conceptions

This section is all about making connections and refining the engineering model to protect the egg in the cart for the *Chapter Challenge*. Placing a visual that emphasizes the two elements of impulse, force and change in time, as large and/or small icons, is a simple anchoring activity for student understanding. Following this with a second visual that connects impulse to change in momentum reinforces the *Physics Talk*: "You can get an identical change in momentum with a large force over a short time or a small force over a longer time." A welcome strategy for students is to encourage them to make a small 'comic' poster to illustrate the connection shown at the right:

$$\text{Impulse} = \mathbf{F} \times \Delta t \qquad \text{Impulse} = \text{change in momentum}$$

$$\mathbf{F} \times \Delta t = m \times \Delta v$$

or,

$$\text{Impulse} = F \times \mathbf{\Delta t} \qquad \text{Impulse} = \text{change in momentum}$$

$$F \times \mathbf{\Delta t} = m \times \Delta v$$

Section 7 Impulse and Changes in Momentum: Crumple Zone

Section 7 Impulse and Changes in Momentum: Crumple Zone

What Do You See?

Learning Outcomes

In this section, you will

- **Design** a device that is able to absorb the energy of a collision and reduce the net force on an object in an automobile.
- **Describe** collisions and crumple zones in terms of momentum, impulse, and force.
- **Apply** the concept of impulse in the analysis of collisions.
- **Use** a computer's motion probe (sonic ranger) to determine the velocity of moving vehicles.
- **Use** a computer's force probe to determine the force exerted during a collision.
- **Compare** the change of momentum of a model vehicle before a collision with the impulse applied during a collision.
- **Explore** ways of using cushions to increase the time that a force acts during a primary collision.

What Do You Think?

When an automobile collides with a wall and is brought to a stop, all of the energy of motion that the automobile had has to go somewhere. Also, a force must act to stop the automobile, and any passengers inside. Automotive engineers design something called a crumple zone to absorb the energy of the collision and to lessen the force on the passengers. (*A crumple zone is part of the body of an automobile that compresses during an impact.*)

- **What are some of the factors that automobile designers and engineers must consider when designing a crumple zone as a safety feature?**

Record your ideas about this question in your *Active Physics* log. Be prepared to discuss your responses with your small group and the class.

Teaching Tip

If you decide to do Part B of this *Investigate* as a class demonstration, be sure to have the students still do Part A. Otherwise, the students may not grasp that allowing the front of the cart to crumple will decrease the force exerted, and this will allow them to test several designs.

Part A: Designing a Crumple Zone

Teaching Tip

Some students may choose to use all of the material available for the crumple zone, and others only part of the material. Let them determine and test their own designs, but put a time limit on the investigation.

1.

Students should work in groups. Allow students to design their crumple zones in whatever manner they choose and allow them to test their design and change it using the *Engineering Design Cycle.*

2.–3.

Students test and revise their designs as necessary. Check that students are following the guidelines.

1. **Collision severity is the same for both automobiles involved in a head-on collision.** Measurement of impulse and momentum to identify the forces involved in the collision and attesting to the equal and opposite nature of those forces serves as a tool for students to organize their knowledge of crumple zones. They learn to assess the severity of the collision by considering the time through which the force acts. Crumple zones allow for more time for the stopping force to act. This observation expands students' conception of work when they associate the force of impact with the distance over which that force is applied. The work done has a consistent value but the change in distance reduces the impact of the force. Energy lost equals work done, and work is force applied multiplied by the distance over which the force is applied.
2. **Students do not recognize the importance of the time factor involved with the concept of impulse.** As students model the work required to stop an object, they become more aware of the change in distance required to stop an object and the time required for the force to act over this distance. This triggers the association with time being critical in the design of safety restraints, and how it relates to the change in the motion and the momentum of the object. As impulse is mathematically related to the change in motion, students will reason about the time required to stop more correctly.

4.

Students should describe why the tape does not keep the block on the cart in terms of Newton's laws. The tape does not apply a net force acting against the direction of motion of the block to stop its motion. Rather, the tape applies a force toward the pivot point on the block when the block tries to continue its forward motion (due to Newton's first law), causing the block to tip forward. If the tape breaks free, the block will continue its motion until a force acts on it.

5.

Students should demonstrate their designs and record their thinking and data.

Before beginning Part B, demonstrate how to open the software and use the probes.

Part B: Cushioning Collisions (Computer Analysis)

Teaching Tip

You may choose to do Part B as a demonstration due to time or equipment considerations. If you do, demonstrate to the students how the apparatus works prior to testing the various cushioning materials. You can also choose one of the class's crumple designs to compare to other cushioning material.

1.a)

Students should sketch the force vs. time graph that is produced by pushing back and forth gently on the force probe.

1.b)

Students should record their actions that made the force vs. time graph. They should note particularly whether pushing on the probe gave a positive or negative reading.

Investigate

Part A: Designing a Crumple Zone

1. Form teams of three to five students. Each team will design a crumple zone that will attach to the front of a cart. You may use one sheet of paper, 30 cm of tape, 2 rubber bands, and 30 cm of string.
2. The task is as follows:
   - Your cart (automobile) will roll down the ramp from a height of 10 cm, cross 20 cm of level ground, then hit a wall of books and come to a complete halt.

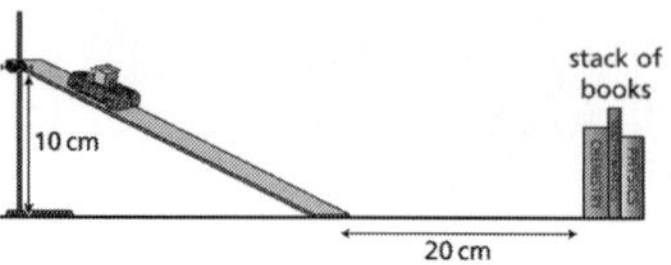

   - A 4 cm by 4 cm by 3 cm block (passenger) will be attached to the cart. The wooden block will be held in the automobile by a single 2-cm length piece of tape attached to the front of the block.

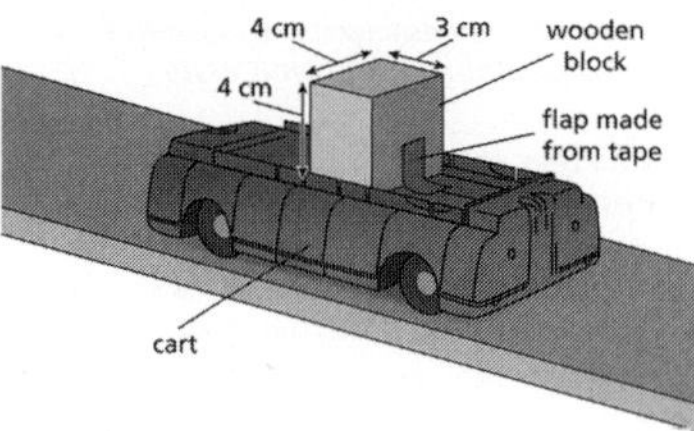

   - The crumple zone which you will design must allow the cart to stop without the block falling over.

If the challenge appears too difficult at first, try heights below 10 cm. You can then demonstrate the height below 10 cm that allows the automobile with the crumple zone to stop without disturbing the wooden block. If the challenge appears too easy, you can try heights above 10 cm.

3. Follow your teacher's guidelines for considering the use of time, space, and materials as you design your crumple zone.
4. Begin by discussing why the tape on the front of the block will not help hold the block in place when the automobile hits the books. (Hint: Use Newton's first law.)
5. Demonstrate your design team's crumple zone for the class. Keep a careful record of your *Engineering Design Cycle*. When you make changes, record the changes and the reason for these changes.

Part B: Cushioning Collisions (Computer Analysis)

1. In this part of the investigation, you will be using a force probe that is attached to a computer to determine the effectiveness of different types of cushions for a cart. A force probe is a device that measures the force of an impact. Before beginning the experiments, investigate how the force probe works. Open the computer files that will display a graph of force versus time. Your teacher will help you locate the program. As the computer is recording data, use your finger to move the lever of the force probe. Investigate how the force of your finger on the lever affects the graph that appears on the computer screen.

a) Draw the graph that you have generated with the detector.

b) Beneath the graph, describe what you were doing to create that graph.

Teaching Tip

If a force probe is not available, a substitute setup can be constructed as follows. In place of the force probe, fill a paper tube, such as that from a roll of paper towels or toilet paper with modeling clay. Insert a sharpened pencil about 1 in. deep into the clay along the axis of the tube. When the cart comes down the ramp, the tube should be aligned so that the center of the cart strikes the end of the pencil perpendicularly, pushing the pencil into the clay. How far the pencil pierces the clay is a measure of the force exerted. After each trial, repair the clay in the tube, and place the pencil back into the tube at the 1-in. depth. Of course, this does not allow you to show that the area under a force vs. time graph equals the change in momentum, but it does give a relative measure of the force. If you assume depth of travel is also proportional to the time required to stop the cart, approximate results may be obtained for $F\Delta t = m\Delta v$. Numerous demonstrations illustrating the change in momentum relationship can be found online.

Section 7 Impulse and Changes in Momentum: Crumple Zone

2. Release a cart at the top of a ramp and measure the force of impact as the cart strikes a barrier at the bottom. A sonic ranger can be mounted at the top of the ramp to measure the speed of the cart prior to the collision. Open the appropriate computer files to prepare the sonic ranger to graph velocity vs. time and the force probe to graph force vs. time.

3. Mount the sonic ranger at the top of a ramp and place the force probe against a barrier at the bottom of the ramp, as shown in the diagram. Attach an index card to the cart to obtain better reflection of the sound waves and improve the readings of the sonic ranger.

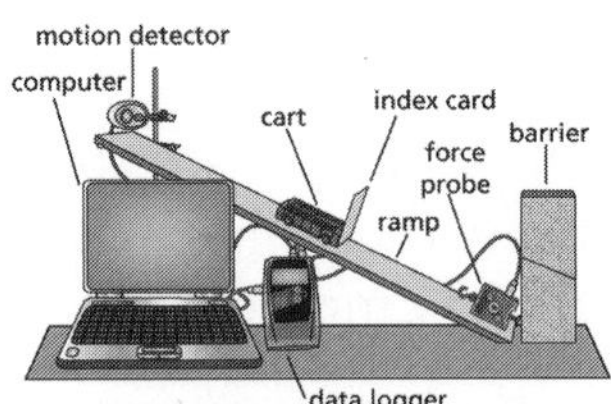

4. To ensure that the data collection equipment is working properly, conduct a few runs of releasing the cart down the ramp so that it collides with the force probe.
   - a) Make copies of the velocity vs. time and force vs. time graphs that are displayed on the computer.
   - b) Some computer probe programs calculate the area between the curve and the x-axis. Record this calculated value. Alternatively, you can use a transparency of graph paper to place over the curve and count the number of boxes that are between the curve and the x-axis.

5. Attach your cushioning material to the front of the cart. Conduct a number of runs with the same type of cushioning. Make sure that the cart is coasting down the same slope from the same position each time.
   - a) Make copies of the velocity vs. time and force vs. time graphs that are displayed on the computer.
   - b) Record the area between the curve and x-axis.

6. Repeat *Step 5* using other types of cushioning materials.
   - a) Record your observations in your log.

7. Using the information from the graphs you obtained in this investigation, answer the following:
   - a) Compare the force vs. time graphs for the cushioned carts with those for the carts without cushioning.
   - b) Compare the areas under the force vs. time graphs for all of the experimental trials.
   - c) Compute the *momentum* of the cart (the product of the mass and the velocity) prior to the collision and compare it with the area under the force vs. time graphs.
   - d) Summarize your comparisons in a chart.
   - e) The area under the force vs. time graph is called the *impulse*. How can impulse be used to explain the effectiveness of cushioning systems?
   - f) Describe the relationship between impulse ($F\Delta t$) and the change in momentum = $\Delta(mv)$.

323

Active Physics

2.

Students set up equipment and computer files.

3.

Check that students have adjusted the equipment to obtain good signals with the sonic ranger.

4.

The students should take several trials before actually testing their crumple zones to ensure the computer-generated graphs are reasonable (see the following sample graphs). Good velocity vs. time and force vs. time graphs are shown.

4.a)

Students should make copies of the velocity vs. time and the force vs. time graphs.

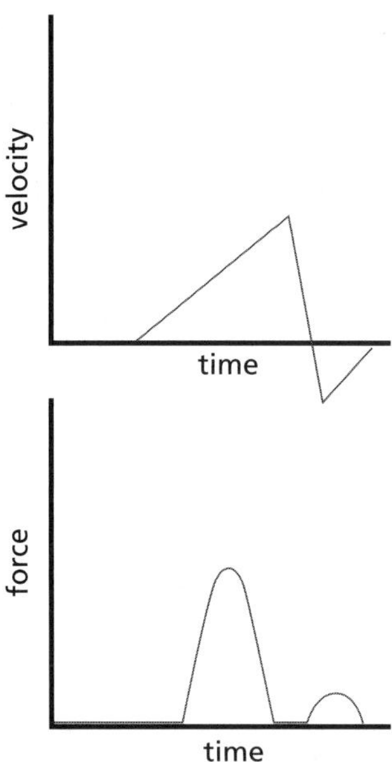

4.b)

Students compute the area under the curve and record the value for the force vs. time graph. Force vs. time graphs with the area under the graph computed should appear similar to the graph shown below.

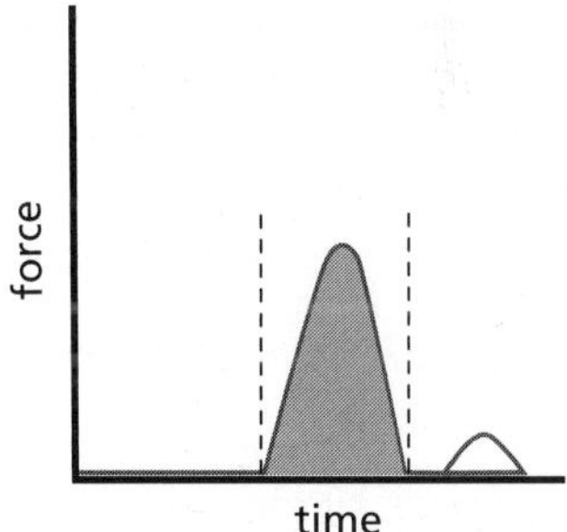

5.a)

Students take data on force vs. time and velocity vs. time. If the cushioning material is working well, the force vs. time graph should be stretched out in the horizontal direction (the time axis) with a lower maximum value for the force. The velocity vs. time graph should have a lower slope indicating a smaller

CHAPTER 3

negative acceleration.

5.b)

Students record the area under the force vs. time graph as shown in *4.b)*. The area under the curve should remain approximately the same if the carts strike the probe with the same speed and come to rest without rebound.

6.a)

Have students try several other types of cushion materials and record values and graphs for these other types of cushion material.

7.a)

Students' data will vary. Students should observe that the force is greater on the collision without cushioning.

7.b)

Students should describe the similarities and differences between the areas under the force vs. time graphs for all the graphs. If the speeds before the collisions are the same, then the areas under the graphs are the same.

7.c)

The mass multiplied by the velocity just before the collision is roughly equal to the area under the force vs. time graphs. The relevant velocity is the velocity at impact, which should be the highest value recorded by the sonic ranger. The area under the graph should equal the change in momentum of the cart because the final momentum is zero. Student answers may vary due to calculation and measurement errors.

7.d)

Students summarize and record their comparisons.

7.e)

The impulse is equal to the change in momentum, and is therefore, related to the mass and the velocity of the cart. Depending on the cushioning, there will be an increase in the time, with the impulse remaining constant; thus, the force acting on the cart and individuals inside will be less.

7.f)

Students' responses should indicate that impulse equals the change in momentum.

NOTES

NOTES

CHAPTER 3

Physics Talk

This *Physics Talk* discusses how force, time, and the change in momentum are related. There is also a discussion on the two approaches used to explain a collision, which are the work and energy approach, and the impulse and momentum approach. Check students' understanding on the factors that play a role in collisions and safety devices.

Have a discussion on how the work-energy theorem applies to students' crumple zones using the information in the student text. Remind the class that the physics is the same as that used when describing air bags. Discuss how the amount of time in which a collision occurs affects the amount of force involved in changing the colliding objects' combined momentum, and how the impulse is equal to the change in momentum.

Go through the derivation in the student text of calculating the force if the change in momentum is known. Discuss how this connects with Newton's second law and show students how the change in momentum over time is equal to the force acting on the object. Let students know that the change in momentum is called the impulse and that it is equal to the force acting on the object multiplied by the time the force acts on the object.

Make connections with students' observations in the *Investigate* and with everyday experiences, such as falling on concrete versus grass, or falling on the floor versus falling on a trampoline or bed. Emphasize that as the time to change the momentum decreases, the force increases. Ask students to provide examples.

As the example of the 50-kg mass is discussed with the class, point out how the force required to stop the object changes as the time of the collision changes. Emphasize that the change in momentum is always equal to the impulse. Describe how the type of surface the object collides with changes the collision time, making connections to the example of the grass and concrete, as well as students' observations from the *Investigate*.

Discuss how crumple zones reduce the force acting on a passenger and the energy transferred to the passenger. Emphasize to students that they can use the work and energy approach (increase the distance over which the force acts) or

Chapter 3 Safety

Physics Talk

FORCES AFFECTING COLLISIONS

By examining the different crumple zones designed by other teams in your class, you can get an insight into the physics of collisions. Probably no team just used the flat sheet of paper. Most teams tried to fold the paper in specific ways. It seemed the goal was to have a "softer" collision rather than have the cart hit the wall without any crumple zone.

By trying to make a "softer" collision, you may have used the physics of work and change in kinetic energy.

$$W = F \cdot d = \Delta KE$$

In this case, as in the air bag, you wanted to decrease the force by increasing the distance to stop the automobile. The work done on the cart then reduced the kinetic energy of the cart.

Impulse and Changes in Momentum

There is an equivalent way of describing the physics of a collision. Rather than focusing on the distance that the force acts, you can look at the amount of time that the force acts. By maximizing the time, you can minimize the force. When creating the crumple zone, you increased the time and thereby minimized the force.

It takes an unbalanced, opposing force to stop a moving automobile. Newton's second law of motion, $F = ma$, lets you find out how much force is required to stop any automobile of any mass with a corresponding *acceleration* (or change in speed with respect to time). For example, if a 1000-kg automobile accelerates at $-2\ m/s^2$, then the force required is

$$F = ma = (1000\ kg)(-2\ m/s^2) = -2000\ N$$

Notice that the acceleration is negative because the automobile's **velocity** decreased. Also, notice that the force is negative to indicate that it produces a negative acceleration as it slows the automobile down to a stop.

The overall idea can be shown using a concept map seen at the top of the next page.

Physics Words

velocity: speed in a given direction; displacement divided by the time interval; velocity is a vector quantity; it has magnitude and direction.

324

Active Physics

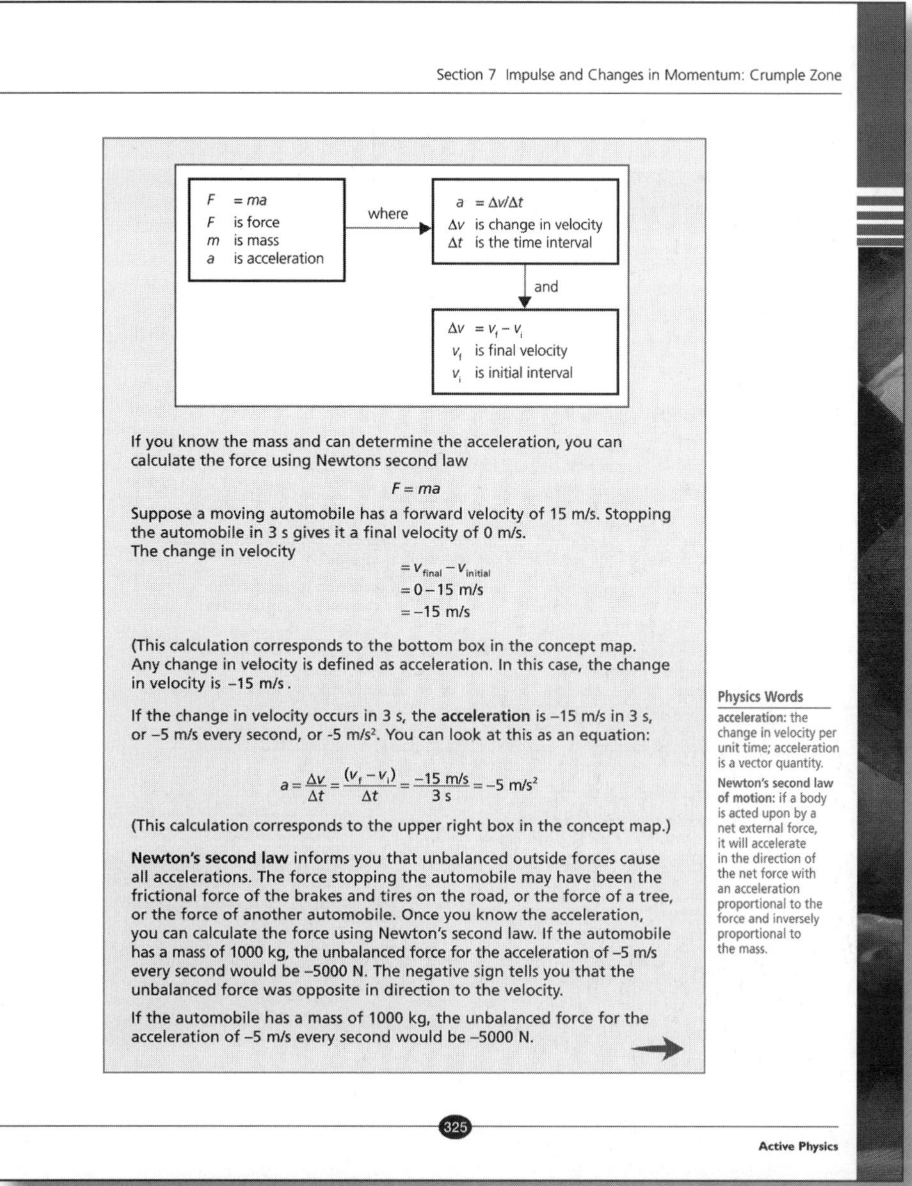
Section 7 Impulse and Changes in Momentum: Crumple Zone

If you know the mass and can determine the acceleration, you can calculate the force using Newtons second law

$$F = ma$$

Suppose a moving automobile has a forward velocity of 15 m/s. Stopping the automobile in 3 s gives it a final velocity of 0 m/s.
The change in velocity

$$= v_{final} - v_{initial}$$
$$= 0 - 15 \text{ m/s}$$
$$= -15 \text{ m/s}$$

(This calculation corresponds to the bottom box in the concept map. Any change in velocity is defined as acceleration. In this case, the change in velocity is −15 m/s.

If the change in velocity occurs in 3 s, the **acceleration** is −15 m/s in 3 s, or −5 m/s every second, or -5 m/s². You can look at this as an equation:

$$a = \frac{\Delta v}{\Delta t} = \frac{(v_f - v_i)}{\Delta t} = \frac{-15 \text{ m/s}}{3 \text{ s}} = -5 \text{ m/s}^2$$

(This calculation corresponds to the upper right box in the concept map.)

Newton's second law informs you that unbalanced outside forces cause all accelerations. The force stopping the automobile may have been the frictional force of the brakes and tires on the road, or the force of a tree, or the force of another automobile. Once you know the acceleration, you can calculate the force using Newton's second law. If the automobile has a mass of 1000 kg, the unbalanced force for the acceleration of −5 m/s every second would be −5000 N. The negative sign tells you that the unbalanced force was opposite in direction to the velocity.

If the automobile has a mass of 1000 kg, the unbalanced force for the acceleration of −5 m/s every second would be −5000 N.

Physics Words

acceleration: the change in velocity per unit time; acceleration is a vector quantity.

Newton's second law of motion: if a body is acted upon by a net external force, it will accelerate in the direction of the net force with an acceleration proportional to the force and inversely proportional to the mass.

325

Active Physics

the impulse and momentum approach (increase the time during which the force acts) when analyzing and discussing collisions. In physics, there are two fundamental approaches to describing phenomena: the force approach and the energy approach. The force approach is dependent on direction, however, the energy approach is not.

Discuss how crumple zones reduce the force acting on a passenger and the energy transferred to the passenger, but no matter what the change in momentum of the vehicle is, the total momentum is always conserved. Ask students to describe how it can be that a cart can change its momentum and yet the total momentum is conserved.

3-7b Blackline Master

Sometimes students have difficulty remembering that momentum is a vector quantity. Ask students if momentum is conserved when two objects of equal mass move toward each other with equal speeds, collide, and come to rest. In this case, both objects initially had momentum and finally do not. Emphasize that the total momentum of the two objects before the collision is equal to the total momentum after the collision. It is zero before and zero after, and it is extremely important to consider the direction when they add together the momentums of each individual cart.

Then ask students to describe what happens when two carts of the same mass are connected together with a compressed spring between them. They are at rest initially, and then they are released from each other. Note that this collision is sometimes called an explosion. Students should describe both carts initially having no momentum, and finally, each cart has momentum, but the total momentum is zero initially and finally.

Return to the collision with a cushion and ask students to describe what happens to the momentum during the collision with the cushion, and how the momentum can be conserved in the collision but changed by a cushion. Ask students to describe what happens to the momentum of the cushion and the energy transferred to it; the momentum of the cart and the energy transferred to it; and the momentum of the force probe and the energy transferred to it. Then, describe what occurs as the cart with the cushion collides with the force probe, using the information in the student text.

Remind students that another approach is the work energy approach, and that the change in kinetic energy of the carts for both cushions was the same. This means that the work done on the cart was the same for the hard cushion and the soft cushion. The forces and distances over which they acted were different, but the work done was the same. Review and summarize how the change of momentum equals the impulse. Review and summarize how the change of kinetic energy equals the work.

$$F = ma$$
$$= (1000\text{ kg})(-5\text{ m/s}^2)$$
$$= -5000\text{ N}$$

The same problem can be solved with the same automobile stopping in 0.5 s. The change in speed would still be –15 m/s, and the acceleration can be calculated.

$$a = \frac{\Delta v}{\Delta t} = \frac{v_i - v_o}{\Delta t} = \frac{-15\text{ m/s}^2}{0.5\text{ s}} = -30\text{ m/s}^2$$

There is another, equivalent picture that describes the same collision:

$$F = ma$$

Multiplying both sides of the equation by Δt, you get

$$F\Delta t = m\Delta v$$

then you can rewrite Newton's second law as

$$F = \frac{m\Delta v}{\Delta t}$$

The term on the right-hand side of the equation is change in **momentum**. The term on the left side of the equation is impulse.

Any moving automobile has momentum. Momentum is represented with a small p. Momentum is defined as the mass of the automobile multiplied by its velocity $p = mv$.

A change in momentum is called **impulse**. The impulse-momentum equation tells you that the momentum of the automobile can be changed by applying a force for a given amount of time. The impulse is the force multiplied by the time. So a small force exerted over a long time produces the same impulse (change in momentum) as a large force exerted over a short time. "The small force exerted over a long time" is what you found in the effective crumple zones.

Impulse-momentum is an effective way in which to describe all collisions.

Physics Words

momentum: the product of the mass and the velocity of an object; momentum is a vector quantity.

impulse: a change in momentum of an object.

Consider this question: "Why do you prefer to land on soft grass rather than on hard concrete?" Soft grass is preferred because the force on your body is less when you land on soft grass. This can be explained by using the impulse-momentum relation.

Whether you land on concrete or soft grass, your change in velocity will be identical. Your velocity may decrease from 3 m/s to 0 m/s.

Section 7 Impulse and Changes in Momentum: Crumple Zone

On concrete, this change occurs very fast, while on soft grass this change occurs in a longer period of time. Your acceleration on soft grass is smaller because the change in velocity occurred in a longer period of time.

$$a = \frac{\Delta v}{\Delta t}$$

When the change in the period of time gets larger, the denominator of the fraction gets larger and the value of the acceleration gets smaller.

When landing on grass, Newton's second law then tells you that the force must be smaller because the acceleration is smaller for an identical mass, $F = ma$. Smaller acceleration on grass requires a smaller force. Smaller forces are easier on your body and that is why you prefer to land on soft grass.

Change in value of momentum $\Delta p = \Delta mv$	Force F	Change in time Δt	Impulse $F\Delta t$
150 kg • m/s	50 N	3 s	150 N/s (150 kg • m/s)
150 kg • m/s	150 N	1 s	150 N/s (150 kg • m/s)
150 kg • m/s	15,000 N	0.01 s	150 N/s (150 kg • m/s)

You can get this change in momentum with a large force over a short time or a small force over a longer time.

If your mass is 50 kg, the amount of your change in momentum will be 150 kg • m/s when you decrease your velocity from 3 m/s to 0 m/s. There are many forces and associated times that can give this change in the value of the momentum.

If you could land on a surface that requires 3 s to stop, it will only require 50 N. A more realistic time of 1 s to stop will require a larger force of 150 N. A hard surface that brings you to a stop in 0.01 s requires a much larger force of 15,000 N.

On concrete, this change in the value of the momentum occurs very fast (a short time) and requires a large force. It hurts. On soft grass this change in the value of the momentum occurs in a longer time and requires a small force that is less painful and is preferred.

Notice that in the chart above, the change in momentum is always equal to the impulse.

CHAPTER 3

"Work and Energy" or "Impulse and Momentum"

The effective crumple zone decreases the force on the passenger. Using the work-energy theorem the force can be minimized by increasing the distance required to stop. Using the impulse-momentum theorem, the force can be minimized by increasing the time required.

As a physics student, it is your job to decide which of these two approaches is best for explaining a particular collision.

$$W = Fd = \Delta KE$$

or

$$F\Delta t = \Delta p$$

Both show that the force can be minimized. In work-energy, force is minimized by an increase in impact distance. In impulse-momentum, force is minimized by an increase in impact time.

Conserving Momentum

In *Part B* of the *Investigate*, a moving cart was brought to rest. The moving cart had momentum. The force probe was able to record the force that was applied to the cart over time to stop the cart. This force over time changed the momentum of the cart.

The force vs. time graphs for two carts with the same initial momentum are sketched below.

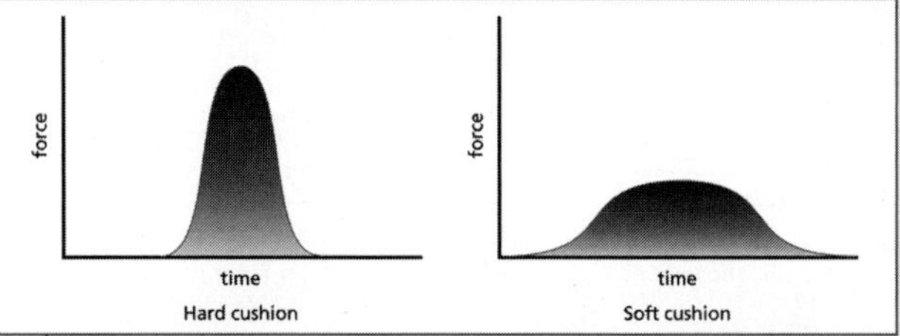

As you can see, the cart with the hard cushion had a larger force acting over a shorter time than the cart with the soft cushion. For the safety of passengers, a small force over a long time is what is needed.

Designing a safety device for an automobile is often determined by finding ways to decrease the force and increase the time during an impact. The change in momentum of a cart or an automobile can be identical, but a smaller force over a larger time will be safer.

Section 7 Impulse and Changes in Momentum: Crumple Zone

You know that during a collision, momentum is always conserved. However, confusion may arise because now you find that a force can change the momentum of a cart by bringing it to rest. How can momentum be conserved in collisions but momentum can be changed by a cushion? It may help to view the collision from two different perspectives. It is true that the total momentum before the collision is equal to the total momentum after the collision. However, during the collision the first cart may lose momentum while the second cart may gain momentum. Each cart changed momentum but the total momentum remained the same. A similar thing happens when the force probe stops the automobile. The automobile loses momentum. The probe gains that same amount of momentum. Since its mass is large (it is connected to the lab table), then this large gain in momentum corresponds to a very small gain in velocity.

You may recall that another way in which to describe the same cushion is to say that the larger force acted over a shorter distance. The work done by the force over the distance was the same for both bumpers, and both bumpers decreased the kinetic energy of the automobile.

Change in Momentum and Impulse

Momentum is the product of the mass and the velocity of an object,where p is the momentum,
m is the mass, and
v is the velocity.

Change in momentum is the change in the product of mass and velocity. If the mass remains the same, the change in the momentum is the product of the mass and the change in velocity.

$$p = \Delta mv$$

Impulse is the change in momentum.

$$F\Delta t = \Delta mv$$

Change in Kinetic Energy and Work

Kinetic energy is the product of ½ the mass and the square of the velocity.

$$KE = \frac{1}{2}mv^2$$

Work is the product of the force and the distance over which the force acts.

Work is equal to the change in kinetic energy.

$$W = \Delta KE$$

Checking Up

1. What is a crumple zone?
2. Why is it safer to collide with a soft cushion than a hard surface?
3. What is momentum?
4. What is the relationship between impulse and momentum?

Checking Up

1.

A crumple zone is a part of a vehicle that compresses during impact to reduce the amount of force acting on passengers during a collision, and it reduces the amount of energy transferred to passengers during collisions.

2.

The softer cushion reduces the force of impact by increasing the time of the collision and the distance over which the force acts.

3.

Momentum is the product of the mass and the velocity of an object, and it is a vector quantity.

4.

The impulse on an object is equal to the change in momentum of the object, which is equal to the net external force acting on the object multiplied by the time this force acts on the object.

Active Physics Plus

This *Active Physics Plus* provides an opportunity for students to apply the concepts presented in this section in a more mathematical context.

1.

Students should calculate the time involved for changing the momentum by 12,000 kg · m/s using a force of 8000 N.

$F\Delta t = \Delta p$

$$\Delta t = \frac{\Delta p}{F} = \frac{12{,}000\ \text{kg}\cdot\text{m/s}}{8000\ \text{N}} = 1.5\ \text{s}.$$

Students should construct a graph that shows the force versus time similar to the graph shown below. The initial value of time is not important, but the time over which the force is applied is, and it should be 1.5 s.

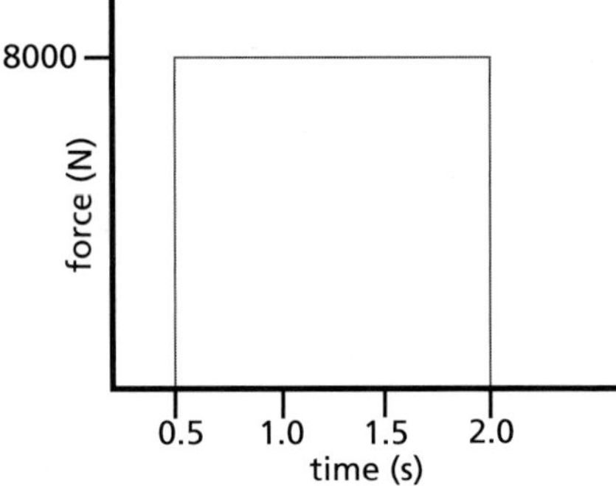

Show students that the area under the curve for this graph is equal to the base (the time interval or 1.5 s) multiplied by the height (the force or 8000 N) and this is equal to the change in momentum.

Area = (base)(height) = ΔtF = (1.5 s)(8000 N) = 12,000 kg · m/s.

Active Physics *Plus*

+Math	+Depth	+Concepts	+Exploration
♦♦			

Graphing Momentum Change

An automobile undergoing a collision has a momentum change of 12,000 kg•m/s during a collision. This could occur with a constant force of 4000 N (newtons) over 3 s (seconds).

The graph of the force vs. time for this collision would look like this:

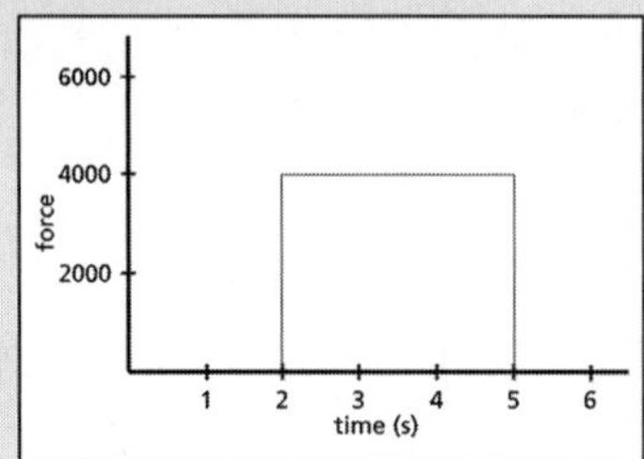

Notice that the area under the graph is the area of the rectangle:

$$A = bh$$
$$A = (3\ \text{s})(4000\ \text{N})$$
$$A = 12{,}000\ \text{N}\cdot\text{s}$$

You can find the units of 1 N by reminding yourself that the newton is defined in Newton's second law:

$$F = ma$$
$$1\ \text{N} = (1\ \text{kg})(1\ \text{m/s}^2)$$
$$1\ \text{N} = 1\ \text{kg}\cdot\text{m/s}^2$$

The area under the graph has the units of N s. Since 1 N is 1 kg•m/s^2, the area of the graph has the identical units of kg•m/s.

1. Create a graph that shows the same momentum change of 12,000 kg•m/s where the force is a constant 8000 N. You will have to calculate the corresponding time for the collision.
2. Create a graph that shows the same momentum change where the force is a constant 4000 N for 1 s and then 8000 N for the remainder of the required time to bring the automobile to rest.
3. Create a graph that shows the same momentum change where the force is a constant 4000 N for 1 s, then 8000 N for 0.5 s, and then 4000 N for the remainder of the required time to bring the automobile to a rest.
4. Create a graph that shows the same momentum change where the force gradually increases from 0 N to 4000 N, remains at 4000 N for 1 s and then gradually decreases to 0 N when the automobile comes to rest.

2.

Students should find the time needed for the second interval.

For the first interval a 4000 N force acts on the object for 1 s. This changes the momentum by $\Delta p = F\Delta t$ = (4000 N)(1 s) = 4000 N · s.

For the vehicle to stop, the momentum has to change a total of 12,000 kg · m/s. Therefore, the second force needs to be applied long enough to reduce the momentum by an additional 12,000 kg · m/s – 4000 kg · m/s = 8000 kg · m/s.

This means the 8000 N force should be applied for an additional time of

$$\Delta t = \frac{\Delta p}{F} = \frac{8000 \text{ kg} \cdot \text{m/s}}{8000 \text{ N}} = 1 \text{ s}.$$

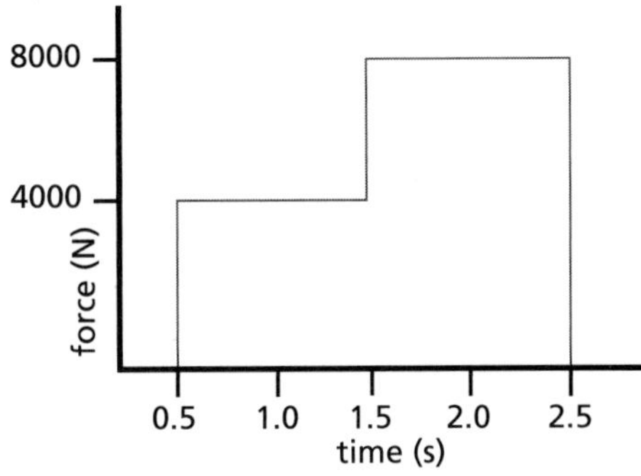

Have students check that the area under the curve (the sum of areas of the two rectangles formed) is equal to the total change in momentum.

3.

For the first interval a 4000 N force acts on the object for 1 s. This changes the momentum by $\Delta p = F\Delta t = (4000 \text{ N})(1 \text{ s}) = 4000 \text{ N} \cdot \text{s}$.

During the second interval an 8000 N force acts on the object for 0.5 s. This changes the momentum by $\Delta p = F\Delta t = (8000 \text{ N})(0.5 \text{ s}) = 4000 \text{ N} \cdot \text{s}$.

For the vehicle to stop, the momentum has to change a total of 12,000 kg · m/s. Therefore, the third force needs to be applied long enough to reduce the momentum by an additional

12,000 kg · m/s – 4000 kg · m/s – 4000 kg · m/s = 4000 kg · m/s.

This means the third applied force of 4000 N force should be applied for an additional

$$\Delta t = \frac{\Delta p}{F} = \frac{4000 \text{ kg} \cdot \text{m/s}}{4000 \text{ N}} = 1 \text{ s}.$$

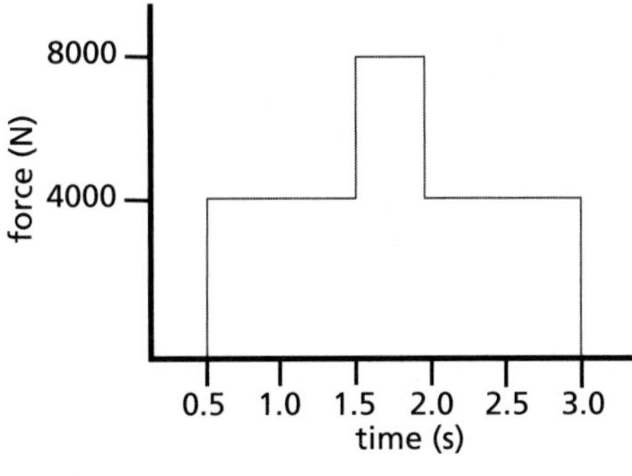

Have students check that the area under the curve (the sum of areas of the three rectangles formed) is equal to the total change in momentum.

4.

If students are having difficulty starting, point out that this problem should be solved in three parts and have them find the area under the curve. They can do this by constructing a graph and figuring out the areas under each part of the curve. For the first and last parts of the curve, the area of a triangle ($A = \frac{1}{2}bh$) should be calculated. The middle part forms the area of a rectangle ($A = l \times h$). The total area under the curve is equal to the change in momentum or 12,000 N · s. This can be found by adding the area under each part of the curve, or the area of the two triangles plus the area of the rectangle. Help students to visualize this by showing them a graph similar to the one below. Note that the two triangles may be the same or different depending on how the gradually changing forces are shown.

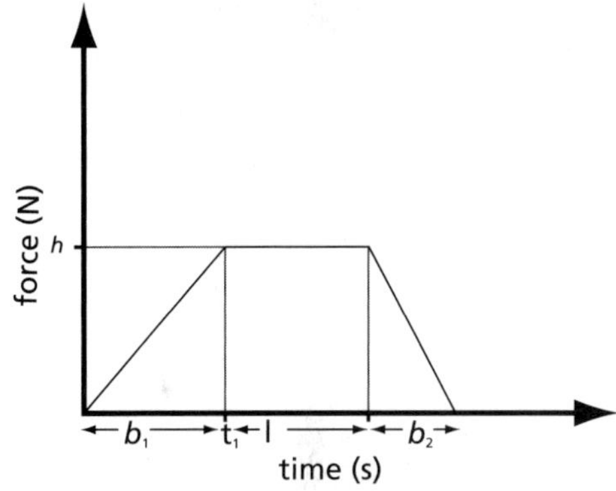

The simplest is the middle part in which a 4000 N force is applied for 1 s. This corresponds to a rectangle of height 4000 N and time 1 s on the graph. This force changed the momentum by $\Delta p = F\Delta t = (4000 \text{ N})(1 \text{ s}) = 4000 \text{ N} \cdot \text{s}$.

An area of 4000 N·s is under this part of the curve.

NOTES

The remaining two sections must add up with the middle section to account for the total change in momentum of 12,000 N. Therefore, together they need to reduce the momentum by 8000 N. If you assume that each gradual force reduces the momentum equally, then the first gradually applied force must reduce the momentum by 4000 N · s, as does the last. This can be found by using the relationship for the area of the triangle

$A = \frac{1}{2}bh$

$\Delta p_{\text{first part}} = \frac{1}{2}(t_1)(4000\text{ N})$

$4000\text{ N} \cdot \text{s} = (2000\text{ N})t_1$

$t_1 = 2\text{ s}.$

This means that the gradually increasing force must increase from 0 to 4000 N over a time period that needs to be calculated by figuring out the area under the curve, which is equal to the change in momentum during the time it is applied. This time is 2 s. An equal area will be needed for the gradually decreasing force applied in the last part, and an equal amount of time.

Point out that if the time the force is applied increases, the area under the curve increases, and if the time it is applied decreases, the area under the curve decreases.

The following image is for a more general case in which the two gradually changing forces applied are not changing equally. As long as the total area under the curve (the sum of the two triangles and the rectangle) are equal to the total change in momentum, then the curve correctly depicts a physical solution.

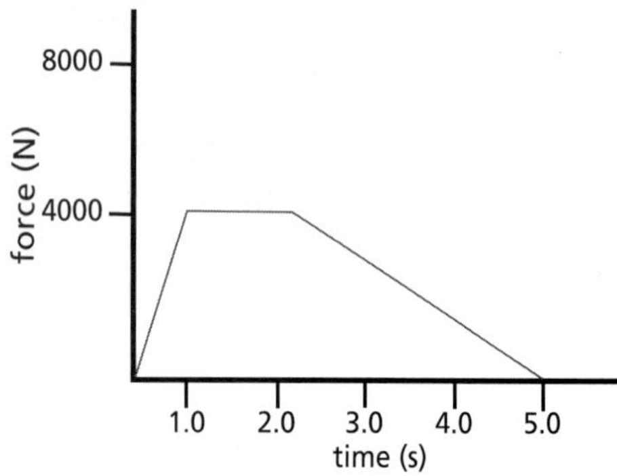

NOTES

Section 7 Impulse and Changes in Momentum: Crumple Zone

What Do You Think Now?

- **What are some of the factors that automobile designers and engineers must consider when designing a crumple zone as a safety feature?**

Now that you have completed this section on crumple zones, what do you think are some of the important considerations in designing a crumple zone? Compare and contrast crumple zones and air bags.

Physics Essential Questions

What does it mean?

What is the difference between impulse and momentum?

How do you know?

What were the key design features of your crumple zone and why were they important? What were the physics principles that you used?

Why do you believe?

Connects with Other Physics Content	Fits with Big Ideas in Science	Meets Physics Requirements
Forces and motion	Systems	* Good clear explanation, no more complicated than necessary

* The same physics principles can be applied to many situations. Newton's second law $(F = ma)$ can describe the effectiveness of air bags or crumple zones. Another form of Newton's second law $(F\Delta t = \Delta mv)$ provides other insights into the design of air bags or crumple zones. Show how the second equation can be derived from the first equation.

Why should you care?

How could the crumple zone concept be used in your safety system for the *Chapter Challenge?*

Reflecting on the Section and the Challenge

In this section, you found that a crumple zone, as you would find in bumpers or an air bag, is able to protect you by extending the time it takes to stop you. Without the air bags, you will hit something and stop in a brief time. This will require a large force, large enough to injure you. With the air bag (or other crumple device), the time to stop is longer and the force required is therefore smaller.

What Do You Think Now?

Ask students to review the previous answers to the *What Do You Think?* question. Ask students how they would answer this question now and survey the class for how many students changed their ideas. Point out that scientists often change their ideas as they gather more information. Consider discussing the information in *A Physicist's Response*.

Reflecting on the Section and the Challenge

Using the information in the student text, review crumple zones and the idea of reducing the force acting on passengers and energy transferred to passengers by having a smaller force act during a greater time during a collision, or having the force act over a greater distance. Emphasize that the crumple device reduces the amount of energy transferred to the passenger because more energy is transformed to heat

Physics Essential Questions

What does it mean?

Impulse and momentum will have the same mathematical value. Impulse is equal to the change in momentum. Impulse is the product of force and time. Momentum is the product of mass and velocity.

How do you know?

Students' crumple zones reduced the amount of force acting on passengers, and the energy transferred to passengers during a collision. Crumple zones decrease the impact felt by passengers during a crash by reducing the force needed to stop a vehicle.

Why do you believe?

For constant mass

$F = ma$

$F = m\dfrac{\Delta v}{\Delta t}$

$F\Delta t = m\Delta v = \Delta p$

Why should you care?

If an object is going to crash, adding a crumple zone will increase the time of the crash and decrease the force exerted during the crash.

energy or energy of deformation during the "crumpling." Remind students that the impulse-momentum and work-energy concepts are important to keep in mind for the *Chapter Challenge*, and that they should consider these concepts as they design their prototype and construct their explanations.

Force and impulse must be considered in designing your safety system. Stopping an object gradually reduces damage. The harder a surface, the shorter the stopping time, and greater the damage. In part, this provides a clue to the use of padded dashboards and sun visors in newer automobiles. Understanding impulse allows designers to reduce damages both to automobiles and passengers.

You can describe the decrease in force by using work (small force and large distance) or impulse (small force and large time). Which explanation to use depends on whether the stopping distance or stopping time is more easily measured. Work relates to changes in kinetic energy ($KE = \frac{1}{2}mv^2$). Impulse relates to changes in momentum ($p = mv$).

An automobile can be stopped in a very short time or over a longer time. In both cases, there is an identical change in momentum of the automobile. The impulse on the automobile is also identical in both cases since impulse is equal to the change in momentum. However, the potential damage to the automobiles and the passengers is not identical. A large force over a short time can produce severe damage to the automobile and injury to the driver and passengers. Therefore, you will want to consider how you can increase the time (and decrease the force) for your automobile to minimize dangers.

Similarly, an automobile can be stopped in a very short distance or over a longer distance. Although an automobile crashing into a snow bank or a highway barrier can lead to damage and injuries, they are not nearly as severe as what is experienced if the same automobile hits a concrete wall. In this case, you describe the work (force•distance) required to change the kinetic energy of the automobile.

When demonstrating your automobile's safety devices, you may use "momentum and impulse" or "work and kinetic energy" to provide a rationale for your design.

Physics to Go

1. How do impulse and Newton's first law (the law of inertia) play a role in your crumple-zone design?
2. Automobiles today have crumple zones designed into the body of the automobile, and they also have air bags inside the automobile. How do these systems work together to protect the passengers?
3. In automobiles built before 1970, the dashboard was made of hard metal. After 1970, the automobiles were installed with padded dashboards like you find in automobiles today. In designing a safe automobile, why is it better to have a passenger hit a cushioned dashboard than a hard metal dashboard?
   a) Using Newton's second law, explain why the padded dashboard is better.
   b) Using impulse and momentum, explain why a padded dashboard is better.
4. Explain why you bend your knees when you jump to the ground.

Physics to Go

1.

Students should describe how Newton's first law, as an object in motion, keeps its motion unless acted upon by a force. Before a collision, the vehicle and its occupants will keep their motion. During a collision, a force is needed to reduce both the vehicle's momentum and that of the occupants to zero. Students should describe how their crumple zone is designed to reduce the amount of force acting on passengers, and energy transferred to passengers during a collision. Then they should explain that the crumple zone increases the amount of time of the collision, but is designed to increase it by "safely" crumpling a part of the vehicle. This reduces the force needed to stop the vehicle. Hence, the impact the occupants feel is also less.

2.

Both the crumple zone and the air bag increase the stopping time of a passenger in a collision. The crumple zone increases the time for the automobile to stop, and the air bag increases the time for the passenger inside the vehicle to stop. The crumple zones reduce the amount of force acting on the vehicle during the collision and on the passenger. They also dissipate energy. The air bags further reduce the amount of force acting on the passenger during a collision and dissipate energy.

3.a)

With a padded dashboard, the increased stopping time means the acceleration is reduced, since the change in velocity is the same, but the time is longer. Using Newton's second law of motion, $F = ma$ means the force is lower if the acceleration is less.

3.b)

Because change in momentum is the same in both cases (the mass, and initial and final velocities, are the same), as the stopping time is increased through the use of a padded dashboard, the required stopping force decreases.

5. Helmets are designed to protect cyclists. How would the designers of helmets make use of the concept of impulse to improve their effectiveness?
6. An automobile has a mass of 1200 kg and an initial velocity of 10 m/s (about 20 mi/h). Calculate the change in momentum required to do the following:
   a) Bring it to rest
   b) Slow it to 5 m/s (approximately 10 mi/h)
7. If the braking force for an automobile is 10,000 N, calculate the impulse if the brake is applied for 1.2 s (seconds). If the automobile has a mass of 1200 kg, what is the change in velocity of the automobile over this 1.2 s time interval?
8. A 1500-kg automobile, traveling at 5.0 m/s after braking, strikes a power pole and comes to a full stop in 0.1 s. Calculate the force exerted by the power pole and brakes required to stop the automobile.
9. For the automobile described in *Question 8*, explain why a breakaway pole that brings the automobile to rest after 2.8 s is safer than the conventional power pole.
10. Compare and contrast the two force vs. time graphs shown below.

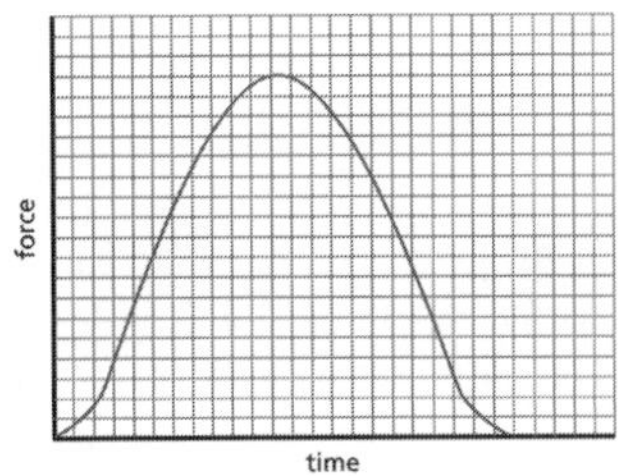

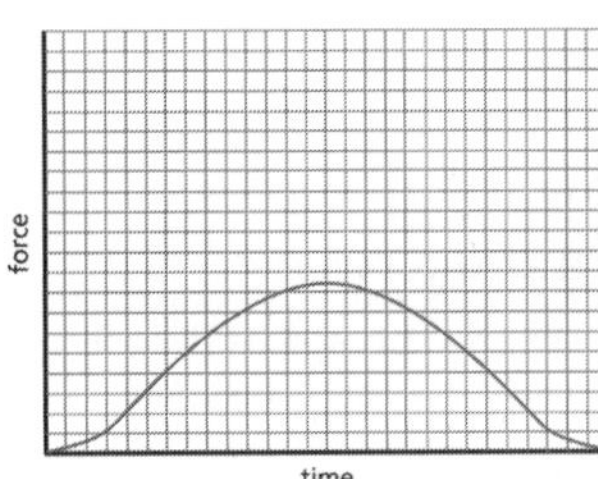

11. ***Preparing for the Chapter Challenge***

    How can your safety device reduce the force experienced during an impulse? What features of your device increase the stopping time for the passenger? Record a description of how these features work in terms of impulse and momentum.

4.

Bending your knees when you land on the ground increases the stopping time as you collide with the ground, and therefore, the stopping force exerted on you by the ground is decreased.

5.

Helmets use cushioning materials that lengthen the time of impact and thus, decrease the force of impact. Note that helmets are designed to break and should not be used again after an accident. A rigid, unbreakable helmet would be useless.

6.a)

$\Delta p = p_{\text{final}} - p_{\text{initial}} =$

$mv_{\text{final}} - mv_{\text{initial}} =$

$12{,}000\ \text{kg}(0) - 12{,}000\ \text{kg}(10\ \text{m/s}) =$

$-120{,}000\ \text{kg} \cdot \text{m/s}.$

6.b)

$\Delta p = p_{\text{final}} - p_{\text{initial}} =$

$mv_{\text{final}} - mv_{\text{initial}} = m\left(v_{\text{final}} - v_{\text{initial}}\right) =$

$12{,}000\ \text{kg}\left(5\ \text{m/s} - 10\ \text{m/s}\right) =$

$-60{,}000\ \text{kg} \cdot \text{m/s}.$

7.

Students should use the impulse-momentum concepts. Here the direction of motion of the vehicle is taken as positive. This means the force acts on the vehicle in the negative direction. Consider asking students what direction the force acts relative to the motion of the vehicle.

$\Delta p = F\Delta t$

$m\Delta v = (-10{,}000\ \text{N})(1.2\ \text{s})$

$(1200\ \text{kg})\Delta v = -12{,}000\ \text{N} \cdot \text{s}$

$\Delta v = -10\ \text{m/s}.$

CHAPTER 3

8.

Using the impulse-momentum concepts,

$F\Delta t = \Delta p$

$F\Delta t = m\Delta v = (1500 \text{ kg})(-5 \text{ m/s})$

$F(0.1 \text{ s}) = (-7500 \text{ kg} \cdot \text{m/s})$

$F = (-75{,}000 \text{ N})$

The negative sign indicates that the force acts opposite to the automobile's direction of motion. Consider asking students the direction of the force relative to the direction of the automobile.

9.

Increasing the time of the collision decreases the force exerted on the vehicle during the collision, even though the change in momentum is the same. This can be verified by calculating the value of the force for this collision.

$F\Delta t = \Delta p$

$F\Delta t = m\Delta v = (1500 \text{ kg})(-5 \text{ m/s})$

$F(2.8 \text{ s}) = (-7500 \text{ kg} \cdot \text{m/s})$

$F = (-2700 \text{ N})$

10.

Students should use the impulse-momentum ideas in their comparison. They should point out that the first graph has a greater force over a shorter time period. The second graph has a smaller force over a longer period of time, and that the area under each graph is equal to the impulse or change in momentum.

11.

Preparing for the Chapter Challenge

Students should develop a description of their safety device, and explain how it will increase stopping time and decrease stopping force to provide the impulse required to stop an automobile. Students should realize that the impulse required to stop the vehicle will be the same since the vehicle's mass and initial velocity are not changing. Their safety device is concerned with decreasing the force required, not the impulse.

NOTES

SECTION 7 QUIZ

3-7c **Blackline Master**

1. Three groups of students are testing their designs for cushioning collisions and obtain the force vs. time graphs shown below. If the areas under all three graphs are equal, which group's design worked best at cushioning the collision?

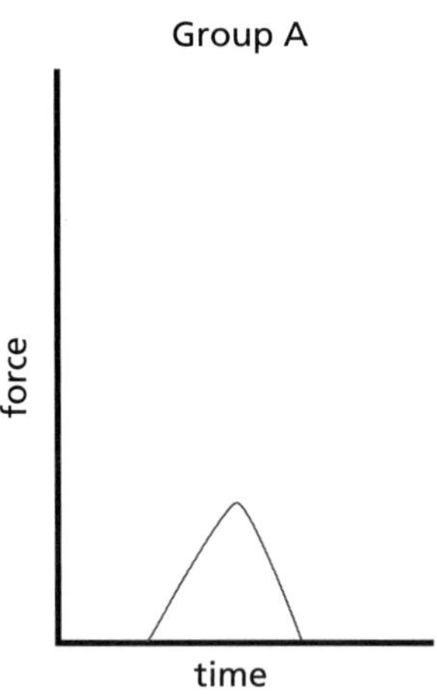

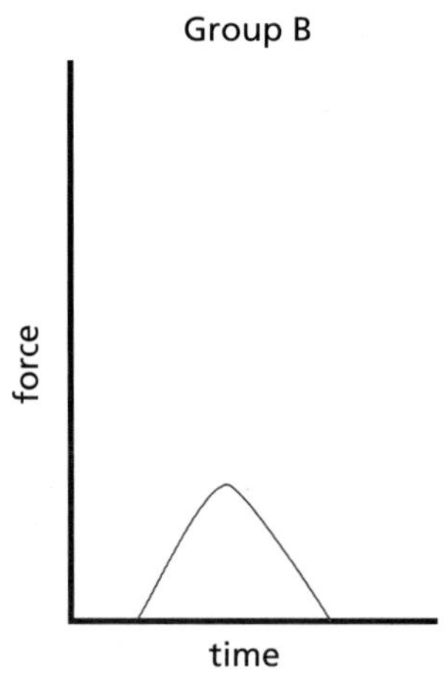

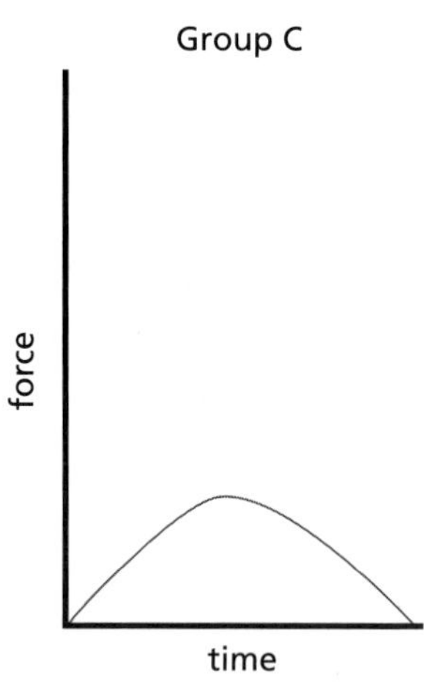

a) Group A
b) Group B
c) Group C
d) All three groups were equal

2. In *Question 1*, which group's cart had the greatest change in momentum when the three carts struck the force probe?

a) Group A
b) Group B
c) Group C
d) All three carts were equal

3. A student drops two eggs of equal mass simultaneously from the same height. Egg A lands on the tile floor and breaks. Egg B lands intact without bouncing on a foam pad lying on the floor. Compared to the impulse exerted to stop egg A as it lands on the tile floor, the impulse exerted to stop egg B is

a) less.
b) greater.
c) the same.

4. In an automobile collision, a 44-kg passenger moving at 15 m/s is brought to rest by an air bag in 0.10 s. What is the average force the air bag exerts on the passenger?

a) 440 N
b) 4400 N
c) 660 N
d) 6600 N

5. A 1.0-kg mass changes speed from 2.0 m/s to 5.0 m/s. What is the mass's change in momentum?

a) 9.0 kg · m/s
b) 21 kg · m/s
c) 3.0 kg · m/s
d) 29 kg · m/s

CHAPTER 3

SECTION 7 QUIZ ANSWERS

1. c) Group C
2. d) All three carts were equal
3. c) the same.
4. d) 6600 N
5. c) 3.0 kg · m/s

NOTES

NOTES

Chapter Assessment

Physics You Learned

A quick reference for the physics content supporting the *Learning Outcomes* of this chapter is provided in the *Physics You Learned*. Consider reviewing and summarizing the material, and evaluating students' understanding. By assessing your students' understanding, you can determine if any content needs to be reviewed before students complete their challenge and practice test. Students can review as a class, in groups, or individually the concepts in the table.

Describe to students how they can use the information in *Physics You Learned* as a study guide. Suggest to students that they quiz and assess each other by preparing questions based on the concepts listed in the table. Consider having student groups work together to review, quiz, and assess their understanding of the concepts presented. Cooperative learning strategies are useful in helping students refresh their knowledge of equations and science vocabulary as well as reinforcing the basic concepts developed in the chapter. Point out to students that reviewing the physics concepts and equations will help them prepare for the *Chapter Challenge*, and prepare for the *Physics Practice Test*.

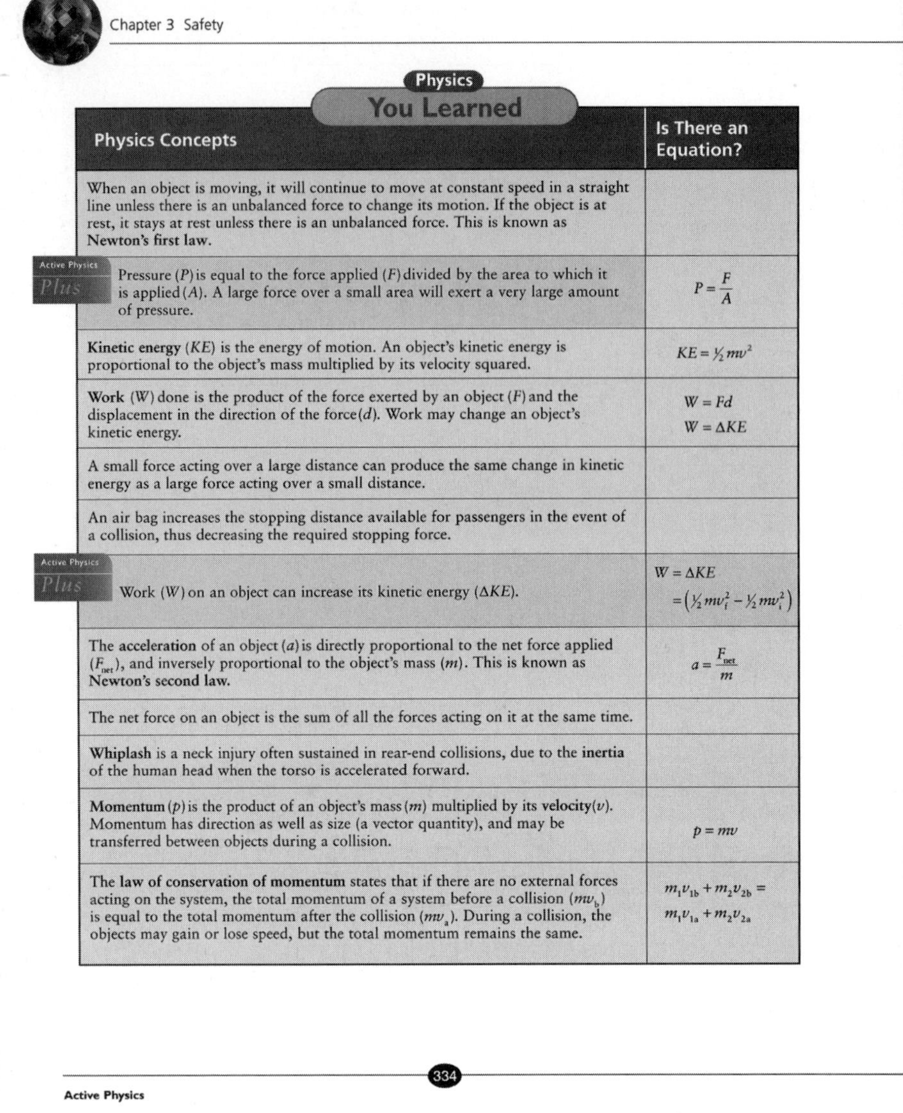

Chapter 3 Safety

Physics You Learned

Physics Concepts	Is There an Equation?
When an object is moving, it will continue to move at constant speed in a straight line unless there is an unbalanced force to change its motion. If the object is at rest, it stays at rest unless there is an unbalanced force. This is known as **Newton's first law.**	
Active Physics Plus: Pressure (P) is equal to the force applied (F) divided by the area to which it is applied (A). A large force over a small area will exert a very large amount of pressure.	$P = \frac{F}{A}$
Kinetic energy (KE) is the energy of motion. An object's kinetic energy is proportional to the object's mass multiplied by its velocity squared.	$KE = \frac{1}{2}mv^2$
Work (W) done is the product of the force exerted by an object (F) and the displacement in the direction of the force (d). Work may change an object's kinetic energy.	$W = Fd$ $W = \Delta KE$
A small force acting over a large distance can produce the same change in kinetic energy as a large force acting over a small distance.	
An air bag increases the stopping distance available for passengers in the event of a collision, thus decreasing the required stopping force.	
Active Physics Plus: Work (W) on an object can increase its kinetic energy (ΔKE).	$W = \Delta KE$ $= \left(\frac{1}{2}mv_f^2 - \frac{1}{2}mv_i^2\right)$
The **acceleration** of an object (a) is directly proportional to the net force applied (F_{net}), and inversely proportional to the object's mass (m). This is known as **Newton's second law.**	$a = \frac{F_{net}}{m}$
The net force on an object is the sum of all the forces acting on it at the same time.	
Whiplash is a neck injury often sustained in rear-end collisions, due to the **inertia** of the human head when the torso is accelerated forward.	
Momentum (p) is the product of an object's mass (m) multiplied by its **velocity** (v). Momentum has direction as well as size (a vector quantity), and may be transferred between objects during a collision.	$p = mv$
The **law of conservation of momentum** states that if there are no external forces acting on the system, the total momentum of a system before a collision (mv_b) is equal to the total momentum after the collision (mv_a). During a collision, the objects may gain or lose speed, but the total momentum remains the same.	$m_1v_{1b} + m_2v_{2b} =$ $m_1v_{1a} + m_2v_{2a}$

334

Active Physics

Encourage students to use this table as a quick reference guide of important concepts and as a checklist of the physics concepts that should be incorporated into their *Chapter Challenge*.

Active Physics Plus — In an elastic collision, both momentum and kinetic energy are conserved. The sum of the kinetic energies after the collision must equal the sum of the kinetic energies before the collision. When two particles of equal mass collide and one is initially at rest, they must travel off at right angles after the collision for this condition to be met.	$p_{before} = p_{after}$ $KE_{before} = KE_{after}$
Impulse equals the product of a force acting on an object, and the time period during which the force acts.	$\text{Impulse} = F\Delta t$
When an impulse acts on an object, the momentum of the object changes by an amount equal to the applied impulse.	$F\Delta t = m\Delta v$
Active Physics Plus — The area under a force vs. time graph is the impulse and therefore, equal to the change in momentum of the object.	
Crumple zones are built into automobiles as cushioning devices. A crumple zone increases the time a force may act to bring an automobile to rest, which allows a smaller force to be exerted during the stopping process. The net impulse required to stop an automobile does not change if the automobile has a crumple zone, but the net force applied is decreased as the time it acts is increased.	

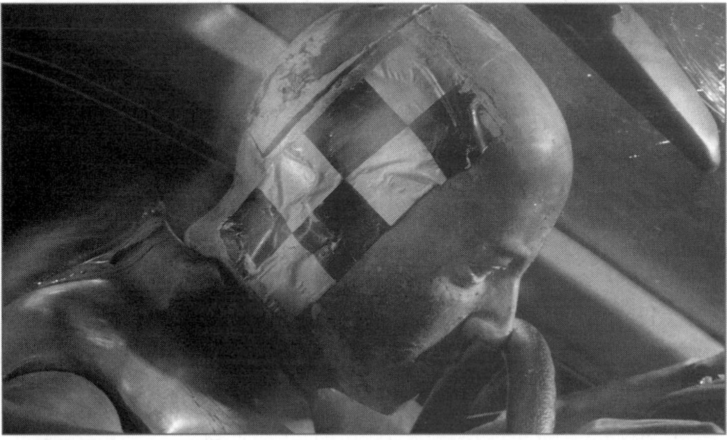

CHAPTER 3

Physics Chapter Challenge

Students review the *Goal* and the additional *Inputs* as they repeat the *Engineering Design Cycle* during the *Physics Chapter Challenge*. Remind students of previous class discussions during the chapter opener and the *Chapter Mini-Challenge*. Review each part of the cycle and how it pertains to this *Chapter Challenge*. Consider using the *Black Line Master* of the *Engineering Design Cycle* as a focal point of the discussion.

Discuss the *Goal* with the class—to create a safety system to protect passengers in a vehicle of their choice, and to build a model of the safety system that will be tested in the classroom using an egg as the passenger. Emphasize that each group must create a written or multimedia report and present an oral presentation to the class containing their research, investigation results, and explanations of the design chosen and how it works using the physics concepts presented in this chapter. Utilize the class's rubric during the discussion and after the review, and revise it if needed.

Chapter 3 Safety

Physics

Chapter Challenge

You will now be completing a second cycle of the *Engineering Design Cycle* as you prepare for the *Chapter Challenge*. The goals and criteria remain unchanged. However, your list of *Inputs* has grown.

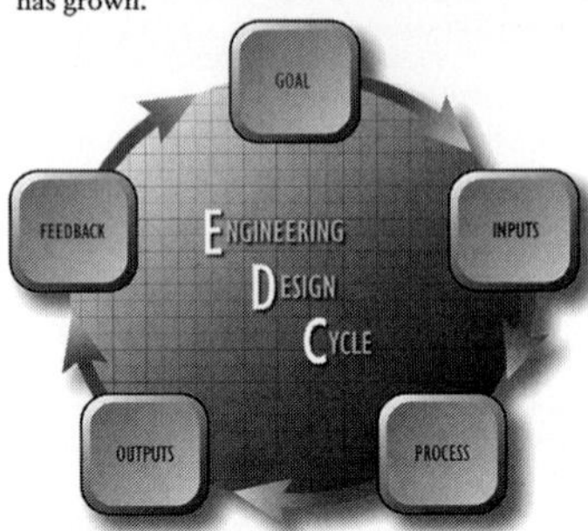

Goal

Your challenge for this chapter is to create a safety system to protect passengers in a vehicle of your choice. As part of your design, you will build a model to protect an egg in a collision that you will enact in the classroom during your oral presentation. You will also submit a written or multimedia report of your research and investigation results.

Inputs

You have completed all the sections of this chapter and learned the physics content you need to complete this challenge. You now have additional physics information to help you optimize the design of your safety system. Remember, you will also be protecting an egg using your new physics knowledge. This is part of the *Inputs* phase of the *Engineering Design Cycle*.

Your group must define the vehicle that your safety device will be used in and apply the appropriate physics concepts to build your presentation. The type of vehicle will determine some key constraints, like the typical speed of the vehicle, the mass and velocity of typical obstructions your vehicle may collide with. The other key *Inputs* for this challenge will be the physics principles you have learned from each section of the chapter.

Section 1 You considered the topic of passenger safety in automobile collisions and explored some of the safety features that are available in modern automobiles. You learned that analyzing safety has led to many safety improvements in vehicle design.

Section 2 You learned about Newton's first law of motion and the role it plays in collisions. You also evaluated different seat belt materials and considered the relationships of force, surface area, and pressure.

Section 3 You built a model of a vehicle air bag. You used the ideas of energy and work to help explain why an air bag can be helpful to passengers in the event of an collision.

Section 4 You explored what happens in rear-end collisions and learned about Newton's second law, which describes the relationship between the force applied to a mass and the acceleration it will experience. You recognized how this law explains the way objects respond when forces are applied to them.

Section 5 You learned about momentum through investigating staged collisions. The momentum of an object depends on both its mass and its velocity, which is why speed and size are both factors in the outcome of an automobile accident.

Active Physics 336

Emphasize to students that they will be analyzing and applying the *Inputs* in the design of their safety system. Review the *Inputs* as a class or in student groups, having students summarize the *Inputs* from each section and recording in their logs the key physics concepts. Remind the class that it is important to also include the *Feedback* they received from the *Mini-Challenge*. As student groups or the class review the *Feedback*, have them provide examples of how it should be used to complete the challenge. This sharing of ideas leads to the *Process* phase, in which students determine their design and the information needed for their reports and presentations.

Discuss the *Process* phase with the class, emphasizing that organization and communication between group members is very important. Ask the class for ideas for keeping groups organized and communicating well, keeping in mind their *Mini-Challenge* experience. Then ask them how they might incorporate these into a procedure that their group can follow. Let the class know that during the *Process* phase, each student in a group should contribute toward the creative process that is developed during this step of

Section 6 You learned about the conservation of momentum, which establishes that the total amount of momentum in a system remains the same. This concept can be used to determine the momentum vehicles have after a collision when no external forces are applied.

Section 7 You compared changes in momentum to the forces applied to objects during a collision. Impulse, or change in momentum, can determine the amount of force applied if you know how much time the change took. You can make the force smaller if you make the time of collision longer!

Process

In the *Process* phase, your group must decide what information you will use to meet the *Goal*. Choose the vehicle that will be the model for your safety system. Your group may brainstorm ideas for many different vehicles. Once you have lots of ideas to choose from, select one that your group agrees is workable and that will also allow you to protect an egg in the classroom collision.

Organization and good communication among your group members will be very important for this challenge. One way to stay organized is to assign roles for each person and make a list of the responsibilities. For instance, one person might have the role of "fact checker" and be responsible for making sure the oral report and written report contain accurate information about the physics principles your safety system applies. Other roles might include model builder, scriptwriter, report writer, and so on. Time constraints are also going to make this challenge difficult, so make sure each person knows when his or her portion of the project must be completed.

When you build your model system to protect an egg, be sure to create a safety system that uses the principles you have chosen to protect human passengers. You may have some constraints, such as time, available materials, or even the shape of the cart your egg will ride on. Engineers are constantly working to meet design goals within constraints.

Make sure that your report or presentation explains why your safety system works and contains the information you want your audience to know. You should include all of the key features of your design along with example calculations and results for all of your safety system. You may also describe differences between your human safety system and your egg safety system. If your class prepares a rubric for this challenge, make sure you refer to it often to ensure that you address each category and include all important information.

Outputs

Presenting your information to the class are your design-cycle *Outputs*. It is very important that your egg safety system be effective since it is the model for your design, so make sure your demonstration is well rehearsed. Each piece of the presentation will have similar information, but it is important to make sure each one is complete.

Feedback

Your classmates will give you *Feedback* on the accuracy and the overall appeal of your presentation based on the criteria of the challenge. This feedback will likely become part of your grade but could also be useful for additional design iterations. Remember that there is always room for some improvement and no design is perfect. From your experience with the *Mini-Challenge*, you should recognize how it is possible to continuously refine any idea by constantly rotating through the design cycle.

Active Physics

the *Engineering Design Cycle*. For a successful implementation of group work, individual tasks should be thoughtfully assigned. Asking students to plan their own tasks and demonstrate how they contributed toward their presentation enables them to be personally responsible. Discuss the constraints that groups may have based on the vehicle they choose for their safety system. Have students consider their *Mini-Challenge* experience, refining the techniques they used in their previous design and presentation. Point out the importance of a rubric and how students can use it to determine whether they are meeting the different criteria of the *Chapter Challenge*.

Reiterate to students that for the *Outputs* phase, an effective presentation depends on how accurately they convey the physics concepts behind their safety system in both their written report and class presentation. Remind students of the time allowed for their presentations and prepare your class for interactive *Feedback* that is constructive.

After reminding students of the design cycle, review and revise the assessment criteria. As a class, develop a comprehensive assessment criteria. Emphasize to the class that the rubric will be used to determine their grade. A *Sample Assessment Rubric* is provided at the end of this *Teacher's Edition*, which may be used as a foundation for developing scoring guidelines and expectations that suit your needs. For example, you might want to ensure that the core concepts and abilities from your local or state science standards also appear in the rubric. However, if you decide to evaluate the *Chapter Challenge*, be sure that the students actively participate in deciding both the criteria for evaluation and the guidelines for scoring. Make sure that students understand how they will be evaluated before they start their work, and that they actively participate in deciding on the criteria for evaluation.

Physics Connections to Other Sciences

Students gain understanding of interdisciplinary interactions when they are actively engaged in thinking about connections between scientific disciplines. This section provides a glimpse of the interconnectedness between the physics concepts presented in this chapter and the scientific disciplines of biology, chemistry, and Earth science. Brief sketches relate students' study of physics concepts to biology, chemistry, and Earth science.

Encourage students to draw analogies with science connections they are familiar with while discussing the science connections. The science concepts discussed in this chapter must hold true for every branch of science, and can provide scientists in other fields with new insights. Emphasize the growing interdisciplinary approach to science and the need for scientists to understand the fundamentals of forces, momentum, and energy. Discuss how scientists try to understand phenomena by studying interactions using either a force approach and applying ideas such as the conservation of momentum law that involve vector quantities, or an energy approach which involves scalar quantities. Encourage appreciation for physics in relation to the broader framework of science by describing the increase in demand for scientists with interdisciplinary backgrounds. (For example, geophysics, biochemistry, and biophysics.)

Consider developing an interdisciplinary lesson plan that investigates how these *Key Physics Concepts* of force, momentum, and energy are applied in different sciences. Groups could select an object and determine how it relates to biology, chemistry, and Earth science based on their reading of *Physics Connections to Other Sciences*, and how it relates to the *Key Physics Concepts* of force, momentum, and energy. A set of questions could be constructed for students to focus on to build a constructive inquiry. Each group member could write down the highlights of their discussion. Once students have recorded the focal points of their discussion, bring the whole class together and have a student volunteer from each group to share their knowledge of important science connections.

Physics Connections to Other Sciences

Here are some examples of how the concepts you studied in this chapter relate to other sciences.

Newton's First Law – Inertia

Biology Just as an object's motion does not change until a force comes along to change it, a biological species does not change until something changes its genetic makeup. This is the basis of natural selection which leads to the biological evolution of species.

Chemistry The electric polarity of a molecule in a liquid can be demonstrated by the ability of an electrically charged rod to deflect the path of a falling stream of the liquid.

Earth Science If gravity were turned off, all the planets presently orbiting the Sun would continue moving with the same speed, along a line in the direction of their present motion.

Energy and Work

Biology The work done by the tail of a fish as it pushes against the water is responsible for the kinetic energy of a fish as it swims.

Chemistry When heat energy is added to a gas, such as the steam in a steam engine, the gas expands and does work moving a piston. The movement of the piston can then be converted into the kinetic energy of a moving train.

Earth Science Stray particles in the Solar System are constantly bombarding Earth. Smaller particles are stopped over long distances by their frictional contact with gases in the atmosphere. Larger objects that eventually collide with Earth must be stopped in much shorter distances by the ground. This requires huge forces that cause terrible damage if the object is sufficiently large, as was the case with the meteor colliding with the Earth 65 million years ago, believed to have been responsible for making dinosaurs extinct.

Newton's Second Law

Biology The ability of all animals to accelerate and change their velocity depends upon their mass and the net force they can exert.

Chemistry The mass of an ionized atom can be determined by the acceleration that it undergoes when it is placed in an electric field.

Earth Science How quickly rock particles that are picked up by a swiftly flowing stream settle out of the water is determined by the net force on each particle, and the particle's mass. This leads to a sorting of rock particles as the stream runs into a lake or ocean.

Momentum

Biology When an eagle grabs a fish swimming near the top of a lake, the eagle will slow down after the catch, due to conservation of momentum. If the eagle slows too much or if the fish is too large, it may not be able to regain momentum, and may land in the water.

Chemistry The momentum of an atom when it collides with another atom is one factor that determines whether the atoms combine to form a molecule, or simply bounce off each other.

Earth Science The tremendous momentum of a large mass of snow in an avalanche allows it to knock down anything that stands in its path.

Impulse

Biology When animals jump from a height down to the ground, they bend their legs. This results in the exertion of a smaller force over a larger time and prevents injury when they land.

Chemistry The impulse delivered to a wall when gas molecules strike is responsible for the pressure the gas exerts upon the wall.

Earth Science The power of erosion of a waterfall is much greater than the power of a stream flowing down a gently sloping hill from the same height. Stream water stops over a much longer time, requiring less force. Water at the bottom of a waterfall stops quickly, requiring a large force with a greater effect.

Physics At Work

Joe Nolan

Senior Vice President, Vehicle Research Center; Ruckersville, VA

Joe Nolan realizes his career, much like automobile accidents, was not planned. While earning his graduate degree in engineering, he realized he needed a research position to graduate. "The only research position available was at the Automobile Safety Lab at the University of Virginia's School of Engineering," said Nolan. Today, he is the senior vice president at the Vehicle Research Center for the Insurance Institute for Highway Safety, located in Ruckersville, Virginia.

The Insurance Institute for Highway Safety is a nonprofit, nongovernmental research organization. It is funded by insurance companies and performs over 100 full-scale crash tests per year. The Vehicle Research Center currently reproduces three different real-world crashes to test a vehicle's safety: frontal offset collisions, side impacts, and rear-end whiplash tests. "In the front and side tests, we observe the way the dummies move in the car, the way the restraint system works, and how the car's structure holds up," he said.

While automobile manufacturers continue to upgrade their safety features, Nolan believes that new drivers still need to be more responsible behind the wheel. "Car crashes are the number one cause of death for people under the age of 18, even though vehicles are getting better and belt use is going up," argues Nolan.

Marjorie Cooke

Marine Safety Expert, Robson Forensic; Fairfax County, VA

Dave Cooke

Professional Engineer, Robson Forensic; Fairfax County, VA

Dave and Marjorie Cooke met at the State University of New York Maritime College, and have been working together for over 30 years. According to the couple, their job is never routine. "We may be inspecting a ship or boat to determine its condition, or we may be interviewing someone to determine what happened during the incident."

The couple's understanding of physics is vital during an investigation. "You have to be able to determine what direction and the amount of force applied to various components. In boating accidents, there are no skid marks to help you out."

Michael Jackson

Emergency Medical Technician (EMT); North Apollo, PA

Michael Jackson is an Emergency Medical Technician (EMT) at Allegheny Valley Hospital, located in Natrona Heights, PA. A typical day involves responding to a variety of emergencies, from shortness of breath and chest pains, to automobile accidents. The most severe automobile accident Jackson responded to in his seven-year career was a vehicle that had gone over a hillside. "The passengers were young adults, and luckily, both survived," said Jackson.

Jackson believes that most car accidents are caused by speeding, especially when road conditions are poor. "So many accidents could be prevented if people would drive at safe and appropriate speeds for the current road conditions."

339

Active Physics

Physics At Work

This section provides examples of how the physics concepts presented in this chapter are applied in the real world. The examples support the relevance of the *Chapter Challenge* as a real-world application. Emphasize how the investigations performed in the chapter provided hands-on experience of the science concepts needed to complete the challenge. From the examples, discuss how physics is embedded in our everyday lives, and how it helps us to analyze accidents, design safety systems to keep us safe, and create driving rules. Emphasize to the class that there are no geographic, ethnic, or gender barriers for people who apply physics in their professions. It is not necessary to have an advanced degree or be a scientist to employ physics. The different profiles presented show how people from diverse backgrounds use the physics concepts from this chapter and vehicle safety in their lives and their jobs. Discuss with the class each profile.

Discuss how Joe Nolan, Senior Vice President of the Vehicle Research Center, was introduced to his career while earning his graduate degree in engineering. Describe how his company conducts over 100 crash tests on vehicles per year with crash-test dummies. Emphasize that the crash tests they perform are analyzed using the same physics principles presented in this chapter. Ask students what some of those principles are and how they are used to analyze a crash. Point out that insurance companies fund the research conducted and ask the class why insurance companies might be interested in crash studies.

Describe how Marjorie and Dave Cooke investigate boat accidents and determine the condition of boats and ships. Emphasize how vital the physics concepts in this chapter are to their work. Discuss the last quote in the student text and ask students what information they might use to help them analyze a boating accident, since no skid marks are left on the water. Ask students to compare what they think the safety features are on a boat to those found in a car.

Connect observations made by the emergency medical technician, Michael Jackson, with the need for safety devices and appropriate driving laws.

Physics Practice Test

The *Physics Practice Test* is provided as a *Blackline Master* on your *Teacher Resources CD*.

3b Blackline Master

Content Review

Have students take the practice test to evaluate their understanding. Students should use their results in conjunction with the checklist to evaluate and review their understanding of the physics concepts.

1. b

2. d

Pushing on a dull and a sharp knife with equal force distributes that force to the place of contact of the edge. The smaller the area of the edge, the greater the pressure (or force per area) at the edge, and hence, the better it cuts.

3. c

4. b

5. d

This question applies the $v^2 = 2ad$ rule, where the stopping distance increases as the square of the velocity. Since the velocity has increased by a factor of 4, from 5 m/s to 20 m/s, the stopping distance will increase by a factor of 4 squared or 16. Thus, the new stopping distance will be 16 times larger than the previous stopping distance or 16 m × 2, which equals 32 m.

Chapter 3 Safety

Physics Practice Test

Before you try the Physics Practice Test, *you may want to review Sections 1–7, where you will find* 24 Checking Up *questions,* 9 What Do You Think Now? *questions,* 28 Physics Essential Questions, 70 Physics to Go *questions, and* 7 Inquiring Further *questions.*

Content Review

1. When an elevator going down comes to a stop, blood tends to rush from the occupants' heads. This phenomenon is best explained by
   a) conservation of energy.
   b) Newton's first law of motion.
   c) action-reaction.
   d) the law of universal gravitation.

2. A dull knife does not cut as well as a sharp knife when pushed equally hard. A sharp knife cuts better because it
   a) requires more energy.
   b) continues in motion with constant speed.
   c) has a longer edge.
   d) exerts greater pressure.

3. Two eggs are dropped from equal heights. Egg A lands on a hard floor, while egg B lands on a soft foam pad. Both eggs stop without bouncing, and egg A breaks while egg B does not. Compared to egg A, egg B has
   a) less work done on it, and requires less force to stop.
   b) more work done on it and requires less force to stop.
   c) the same amount of work done on it and requires less force to stop.
   d) less work done on it and requires more force to stop.

4. An egg thrown at a sheet at high speed is stopped and does not break. The best explanation is that
   a) the sheet is made of a soft material and cannot break the egg.
   b) the sheet exerts a small force over a large distance to stop the egg.
   c) the sheet is not held tightly.
   d) the kinetic energy of the thrown egg is not enough to cause the egg to break.

5. An automobile traveling at a speed of 5 m/s brakes to a stop at a distance of 2 m. What distance would be required by the same braking force to stop the automobile at a speed of 20 m/s?
   a) 8 m b) 16 m
   c) 24 m d) 32 m

6. As the distance an automobile has available to stop increases, the force required to bring it to a stop decreases. Which physics principle best explains this?
   a) work and energy
   b) conservation of momentum
   c) Newton's first law
   d) pressure depends upon the area over which a force is exerted

7. In the design of some modern automobiles, each headrest automatically snaps forward against the back of a passenger's head if the vehicle is struck from behind, pushing forward with the same acceleration as the rest of the vehicle. This safety device protects a passenger's head by overcoming its
   a) inertia. b) kinetic energy.
   c) momentum. d) impulse.

8. Which best describes the forces acting on an automobile moving at constant speed?
   a) There is a constant unbalanced force on the automobile, pushing it forward.
   b) There are no forces acting on the automobile.
   c) All the forces acting on the automobile add up to zero.
   d) There is a small net force on the automobile to keep it moving.

9. Which object would have the greatest momentum?
   a) a 5-kg bowling ball moving at 7 m/s
   b) a 0.4-kg bird flying at 30 m/s
   c) a 20-kg wheelbarrow moving at 2 m/s
   d) a 0.3-kg baseball batted at 60 m/s

Active Physics 340

6. a

7. a

8. c

9. c

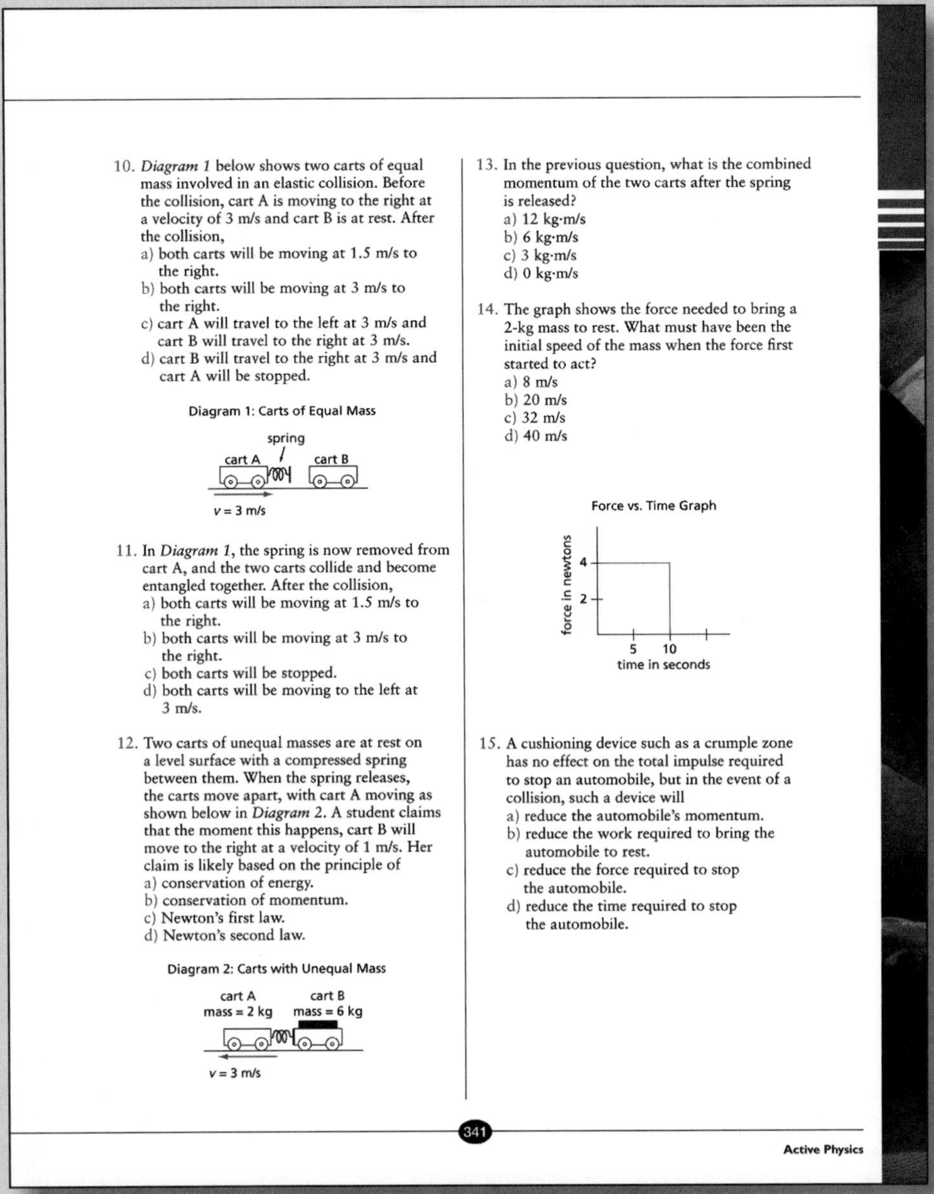

10. *Diagram 1* below shows two carts of equal mass involved in an elastic collision. Before the collision, cart A is moving to the right at a velocity of 3 m/s and cart B is at rest. After the collision,
a) both carts will be moving at 1.5 m/s to the right.
b) both carts will be moving at 3 m/s to the right.
c) cart A will travel to the left at 3 m/s and cart B will travel to the right at 3 m/s.
d) cart B will travel to the right at 3 m/s and cart A will be stopped.

Diagram 1: Carts of Equal Mass

spring
cart A
cart B
v = 3 m/s

11. In *Diagram 1*, the spring is now removed from cart A, and the two carts collide and become entangled together. After the collision,
a) both carts will be moving at 1.5 m/s to the right.
b) both carts will be moving at 3 m/s to the right.
c) both carts will be stopped.
d) both carts will be moving to the left at 3 m/s.

12. Two carts of unequal masses are at rest on a level surface with a compressed spring between them. When the spring releases, the carts move apart, with cart A moving as shown below in *Diagram 2*. A student claims that the moment this happens, cart B will move to the right at a velocity of 1 m/s. Her claim is likely based on the principle of
a) conservation of energy.
b) conservation of momentum.
c) Newton's first law.
d) Newton's second law.

Diagram 2: Carts with Unequal Mass

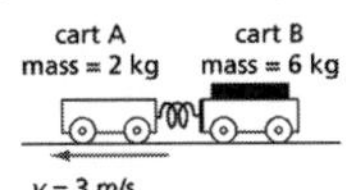

13. In the previous question, what is the combined momentum of the two carts after the spring is released?
a) 12 kg·m/s
b) 6 kg·m/s
c) 3 kg·m/s
d) 0 kg·m/s

14. The graph shows the force needed to bring a 2-kg mass to rest. What must have been the initial speed of the mass when the force first started to act?
a) 8 m/s
b) 20 m/s
c) 32 m/s
d) 40 m/s

Force vs. Time Graph

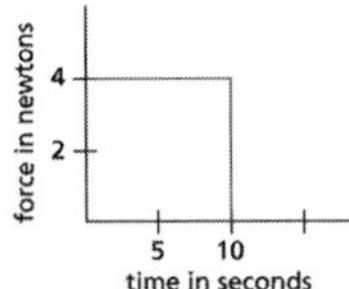

15. A cushioning device such as a crumple zone has no effect on the total impulse required to stop an automobile, but in the event of a collision, such a device will
a) reduce the automobile's momentum.
b) reduce the work required to bring the automobile to rest.
c) reduce the force required to stop the automobile.
d) reduce the time required to stop the automobile.

10. d

11. a

This is a perfectly inelastic collision. Momentum is conserved, but energy is not. Students should realize that the final velocity of the carts must be the same as the initial. Below, the positive direction is chosen to the left.

$$p_{\text{initial}} = p_{\text{final}}$$

$$mv_{\text{initial A}} + mv_{\text{initial B}} = 2mv_{\text{final}}$$

$$m(3 \text{ m/s}) + 0 = 2mv_{\text{final}}$$

$$v_{\text{final}} = \frac{m(3 \text{ m/s})}{2m} = 1.5 \text{ m/s}.$$

12. b

13. d

Students should reason this by noting that the total momentum of the system is conserved. Because the momentum of the system before the event was zero, it must be zero after the event.

14. b

Using the impulse-momentum theory, students should be able to calculate the initial speed.

$$F\Delta t = m\Delta v$$

$$(4 \text{ N})(10 \text{ s}) = (2 \text{ kg})\Delta v,$$

$\Delta v = 20$ m/s, indicating the initial velocity points in the direction opposite to the applied force.

15. c

CHAPTER 3

Practice Test *(continued)*

Critical Thinking

16. In an "elastic collision," the total kinetic energy as well as the total momentum of colliding objects before and after a collision is the same. Design an experiment to determine if momentum is conserved in an elastic collision between two carts of unequal mass that have springs between them. You should include a description of the equipment setup, the procedure to follow, and the equipment needed to complete the investigation.
    a) What measurements should you take to determine if momentum has been conserved during the collision?
    b) How would you analyze the data to confirm that momentum was conserved?
    c) Would you need to take any additional measurements to determine if the kinetic energy before the collision was equal to the kinetic energy after the collision? If so, what would those measurements be?

17. Imagine you are riding in an automobile and the following events occur. Describe which of Newton's laws applies to each event.
    a) A large box in the trunk slides forward each time the brakes are applied.
    b) You feel you are "pushed back" into the seat when the automobile accelerates.
    c) When you come to a quick stop, your seat belt stops you from moving forward.

18. Automobiles have crumple zones and air bags to protect passengers. Using the concept of impulse equals the change in momentum, explain how these two systems work together to decrease the forces exerted on a passenger in the event of a collision.

19. Two identical cars are equipped with seat and shoulder belts. The first car has narrow belts, and the second car has wider belts.
    a) Which set of belts would be safer for passengers and why?
    b) The seat belts are designed to stretch slightly when a sudden, severe collision occurs and the belts are needed to hold passengers in place. Why would a seat belt that stretches be better than one that does not stretch?

20. The following data was taken when two different amounts of mass were dropped into separate containers of sand and kitty litter. In both cases, the mass did not reach the bottom of the container.

	Trial 1 (sand)	Trial 2 (kitty litter)
Mass	2.0 kg	1.0 kg
Drop height	0.30 m	0.50 m
Leaves an indentation	Yes	Yes
Depth	0.04 m	0.03 m

    a) Calculate the *GPE* of the two masses above the container before they are dropped.
    b) What is the kinetic energy of each mass as it strikes the surface of the sand or kitty litter?
    c) How much work was done on each mass to bring it to rest?
    d) Show your calculations to determine which mass required the greater force to stop once it struck the surface.

21. An automobile with a mass of 1500 kg is traveling at a speed of 20 m/s when the brakes are applied for a distance of 100 m. If the average braking force during this time is 2500 N, what is the automobile's final speed after braking?

22. In a football game on a muddy field, a 120-kg linebacker running north at 5 m/s tackles an 80-kg running back running west at 10 m/s. If the two players slide off together, what is the speed of the combination?

23. A baseball with a mass of 0.160 kg is thrown at a speed of 40 m/s toward a batter. The batter hits the ball back at 60 m/s. What impulse was provided on the ball by the bat?

Critical Thinking

16.a)

Measure the masses of each cart before and after the collision, and the velocities of each cart immediately before and immediately after the collision.

16.b)

Calculate the total momentum before the collision and after the collision. These should be equal to each other, within experimental uncertainty, if momentum is conserved. Otherwise it is not conserved.

$p_{\text{initial}} = p_{\text{final}}$

$mv_{\text{initial A}} + mv_{\text{initial B}} =$

$mv_{\text{final A}} + mv_{\text{final B}}$

16.c)

No additional measurements would be needed. The measured quantities could be used to determine the kinetic energy by using the formula

$KE_{\text{initial}} = \frac{1}{2} mv^2_{\text{initial A}} + \frac{1}{2} mv^2_{\text{initial B}}$

$KE_{\text{final}} = \frac{1}{2} mv^2_{\text{final A}} + \frac{1}{2} mv^2_{\text{final B}}$

17.a)

Newton's first law: An object in motion stays in motion until a force acts on it. The box keeps moving forward because the force of friction between the box and the truck is not great enough to hold it in place when the forward moving truck applies its brakes.

17.b)

Newton's first law and second law: An object at rest stays at rest and an object in motion stays in motion unless a net external force acts on it. Before the automobile accelerates (in the forward direction) you are moving with the automobile at a constant speed and direction (the speed can be zero). When the automobile begins to move faster, you do not move until an external force acts on you to push you forward. The frictional force from the bottom of the seat is not great enough to do this, and is not directly applied to your upper torso. Your torso is pushed forward by the seat back (Newton's second law). The organs within you are pushed forward by your body cavity. This creates the sensation of being pushed back in your seat.

17.c)

Newton's first law and second law. When you move forward in your seat, it is due to Newton's first law because no net external force is acting on you. When your seat belt stops you, it is due to Newton's second law as it exerts a force on you to change your motion (it brings your forward motion relative to the vehicle) to rest.

18.

Both air bags and crumple zones increase the distance and time over which a force is exerted. The impulse momentum equation, $Ft = m\Delta v$ says that since the automobile will be stopped in the collision, increasing the time available to stop the automobile means less force will be required to stop the automobile and its occupants. The smaller the force needed, the less harm will come to the passengers.

19.a)

The wider seat belts are safer for passengers because they spread out the applied force from the seat belt over a greater area, reducing the amount of pressure at any one point applied from the seat belt to the passenger. This avoids the seat belt from cutting in to the passenger.

19.b)

A seat belt that stretches is safer because it supplies a force on the passenger over a greater distance, reducing the force necessary to stop the passenger, and hence, reducing the risk of possible injury.

20.a)

$GPE = mgh$

$GPE_{sand} =$

$(2.0 \text{ kg})(9.8 \text{ m/s}^2)(0.30 \text{ m}) =$

5.9 J

$GPE_{litter} =$

$(1.0 \text{ kg})(9.8 \text{ m/s}^2)(0.50 \text{ m}) =$

4.9 J

20.b)

$\Delta KE = -\Delta GPE$

$KE_{final} = GPE_{initial}$

$KE_{final\ sand} = 5.9 \text{ J}$

$KE_{final\ litter} = 4.9 \text{ J}$

20.c)

$W = \Delta KE$

$W_{sand} = \Delta KE_{sand} = 5.9 \text{ J}$

$W_{litter} = \Delta KE_{litter} = 4.9 \text{ J}$

20.d)

$F \cdot d = \Delta KE$

$F_{sand} = \dfrac{\Delta KE_{sand}}{d_{sand}} = \dfrac{5.9 \text{ J}}{0.04 \text{ m}} = 148 \text{ N}$

$F_{litter} = \dfrac{\Delta KE_{litter}}{d_{litter}} = \dfrac{4.9 \text{ J}}{0.03 \text{ m}} = 163 \text{ N}$

The litter trial required more force to stop the mass once it struck the surface.

21. Active Physics Plus

Students should use the work-energy theorem.

$W = F \cdot d = \Delta KE$

$F \cdot d = \frac{1}{2} mv^2_{final} - \frac{1}{2} mv^2_{initial}$

$(-2500 \text{ N})(100 \text{ m}) =$

$\frac{1}{2}(1500 \text{ kg})v^2_{final} -$

$\frac{1}{2}(1500 \text{ kg})(20 \text{ m/s})^2$

$-250{,}000 \text{ J} =$

$(750 \text{ kg})v^2_{final} - 300{,}000 \text{ J}$

$50{,}000 \text{ J} = (750 \text{ kg})v^2_{final}$

$v_{final} = \sqrt{\dfrac{50{,}000 \text{ J}}{750 \text{ kg}}} = 8 \text{ m/s}$

22. Active Physics Plus

Students should use the conservation of momentum.

$p_{initial} = p_{final}$

$m_A v_{initial\ A} + m_B v_{initial\ B} = m_{A+B} v_{final}$

(120 kg)(5 m/s, north) +

(80 kg)(10 m/s, west) =

$(200 \text{ kg})v_{final}$

$v_{final} = \dfrac{(120 \text{ kg})(5 \text{ m/s, north})}{(200 \text{ kg})} +$

$\dfrac{(80 \text{ kg})(10 \text{ m/s, west})}{(200 \text{ kg})} =$

3 m/s, north + 4 m/s, west.

Students can then add these components together to find the magnitude and direction of the final velocity. The magnitude (size) of the final velocity is 5 m/s at an angle of 53° west of north.

23. Active Physics Plus

$F\Delta t = m\Delta v = m(v_{final} - v_{initial})$

Assuming the direction of the incoming ball is positive, you have

$F\Delta t =$

$(0.160 \text{ kg})(-60 \text{ m/s} - 40 \text{ m/s}) =$

$-16 \text{ kg} \cdot \text{m/s}.$

CHAPTER 3

Sample Assessment Rubric

Introduction

Development of rubrics as a template for identifying performance criteria has been shown to increase student achievement. A typical rubric clearly denotes each category to be evaluated and provides specific, required criteria for defining excellence, proficiency, and below-proficiency levels of performance. The sample rubrics for each chapter are intended to serve as guidelines. It should be understood that assessment is more effective when students and teachers tailor it to fit their needs. You are encouraged to work with your colleagues and especially with students to customize the rubric and the criteria. Decisions should be made together with respect to the curricular goals of the project within the particular context. For example, a class may choose to add one requirement in lieu of another, or to change the relative weighting of categories. It is helpful to remember the following recommendations:

1. Assessment should directly address the goals of the *Chapter Challenge*.

   Attention has been paid to the suggested rubrics in addressing the goals of the chapter, and the *Physics You Learned* section should serve as a guide for students and teachers working with the challenge. You may choose to make changes to the rubric in order to emphasize goals important to their context.

2. Students should participate in the assessment of their own performance.

   Students submit their rubric along with the grade they have given themselves. This not only encourages students to take ownership of the project, but it becomes a useful assessment tool for the teacher. If a student earns a "C" and gives himself or herself a "C," the conversation is very different than if a student were to earn a "C" and give himself or herself an "A." The question you might have for the first student is, "Why didn't you choose to do more?" While you might need to review the criteria with the second student and help understand what it takes to get an "A." After the teacher has graded the assignment, students have an opportunity to revise their work, and resubmit it for the "revision" grade. Emphasis should be placed not only on the finished project, but on progress with the rubric during revision.

3. Assessment should begin from a foundation of a proficient level of performance, providing ladders for students to achieve higher orders of thinking.

Finally, the rubric is built from a foundation of proficiency (meets standards). An analogy for this level is that in the real world their is a minimum acceptable standard for performance. A CD must play without skipping, and a shirt must have all of its buttons. Anything less, is substandard. This rubric works the same way. To get a "C" or better, the work must meet all the standards, and fall into the "proficient" category. Work not meeting all standards must be revised and resubmitted. Beyond proficiency, students can do work which shows mastery and may therefore earn a "good" or "excellent" rating (a "B" or an "A"). The scoring column of the rubric includes suggested point ranges for each level of mastery.

For further discussion, see: "Assessment of Laboratory Investigations," Eisenkraft, Arthur and Anthes-Washburn, Matthew. Assessment: Research and Practical Approaches, eds. Coffey, Douglas and Stearns. NSTA Press 2008.

Guide to the Sample Assessment Rubric

Assessment via this rubric will assign students to one of three major groups:

Excellent: Work meets all standards and demonstrates extensive evidence of mastery.

Good: Work meets all standards and demonstrates moderate evidence of mastery.

Proficient: Work meets standards without further evidence of mastery.

Please note that these groups are written at the top of the rubric page as a reminder to students.

In the table, there are three main groups of criteria—Mastery, Meets Standards, and Interventions. A student or team of students should achieve all of the criteria in order to satisfactorily complete the project. Anything less than this will require that the student make another attempt using the Interventions listed in the last column. This is the foundation, or floor of expectations. As teachers, we have to beware that our floor of expectations does not become a ceiling for some students.

1. In the first column, there are suggestions for demonstrating mastery. Completing one or more of these may raise a student or team from Proficient, to Good or Excellent.
2. In the second column, the criteria to meet the standards for the assignment are listed.
3. Some students may have trouble meeting the standards in the Meeting Standards column. The last column provides Interventions, or suggestions for how a student might meet the requirements of the project.
4. In the Scoring column, students submit their own grade, and you respond with a grade and feedback. Students receive a final grade after a revision is submitted. The range of scores for Excellent, Good, and Proficient allow you to assign points that match a student or team's demonstrated mastery. Thus, a student who barely meets standards can receive a different score than one who shows a higher level of mastery.

Implementing the Sample Assessment Rubric

- Modify the rubric with discussions from students.
- Hand out the rubric.
- Review the rubric so that you are confident that students understand each component.
- Have students complete the Scoring column for their work in the chapter by placing checks in each of the boxes. Have students assign themselves a point value for each component.
- Have students total their score for the rubric.
- Collect the student self-appraisal of their work.
- Use the rubric to grade the student work.
- Grade students' work after you and the student agree on the grade. Encourage the student to improve their work for the next chapter. If you and a student disagree, have an appropriate conversation with the student about his or her work and how it could be improved.

The *Sample Assessment Rubric* on the following page is provided as a *Blackline Master* on your *Teacher Resources CD*.

3c **Blackline Master**

CHAPTER 3

Sample Assessment Rubric

Mastery (Students may show mastery through these or other ideas provided by students and teachers.)	**Meets Standards**	**Scoring** (To be discussed by students and teacher)	**Interventions** (Guiding questions and instructions for students falling short of the Standards)
• Design your own safety test for your device, discussing safety limits and their justification. • Create a demonstration comparing your design to standard devices, explaining the physics principles involved.	**1. Physics Principles** • Explain how your safety device works from three points of view: Forces, Energy and Momentum. • Discuss consumer acceptance and market potential. • Use expert information and document sources. • Use scientific vocabulary consistently and precisely. • Use appropriate scientific symbols for units of measurement.	**Maximum:** 40 Points ***Excellent:** 36–40 **Good:** 32–35 **Proficient:** 28–31 *Student Self Grade:* *Teacher Grade:* *Revision:*	• **Forces:** • Why do people go flying forward in a head-on crash? • **Energy**: • How do we protect people from the energy in a collision? • **Momentum**: • Why are seat belts more important in car than in a bus?
• Produce a commercial for the safety device. • Write a letter to a public official regarding the need for the safety device.	**2. Quality of the Oral Report** • Discuss the physics principles of how the safety device works as well as the need for the safety device. • Discuss the development and design of the prototype, and describe a next-generation version of the safety device. • Prepare and practice your presentation with the team; cooperate with your team to ensure that all members participate. • Keep your presentation within the agreed-upon time limit. (Recommended—5 minutes) • Answer questions presented by the audience.	**Maximum:** 20 Points ***Excellent**: 18–20 **Good:** 16–17 **Proficient:** 14–15 *Student Self Grade:* *Teacher Grade:* *Revision:*	• First, write out your presentation. • Then, practice with your team.
• Present a poster for the safety device with three independent explanations of how it works, using force, energy and momentum. • Create a brochure or user's manual for the device, with force, energy and momentum explanations of how it works.	**3. Quality of the Written Report** • Organize the report so it is easy to follow and understand. • Use correct sentence structure. • Use correct spelling, punctuation, and grammar. • Use the correct number of pages (determined by class and teacher). Suggested–2-3 pages double spaced.	**Maximum:** 20 Points ***Excellent:** 18–20 **Good:** 16–17 **Proficient:** 14–15 *Student Self Grade:* *Teacher Grade:* *Revision:*	• Follow the suggestions of your teacher and submit a revised script.
• The prototype with the best performance in the class will receive a bonus!	**4. Performance of Working Model** (prototype) • Prepare your prototype to survive the minimum test conditions agreed upon by the class.	**Maximum:** 20 Points **Egg survives:** 20 points **Best-in-class bonus:** 5 *Student Self Grade:* *Teacher Grade:* *Revision:*	• Devise a way to test your prototype to perfect it before the challenge.
		TOTAL: ***Excellent:** 90–100 **Good:** 80-89 **Proficient:** 70-79	

* **Excellent:** Work meets all standards and demonstrates extensive evidence of mastery.
Good: Work meets all standards and demonstrates moderate evidence of mastery.
Proficient: Work meets standards without further evidence of mastery.

84 Business Park Drive, Armonk, NY 10504
Phone (914) 273-2233 Fax (914) 273-2227
www.its-about-time.com

Publishing Team

President
Tom Laster

Director of Product Development
Barbara Zahm, Ph.D.

Managing Editor
Maureen Grassi

Project Development Editor
Ruta Demery

Quality Control
Alexander Mari

Editors
Danielle Bouchat-Friedman
Heidi Doss
Tamara Kathwari
Zhiren Qin
Sampson Starkweather
Daniel M. Wolff

Writer – Physics At Work
Danielle Bouchat-Friedman

Editorial Coordinator
Susan Gibian

Creative Director
John Nordland

Assistant Art Director
Mauricio Gonzalez

Equipment Kit Developers
Dana Turner
Joseph DeMarco
Henry Garcia

Staff Photographer
Jason Harris

Creative Artwork
Thomas Bunk

Illustrators
Sean Campbell
Richard Ciotti
Doreen Flaherty
Fredy Fleck
Michael Hortens
Marie Killoran
Louise Landry
MaryBeth Schulze
Jason Skinner

Production/Studio Manager
Robert Schwalb

Production Studio Coordinator
Marie Killoran

Layout Artists
Robert Aleman
Sean Campbell
Richard Ciotti
Doreen Flaherty
Fredy Fleck
Mauricio Gomez
Marie Killoran
Louise Landry
MaryBeth Schulze

NOTES

NOTES

NOTES